Agriculture, Hydrology and Water Quality

Agriculture, Hydrology and Water Quality

Edited by

P.M. Haygarth and S.C. Jarvis

Institute of Grassland and Environmental Research,
North Wyke Research Station, Devon, UK

CABI *Publishing*

CABI is a trading name of CAB International

CABI Head Office
Nosworthy Way
Wallingford
Oxfordshire OX10 8DE
UK

Tel: +44 (0)1491 832111
Fax: +44 (0)1491 833508
Email: cabi@cabi.org
Web site: www.cabi.org

CABI North American Office
875 Massachusetts Avenue
7th Floor
Cambridge, MA 02139
USA

Tel: +1 617 395 4056
Fax: +1 617 354 6875
Email: cabi-nao@cabi.org

A catalogue record for this book is available from the British Library, London, UK

Library of Congress Cataloging-in-Publication Data
Agriculture, hydrology and water quality / edited by P.M. Haygarth & S.C. Jarvis
p. cm.
Includes bibliographical references.
ISBN 0-85199-545-4 (alk. paper)
1. Agricultural pollution—Environmental aspects. 2. Water—pollution. 3. Hydrology. 4. Water quality. I. Haygarth, P.M. II. Jarvis, S.C.
TD428.A37 A366 2002
628.1´684—dc21

2001005591

ISBN-13: 978-0-85199545-8
ISBN-10: 0-85199-545-4

First published 2002
Transferred to print on demand 2009

Printed and bound in the UK by CPI Antony Rowe, Chippenham and Eastbourne

Contents

Contributors

S. Anthony, *ADAS Research and Development, Woodthorne, Wergs Road, Wolverhampton WV16 8TQ, UK*

D.S. Baldwin, *Cooperative Research Centre for Freshwater Ecology and Murray–Darling Freshwater Research Centre, PO Box 921, Albury, NSW 2640, Australia*

D.B. Beegle, *Department of Crop and Soil Sciences, The Pennsylvania State University, 116 ASI Building, University Park, PA 16802, USA*

K. Beven, *Institute of Environmental and Natural Sciences, Lancaster University, Lancaster LA1 4YQ, UK*

M.S.A. Blackwell, *Royal Holloway Institute for Environmental Research, Department of Geography, Royal Holloway University of London, Surrey GU25 4LN, UK*

J.J.B. Bronswijk, *National Institute of Public Health and Environment (RIVM), PO Box 1, 3720 BA, Bilthoven, The Netherlands*

K.C. Cameron, *Centre for Soil and Environmental Quality, PO Box 84, Lincoln University, Canterbury, New Zealand*

G. Campbell, *Department of Plant and Soil Science, University of Aberdeen, Aberdeen AB24 3UU, UK*

D.R. Chadwick, *Institute of Grassland and Environmental Research, North Wyke Research Station, Okehampton, Devon EX20 2SB, UK*

P.A. Chambers, *National Water Research Institute, 867 Lakeshore Boulevard, PO Box 5050, Burlington, Ontario L7R 4S6, Canada*

S. Chen, *Centre for Land and Water Stewardship, University of Guelph, Guelph, Ontario N1G 2W1, Canada*

F.J. Coale, *Department of Natural Resource Sciences and Landscape Architecture, University of Maryland, College Park, MD 20742, USA*

L.M. Condron, *Centre for Soil and Environmental Quality, PO Box 84, Lincoln University, Canterbury, New Zealand*

J. Cox, *CSIRO Land and Water, PMB 2, Glen Osmond, SA 5064, Australia*

H.J. Di, *Centre for Soil and Environmental Quality, PO Box 84, Lincoln University, Canterbury, New Zealand*

A.C. Edwards, *Macaulay Land Use Research Institute, Craigiebuckler, Aberdeen AB15 8QH, UK*

G. Fishburn, *New South Wales Department of Land and Water Conservation, PO Box 10, Wagga Wagga, NSW 2650, Australia*

R.H. Foy, *Agricultural and Environmental Science Division, Department of Agriculture and Rural Development, Newforge Lane, Belfast BT9 5PX, UK*

A.I. Fraser, *National Soil Resources Institute, North Wyke Research Station, Okehampton, Devon EX20 2SB, UK*

W.J. Gburek, *United States Department of Agriculture – Agricultural Research Service, Pasture Systems and Watershed Management Research Unit, Curtin Road, University Park, PA 16802-3702, USA*

B. Gevao, *Institute of Environmental and Natural Sciences, Lancaster University, Lancaster LA1 4YQ, UK*

K.E. Giller, *Department of Soil Science and Agricultural Engineering, University of Zimbabwe, PO Box MP167, Mount Pleasant, Harare, Zimbabwe*

K. Goulding, *Soil Science Department, Institute of Arable Crop Research, Rothamsted, Harpenden, Hertfordshire AL5 2JQ, UK*

R. Grant, *National Environmental Research Institute, Department of Streams and Riparian Areas, Vejlsøvej 25, DK-8600, Silkeborg, Denmark*

D. Halliwell, *Agriculture Victoria Ellinbank, RMB 2460, Ellinbank, VIC 3821, Australia*

G. Harris, *CSIRO Land and Water, Canberra, Australian Capital Territory, Australia*

T.R. Harrod, *National Soil Resources Institute, North Wyke Research Station, Okehampton, Devon EX20 2SB, UK*

D. Hatch, *Institute of Grassland and Environmental Research, North Wyke Research Station, Okehampton, Devon EX20 2SB, UK*

T.J. Hatton, *CSIRO Land and Water GPO Box 1666, Canberra, ACT 2601, Australia*

P.M. Haygarth, *Institute of Grassland and Environmental Research, North Wyke Research Station, Devon, UK*

D.V. Hogan, *Royal Holloway Institute for Environmental Research, Department of Geography, Royal Holloway University of London, Surrey GU25 4LN, UK*

S.C. Jarvis, *Institute of Grassland and Environmental Research, North Wyke Research Station, Devon, UK*

D.L. Jones, *School of Agricultural and Forest Sciences, University of Wales, Bangor, Gwynedd LL57 2UW, UK*

K.C. Jones, *Institute of Environmental and Natural Sciences, Lancaster University, Lancaster LA1 4YQ, UK*

C.W. Kaspar, *Department of Food Microbiology and Toxicology, Food Research Institute, University of Wisconsin, Madison, WI 53706-1187, USA*

P.J.A. Kleinman, *United States Department of Agriculture – Agricultural Research Service, Pasture Systems and Watershed Management Research Unit, Curtin Road, University Park, PA 16802-3702, USA*

B. Kronvang, *National Environmental Research Institute, Department of Streams and Riparian Areas, Vejlsøvej 25, DK-8600, Silkeborg, Denmark*

L.E. Lanyon, *Department of Crop and Soil Sciences, The Pennsylvania State University, 116 ADI Building, University Park, PA 16802, USA*

A.R. Laubel, *National Environmental Research Institute, Department of Streams and Riparian Areas, Vejlsøvej 25, DK-8600, Silkeborg, Denmark*

T. van Leeuwen, *Agricultural Economic Research Institute (LEI), PO Box 29703, 2502 LS, The Hague, The Netherlands*

P. Leinweber, *Institute of Soil Science, University of Rostock, Justus-von-Liebig-Weg 6, D-18059 Rostock, Germany*

E. Maltby, *Royal Holloway Institute for Environmental Research, Department of Geography, Royal Holloway University of London, Surrey GU25 4LN, UK*

E. Mangwayana, *Department of Soil Science and Agricultural Engineering, University of Zimbabwe, PO Box MP167, Mount Pleasant, Harare, Zimbabwe*

R.W. McDowell, *United States Department of Agriculture – Agricultural Research Service,*

Pasture Systems and Watershed Management Research Unit, Curtin Road, University Park, PA 16802-3702 USA. Present address: AgResearch Invermay, Private Bag 50034, Mosgiel, New Zealand

R. Meissner, *UFZ Centre for Environmental Research, Leipzig-Halle, Lysimeter Station, D-39615 Falkenberg, Germany*

A.M. Mitchell, *Cooperative Research Centre for Freshwater Ecology and Murray–Darling Freshwater Research Centre, PO Box 921, Albury, NSW 2640, Australia*

K.E. Motsi, *Department of Soil Science and Agricultural Engineering, University of Zimbabwe, PO Box MP167, Mount Pleasant, Harare, Zimbabwe*

D. Murphy, *Centre for Land Rehabilitation, Soil Science and Plant Nutrition, The University of Western Australia, Nedlands 6009, Western Australia*

D. Nash, *Agriculture Victoria Ellinbank, RMB 2460, Ellinbank, VIC 3821, Australia*

W.C.K. O'Connor, *Agricultural and Environmental Science Division, Department of Agriculture and Rural Development, Newforge Lane, Belfast BT9 5PX, UK*

J.M. Olley, *CSIRO Land and Water, PO Box 1666, Canberra, ACT 2601, Australia*

M.L. Pedersen, *National Environmental Research Institute, Department of Streams and Riparian Areas, Vejlsøvej 25, DK-8600, Silkeborg, Denmark*

S. Pengelly, *New South Wales Department of Land and Water Conservation, PO Box 365, Queanbeyan, NSW 2620, Australia*

P.F. Quinn, *Water Resources Systems Research Laboratory, Newcastle University, Newcastle-upon-Tyne, UK*

P.F.A.M. Römkens, *Alterra Green World Research, PO Box 44, 6700 AA, Wageningen, The Netherlands*

A.N. Sharpley, *United States Department of Agriculture – Agricultural Research Service, Pasture Systems and Watershed Management Research Unit, Curtin Road, University Park, PA 16802-3702, USA*

J.T. Sims, *Department of Plant and Soil Science, 163 Townsend Hall, University of Delaware, Newark, DE 19717-1303, USA*

F.D. Theurer, *National Water and Climate Center, Natural Resources Conservation Service, United States Department of Agriculture, 7413 Cinnabar Terrace, Gaithersburg, MD 20879-4575, USA*

P.A. Troch, *Department of Water Management, Wageningen Agricultural University, Wageningen, The Netherlands*

B.L. Turner, *United States Department of Agriculture – Agricultural Research Service, Northwest Irrigation and Soils Research Laboratory, 3793 N. 3600 E. Kimberly, ID 83341, USA*

Y.J.P. Van Herpe, *Laboratory of Hydrology and Water Management, Ghent University, Ghent, Belgium*

W. de Vries, *Alterra Green World Research, PO Box 44, 6700 AA, Wageningen, The Netherlands*

G.R. Walker, *CSIRO Land and Water, GPO Box 1666, Canberra, ACT 2601, Australia*

J. Williams, *CSIRO Land and Water, GPO Box 1666, Canberra, ACT 2601, Australia*

Acknowledgements

Special thanks are due to Trisha Butler for extensive editorial assistance and support, also to Anne Roker, Linda Jewell, Jo Chisholm and Adrian Joynes all at the Institute of Grassland and Environmental Research. Thanks also to Louise Heathwaite at Sheffield University, Leo Condron and Neil Smith at Lincoln University, Ben Turner at USDA Idaho and Anne Haygarth for help, discussions and inspiration in producing this volume.

Note on Terminology and Abbreviations

Terminology

At the start of the editorial process, we referred authors to three aspects of our published work, which were intended be used for guidance on use of terminology (Haygarth and Jarvis, 1999; Haygarth *et al.*, 2000; Haygarth and Sharpley, 2000). Despite these efforts, there remain five key areas of terminology that were vulnerable to ambiguity and therefore worthy of special note:

1. 'Runoff' is a term to be avoided as it is not pathway specific and is ambiguous. Instead, we encouraged use of specific pathway terms such as overland flow and subsurface lateral flow. Our policy was that runoff is only acceptable in a general context and where this is made clear.
2. 'Leaching' was often incorrectly referred to as a pathway, while it is actually a mechanism that describes the elluviation of solutes through soil, and is not pathway specific. Preferred alternative terms for pathways are preferential flow, saturated flow, macropore flow, etc.
3. Hydrological scale is often difficult to conceptualize. Haygarth *et al.* (2000) attempted a definition of 'soil profile', 'slope/field' and 'catchment' scale, but it is clear that contributors have individual views about these boundaries. This highlights the need for caution and clarity when interpreting issues of scale.
4. Terms such as 'dissolved' and 'particulate' need careful interpretation as some researchers use different filter sizes to define the thresholds.
5. The use of 'phosphate' is unacceptable if describing forms determined by Mo-reaction chemistry, because the reaction does not exclusively determine free phosphate. Reactive P is operationally defined and thus more correct.

Abbreviations

Below is a list of selected 'common' abbreviations used throughout the book. The abbreviated form is always used, except at the start of sentences or in headings. Any abbreviation not listed here is defined at the first time of using in the specific chapter; thereafter it follows the style of common abbreviations.

Al	aluminium
BMP(s)	best management practice(s)
BOD	biological oxygen demand
C	carbon
Ca	calcium
Cd	cadmium
Co	cobalt

CO_2	carbon dioxide
COD	chemical oxygen demand
CSA	critical source area
Cu	copper
DDD	1,1-dichloro-2,2-*bis* (*p*-chlorophenyl)ethane
DDE	1,1-bis-(*p*-chlorophenyl)-2,2-dichloroethane
DDT	1,1,1-trichloro-2,2-*bis* (*p*-chlorophenyl)ethane
DO	dissolved oxygen
DOC	dissolved organic carbon
DOM	dissolved organic matter
DON	dissolved organic nitrogen
DRP	dissolved reactive phosphorus
DW	dry weight
Fe	iron
GIS	geographical information systems
H	hydrogen
ha	hectare
HCB	hexachlorobenzene
Hg	mercury
K	potassium
LU	livestock units
Mn	manganese
Mo	molybdenum
MRP	molybdate reactive phosphorus
N	nitrogen
NH_3	ammonia
NH_4^+	ammonium
Ni	nickel
NO_3^-	nitrate
O_2	oxygen
OM	organic matter
P	phosphorus
Pb	lead
PCP	pentachlorophenol
PO_4^{3-}	phosphate
PP	particulate phosphorus
RP	reactive phosphorus
SO_4^{2-}	sulphate
SRP	soluble reactive phosphorus
SS	suspended solids
TP	total phosphorus
UP	unreactive phosphorus
Zn	zinc

References

Haygarth, P.M. and Jarvis, S.C. (1999) Transfer of phosphorus from agricultural soils. *Advances in Agronomy* 66, 195–249.

Haygarth, P.M. and Sharpley, A.N. (2000) Terminology for phosphorus transfer. *Journal of Environmental Quality* 29, 10–15.

Haygarth, P.M., Heathwaite, A.L., Jarvis, S.C. and Harrod, T.R. (2000) Hydrological factors for phosphorus transfer from agricultural soils. *Advances in Agronomy* 69, 153–178.

Introduction: an Interdisciplinary Approach for Agriculture, Hydrology and Water Quality

In the short time since we started the production of *Agriculture, Hydrology and Water Quality*, interest in the subject has already increased. In the USA, Congress has been called to strengthen the Clean Water Act, with polluted runoff being described as the 'most pervasive problem in the coastal marine environment' (Schlein and Yum, 1999). Most recently, the European Commission and the European Parliament have confirmed the Water Framework Directive, which aims to achieve 'good water status' in all European countries by 2010.

This volume is aimed at undergraduates, postgraduates, researchers and policy makers in all related areas: the science is complex and, by its nature, interdisciplinary. Herewith was the challenge, with the aim of drawing in researchers from different fields and different spheres of understanding. Agronomists, aquatic ecologists, chemists, hydrologists and soil scientists all have perspectives that are different, but of equal value. The aim of this book has been to bring these disciplines together, to work towards an improved understanding of the processes involved and impacts of agriculture on water. Our first priority was to establish a logical way of developing a structure. The result is a separation into the three sections: (i) agriculture as a potential source; (ii) hydrology as the transport; and (iii) impacts and case studies from around the world.

During the editorial process we were able to identify some emerging areas of interest and importance:

1. The increasing interest in soil biology and organic forms of particularly N and P in soil and water; traditional thinking has been dominated by inorganic chemistry (Chapters 1 and 2).
2. The terrible cost of salinization in Australia (Chapter 21).
3. Discrete soil erosion problems contributing to long-term sediment transfers and river sedimentation, presenting dual problems of: (i) impaired freshwater ecology; and (ii) 'carrying' other contaminants such as pathogens, persistent organic pollutants and P (Chapters 2, 4, 6 and 7).
4. Problems of manure production, storage and land application, confounded by conflicts of N versus P management priorities (Chapters 3 and 8).
5. The rise in farm soil P levels in some intensive agricultural regions, representing a potential reservoir that may contribute to water problems in the future (Chapter 19).

Uncertainty will always be a factor in the complexity of the various features of the system. We are students of a 'real world' where climates, rainfall and soils, among others, are varied; controlled and replicated experimental catchments do not exist. Because of these difficulties, the subject is especially vulnerable to the proliferation of 'anecdotal' at the expense of 'scientific' evidence, which can become self-perpetuating and lead to misunderstandings. Take, for example, the problems of P transfer from land to water: the impact may be universal (i.e. eutrophication and the associated problems in waters) but it would be a mistake to assume there was a single global solution. Indeed, the science may vary depending on location, from physical detachment of soil (and attached P) (e.g. extensively farmed catchments of south-east Australia) to the leaching of P following a history of high fertilizer inputs (e.g. intensively farmed pastures of Northern Ireland). In these cases the available mitigation options are completely different; thus a solution in one part of the world cannot be transposed to another region. We need to become more aware of such differences and establish scientific principles that transcend regional and causal differences.

Therefore, without an interdisciplinary approach, these issues become overlaid and muddled. The popular wisdom has been that reducing the inputs to a farm will reduce water problems: indeed there are some powerful examples where this is the case (see, for example, Chapters 1, 9, 16, 17 in relation to N effects and Chapters 2, 9, 16, 17, 19 in relation to P effects). In the practical sense, it is easy to understand why reducing fertilizer inputs is thought to be a more effective mitigation strategy than, for example, controlling irrigation water flow on to and over the soil. However, while this may be appropriate for some issues (N for example) a concern is that in other instances, such 'source control' wisdom may exist because it is convenient, rather than because of any objective scientific evaluation. A large source may not necessarily equate to a high impact, especially if issues of scale and connectivity are taken into consideration (Haygarth *et al.*, 2000).

For these reasons, there is a pressing need to rationalize the scientific basis of land management and its impacts downstream, through a holistic perspective (Haygarth and Jarvis, 1999). We hope that the assembly of these 22 chapters in *Agriculture, Hydrology and Water Quality* has made a contribution to this process.

Philip Haygarth and Steve Jarvis
Institute of Grassland and Environmental Research, North Wyke Research Station, Devon
July 2001

References

Haygarth, P.M., Heathwaite, A.L., Jarvis, S.C. and Harrod, T.R. (2000) Hydrological factors for phosphorus transfer from agricultural soils. *Advances in Agronomy* 69, 153–178.

Haygarth, P.M. and Jarvis, S.C. (1999) Transfer of phosphorus from agricultural soils. *Advances in Agronomy* 66, 195–249.

Schlein, S. and Yum, S. (1999) Polluted runoff called 'Most Pervasive problem in Coastal Marine Environment'. Press release (http://www.cmc-ocean.org), Centre for Marine Conservation, Washington DC.

Section 1

Agriculture:
Potential Sources of Water Pollution

Introduction: Agriculture as a Potential Source of Water Pollution

Andrew N. Sharpley
USDA–ARS, Pasture Systems and Watershed Management Research Unit, Curtin Road, University Park, PA 16802-3702, USA

The production of grain and protein has increased so dramatically since the industrialization of agricultural systems in the 1950s, that supply now exceeds demand in most developed countries. This has drawn public, political and ultimately research attention from issues of food supply to those of maintaining or improving environmental quality (National Research Council, 2000). With the advent of point source pollutant controls in most European and North American countries in the 1970s, came some improvement in water quality, most notably in the Baltic and North Seas, Rhine River (Germany), as well as the Chesapeake Bay and Great Lakes (USA) (Sharpley, 2000). Even so, many water quality problems still exist, such as eutrophication, elevated groundwater NO_3^-, siltation of navigational waterways, and contamination of waters with pesticides, heavy metals and pathogens. Thus, the relative importance and role of agriculture, along with urban development, as contributors to water quality degradation are now being questioned (Carpenter *et al.*, 1998).

In developed countries, agricultural systems have been developed to produce cheaper grain and protein on a smaller land area. Most agricultural produce (> 80%) from these countries now comes from a few farms (< 10% of farms) (Withers *et al.*, 1999). The result has been greater use of fertilizers (N and P) and pesticides, and a greater potential for erosion and runoff with current land cultivation methods. Studies have shown that modern agricultural systems accumulate nutrients (especially from animal feed in manure), increasing their potential for loss to surface and groundwaters (Haygarth *et al.*, 1998).

The controlling factors influencing loss include physical location, chemical speciation, fate and environmental availability. These must be combined with hydrologic processes transporting chemicals from a field (where they are perceived as a resource) to point of impact (where they are perceived as a pollutant) (Gburek *et al.*, 2000). Overlaying these are the complex interactions of land management, topography and climate; these act in combination with market forces that can dictate on-farm management decisions. Even so, an awareness of agricultural or environmental problems and potential solutions does not necessarily cause people to change their behaviour to correct such problems. Solutions have to be adapted in practical ways to individual circumstances, as recognized in Australia, where action agencies are investigating the capacity of rural communities to implement changes to help protect soil and water resources (National Land and Water Resources Audit, 1998).

The following chapters in this section detail many of the factors and processes determining the fate of nutrients, pesticides, heavy metals, manure, pathogens and sediment in agriculture. Only when these processes are more clearly understood, will we be able to develop effective remedial strategies that can be targeted to reduce the impact of critical pollutant sources.

References

Carpenter, S.R., Caraco, N.F., Correll, D.L., Howarth, R.W., Sharpley, A.N. and Smith, V.H. (1998) Nonpoint pollution of surface waters with phosphorus and nitrogen. *Ecological Applications* 8, 559–568.

Gburek, W.J., Sharpley, A.N., Heathwaite, A.L. and Folmar, G.J. (2000) Phosphorus management at the watershed scale: a modification of the phosphorus index. *Journal of Environmental Quality* 29, 130–144.

Haygarth, P.M., Chapman, P.J., Jarvis, S.C. and Smith, R.V. (1998) Phosphorus budgets for two contrasting grassland farming systems in the UK. *Soil Use and Management* 14, 160–167.

National Land and Water Resources Audit (1998) *National Land and Water Resources Audit: Draft Strategic Plan – 1998–2001.* National Land and Water Resources Audit, Canberra, Australia.

National Research Council (2000) Clean Coastal Waters: Understanding and Reducing the Effects of Nutrient Pollution. National Academy Press, Washington, DC.

Sharpley, A.N. (2000) *Agriculture and Phosphorus Management: the Chesapeake Bay.* CRC Press, Boca Raton, Florida.

Withers, P.J.A., Peel, S., Mansbridge, R.M., Chalmers, A.C. and Lane, S.J. (1999) Transfers of phosphorus within three dairy farming systems receiving varying inputs in feeds and fertilizers. *Nutrient Cycling in Agroecosystems* 55, 63–75.

1 Nitrogen

David Hatch,[1] Keith Goulding[2] and Daniel Murphy[3]
[1] *Institute of Grassland and Environmental Research, North Wyke Research Station, Okehampton, Devon EX20 2SB, UK;* [2] *Soil Science Department, IACR-Rothamsted, Harpenden, Hertfordshire AL5 2JQ, UK;* [3] *Centre for Land Rehabilitation, Soil Science and Plant Nutrition, The University of Western Australia, Nedlands 6009, Western Australia*

Key words: ammonium, dissolved organic nitrogen, eutrophication, fertilizer, leaching, manure, nitrate, nitrite, nitrogen, organic farming

Introduction

As a major plant nutrient, N is often applied in large amounts to agricultural land to maintain optimal yields. Nitrogen can also enter the soil through atmospheric deposition in regions influenced by high levels of industrial pollution (Harrison, 1999; Langan, 1999) or, in the special case of legume plant species, through their association with dinitrogen (N_2) fixing bacteria. Considerable amounts of plant-available N are released during the decomposition of soil organic matter, but the rate at which this occurs is difficult to predict and varies throughout the year. Plants compete with soil microbial populations for this available N; some is utilized by the microorganisms for their own growth and metabolism, while the remainder, if not utilized by plants, is vulnerable to being lost from the surface layers of the soil. The charge on the inorganic N ion affects the loss processes. Nitrate carries a negative charge and the amount of NO_3^- present tends to increase in most fertilized soils and, in this form, is very susceptible to loss via leaching and runoff. Nitrate pollution in drinking water has been considered as a hazard to human health for many years, although recent medical research has questioned some of the reasoning behind this established view (Wilson *et al.*, 1999). As an environmental pollutant, NO_3^- causes eutrophication of surface waterways. Dissolved molecules of organic N (DON) can also be lost through water movement and may be of concern for surface water quality. Now, farmers, legislative bodies and the general public are increasingly concerned with reducing N pollution to the environment. A number of management options are available to reduce N losses from the farm: these are based on the strategic use of N fertilizer and improved N recycling within farming systems (Burt *et al.*, 1993; MAFF, 1999). In some regions of the world, it has also been necessary to impose government restrictions on the application of N fertilizers to agricultural land (MAFF/WOAD, 1998). The key for the future is to develop sustainable farming systems which maintain acceptable yields, while causing minimal pollution to the environment.

Nitrogen: an Essential Plant Nutrient but Major Environmental Pollutant

Nitrogen is an essential nutrient to both plants and animals, being a vital component of amino acids, proteins and nucleic acids (i.e. the building blocks of life). For example, on a dry weight basis, plant material contains 2–4% N and 40% C. Although N makes up 78.1% (by volume) of the atmosphere and is a significant component of all soils, it is often a major limitation to the growth of plants (along with sunlight and water). This is because only legume species of plants (i.e. pea family: beans, lupins, clover, acacia) can obtain N that has been 'fixed' from atmospheric N_2 via a symbiotic relationship with specialized organisms (*Rhizobia*) that colonize their roots. Also, most of the N contained in soil is complexed within soil organic matter and is not immediately available for plant uptake. Microbial decomposition of soil organic matter (i.e. ammonification, often termed gross mineralization) occurs in all soils and results in the release of plant-available mineral (i.e. inorganic) forms of N. Plants compete with soil microbial processes for the available N in the soil surrounding their roots. Plants generally take up N as NO_3^-, or NH_4^+ from the soil solution; NO_3^- is favoured, although some plants (e.g. Arctic sedges) are also able to utilize N contained within small dissolved organic molecules (Chapin *et al.*, 1993). Except in some natural systems, where soil N supply is in equilibrium with plant demand, most plants will increase growth if additional N is supplied.

To ensure that plant N availability does not limit crop yields, additional N is generally added to agricultural crops as inorganic fertilizer, or in organic forms (e.g. farmyard manure, slurry). In developed countries (e.g. in the UK), where high crop yields are achievable and commercial sources of N are readily available, fertilizer application rates per annum of up to 200 kg N ha^{-1} for cereal crops and 400 kg N ha^{-1} for grass cut for silage are common. Any remaining N that is not immobilized by microorganisms, or utilized by plants, is a potential source of N pollution. Understanding the timing and amount of N supplied from mineralization, while adjusting application rates of additional inorganic or organic N fertilizer to ensure plant N demand is met, is one of the major challenges to farming throughout the world. If this balance can be maintained, then optimal crop yields can be reached with minimal fertilizer wastage, financial benefits and a reduction in environmental impact. However, when N fertilizer is applied in excess of plant requirements, it undergoes a series of transformations and transfers in soil, which can lead to pollution of the waterways and gaseous emissions from the soil.

The Forms of Nitrogen that are Potential Pollutants in Water

The inorganic forms of N that are 'reactive' and therefore, potential pollutants, are likely to cause problems only when present in the wrong place, or in excess of local requirements. In the context of water pollution, forms of reactive N of known concern are NH_3 (which dissolves to form NH_4^+), NO_2^- and NO_3^-. Ammonium is rapidly nitrified to NO_3^- by soil microorganisms and is also held tightly on the negative charges of clay minerals and soil organic matter, and so is relatively immobile and harmless. However, it can be removed occasionally in overland flow, especially in poorly structured soils subject to erosion.

Nitrite is an intermediary product in the process of nitrification (Fig. 1.1); it has a short half-life and, therefore, does not usually pose a problem. However under conditions of high temperature and poor aeration, NH_4^+ oxidation exceeds NO_2^- oxidation and the latter can accumulate. Other factors, including high NO_3^- concentrations and pH > 7.5, or combinations of these factors, can also lead to NO_2^- accumulation and subsequent leaching. Nitrite is very

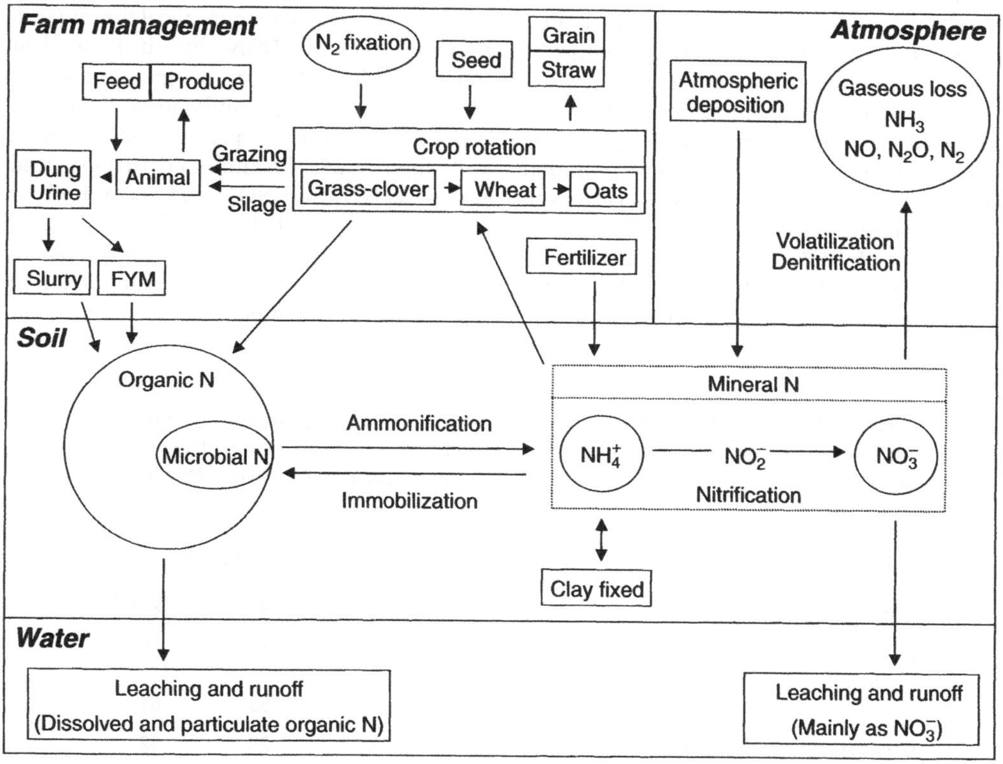

Fig. 1.1. A simplified diagram of the farm nitrogen cycle.

reactive, toxic to aquatic life and usually present in soils and waters in only small quantities. The maximum acceptable concentration in drinking water is 30 μg l^{-1}, but may only be allowed to reach up to 3 or 9 μg l^{-1} to protect salmonids and coarse fish, respectively (European Economic Community, 1980). Typical values in unpolluted streams are generally between 1 and 3 μg l^{-1}, but may range between 8–17 μg l^{-1} in highly contaminated rivers (e.g. Rhine and Rhone). Higher NO_2^- concentrations of 100–150 μg l^{-1} have been reported in major rivers entering Lough Neagh, Ireland (Smith *et al.*, 1995).

The conversion of NH_4^+ to the intermediate NO_2^- and then through to NO_3^- by nitrifying bacteria (i.e. nitrification) is a key process which mobilizes N and promotes losses to watercourses. Within agricultural soils, the rate of NO_3^- production is usually non-limiting, meaning that NH_4^+ levels are generally kept low (typically 1–3 mg l^{-1}), while NO_3^- pool sizes can be considerable. Nitrate is relatively stable, very soluble and does not become fixed on clays or organic matter because of its negative charge; it therefore remains highly mobile.

Until recently, the problem of N leaching from soils to waters was thought to be mainly associated with NO_3^-, but included small amounts of NH_4^+ and NO_2^-. However, recent attention has focused on the large quantities of DON present in agricultural soils (Murphy *et al.*, 2000). Table 1.1 shows some data collected from the Broadbalk Continuous Wheat Experiment (since 1843) at the Rothamsted Experimental Station, England. Each plot is drained and the DON collected from the drains is compared with DON measured in 2 M KCl extracts from the soil, from which drainage is collected. The data show that the DON leached from the soil was equivalent to only 2–10% of the total DON pool held within the soil profile. Although leaching losses of DON were small under treatments receiving inorganic N fertilizer, significant amounts of DON were leached from plots receiving farmyard manure (FYM).

Large concentrations of DON have also been found in drainage waters leaving grassland lysimeters in Devon, England. Within a high input beef grazing system approximately 30% of the 89–272 kg N ha^{-1} leached was as DON (Scholefield and Hawkins, personal communication cited in Jarvis, 1999). Certainly, the potential exists for large leaching losses of DON. The microbially available components of DON are probably decomposed before leaving the surface soil, thus leaching as NO_3^-. The more recalcitrant components leach as DON, but it is unclear whether this DON can be transformed to NO_3^- in surface or groundwaters. If DON is mineralized during percolation through soil, then it will be included in measured losses (as NH_4^+ and NO_3^-) and accounted for in the overall N balance. However, DON may be sorbed in lower soil horizons, in which case it will not be accounted for in current measurements.

Table 1.1. Mineral nitrogen (N) and dissolved organic N (DON) in 2 M KCl extracts from the 0–75 cm soil layer and in the drainage waters collected from tile drains at 65 cm on the Broadbalk Experiment, Rothamsted, south-east England, September–November 1998. (Source: Murphy *et al.*, 2000.)

		N applied (kg ha^{-1} year^{-1})			
Source	Analysis	0	144	288	240 (as FYM)
Soil extracts in 2 M KCl	Mineral N (kg ha^{-1})	19	36	49	83
	DON (kg ha^{-1})	47	24	55	61
Drainage water	Mineral N (kg ha^{-1})	10	6	29	52
	DON (kg ha^{-1})	1	1	3	7

The Nitrogen Cycle within a Farming System

Figure 1.1 shows a simplified farm N cycle. It summarizes the movements of N into and out of a mixed farm (one with arable crops and livestock). A typical conventional, mixed farm of 250 ha might have 250 kg N ha^{-1} (> 60 t N) moving into, through, and out from the farm each year. Nitrogen enters the farm from the atmosphere, as fertilizer, in irrigation water, bought-in animals, seed, bought-in feed and manure (i.e. not produced on the farm), in sewage sludge (i.e. human waste) and by N_2 fixation. Inputs in animals and seed are easily quantified. Other inputs are spatially and temporally variable. Even the above ground component of the N cycle is complicated, with many inputs and outputs, and it ignores the numerous microbially mediated processes occurring within the soil.

Atmospheric nitrogen deposition

Current rates of atmospheric N deposition in Europe and North America are now 5–20 times more than in pre-industrial times (25–100 kg N ha^{-1} $year^{-1}$ compared with 5–10 kg N ha^{-1} $year^{-1}$). Atmospheric deposition can have a significant impact on NO_3^- leaching. Computer models have shown that up to 40% of the NO_3^- leached from some arable crops originated directly from atmospheric deposition (Goulding *et al.*, 1998). Deposited N can also increase crop yield, suggesting that fertilizer recommendations should be adjusted to take account of regional data on atmospheric N deposition rates.

Biological dinitrogen fixation

In the special case of legumes, these plants can also 'acquire' additional N through a symbiotic relationship with certain soil microorganisms (*Rhizobia*) located in their roots which can 'fix' N directly from atmospheric N_2. The amount of transfer between microorganism and plant varies with environmental conditions, but can be as much as 90%. For this reason legumes such as lupins, lucerne and clover are commonly included within agricultural crop rotations in soils of naturally low fertility, such as in parts of Australia, so as to reduce the need to apply N fertilizers.

The fixed N_2 gained by legumes may benefit subsequent non-leguminous crops, since the N excreted in root exudates, and that which remains in the legume roots after harvest, decomposes to produce mineral N. Within grazed pastures containing legumes, the N is rapidly recycled to non-legumes via excretal returns. Also, by ploughing the whole crop back into the ground as a green manure, considerable N can be released over the following season. However, while this generally results in a benefit to the next crop (in terms of N), it is possible for this cropping phase to contribute to groundwater pollution, since decomposing plant residues can release as much as 200 kg N ha^{-1} $year^{-1}$.

Soil nitrogen cycle

Considerable N is stored below ground within plant root systems and in the soil organic matter. In cereal plants, approximately 50% of the total plant N is located below ground. This figure is even higher for pastures which can accumulate large amounts of OM. The soil therefore contains a significant amount of N, perhaps 1–10 Mg ha^{-1} in the surface layer of grass soils (0–15 cm), or in the plough layer of arable soils (0–23 cm). Most of this N occurs in complex organic compounds, which may include organic N that is virtually inert, as well as N present in the bodies of living soil microorganisms (bacteria, fungi). Mineral N (NO_3^- + NH_4^+) comprises only a small part of the total N in the soil, usually about 1% in arable soils, except after fertilizer application (Jarvis *et al.*, 1996). Mineral N is added to this pool in the soil as fertilizer, in manure and by atmospheric deposition, and is depleted by plant uptake and immobilization by microorganisms, by denitrification and by leaching (Fig. 1.1). In grassland soil receiving fertilizer, the mineral N component represents < 10% of the amount of N contained in the soil microorganisms. However, the soil mineral N pool is under constant flux and is supplemented by the decomposition of soil organic N by the process of net mineralization (i.e. the net result of the opposing processes of ammonification and immobilization). Net N mineralization ranges from 50–130 kg N ha^{-1} $year^{-1}$ for low organic matter cropped soils to > 400 kg N ha^{-1} $year^{-1}$ under fertile pastures (Jarvis *et al.*, 1996).

Competition for soil mineral nitrogen

As can be seen in Fig. 1.1, there are a number of competing processes for the consumption of mineral N from the soil. Immobilization of NH_4^+ and NO_3^- by soil microorganisms regulates the net production of mineral N from the mineralization of soil organic matter. When microbial demand for N is high (e.g. when high C:N ratio plant material such as wheat straw is added to soil), the net release of N from soil is low, or even negative (i.e. when immobilization > ammonification). Under these conditions, soil-derived N loss is minimal. Together with gaseous forms of N loss, these microbial processes determine the amount of soil mineral N available for plant uptake and any surplus will be vulnerable to leaching.

Denitrification involves the microbial reduction of oxides of N to N_2. These microorganisms only reduce NO_3^- when oxygen is unavailable, so that the process is confined to anaerobic micro-sites within the soil. Denitrification can be considered as either a beneficial or a detrimental soil process. When soils have too much N, denitrification can assist by protecting groundwater and streams from excessive NO_3^- loadings. This approach can be utilized to remove NO_3^-, by re-processing wastewater and encouraging denitrification before water percolates into subsoil layers, or streams. The introduction of buffer strips between agricultural land and watercourses or boreholes is an 'end of pipe' solution that can prevent the movement of NO_3^- (and also P and pesticides) into watercourses at some sites. Zones of denitrification (denitrification walls) can also be constructed within agricultural fields to reduce NO_3^- contamination of agricultural streams (see Blackwell *et al.*, Chapter 15, this volume).

Annual denitrification rates reported in the literature range from 0–40 kg N ha^{-1} $year^{-1}$ under forest systems and 0–239 kg N ha^{-1} $year^{-1}$ under agricultural systems (Barton *et al.*, 1999). The majority of these data have been collected from the northern hemisphere. In general, the application of inorganic N fertilizer causes a larger increase in annual denitrification rates compared with organic-N additions, due to greater NO_3^- availability. However, the denitrification process may result in the emission of N_2O, a global warming gas (some 200 times more reactive than CO_2), thus swapping one pollutant for another. At best, these are only short-term solutions and are better replaced by actions that reduce all emissions at source.

Agricultural Management Practices as a Source of Pollution

Nitrogen as a pollutant

More reactive N comes from anthropogenic (human-driven) sources than is produced by natural processes. Much of this is associated with agriculture: worldwide 80 Tg N $year^{-1}$ is fixed in fertilizer production, 40 Tg N $year^{-1}$ fixed by the cultivation of legumes and other crops, and 20 Tg N $year^{-1}$ is emitted from energy production; a total of 140 Tg N $year^{-1}$. This compares with 90–130 Tg N $year^{-1}$ by natural processes (lightning and biological fixation). In addition, N is being mobilized from long-term storage pools through the burning of biomass, land clearance and conversion, and the drainage of wetlands. Reactive N is readily transported in solution, or via the atmosphere and spreads regionally and globally and accumulates in the atmosphere, soils, groundwater, land vegetation, oceans and marine sediments (Galloway *et al.*, 1995; Mansfield *et al.*, 1998).

Nitrogen pollution from agricultural sources is now considered to be a major problem in many regions of the world. This has arisen from agricultural management practices that have promoted high N fertilizer application rates to ensure that maximum yields are achieved. There are also increasing levels of soil organic matter in some farming systems. Since the pattern and rate of soil organic matter breakdown may be out of sequence with plant N demand, losses may increase when these sources are not considered. Within regions of Europe and North America there is the additional problem of atmospheric N deposition that has caused increased annual loadings above plant N requirements (Goulding *et al.*, 1998).

High application rates of nitrogen fertilizer

The application of N fertilizer to arable and horticultural crops has, until recently, been very cost-effective; that is, the extra value of the crops produced has far outweighed the cost of applying it. This has tended to encourage farmers to apply plenty of N to ensure they get the most economic (usually the maximum) yield and this creates a surplus of inputs over outputs in the crop. Figure 1.2 shows response of crop yield against N fertilizer applied, and also the quantity of N leached for each amount of N applied. Applying more N than is needed for the optimum yield greatly increases losses by leaching. Research has shown that surpluses of N on well-managed arable land can be as little as 20 kg N ha^{-1} $year^{-1}$, but even where no fertilizer N is applied there is always some leaching loss: 5–10 kg ha^{-1} $year^{-1}$ under winter wheat in the example shown in Fig. 1.2.

However, the economics in favour of excessive commercial fertilizer applications are changing. At the time of writing, 1 kg of N costs 25p. Thus, the above optimal application rate of 192 kg N ha^{-1} applied to the cereal crop (shown in Fig. 1.2) costs under £50. With average yields at 8 t ha^{-1} and at cereal prices of > £100 t^{-1}, applying extra fertilizer N to avoid yield loss made good economic sense. With cereal prices now nearer world averages at £60–70 t^{-1}, fertilizer N must be used more carefully, as must all other inputs.

Table 1.2 shows the increases in the use of N fertilizer around the world in the last 30 years. Some of these are very large. The UK's use of N fertilizer has been relatively stable for the last 10 years because of: (i) the increased awareness among farmers of pollution arising from farming; (ii) the European Community limit for NO_3^- in waters; and (iii) a depressed market for

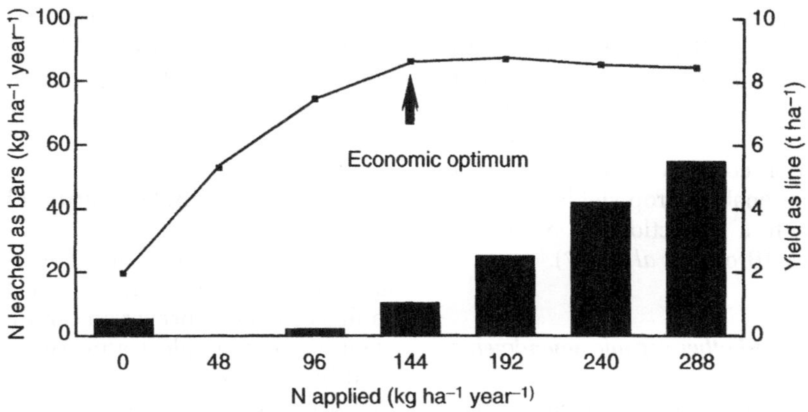

Fig. 1.2. A nitrogen (N) response curve and corresponding leaching losses. Data from the Broadbalk Experiment, Rothamsted, UK (Goulding, 2000).

Table 1.2. Nitrogen fertilizer use by region (Mt) (FAO, 1996).

Year	North America	Western Europe	Eastern Europe	Eurasia	Oceania	Africa	Latin America	Asia	World
1959/60	2.55	3.26	0.6	0.71	0.03	0.22	0.29	1.87	9.54
1989/90	11.2	11.2	4.56	9.92	0.50	2.04	3.80	35.9	79.1
1994/95	11.9	93.8	2.03	2.80	0.72	2.02	3.96	40.4	73.6

agricultural produce. In developing nations, fertilizer costs are often prohibitive and consumption of N fertilizer is, therefore, below optimal for maximal yield. However, fertilizer use in many other countries is still increasing rapidly. For example in India, N fertilizer use has been increasing exponentially; 15 Mt were expected to be applied in 2000 (Pathak, 1999). This is, however, still approximately 10 Mt less than the N exported in crops. Clearly, N application must increase if world food production is to be sustained.

Inefficiency of fertilizer use is essentially due to a lack of synchronization between the release of N from mineralization and uptake by the growing plant. Any surplus remaining at the end of the growing season is then likely to lead to losses in autumn and winter. The large amounts of N applied as fertilizer, or manure, often exceed the immediate and short-term requirements of the crop and so are at risk to leaching, or other forms of loss. The situation is further complicated by the unpredictable supply of N mineralized from the soil. One approach (see Fig. 1.3) is to assess the mineral N available in the soil and then to add only sufficient fertilizer to meet the predicted demands of the crop at the various stages of development. This method, based on a computer simulation model (NFERT), has enabled crop yields to be maintained with a reduction in N input of nearly 30% (Brown *et al.*, 1997).

Manures and other organic amendments

Organic manures are a valuable source of major plant nutrients that can replace inorganic fertilizers in supplying crop nutrients and maintaining soil fertility. Nitrogen consumed by animals usually far exceeds the daily requirement and some 80% of intake is excreted by herbivores and humans alike. The excretal wastes of animals are relatively high in N content (e.g. cattle have a faecal N content of about 4% and a urinary N content of 0.5%). The disposal of such wastes, therefore, offers a potential source of nutrient supply to farm land (if applied in the right amount), or pollution (if applied in excess of requirements). Currently, some 67 Mt of animal manure is collected annually in the UK, of which approximately half is solid based manures and the remainder is liquid slurries (Chambers *et al.*, 1999). In 1996/97, 520 kt of sewage sludge dry solids were applied to 80,000 ha of agricultural land (approximately 47% of the UK sludge production).

Irrigation

Amounts of N applied in irrigation water are negligible, with the exception of 'dirty water' (washings from livestock units). A typical farm in East Anglia, England, would be adding < 0.5 kg N ha^{-1} in every 10 mm irrigation.

Nitrogen Leaving the Farm in Water

Exchanges of N can occur between air, land and water. Losses to air are not discussed in this chapter. However, in certain circumstances, for example warm/wet soils with readily available sources of C and NO_3^-, denitrification losses can exceed leaching losses (Hatch *et al.*, 1998). Experiments in the UK have shown that most N is lost

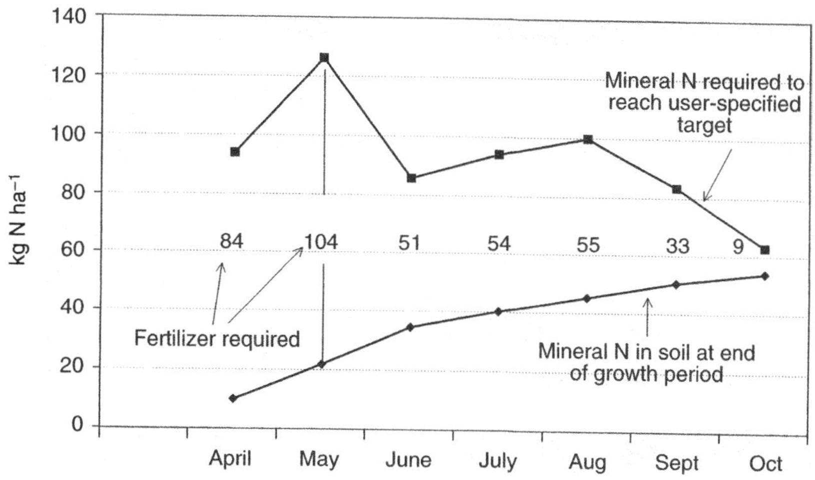

Fig. 1.3. Example of a soil mineral nitrogen (N) profile for a grass crop yield target of 10.3 t dry matter on a moderately drained sandy loam soil (Brown *et al.*, 1997).

from arable farms by leaching during the autumn and winter, but most by denitrification in the spring.

Transfer of nitrogen to water from the farm

There are two general hydraulic pathways by which mobile forms of N can be leached, or transferred into waterways (see Goulding and Webster, 1992). Horizontal flow occurs in soils with poor drainage. This may occur on the soil surface, or above impermeable layers within the soil profile, such as the water movement in duplex soils (sand over clay). Losses of N by surface flow in eroded soils may contribute to NO_3^- leaching, if the organic N present in the soil particles released into watercourses then decomposes during, or after, the erosion process. This may happen many years after the erosion event. Sediments in lakes can release nutrients when disturbed by water movements in the lake, temperature changes and, of course, if the lake is dredged.

Vertical flow through the soil profile can be either via matrix flow through the whole soil body in light textured soils, or through bypass flow in large macropores and cracks in heavy textured soils. Retention times of mobile N within a soil is therefore, dependent on many soil properties: for example, texture, porosity, regolith (the mantle of material overlying bedrock) and slope, or position in the landscape, as well as the amount and frequency of rainfall events. For these reasons, the prediction of N leaching from soils is difficult (see review by Stockdale, 1999), and in heavy textured soils, conventional sampling using ceramic suction cups may give misleading results, since the soil solution extracted may not be representative of the mobile water phase (Hatch *et al.*, 1997). However, Ryden *et al.* (1984), by sampling deep soil profiles, were able to show that the potential for leaching losses from grazed grassland could exceed losses from arable systems. Using field-scale lysimeters, Scholefield *et al.* (1993) obtained leaching data over 7 years, which averaged 39 and 134 kg N ha^{-1} $year^{-1}$ from grazed swards receiving fertilizer applications of 200 and 400 kg N ha^{-1} $year^{-1}$, respectively.

Vulnerable times for transfer of nitrogen to water from the farm

Farming systems (e.g. arable, livestock, mixed) each have key periods in the year and certain management practices during which NO_3^- is most likely to leach. For arable farming, the weather dominates N loss through the impact of rainfall and temperature on drainage, crop growth and N utilization; rainfall in the weeks following N application is especially important.

Besides the weather, losses from agricultural lands are determined by:

- Residual mineral N concentrations present in the soil and the amount of soil-derived N produced through the autumn; the more there is and the less taken up by a crop, the greater the leaching loss.
- Cultivation, especially in ploughed grassland, or with legume crops, opens up the soil making previously inaccessible soil organic matter available to microbial decomposition. This releases large quantities of mineral N which can be used by growing crops (if they are planted quickly and grow well) but are at considerable risk to leaching, depending on the weather.
- Drilling date: the earlier a crop emerges the quicker it covers the ground and the more N is taken up.
- Amount and timing of N fertilizer and manures applied. These must be applied to growing crops when required, otherwise there is a great risk of leaching as a result of poor N fertilizer use efficiency. Fertilizer research has shown that leaching losses are directly related to rainfall in the 3–4 weeks following fertilizer application (Addiscott *et al.*, 1991). The N fertilizer must also be applied correctly. This means having a fertilizer spreader that has been calibrated to apply the correct rate and to apply it evenly across a field. An ineffective spreader can give a very uneven spread and deliver double the amount required in some areas and none in others. Finally, applying N fertilizer in autumn to winter cereal crops is inefficient because of under-developed root systems and low nutrient demand by the crop.
- A bare, fallow soil is more likely to leach N than a cropped soil, because there are no plants to take up any mineral N present. However the maintenance of a 'green cover' on soil increases the risk of pest and disease problems and thus increases the potential need for more herbicides and pesticides. This is one of the many conflicting needs that a farmer has to balance.
- Legume rotations are likely to increase soil organic matter reserves and subsequent net N mineralization during the following crops. However, poor timing of release can result in large losses of NO_3^-.
- Grazing can dramatically increase losses due to the impact of excretal deposits on pasture plants, since the concentrated N input can greatly exceed the uptake requirements within the affected area (Ryden *et al.*, 1984).

Nitrogen leaving the farm

The amounts of N leaving a farm by different routes can be summarized in a budget. As an example, N inputs, outputs and transfers have been measured at Coates Manor Farm at the Royal Agricultural College, England, to identify key parts of the system where losses may occur (Table 1.3). The farm has three enterprises: 160 Friesian Holstein dairy cows, 300 ewes and 120 ha of cereals. The system has a 7-year rotation: first wheat, second wheat, first barley, second barley, or forage crop then spring cereal/forage maize, and either 2 or 3 years of grass ley. In addition, long leys and permanent pastures are grazed by the dairy herd and by the sheep flock that are brought in from other farms. Measurements were made within ten 120 m × 20 m areas in components of the system and a series of budgets for Coates Farm for 1994/95 to 1997/98 are shown in Table 1.3. The dominant input was fertilizer, comprising *c.* 60% of the total, but feed (20%) and atmospheric deposition (9%) were also significant. Atmospheric N deposition near the dairy complex was almost twice that found elsewhere on the farm (183%) because of localized NH_3 volatilization and re-deposition.

Table 1.3. Nitrogen (N) budgets for Coates Farm, Cirencester, UK, 1994–1998. Total for the farm (kg N) (Goulding *et al.*, 1999).

	kg N				% of inputs 1997/98
	1994/95	1995/96	1996/97	1997/98	
Inputs:					
Fertilizer	36,398	33,267	34,530	38,746	61
Deposition[a]	5,002	5,002	5,368	4,636	6
Seed	510	443	443	537	1
Legumes	3,358	3,358	3,358	4,997	8
Feed	12,258	12,233	12,392	15,408	24
Total	57,526	54,303	56,091	64,324	100
Outputs:					
Grain	14,470	17,228	14,069	15,483	24
Milk	5,106	5,365	5,369	5,183	8
Calves, lambs, wool	1,004	988	1,118	1,339	2
Leaching	10,407	10,222	6,742	17,508	27
Volatilization	12,833	13,904	13,647	25,017	39
Denitrification	5,649	5,094	3,192	4,461	7
Total	49,469	52,801	44,137	68,991	107
Unaccounted for	8,057	1,502	11,954	−4,667	−7

[a] Deposition of N from the atmosphere in precipitation and as nitrogen dioxide.

Of the N entering the system, *c.* 38% was exported in saleable product. This is typical of effective, productive management from a mixed farm for this region. Of the remaining N, some 14, 3 and 21% were unaccounted for in the drier years of 1994/95 to 1996/97. By contrast, in the wet year of 1997/98, exports were 7% larger than inputs. This reflects net immobilization of N in dry years and net mineralization in wet years. Leaching losses were much larger in the wet 1997/98 season; Table 1.4 illustrates some of the practices that increased leaching and which did not feature in the drier years. For example, the grazing of stubble turnips by sheep is a very common practice for this region, but also causes much NO_3^- leaching in wet seasons.

Table 1.4. Nitrogen (N) leaching losses (kg N ha^{-1}) from Coates Farm, Cirencester, UK, 1996–1998 (Goulding *et al.*, 1999).

Field name	Crop grown, 1996/97	N leached, 1996/97 (kg ha^{-1})	Crop grown, 1997/98	N leached, 1997/98 (kg ha^{-1})
Field 1	Stubble turnips/forage maize	12	Winter wheat	55
Field 2	Winter wheat	40	Winter barley	26
Jarvis Quarry	Winter barley	6	Stubble turnips/Linola	181
Gooseacre	Stubble turnips/Linola	39	Winter wheat	115
Shepherd's Piece	Winter barley	7	Stubble turnips/ Peas	129
Paddimore	Grass ley	24	Winter wheat	42
Peel's Piece	Lucerne	32	Winter wheat	77
Oathills South	Kale/forage maize	48	Short ley	184
Oathills North	Winter wheat	8	Winter barley	41
Cricket Field	Permanent pasture	58	Permanent pasture	78

Losses of N to the environment at Coates Farm are most likely to be reduced by increasing the efficiency of the livestock part of the system, especially the use of manures and slurries. The sale of half the dairy herd in 1998, together with an increase in arable cropping, increased the N efficiency to 48%.

Nitrogen Pollution Causing Problems

Human health issues

Agricultural soils are a major source of NO_3^- in potable waters (used for drinking, but also for washing and general domestic purposes), which has been considered a hazard for drinking water quality for many years. This led to the introduction of the European Community limit on NO_3^- in potable waters of 50 mg l^{-1} in 1980. The concern over NO_3^- in drinking water was twofold. First, water containing NO_3^- concentrations of > 100 mg l^{-1} was associated with methaemoglobinaemia ('blue baby syndrome'). Second, the ingestion of large amounts of NO_3^- was thought by some to cause stomach cancer (Addiscott *et al.*, 1991; Emsley, 1994).

The 'blue baby syndrome' was thought to be caused when NO_3^- in the stomach is converted by naturally present microorganisms into NO_2^-, which enters the blood and reacts with haemoglobin. Normal oxyhaemoglobin (i.e. oxygenated haemoglobin), which contains Fe as Fe^{2+} becomes methaemoglobin (with Fe as Fe^{3+}) which does not transport O_2, and the O_2 carrying capacity of the blood is decreased. Young children (generally under 1-year old) are susceptible because: (i) fetal haemoglobin (which has a greater affinity for NO_2^-) persists for a while in their bloodstream; and (ii) their stomachs are not acid enough to inhibit the microorganisms that convert NO_3^- to NO_2^- (Addiscott *et al.*, 1991). The infants suffer what can be described as 'chemical suffocation' and turn blue. However, the last death in the UK was in 1950 and the last confirmed case in 1972. Recent French research (L'Hirondel and L'Hirondel, 1996) has instead attributed all cases of methaemoglobinaemia in France to poor kitchen hygiene: bacteria on feeding bottles convert NO_3^- to NO_2^- when infant feeds are warmed before ingestion. Methaemoglobinaemia is also usually associated with well water. Local wells are often near housing and can be polluted with human or animal excreta, suggesting that the excreta and not NO_3^- are the prime cause of illness.

There were good theoretical reasons for thinking that NO_3^- may cause stomach cancer. Nitrite produced in the stomach could react with secondary amines from the digestion of proteins to produce carcinogenic N-nitroso compounds. These compounds are found in urine and also in saliva. It therefore seems unlikely that they should prove to be especially toxic when they occur naturally in the mouth. Also, many epidemiological studies of stomach cancer (Golden and Leifert, 1999) have found no firm links between NO_3^- and cancer among people at higher risk (e.g. fertilizer factory workers) who may ingest large amounts of NO_3^-. In contrast, it is now clear that a NO_2^- supply to the stomach is essential for the formation of nitric oxide (NO) which provides protection against gastroenterinal pathogens: there are 3–5 million deaths worldwide from gastroenterinal diseases each year (Golden and Leifert, 1999). In addition, NO_3^- reduces platelet function and hence the risk of heart attack. Losing NO_3^- in sweat creates NO_2^- and NO on the surface of the skin, preventing skin infections. Thus, researchers are beginning to think that NO_3^-, far from being toxic, is an essential part of the human diet and that the European Community limit of 50 mg l^{-1} is not necessarily based on a firm, medically and scientifically sound basis (Addiscott, 1999). The emphasis concerned with NO_3^- in water is now shifting to the recognized environmental problems, rather than unproven issues of human health.

Environmental pollution

Reactive forms of N, collectively termed NO_x ($NO + NO_2$ + any other single N species with an oxygen atom) and NH_x (NH_3 + aerosol NH_4^+), contribute to acid deposition and soil acidification. Ammonium-based fertilizers, along with legume-based crops, tend to acidify soils: this can lead to the leaching of Al and other toxic metals into waters.

Excess N in ecosystems can cause serious damage; for example, algal blooms in water bodies, changing heathland to grassland and reducing species diversity in grassland generally (Mansfield *et al.*, 1998). As the concentration of N in waters increases and the water becomes nutrient rich (eutrophic), there is a gradual reduction in the number of species of plankton and diatoms, while epiphytes increase. Cyanobacteria and unicellular blue-green algae (both are toxic to animals and humans) increase as food webs in water bodies move up in trophic levels. As phytoplankton and macrophyte debris accumulate, so O_2 concentrations decrease, leading ultimately to the death of fish (Heathwaite, 1993).

Nitrogen Pollution in Agroecosystems of the World

Fertilizers are a very large part of the total reactive N, especially in less developed countries (Table 1.5). The excessive use of fertilizer, above the amount that can be used by growing plants, has created a large N surplus in some European countries (Table 1.6). Generally, the more farm animals per hectare of land, the greater the surplus. The efficiency of conversion of N inputs into saleable product for arable crops can be 60–70%, or even more; for livestock-only systems a 20% efficiency is to be expected. These are average surpluses for each country, but some farms may have N surpluses of > 1000 kg ha^{-1} $year^{-1}$.

For the Broadbalk Experiment at Rothamsted, SE England, amounts of N leached are directly proportional to the magnitude of N surplus (Fig. 1.4). The relationship may not be the same, or as close on other farms, systems or countries, but there is a clear link between N surpluses and losses to the environment. Nitrogen pollution is widespread in countries with a more developed, intensive agriculture.

Table 1.5. Conversion of unreactive nitrogen (N) to reactive N by energy and fertilizer production as Tg N $year^{-1}$ (Galloway *et al.*, 1995).

	Energy		Fertilizer	
	Present	2020	Present	2020
USA & Canada	7.6	10.1	13.3	14.2
Europe	4.9	5.2	15.4	15.4
Australia	0.3	0.4	—	—
Japan	0.8	0.8	—	—
Asia	3.5	13.2	35.9	85.1
Mexico, Latin & S. America	1.5	5.9	1.8	4.5
Africa	0.7	4.2	2.1	5.2
Former Soviet Union	2.2	5.7	10.0	10.0
Total	21.2	45.6	78.5	134.4

Table 1.6. Nitrogen (N) surpluses in some European Union (UN) countries (kg N ha^{-1} $year^{-1}$) in 1990/91 (Brouwer and Hellegers, 1997).

Netherlands	Belgium	Germany	France	UK	Portugal
321	170	121	73	59	6

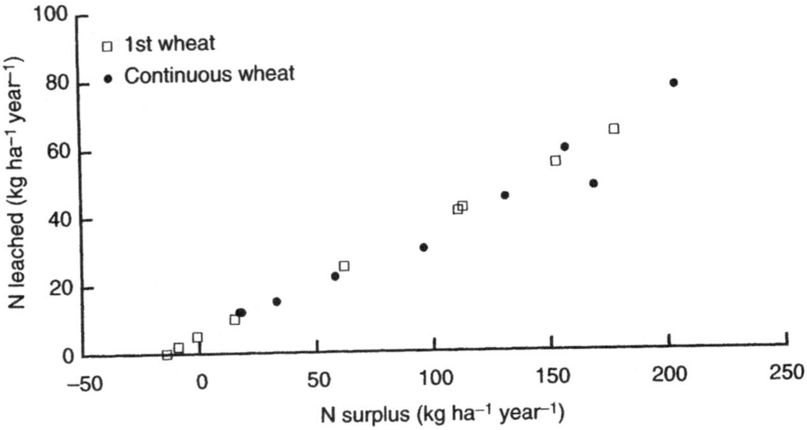

Fig. 1.4. Amounts of nitrogen (N) leached in relation to N surplus. Data from the Broadbalk Experiment, Rothamsted, UK.

Management Options to Minimize Nitrogen Pollution to Waterways

As noted above, the weather dominates N loss through the impact of rainfall and temperature on drainage, crop growth and N utilization. Maximizing the efficiency of N use for agricultural crops involves a series of difficult tasks for the farmer. The farmer has to determine the amount of residual mineral N present in the soil, estimate soil N supply through the autumn, assess the appropriate cultivation method, determine the drilling date and the amount and timing of N fertilizer applied. The amount of organic manures or slurries to be applied must be calculated; then the balance needed as fertilizer is obtained. Finally, the fertilizers and manures must be applied correctly and evenly, when and where the crop needs them. Accurate prediction of the N requirement is especially important, as is the timing of the application. Applying the required amount in several doses ('split applications') is usually most effective at getting N into the crop and minimizing leaching loss; this is especially the situation in poorly buffered soils that leach easily. To aid farmers with these calculations, N fertilizer recommendation models have been developed for agricultural, horticultural and grassland systems from the results of over 100 years of N research (Dampney *et al.*, 2000).

Applying N fertilizer is both agronomically and economically beneficial, but it does cause environmental problems. To counter this, strategies for reducing inputs are being developed. These are considered in detail below, but certain farm systems intrinsically use less N, notably integrated and organic systems.

Integrated farming systems

The aim of integrated farm systems is to maintain farm income by balancing reduced outputs with a reduction in inputs and so limit the overall environmental impact. LIFE (Less Intensive Farming and the Environment) is one such system that aims to be environmentally benign. The system utilizes a 7-phase rotation to minimize pest, disease and weed problems and to improve soil structure and fertility. Minimum cultivation techniques are used to counteract N losses from late sowing. For example, immediately after harvest, crop residues are incorporated into the topsoil to encourage weed and 'volunteer' cereal growth to mop up excess soil N. The herbage is killed with glyphosate, prior to drilling with one-pass, non-inversion tillage. This creates an adequate seedbed while minimizing autumn mineralization, reducing erosion and energy input and encouraging soil biota; it can however, exacerbate weed and slug problems. Late sowing, correct crop rotation and natural predators reduce the need for pesticides, since limited chemical application (at minimum dose rates) is the target. Nitrogen is applied to attain a sub-optimal yield; other fertilizers are applied to balance removals.

Integrated farming systems like this can reduce N losses considerably. A LIFE experiment on a deep loam soil at Long Ashton Research Station, England, showed that NO_3^- loadings from non-inversion tillage, were 80% less than those from conventionally tilled land (Fig. 1.5). However, a similar experiment on a shallow (30 cm deep or less) stony clay loam soil over limestone at the Royal Agricultural College, England, had N losses which were identical to those from the same crops grown conventionally. Thus, for each farm, a complex assessment must be tailored to optimize N use and all other inputs and so minimize the total environmental impact (Goulding *et al.*, 1999).

Organic farming

Organic farming systems seek to minimize all inputs (including nutrients) by maximizing recycling. Water-soluble fertilizers containing NH_4^+ and NO_3^- salts, are not permitted, so a major N input present in conventional farming is thereby removed. This would be expected to reduce losses. The Organic Farming Study (at the Duchy of Cornwall's Home Farm, Gloucestershire, England) determined the N budgets, economics and environmental benefits of this organically based farming system. On this, as on most organic farms with livestock, inputs of N from bought-in feed partly compensate for the absence of inorganic fertilizers. There was also good evidence of a net release of N from soil organic matter decomposition.

On this organic farming study, leaching was estimated across the whole rotation and also for a comparable conventional system using a whole farm, nutrient budget model. On average, NO_3^- leaching from the organic system was 66% of that in the conventional system (Fig. 1.6). Calculations from the whole farm N budget showed that 55% of the N entering the organic system was exported in saleable products. However, measurements of internal N flows showed that when the supply of N

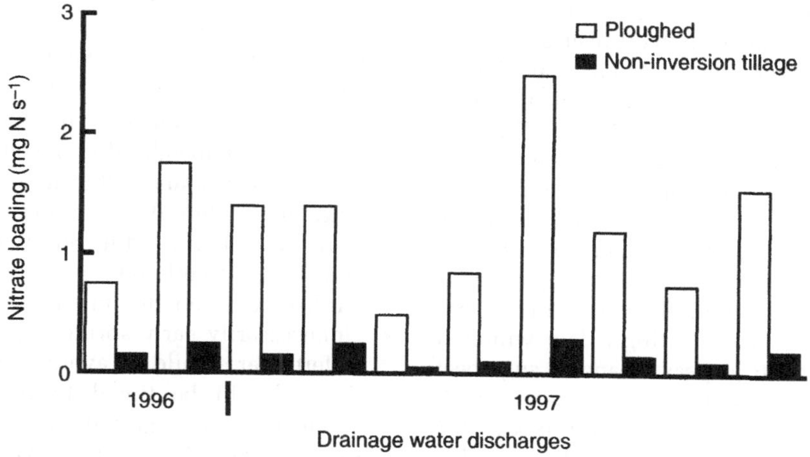

Fig. 1.5. Nitrate loadings in drainwater discharges from the LIFE experiment at Long Ashton, Bristol, UK (Goulding *et al.*, 1999).

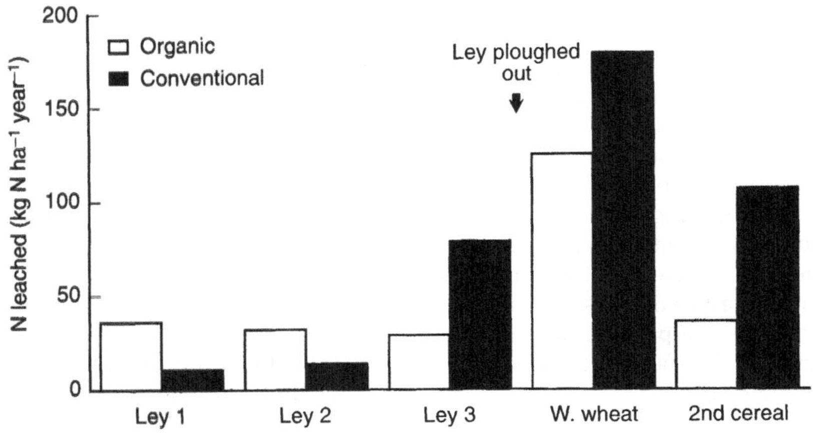

Fig. 1.6. Leaching from fields in organic and conventional 5-year rotations, estimated using a whole farm nutrient budget model and calibrated from parts of the organic system (Goulding *et al.*, 1999).

from soil organic matter decomposition was included, the efficiency of N use was only 34%, about the same as that on a conventionally managed farm in this region (Goulding *et al.*, 1999).

Key farm management practices controlling leaching

Several farm management practices are key to controlling losses. A major contribution to NO_3^- leaching in the UK comes from ploughing out permanent grassland, leys (Fig. 1.6) and set-aside. Set-aside and leys can release 200–300 kg N ha^{-1} over 1–2 years, reaching 4 t N ha^{-1} over 20 years in permanent grassland. Model calculations suggest that ploughing of grassland during and after the Second World War contributed > 50 mg l^{-1} NO_3^- to the water supplies of much of central England in the 1950s, 1960s and 1970s (Addiscott *et al.*, 1991). Experiments have shown that planting a cover crop (e.g. arable crops such as rye, white mustard, phacelia) between harvest and before planting a late winter or spring crop (e.g. peas, beans or sugarbeet) is the most effective way of retaining N. However, when the cover crop is killed, N is released back into the soil from plant residues at a rate that depends on climate and management. This N can be used effectively by the following crop if conditions are conducive, but if not, may be leached in subsequent seasons (Goulding, 2000).

The best management practices to minimize leaching from land growing arable and horticultural crops can be defined as follows:

- Farmers should choose the highest-yielding variety appropriate to maximize the use of the available N (bearing in mind quality, e.g. for milling).
- A green cover should be maintained as much as is practicable with a cover crop if necessary and by drilling autumn-sown crops early. However, this must be balanced against effective weed, pest and disease control.
- Fertilizer requirements should be calculated using a recommendation system, allowing for soil mineral N residues and N that will be mineralized from soil organic matter, crop residues and manures.
- Autumn N should be applied only to those crops that need it, certainly not to cereals, and to oilseed rape only if straw has been incorporated previously.
- Unnecessarily early spring applications of fertilizer should be avoided; applications should be timed to provide N when the crop is growing quickly. The use of a starter fertilizer and banding for vegetables can greatly increase the efficiency with which the N is used.

- Leaching losses may be reduced by splitting spring fertilizer applications, but yield benefits should not be expected.
- Fertilizers and manures should be applied evenly, and well away from watercourses, using a properly calibrated spreader.
- Appropriate controls to minimize pest, disease and weed infestation are essential because a diseased crop is less able to use N.
- If irrigation is required, this should be done carefully; that is, only to support crop yield, and using a scheduling system that takes account of the weather.

These strategies have been proved to work. For example, N fertilizer applications to *Brassica* rotations can be reduced by 50% without loss of yield, if residual N is taken into account. Using starter, or banded fertilizers on vegetable crops can reduce leaching losses by up to 75% (Rahn *et al.*, 1993).

Legislative Policy: Reducing Nitrogen Pollution to Waterways

There have been some suggestions (Romstad *et al.*, 1997) that a tax should be placed on fertilizer N to reduce its use and obtain funds for environmental protection. However, money raised in taxes is often not targeted at specific problems, and calculations suggest that the tax would need to be > 100% of price to influence fertilizer use significantly.

Since 1998, the application of untreated sludge to grassland in the UK has been prohibited, and from the end of 1999 this regulation was extended to include land growing combinable and animal feed crops. With the banning of sea disposal for sewage, the amounts applied to land are set to rise. Unless careful account of the N fertilizer value of these organic wastes is taken, the losses from land receiving organic amendments will also increase considerably.

Table 1.7 lists N in potable waters for some countries. Fresh waters that are not intended to be potable are classified by the OECD according to nutrient and chlorophyll *a* contents and the degree of light penetration. These range from oligotrophic (661 μg N l^{-1}) to mesotrophic (753 μg N l^{-1}) up to eutrophic (1875 μg N l^{-1}).

To counter large surpluses, The Netherlands has introduced a compulsory nutrient accounting policy. This requires obligatory nutrient budgets to be made on all farms with > 2.5 livestock units ha^{-1} and specifies surplus allowances (Table 1.8). If these values are exceeded, farmers are taxed at 50p kg^{-1} for each kg N above the limit. However, it should be noted that farmers do not have to include atmospheric N deposition, or N_2 fixation by legumes in their calculations of inputs, and some NH_3 losses are allowable. Despite these relatively generous regulations, Dutch farmers are not happy with the arrangements and will have great difficulty meeting the requirements because of the focus on livestock production systems in The Netherlands.

Table 1.7. Guide and maximum admissible concentrations of various forms of nitrogen (N) in potable waters (ECE, 1993).

Country/organization	Form of nitrogen	Guide (mg l^{-1})	Maximum admissible concentration (mg l^{-1})
European Community	Nitrate	25.0	50.0
	Nitrite	—	0.1
	Ammonium	0.05	0.5
USA	Nitrate	—	45.0[a]
World Health Organization	Nitrate	—	45.0[a]

[a] Limit defined as 10 mg N l^{-1}; recalculated to mg NO_3^- l^{-1} = 44.3, rounded up to 45.

Table 1.8. Allowed nitrogen (N) surpluses in The Netherlands using the MINAS Nutrient Accounting Scheme. (kg N ha^{-1} $year^{-1}$). Figures in parentheses are not yet agreed (Oenema *et al.*, 1997).

Year	1998	1999	2000	2002	2005	2008
Arable	175	175	150	(125)	(110)	(100)
Grassland	300	300	275	(250)	(200)	(180)

Within the UK, the main practical ideas for reducing NO_3^- leaching have been tested voluntarily in Nitrate Sensitive Areas (NSAs) and, from December 1998, enforced in 68 Nitrate Vulnerable Zones (NVZs) covering 600,000 ha. Farmers in these areas must adhere to the Codes of Good Agricultural Practice that have been prepared by a government regulatory body (Davies, 2000; Dampney *et al.*, 2000). Limits for N application rates and on the timing of manure applications have been defined and are monitored for these zones. Results from measurements and modelling studies in NSAs show a significant reduction in N usage (Table 1.9) and losses (Table 1.10). Generally, however, large reductions can only be achieved by a change from intensive arable and horticultural cropping to extensive, unfertilized grass. Also, in the driest parts of the UK (East Anglia) where through drainage is less than 150 mm $year^{-1}$, leaching of no more than 15 kg N ha^{-1} $year^{-1}$ will breach the European Community limit in water draining from land. This is almost certainly impossible to achieve.

Table 1.9. Nitrogen fertilizer used (kg N ha^{-1}) on crops grown in the Nitrate Sensitive Areas (NSAs) scheme and non-scheme areas (Dampney *et al.*, 2000).

Crop	Non-scheme areas	Scheme areas
Winter wheat	194	143
Winter barley	127	120
Spring barley	98	97
Potatoes	218	166
Sugar beet	120	123
Oilseed rape	173	139

Table 1.10. Modelled changes in mean nitrate concentrations (mg NO_3^- l^{-1}) in drainage waters within the Pilot Nitrate Sensitive Areas (NSAs) scheme (Dampney *et al.*, 2000).

Area	Concentration before scheme introduced	Concentration following introduction of scheme
Broughton	95	70
Branston	106	81
Sleaford	82	52
Wellings	61	39
Wildmoor	77	59

Future legislation

In addition to the legislation on NO_3^-, other controls are likely to be imposed in the near future:

1. The UK is committed to the aims of the Second North Sea Conference 1987 and the Oslo/Paris Commission (OSPARCOM), both of which require action to reduce nutrient inputs to the North Sea in the event that they are deemed to be causing a problem.
2. The EU Directive (96/61/EU) on Integrated Pollution Prevention and Control (IPPC), implemented in October 1999, aims to reduce a range of emissions to air, soils and water from industry (including larger pig and poultry units) by requiring the owners to have a permit to carry out their operations. For agriculture, the list of relevant pollutants includes NH_3, so that implementation of IPPC in the UK could be expensive.
3. The UNECE Convention on 'Long-Range Transboundary Air Pollution' requires reductions in NH_3 emissions to abate acidification and eutrophication. The main measure for implementing this is the proposed EU National Emissions Ceilings Directive, which is currently being negotiated and which will also set legally binding emission levels for member states.
4. Under the Kyoto Protocol, the UK is legally bound to reduce its greenhouse gas emissions by 12.5% over the period from 2008 to 2012. The main gases of concern for agriculture include N_2O.
5. The EU Water Framework Directive sets a new standard for 'good ecological status' of receiving waters and will have a far reaching and as yet unmeasureable effect on N pollution.

Conclusion

An adequate N supply is essential to all productive agriculture, including organically farmed systems, but an excess of supply over demand is causing environmental pollution. Legislation to reduce NO_2^- leaching is likely to be followed by further statutory limits on NH_3 and N_2O emissions. The sustainability of profitable farming with minimal environmental impact is, therefore, a considerable challenge to the agricultural industry.

Acknowledgements

IGER and IACR are is supported by the Biotechnology and Biological Research Council (BBSRC), Swindon, UK.

References

Addiscott, T.M. (1999) Nitrate and health. In: Wilson, W.S., Ball, A.S. and Hinton, R.H. (eds) *Managing Risks of Nitrates to Humans and the Environment.* Royal Society of Chemistry, Cambridge, pp. 247–249.

Addiscott, T.M., Whitmore, A.P. and Powlson, D.S. (1991) *Farming, Fertilizers and the Nitrate Problem.* CAB International, Wallingford, UK.

Barton, L., McLay, C.D.A., Schipper, L.A. and Smith, C.T. (1999) Annual denitrification rates in agricultural and forest soils: a review. *Australian Journal of Soil Research* 37, 1073–1093.

Brouwer, F. and Hellegers, P. (1997) Nitrogen flows at farm level across European Union agriculture. In: Romstad, E., Simonsen, J. and Vatn, A. (eds) *Controlling Mineral Emissions in European Agriculture. Economics, Policies and the Environment.* CAB International, Wallingford, UK, pp. 11–26.

Brown, L., Scholefield, D., Jewkes, E.C. and Preedy, N. (1997) Integrated modelling and soil testing for improved fertilizer recommendations for grassland. In: van Cleemput, O., Haneklaus, S., Hofman, G., Schnug, E. and Vermoesen, A. (eds) *Fertilization for Sustainable Plant Production and Soil Fertility. 11th International World Fertilizer Congress, Gent, Belgium. Proceedings* Vol. III. CIEC editorial board in co-operation with FAL, Braunschweig-Volkenrode, Germany, pp. 248–255.

Burt, T.P., Heathwaite, A.L. and Trudgill, S.T. (eds) (1993) *Nitrate. Processes, Patterns and Management.* John Wiley & Sons, Chichester, UK.
Chambers, B., Smith, K. and Pain, B. (1999) Strategies to encourage better use of nitrogen in animal manures. In: *Tackling Nitrate from Agriculture. Strategy from Science.* MAFF Publications, London.
Chapin III, F.S., Moilanen, L. and Kielland, K. (1993) Preferential use of organic nitrogen for growth by a non-mycorrhizal arctic sedge. *Nature* 316, 150–153.
Dampney, P.M.R., Lord, E.I. and Chambers, B.J. (2000) Development of improved advice for farmers and advisers. *Soil Use and Management* 16, 162–166.
Davies, B. (ed.) (2000) Tackling nitrate. Strategies from science. *Soil Use and Management* 16, 141–174.
ECE (Economic Commission for Europe) (1993) *Protection of Water Resources and Aquatic Ecosystems* (Water Series No. 1) ECE/ENVWA/31. United Nations, New York.
Emsley, J. (1994) *The Consumer's Good Chemical Guide.* W.H. Freaman Spektrum, Oxford, pp. 204–233.
European Economic Community (1980) EEC 80/77, Brussels. Council directive of 15 July 1980 relating to the quality of water intended for human consumption (80/778/EEC). *Official Journal of the European Communities* L 229, 11–29.
FAO (Food and Agriculture Organisation) (1996) Production data diskette. FAO, Rome.
Galloway, J.N., Schlesinger, W.H., Levy, Hiram II, Michaels, A. and Schnoor, J.L. (1995) Nitrogen fixation: anthropogenic enhancement–environmental response. *Global Biogeochemical Cycles* 9, 235–252.
Golden, M. and Leifert, C. (1999) Potential risks and benefits of dietary nitrate. In: Wilson, W.S., Ball, A.S. and Hinton, R.H. (eds) *Managing Risks of Nitrates to Humans and the Environment.* Royal Society of Chemistry, Cambridge, pp. 269–280.
Goulding, K.W.T. (2000) Nitrate leaching from arable and horticultural land. *Soil Use and Management* 16, 145–151.
Goulding, K.W.T. and Webster, C.P. (1992) Methods for measuring nitrate leaching. *Aspects of Applied Biology* 30, 63–69.
Goulding, K.W.T., Bailey, N.J. and Bradbury, N.J. (1998) Model study of the fate of nitrogen deposition on arable land. *Soil Use and Management* 14, 70–77.
Goulding, K.W.T., Jordan, V., Leake, A., Warman, K., Donaghy, D., Reynolds, F., Stockdale, E., Crew, A., Webster, C.P., Conway, J. and Jarvis, S.C. (1999) Interactions between agricultural emissions to the environment: the value of system studies in minimizing all emissions. In: Turner, S.T.D. and Alford, D. (eds) *Agriculture and the Environment. Challenges and Conflicts for the New Millennium*, Proceedings of an International Conference, 14–16 April 1999, University of Warwick. ADAS, Wolverhampton, pp. 55–63.
Harrison, R.M. (ed.) (1999) *Understanding Our Environment. An Introduction to Environmental Chemistry and Pollution*, 3rd edn. Royal Society of Chemistry, Cambridge.
Hatch, D.J., Jarvis, S.C., Rook, A.J. and Bristow, A.W. (1997) Ionic contents of leachate from grassland soils: a comparison between ceramic suction cup samplers and drainage. *Soil Use and Management* 13, 68–74.
Hatch, D.J., Jarvis, S.C. and Parkinson, R.J. (1998) Concurrent measurements of net mineralization, nitrification, denitrification and leaching from field incubated soil cores. *Biology and Fertility of Soils* 26, 323–330.
Heathwaite, A.L. (1993) Nitrogen cycling in surface waters and lakes. In: Burt, T.P., Heathwaite, A.L. and Trudgill, S.T. (eds) *Nitrate. Process Patterns and Management.* John Wiley & Sons, Chichester, UK, pp. 99–140.
Jarvis, S.C. (1999) Nitrogen dynamcs in natural and agricultural ecosystems. In: Wilson, W.S., Ball, A.S. and Hinton, R.H. (eds) *Managing Risks of Nitrates to Humans and the Environment.* Royal Society of Chemistry, Cambridge, pp. 2–20.
Jarvis, S.C., Stockdale, E.A., Shepherd, M.A. and Powlson, D.S. (1996) Nitrogen mineralization in temperate agricultural soils: processes and measurements. *Advances in Agronomy* 57, 187–235.
Langan, S.J. (ed.) (1999) *The Impact of Nitrogen Deposition on Natural and Semi-natural Ecosystems*, Environmental Pollution, Vol. 3. Kluwer, Dordrecht.
L'Hirondel, J. et L'Hirondel, J.-L. (1996) *Les Nitrates et L'homme. Le Mythe de leur Toxicité.* Les editions de l'Institute de l'Environment, Liffré, France (English translation available).
MAFF (1999) *Tackling Nitrate. Strategies from Science*, PB4401. MAFF Publications, London.

MAFF/WOAD (1998) *Code of Good Agricultural Practice for the Protection of Water*. PB0585. MAFF Publications, London.
Mansfield, T.M., Goulding, K.W.T. and Sheppard, L. (eds) (1998) Disturbance of the nitrogen cycle. *New Phytologist* 139, 1–234.
Murphy, D.V., Macdonald, A.J., Stockdale, E.A., Goulding, K.W.T., Fortune, S., Gaunt, J.L., Poulton, P.R., Wakefield, J.A., Webster, C.P. and Wilmer, W.S. (2000) Soluble organic nitrogen in agricultural soils. *Biology and Fertility of Soils* 30, 374–387.
Oenema, O., Boers, P.C.M., van Eerdt, M.M., Fraters, B., van der Meer, H.G., Roest, C.W.J., Schroder, J.J. and Willems, W.J. (1997) The nitrate problem and nitrate policy in The Netherlands. Nota 88, ab-dlo/cbs/rivm/sc-dlo, Haren, The Netherlands.
Pathak, H. (1999) Fertilizer and nitrate pollution in India. In: Wilson, W.S., Ball, A.S. and Hinton, R.H. (eds) *Managing Risks of Nitrates to Humans and the Environment*. Royal Society of Chemistry, Cambridge, pp. 228–239.
Rahn, C.R., Paterson, C.D. and Vaidyanathan, L.V.V. (1993) Improving the use of nitrogen in brassica rotations. *Acta Horticulturae* 339, 207–218.
Romstad, E., Vatn, A., Bakken, L.R. and Botterweg, P. (1997) Economics–ecology modelling – the case of nitrogen. In: Romstad, E., Simonsen, J. and Vatn, A. (eds) *Controlling Mineral Emissions in European Agriculture. Economics, Policies and the Environment*. CAB International, Wallingford, UK, pp. 225–248.
Ryden, J.C., Ball, P.R. and Garwood, E.A. (1984) Nitrate leaching from grassland. *Nature* 311, 50–53.
Scholefield, D., Tyson, K.C., Garwood, E.A., Armstrong, A.C., Hawkins, J. and Stone, A.C. (1993) Nitrate leaching from grazed grassland lysimeters: effects of fertilizer input, field drainage, age of sward and patterns of weather. *Journal of Soil Science* 44, 601–613.
Smith, R.V., Foy, R.H., Lennox, S.D., Jordon, C., Burns, L.C., Cooper, J.E. and Stevens, R.J. (1995) Occurrence of nitrite in the Lough Neagh River System. *Journal of Environmental Quality* 24, 952–959.
Stockdale, E.A. (1999) Predicting nitrate losses from agricultural systems: measurements and models. In: Wilson, W.S., Ball, A.S. and Hinton, R.H. (eds) *Managing Risks of Nitrates to Humans and the Environment*. Royal Society of Chemistry, Cambridge, pp. 21–41.
Wilson, W.S., Ball, A.S. and Hinton, R.H. (eds) (1999) *Managing Risks of Nitrates to Humans and the Environment*. Royal Society of Chemistry, Cambridge.

2 Phosphorus

Peter Leinweber,[1] Benjamin L. Turner[2] and Ralph Meissner[3]
[1]*Institute of Soil Science, University of Rostock, Justus-von-Liebig-Weg 6, D-18059 Rostock, Germany;* [2]*USDA – ARS, Northwest Irrigation and Soils Research Laboratory, 3793 N. 3600 E. Kimberly, ID 83341, USA;* [3]*UFZ Centre for Environmental Research, Leipzig-Halle, Lysimeter Station, D-39615 Falkenberg, Germany*

Key words: detachment, fertilizer, inorganic phosphorus, leaching, manure, organic phosphorus, phosphorus runoff, soil solution, solubilization

Introduction

Phosphorus (P) is one of the most important mineral nutrients for biological systems, yet it is also one of the most scarce nutrients in terms of its demand in both terrestrial and aquatic environments (Moss, 1988). In natural systems, P is tightly cycled through the plant–soil continuum, but in agricultural systems soil P is removed in the crop or animal products and must be replaced if P deficiency is to be avoided. Therefore, mineral PO_4^{3-} fertilizers and animal manures are applied to agricultural land to raise soil

P levels and maintain crop yields (Sibbesen and Sharpley, 1997).

Phosphorus fertilizer has traditionally been applied to agricultural land with no concern over losses to water, because inorganic P is strongly fixed in the soil and prevented from leaching (Sample *et al.*, 1980). However, while this is true from an agronomic standpoint, the small amounts of P lost from the soil can have severe impacts on water quality in receiving catchments, especially in terms of eutrophication and the growth of undesirable algae (Moss, 1996). In aquatic ecosystems, P can contribute to eutrophication, toxic algal blooms and a general deterioration of water quality. Known far-reaching consequences of these aquatic processes include the disappearance of sensitive ichthyofauna, the increased cost of drinking water purification and loss of income from recreational tourism. In particular, accumulations of P can occur under heavy manure or fertilizer application (Breeuwsma *et al.*, 1995), which can eventually reduce the capacity of the soil to retain P and accelerate the loss of P in runoff to watercourses. Examples of areas with severe eutrophication problems that are directly attributable to agricultural P include the Chesapeake bay area of north-eastern USA and the Gippsland Lakes area in south-eastern Australia (Burkholder *et al.*, 1992; EPA, 1995).

In this chapter we discuss the role of agricultural management of P in contributing to water quality problems. The main areas of consideration are the imbalances in P use at the farm scale, the dynamics of soil P and the transfer of P from soils to watercourses. Emphasis is placed on understanding the mechanisms and pathways of losses from soil to water and forms of P in surface waters. Finally, options for improved management of agricultural P for the prevention of surface water pollution are considered.

Phosphorus Use on Farms

Agricultural soils receive P from livestock (direct excreta and manures), mineral fertilizers and municipal sludges and wastewaters. Imbalances originate because P is applied at levels that are in excess of the amounts required for optimum crop yields. This is often exacerbated where areas requiring P fertilizer (e.g. crop producing areas) are spatially separated from areas with large P surpluses (e.g. areas of intensive livestock production). The following sections examine the basic sources and pathways of P in agricultural systems that can lead to the over-application and consequent acceleration of transfer to watercourses.

Livestock feeding and manure production

Dietary P supplements are often used to maintain the optimum P requirements of livestock, because P deficiency can exert a number of negative effects on animal growth, productivity and fertility. The optimal requirement of digestible P in animal feeds ranges from 0.12–0.26% dry matter for pigs to 0.4–0.8% dry matter for poultry; ruminants have an intermediate requirement of 0.2–0.6% dry matter (Jongbloed and Everts, 1992). The digestibility of plant P ranges between 15 and 35% depending on the type of plant, while the availability of P in mixed feed for ruminants is *c.* 45–50% (Lynch and Caffrey, 1997). Balancing plant P and animal requirements for digestible P can be achieved by the supplementation of diets, as Ca-, NH_4- or Na-phosphate or phosphoric acid. Alternatively, deficiency can be avoided by the use of enzyme supplements to animal diets to increase the digestibility. Between 60 and 90% of the P in animal feeds occurs as phytic acid (inositol hexabisphosphate), an organic P compound found widely in the environment (Turner *et al.*, 2002). Phytic acid is poorly available to monogastrics (pigs, poultry), which have low levels of intestinal phytase enzymes. Phytase catalyses the release of PO_4^{3-} from phytic acid. Therefore, the uptake of P from phytic acid can be improved by the use of phytase as a dietary supplement. Phytase supplements are also used for ruminants, although these animals can use phytic acid to a larger extent because of a greater level of intestinal phytase. Phosphorus uptake efficiency can be greatly improved through the use of phytase supplements, reducing P contents in excrement by up to 25–35% (Oloffs, 1998). This has environmental as well as financial benefits, because increased efficiency in livestock production reduces the P returns to land. It should be noted, however, that P in manures from animals fed with phytase supplements may be present in more soluble forms and, hence, may actually increase the likelihood of manure-derived P transfer to watercourses.

Mineral fertilizers

The addition of mineral P fertilizer to crops is commonly based on soil tests and the resulting fertilizer recommendations. More than ten different soil tests are used in European countries (Tunney *et al.*, 1997). These include various lactate methods such as double lactate (DL), calcium-ammonium lactate (CAL) and ammonium acetate (AL) (Austria, Belgium, Germany, Hungary, The Netherlands, Norway, Poland, Romania, Sweden), the Olsen method (sodium bicarbonate ($NaHCO_3$): Denmark, Greece, Italy, Spain, UK) and the water-extractable P method (Austria, Netherlands, Switzerland). A more recent development is the extraction of soil with 0.01 M calcium chloride ($CaCl_2$), which is used for multi-element determination and evaluation for various nutrients and contaminants. The critical soil value to obtain the desired high crop yield varies widely, even if the same soil P test is used (Table 2.1). This results from differences in soil and climatic conditions, P requirements and uptake by crops, and also from differences in the experimental basis used to derive these critical levels. Therefore, a re-evaluation of experimental data to match P requirements of crops and minimize the risk of P losses from soils is urgently needed.

Clearly, the differences in soil P test methods and their evaluations result in widely different P fertilizer recommendations in various regions (Sibbesen and Sharpley, 1997; Tunney *et al.*, 1997). Mineral P applications are typically in the range of 10–20 kg P ha^{-1} $year^{-1}$. Common types of P fertilizer are superphosphate (8–10% P), triple superphosphate (20% P), ammonium phosphates (21–24% P) and rock phosphates (apatites, 13% P) (Mengel and Kirkby, 1987). Examples of current mean fertilizer P applications in European countries are given in Table 2.2. In calculating fertilizer application rates, it is necessary to consider the manure-P inputs so that the farmer can reduce mineral fertilizer application, thereby saving money, avoiding undesirable soil P enrichments and reducing the potential impacts of P transfer to aquatic ecosystems. However, this is not always the case, especially when manure applications are calculated on the basis of N, rather than P, content, which can result in disproportionate quantities of P being applied to the land.

Table 2.1. Examples of soil-test phosphorus (P) levels for 95% relative yield or no yield response (data from Sibbesen and Sharpley, 1997).

Soil P test method	Critical level (mg P kg^{-1})	Crops	Number and duration of trials	Region, soils
H_2O	5	Rotation	105 trials, 1 year	Germany, loess
H_2O	14	Potatoes	37 trials, 1 year	The Netherlands
DL	26	Rotation	150 trials, 1–3 years	Germany, various
DL	50–60	Rotation	1 trial, 42 years	Germany, Pheozem
Olsen	5	Maize	25 trials, 1 year	Iowa, various
Olsen	21–35	Rotation	7 trials, 20 years	Denmark, various
Bray-1	11	Maize	25 trials, 1 year	Iowa, various
Bray-1	22	Wheat	4 trials, 12 year	Nebraska, Mollisols

Table 2.2. Phosphorus field balances in some European Countries (Isermann, 1999).

	The Netherlands (1997)	Denmark (1996)	Poland (1997/98)	Greece (1988)
Input (kg ha^{-1} year^{-1})	58	28	14	> 22
Animal manure	41	18	6	n.d.
Fertilizer	14	8	8	22
Deposition	1	n.d.	n.d.	n.d.
Other	2	2	n.d.	n.d.
Output (kg ha^{-1} year^{-1})				
Plant products	31	20	14	9
Surplus	27	8	> 0	> 13
Efficiency of use (%)	53	71	98	< 41

n.d. = Not determined.

Soil phosphorus

Soils are the source of P for plant uptake and also for transfer to watercourses, so an understanding of soil P dynamics underpins the two fundamental aspects of agricultural P research.

Phosphorus contents and forms

Total P contents of topsoils in temperate regions typically range between 300 and 2000 mg kg^{-1} depending on parent material, soil type, land use and fertilizer/manure application method (Frossard *et al.*, 1995; Leinweber, 1996). In contrast, some soils in densely stocked areas of Western Europe and the USA are greatly enriched in P (Barberis *et al.*, 1996). The proportions of organic P can range between 25 and 50% of total P in mineral soils and up to 60–90% in organic soils (Harrison, 1987). Soil management (e.g. arable vs. permanent grassland and biological vs. conventional farming) and rates of P fertilizer/manure application all affect soil P contents and the ratio of inorganic to organic compounds.

Inorganic P-containing minerals include the various calcium phosphates, such as $Ca(H_2PO_4)_2$, $CaHPO_4$, $Ca_4(HPO_4)_3$ and $Ca_5(PO_4)_3OH$. The solubility of calcium phosphates depends on pH and the Ca^{2+} concentrations in soil solution. In alkaline soils, the availability of freshly applied fertilizer P gradually decreases through precipitation of less soluble calcium phosphates. In neutral and acidic soils, the dynamics of inorganic P are dominated by PO_4^{3-} sorption on to the edges of clay minerals and to pedogenic Al and Fe oxides. Sorbed PO_4^{3-} can diffuse inside pedogenic oxides and become less soluble. These sorbed and occluded insoluble phosphates of Al and Fe predominate in acid soils. This presents a serious problem in highly weathered soils rich in Al and Fe oxides, like Ferralsols and Sesquisols, especially in tropical regions. Low molecular organic anions and higher molecular organic acids such as fulvic and humic acids compete with PO_4^{3-} and, possibly, organic P compounds, for sorption sites. Furthermore, chelating organic compounds may also contribute to the desorption of PO_4^{3-}. Hence, agronomic treatments that enhance soil organic matter content often also improve PO_4^{3-} availability. There is substantial evidence that redox dynamics are closely linked to the dynamics of sorbed and dissolved PO_4^{3-} in soil which can directly affect the amounts and forms that leach from soils.

Quantitatively, the most important organic P compound in most soils is phytic acid, which can form up to 80% of the organic P (Turner *et al.*, 2002). Other measurable organic P compounds include phospholipids and nucleotide phosphates (Harrison, 1987). Despite their prevalence, soil organic P compounds remain poorly understood, especially in terms of their availability to plants and transfer to watercourses (Frossard *et al.*, 2000). Indeed, a large proportion of the soil organic P has eluded even basic chemical characterization (Harrison, 1987). This is partly attributable to the fact that recoveries of organic P compounds are often < 50% in common extractants such as sodium hydroxide (NaOH) (Cade-Menun and Preston, 1996). This may result from the 'trapping' of organic P compounds by more complex organic macromolecules, such as fulvic and humic acids, which makes their direct determination difficult.

The soil microbial biomass can contain a large and active P fraction, which can be determined by the fumigation–extraction method (Brookes *et al.*, 1982). Typical proportions of microbial P range from 2 to 5% of organic P in arable soils (Brookes *et al.*, 1982) and up to 24% in grassland soils (Turner *et al.*, 2001). Estimated annual fluxes of P through the microbial biomass of 5 (arable) to 23 (grassland) kg ha^{-1} $year^{-1}$ (Brookes *et al.*, 1984) indicate that microbial biomass P is an important pool for plant nutrition and possible losses to water.

Advances in the characterization of organic P in soils have been achieved by the application of solution-state ^{31}P nuclear magnetic resonance (^{31}P NMR) spectroscopy to soil extracts (Newman and Tate, 1980). This technique enables the characterization of functional classes of P in alkaline extracts with NaOH (Table 2.3). In general, these studies have shown that, as well as PO_4^{3-}, monoester P is the dominant class of compounds in most soils. Individual studies have shown the effects of land use and manuring or fertilizer practice on the relative abundance of P compounds. However, the data in Table 2.3 do not show a significant influence of P applications on the relative proportions of P compounds. Direct comparison between studies is often difficult because of the differences in P forms extracted by the various extraction methods employed in these studies (Cade-Menun and Preston, 1996). Solid-state ^{31}P NMR spectroscopy has been used to investigate Ca-P and Al-P compounds in soils, but bonds between P and Fe cannot be observed because of the paramagnetic character of Fe (Frossard *et al.*, 1994; Lookman *et al.*, 1994).

Sequential extractions have been widely used to distinguish P forms of different solubilities and, thus, with different ecological significance (Hedley *et al.*, 1982). The basic sequential extraction starts with the removal of PO_4^{3-} ions by an anion exchange resin (resin P). The following $NaHCO_3$ and NaOH extracts usually contain inorganic and organic P, which are considered as labile and moderately labile fractions, respectively (Cross and Schlesinger, 1995). The sulphuric acid (H_2SO_4) or hydrochloric

Table 2.3. Ranges (and median) percentage of phosphorus (P) compounds from total signal area as compiled from investigations by ^{31}P NMR spectroscopy. The samples included humic acids (HA), fulvic acids (FA) and NaOH extracts of whole soils, and particle-size fractions which were either native (Soil – P) or received organic or mineral P applications (Soil + P) (Leinweber *et al.*, 1997b).

Sample	No. of studies	Phosphonate	Ortho-P	Monoester-P	Teichoic acid P	Diester-P	Pyro-P	Poly-P	Unknown P
HA	3	3–7 (4)	3–4 (3)	51–59 (53)	1–3 (1)	30–38 (37)	+ –2 (1)		
FA	3	n.d.[2]	15–18 (16)	54–64 (54)	20–24 (21)	+	+ –8 (6)		
Soil + P	24	n.d.–0.6 (n.d.)	47–89 (81)	10–39 (15)	n.d.–20 (n.d.)	n.d.–10 (3.0)	n.d.–3.0 (1.6)	n.d.–1.4 (n.d.)	n.d.–10 (5.4)
Soil – P	42	n.d.–3.0 (0.69)	9.0–76 (36)	22–72 (49)	n.d.–9.7 (6.0)	n.d.–34 (12)	n.d.–18 (7.0)	n.d.–3.9 (0.84)	2.9–13 (5.0)

+ = Below quantification limit.
n.d. = Below detection limit.

acid (HCl) soluble P is assumed to be linked with Ca. After this strong acid extraction, a residual P fraction remains comprising stable inorganic and organic P. Examples of the differences in P forms among soil types are shown in Table 2.4. Heavily manured soils in an area of intensive livestock production in north-west Germany contained P contents an order of magnitude larger and also a disproportional enrichment in PO_4^{3-} ions and labile P fractions compared with the other soils.

The distribution of P compounds between soil particle size fractions is an important property with regard to P transfer in runoff, because of the different susceptibilities of physical fractions to detachment, erosion and transport. The P content of soil particles often increases with decreasing particle size. For example, the contents of PO_4^{3-} (resin-P), $NaHCO_3$-P, NaOH-P, H_2SO_4-P and residual-P decreased from the clay fraction to the sand fraction of two loam soils under originally native prairie (Tiessen *et al.*, 1983). Despite the decreasing P contents with increasing particle size, large proportions of soil P are associated with the sand fraction. For example, in sandy soils receiving large amounts of manure, the proportions of $NaHCO_3$-P and residual-P were larger in sand than in clay (86–87% sand) (Fig. 2.1). This P probably comes from manure-derived specific light particles, which enter the sand fraction. Under the geographical conditions of north-west Germany, this P may be especially susceptible to losses by wind erosion.

Soil phosphorus dynamics and transformations

According to turnover rates, slow and rapid cycling organic and inorganic P pools can be distinguished (Fig. 2.2). Inorganic and organic P compounds can be assigned to these pools on the basis of sequential soil P fractionation methods. For instance, resin-P, $NaHCO_3$-P and microbial biomass P represent most of the rapid cycling P. These fractions, along with organic NaOH-P (which is interpreted as moderately labile organic P), can be involved in biotic processes controlling P release to the soil solution (Frossard *et al.*, 2000).

Determination of the concentrations and forms of P in soil solution is important for the investigation of P uptake by plants and of leaching losses. This requires soil solution collection by zero-tension lysimeters or suction cups. However, the exchange of P ions between the solid phase and soil solution is time-dependent, so isotopic exchange methods can be used to quantify P transfer and determine the effects of soil type, P concentration in solution and time on the kinetics of P transfer. Morel *et al.* (2000) showed that more reliable predictions of the crop response to soil P were obtained with isotopic exchange methods compared with traditional soil extraction methods and concluded that the isotopic exchange approach could be helpful in estimating P loss to water. Applications of P to soils in fertilizer or manures initially reach the rapidly cycling fractions. The P that is not taken up into the biomass or removed

Table 2.4. Typical contents of total phosphorus (P) (g kg^{-1}) and proportions of sequentially extracted forms of phosphorus in topsoils of different origin (% of total P).

Material	Total P	Resin P	$NaHCO_3$-P	NaOH-P	H_2SO_4-P
Mollisol[a]	0.3–0.7	1–9	1–17	1–12	10–42
Vertisol[a]	0.3–0.7	1–6	1–2	3–7	22–46
Spodosol[b]	0.4	3	15	43	3
Histosol	0.5–0.9	1–9	3–8	11–28	0–2
Gleysol	0.5–0.8	2–9	4–8	4–25	16–24
Manured soils[c]	0.5–8.1	2–17	4–33	2–62	1–41

[a] Sharpley *et al.* (1985).
[b] Beck and Elsenbeer (1999).
[c] Leinweber (1996).

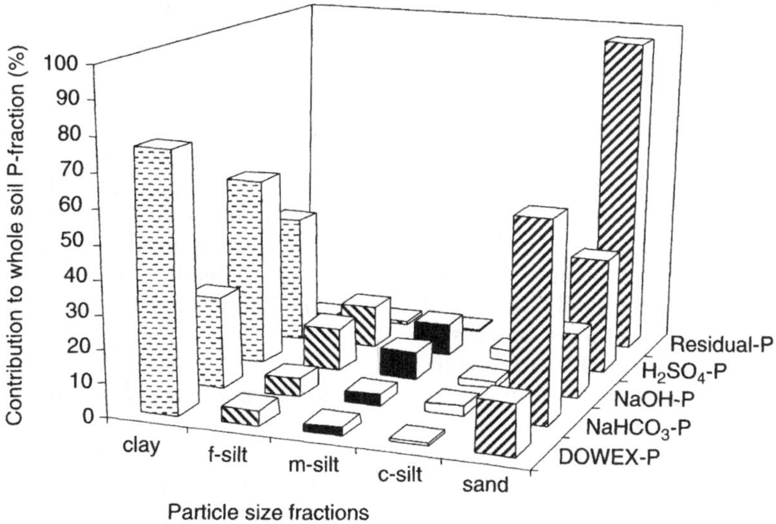

Fig. 2.1. Distribution of sequentially extracted P fractions in organic-mineral particle size fractions averaged for two soils.

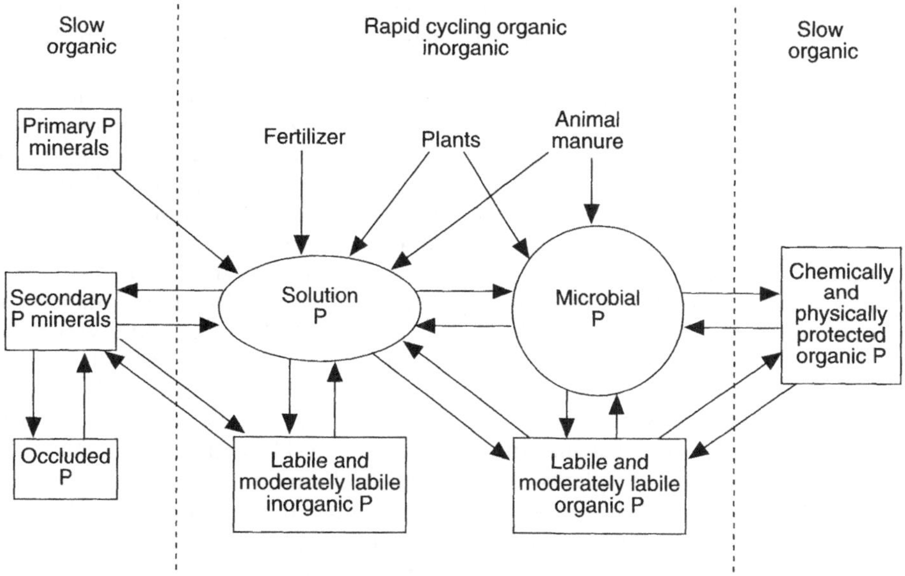

Fig. 2.2. The soil phosphorus cycle: its components and measurable fractions (adapted from Stewart and Sharpley, 1987).

from the soil can be sorbed, occluded or incorporated into soil organic matter. These reactions reduce the rate of P turnover and explain why soils can be a sink for P. This is especially valid for soils with large amounts of pedogenic oxides or Ca, which tend to adsorb PO_4^{3-} strongly or form secondary P minerals. Processes such as mineralization of soil organic matter, mineral solubilization and PO_4^{3-} desorption, result in P inputs into the labile and mobile pool. This may be one reason why long-term cessation of P fertilizer application does not necessarily lead to decreasing yields.

Phosphorus uptake by crops

Soil P is taken up by crops as dihydrogen phosphate ions ($H_2PO_4^-$) under acid conditions and as hydrogen phosphate ions (HPO_4^{2-}) under neutral and alkaline conditions. To maintain equilibration, crop roots release anions such as OH^- or HCO_3^-. In plant physiology, P is essential to the production of a variety of organic P compounds that are building blocks or part of coenzymes in many synthesis processes. Adenosine diphosphate (ADP) and adenosine triphosphate (ATP) take part in energy metabolism, which is involved in all syntheses in plant cells. The hydrogen-transferring coenzymes, such as nicotinamide adenine dinucleotide (NAD) and nicotinamide adenine dinucleotide phosphate (NADP), are involved in oxidation and reduction reactions. Phosphates are also building blocks of nucleic acids and phospholipids; the latter are important cell wall components. In addition to these and other organic P containing compounds, inorganic P occurs in larger quantities in the vegetative organs of crops. Total P contents of plant materials range from 2 to 8 g kg^{-1}. They tend to be smaller in root crops than in grain crops and smaller in the vegetative than in the generative organs. Phosphorus uptake from the soil (kg ha^{-1} $year^{-1}$) at mean crop yields follows the order of cereals and potatoes (25) < forage maize (30) < sugarbeets (35) < oilseed rape (38).

Agricultural crops have different strategies to acquire soil PO_4^{3-}. Some grasses and cereals increase their root systems under P-deficient conditions. Other plants, such as maize, oilseed rape and some legumes and inter-cultures (e.g. buckwheat), exude protons, organic acids and phosphatase enzymes into the rhizosphere to enhance P availability. In this way, inter-cultures can mobilize P from insoluble fractions and provide this in plant-available form for use by the main crops (Eichler and Köppen, 1998). Deficiency of P results in a general reduction of most metabolic processes and, therefore, in reduced crop yields. The large concentrations of Al and Fe oxides in the soil present problems for the supply of P to plants, because they strongly adsorb PO_4^{3-} and prevent availability for biological uptake. Traditionally, P fertilizer recommendations have aimed to maintain large amounts of P in the soil, but the current concerns over P loss to water from over-fertilized soils has challenged this approach. However, the experimental bases for target levels of soil P ensuring the economical yield optimum and protecting aquatic ecosystems from inevitable P inputs are not well established.

Imbalances of agricultural phosphorus

Imbalances in agricultural P use can result from the use of P in livestock feeding, P excretion and manure disposal, mineral P fertilizers and P turnover in soils, which together provide total and available P above the P requirements of crops. The magnitude of these imbalances can be calculated by various approaches and at different scales. These scales can be the field, farm or catchment scale, or for political units they can be districts or nations.

Tables 2.2 and 2.5 show examples of P balances in European countries. From the farmer's point of view, field or farm balances are most important. Mean values show a P surplus ranging from > 0 (Poland) to 27 kg ha^{-1} $year^{-1}$ (The Netherlands). However, even at low field balances, such as in Poland (Table 2.2), individual farms with a large stocking density (e.g. *c.* 5 animal units ha^{-1}) can have large surplus values of up to 50 kg P ha^{-1} $year^{-1}$ (Marcinkowski and Sapek, 1998). For 596,000 farms in North Rhine-Westphalia, the median P surplus value did not indicate a P surplus, but for 24,800 farms with intensive livestock feeding, a median of 35 kg P surplus ha^{-1} $year^{-1}$ was calculated (Werner and Brenk, 1997). Maximal P surplus values of > 80 kg P ha^{-1} $year^{-1}$ were reported in this study. Often, regional P balances disclose such great surplus values, especially if stocking density on the basis of imported feeds contributes to excessive manure P. For the county of Vechta (Lower Saxony, north-west Germany), one of the areas with the largest

Table 2.5. Mean values of phosphorus farm balances in European countries (Isermann, 1999 and references cited therein).

	Austria (1992)	Germany (1995)	Switzerland (1995)	UK (1993)
Input (kg ha^{-1} $year^{-1}$)	11	21	20	13
Mineral fertilizer	8	13	10	9
Imported feeds	2	7	5	3
Deposition	<1	n.d.	<1	<1
Sludge, compost	<1	1	4	<1
Output (kg ha^{-1} $year^{-1}$)	5	9	6	3
Plant products	3	6	2	<1
Animal products	2	3	4	>2
Surplus	6	12	14	10
Efficiency of utilization (%)	45	43	30	23

n.d. = Not determined.

stocking density in Europe, the mean P surplus originating from manure is 37 kg ha^{-1} $year^{-1}$ (von Hammel, 1994).

National P balances have revealed surplus values ranging from 0.8 (Lithuania) to 6.4 (Poland) and *c.* 7 kg ha^{-1} $year^{-1}$ (Sweden and Denmark) (data for 1995 compiled by Marcinkowski and Sapek, 1998). A national P balance for Germany showed that the contribution of animal excrement is on average *c.* 17 kg P ha^{-1} agricultural area $year^{-1}$. This confirms that livestock feeding is an important source of P inputs to soils (Fleischer, 1997).

Impact of excess fertilizer application on the amounts and forms of soil phosphorus

The application of excess mineral fertilizer, manures and sewage sludges to soil increases the total P concentration. In a European Community research project, over-fertilized soils had total P concentrations of 890–1168 mg kg^{-1} (Germany), 333–1400 mg kg^{-1} (Spain), 1331–1920 mg kg^{-1} (UK) and 709–1067 mg kg^{-1} (Italy) (Barberis *et al.*, 1996). In a region with intensive livestock feeding in north-west Germany, total P contents ranging from 565 to 8173 mg kg^{-1} with mean values of 1335–2693 mg P kg^{-1} clearly indicated large P enrichments (Leinweber, 1996). Three farms with alternative agricultural management systems in that area had smaller total P concentrations (1580–1690 mg kg^{-1}) than farms with known high livestock densities (mean 3113 mg P kg^{-1}).

Soil P test values are often unequivocal indicators of over-fertilization and P accumulation from manure application and waste disposal. In Delaware, north-eastern USA, which has a large and growing poultry industry, even the maximum values of P soluble in 0.05 M HCl + 0.0125 M H_2SO_4 of up to 632 mg P kg^{-1} were in the range of normal total P concentrations of other soils (Mozaffari and Sims, 1994; and see Sims and Coale, Chapter 16, this volume). Figure 2.3 shows the frequency distribution of double lactate extractable P (DL-P) contents in soils from Vechta, north-west Germany. Forest soils, which serve as a reference, contained low to medium DL-P contents. Soils under agricultural use had much higher DL-P concentrations: about 95% of samples from grassland, horticulture (raspberries, asparagus) and arable land had DL-P concentrations above the levels for crop response to P fertilizer addition. Soils in counties with a large stocking density in North Rhine-Westphalia had mean calcium-ammonium lactate extractable P (CAL-P) contents of 100–115 mg P kg^{-1}. About 60–80% of soils in these counties had CAL-P contents classified as 'high' or 'very high' (Werner and Piehl, 1998).

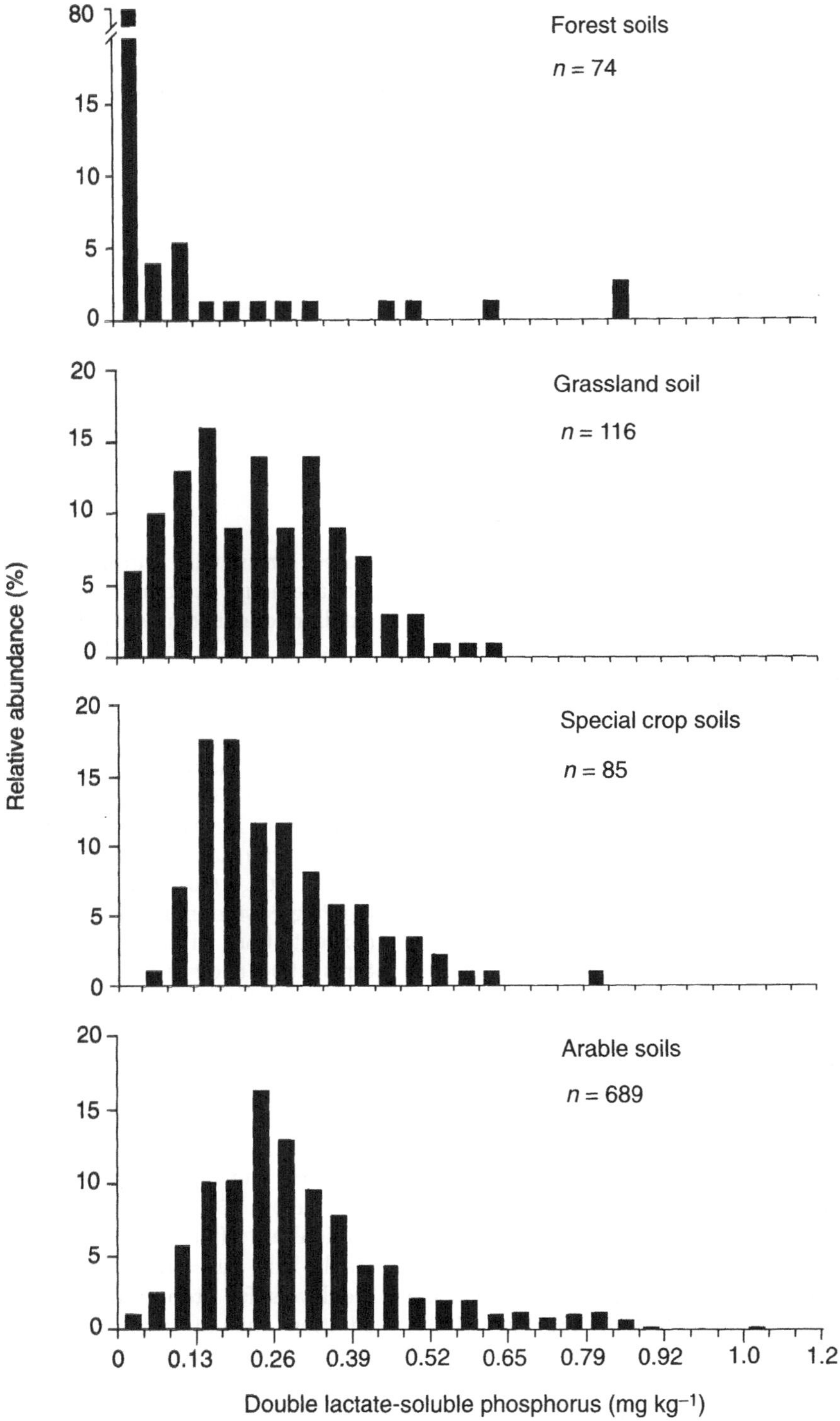

Fig. 2.3. Frequency distribution of soil phosphorus test values in a densely stocked area in north-west Germany (Leinweber *et al.*, 1993).

In The Netherlands, soil P accumulations were recognized in the early 1980s. Studies of the regional P distribution were based on PO_4^{3-} saturation characteristics (Breeuwsma *et al.*, 1995). According to these authors, the degree of PO_4^{3-} saturation is defined as:

$$DPS\ (\%) = P_{ox} / PSC \times 100$$

where DPS = degree of phosphate saturation, P_{ox} = sorbed P (extracted by oxalate) (mol kg^{-1}) and PSC = PO_4^{3-} sorption capacity, often determined by $0.5 \times (Al_{ox} + Fe_{ox})$ (mol kg^{-1}).

These detailed studies revealed that 70% of farmland soils were P saturated, with values of 88% for maize and 69% for grassland. More than 21% of farmland was strongly (DPS > 50%) and 6% very strongly (DPS > 75%) P saturated. In north-west Germany, Leinweber *et al.* (1997a) determined DPS values from 14 to 145% in soils from a catchment with relatively low stocking density. In contrast, in an area with high stocking densities the DPS ranged from 13 to 179%, with most frequent values in the range of 70 to 90%. The differences between the two areas clearly originated from the contents of P_{ox} and not differences in the PSC, hence they could be assigned to the P input with manure rather than to differences in the P sorption.

Besides these effects on total P contents, soil P test values and DPS, excessive fertilizer application and waste disposal also alter the amounts and proportions of soil P fractions, as shown by sequential extraction and spectroscopic investigations. Mineral fertilizer application usually increases the proportions of inorganic P in labile and moderately labile fractions. Repeated manure applications can increase the amounts and percentages of Ca-P phases and residual P.

Combinations of inorganic and manure P increase the P proportions over a wider range of differently soluble fractions (Zhang and MacKenzie, 1997). Extensive soil management can lead to reduced proportions of resin-P, $NaHCO_3$-P and NaOH-P, whereas intensively used arable soils with organic and mineral fertilization contain high levels of resin-P, H_2SO_4-P and residual-P (Leinweber *et al.*, 1999). Generally, there are various interactions between amounts and forms of P application, fertilization with other nutrients, soil type and properties, crop rotation and the contents and the proportions of inorganic and organic P forms. Sequential fractionation of P forms in 69 soils from the densely stocked area in north-west Germany yielded results in the order: resin-P (2–17%) < $NaHCO_3$-P (4–33%) < NaOH-P (2–62%). Soil management influenced the percentages of the different P fractions. Grouping of the samples according to different soil use gave the order, arable > horticultural crops and grassland for the mean concentrations of resin-P and $NaHCO_3$-P. This was in accordance with the large proportions of labile P in manure and the particular use of arable soils for its disposal. The large proportions of resin-P and $NaHCO_3$-P show that, as well as greater total P concentration, the soils of this densely stocked area have more soluble P (Leinweber, 1996). In particular, liquid pig manure and poultry litter contain large amounts of P, of which large proportions occur as PO_4^{3-} (Leinweber, 1996). Therefore, it can be expected that soils receiving large amounts of these materials will be enriched in these labile P compounds. This is a serious problem in regions with high livestock densities, such as north-west Germany (Vechta, Lower Saxony), where it has been calculated that the P contents in some soils could only be explained by annual manure-P inputs of up to 180 kg P ha^{-1} $year^{-1}$ over the last 40 years (Leinweber *et al.*, 1993). Such high P inputs may increase the potential for losses of P to water.

Environmental Consequences of Excessive Phosphorus Enrichment in Soils

Introduction and general problem

As described above, large stocking densities and excessive manure applications are reasons for surplus P inputs to agricultural soils. The perception that fertilizer P is strongly held in soil and prevented from leaching has meant that P fertilizer recommendations have often been above the requirements of plants and have not adequately considered the P inputs from manures. As a consequence, P contents have gradually increased in many European soils, which can accelerate the transfer of P to watercourses in runoff. In the following section we describe the effects of oversupply on the contents and forms of soil P. The movement of P from soil to surface waters is described, including the mechanisms of P release and pathways of P transfer from soil to water: this is not only affected by the soil P contents and forms, but also by land use. Finally, an assessment of the environmental consequences of excessive P enrichment of soils requires basic information on the forms of P in soil solution and runoff and on critical concentrations and loads in drainage waters.

Forms of phosphorus in soil solution and runoff waters

Phosphorus forms in water are conventionally determined on the basis of operational procedures, which are used throughout the world. These procedures involve: (i) chemical fractionation, based on the reaction of inorganic P with molybdate followed by subsequent reduction and spectrophotometric detection; and (ii) physical fractionation, based on filtration through a sub-micron filter. The spectrophotometric procedures are based on an original method described by Murphy and Riley (1962) for the determination of dissolved PO_4^{3-}. The P determined by this procedure is termed reactive P (RP) rather than inorganic P (Haygarth and Sharpley, 2000) because the procedure can overestimate true PO_4^{3-}. The sample preparation and reaction conditions can result in other forms of P, such as labile organic P species, being hydrolysed in the strongly acid conditions required for colour development (Tarapchak, 1993), while interference from silica (Ciavatta *et al.*, 1990) and sample turbidity has also been noted. Recent studies using alternative methods of detecting PO_4^{3-}, such as ion chromatography, have indicated the extent of this problem in soil runoff and lake waters (Baldwin, 1998; Halliwell *et al.*, 2000).

Total P can be determined in the sample using an appropriate digestion procedure (recently reviewed by Rowland and Haygarth, 1997), which allows the UP fraction to be determined by difference. This fraction generally represents organic P compounds, including monoester P (e.g. inositol phosphates, sugar phosphates or mononucleotides), diester P forms (e.g. nucleic acids and phospholipids), phosphonates (e.g. aminoethylphosphonic acid) and condensed organic P forms (e.g. ATP), although only a few of these have been directly identified in soil waters (Espinosa *et al.*, 1999). However, UP (< 0.45) cannot strictly be termed organic P, because inorganic condensed forms that are not Mo-reactive, such as pyrophosphate, are included in the UP fraction (Ron Vaz *et al.*, 1993). Alternative methods for the determination of the true 'organic' P fraction using ultraviolet light induced persulphate oxidation are available, but are not widely employed (Ron Vaz *et al.*, 1992; Benson *et al.*, 1996). However, some studies have indicated that condensed inorganic P can represent a large part of the soil solution P (Ron Vaz *et al.*, 1993).

Samples can be further fractionated by filtration through a sub-micron filter, most commonly of pore size 0.45 μm, although 0.2 μm filters are used in some countries where a sterile sample is required. Filtration is often used to define 'dissolved' (< 0.45 μm) and 'particulate' (> 0.45 μm) P fractions, but this is incorrect, because soil waters have been shown to contain a continuum of particles in the < 0.45 μm size range (Haygarth *et al.*, 1997; Sinaj *et al.*, 1998). Particles of size < 0.45 μm are often referred to as colloidal, although this term strictly defines particles ranging in size from 1 nm to 1 μm, composed of specific mineral forms (Kretzschmar *et al.*, 1999).

Therefore, four fractions are directly determined on the basis of Mo reaction with or without digestion and on filtered and unfiltered samples (TP, TP (< 0.45), TRP and RP (< 0.45)). From these, five additional fractions can be determined by difference (Fig. 2.4).

Identification of the various P forms is complicated by changes within the solution after release from the soil (Sharpley *et al.*, 1981). For example, rapid sorption or desorption of orthophosphate between the solid soil, solution particles and solution can occur, in addition to the hydrolysis of organic forms by enzymes and immobilization by biomass uptake. These changes can occur rapidly, as noted by Haygarth *et al.* (1995), who discovered that RP (< 0.45) concentrations in soil leachate were substantially reduced within 8 hours of sampling. Clearly, it is strictly wrong to describe RP as 'inorganic', UP as 'organic', filtered P as 'dissolved' and unfilterable P as 'particulate'. These ambiguities mean that is only correct to use strict operational definitions for the fractions determined.

Movement of phosphorus from soils to watercourses

The transfer of P from agricultural soils to watercourses is now regarded as a serious problem in many areas of the world, highlighted by recent major conferences on the subject (Tunney *et al.*, 1997; Foy and Dils, 1998). Phosphorus transfer is a complex phenomenon, but can be defined as a function of the forms of P, the mechanisms of their release from soil to water and the hydrological pathways by which the mobilized P moves from the land (Haygarth and Jarvis, 1999). The following sections present brief reviews of these three aspects of P transfer from the soil shown conceptually in Fig. 2.5. Each aspect is important for a comprehensive understanding of P transfer.

Mechanisms of phosphorus release from soil to water

The critical first step in the transfer process is release from the soil solid phase to soil

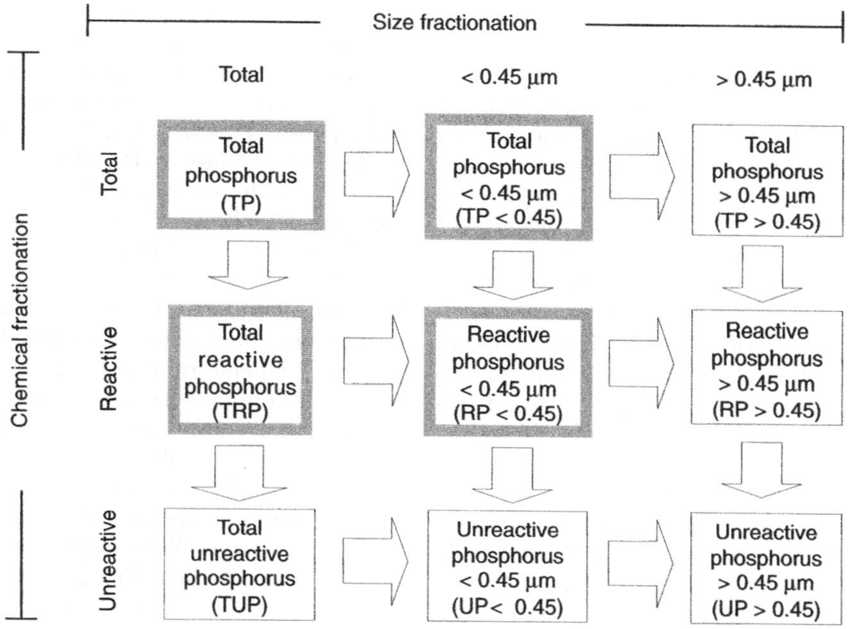

Fig. 2.4. Operationally defined phosphorus fractions determined in water, showing the directly determined fractions (in boxes with shaded borders) and fractions determined by difference.

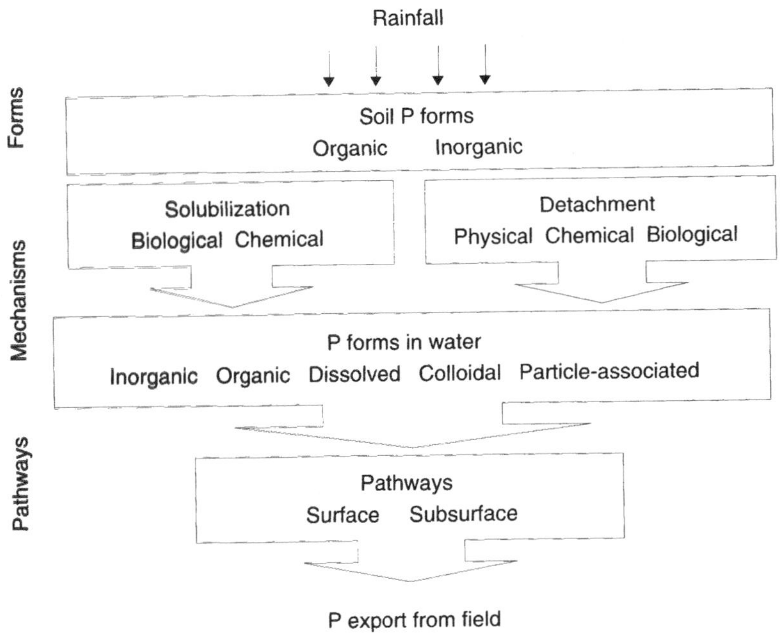

Fig. 2.5. A conceptual illustration of the transfer process (adapted from Haygarth and Jarvis, 1999).

water. This occurs through various mechanisms of P release. These can be classified into solubilization and detachment mechanisms of solid colloids and particles.

The processes controlling the solubilization of soil inorganic P have received considerable attention, mainly in terms of the cycling and plant availability of inorganic P (Barrow, 1978; Frossard *et al.*, 1995; Barberis *et al.*, 1996), and a limited amount of information also exists on organic P sorption by chemical mechanisms (Anderson *et al.*, 1974; Frossard *et al.*, 1989). However, relatively little attention has been paid to the implications of these processes for P transfer (Haygarth and Jarvis, 1999). Solubilization of inorganic P results primarily from desorption of PO_4^{3-} from solid surfaces. Similar processes apply to sorbed organic P. Release of PO_4^{3-} to solution is largely controlled by the amount of readily desorbable P in the soil. Desorption of inorganic P from soil to solution appears to involve a readily exchangeable pool of inorganic P, which can move rapidly between the solid–liquid phases, at timescales of seconds to minutes (Cole *et al.*, 1977; Lopez-Hernandez *et al.*, 1998). The steady-state concentration of PO_4^{3-} in solution is termed the equilibrium P concentration (EP_0). Removal of inorganic P from the soil solution, for example by plant uptake, results in the movement of P ions into solution. Conversely, an increase of inorganic P in solution, for example through fertilizer application, results in P movement on to the solid phase. Whether equilibrium will be reached in drainage water will depend on the rate of water movement through the soil and the equilibrium P concentration, but should be fairly rapid where the amounts of P in the soil are low (Cole *et al.*, 1977; Fardeau, 1996). Other factors that affect desorption from the soil include ionic strength, pH and temperature. Effects of pH, which can result from the release of plant root exudates or lime additions, can be variable, because increasing pH has been shown to both increase (Barrow, 1984) and decrease (Wild, 1988) the inorganic P concentration of the soil solution. Waterlogging of soil can release P through the reduction of Fe^{3+} compounds and the mobilization of clay–humic complexes (and associated P) through reduction of the metal (Fe) bridge (Ryan and Gschwend, 1990; Frossard *et al.*, 1995).

Biological mechanisms can exert strong controls on the solubilization of P, but represent the most under-investigated aspect of P transfer. Most studies have centred on the role of the soil microbiology in soil P cycling and the availability of P to plants, although even this information has been fairly scarce until relatively recently. This is highlighted by the fact that reliable techniques for measuring P in the microbial biomass were only developed as late as the 1980s (Brookes *et al.*, 1982; Hedley and Stewart, 1982). However, biological mechanisms appear to be critically important in determining soil solution P concentrations (Cole *et al.*, 1977; Lee *et al.*, 1990; Seeling and Zasocki, 1993; Turner and Haygarth, 2002) and, therefore, have the potential to exert a considerable influence on P solubilization. Indeed, recent studies of leaching indicate that seasonal patterns of microbial release in the soil translate into P concentrations in leachate water at depth (Turner and Haygarth, 2000).

Growing plants and microorganisms immobilize inorganic P from the soil solution in their biomass. The immobilized P is released on cell death and decay, mostly as organic P, particularly through lysis of microbial biomass cells. Therefore, these biological mechanisms transform inorganic forms into organic P. These processes maintain a low soil solution inorganic P concentration, with a relatively large organic P concentration (Lee *et al.*, 1990; Seeling and Zasocki, 1993). The cycle is completed by phosphatase enzymes, released into the soil from plants and microorganisms, which catalyse the release of inorganic forms from organic compounds (Spier and Ross, 1978; Turner and Haygarth, 2002). Phosphatase enzymes appear to have a fundamental, but poorly understood, role in the P transfer process, by regulating the availability of P to the microbial biomass and the forms of P available for transfer in runoff. In addition to microbial controls, macrofauna such as earthworms play a major role through the mineralization and translocation of P in the soil profile (Sharpley *et al.*, 1979), while plants release substantial amounts of water-soluble P after defoliation (Bromfield and Jones, 1972).

Soil detachment and transport processes are well described in standard texts (e.g. Hudson, 1995; Morgan, 1995). Detachment of P primarily involves the erosion of soil particles by the hydraulic action of flowing water or by raindrop impact. However, chemical mechanisms, such as changing ionic strength or pH, can also result in the detachment of particle-associated P forms sorbed to the soil. In addition, the release of some biological P may be classed as detachment, such as the cell fragments released by cell lysis.

Erosion mechanisms have been relatively well investigated, partly because of the traditional perception of P transfer being dominated by P export in overland flow (see Sharpley and Smith, 1990, for review). Erosion is frequently associated with P transfer by overland flow, especially from arable land, where losses of particulate P can be extremely high and constitute most of that transferred (e.g. Burwell *et al.*, 1977; Sharpley and Syers, 1979; Haith and Shoemaker, 1987). This is often associated with the preferential transport of fine clay-sized materials, which are relatively enriched in P because of their high P sorption capacities (Sharpley, 1985; Kretzschmar *et al.*, 1999). Erosion can also result from rapid subsurface drainage, with particles eroded from the edges of wormholes, cracks and underground fissures. This can contribute to high concentrations of particulate P in subsurface drainage through macropores (Dils and Heathwaite, 1996) and in tile drainage (Turtola and Jaakkola, 1995; Grant *et al.*, 1996; Ulen, 1999). Fine colloidal sized (1 nm–1 μm) particles may be particularly efficient at transporting P through subsurface pathways, because they have high capacities to sorb P but are not easily stabilized in the soil profile (Kretzschmar *et al.*, 1999). Therefore, although physical release of P has traditionally been associated with surface pathways, there is the potential for particle-associated P forms to make a significant contribution to P transfer through subsurface pathways.

Chemical detachment can result from changes in ionic strength, which controls the release of particle-associated P in terms of clay and organic matter. For example, $CaCl_2$ solution suppresses the release of these particles to solution (Wild and Oke, 1966), while high Na^+ and low ionic strength can result in the release of particulate P through dispersion of aggregates (Kaplan *et al.*, 1996). These processes have been suggested as particularly important in P movement through soils (Chardon *et al.*, 1997).

Surface applied P in animal manure or mineral fertilizer can be directly lost from the field without ever becoming part of the soil P. This process cannot be classed by a mechanism of release, because the P never actually becomes part of the soil matrix. This type of P loss has recently been termed 'incidental' (Haygarth and Jarvis, 1999) and can be especially important in grassland systems where surface applied manure is used, especially when 'best management practices' are not employed. Incidental losses may account for substantial losses of P from the land (Harris *et al.*, 1995), but little information currently exists on this potentially important aspect of P transfer.

Pathways of phosphorus transfer from soil to water

The driving force behind P transfer from the land is the hydrology, because water provides the energy and the carrier for P movement (Haygarth and Jarvis, 1999). The route taken by water as it carries P off the land describes the pathway of P transfer. Pathways can be broadly separated into surface (overland flow) and subsurface pathways and the pathways taken are of critical importance in determining the extent of P loss from agricultural land. Comprehensive details and definitions of the pathways within these broad separations can be found in Ryden *et al.* (1973) and Haygarth and Sharpley (2000).

The traditional perception of P being strongly fixed in the subsoil has meant that research into transfer traditionally has focused on surface pathways (Baker *et al.*, 1975; Burwell *et al.*, 1977; Sharpley and Syers, 1979). Detachment mechanisms and loss as particulates are considered to dominate transfer in overland flow, but all forms and release mechanisms can contribute to loss via overland flow. Overland flow is often associated with catastrophic events, which are considered to be the most severe in terms of loss, although there is evidence that lower frequency events that occur for a much greater time throughout the drainage period are equally important in P transfer (Fraser *et al.*, 1999). This is because overland flow preferentially erodes and transports clay-sized material (Sharpley, 1985). As shown in Fig. 2.1, total P and desorbable PO_4^{3-} are enriched in clay fractions. Using ^{32}P studies, overland flow has been found to interact with only the top few mm of soil, termed the effective depth of interaction (Ahuja *et al.*, 1981). This coincides with the depth of soil that often contains the greatest amount of P and can result in especially high concentrations in overland flow.

Subsurface pathways include lateral flow through the soil, vertical drainage, preferential flow through macropores and artificial drainage channels. Until recently, subsurface P losses were considered to be negligible, because of the propensity for P to be retained at depth in the soil profile (Baker *et al.*, 1975; Burwell *et al.*, 1977; Sharpley and Syers, 1979). This was exacerbated by the fact that measured losses of P in subsurface flows were small compared with the total amount of P in the soil (Sample *et al.*, 1980). However, the change from an agronomic to an environmental perspective on soil P has refocused attention on subsurface P transfer, because small concentrations of P can cause significant problems for water quality (Tunney *et al.*, 1997; Turner and Haygarth, 2000). This phenomenon is not restricted to waterlogged soils (Khalid *et al.*, 1977) or sandy-textured soils under heavy fertilizer addition as traditionally thought (Ozanne *et al.*, 1961; Breeuwsma and Silva, 1992), but includes many soil types, especially clay soils that are susceptible to cracking and preferential flow (Stamm *et al.*, 1998; Simard *et al.*, 2000; Turner and Haygarth, 2000).

Studies using lysimeters have revealed that P concentrations in leachate water from a range of soil types occur at levels that could present problems for water quality. For example, Leinweber *et al.* (1999) and Turner and Haygarth (2000) measured TP concentrations routinely in excess of 100 μg l^{-1} in leachate from a range of contrasting grassland soil types and management conditions. Organic P constitutes a large proportion of the total P transported in leachate (Magid and Nielsen, 1992; Espinosa *et al.*, 1999; Turner and Haygarth, 2000), where its mobility in the soil compared with inorganic P forms makes it an important mechanism by which P may escape from the soil. Inorganic P is readily fixed in the soil and prevented from leaching by precipitation with Ca, Fe and Al, or sorption to clays and other soil particles (Frossard *et al.*, 1995). However, some organic P is not strongly sorbed and can move easily through the soil to surface waters (Rolston *et al.*, 1975; Frossard *et al.*, 1989; Chardon *et al.*, 1997), where it can contribute to algal growth after the release of inorganic P by phosphatase enzymes (Whitton *et al.*, 1991).

Preferential flow pathways may be important in subsurface P transfer. The high concentrations of > 0.45 μm P forms in leachate water indicates the importance of preferential flow even in sandy-textured soils (Turner and Haygarth, 2000). The potential for P transport through macropores was demonstrated by Simard *et al.* (2000), while Dils and Heathwaite (1996) detected P concentrations in macropore flow through agricultural grassland soils of > 1 mg l^{-1}. The large amounts of TP (> 0.45) present in subsurface drainage water indicate that erosion mechanisms within macropores may contribute to P transfer. The importance of TP (> 0.45) transport confirms that process terms such as leaching (which is synonymous with all subsurface transport, but actually means the release and movement of exclusively dissolved P) can be very misleading (Haygarth and Sharpley, 2000).

Colloidal P can facilitate export through soil, through protection by sorption in the soil profile, and is a large component of P in overland flow (Matthews *et al.*, 1998) and in river waters (Haygarth *et al.*, 1997). Colloidal P may be an especially important component of P transfer through preferential flow pathways, because the low ionic strength water that results from the minimal contact between rainfall and soil results in the mobilization of colloids and associated P (Chardon *et al.*, 1997).

Subsurface export from artificially drained land can be responsible for considerable amounts of P transfer (Turtola and Jaakkola, 1995; Grant *et al.*, 1996; Haygarth *et al.*, 1998; Ulen, 1999), despite the fact that artificially drained soils are mainly loams and clays that are low in P. Much of the P can occur in dissolved forms, although large amounts also occur as particulates, probably by detachment from the walls of drainage channels. High P concentrations in artificial drainage water are probably a result of the rapid movement of water to depth, thereby bypassing the soil matrix and minimizing removal from solution by sorption. However, artificial drainage has the overall effect of reducing P transfer from the land, by as much as 30% compared with undrained plots (Haygarth *et al.*, 1998).

Effects of land use on phosphorus losses

Arable land can be susceptible to substantial P losses, especially by erosion and overland flow, because the land is left bare at certain times of the year. An increase in winter cereal cropping has exacerbated this problem, because it combines the period of maximum rainfall with long periods of bare soil. Under these circumstances, vast amounts of P can be lost in 'catastrophic' events, associated with the removal of large amounts of soil by rill and gully erosion. For example, Haith and Shoemaker (1987) reported exports of up to 24 kg ha^{-1} $year^{-1}$ from maize fields in Delaware, USA, mainly as > 0.45 μm P. The amounts of P lost depend greatly on factors such as slope, rainfall intensity and soil type. In addition, arable land often receives large inputs of mineral fertilizer, which results in large amounts of P available for loss through leaching (Heckrath *et al.*, 1995). However, higher rates of mineral fertilizer applications do not necessarily result in larger P losses. Experience from long-term lysimeter studies showed that P concentrations in leachates and leaching losses depend on soil texture, cropping, mineral fertilizer applications and intensity of soil management. Phosphorus losses were larger in sands than in loamy sands and were greater in arable soils than in grassland soils (Meissner *et al.*, 1995). In lysimeters of a loamy sand under a winter barley–oats rotation, the mean leaching losses were largest (657 g P ha^{-1} $year^{-1}$) with the smallest fertilizer rate (12.5 kg P ha^{-1} $year^{-1}$) and decreased by more than 100 g P ha^{-1} $year^{-1}$ if the fertilizer rates used were up to three times larger. Interestingly, intensively used arable soils with organic and mineral P fertilizer inputs had the smallest leaching losses (Leinweber *et al.*, 1999).

Even if arable land is found on light-textured soils on relatively flat land, it can still be susceptible to overland flow and erosion through soil 'capping' (Fraser *et al.*, 1999). Sandy soils under heavy manure applications can also result in high concentrations of P in leachate water (Breeuwsma and Silva, 1992; Chardon *et al.*, 1997; Meissner *et al.*, 1997).

Grasslands appear to transfer a smaller proportion of soluble inorganic P than arable land, especially on soils with low extractable P contents. However, this does not account for organic and particulate P exports, which can be substantial from grasslands. For example, exports of P of between 2 and 3 kg ha^{-1} $year^{-1}$ have been recorded from 1 ha grassland lysimeters in south-west England (Haygarth and Jarvis, 1997; Haygarth *et al.*, 1998). Because grasslands are often located on heavy, poorly draining soils on steeply sloping ground, they can lose large amounts of > 0.45 μm P by detachment, erosion and overland flow; in one extreme storm event, in excess of 0.5 kg P ha^{-1} was lost by this pathway (Haygarth and Jarvis, 1997). Large P concentrations have also been observed in leachates from grassland established in sandy soils, although mean annual concentrations did not reflect the rate of mineral P fertilizer applications (Leinweber *et al.*, 1999). Grasslands are also susceptible to incidental losses of P from ill-timed manure applications; for example, Harris *et al.* (1995) measured RP (< 0.45) concentrations of 267 μg l^{-1} from plots that had received slurry applications just prior to heavy rainfall, compared with 30 μg l^{-1} from control plots.

Compaction of soil by farm machinery and poaching by livestock can further promote the occurrence of overland flow. The high P concentrations in overland flow from grasslands can be exacerbated by an intense surface accumulation (Haygarth *et al.*, 1998), because of the permanent (untilled) nature of the soil and high inputs from fertilizer, slurry applications and excretal returns from grazing animals. In one extreme case, concentrations of RP (< 0.45) of > 7 mg l^{-1} have been reported in overland flow from high input pasture plots in south-eastern Australia (Nash and Murdoch, 1997).

Pleistocene landscapes covering the northern parts of central Europe have meltwater valleys that are nowadays filled in by peat and were used in the past for intensive grassland. This practice led to the oxidation of peat, accelerated especially if the soils were ploughed to establish a closed new grass population. In more recent years, the restoration of fenlands, set-aside and wetland programmes have been started to prevent the Histosols from further severe degradation. However, a novel problem emerged: the undesired mobilization of dissolved organic matter and PO_4^{3-} if the groundwater level was elevated. For example, Kalbitz *et al.* (1999) measured PO_4^{3-} concentrations in the soil solution and near-surface groundwater ranging from 0.001 to 0.622 mg P l^{-1} which were negatively correlated to the redox potential. Thus, dramatic changes in land use or setting aside land should not be planned and carried out without consideration of possible undesired side-effects and consultation of the pertinent literature.

Critical phosphorus concentrations and loads

The pollution of watercourses with P is a serious problem in many parts of the developed world. In unpolluted freshwaters, TP concentrations are typically below 25 μg P l^{-1}. In water management, it is generally assumed that concentrations above 50 μg P l^{-1} are the result of anthropogenic influences. In Germany, the quality classification in Table 2.6 is applied to running surface waters.

A survey of rivers in Europe (European Environment Agency, 1998) revealed that a large proportion of *c.* 1000 monitoring stations observed TP concentrations exceeding 50 μg l^{-1}. Only *c.* 10% of the monitoring stations reported mean TP concentrations below 50 μg P l^{-1}. The lowest concentrations of P were found in the streams and rivers of the Nordic countries where 91% of the monitoring stations reported mean concentrations below 30 μg P l^{-1} and 50% below 4 μg P l^{-1}. This pattern results from a low human population and the presence of slowly weathering bedrock poor in nutrients. Higher P concentrations were found in a band stretching across central Europe, from southern England to Romania.

A similar pattern of P levels can also be found in lakes with low concentrations in the Nordic and Alpine regions. Lakes can be severely affected by eutrophication after P enrichments. This can be a particularly difficult problem because P may accumulate in lake sediments and be released (for example, by climatic events, changes of redox-conditions or the action of fish) into the water column, maintaining a eutrophic status.

Phosphorus levels in surface waters generally declined across Europe between 1987 and 1996. This trend occurred for annual average concentrations of both total and dissolved P, but it appears that high peak concentrations still occur even where improvements have been achieved. Such improvements originate to a great extent from a more intensive treatment of urban and industrial wastewater and some decline in fertilizer use. In recent years, great efforts have been undertaken in Europe to reduce and control aquatic P pollution and eutrophication in general. These activities (policies, legislation, conventions, etc.) were started at the national and international scale, especially within the framework of the European Union.

It is important to note that some pressure for action comes from international conventions to protect marine environments. These include HELCOM (The Helsinki Convention to protect the Baltic Sea), OSPAR (The Paris and Oslo Conventions to protect the North Sea and Atlantic) and The Barcelona Convention to protect the Mediterranean Sea. The HELCOM and OSPAR conventions focused particularly on reductions of nutrient pollution (N and P). However, Table 2.7 shows that this was not achieved by all countries. Great efforts are necessary to solve this problem. The proposed Water Framework Directive of the European Union is currently under intensive debate within the legislative process. It will provide the basis for an overall management framework for surface (including estuarine and coastal) waters and groundwaters. This is a major

Table 2.6. Threshold values for total phosphorus (TP) in running surface waters as used in Germany (Umweltbundesamt, 1997).

Quality class	Degree of pollution	TP-concentration (μg l^{-1}]
I	Not anthropogenic polluted	≤ 50
I–II		≤ 80
II (target class)	Moderately polluted	≤ 150
II–III		≤ 300
III	Raised polluted	≤ 600
III–IV		≤ 1200
IV	Highly polluted	≥ 1200

Table 2.7. Annual discharges of total phosphorus (106 kg year^{-1}) to the North Sea, the north-east Atlantic Ocean and to the Baltic Sea (European Environment Agency, 1998).

Country	1991	1995
North Sea, Atlantic Ocean		
Belgium[a]	2.0	4.0/5.0
Denmark	2.3	2.0
France	n.d.	n.d.
Germany	11.6	11.5
Ireland	6.3	7.3
The Netherlands	17.0	34.1
Norway	3.3	3.9
Portugal	3.1	3.1
Sweden	0.2	1.3
UK[a]	39/40	36
Baltic Sea		
Denmark	5.3	2.3
Estonia	2.8	1.3
Finland	3.4	3.6
Germany	1.2	0.6
Latvia	3.2	2.2
Lithuania	1.7	1.4
Poland	15.0	14.2
Russia	9.5	7.1
Sweden	4.0	4.7

[a] Low/high estimate.
n.d. = Not determined.

new development for water legislation and a fundamental basis for sustainable protection of water resources against pollution.

Options for Mitigation and Conclusions

To establish successful management options for reducing P losses to watercourses, it is necessary that farmers, other stakeholders and politicians are aware of the problem and accept that improper P use in agriculture is a main reason for surface water pollution with P. Much evidence has been compiled to show that uneven distribution of livestock results in particular enrichments in total and labile P in certain regions. Hence, political and economic measures are necessary to achieve a more even distribution of animal production. However, the socio-economic impacts of such measures on rural and urban communities need to be considered, as emphasized by Sharpley and Tunney (2000).

Besides this international or national level, a lot can be done locally to improve agricultural P management. Phosphorus use in animal production can be reduced by improving utilization (e.g. precise recommendations, improved P utilization, better choice of inorganic P supplements, plant breeding, etc.; Damgaard-Poulsen, 2000), so that better farm balances can be achieved without loss in productivity (Valk *et al.*, 2000). Improved P use in animal production offers the opportunity to reduce inputs to soils, provided it is not compensated for by increases in stocking densities. In any case, all P entering the soil, including that in manure, must be considered in the calculation of fertilizer recommendations. A European-wide re-evaluation of the experimental data that fertilizer recommendations are traditionally based on, is being carried out in the framework of the European Union. More precise recommendations with lower safety margins will help farmers to avoid inefficient fertilizer applications and undesired P enrichments in soil.

Novel techniques such as isotopic exchange (Morel *et al.*, 2000), ^{31}P NMR spectroscopy, or microbiological measurements, will improve the general understanding of P transformations in soil and thus potential for P solubilization. This will enable a closer matching of fertilizer applications with plant requirements. In plant production, various measures can reduce the risk of losses. These include: (i) reducing P availability in excessively fertilized soils by lowering inputs and either increasing removal in produce or increasing the sorption capacity of soil; and (ii) modification of the cropping systems to achieve optimum utilization of soil and fertilizer P (Frossard *et al.*, 2000). One example for increasing the sorption capacity of soil is the application of aluminium sulphate ($Al_2(SO_4)_3.14H_2O$) to poultry litter (Moore *et al.*, 2000). This amendment reduces P in runoff and NH_3 loss from poultry litter broadcast on to grassland. However, from an environmental perspective, all options for avoiding surplus P balances and improving utilization are preferable to 'technical solutions' for temporal immobilization.

The identification of critical source areas and assessment of the risks of eutrophication requires a comprehensive understanding of movement from soil to water, and of the P forms and critical concentrations and loads in water bodies. A first step is to evaluate how agronomic and environmental soil tests predict P losses to water. Studies using lysimeters have revealed that DPS predicts P in leachate better than a range of agronomic tests (Leinweber *et al.*, 1999). For acid sandy soils in The Netherlands, Schoumans and Groenendijk (2000) concluded from model calculations that the water-extractable P used for fertilizer recommendations was a good predictor of pollution risk. They also demonstrated that it will take several decades to reduce P contents in over-fertilized soils to levels that are acceptable from an environmental perspective. This calls for an immediate reduction of inputs along with increasing removal by crops and stabilizing mobile fractions in P enriched acid sandy soils.

If critical source areas within catchments are identified, the risk of P mobilization and transport to surface waters can be minimized. Surface application of manure can be followed by substantial losses in overland flow, so manure should immediately be incorporated into soil. The method of tillage is also important: mouldboard ploughing can exacerbate erosion by creating pathways for water flow, while least damage occurs from direct drilling or no-tillage systems, which are now common in the grain belt of the USA. Contour ploughing and other practices that minimize erosion are equally valid for minimizing P loss from arable land. For example, McDowell and McGregor (1984) found that no-till cropping reduced TP loss ninefold compared with conventional maize cropping practices, although this was at the expense of an eightfold increase in dissolved P loss.

Possible effects of land use on losses must be considered in conjunction with overall P management. These call for a critical evaluation of possible effects on P losses in advance of large-scale land use changes. The same need applies for setting aside land and Histosol re-wetting measures. However, research in Nordic countries has shown that small-scale measures such as buffer zones, constructed wetlands and ponds may be very efficient in retaining P from agricultural runoff. For example, Uusi-Kämppä *et al.* (2000) reported that total P concentrations decreased by 27–97% (0.24–0.67 kg P ha^{-1} $year^{-1}$), depending on the width of buffer zones. Constructed wetlands and ponds retained 17 and 21% of their P loads, respectively. However, the retention of RP (< 0.45) was inconsistent. Therefore, more data are needed on the availability of P, influences of soil types, and effects of cutting and harvesting.

Finally, the setting of scientifically based threshold values for algal-available P in waters of various quality classes, and the continuation and fulfilment of international conventions such as HELCOM and OSPAR will initiate and assist efforts to reduce agricultural P losses. The proposed Water Framework Directive of the European Union is a novel approach in the right direction. Last but not least, more interdisciplinary and international research is needed to fill in the gaps in our knowledge concerning agricultural P cycles, and the forms, pathways and consequences of P losses from land to water.

Acknowledgements

Part of this work was financially supported by the European Community within the network COST 832 Quantifying the Agricultural Contribution to Eutrophication and within the PROWATER project (contract EVKI-1999–00212). Peter Leinweber thanks his research associates K.-U. Eckhardt and A. Schlichting for their excellent cooperation.

References

Ahuja, L.R., Sharpley, A.N., Yamamoto, M. and Menzel, R.G. (1981) The depth of rainfall–runoff interaction as determined by ^{32}P. *Water Resources Research* 17, 969–974.

Anderson, G., Williams, E.G. and Moir, J.O. (1974) A comparison of the sorption of inorganic orthophosphate and inositol hexaphosphate by six acid soils. *Journal of Soil Science* 25, 51–62.

Baker, J.L., Campbell, K.L., Johnson, H.P. and Hanway, J.J. (1975) Nitrate, phosphorus, and sulfate in subsurface drainage water. *Journal of Environmental Quality* 4, 406–412.

Baldwin, D.S. (1998) Reactive 'organic' phosphorus revisited. *Water Research* 32, 2265–2270.

Barberis, E., Marsan, F.A., Scalenghe, R., Lammers, A., Schwertmann, U., Edwards, A.C., Maguire, R., Wilson, M.J., Delgado, A. and Torrent, J. (1996) European soils overfertilized with phosphorus: Part 1. Basic properties. *Fertilizer Research* 45, 199–207.

Barrow, N.J. (1978) The description of phosphate adsorption curves. *Journal of Soil Science* 29, 447–462.

Barrow, N.J. (1984) Modelling the effects of pH on phosphate adsorption by soils. *Journal of Soil Science* 35, 283–297.

Beck, M.A. and Elsenbeer, H. (1999) Biogeochemical cycles of soil phosphorus in southern alpine spodosols. *Geoderma* 91, 249–260.

Benson, R.L., McKelvie, I.D., Hart, B.T., Truong, Y.B. and Hamilton, I.C. (1996) Determination of total phosphorus in waters and wastewaters by on-line UV/thermal induced digestion and flow injection analysis. *Analytica Chimica Acta* 326, 29–39.

Breeuwsma, A. and Silva, S. (1992) *Phosphorus Fertilisation and Environmental Effects in The Netherlands and the Po Region (Italy)*, Report No. 57, DLO, The Winard Staring Centre for Integrated Land, Soil and Water Research (SC-DLO), Wageningen, The Netherlands.

Breeuwsma, A., Reierink, J.G.A. and Schoumans, O.F. (1995) Impact of manure on accumulation and leaching of phosphate in areas of intensive livestock farming. In: Steele, K. (ed.) *Animal Waste and the Land Water Interface*. Lewis Publishers, New York, pp. 239–249.

Bromfield, S.M. and Jones, O.L. (1972) The initial leaching of hayed-off pasture plants in relation to the recycling of phosphorus. *Australian Journal of Agricultural Research* 23, 811–824.

Brookes, P.C., Powlson, D.S. and Jenkinson, D.S. (1982) Measurement of microbial biomass phosphorus in soil. *Soil Biology and Biochemistry* 14, 319–329.

Brookes, P.C., Powlson, D.S. and Jenkinson, D.S. (1984) Phosphorus in the soil microbial biomass. *Soil Biology and Biochemistry* 16, 169–175.

Burkholder, J.M., Noga, E.J., Hobbs, C.W., Glasgow, H.B. Jr and Smith, S.A. (1992) New 'phantom' dinoflagellate is the causative agent of major estuarine fish kills. *Nature* 358, 407–410.

Burwell, R.E., Schuman, G.E., Heinemann, H.G. and Spomer, R.G. (1977) Nitrogen and phosphorus movement from agricultural watersheds. *Journal of Soil and Water Conservation* 32, 226–230.

Cade-Menun, B.J. and Preston, C.M. (1996) A comparison of soil extraction procedures for ^{31}P NMR spectroscopy. *Soil Science* 161, 770–785.

Chardon, W.J., Oenema, O., del Castilho, P., Vriesema, R., Japenga, J. and Blaauw, D. (1997) Organic phosphorus solutions and leachates from soils treated with animal slurries. *Journal of Environmental Quality* 26, 372–378.

Ciavatta, C., Antisari, L.V. and Sequi, P. (1990) Interference of soluble silica in the determination of orthophosphate-phosphorus. *Journal of Environmental Quality* 19, 761–764.

Cole, C.V., Innis, G.S., and Stewart, J.W.B. (1977) Simulation of phosphorus cycling in semi-arid grassland. Ecology 58, 1–15.

Cross, A.F. and Schlesinger, W.H. (1995) A literature review and evaluation of the Hedley fractionation: Applications to the biogeochemical cycle of soil phosphorus in natural ecosystems. *Geoderma* 64, 197–214.

Damgaard-Poulsen, H. (2000) Phosphorus utilization and excretion in pig production. *Journal of Environmental Quality* 29, 24–27.

Dils, R.M. and Heathwaite, A.L. (1996) Phosphorus fractionation in hillslope hydrological pathways contributing to agricultural runoff. In: Anderson, M. and Brookes, S. (eds) *Advances in Hillslope Processes*. John Wiley & Sons, Chichester, UK, pp. 229–252.

Eichler, B. and Köppen, D. (1998) Interculture – an opportunity to utilize phosphorus reserves in soil. In: Foy, R.H. and Dils, R. (eds) *Practical and Innovative Measures for the Control of Agricultural Phosphorus Losses to Water*. Greenmount College of Agriculture and Horticulture, Northern Ireland, pp. 70–71.

EPA (1995) *Protecting Water Quality in Central Gippsland*, Rep. No. 144. Environment Protection Authority, Melbourne, Australia.

Espinosa, M., Turner, B.L. and Haygarth, P.M. (1999) Pre-concentration and separation of trace phosphorus compounds in soil leachate. *Journal of Environmental Quality* 28, 1497–1504.

European Environment Agency (1998) *Europe's Environment: the Second Assesment*. EEA, Copenhagen.

Fardeau, J.C. (1996) Dynamics of phosphate in soils. An isotopic outlook. *Fertilizer Research* 45, 91–100.

Fleischer, E. (1997) Nutztierhaltung und nationale Phosphorbilanz des Bereiches Landwirtschaft. *Archiv Acker- und Pflanzenbau und Bodenkunde* 41, 1–73.

Foy, R.H. and Dils, R. (1998) Practical and innovative measures for the control of agricultural phosphorus losses to water. An OECD sponsored workshop, Antrim, Dept. of Agriculture for Northern Ireland.

Fraser, A.I., Harrod, T.R. and Haygarth, P.M. (1999) The effect of rainfall intensity on soil erosion and particulate phosphorus transfer from arable soils. *Water Science and Technology* 39, 41–45.

Frossard, E., Stewart, J.W.B. and St Arnaud, R.J. (1989) Distribution and mobility of phosphorus in grassland and forest soils of Saskatchewan. *Canadian Journal of Soil Science* 69, 401–416.

Frossard, E., Tekeley, P. and Grimal, J.Y. (1994) Characterization of phosphate species in urban sewage sludges by high-resolution solid-state ^{31}P-NMR. *European Journal of Soil Science* 45, 403–408.

Frossard, E., Brossard, M., Hedley, M.J. and Metherell, A. (1995) Reactions controlling the cycling of P in soils. In: Tiessen, H. (ed.) *Phosphorus in the Global Environment*. John Wiley & Sons, New York, pp. 107–138.

Frossard, E., Condron, L., Oberson, A., Sinaj, S. and Fardeau, J.C. (2000) Processes governing phosphorus availability in temperate soils. *Journal of Environmental Quality* 29, 15–23.

Grant, R., Laubel, A., Kronvang, B., Anderson, H.E., Svendsen, L.M. and Fugslang, A. (1996) Loss of dissolved and particulate phosphorus from arable catchments by subsurface drainage. *Water Research* 30, 2633–2642.

Haith, D.A. and Shoemaker, L.L. (1987) Generalized watershed loading functions for stream flow nutrients. *Water Resources Bulletin* 23, 471–478.

Halliwell, D., Coventry, J. and Nash, D. (2000) Inorganic monophosphate determination in overland flow from irrigated grazing systems. *International Journal of Environmental Analytical Chemistry* 76, 77–78.

Harris, R.A., Heathwaite, A.L. and Haygarth, P.M. (1995) High temporal resolution sampling of P exported from grassland soil during a storm, and the impact of slurry additions. In: Johnston, A.E. (ed.) *Proceedings of the International Workshop on Phosphorus Loss to Water from Agriculture*. TEAGASC, Johnstown Castle, Wexford, Ireland, pp. 25–26.

Harrison, A.F. (1987) *Soil Organic Phosphorus – a Review of World Literature*. CAB International, Wallingford, UK.

Haygarth, P.M. and Jarvis, S.C. (1997) Soil derived phosphorus in surface runoff from grazed grassland lysimeters. *Water Research* 31, 140–148.

Haygarth, P.M. and Jarvis, S.C. (1999) Transfer of phosphorus from agricultural soils. *Advances in Agronomy* 66, 195–249.

Haygarth, P.M. and Sharpley, A.N. (2000) Terminology for phosphorus transfer. *Journal of Environmental Quality* 29, 10–15.

Haygarth, P.M., Ashby, C.D. and Jarvis, S.C. (1995) Short-term changes in the molybdate reactive phosphorus of stored soil waters. *Journal of Environmental Quality* 24, 1133–1140.

Haygarth, P.M., Warwick, M.S. and House, W.A. (1997) Size distribution of colloidal molybdate reactive phosphorus in river waters and soil solution. *Water Research* 31, 439–442.

Haygarth, P.M., Hepworth, L. and Jarvis, S.C. (1998) Forms of phosphorus transfer in hydrological pathways from soil under grazed grassland. *European Journal of Soil Science* 49, 65–72.

Heckrath, G., Brookes, P.C., Poulton, P.R. and Goulding, K.W.T. (1995) Phosphorus leaching from soils containing different phosphorus concentrations in the Broadbalk Experiment. *Journal of Environmental Quality* 24, 904–910.

Hedley, M.J. and Stewart, J.W.B. (1982) Method to measure microbial phosphate in soils. *Soil Biology and Biochemistry* 14, 377–385.

Hedley, M.J., Stewart, J.W.B. and Chauhan, B.S. (1982) Changes in inorganic and organic soil phosphorus fractions induced by cultivation practices and by laboratory incubations. *Soil Science Society America Journal* 46, 970–976.

Hudson, N.W. (1995) *Soil Conservation.* Batsford, London.

Isermann, K. (1999) *Phosphor (P)-Bilanzen der Landwirtschaft Europas vor dem Hintergrund der nachhaltigen Entwicklung ihrer P-Haushalte (COST Action 832).* VDLUFA-Kongressband 111, Darmstadt, Germany.

Jongbloed, A.W. and Everts, H. (1992) Apparent digestible phosphorus in the feeding of pigs in relation to availability requirement and environment. 1. Digestible phosphorus in feedstuffs of plant and animal origin. *Netherlands Journal of Agricultural Science* 40, 123–136.

Kalbitz, K., Rupp, H., Meißner, R. and Braumann, F. (1999) Folgewirkungender Renaturierung eines Niedermoores auf die Stickstoff-, Phosphor- und Kohlenstoffgehalte im Boden und Grundwasser. *Zeitschrift für Kulturtechnik und Landentwicklung* 40, 22–28.

Kaplan, D.I., Sumner, M.E., Bertsch, P.M. and Adriano, D.C. (1996) Chemical conditions conducive to the release of mobile colloids from Ultisol profiles. *Soil Science Society of America Journal* 60, 269–274.

Khalid, R.A., Patrick, W.H. Jr and DeLaune, R.D. (1977) Phosphorus sorption characteristics of flooded soils. *Soil Science Society of America Journal* 41, 305–310.

Kretzschmar, R., Borkovec, M., Grolimund, D. and Elimelech, M. (1999) Mobile subsurface colloids and their role in contaminant transport. *Advances in Agronomy* 66, 121–193.

Lee, D., Han, X.G. and Jordan, C.F. (1990) Soil phosphorus fractions, aluminium, and water retention as affected by microbial activity in an Ultisol. *Plant and Soil* 121, 125–136.

Leinweber, P. (1996) Phosphorus fractions in soils from an area with high density of livestock population. *Zeitschrift für Pflanzenernährung und Bodenkunde* 159, 251–256.

Leinweber, P., Jordan, E. and Geyer-Wedell, K. (1993) *Phosphorversorgung der Böden im agrarischen Intensivgebiet Südoldenburg.* Vechtaer Druckerei und Verlag, Vechta, Germany.

Leinweber, P., Lünsmann, F. and Eckhardt, K.U. (1997a) Phosphorus sorption capacities and saturation of soils in two regions with different livestock densities in northwest Germany. *Soil Use and Management* 13, 82–89.

Leinweber, P., Haumaier, L. and Zech, W. (1997b) Sequential extractions and ^{31}P NMR spectroscopy of phosphorus forms in animal manures, whole soils and particle-size fractions from a densely populated livestock area in northwest Germany. *Biology and Fertility of Soils* 25, 89–94.

Leinweber, P., Meissner, R., Eckhardt, K.U. and Seeger, J. (1999) Management effects on forms of phosphorus in soil and leaching losses. *European Journal of Soil Science* 50, 413–424.

Lookman, R., Grobet, P., Merckx, R. and Vlassak, K. (1994) Phosphate sorption by synthetic amorphous aluminum hydroxides: a ^{27}Al and ^{31}P solid-state MAS NMR spectroscopy study. *European Journal of Soil Science* 45, 37–44.

Lopez-Hernandez, D., Brossard, M. and Frossard, E. (1998) P-isotopic exchange values in relation to Po mineralization in soils with very low P-sorbing capacities. *Soil Biology and Biochemistry* 30, 1663–1670.

Lynch, P.B. and Caffrey, P.J. (1997) Phosphorus requirements for animal production. In: Tunney, H., Carton, O.T., Brookes, P.C. and Johnston, A.E. (eds) *Phosphorus Loss from Soil to Water.* CAB International, Wallingford, UK, pp. 283–296.

Magid, J. and Nielsen, N.E. (1992) Seasonal variation in organic and inorganic phosphorus fractions of temperate-climate sandy soils. *Plant and Soil* 144, 155–165.

Marcinkowski, T. and Sapek, A. (1998) Phosphorus balances in selected farms of Elblag province. In: *Phosphorus in Agriculture and Water Quality Protection.* Institute for Land Reclamation and Grassland Farming (IMUZ) Publisher, Falenty, Poland, pp. 35–43.

Matthews, R.A., Preedy, N., Heathwaite, A.L. and Haygarth, P.M. (1998) Transfer of colloidal forms of phosphorus from grassland soils. In: *Proceedings of the International Association on Water Quality 3rd International Conference on Diffuse Pollution,* Vol. II. SEPA, Edinburgh, Scotland, pp. 11–16.

McDowell, L.L. and McGregor, K.C. (1984) Plant nutrient losses in runoff from conservation tillage corn. *Soil Tillage Research* 4, 79–91.

Meissner, R., Rupp, H., Seeger, J. and Schonert, P. (1995) Influence of mineral fertilizers and different soil types on nutrient leaching: results of lysimeter studies in East Germany. *Land Degradation and Rehabilitation* 6, 163–170.

Meissner, R., Rupp, H. and Seeger, J. (1997) Threat of phosphorus leaching from intensively farmed agricultural soils in the central reaches of the River Elbe. In: Tunney, H., Carton, O.T., Brookes, P.C. and Johnston, A.E. (eds) *Phosphorus Loss from Soil to Water*. CAB International, Wallingford, UK, pp. 430–433.

Mengel, K. and Kirkby, E.A. (1987) *Principles of Plant Nutrition*. International Potash Institute, Worblaufen-Bern, Switzerland.

Moore, P.A. Jr, Daniel, T.C. and Edwards, D.R. (2000) Reducing phosphorus losses and inhibiting ammonia loss from poultry manure with aluminum sulphate. *Journal of Environmental Quality* 29, 37–49.

Morel, C., Tunney, H., Plénet, C. and Pellerin, S. (2000) Transfer of phosphate ions between soil and solution: perspectives in soil testing. *Journal of Environmental Quality* 29, 50–59.

Morgan, R.P.C. (1995) *Soil Erosion and Conservation*. Longman Scientific and Technical, Harlow, UK.

Moss, B. (1988) *Ecology of Fresh Waters: Man and Medium*. Blackwell Scientific Publications, London.

Moss, B. (1996) A land awash with nutrients – the problem of eutrophication. *Chemistry and Industry* 11, 407–411.

Mozaffari, M. and Sims, J.T. (1994) Phosphorus availability and sorption in an Atlantic coastal plain watershed dominated by animal-based agriculture. *Soil Science* 157, 97–107.

Murphy, J. and Riley, J.P. (1962) A modified single solution method for the determination of phosphate in natural waters. *Analytica Chimica Acta* 27, 31–36.

Nash, D. and Murdoch, C. (1997) Phosphorus in runoff from a fertile dairy pasture. *Australian Journal of Soil Research* 35, 419–429.

Newman, R.H. and Tate, K.R. (1980) Soil phosphorus characterization by ^{31}P nuclear magnetic resonance. *Communications in Soil Science and Plant Analysis* 11, 835–842.

Oloffs, K. (1998) Possibilities for reducing phosphorus excretion in poultry feeding. In: *Phosphorus in Agriculture and Water Quality Protection*. Institute for Land Reclamation and Grassland Farming (IMUZ) Publisher, Falenty, Poland, pp. 77–84.

Ozanne, P.G., Kirton, D.J. and Shaw, T.C. (1961) The loss of phosphorus from sandy soils. *Australian Journal of Experimental Agriculture* 12, 409–423.

Rolston, D.E., Rauschkolb, R.S. and Hoffman, D.L. (1975) Infiltration of organic phosphate compounds in soil. *Soil Science Society of America* 39, 1089–1094.

Ron Vaz, M.D., Edwards, A.C., Shand, C.A. and Cresser, M. (1992) Determination of dissolved organic phosphorus in soil solutions by an improved automated photo-oxidation procedure. *Talanta* 39, 1479–1487.

Ron Vaz, M.D., Edwards, A.C., Shand, C.A. and Cresser, M.S. (1993) Phosphorus fractions in soil solution: Influence of soil acidity and fertilizer additions. *Plant and Soil* 148, 179–183.

Rowland, A.P. and Haygarth, P.M. (1997) Determination of total dissolved phosphorus in soil solutions. *Journal of Environmental Quality* 26, 410–415.

Ryan, J.N. and Gschwend, P.M. (1990) Colloid mobilization in two Atlantic coastal plain aquifers: Field studies. *Water Resources Research* 26, 307–322.

Ryden, J.C., Syers, J.K. and Harris, R.F. (1973) Phosphorus in runoff and streams. *Advances in Agronomy* 25, 1–45.

Sample, E.C., Soper, R.J. and Racz, G.J. (1980) Reactions of phosphate fertilizers in soils. In: Khasawneh, F.E. *et al.* (eds) *The Role of Phosphorus in Agriculture*. ASA, Madison, Wisconsin, pp. 263–310.

Schoumans, O. and Groenendijk, P. (2000) Modeling soil phosphorus levels and phosphorus leaching from agricultural land in The Netherlands. *Journal of Environmental Quality* 29, 111–116.

Seeling, B. and Zasocki, R.J. (1993) Microbial effects in maintaining organic and inorganic solution phosphorus concentrations in a grassland topsoil. *Plant and Soil* 148, 277–284.

Sharpley, A.N. (1985) The selective erosion of plant nutrients in runoff. *Soil Science Society of America Journal* 49, 1527–1534.

Sharpley, A.N. and Smith, S.J. (1990) Phosphorus transport in agricultural runoff: the role of soil erosion. In: Boardman, J., Foster, L.D.L. and Dearing, J.A. (eds) *Soil Erosion on Agricultural Land.* John Wiley & Sons, Chichester, UK, pp. 351–366.

Sharpley, A.N. and Syers, J.K. (1979) Phosphorus inputs into a stream draining an agricultural watershed II. Amounts and relative significance of runoff types. *Water, Air and Soil Pollution* 9, 417–428.

Sharpley, A.N. and Tunney, H. (2000) Phosphorus research strategies to meet agricultural and environmental challenges of the 21th Century. *Journal of Environmental Quality* 29, 176–181.

Sharpley, A.N., Syers, J.K. and Springett, J.A. (1979) Effect of surface casting earthworms on the transport of phosphorus and nitrogen in surface runoff from pasture. *Soil Biology and Biochemistry* 11, 459–462.

Sharpley, A.N., Menzel, R.G., Smith, S.J., Rhoades, E.D. and Olness, A.E. (1981) The sorption of soluble phosphorus by soil material during transport in runoff from cropped and grassed watersheds. *Journal of Environmental Quality* 10, 211–215.

Sharpley, A.N., Jones, C.A., Gray, C., Cole, C.V., Tiessen, H. and Holzhey, C.S. (1985) A detailed phosphorus characterization of seventy eight soils. USDA Agricultural Research Services, ARS 31.

Sibbesen, E. and Sharpley, A.N. (1997) Setting and justifying upper critical limits for phosphorus in soils. In: Tunney, H., Carton, O.T., Brookes, P.C. and Johnston, A.E. (eds) *Phosphorus Loss from Soil to Water.* CAB International, Wallingford, UK, pp. 151–176.

Simard, R., Beaucheim, S. and Haygarth, P.M. (2000) Potential for preferential pathways of phosphorus transport. *Journal of Environmental Quality* 29, 97–105.

Sinaj, S., Mächler, F., Frossard, E., Fäisse, C., Oberson, A. and Morel, C. (1998) Interferences of colloidal particles in the determination of orthophosphate concentrations in soil water extracts. *Communications in Soil Science and Plant Analysis* 29, 1091–1105.

Spier, T.W. and Ross, D.J. (1978) Soil phosphatase and sulphatase. In: Burns, R.G. (ed.) *Soil Enzymes.* Academic Press, London, pp. 197–250.

Stamm, C., Flühler, H. and Gächter, R. (1998) Preferential transport of phosphorus in drained grassland soils. *Journal of Environmental Quality* 27, 515–522.

Stewart, J.W.B. and Sharpley, A.N. (1987) Controls on dynamics of soil and fertilizer phosphorus and sulphur. In: Follett, R.F., Stewart, J.W.B. and Cole, C.V. (eds) *Soil Fertility and Organic Matter as Critical Components of Production.* SSSA Special Publication 19, American Society of Agronomy, Madison, Wisconsin, pp. 102–121.

Tarapchak, S.J. (1993) Soluble reactive phosphorus in lake water: evidence for molybdate enhanced hydrolysis. *Journal of Environmental Quality* 12, 105–108.

Tiessen, H., Stewart, J.W.B. and Moir, J.O. (1983) Changes in organic and inorganic phosphorus composition of two grassland soils and their particle size fractions during 60–90 years of cultivation. *Journal of Soil Science* 34, 815–823.

Tunney, H., Carton, O.T., Brookes, P.C. and Johnston, A.E. (eds) (1997) *Phosphorus Loss from Soil to Water.* CAB International, Wallingford, UK.

Turner, B.L. and Haygarth, P.M. (2000) Phosphorus forms and concentrations in leachate under four grassland soil types. *Soil Science Society of America Journal* 64, 1090–1097.

Turner, B.L. and Haygarth, P.M. (2001) Phosphorus solubilisation in rewetted soils. *Nature* 411, 258.

Turner, B.L. and Haygarth, P.M. (2002) Organic phosphorus characterization by phosphatase hydrolysable phosphorus techniques: applications to soil extracts and runoff waters. In: Whitton, B.A. and Hernández, I. (eds) *Phosphatases in the Environment.* Kluwer Academic Press, The Netherlands (in press).

Turner, B.L., Papházy, M., Haygarth, P.M. and McKelvie, I.D. (2002) Inositol phosphates in the environment. *Philosophical Transactions of the Royal Society, London,* p. 357.

Turner, B.L., Bristow, A.W. and Haygarth, P.M. (2001) Rapid estimation of soil microbial biomass in grassland soils by ultra-violet absorbance. *Soil Biology and Biochemistry* 33, 913–919.

Turtola, E. and Jaakkola, A. (1995) Loss of phosphorus by surface runoff and leaching from a heavy clay soil under barley and grass ley in Finland. *Acta Agriculture Scandinavia, Section B, Soil and Plant Science* 45, 159–165.

Ulen, B. (1999) Leaching and balances of phosphorus and other nutrients in lysimeters after application of organic manures or fertilizers. *Soil Use and Management* 15, 56–61.

Umweltbundesamt (1997) *Daten zur Umwelt – Zustand der Umwelt in Deutschland.*

Uusi-Kämppä, J., Braskerut, B., Jansson, H., Syversen, N. and Uusitalo, R. (2000) Buffer zones and constructed wetlands as filters for agricultural phosphorus. *Journal of Environmental Quality* 29, 151–158.

Valk, H., Methcalf, J.A. and Withers, P.J.A. (2000) Prospects for minimizing phosphorus excretion in ruminants by dietary manipulation. *Journal of Environmental Quality* 29, 28–36.

Von Hammel, M.-L. (1994) *Mengen- und Nährstoffbilanz organischer Rest- und Abfallstoffe für die Landkreise Cloppenburg und Vechta und das Oldenburger Münsterland.* Vechtaer Druckerei und Verlag, Vechta, Germany.

Werner, W. and Brenk, C. (1997) *Regionalisierte und einzelbetriebliche Nährstoffbilanzierung als Informationsgrundlage zur gezielten Quantifizierung der Wirkungspotentiale von Maßnahmen zur Vermeidung auftretender Überschüsse.* Forschungsberichte, Nr. 46, Rheinische Friedrich-Wilhelms-Universität Bonn.

Werner, W. and Piehl, U. (1998) *Erhebungsuntersuchungen zu Phosphatgehalten, Phosphatsorptionskapazität und relativer Phosphatsättigung der Böden in den veredlungsstarken Regionen von NRW als Prognosekriterien des potentiellen P-Austrages in Drain- und Grundwasser.* Forschungsberichte, Nr. 64, Rheinische Friedrich-Wilhelms-Universität Bonn.

Whitton, B.A., Grainger, S.L.J., Hawley, G.R.W. and Simon, J.W. (1991) Cell-bound and extracellular phosphatase activities of cyanobacterial isolates. *Microbial Ecology* 21, 85–98.

Wild, A. (1988) Plant nutrients in soil: phosphate. In: Wild, A. (ed.) *Russell's Soil Conditions and Plant Growth*, 10th edn. Longman Scientific and Technical, Harlow, UK, pp. 695–742.

Wild, A. and Oke, O.L. (1966) Organic phosphate compounds in calcium chloride extracts of soils: Identification and availability to plants. *Journal of Soil Science* 17, 356–371.

Zhang, T.Q. and MacKenzie, A.F. (1997) Changes of soil phosphorus fractions under long-term corn monoculture. *Soil Science Society of America Journal* 61, 485–493.

3 Manures

D.R. Chadwick[1] and S. Chen[2]
[1] *Institute of Grassland and Environmental Research, North Wyke Research Station, Okehampton, Devon EX20 2SB, UK;* [2] *Centre for Land and Water Stewardship, University of Guelph, Guelph, Ontario N1G 2W1, Canada*

Key words: biological oxygen demand, dirty water, farmyard manure, heavy metals, nitrogen, phosphorus, pathogens, slurry

Introduction

The increase in agricultural production over the last five decades has resulted in nutrient surpluses in some areas (Brouwer and Hellegers, 1997), which are usually associated with intensive livestock production. Pre-war livestock farming was more extensive and self-sufficient with the emphasis on recycling nutrients and organic matter to the land via animal manures. However, this recycling philosophy has been replaced by one that resembles a disposal attitude, and is the product of very successful, yet regionalized livestock rearing industries.

Whilst the nutrient and organic matter content of a manure is desirable for plant growth, long-term fertility and soil structure in terrestrial systems, transport into aqueous systems can ultimately result in those nutrients, particularly N, P and C, being utilized by aquatic organisms. The consequences of this are eutrophication of watercourses, which manifests itself as toxic algal blooms that are poisonous to livestock and a serious health risk to humans (Kotak *et al.*, 1993), and the depletion of O_2. The result is loss of desirable plants and animals. Other potential pollutants, such as heavy metals and pathogens, can reach ground- and surface water and affect drinking water quality for both livestock and humans (see Jones *et al.* Chapter 6 and De Vries *et al.* Chapter 5, this volume).

Mis-management of manures can result in both point source pollution events (NRA, 1995) (e.g. catastrophic leaks from slurry stores) and diffuse transfers of nutrients and organic matter to watercourses (Sharpley and Withers, 1994; Brenner and Mondok, 1995). Point source pollution events can be controlled with increased farmer vigilance and additional or improved manure storage facilities. In the UK, the total number of pollution incidents from organic manures declined by *c.* 20% between 1985 and 1989. The number of incidents resulting from land runoff increased significantly during the same period, suggesting that incidents of pollution from point sources such as over-spilling slurry stores had declined (NRA, 1992). Therefore, the focus and challenge of managing manures for water quality has now shifted to control of transfers of nutrients from land receiving manures.

In this chapter, we outline the problems associated with managing manures at land spreading and the effect that manures can have on water quality. Other chapters deal with some of these problems in greater detail, but 'manures' themselves also warrant special consideration. We highlight problems with specific examples from the literature and, where appropriate, offer potential solutions. It is appropriate to begin by defining manure.

Definition of manure

Manure is the generic term used to describe animal faeces and urine that are deposited in buildings or on collection yards and then require handling; for example storage and spreading to land (as opposed to dung and urine voided while grazing). Manure can be subdivided into three broad groups according to its physical nature. Some fundamental characteristics are summarized in Table 3.1.

'Solid manure', is a heterogeneous mixture of dung, urine and bedding material, such as straw or sawdust and is commonly called farmyard manure (FYM). Poultry manures may be produced in the absence of bedding. Typically solid manures have a higher dry matter content than slurries and are stacked in heaps until opportunity arises for spreading.

'Slurries' are relatively uniform mixes of dung and urine produced by cattle or pigs kept in buildings with slatted floors or scraped passages in the absence of bedding material. Slurries can be pumped to storage tanks before being spread by a number of different application techniques.

'Dirty water' is generally a product of cattle-farms and is a dilute effluent with low nutrient content. It comprises dairy parlour washings and yard runoff, but it can also contain silage effluent and some animal excreta. It is produced in large

Table 3.1. Typical nutrient content of different manure types (on a fresh weight basis). (Derived from MAFF 2000.)

	Dry matter (%)	Total N (kg t^{-1} FYM or kg m^{-3} slurry)	Total P (kg t^{-1} FYM or kg m^{-3} slurry)	Total K (kg t^{-1} FYM or kg m^{-3} slurry)
Dirty water	< 1	0.3	trace	0.2
Cattle slurry	6	3.0	0.5	2.9
Pig slurry	6	5.0	1.3	2.5
Cattle FYM	25	6.0	1.5	6.6
Pig FYM	25	7.0	3.0	4.1
Layer manure	30	16.0	5.7	7.5
Broiler litter	60	30.0	10.9	14.9

quantities and generally creates a disposal problem. Because of its low dry matter content it is 'flowable' and is often applied to land through low rate irrigation systems.

Potential pollutants associated with manures

Nitrogen

Different animal types are fed with different diets as they have different abilities to assimilate nutrients from their feed (McDonald *et al.*, 1995). Utilization of N in feed ranges from 20 to 25% for dairy cattle, 5 to 10% for beef cattle and 3 to 15% for sheep and lambs (Holmes, 1970; Henzell and Ross, 1973); excess N is voided in dung and urine. Hence, manures from different animal groups have different total N contents (MAFF, 2000; Smith and Frost, 2000; Smith *et al.*, 2000) (Table 3.1).

Nitrogen is present in manures in two principal forms, organic-N and inorganic-N. Ammonium is the main inorganic form of N, although NO_3^- is present in composted solid manure; both forms are generally considered to be plant available, although grasses often show a greater uptake of NH_4^+-N than of NO_3^--N (Clarkson *et al.*, 1986). Organic-N has to be mineralized to inorganic form before it is available to crops. Over-application of manure and hence over-application of N can result in excess N remaining in the soil at times of low crop uptake. The excess N is then at risk of transfer to watercourses (Chambers *et al.*, 2000) and the atmosphere (see Fig. 3.1). Because of the deemed health risks associated with high NO_3^- levels in potable water supplies, legislation within Europe via the European Drinking Water Directive (EC, 1991) and in other non-European countries such as the USA, has set maximum acceptable concentrations of NO_3^--N (10–11.3 mg l^{-1}) for recharging groundwater.

Nitrogen runoff into surface waters is problematic for the environment. Concentrations of 0.3 mg inorganic N l^{-1} have been identified as levels above which excessive algal growth or eutrophication can occur (Nichols *et al.*, 1994). In the case of NH_4^+, the tolerance limit may be as low as 0.1 mg N l^{-1} (Gangbazo *et al.*, 1995). Figure 3.1 illustrates the potential of manures to pollute both the gaseous and aqueous environments with N species.

Phosphorus

Inorganic P is present in animal feed and only a small percentage (30%) is utilized by the animal (Tamminga, 1992). Consequently, animal manures contain significant quantities of P, some of which is in organic forms (Gerritse and Vriesema, 1984). Some of the organic P compounds are relatively stable and mobile in soils (Chardon *et al.*, 1997). Other forms display sorption/desorption kinetics with clay minerals and can be rapidly immobilized in the upper few centimetres of soil (Haygarth and Jarvis, 1999). It is well documented that the P content of soil increases with manure applications (Leinweber *et*

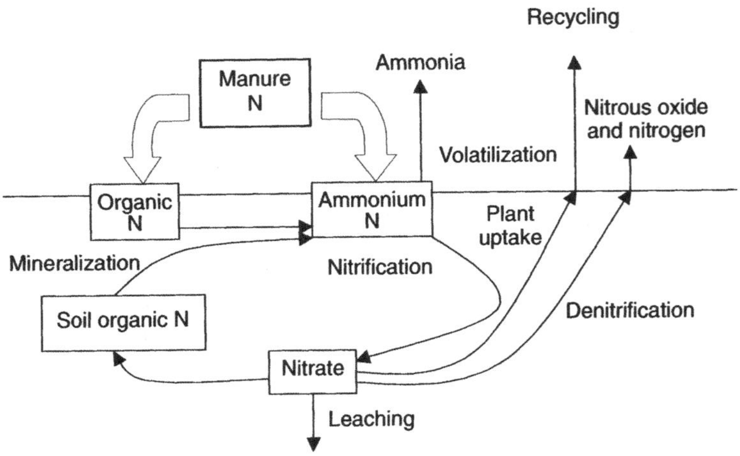

Fig. 3.1. The potential of manures to pollute the aqueous and gaseous environments with nitrogen (adapted from Pain, 2000).

al., 1997). Therefore, potential surface water contamination (eutrophication) exists in areas where surface runoff is likely. Concentrations of > 10 μg inorganic-P l^{-1} can result in excessive algal growth (Nichols *et al.*, 1994).

Organic matter

Agricultural livestock are fed forage and grain-based diets, with high organic matter contents. Hence their manures also contain a large proportion of organic material. Many organic compounds present in manure are readily subjected to microbial breakdown in the soil, or if they enter directly adjacent surface water, they can create a large O_2 demand (MAFF, 1998a) resulting in asphyxiation of aquatic life forms. The magnitude of this BOD varies with manure type and can be measured using a 5 day incubation assay in the laboratory (HMSO, 1980). The measurement is expressed in milligrams of O_2 that can be removed from a litre of water at 25°C over 5 days. Table 3.2 shows typical BOD values for several manure types and the recommended maximum BOD level for discharge to water systems in the UK.

Levels of BOD will depend on the collection system for the dirty water (Cumby *et al.*, 1999) or slurry. If silage effluent is allowed into the storage facility then BOD levels will be increased significantly, as silage effluent has a typical BOD of 80,000 mg l^{-1}. In contrast, ingress of clean rainwater into stores will dilute the effluent.

Heavy metals

Metals such as Cu and Zn are added to animal feed or given as supplements to aid nutrient assimilation and prevent scouring period. As a consequence animal manures can contain high concentrations of heavy metals (Cu, Zn, Fe, Mn, Co, Cd) (L'Herroux *et al.*, 1997; Nicholson *et al.*, 1999).

Pathogens

Livestock manures contain large numbers of microorganisms, including some potential pathogens (Kearney *et al.*, 1993) and recently concern has been expressed about possible distribution of these pathogens to the human population. Many pathogens are excreted in the faeces of infected, and in some cases, healthy 'carrier' animals. Pathogens can include bacteria (Hinton and Bale, 1991) such as *Escherichia coli* O157 and *Salmonella* spp., viruses like *Rotavirus* spp. and protozoa (Mawdsley *et al.*, 1995), for example *Cryptosporidium* and *Giardia* spp. Some pathogens are

Table 3.2. Typical biological oxygen demand and nutrient concentrations of slurries and dirty water.

	BOD ($mg\ l^{-1}$)	MRP ($mg\ l^{-1}$)	NH_4^+-N ($mg\ l^{-1}$)	NO_3^--N ($mg\ l^{-1}$)
Dirty water	500	30	200	< 10
Cattle slurry	10,000	100	1500	< 10
Pig slurry	25,000	150	3000	< 10
Recommended max. concs	20[a]	0.1[b]	–	11.3 NO_3-N[c] or 50 NO_3^-

[a] HMSO (1980).
[b] Sharpley and Smith (1989).
[c] EC (1991).

obligate parasites and are of limited concern; others can survive in the open environment for long periods, sometimes in the form of transitive cysts or spores. Specific factors influencing the movement of microorganisms through and across soils are related to sorption properties and size of the pathogen, continuous water films, pore size and hydrophobicity.

Liquid manure tends to be more uniformly contaminated than the solid manure, because of the greater mobility of bacteria in the liquid phase compared with the solid phase. Pathogens may be present in manure even if the animals present no symptoms, and a small number of infected animals can contaminate a whole source of manure (Strauch, 1987). Consequently, the greater the number of animals on a farm, the greater the likelihood of pathogens being present in the manure.

The bacterial groups present in greatest number in manure are faecal coliforms and streptococci. While bacteria species from these two groups are always present in manure, salmonella (another important group of pathogens) is present occasionally, mostly in pig and poultry manure.

Problems

The principal reasons for contamination of watercourses by manures are untimely applications at rates of nutrient loading in excess of crop requirements. Land availability is one factor which dictates nutrient loading rate. Another is nutrient content and availability.

Rates of application

In Europe, there are in excess of 100 million dairy cattle and there is a similar number in the USA (Eurostat, 1997). On average, an adult dairy cow produces *c.* 20 t of slurry each year (Smith and Frost, 2000). On an annual basis, *c.* 50% of this may require storage and spreading: a similar amount is voided in the field while grazing. The manure output from European cattle that requires management is therefore estimated to be in the region of 1000 Mt $year^{-1}$. Other classes of livestock also produce significant quantities of manure.

If manure were spread to all available grass and arable land, application rates would be much lower than surveys suggest (Smith *et al.*, 1998). For example, in the UK manure is only applied to 18% of the arable land and 48% of the grassland (Burnhill *et al.*, 1994). This results in average manure-P loadings of 80 and 16 kg ha^{-1} to arable and grassland, respectively, instead of 14 and 8 kg ha^{-1} if manure were spread on all the arable and grassland available in the UK (Smith *et al.*, 1998). This nutrient accumulation increases the risk of transfer to watercourses. There are several reasons why manure applications are limited to a fraction of the available area of land and these are outlined below.

Livestock production is often regionalized within a country so that transport costs for manure to other areas are the limiting factor, hence application rates can be high in the region of production. For example, Brittany is the main region in France for the pig and poultry industry. It produces > 50% of the pigs and > 41% of the poultry, within 7% of the agricultural area of France (Dourmad and Aumaitre, 1990), resulting in a density of 337 pigs ha^{-1} in Brittany compared with 40 in the rest of France. At the farm-scale, not all the land receives equal application rates of manure. There can be several reasons why some fields receive more manure than others: for example, proximity to the manure store (transport costs), and land used for growing silage and hay is often avoided as a precautionary measure to reduce the possibility of poor silage or hay quality. The result is that some fields, particularly maize stubbles in autumn, receive large quantities of manure each year. Repeated applications of manure increase the nutrient status of the soil (Leinweber *et al.*, 1997) and increase the risk of transport of potential pollutants to watercourses.

At the field-scale, manures are inherently more difficult to spread evenly than inorganic fertilizers in granular or liquid form. Not only can this result in an uneven yield of crop but excess nutrients can remain in the soil after harvest. Recent advances in spreading efficiency have decreased the coefficient of variation in spread pattern (Smith and Baldwin, 1999).

There are a limited number of occasions during the winter period when soil conditions are suitable to allow spreading machinery to travel on to fields without risk of compaction. Because of this, application rates tend to be greater than those recommended in order to empty ever-filling storage tanks. Inadequate storage capacity often results in manure applications at times when there is a low crop demand, for example to frozen soil. Winter application is not recommended because of the risk of pollutant transfer to water (MAFF, 1998a).

Composition of manures

The chemical and physical composition of manure varies with animal type and growth stage, diet, manure collection design (e.g. amount of clean water allowed into the collection system) and length of time in storage. Some examples of the nutrient content of commonly produced manures are shown in Table 3.1.

One of the major reasons that farmers do not take full account of the nutrients in manures is the uncertainty surrounding the nutrient content. Most of these manures represent a substantial fertilizer resource. Based on UK agricultural census data for 1993 (HMSO, 1993) approximately 80 Mt of excreta were collected from buildings and yards. The potential value of the N, P and K in these manures at current fertilizer prices (Smith and Chambers, 1995) is estimated to be £200 million $year^{-1}$. Annual statistics on fertilizer use, derived from the British Survey of Fertilizer Practice (Burnhill *et al.*, 1994) showed that approx. 1.1 Mt of N, 207,000 t of P and 342,000 t of K were purchased as 'artificial' fertilizer in 1993. This compares with the estimated nutrient content of managed manures for the same year: 447,000 t N, 119,00 t P and 334,000 t K.

There is considerable variation about the typical nutrient contents (Table 3.1), and farmers risk under-application of nutrients and poor crop response or over-application and risk of transfer of nutrients to watercourses. On-farm tests are now available for estimating N (Kjellerup, 1996), P and K contents of slurries.

Availability of nutrients

Knowledge of the total nutrient content is only part of the equation in matching nutrient application to crop demand. An additional complication is that only a percentage of the total nutrient content of manure is in a form that is readily available to plants. So, the form of nutrient and its rate of supply to the crop are also important. For example, inorganic N is required for crop uptake, yet up to 90% of the total N in cattle FYM may be in the organic form and will need to be mineralized before it is available to plants (Chadwick *et al.*, 2000a). Mineralization and release of organically bound nutrients at times of low crop demand can result in increased risk of transfer to watercourses. The availability of the nutrient depends on manure composition, for example the C:N ratio (Chadwick *et al.*, 2000a), and temperature (Sierra, 1997). Chadwick *et al.* (2000a) found that 56% of the organic N in poultry litter was mineralized within 199 days of application. Also, Bitzer and Sims (1988) observed a similar rate of mineralization. For other manure types, significantly lower rates of mineralization have been found. The overlying control is the C to organic N ratio; generally, organic materials with a ratio > 15 will immobilize N and those with ratios < 15 will mineralize organic N. The lower the C:N ratio, the greater will be the rate of mineralization (Kirchmann, 1985). Chadwick *et al.* (2000a) demonstrated that the order of mineralization was poultry manure > pig manures > cattle manures.

The physical and chemical characteristics of some manures mean that they have an inherent potential to pollute watercourses, if mismanaged. In Table 3.2 we compare the typical concentration of several nutrients and BOD of the lower dry matter manures and compare these with the maximum recommended concentrations for watercourses. Some of these recommendations are fixed in national legislation; for example, the EC Nitrate Directive (EC, 1991) limits NO_3^--N concentrations to 11.3 mg l^{-1}. For other nutrients (e.g. P) there is no statutory legislation, but the concentration in surface water at which algae will grow is 0.1 mg l^{-1} (Sharpley and Smith, 1989).

Pollutant Transfers from Manures to Watercourses

How and where are pollutants most vulnerable for transfer?

There are two categories of transport of nutrients from manures to watercourses: (i) point source; and (ii) diffuse transfers. Point source pollution of watercourses by manures can occur as a result of catastrophic leaks and spills from poorly maintained stores. Over the last 10 to 15 years, grant aid has encouraged UK farmers to update and build new, safer stores and this point source of pollution has decreased (NRA, 1995). The leaky part of manure management is now the diffuse transfers of nutrients, pathogens and organic material to watercourses after application to land.

Diffuse pollution can be further subdivided into: (i) incidental transfers of nutrients (Haygarth and Jarvis, 1999), for example, associated with surface runoff events or contaminated drain-flow if manure application is shortly followed by intense rainfall; and (ii) underlying transfers of nutrients which are inherently associated with the previous manure, fertilizer and land management.

Pathways and processes influencing nutrient transfers

The two major hydraulic pathways associated with nutrient transfer from manures to watercourses are: (i) subsurface, either vertically (matrix or macro-pore flow) or laterally (interflow); and (ii) surface (runoff in solution, attached to colloids and soil particles). Transport of potential pollutants only occurs when there is sufficient energy in the system to move pollutants away from their source (see Fig. 3.2). This may occur if there is a sufficiently high hydraulic loading with the manure, or if heavy rainfall occurs shortly after manure application (Sharpley, 1997). If application rate is greater than infiltration rate then surface flow will be induced and surface waters will be at risk. Rapid infiltration through preferential flow pathways (e.g. cracked clay-rich soils, worm burrows or structured soils) may lead to contamination of drains and groundwater (Chadwick and Pain, 1999).

As effluent moves via any pathway, the manure-derived nutrients, organic matter and pathogens will undergo processes within, or on, the soil. The residence time of effluent in contact with soil particles

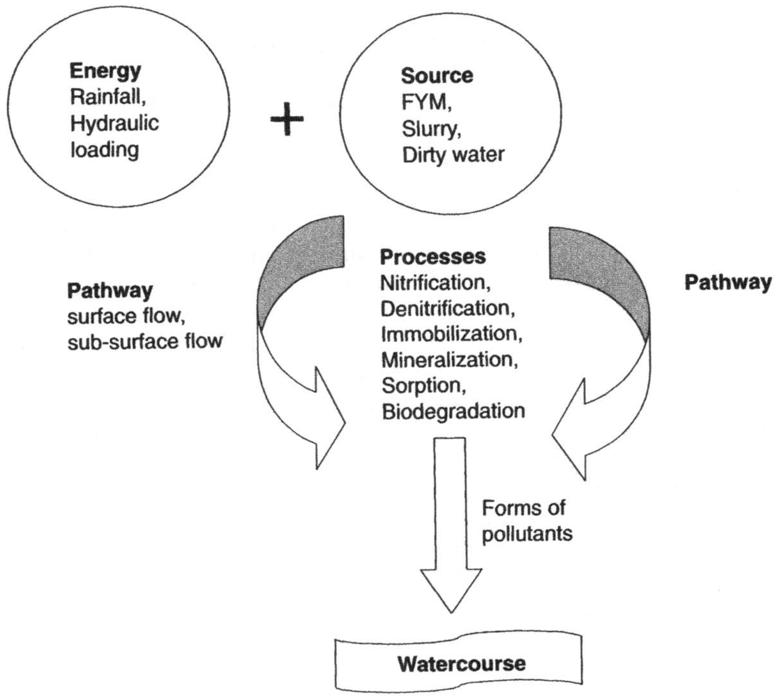

Fig. 3.2. Schematic view of the transport of manure nutrients to watercourses.

will influence the extent to which any process will act on a potential pollutant. These processes will control the forms and concentrations of potential pollutants when they reach the watercourse (Fig. 3.2) and are described briefly below.

Nitrification

The major inorganic-N form in slurries solid manures and dirty water is NH_4^+ (Chadwick et al., 2000a). Excreted urea is rapidly hydrolysed to NH_4^+ mediated by the enzyme urease (Pain *et al.*, 1999) and remains in this alkaline form in anaerobic conditions. Conventional slurry and dirty water management systems are anaerobic in nature so there is little opportunity for nitrification to NO_3^-. However, nitrification occurs in FYM heaps so NO_3^- can be applied to land via FYM. Nitrification of manure-NH_4^+ in soils can result in high concentrations of NO_3^- in leachate, particularly at times of low crop demand (Chambers *et al.*, 2000).

Denitrification

Once NO_3^- is formed in the soil after mineralization and subsequent nitrification, it may be subjected to denitrification losses. The emission of gases such as nitrous oxide (N_2O) (an important greenhouse gas), nitric oxide (NO), nitrogen dioxide (NO_2) and dinitrogen gas (N_2) as a result of nitrification and/or denitrification is likely to be greater than that from soils receiving mineral fertilizers (Thompson and Pain, 1990). Paul *et al.* (1993) showed that manure-amended soil produced N_2O and NO from both nitrification and denitrification processes. Some of the C compounds present in manures are readily available energy sources for denitrifiers (Paul and Beauchamp, 1989).

Volatilization

Ammonia is volatilized from the surface of urine, faeces, slurry or FYM in the animal house, the manure store and after land application (Pain *et al.*, 1998). Ammonia emissions can easily account for 50% of the total NH_4^+ applied in slurry; much of this loss occurs in the first few hours of application (Pain *et al.*, 1989). This reduces the risk of N transfer to watercourses but is undesirable because of the adverse effects that NH_3 has after deposition to natural ecosystems (Schulze *et al.*, 1989). New slurry application techniques have been developed to reduce NH_3 losses from slurries (Pain *et al.*, 1999).

Immobilization/mineralization

Microbial immobilization of N, P and C can occur, thus reducing the risk of nutrient loss to watercourses. Immobilization of soil nutrients by organic amendments with high C:N ratios has been shown by Powlson *et al.* (1994). In contrast, mineralization of organic compounds to inorganic forms at times of low crop demand increases the risk of transport to watercourses.

Biodegradation

Microorganisms will utilize readily degradable carbon compounds as a source of energy and thus reduce the oxygen demand of any effluent that reaches a watercourse.

Sorption

Ammonium and PO_4^{3-} ions and pathogens can become physically or chemically sorbed on to exchange sites on clay lattices. Phosphorus can also bind to or precipitate with Al, Fe or Ca in the soil (Yuan and Lavkulich, 1994).

Examples

Sub-surface transport via leaching

Nitrogen

Nitrate leaching following manure applications to soil arises after nitrification of the applied NH_4^+ (Chambers *et al.*, 2000). Leaching of NO_3^- will also occur after the mineralization of manure-organic-N to NH_4^+ and subsequent nitrification to NO_3^-.

The rate of manure application has a considerable influence on the N leaching loss (Cameron *et al.*, 1995), and excessive application rates can pose a significant risk to water quality. Agricultural land, particularly if it is under row crops, or is used for intensive animal production, is often associated with groundwater, with NO_3^--N concentrations near or above 10 mg l^{-1} (Goss *et al.*, 1998; Toth and Fox, 1998). In contrast, percolation from soil below forests and suburban areas generally has NO_3^--N concentration < 1 mg l^{-1}. Lysimeter studies (Cameron *et al.*, 1995) have shown that high application rates of manure (600 kg N ha^{-1}) can result in very high NO_3^- concentrations in leachates (max. 150 mg N l^{-1}). Lower application rates (200 kg N ha^{-1}) resulted in much lower concentrations (max. < 50 mgN l^{-1}). Increased concentrations of NO_3^- were also measured at a depth of 1.2 m following application of poultry litter at 440 and 880 kg N ha^{-1} in a silt loam soil (Adams *et al.*, 1994).

Repeated applications of manure also result in increased NO_3^- leaching. In an extreme example, concentrations up to 1500 mg l^{-1} have been found in effluent draining from a soil treatment system in Brittany following repeated applications of slurry at rates of up to 200 m^3 ha^{-1} (Martinez, 1997). In contrast, the total P concentration in the leachates from this same plot had not increased significantly as the majority of the P was immobilized in the top 20 cm of the soil (Martinez and Peu, 2000). It should be noted that once NO_3^- enters a preferential flow path, there is little chance of it being removed from solution since it is negatively charged and there will be no adsorption on to soil particles. In contrast, NH_4^+ and PO_4^{3-} can be removed from the water through interaction with the soil matrix (Goss *et al.*, 2000).

Phosphorus

Phosphorus retention mechanisms, such as sorption of PO_4^{3-} on surfaces of reactive minerals or chemical precipitation in most soils are likely to result in a low risk of P leaching (Haygarth and Jarvis, 1999). On the other hand, surface water contamination by P associated with the runoff events is not uncommon. Calcareous soils have been promoted as being suitable for manure recycling disposal because of their ability to sorb P and reduce risk of P transfer to water (Lance, 1977). There are indications that there is only a finite capacity of soils to sorb P. In The Netherlands, under continued high manure loading, the sorption capacity of some soils is diminishing at such a rate that van Riemsdijk *et al.* (1987) predict that a breakthrough of high P concentrations to groundwater is likely to occur in the next 20–30 years. Heckrath *et al.* (1995) defined a 'change point' for Olsen P of 60 mg kg^{-1} above which P losses during drainage events increase markedly (see Leinweber *et al.*, Chapter 2 and McDowell *et al.*, Chapter 9, this volume).

Manure applications based on N loading can result in a significant enrichment of the soil P content because of the N:P ratio (Smith *et al.*, 1998). Table 3.3 shows P loadings with manure applications targeting a N application of 250 kg N ha^{-1}. This N loading is the maximum rate recommended in the UK *Code of Good Agricultural Practice for the Protection of Water* (MAFF, 1998a), but could result in significant accumulation of P in the soil.

Organic matter

Repeated applications of dirty water are often required to dispose of the large volumes concerned. Chadwick *et al.* (2000b) conducted experiments to determine the influence of repeated applications of dirty water on leachate chemistry from large lysimeters, 2 m × 2 m × 1.2 m deep, of an intact sandy loam soil. Figure 3.3 summarizes the effect of application rate on BOD removal in leachates. These lysimeters have shown that repeated applications of dirty water of < 75 mm $week^{-1}$ have resulted in $> 90\%$ reductions in the initial BOD level of dirty water. However, at higher application rates, BOD levels, NH_4^+-N, NO_3^- and MRP concentrations can reach levels greater than those recommended for watercourses.

Metals

Most metals, such as Cu and Zn, applied to soils in manures and other organic materials accumulate in the upper layers of soil (L'Herroux *et al.*, 1997). There appears to be little movement of metals below the zone of incorporation as metals are strongly bound to organic matter particles, which become trapped in the upper soil layers. There can be increased risk of leaching from acidic sandy soils with low organic matter content. Migration through soils greatly depends on the metal physico-chemical forms (Alloway and Jackson, 1991). Other metals, such as Co and Mn, have been shown to leach (L'Herroux *et al.*, 1997) (see also De Vries *et al.*, Chapter 5, this volume, for details of this).

Table 3.3. The equivalent phosphorus applied with a nitrogen loading of 250 kg ha^{-1} of manure.

	Dry matter (%)	P applied with N @ 250 kg ha^{-1} [a]
Cattle slurry	6	42
Pig slurry	6	65
Cattle FYM	25	62
Pig FYM	25	107
Broiler litter	60	91
Layer manure	30	89

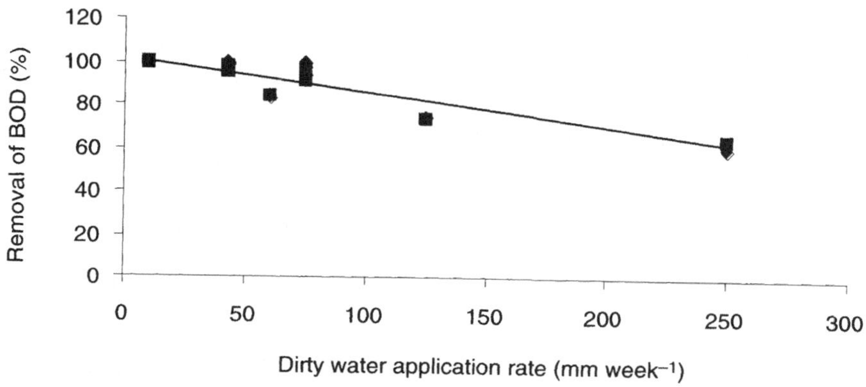

Fig. 3.3. Influence of application rate of dirty water on removal of biological oxygen demand eg by soil.

Pathogens

Microorganisms are transported in the same way as colloids and their movement is likely to be affected by the flow associated with the size of pores in which they are transported. Microbial movement is generally greater in coarse soils with large pores than in fine textured soils (Tan *et al.*, 1991). Microorganisms have a variable surface charge which allows strong adsorption of the bacteria to soil particles. Soil type is also a major factor influencing transport of pathogens in soils (Paterson *et al.*, 1993); the organic matter and clay content have the greatest effect on availability of negatively charged surfaces. Microorganisms are also influenced by factors that affect their survival, for example temperature (Kemp *et al.*, 1992) and soil water content (Labeda *et al.*, 1976). Many are also motile (Goss *et al.*, 2000; see also Jones *et al.*, Chapter 6, this volume for more detail on this subject).

Sub-surface transfers via preferential flow

Movement of metals, pathogens and nutrients to depth can be via rapid transport through preferential flow pathways. Soil pore characteristics are important for water and solute transport. However, the transport of contaminants in soils with strongly aggregated structure, or with large and continuous pores, is not well described by traditional convection–dispersion equations because preferential flow occurs (Wagenet, 1990). Preferential flow is the process whereby water and its constituents move by preferred pathways through an otherwise impermeable medium. It means that part of the matrix is effectively bypassed. The term preferential flow does not itself convey a mechanism for the process (Helling and Gish, 1991), whereas the often-used term macropore flow implies transport through relatively large pores, worm channels (Ehlers, 1975), root channels (Barley, 1954), fissures or other semi-continuous voids within the soil.

As much as 70–90% of applied chemicals may move preferentially through macropores (Ahuja *et al.*, 1991). Quisenberry *et al.* (1993) found that 50% of applied water and Cl^- was transported through $< 20\%$ of the soil matrix. Macropore flow can either increase or decrease the residence time of solutes in soil, depending on the location of solutes relative to macropores (Chen *et al.*, 1997; 1999). Macropore flow is a major cause of groundwater contamination and transport uncertainty. Contamination of drains with high BOD levels was observed following low application rates of dirty water to a clay grassland soil under soil moisture deficit (Chadwick and Pain, 1999). The speed of the contamination was attributed to preferential flow through fissures.

Mouldboard ploughing may destroy the continuity of pores between the plough layer and the deep horizons. Long-term no-tillage plots, on the other hand, often develop a high density of continuous, relatively large vertical channels. Manure application may encourage the activity of earthworms which may result in the greater continuity of macropores. Hence, there may be a faster breakthrough of contaminants than predicted (Munyankusi *et al.*, 1994). Greater transport of pathogenic organisms has been observed in intact soils compared with repacked soils. This increased transfer is considered to be due to water flow bypassing the filtering effect of the soil matrix (White, 1985; van Elsas *et al.*, 1991).

Sub-surface transfers via lateral flow (interflow)

Subsurface flow may occur in the case of tile drains or as a result of a shallow hardpan (Lowrance, 1992; Goss *et al.*, 1993). Most of the losses of NO_3^- in the lower Southern Coastal Plain of the USA were from subsurface flow taking place in the top 30 cm of soil rather than from surface flow. Over a 10 year period, 20% of the N in the applied fertilizer was lost via surface and subsurface flow (Hubbard and Sheridan, 1983). Such results suggest that tile drainage systems can greatly reduce groundwater contamination at the expense of surface water contamination. However, not all drainage water may be intercepted by pipe drains, even during major flow events so that groundwater contamination is still likely.

Liquid manure adversely affected tile water quality when applied to the land following the farming guidelines current in Ontario (Dean and Foran, 1992). These authors found that 75% of the manure-spreading events investigated resulted in water quality impairment. The difficulty of determining an acceptable rate of application of liquid manure because of the numerous factors which affect the contamination of watercourses is apparent (Foran and Dean, 1993).

There are contrasting effects of drainage on NO_3^- and P transport to watercourses. Nitrate can move rapidly towards drains reducing the residence time in the soil for plant uptake and denitrification. However, P is drawn into the soil towards the drains and sorbed/fixed in the soil matrix reducing the risk of P transport over the soil surface (Haygarth *et al.*, 1998). Heckrath *et al.* (1995) found a critical concentration of soluble-P in the ploughed layer that, if exceeded, resulted in an enhanced contribution of P losses through tile drains in clay loam soils. Hergert *et al.* (1981) found P losses in field tile drain effluent to be increased where manure was applied compared with unfertilized control plots. Dils and Heathwaite (1997) found that subsurface transport of P may occur as water-soluble P and as particulate P for both undrained and tile-drained plots.

The importance of preferential flow in P transport is confirmed by the amounts of P leached despite the large sorption potential of some subsoils (Eghball *et al.*, 1996; Thomas *et al.*, 1997). It is important to note that there is strong evidence to show that as nutrients move through the soil in interflow, there is greater opportunity for them to take part in soil processes (e.g. sorption of P and NH_4^+), than if they were to run over the soil surface (Smith *et al.*, 1998; Chadwick and Pain, 1999). For this reason, P losses in interflow are much lower than those in runoff. Hence, increasing infiltration of effluent on sloping land should be encouraged as a means of reducing the risk of nutrient transfer to surface water (Unger, 1984).

Surface transfers

Manure-derived pollutants can be transported over the surface of the land in either dissolved or particulate forms. The soil surface can be sufficiently unstable and the hydraulic energy sufficiently great to cause erosion of soil particles that may also be transferred down-slope in overland flow.

Nitrogen and phosphorus

Overland flow generally accounts for only a small portion of applied N compared with the portion transferred in percolates through the profile (Burgoa *et al.*, 1993). Blevins *et al.* (1996) found that after two growing seasons less than 2% of fertilizer-N was lost to overland flow whereas 30% had moved below 1 m in the soil.

Nitrogen in any manure solids left on the soil surface, or associated with fine particles that are readily moved during soil erosion, can be lost through surface runoff to a watercourse. The factors that determine N loss by erosion are: the amount of sediment moved, the N content of the soil moved, and the N content of the manure solids. The other material subject to loss to surface water is N dissolved in runoff water. This portion is usually small (Meisinger and Randall, 1991; Blevins *et al.*, 1996); however, it is variable, and depends on a number of factors, such as the degree of soil cover, source of N applied, application rate, timing and duration of the application. Surface conditions are also important, and affected by slope, soil characteristics, and land management. Finally, runoff is highly dependent on the intensity of rainfall after application. The largest losses occur if a soluble N source is applied to a bare soil surface and a significant rainfall event occurs soon after application (Edwards and Daniel, 1993; Sharpley, 1997). In many cases, dissolved N is transported into the soil with the initial infiltration that precedes runoff (Meisinger and Randall, 1991). Incorporating the N source such as manure dramatically reduces runoff losses. In most cases, runoff N losses are small, being 3 kg-N ha^{-1} annually or less (Meisinger and Randall, 1991; Nichols *et al.*, 1994; Blevins *et al.*, 1996).

Sharpley (1997) investigated N and P in overland flow on ten Oklahoma (USA) soils amended with poultry litter. Nitrogen and P concentrations decreased with ten successive rain events, starting 7 days after litter application. Increasing the time between litter application and rainfall from 1 to 35 days reduced total N from 7.5 to 2.3, NH_4^+-N from 5.5 to 0.1, dissolved P from 0.7 to 0.4, and bioavailable P from 1.0 to 0.6 mg l^{-1}. However, NO_3^--N concentrations were unaffected by rainfall frequency and timing. Hooda *et al.* (1996) measured up to 42% of the annual MRP loss in drain water within 1 week of an application of cattle slurry at 50 m^3 ha^{-1}. Peak concentrations in drainage water exceeded 10 mg l^{-1}.

Runoff from agricultural land has been shown to be responsible for *c.* 70% of the P reaching Lake Erie from the tributaries in Ontario (Miller *et al.*, 1982). About 20% of this amount was estimated to be from direct inputs from livestock operations, including runoff from storage areas and surface runoff from manure applied close to streams. The remainder was due largely to P associated with eroded sediment. Manure application would have two opposing effects on this latter contribution. Firstly, it would increase the P content of the soil and hence the concentration on the eroded sediment (Sharpley and Halvorson, 1994). On the other hand, manure would tend to improve soil structural stability (Sommerfeldt and Chang, 1985) and hence reduce erosion. Where liquid or solid manure was not incorporated after application, the loss of P in surface runoff was greater from land under mouldboard cultivation than under no-till. Loss from no-till land was similar to that from land where the manure had been incorporated after application (King *et al.*, 1994).

The application rate of total solids in slurries has a major influence on the magnitude of P loss in runoff (Smith *et al.*, 1998). An application rate of 80 m^3 ha^{-1} with 8% solids content resulted in surface capping and increased runoff volume compared with runoff from a similar application rate of dairy slurry with a 4% solids content. Total P losses of 1.8 kg ha^{-1} were measured from the 8% solid content slurry application, and 0.2 kg ha^{-1} for the low solid content slurry. Losses from the control area were 0.02 kg ha^{-1}. It is important to note that losses of P in runoff may only represent a small percentage of that applied in the manure application, 3% in the study by Smith *et al.* (1998), but peak concentrations of MRP were 11 mg l^{-1} and could have major implications for water quality.

One aspect of P in runoff that is still poorly understood is the bioavailability of the different forms of P, for example to algae (Sharpley *et al.*, 1992). While on the farm manure application may not increase the total P in runoff, it is more likely to increase the amount of bioavailable P. In the State of Delaware, continual land application of animal manures has resulted in an accumulation of P in the surface soil (Sharpley and Halvorson, 1994). These authors concluded that there would be an increase in the bioavailability of P in runoff water from land following the application of manure because of the increased transport of low-density organic material together with the high solubility of manure P. The magnitude of the increase would be expected to vary dependant on the density of the manure, the water and P content, for different animal sources (Sharpley and Halvorson, 1994).

Pathogens

Runoff water from grazed and ungrazed grass pastures can contain large numbers of bacteria (Kirchmann, 1994). Bacteria from poultry manure were not detected in runoff when the manure was applied to bare soil, but were present when the manure was applied to grassland (Giddens and Barnett, 1980). More bacteria may be lost in overland flow from no-till land than from ploughed land within 24 h of manure application, but the rate of decline in the concentration of bacteria in the runoff water can also be greater (King *et al.*, 1994). This issue is discussed further in Jones *et al.*, Chapter 6, this volume.

Options

The key to reducing nutrient transfer from manures to watercourses is to move from an attitude of disposal to one of utilization by matching nutrient application with crop demand. This requires: (i) adequate land availability; (ii) adequate storage capacity; and (iii) improved knowledge of manure nutrient content and availability.

Land availability

Some livestock enterprises produce more animal manure than can be applied safely to land, and manure has to be exported from the farm of origin to neighbouring farms. In some circumstances, most farms in a given area have the same problem, requiring manure to be transported large distances to alternative agricultural land. An extreme example is Belgium, where an agreement was reached to limit N and P production, as well as animal numbers, to 1991 levels and a P production level was designated for each community (Vlassak, 1998).

A restriction in spreading manures on vulnerable land has resulted in the introduction of Farm Waste Management Plans in the UK. These have been used to educate farmers about which parts of their farms are suitable for manure applications and those areas which are specifically high risk areas where no manure should be applied. New large pig and poultry enterprises in the UK are required to submit Farm Waste Management Plans under the Integrated Pollution Prevention and Control (IPPC, 1997) Directive to demonstrate that all the manure produced can be applied safely without the risk of nutrient transfers to watercourses. Specific areas of land have been designated as 'vulnerable' or 'at risk'. Following the EU Nitrate Drinking Water Directive (EC, 1991), 66 zones in England have been designated as Nitrate Vulnerable Zones (NVZs). Further details about restriction of manure use in NVZs are discussed in the section on legislation below (MAFF, 1998d).

Storage capacity

Farmers with adequate storage capacity have greater flexibility in terms of when they can apply manures to the land. Different countries have different recommendations for what is considered adequate storage. They range from 4 months for slurry in England and Wales (for the UK) to 6 months in Canada and 7–9 months in Denmark

(Ingénieries-EAT, 1996). In some countries this is laid down in legislation whereas in others (e.g. UK) this is recommended in the *Code of Good Agricultural Practice for the Protection of Water* (MAFF, 1998a) but is not mandatory. Because of the increased risk of nutrient transfer after applications of manure in autumn and increased risk of mineralization and leaching losses as a result of autumn applications, some countries have imposed curfews on land application during late autumn/winter; for example The Netherlands, Denmark and Belgium (Vlassak, 1998).

Options to maximize current storage capacity include reducing the clean water entry into stores by channelling clean rainwater into separate drains, fixing broken roof gutters and covering stores. An extended grazing season will also reduce the volume of manure requiring storage. The extreme option is to graze continuously and have no housing period, a popular choice in areas where climatic and edaphic factors are favourable, e.g. New Zealand.

Manure nutrient content and availability

Increased farmer confidence and knowledge of manure nutrient content is critical in moving from waste management to nutrient management. At worst, farmers are encouraged to use average values of N, P and K contents published in national handbooks and literature, (e.g. see RB209, MAFF, 1994). A better alternative is to persuade farmers to submit representative slurry and manure samples to commercial laboratories for nutrient analysis. If feeding and manure management regimes remain similar then manure nutrient composition should remain fairly constant and a limited number of samples would require analysis.

A further option is for farmers to analyse manures on their farm with portable test kits. There are several N meters available for estimating the plant available N content of manures. Nitrogen meters are used on sub-samples of slurry prior to application to the field (Kjellerup, 1996). These tests take approximately 10 min to complete and are accurate and reliable at estimating the NH_4^+-N content of pig and dairy slurries (Williams *et al.*, 1999). Hydrometers can be used to indicate total N, P and K content of slurries as there is a good relationship between dry matter and nutrient contents (Piccini and Bortone, 1991). Significant progress has also been made in quantifying nutrient contents in-line on slurry tankers (Lenehan *et al.*, 1999).

The availability of nutrients, for example as determined by the mineralization rates of organic N from different manure types applied to land, requires an understanding of the biological processes involved. Promising research has revealed useful indicator relationships for mineralization rates of organic matter such as the C:organic N ratio of the material (Chadwick *et al.*, 2000a) and soil thermal units (Honeycutt and Potaro, 1990). Some information is made available to farmers in 'look-up' tables (MAFF, 2000), but increasingly PC-based decision support systems (DSS) will be required to assimilate all the information and provide advice to farmers on the best time of year to apply manures to gain maximum utilization of manure nutrients. Examples of such DSS are MANNER (Chambers *et al.*, 1999) and MCLONE4 (University of Guelph, 1999).

Other options to reduce risk of nutrient transfer to watercourses include treatment of manures prior to land application and use of vegetative buffer strips between sources of nutrients and watercourses (see below).

Treatment

Excess manure can be treated to reduce potential pollutant content prior to land spreading (Burton, 1999). There are a wide range of available technologies for processing manures, some of which are better developed than others. Treatment systems rely on one or a combination of biological, chemical, physical and thermal processes and some can be costly to set up and maintain. Table 3.4 summarizes the more common treatment processes and potential effects on water quality when the treated

Table 3.4. Processes for treating manures.

Treatment	Process	Principal benefit	Other comments
Separation Screening, Centrifugation	Physical	Results in a uniform liquid fraction and a solid compostable fraction	Easier handling, increased storage capacity. Phosphate in liquid fraction reduced
Thermal Drying Incineration	Physical	Disinfection, reduces odours and N content	Produces energy, produces a dry concentrated product
Aeration Slurry aeration FYM composting	Biological	Effective at reducing BOD, odour and removing N	Reduces pathogen numbers, particularly during composting
Anaerobic	Biological	Reduces BOD, odours and thermophilic treatment reduces pathogen numbers	Option to generate biogas and generate electricity
Additives	Physical Chemical Biological	Use of zeolites, acid, floating oil cover to reduce NH_3 volatilization. Addition of lime to precipitate P. Addition of microbes or microbial substrates	Precipitation of P and separation to reduce P content of liquid fraction. Immobilization of mineral N and P by microbial biomass. Lowering of pH, e.g. by *Lactobacillus*
Recovery	Biological Chemical	Reduces P content of slurry	Refinement of recovered P for further use in agriculture
Wetlands Reedbeds Soil	Chemical Biological	Reduces BOD, phosphate and organic matter	Biological nature of these systems restricts efficient treatment to warmer conditions. High nitrate concentrations can result

manure is spread to land. Reed beds (Sun *et al.*, 1999), constructed wetlands (Tanner *et al.*, 1995a, b) and soil-based systems (Martinez, 1997; Chadwick *et al.*, 2000b) are alternative treatment strategies if sufficient land is available.

Additional benefits of some of these treatment processes are the generation of income through conversion of manures to fertilizer products and energy (Amon *et al.*, 1999). Although a relatively new process, nutrient recovery of P, is worthy of additional thought (Greaves *et al.*, 1999). Slurries, in particular, are a source of compounds other than key plant nutrients. They are also a potentially valuable source of heavy metals and organic compounds such as acetic acid (Hobbs *et al.*, 1996), which could be recovered with suitable technology.

Buffer strips

Research has shown that soil can act as a sink for slurry-derived P (Chadwick and Pain, 1999). Biochemical oxygen demand levels and NH_4^+-N concentrations can also be reduced as effluent travels over the soil surface and particularly through the soil matrix (Chadwick and Pain, 1999; Chadwick *et al.*, 2000b). Thus, vegetated buffer strips or constructed buffer wetlands between agricultural fields and watercourses or natural wetlands offer additional protection from nutrient transfers (Fennessy and Cronk, 1997; Hefting and de Klein, 1998). The UK *Code of Good Agricultural Practice for the Protection of Water* (MAFF, 1998a) recommends that manure applications should not take place within 10 m of a watercourse, although this may not be sufficiently wide to offer adequate protection under all soil and weather conditions (Chadwick and Pain, 1999).

Aiding infiltration into the soil matrix offers a greater degree of *purification* than overland flow (Chadwick and Pain, 1998), as the effluent is in greater contact with soil surfaces and residence time within the soil is longer than effluent travelling over the soil surface, assuming no preferential flow. Cultivation practices such as contour ploughing will reduce the risk of runoff and any management technique to increase the storage capacity of land will have a similar effect (Unger, 1984).

Feeding strategies to reduce nutrient excretion

A final option that would help to reduce overall nutrient loading through manure applications would be to alter feeding strategies to reduce nutrient excretion. Nutrient content and availability is affected by manure composition, which is influenced by the nutrients in feed and assimilation rates by different animal groups. There is considerable scope to decrease the nutrient content of manures by tailoring feeds more closely to the nutritional requirements of the animals. Hobbs *et al.* (1996) demonstrated that N excretion by pigs could be reduced by decreasing the crude protein content of the feed and adding specific amino acids to maintain the nutritional balance. The N content of cattle excreta can also be reduced through modification of the feed (Smits *et al.*, 1997).

Manipulation of animal diets also has the potential to reduce P excretion without affecting animal performance (Morse *et al.*, 1992; Brintrup *et al.*, 1993). Increasing the availability of P in maize through genetic manipulation decreases P excretion of poultry (Ertl *et al.*, 1998). Although these studies have shown the potential to reduce nutrient excretion through modification of animal feeds, implementation remains costly.

The role of plant and animal breeders should not be ignored in targeting future requirements for managing manures for water quality. Breeding forages with lower crude protein contents and composition which is more readily assimilated by animals will reduce nutrient excretion. Equally, the selection of animals with increased assimilation rates will also have a similar effect. Matching diets to growth stage is also a potential method of reducing nutrient and heavy metal excretion.

Regulations

Legislation is an option open to governments to impose management practices on farmers as a means to reduce risks of water pollution. There are varying degrees of legislation and guidelines for managing manures for water quality throughout different countries of the world. For example, there are the voluntary guidelines for Good Agricultural Practice for the Protection of Water (MAFF, 1998a), Soil (MAFF, 1998b) and Air (MAFF, 1998c) in the UK. These are recommendations and are not enforced. In contrast, The Netherlands and Belgium have enforced legislation and nutrient budgeting plans to reduce risks of N and P transfers to watercourses.

For example, the Manure Action Plan (MAP) was introduced in Belgium in 1995 to control the use of manures. Manure applications are prohibited from 21 September to 21 January (22 October to 21 January for grass and maize) and there is restricted application in surface and groundwater producing areas, ecologically sensitive areas and in P risk and P saturated zones. Also, a levy system on the production of surplus manures was introduced. Any farmer producing > 655 kg P or > 3000 kg N year^{-1} has to pay the levy of 1.25 Belgium francs (37BEF = US$1, October 1997) per kg production of P or N (Vlassak, 1998).

The mineral balance approach has been introduced into other European countries (e.g. The Netherlands) and aims to reduce excess nutrient production on farms, specifically for N and P (Derikx, 1999). Farmers are required to balance the nutrients coming into the farm with those leaving it in products, as milk, meat, grain and straw. However, it should be noted that even if the inputs match the outputs not all the nutrients are utilized and there will still be some risk of diffuse pollution.

In Denmark, there is a comprehensive Action Plan for the Aquatic Environment and a Programme for Sustainable Development. Farmers must have sufficient areas for the spreading of manures, and manure spreading is formally prohibited between September and February.

In the USA, the Clean Water Act of 1972 was responsible for significant progress in controlling point source pollution incidents. Since then attention has been placed on targeting controls of non-point sources using a watershed/catchment based approach (US EPA, 1993). Clearly P management is of considerable environmental concern in certain states of the USA, particularly those located close to sensitive water bodies, such as the Great Lakes and Chesapeake Bay (Sharpley *et al.*, 1994; see also Sims and Coale, Chapter 6, this volume). A P indexing system (Lemunyon and Gilbert, 1993) has been introduced to locate soils with P levels in excess of requirements for crop growth and in areas where the risk of transport phenomena (detachment and erosion, overland flow) could deliver P to surface waters. Most states have adopted policies designed to restrict increased nutrient loading.

EU Nitrate Directive

The one directive that has resulted in common regulations in all Member States of the European Union is the EU Nitrate Directive (EC, 1991). This directive has required Member States to identify waters where NO_3^- levels exceed or are at risk of exceeding 50 mg l^{-1}. These areas of land are then designated as Nitrate Vulnerable Zones (NVZs) and then subject to a mandatory action programme to reduce NO_3^- pollution or to avoid further pollution from occurring.

The key areas of the action programme for managing NVZs in the UK are (MAFF, 1998d);

1. Use of inorganic fertilizers.
2. Use of organic manures.
3. Storage of slurry.
4. Farm records.

Maximum N loadings across the farmed area through manure applications are 210 kg ha^{-1} for the first 4 years of the action programme and 170 kg ha^{-1} thereafter. This loading includes N voided by grazing animals. Grassland is given a higher N loading via manures (including grazing) of 250 kg ha^{-1} because NO_3^- leaching is deemed to be less of a risk than from arable land. On freely draining and shallow soils the action programme also defines a period during the autumn/winter when high available N content manures (slurry, poultry manures and liquid digested sewage sludge) cannot be applied.

Therefore, for the majority of cattle farmers, manure management to satisfy the NVZ Action Programme is not practically prohibitive if stocking rates can be kept below *c.* 2.6 cows ha^{-1} (550 kg liveweight). However, managing the manure N loadings on pig and poultry units within NVZs presents farmers with real problems as many do not have sufficient land area available on the farm of origin.

Water Framework Directive

The EC Water Framework Directive is expected to receive final agreement by the Council of Ministers in the near future. It will provide a framework for a coordinated approach to water management within catchments. Assessments of the state of watercourses will be made and plans implemented to improve water quality and maintain such quality by sustainable agricultural management practices. The aim of the directive is to achieve good water quality with an emphasis on ecological quality and will include prevention of eutrophication. As such, the directive will focus on the control of diffuse losses of not only N, but also P and organic materials from agricultural sources such as manures.

Integrated Pollution Prevention and Control (IPPC)

The European Directive on Integrated Pollution Prevention and Control (IPPC, 1997) applies to installations for intensive rearing of pigs (> 2000 sows or 750 sow places) and poultry (> 40,000 birds). Farmers managing such units will need to submit plans to demonstrate adequate manure storage facilities and sufficient land for spreading manures safely before licences are agreed and production commences.

Discussion

Farmers do not wish to pollute watercourses, but manures that are generated on farms have an inherent ability to do so. Pollution is often a result of ignorance of the problem. Increased awareness, availability of tools such as on-farm N test meters and decision support systems will help to ensure that manure nutrients are applied at suitable rates and times of the year. The development of automatic nutrient monitoring systems coupled with high precision application techniques will, in the future, result in more accurate application rates.

Further research is required to reduce risks of overland flow from applied manures in the field, principally via effective cultivation techniques. However, the most significant influence on management of manures for water quality will be the recognition of manures as a resource and not a disposal problem. There is the need to move from an attitude of waste disposal to environmental management (see Fig. 3.4), and some countries are further down this route than others. In general, those countries that have introduced legislation are closer to the desired goal than those countries which use recommended guidelines.

Education to increase awareness of the nutrient content of manures is essential with a strong emphasis on the cost benefits of utilizing these as a primary source of nutrients. The low cost of inorganic fertilizers often means that farmers use them as their principal form of nutrients. Allowance must be made for nutrients in manures by reducing the use of inorganic fertilizers. The costs of fines for polluting watercourses are low relative to the investment required to increase manure storage capacity and safe manure management. Education may not be sufficient on its own and legislation may be required to achieve this goal of managing manures as a nutrient resource.

Because of the diverse nature of agriculture throughout the world, a flexible

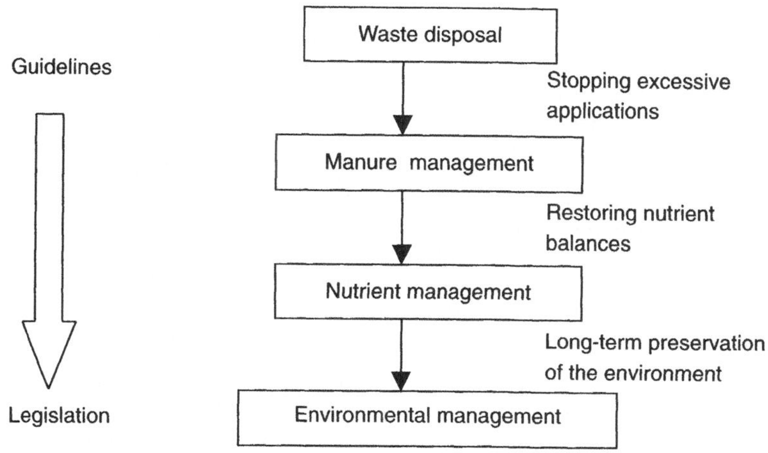

Fig. 3.4. Manure management continuum from waste disposal to management of the environment.

approach to management of manures is required. It is not the excessive manure production that is the problem, rather their excessive nutrient contents. There are countries with overall nutrient surpluses, such as The Netherlands, and others with soils with low organic matter contents, such as the southern Mediterranean countries which require additional organic manures for soil conditioning qualities. Blanket policy making for manure management will not deal with these extremes.

Conclusions

There are several specific measures for managing manures to reduce rapid 'incidental' transfers of pollutants to watercourses:

- avoid applying manures to cracked and waterlogged soils;
- avoid application to frozen land;
- avoid under-drained land;
- avoid steeply sloping land;
- increase infiltration/storage capacity of land;
- use a buffer zone of at least 10 m between the edge of the applied area and a watercourse;
- restrict application rate/hydraulic loading rate to reduce risk of generating runoff.

Specific measures for managing manures to reduce 'diffuse' transfers of pollutants to watercourses require farmers to:

- match rate of nutrient application to agronomic needs;
- reduce accumulation of nutrients by utilizing all available land for application.

A combined effort by farmers, advisers, policy makers, researchers and agricultural suppliers is required to: (i) recognize the practical problems; (ii) commission the research to provide the information and tools to tackle the problems: (iii) communicate research findings and advice effectively to the farmers; and (iv) provide effective legislation/recommendations that can be monitored. Through this process manures will be managed as a resource with the environment in mind.

Acknowledgements

IGER is supported by the Biotechnology and Biological Research Council (BBSRC), Swindon, UK.

References

Adams, P.L., Daniel, T.C., Edwards, D.R., Nichols, D.J., Pote, D.H. and Scott, H.D. (1994) Poultry litter and manure contributions to nitrate leaching through the vadose zone. *Soil Science Society of America Journal* 58, 1206–1211.

Ahuja, L.R., DeCoursey, D.G., Barnes, B.B. and Rojas, K.W. (1991) Characteristics and importance of preferential macropore transport studied. In: Gish, T.J. and Shrimohammadi, A. (eds) *Preferential Flow.* American Society of Agricultural Engineers, St Joseph, Michigan, pp. 32–49.

Alloway, B.J. and Jackson, A.P. (1991) The behaviour of heavy metals in sewage-amended soils. *The Science of the Total Environment* 100, 151–176.

Amon, T., Boxberger, J., Lindworsky, J. and Pfeffer, M. (1999) Biogas production in agriculture: safety guidelines, co-fermentation and emissions from combined heat and power couplings. In: Martinez, J. and Maudet, M.M. (eds) *Proceedings of the Eighth International Conference on Management Strategies for Organic Waste Use in Agriculture.* CEMAGREF, Rennes, France, pp. 405–416.

Barley, K.P. (1954) Effects of root growth and decay on the permeability of a synthetic sandy loam. *Journal of Soil Science* 78, 205–210.

Bitzer, C. and Sims, J.T. (1988) Estimating the availability of nitrogen in poultry manure through laboratory and field studies. *Journal of Environmental Quality* 17, 47–54.

Blevins, D.W., Wilkison, D.H., Kelly, B.P. and Silva, S.R. (1996) Movement of nitrate fertilizer to glacial till and runoff from a claypan soil. *Journal of Environmental Quality* 25, 584–593.

Brenner, F.J. and Mondok, J.J. (1995) Nonpoint source pollution potential in an agricultural watershed in NW Pennsylvania. *Water Resources Bulletin* 31, 1101–1112.

Brintrup, R., Mooren, T., Meyer, U., Spiekers, H. and Pfeffer, E. (1993) Effect of two levels of phosphorus intake on performance and faecal phosphorus excretion of dairy cows. *Journal of Animal Physiology and Animal Nutrition* 69, 29–36.

Brouwer, F. and Hellegers, P. (1997) Nitrogen flows at farm level across European Union agriculture. In: Romstad, E., Simonsen, J. and Vatn, A. (eds) *Controlling Mineral Emissions in European Agriculture, Economics, Policies and the Environment.* CAB International, Wallingford, UK, pp. 11–26.

Burgoa, B., Hubbard, R.K., Wauchope, R.D. and Davis-Carter, J.G. (1993) Simultaneous measurement of runoff and leaching losses of bromide and phosphate using tilted beds and simulated rainfall. *Communications in Soil Science and Plant Analysis* 24, 2689–2699.

Burnhill, P., Chalmers, A.G. and Fairgreave, J. (1994) *Fertiliser use on Farm Crops 1993.* British Survey of Fertiliser Practice 1994. FMA, MAFF, SOAFD, HMSO, Edinburgh, Scotland.

Burton, C.H. (1999) Process strategies for organic wastes. In: Martinez, J. and Maudet, M.M. (eds) *Proceedings of the Eighth International Conference on Management Strategies for Organic Waste Use in Agriculture.* CEMAGREF, Rennes, France, pp. 349–363.

Cameron, K.C., Rate, A.W., Carey, P.L. and Smith, N.P. (1995) Fate of nitrogen in pig effluent applied to a shallow stony pasture soil. *New Zealand Journal of Agricultural Research* 38, 533–542.

Chadwick, D.R. and Pain, B.F. (1998) Transport of nutrients and organic material in soils following applications of animal wastes. In: Matsunaka, T. (ed.) *Environmentally Friendly Management of Farm Animal Waste.* Proceedings of an International Workshop, Sapporo, Japan, 25–29 November 1997. Kikanshi, Insatsu Co. Ltd, Sapporo, Japan, pp. 103–107.

Chadwick, D. and Pain, B. (1999) Transport of nutrients and organic material following applications of animal wastes to grassland soils. In: Corrall, A.J. (ed.) *Accounting for Nutrients – A Challenge for Grassland Farmers in the 21st Century.* BGS Occasional Symposium No. 33, Malvern, UK. Antony Rowe Ltd, Reading, UK, pp. 87–92.

Chadwick, D.R., John, F., Pain, B.F., Chambers, B.J. and Williams, J. (2000a) Plant uptake of nitrogen from the organic nitrogen fraction of animals manures: a laboratory experiment. *Journal of Agricultural Science, Cambridge* 134, 159–168.

Chadwick, D.R., Brookman, S., Leeds-Harrison, P., Tyrrel, S., Fleming, S. and Pain, B. (2000b) Soil based treatment systems for dirty water. In: Moore, J.A. (ed.) *Proceedings of the Eighth International Symposium on Animal, Agricultural and Food Processing Wastes.* ASAE, Michigan, USA. Des Moines, Iowa, pp. 728–737.

Chambers, B.J., Lord, E.I., Nicholson, F.A. and Smith, K.A. (1999) Predicting nitrogen availability and losses following application of organic manures to arable land: MANNER. *Soil Use and Management* 15, 137–143.

Chambers, B.J., Smith, K.A. and Pain, B.F. (2000) Strategies to encourage better use of nitrogen in animal manures. *Soil Use and Management* 16, 157–161.

Chardon, W.J., Oenema, O., del Castilho, P., Vriesema, R., Japenga, J. and Blaauw, D. (1997) Organic phosphorus in solutions and leachates from soils treated with animal slurries. *Journal of Environmental Quality* 26, 372–378.

Chen, S., Franklin, R.E. and Johnson, A.D. (1997) Clay film effects on ion transport in soil. *Journal of Soil Science* 162, 91–96.

Chen, S., Franklin, R.E., Quisenberry, V. and Dang, P. (1999) The effect of preferential flow on the short and long-term spatial distribution of surface applied solutes in the structured soil. *Geoderma* 90, 229–241.

Clarkson, D.T., Hopper, M.J. and Jones, L.H.P. (1986) The effect of root temperature on the uptake of nitrogen and the relative size of the root system in *Lolium perenne*. I. Solutions containing both NH_4^+ and NO_3^-. *Plant, Cell and Environment* 9, 535–545.

Cumby, T.R., Brewer, A.J. and Dimmock, S.J. (1999) Dirty water from dairy farms, 1: biochemical characteriztics. *Bioresource Technology* 67, 155–160.

Dean, D.M. and Foran, M.E. (1992) The effect of farm liquid waste application on tile drainage. *Journal of Soil and Water Conservation* 47, 368–369.

Derikx, P. (1999) Mineral book-keeping, new manure legislation in The Netherlands. In: Martinez, J.

and Maudet, M.M. (eds) *Proceedings of the Eighth International Conference on Management Strategies for Organic Waste Use in Agriculture.* CEMAGREF, Rennes, France, pp. 49–58.

Dils, R.M. and Heathwaite, A.L. (1997) Phosphorus fraction in grassland hill-slope hydrological pathways. In: Tunney, H., Carton, O.T., Brookes, P.C. and Johnston, A.E. (eds) *Phosphorus Loss from Soil to Water.* CAB International, Wallingford, UK, pp. 349–351.

Dourmad, J.Y. and Aumaitre, A. (1990) New trend in pig production in France. *Pig News and Information* 11, 39–42.

EC, European Community (1991) Council Directive on Nitrate in Drinking Water. *Official Journal of the European Communities,* (92/676/EEC), Legislation 1375/1–1375/8. European Community, Brussels.

Edwards, D.R. and Daniel, T.C. (1993) Effects of poultry litter application rate and rainfall intensity on quality of runoff from fescue grass plots. *Journal of Environmental Quality* 22, 361–365.

Eghball, B., Binford, G.D. and Baltensperger, D.D. (1996) Phosphorus movement and adsorption in a soil receiving long-term manure and fertilizer application. *Journal of Environmental Quality* 25(6), 1339–1343.

Ehlers, W. (1975) Observations on earthworm channels and infiltration on tilled and untilled loess soil. *Journal of Soil Science* 119, 242–249.

Ertl, D.S., Young, K.A. and Raboy, V. (1998) Plant genetic approaches to phosphorus management in agricultural production. *Journal of Environmental Quality* 27, 299–304.

Eurostat (1997) *Agriculture Statistical Yearbook 1996.* European Commission, Brussels.

Fennessy, M.S. and Cronk, J.K. (1997) The effectiveness and restoration potential of riparian ecotones for the management of non-point source pollution, particularly nitrate. *Critical Reviews in Environmental Science and Technology* 27, 285–317.

Foran, M.E. and Dean, D.M. (1993) The land application of liquid manure and its effect on tile drain water and groundwater quality. In: *Agricultural Research to Protect Water Quality. Proceedings of the Conference February 21–26, 1993, Minneapolis, Minnesota.* Soil and Water Conservation Society, Ankheny, Iowa, USA, pp. 279–281.

Gangbazo, G., Pesant, A.R., Barnett, G.M., Charuest, J.P. and Cluis, C.C. (1995) Water contamination by ammonium nitrogen following the spreading of hog manure and mineral fertilizers. *Journal of Environmental Quality* 24, 420–425.

Gerritse, R.G. and Vriesema, R. (1984) Phosphorus distribution in animal waste slurries. *Journal of Agricultural Science* 102, 159–161.

Giddens, J.D. and Barnett, A.P. (1980) Soil loss and microbiological quality of runoff from land treated with poultry litter. *Journal of Environmental Quality* 9, 518–520.

Goss, M.J., Howse, K.R., Lane, P.W., Christian, D.G. and Harris, G.L. (1993) Losses of nitrate-nitrogen in water draining from under autumn crops established by direct drilling or mouldboard ploughing. *Journal of Soil Science* 44, 35–48.

Goss, M.J., Barry, D.A.J. and Rudolph, D.L. (1998) Groundwater contamination in Ontario farm wells and its association with agriculture. 1. Results from drinking water wells. *Journal of Contaminant Hydrology* 32, 267–293.

Goss, M.J., Unc, A. and Chen, S. (2000) Transport of nitrogen, phosphorus and microorganisms from manure into surface- and ground-water. In: Balázs, E., Galaste, E., Lynch, J.M., Schepers, J.S., Toutant, J.-P., Werner, D. and Werry, P.A.Th.J. (eds) *Biological Resource Management – Connecting Science and Policy.* Springer-Verlag, New York, pp. 31–55.

Greaves, J., Hobbs, P., Chadwick, D. and Haygarth, P.M. (1999) Prospects for the recovery of phosphorus from animal manures: a review. *Environmental Technology* 20, 697–708.

Haygarth, P.M. and Jarvis, S.C. (1999) Transfer of phosphorus from agricultural soils. *Advances in Agronomy* 66, 195–249.

Haygarth, P.M., Hepworth, L. and Jarvis, S.C. (1998) Forms of phosphorus transfer in hydrological pathways from soil under grazed grassland. *European Journal of Soil Science* 49, 65–72.

Heckrath, G., Brookes, P.C., Poulton, P.R. and Goulding, K.W.T. (1995) Phosphorus leaching from soils containing different phosphorus concentrations in the Broadbalk experiment. *Journal of Environmental Quality* 24, 904–910.

Hefting, M.M. and de Klein, J.J.M. (1998) Nitrogen removal in buffer strips along a lowland stream in The Netherlands: a pilot study. *Environmental Pollution* 102, 521–526.

Helling, C.S. and Gish, T.J. (1991) Physical and chemical processes affecting preferential flow. In: Gish, T.J. and Shrimohammadi, A. (eds) *Preferential Flow.* American Society of Agricultural Engineers, St Joseph, Michigan, pp. 77–86.

Henzell, E.F. and Ross, P.J. (1973) The nitrogen cycle of pasture ecosystems. In: Butler, G.W. and Bailey, R.W. (eds) *Chemistry and Biochemistry of Herbage*, Vol. 2. Academic Press, London, pp. 227–246.

Hergert, G.M., Bouldin, D.R., Klausner, S.D. and Zwerman, P.J. (1981) Phosphorus concentration-water flow interactions in tile effluent from manured land. *Journal of Environmental Quality* 10(3), 338–344.

Hinton, M. and Bale, M.J. (1991) Bacterial pathogens in domesticated animals and their environment. *Journal of Applied Bacteriology Symposium Supplement* 70, 81S-90S.

HMSO (1980) 5 day biochemical oxygen demand (BOD_5). In: *Methods for Examination of Waters and Associated Materials*, 2nd edn. Royal Commission, HMSO, London.

HMSO (1993) *Agriculture in the UK* 1993. MAFF, SOAFD, DANI, WO, HMSO, London.

Hobbs, P.J., Pain, B.F., Kay, R.M. and Lee, P.A. (1996) Reduction of odorous compounds in fresh pig slurry by dietary control of crude protein. *Journal of Science of Food and Agriculture* 71, 508–514.

Holmes, W. (1970) Animals for food. *Proceedings of the Nutrition Society* 29, 237–244.

Honeycutt, C.W. and Potaro, L.J. (1990) Field evaluation of heat units for predicting crop residue carbon and nitrogen mineralization. *Plant and Soil* 125, 213–220.

Hooda, P.S., Moynagh, M. and Svoboda, I.F. (1996) A comparison of phosphorus losses in drainage water from two different grassland systems. *Soil Use and Management* 12, 224.

Ingénieries-EAT (1996) *Animal Manures and Environment in Europe*, special issue. CEMAGREF, Paris, France.

IPPC (1997) BC Directive 96/61 concerning integrated pollution prevention and control (IPPC) – implementation in the pig and poultry sectors. Consultation paper, Ministry of Agriculture, Fisheries and Food.

Kearney, T.E., Larkin, M.J. and Levett, P.N. (1993) The effect of slurry storage and anaerobic digestion on survival of pathogenic bacteria. *Journal of Applied Bacteriology* 74, 86–93.

Kemp, J.S., Paterson, E., Gammack, S.M., Cresser, M.S. and Killham, K. (1992) Leaching of genetically modified *Pseudomonas fluorescens* through agricultural soils: Influence of temperature, soil pH and roots. *Biology and Fertility of Soils* 13, 218–224.

King, D.J., Watson, G.C., Wall, G.J. and Grant, B.A. (1994) *The Effects of Livestock Manure Application and Management on Surface Water Quality.* Summary Technical Report GLWQP-AAFC Pest. Canada, Agriculture and Agrifood Canada, Management Research Centre, London, Ontario.

Kirchmann, H. (1985) Losses, plant uptake and utilisation of manure nitrogen during a production cycle. *Acta Agriculturae Scandinavica Supplement* 24, 75 pp.

Kirchmann, H. (1994) Animal and municipal organic wastes and water quality. In: Lal, R. and Stewart, B.A. (eds) *Advances in Soil Science: Soil Processes and Water Quality.* CRC Press, Boca Raton, Florida, pp. 163–232.

Kjellerup, V. (1996) Agros nitrogen meter for estimation of ammonium nitrogen in slurry and liquid manure. In: Dam Kofooed, A., Williams, J.H. and L'Hermite, P. (eds) *Land Use of Sludge and Manure.* Elsevier Applied Science, London, pp. 216–223.

Kotak, B.G., Kenefick, S.L., Fritz, D.L., Rousseaux, C.G., Prepas, E.E. and Hrudey, S.E. (1993) Occurrence and toxicological evaluation of cyanobacterial toxins in Alberta lakes and farm dugouts. *Water Research* 27, 495–506.

Labeda, D.P., Kang-Chien, J. and Casida, L.E. Jr (1976) Colonisation of soil by Arthrobacter and Pseudomonas under varying conditions of water and nutrient availability as studied by plate counts and transmission electron microscopy. *Applied Environmental Microbiology* 31, 551–561.

Lance, J.C. (1977) Phosphate removal from sewage water by soil columns. *Journal of Environmental Quality* 6, 279–284.

Leinweber, P., Lunsmann, F. and Eckhardt, U. (1997) Phosphorus sorption capacities and saturation in soil in two regions with different livestock densities in north-west Germany. *Soil Use and Management* 13, 82–89.

Lemunyon, J.L. and Gilbert, R.G. (1993) Concept and need for a phosphorus assessment tool. *Journal of Production Agriculture* 6, 483–486.

Lenehan, J.J., Cumby, T., Huijsmans, J., Lorenz, F., Provolo, G., Perros, J. and McGechan, M. (1999) SWAMP: optimising the use of slurry. In: Martinez, J. and Maudet, M.M. (eds) *Proceedings of the*

Eighth International Conference on Management Strategies for Organic Waste Use in Agriculture. CEMAGREF, Rennes, France, pp. 37–48.
L'Herroux, L., Le Roux, S., Appriou, P. and Martinez, J. (1997) Behaviour of metals following intensive pig slurry applications to a natural field treatment process in Brittany (France). *Environmental Pollution* 97, 119–130.
Lowrance, R. (1992) Nitrogen outputs from a field-size agricultural watershed. *Journal of Environmental Quality* 21, 602–607.
MAFF (1998a) *Code of Good Agricultural Practice for Protection of Water.* MAFF Publications, London.
MAFF (1998b) *Code of Good Agricultural Practice for Protection of Soil.* MAFF Publications, London.
MAFF (1998c) *Code of Good Agricultural Practice for Protection of Air.* MAFF Publications, London.
MAFF (1998d) *Guidelines for Farmers in Nitrate Vulnerable Zones.* Booklet PB 2377, MAFF. London.
MAFF (2000) *Fertiliser recommendations for Agricultural and Horticultural Crops.* Reference Book 209, 7th edn. Ministry of Agriculture, Fisheries and Food, HMSO, London.
Martinez, J. (1997) Solepur: a soil treatment process for pig slurry with subsequent denitrification of drainage water. *Journal of Agricultural Engineering Research* 66, 51–62.
Martinez, J. and Peu, P. (2000) Nutrient fluxes from a soil treatment process for pig slurry. *Soil Use and Management* 16, 100–107.
Mawdsley, J.L., Bardgett, R.D., Merry, R.J., Pain, B.F. and Theodorou, M.K. (1995) Pathogens in livestock waste, their potential for movement through soil and environmental pollution. *Applied Soil Ecology* 2, 1–15.
McDonald, P., Edwards, R.A., Greehalgh, J.F.D. and Morgan, C.A. (1995) *Animal Nutrition*, 5th edn. Longman, London.
Meisinger, J.J. and Randall, G.W. (1991) Estimating nitrogen budgets for soil-crop system. In: Follet, R.F., Keeney, D.R. and Cruse, R.M. (eds) *Managing Nitrogen for Groundwater Quality and Farm Profitability.* SSSA, Madison, Wisconsin, pp. 85–124.
Miller, M.H., Robinson, J.B., Coote, D.R., Spires, A.C. and Wraper, D.W. (1982) Agricultural and water quality in the Canadian Great Lakes Basin: III. Phosphorus. *Journal of Environmental Quality* 11(3), 487–493.
Morse, D., Head, H.H., Wilcox, C.J., van Horn, H.H., Hissem, C.D. and Harris, B. Jr (1992) Effects of concentration of dietary phosphorus on amount and route of excretion. *Journal of Dairy Science* 75, 3039–3049.
Munyankusi, E., Gupta, S.C., Moncrief, J.F. and Berry, E.C. (1994) Earthworm macropores and preferential transport in a long-term manure applied Typic Hapludalf. *Journal of Environmental Quality* 23, 733–784.
Nichols, D.J., Daniel, T.C. and Edwards, D.R. (1994) Nutrient runoff from pasture after incorporation of poultry litter of inorganic fertilizer. *Soil Science Society of America Journal* 58, 1224–1228.
Nicholson, F.A., Chambers, B.J., Williams, J.R. and Unwin, R.J. (1999) Heavy metal contents of livestock feeds and animal manures in England and Wales. *Bioresource Technology* 70, 23–31.
NRA (National Rivers Authority) (1992) *The Influence of Agriculture on the Quality of Natural Waters in England and Wales,* NRA Water Quality Series Report No. 6. HMSO, London.
NRA (National Rivers Authority) (1995) *Water Pollution Incidents in England and Wales 1994,* NRA Water Quality Series Report No. 25. HMSO, London.
Ogilvie, J.R., Goss, M.J. and Stonehouse, D.P. (1999) *MCLONE4 – an Integrated Systems Approach to Manure Handling Systems and Nutrient Management,* University of Guelph, Ontario, Canada.
Pain, B.F. (2000) Control and utilization of livestock manures. In: Hopkins, A. (ed.) *Grass Its Production and Utilization,* 3rd edn. BGS, Blackwell Science, Oxford, pp. 343–364.
Pain, B.F., Phillips, V.R., Clarkson, C.R. and Klarenbeek, J.V. (1989) Loss of nitrogen through ammonia volatilization during and following the application of pig or cattle slurry to grassland. *Journal of the Science of Food and Agriculture* 47, 1–12.
Pain, B.F., van der Weerden, T.J., Chambers, B.J., Phillips, V.R. and Jarvis, S.C. (1998) A new inventory for ammonia emissions from U.K. agriculture. *Atmospheric Environment* 32, 309–313.
Pain, B.F., Misselbrook, T.H. and Chadwick, D.R. (1999) Controlling losses of ammonia from manures. In: Corrall, A.J. (ed.) *Accounting for Nutrients – a Challenge for Grassland Farmers in the 21st Century,* BGS Occasional Symposium No. 33, Malvern, UK. Antoney Rowe, Reading, UK, pp. 93–98.
Paterson, E., Kemp, J.S., Gammack, S.M., Fitzpatrick, E.A., Cresser, M.S., Mullins, C.E. and Killham, K.

(1993) Leaching genetically modified *Pseudomonas fluorescens* through intact soil microcosms – Influence of soil type. *Biology and Fertility of Soils* 15, 308–314.

Paul, J.W. and Beauchamp, E.G. (1989) Effect of carbon constituents in manure on denitrification in soil. *Canadian Journal of Soil Science* 69, 49–61.

Paul, J.W., Beauchamp, E.G. and Zhang, X. (1993) Nitrous and nitric oxide emissions during nitrification and denitrification from manure-amended soil in the laboratory. *Canadian Journal of Soil Science* 73, 539–553.

Piccini, S. and Bortone, T. (1991) The fertilizer value of agricultural manure: simple rapid methods of assessment. *Journal of Agricultural and Engineering Research* 49, 197–208.

Powlson, D.S., Stockdale, E.A., Jarvis, S.C. and Shepherd, M.A. (1994) *A Review of Nitrogen Mineralisation and Immobilization in UK Soils.* Report for the Ministry of Agriculture, Fisheries and Food, Environmental Protection Unit, London.

Quisenberry, V.L., Smith, B.R., Phillips, R.E., Scott, H.D. and Nortcliff, S. (1993) A soil classification system for describing water and chemical transport. *Journal of Soil Science* 156(5), 306–315.

Schulze, E.D., de Vries, W., Hauks, M., Rosen, K., Rasmussen, L., Tamm, S.O. and Nillson, J. (1989) Critical loads for nitrogen deposition on forest ecosystems. *Water, Air and Soil Pollution* 48, 451–456.

Sharpley, A.N. (1997) Rainfall frequency and nitrate and phosphorus runoff from soil amended with poultry litter. *Journal of Environmental Quality* 26, 1127–1132.

Sharpley, A.N. and Halvorson, A.D. (1994) The management of soil phosphorus availability and its transport in agricultural runoff. In: Lal, R. (ed.) *Soil Processes and Water Quality.* Advances in Soil Science. Lewis Publishers, Boca Raton, Florida, pp.1–84.

Sharpley, A.N. and Smith, S.J. (1989) Prediction of soluble phosphorus transport in agricultural runoff. *Journal of Environmental Quality* 18, 131–316.

Sharpley, A.N. and Withers, P.A. (1994) The environmentally sound management of agricultural phosphorus. *Fertiliser Research* 39, 133–146.

Sharpley, A.N., Smith, S.J., Jones, O.R., Berg, W.A. and Coleman, G.A. (1992) The transport of bioavailable phosphorus in agricultural runoff. *Journal of Environmental Quality* 21, 30–35.

Sharpley, A.N., Chapra, S.C., Wedepohl, R., Sims, J.T., Daniel, T.C. and Reddy, K.R. (1994) Managing agricultural phosphorus for protection of surface waters: issues and options. *Journal of Environmental Quality* 23, 437–451.

Sierra, X.X. (1997) Temperature and soil moisture dependence of N mineralization in intact soil cores. *Soil Biology and Biochemistry* 24, 1557–1563.

Smith, K.A. and Baldwin, D.J. (1999) A management perspective on improved precision manure and slurry application. In: Martinez, J. and Maudet, M.M. (eds) *Proceedings of the Eighth International Conference on Management Strategies for Organic Waste Use in Agriculture.* CEMAGREF, Rennes, France, pp. 89–96.

Smith, K.A. and Chambers, B.J. (1995) Muck: from waste to resource. Utilisation: the inputs and implications. *Agricultural Engineer* 50, 33–38.

Smith, K.A. and Frost, J.P. (2000) Nitrogen excretion by farm livestock with respect to land spreading requirements and controlling nitrogen losses to ground and surface waters. Part 1: cattle and sheep. *Bioresource Technology* 71, 173–181.

Smith, K.A., Chalmers, A.G., Chambers, B.J. and Christie, P. (1998) Organic manure phosphorus accumulation, mobility and management. *Soil Use and Management* 14, 1–7.

Smith, K.A., Charles, D.R. and Moorhouse, D. (2000) Nitrogen excretion by farm livestock with respect to land spreading requirements and controlling nitrogen losses to ground and surface waters. Part 2: pigs and poultry. *Bioresource Technology* 71, 183–194.

Smits, M.C.J., Valk, H., Monteney, G.J. and van Vuuren, A.M. (1997) Effect of protein nutrition on ammonia emissions from cow houses. In: Jarvis, S.C. and Pain, B.F. (eds) *Gaseous Nitrogen Emissions from Grassland.* CAB International, Wallingford, UK, pp. 101–108.

Sommerfeldt, T.G. and Chang, C. (1985) Changes in soil properties under annual applications of feedlot manure and different tillage practices. *Soil Science Society of America Journal* 49, 983–987.

Strauch, D. (1987) Hygiene of animal waste management. In: Strauch, D. (ed.) *Animal Production and Environmental Health.* Elsevier Science Publisher, Amsterdam, pp. 155–202.

Sun, G., Gray, K.R., Biddlestone, A.J. and Cooper, D.J. (1999) Treatment of agricultural wastewater in a combined tidal flow-downflow reed bed system. *Water Science Technology* 40, 139–146.

Tamminga, S. (1992) Nutrition management of dairy cows as a contribution to pollution control. *Journal of Dairy Science* 75, 345–357.
Tan, Y., Bond, W.J., Rovira, A.D., Brisbaine, P.G. and Griffin, D.M. (1991) Movement through soil of a biological control agent, *Pseudomonas fluorescens. Soil Biology and Biochemistry* 23, 821–825.
Tanner, C.C., Clayton, J.S. and Upsell, M.P. (1995a) Effect of loading rate and planting on treatment of dairy farm wastewaters in constructed wetlands-I. Removal of oxygen demand, suspended solids and faecal coliforms. *Water Research* 29, 17–26.
Tanner, C.C., Clayton, J.S. and Upsell, M.P. (1995b) Effect of loading rate and planting on treatment of dairy farm wastewaters in constructed wetlands-II. Removal of nitrogen and phosphorus. *Water Research* 29, 27–34.
Thomas, D., Heckrath, G. and Brooks, P.C. (1997) Evidence of phosphorus movement from Broadbalk soils by preferential flow. In: Tunney, H., Carton, O.T., Brooks, P.C. and Johnston, A.E. (eds) *Phosphorus Loss From Soil to Water.* CAB International, Wallingford, UK, pp. 369–370.
Thompson, R.B. and Pain, B.F. (1990) The significance of gaseous losses of nitrogen from livestock slurries applied to agricultural land. In: Merckx, R., Vereecken, H. and Vlassak, K. (eds) *Fertiliser and Environment.* Leuven University Press, Leuven, Belgium, pp. 290–296.
Toth, J.D. and Fox, R.H. (1998) Nitrate losses from a core-lucerne rotation: lysimeter measurement of nitrate leaching. *Journal of Environmental Quality* 27, 1027–1033.
Unger, P.W. (1984) Tillage systems for soil and water conservation. *FAO Soils Bulletin* 54, 278 pp.
US EPA (1993) *A Watershed Approach.* Office of Water, US Government Print Office, Washington, DC.
Van Elsas, J.D., Trevors, J.T. and van Overbeek, L.S. (1991) Influence of soil properties on the vertical movement of genetically marked *Pseudomonas fluorescens* through large soil microcosms. *Biology and Fertility of Soil* 10, 249–255.
Van Riemsdijk, W.H., Lexmond, T.H., Enfield, C.G. and van der Zee, S.E.A.T. (1987) Phosphorus and heavy metals: accumulation and consequences. In: van der Meer, H.E., Unwin, R.J., van Dijk, T.A. and Ennick, G.C. (eds) *Animal Manures on Grassland and Forage Crops.* Martinus Nijhoff, Dordrecht, pp. 213–227.
Vlassak, K. (1998) Environmental problems, current recommendations and regulations in Flanders (Belgium). In: Matsunka, T. (ed.) *Environmentally Friendly Management of Farm Animal Waste.* Proceedings of an International Workshop, Sapporo, Japan, 25–29 November 1997. Kikanshi Insatsu Co. Ltd, Sapporo, Japan, pp. 13–23.
Wagenet, R.J. (1990) Quantitative prediction of the leaching of organic and inorganic solutes in soil. *Philosophical Transactions of the Royal Society of London B* 329, 321–330.
White, R.E. (1985) The influence of macropores on the transport of dissolved and suspended matter through soil. *Advances in Soil Science* 3, 95–120.
Williams, J.R., Hurst, C.L., Chambers, B.J., Brookman, S. and Chadwick, D. (1999) Rapid methods for the analysis of readily available nitrogen in manures. In: Corrall, A.J. (ed.) *Accounting for Nutrients – a Challenge for Grassland Farmers in the 21st Century,* BGS Occasional Symposium No. 33, Malvern, UK. Antony Rowe, Reading, UK, pp. 171–172.
Yuan, G. and Lavkulich, L.M. (1994) Phosphate sorption in relation to extractable iron and aluminium in spodosols. *Soil Science Society of America Journal* 58, 343–346.

4 Pesticides and Persistent Organic Pollutants

Bondi Gevao and Kevin C. Jones
Institute of Environmental and Natural Sciences, Lancaster University, Lancaster LA1 4YQ, UK

Key words: degradation, environmental fate, leaching, persistent organic pollutants, pesticides, runoff, sorption, spray drift, volatilization, water pollution

Introduction

The term 'pesticide' encompasses a great range of diverse substances, falling into a number of broad groupings such as herbicides, insecticides, molluscicides and fungicides. The use of pesticides has substantially increased since the introduction of synthetic chemicals in the 1940s, revolutionizing agriculture. Specifically, pesticides have been used in: (i) agriculture to increase productivity and the quality and quantity of food, fibre and feed; (ii) forestry for pest control; (iii) the industrial, commercial, municipal and military sectors for rodent and weed/brush control around plant sites and rights-of-way; (iv) medical vector sector for control of mosquitoes and rodents etc.; and (v) urban environments for termite control around structures and pest control in gardens (Garthwaite *et al.*, 1995; Centner, 1998; Falconer, 1998). Agriculture is the principal market for pesticides (Himel *et al.*, 1990; Bower *et al.*, 1993).

Hundreds of different active ingredients are now available. Selective herbicides were particularly instrumental in transforming arable production methods, by enabling farmers to move from traditional systems of crop rotation to continuous cereal cropping (Ward *et al.*, 1993; Falconer, 1998). In the UK, for instance, over 99% of arable crops receive at least one pesticide application annually, with *c.* 4.5 Mha of land treated in total (Ward *et al.*, 1993; Falconer, 1998). It is not uncommon for six or seven different chemicals to be applied to a cereal crop in a growing season (Garthwaite *et al.*, 1995).

The generic term 'persistent organic pollutants (POPs)' encompasses many hundreds of chemicals which: (i) are persistent in the environment, having long half-lives in the soils, sediments, air and biota; (ii) are typically hydrophobic and lipophilic; (iii) have the propensity to enter the gas phase under environmental temperatures and are subject to long range transport; and (iv) are distributed globally and even found in pristine environments such as the Arctic. The combination of their resistance to metabolism and lipophilicity means that they will bioaccumulate and be transported through food chains. Among the important classes of POPs are many families of chlorinated (and brominated) aromatics, including polychlorinated biphenyls (PCBs), polychlorinated naphthalenes (PCNs), polychlorinated dibenzo-*p*-dioxins and -furans (PCDD/Fs) and organochlorine pesticides (e.g. DDT, chlordane, toxaphene, lindane). Many of these chemicals have been synthesized for industrial applications (e.g. PCBs, PCNs) and as agrochemicals (e.g. lindane, chlordane). Apart from the agrochemicals that are deliberately applied to land, most POPs reach agricultural land predominantly from atmospheric transport and deposition.

Types of pesticides found on agricultural land

There are a large number of pesticides currently in use, with a wide range of physico-chemical properties and belonging to a wide variety of chemical classes. Clearly, the physico-chemical properties of a given pesticide will govern its behaviour and biological activity in the soil. Molecular size, ionizability, water solubility, lipophilicity, polarizability and volatility are all key properties, but generally one or two properties have a dominating influence (Bailey and White, 1970; Weber, 1971, 1972; Stevenson, 1976; White, 1976). Figure 4.1 is a classification scheme for selected pesticides on the basis of their significant chemical properties and reported behaviour in soils and water.

Sources of Pesticides and Persistent Organic Pollutants on Agricultural Farms

Direct applications

Spray applications

Pesticides are voluntarily spread on soils or sprayed on to crops to control pests in agriculture. Application rates for the active ingredients vary tremendously.

Accidental spillage or irresponsible disposal

Significant quantities of unwanted pesticides may be held by agricultural producers in barns and other storage buildings. A survey carried out in the USA, for instance, estimated that millions of tonnes of unwanted products have accumulated over the past 60 years in thousands of barns across the country (Jones, 1993). Pesticides are unwanted for various reasons: (i) some are products whose registration has been cancelled; (ii) they did not perform as

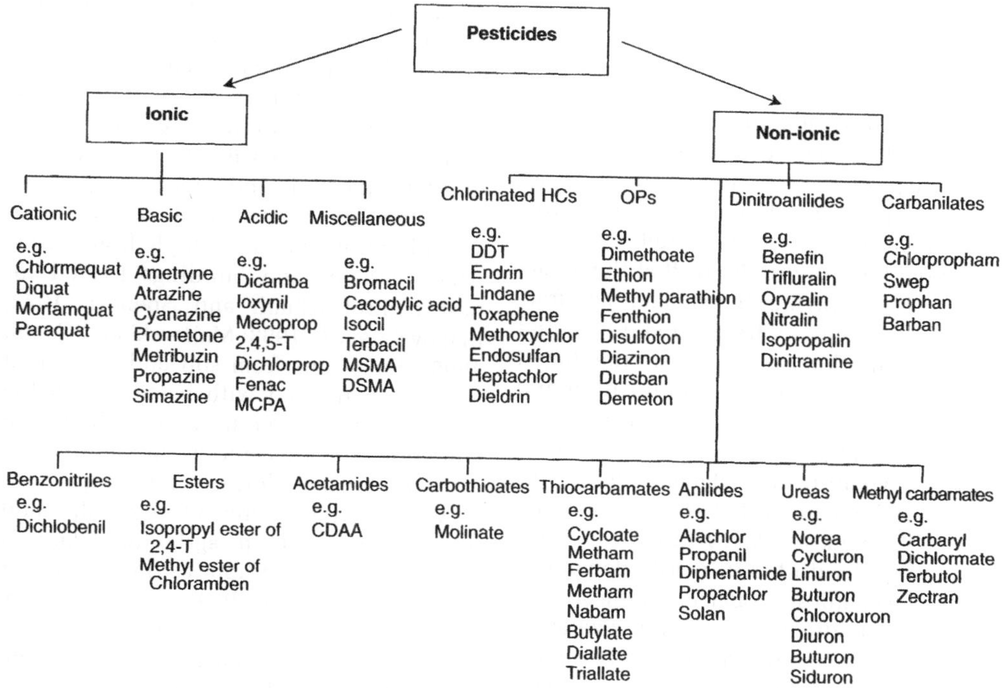

Fig. 4.1. Classification of pesticides (OPs = organo-phosphates).

desired or intended; (iii) others have been replaced by alternative products; or (iv) pests became resistant to the pesticides (Centner, 1998; Centner and Gunter, 1999).

Surveys suggest that most people with quantities of unwanted pesticides continue to store them because there is no viable disposal option (Karnatz, 1991; Centner, 1998; Centner and Gunter, 1999). One impediment to the disposal of pesticides is cost. Usually, farmers are unwilling to incur the full cost of legal disposal. Some end up being disposed of in landfill or dispersed into the environment, where they may cause pollution and other problems (Karnatz, 1991; Centner, 1998; Centner and Gunter, 1999).

Sewage sludge

Sewage sludges are by-products of wastewater treatment, so they are an inevitable sink for a wide range of organic chemicals, depending on the origin of the sewage. Organic chemicals may enter the wastewater through a variety of sources: (i) point source discharge by industry; (ii) diffuse discharge from industrial, commercial and domestic properties; (iii) in runoff following atmospheric deposition; and (iv) addition at the wastewater treatment works (Jones and Sewart, 1997).

During wastewater treatment, many contaminants are preferentially removed from the water and concentrated in the sludges. Many potentially hazardous organic compounds can be present in sludge that is applied to agricultural soils. Although the recycling of organic matter and nutrients contained in wastewater through land application is a worthwhile objective, it can lead to an increase in the concentrations of organic chemicals in soils, which has raised concerns about their potential to transfer to food chains (Wild and Jones, 1992; Wang and Jones, 1994b; Fries, 1996) and groundwater (Wilson *et al.*, 1997). In the UK, for example, 50% of the 1 Mt of sludge produced each year is currently applied to land (Bruce and Davis, 1989).

Indirect applications

Spray drift

Whenever pesticides are applied to agricultural fields as spray from either ground-based equipment or aircraft, there is an initial migration of some of the active ingredient away from the target area. This takes the form of droplet drift at the actual time of the spraying; the smaller size fraction is preferentially transferred downwind rather than deposited out in the target area. Then, following the application and extending over several hours at least, more pesticide may be carried downwind through volatilization depending on the compound volatility (Akesson and Yates, 1964; Maybank *et al.*, 1978; Grover *et al.*, 1997). Droplet drift is primarily dependent on the mechanical properties of the sprayer that controls dissemination and size spectrum, while vapour drift depends on the pesticide vapour pressure and the soil surface temperature of the target area. In addition, both factors will vary with wind speed and atmospheric turbulence conditions and chemical formulation of the pesticide (Akesson and Yates, 1964; Maybank *et al.*, 1978; Elliott and Wilson, 1983; Grover *et al.*, 1997). The processes of spray drift and its potential impact have been discussed in detail elsewhere (Akesson and Yates, 1964; Maybank *et al.*, 1978; Parkin and Merritt, 1988).

Atmospheric transport and deposition

Global pesticide use is currently estimated at *c.* 2.5 billion kg year^{-1} (Bergstrom and Stenstrom, 1998; Centner, 1998). To be effective, pesticides need to persist for a certain period of time. However, the greater their persistence, the greater their potential for transport away from the target area. Agricultural pesticides therefore enter the atmosphere by spray drift, post-application volatilization and wind erosion of soil. Volatilization can remove a large fraction of the pesticide initially applied to the field. Willis *et al.* (1983), for example, followed losses of toxaphene and DDT after spray application to cotton plants and found that 53–58% volatilized within a month. Measured trifluralin and triallate volatilization fluxes accounted for 57–74% of total losses from soil within 5 days of application (Cliath and Spencer, 1972; Majewski *et al.*, 1993; Majewski and Capel, 1995).

Pesticides are also emitted from non-agricultural usage; for example from lawns, parks, gardens and buildings treated for pest control (Cliath and Spencer, 1972; Majewski *et al.*, 1993; Majewski and Capel, 1995). Emissions from soil and water that were contaminated with pesticides in the past may be a significant contributor to contemporary atmospheric burdens, especially for organochlorine compounds.

Whereas the source of pesticides and agrochemical POPs to agricultural land is the deliberate addition as herbicides or insecticides, other POP chemicals are delivered to agricultural land principally or exclusively by atmospheric deposition. Pesticides and POPs will be present in air either in the gaseous phase or associated with particles, a function of their physico-chemical properties (e.g. water solubility and vapour pressure), environmental variables (e.g. temperature, humidity) and the nature and concentration of dust particles in air. The gas/particle distribution influences atmospheric removal by precipitation and dry particle deposition and, consequently, the residence time in the atmosphere (Ligocki *et al.*, 1985; Bidleman, 1988; Ballschmiter and Wittlinger, 1991; Pankow *et al.*, 1993). Clearly, the longer a substance persists in the atmosphere, the further it can be transported. Pesticides with low Henry's law constants (e.g. dieldrin, hexachlorocyclohexanes (HCHs)) are removed by gas scavenging, while those with high constants (e.g. DDT, chlordane) are removed mainly by washout of particles. Once they have been deposited on soil or vegetation, persistent compounds can re-enter the atmosphere and recycle between the atmosphere and the surface, depending on temperature (Majewski and Capel, 1995; Muir *et al.*, 1996; Wania and Mackay, 1996). This subject has received great coverage in the literature in recent years (Bidleman, 1988; Ligocki and Pankow, 1989; Pankow *et al.*, 1993; Hoff *et al.*, 1996; Bidleman, 1999).

Estimated deposition of pesticides based on the analysis of precipitation collected with bulk or wet-only samplers, is generally of the order of a few mg ha^{-1} $year^{-1}$ up to more than 1 g ha^{-1} $year^{-1}$ for individual pesticides (Buser, 1990; Waite *et al.*, 1995; Bucheli *et al.*, 1998). Considerable between-year variations may exist in the deposition of a certain pesticide at a specific sampling site (McConnell *et al.*, 1998a,b; Rawn *et al.*, 1999a,b). This may be explained by differences in application amounts and atmospheric conditions, including the timing of rainfall events relative to application (Rawn *et al.*, 1999a). In general, the occurrence of many pesticides in air and rainwater shows a seasonal variation, with peak concentrations occurring during application periods (Waite *et al.*, 1995; Rawn *et al.*, 1999a).

Pesticides do not only occur in the atmosphere in agricultural and urban areas where they are applied. They are also found in areas where no pesticides are or ever have been used, such as mountains, boreal forests and the Arctic (Buser, 1990; Waite *et al.*, 1995; Wania and Mackay, 1996; Rawn *et al.*, 1999a,b). In general, atmospheric deposition of pesticides is lower in these pristine environments, but there is evidence that the more environmentally stable pesticides, such as lindane, are accumulating in the cooler regions of the globe (Wania and Mackay, 1996). Hence agroecosystems can receive inputs of pesticides indirectly, without deliberate application.

Fate of Pesticides in Soil

Once pesticides and POPs have entered the soil (by the processes discussed in the previous section) they can undergo several different fates in the soil–groundwater system. These are summarized in Fig. 4.2. Whether they are retained by the soil and the extent to which they volatilize, leach, degrade, etc., is controlled, fundamentally, by their interaction with the soil itself.

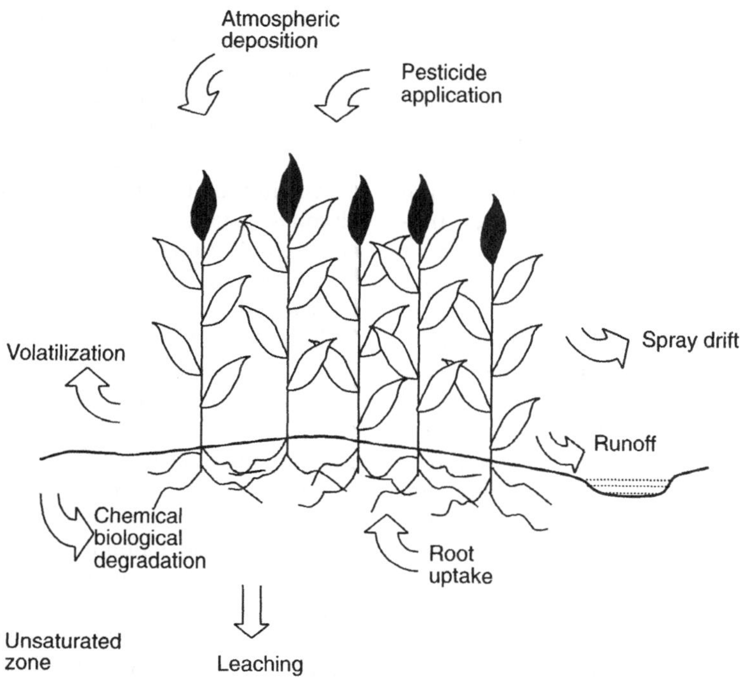

Fig. 4.2. Pathways of pesticide degradation and transportation.

Degradation

Pesticide degradation is one of the major loss processes in the environment and can proceed through biological and non-biological pathways.

Biotic degradation

The microbially mediated breakdown of pesticides has been identified to be more important in degradation compared with chemical and physical means. The microbial influence either proceeds directly (through mineralization, co-metabolism, polymerization or conjugation/accumulation) or indirectly through secondary effects of microbial activity altering soil pH and redox conditions (Bollag and Liu, 1990). The microbial transformation of a pesticide molecule may involve one or a combination of the mechanisms outlined above. The transformation process can be mediated by one or several organisms. A huge volume of literature exists on the microbial degradation of pesticides and other xenobiotics, including numerous books and reviews (Moorman, 1993; Alexander, 1994; Skipper and Turco, 1995).

The rate of degradation of pesticides is a function of the chemical's properties (i.e. structure, toxicity, solubility, concentration, etc.) and the soil and environmental conditions affecting microbial activity in the system. Even if soil conditions are favourable for microbial activity, biodegradation can be hampered by poor compound bioavailability. Mass transfer limitations via the processes of sorption, desorption and solubilization (Pignatello, 1990; Scow, 1993; Beck and Jones, 1995) can potentially reduce the amount of chemical available for microbial uptake/mineralization. The rate of degradation is also a function of the prevailing temperature and moisture regime, and the content and type of organic matter and clays present (Kookana *et al.*, 1998).

Since soil microbial activities are strongly modulated by temperature, pesticide degradation would be expected to be greater in tropical soils, which experience higher year-round temperatures, than in temperate soils. This explanation would be consistent with observations of the elevated rates of soil organic matter turnover that characterize tropical environments (Sanchez, 1976). In an excellent review of microbial pesticide degradation in tropical soils, Sethunathan *et al.* (1982) concluded that acceleration of microbial activities because of elevated temperatures was the major factor responsible for observations of increased degradation of pesticides under tropical rice paddy soil conditions.

Organic matter, temperature, moisture and pH generally vary substantially with depth in the soil profile and so does the associated microbial activity. Therefore, microbially mediated degradation may markedly decline with depth in the soil profile not only because of decreasing microbial activity but also through faster movement of pesticide as a result of decreasing organic matter content with depth (Kookana and Aylmore, 1994; Kookana *et al.*, 1998). It is therefore imperative to consider profile distribution of these soil properties in evaluating the fate and behaviour of a pesticide in soil.

Furthermore, the history of pesticide use can also influence the rate of degradation in soil. Pesticide degradation in soils is chiefly mediated by microorganisms, and their involvement is generally established by an initial lag in degradation while microbes adapt. Prior exposure to, or repeated additions of, a pesticide shortens the lag period, because of the enrichment of pesticide degrading micro-organisms (Sethunathan *et al.*, 1982; Roeth, 1986; Smith and Lafond, 1990). The phenomenon of cross-enhancement, or the ability of soils pre-treated with specific herbicides to degrade other structurally related chemicals more rapidly than in untreated soils, has been known for many years (Roeth, 1986; Smith and Lafond, 1990). This phenomenon has been reported in field situations for phenoxyalkanoic acid herbicides (Kirkland and Fryer, 1972; Tortensson *et al.*, 1975; Roeth, 1986; Smith and Lafond, 1990). Instances of such enhanced degradation of pesticides following repeat applications have been reported previously (Tal *et al.*, 1989; Suett and Jukes, 1990).

Abiotic degradation

Although biotic degradation is often seen as the dominant process for the majority of pesticides, under certain conditions the abiotic pathway for some pesticides can be the key breakdown mechanism in soils. For example, under conditions of low microbial activity, which may result at depths in a soil profile or in groundwater, abiotic transformations may dominate. The major abiotic transformation or degradation processes include hydrolysis, oxidation–reduction, dehydrohalogenation and photolysis (Wolfe *et al.*, 1990).

Hydrolysis

Hydrolytic degradation of pesticides in soil occurs in soil pore water or on the surface of clay minerals. Although investigations have been carried out on the significance and mechanism of hydrolysis for many pesticides (Chapman and Cole, 1982; Lehmann and Miller, 1989; Racke *et al.*, 1996), it has not been as thoroughly examined as other degradative processes in soils, probably because of experimental limitations in studying photolysis in the absence of competing processes and in the complex soil matrix. Hydrolysis has been identified as an important if not primary route of degradation for members of several classes of pesticides (e.g. organo-phosphates (OPs), carbamates, phenoxy herbicide esters) (Chapman and Cole, 1982; Laskowski *et al.*, 1983; Racke *et al.*, 1996).

Temperature is an important factor governing the rate of hydrolysis in soil pore water. The acceleration of hydrolytic reactions can be described by the Arrhenius equation and may be used to predict the behaviour of pesticides in soil (Getzin, 1981). The effect of temperature on the rate of hydrolysis is exemplified by the work of Getzin (1981) who reported half-lives of chlorpyrifos of > 20 days to 1 day for temperatures of 5 to 45°C, respectively. The rate of hydrolysis increases by a factor of two for every 10°C rise in temperature (Getzin, 1981). It is therefore expected that hydrolysis of pesticides occurs to a larger extent in warmer countries compared with rates in colder regions. Soil pH is also an important property influencing hydrolytic reactions of pesticides. The effects of soil pH on degradation of a given pesticide depend greatly on whether the compound is most susceptible to alkaline- or acid-catalysed hydrolysis (Laskowski *et al.*, 1983; Korpraditskul *et al.*, 1993a,b).

Redox reactions

Some pesticides are susceptible to oxidation or reduction reactions, which occur predominantly in aerobic and anaerobic soils, respectively. Some OP and carbamate insecticides (e.g. terbufos, phorate, isofenphos, aldicarb), for example, may undergo rapid oxidation in aerobic soils. The organochlorine insecticides and various pesticides with free nitro- groups (e.g. parathion, fenitrothion, chlomethoxynil), undergo anaerobic degradation. Yoshida and Castro (1970), for example, reported that although very little lindane exists in upland tropical soils, significant degradation occurred within a month under flooded conditions. The extent of degradation was dependent on the OM content, presumably due to the rapid onset of reducing conditions in soils with higher OM contents. DDT was rapidly converted to DDD via reductive dechlorination in flooded soils, the rate being dependent on the OM contents of the soils (Gohre and Miller, 1986). These degradative mechanisms are particularly important in the tropics, given the magnitude of pesticide use that is usually associated with flooded rice paddy agriculture.

Photolysis

Photolysis is the transformation of pesticides due to their exposure to radiation. Ultraviolet radiation in sunlight can break down pesticide molecules. Photolysis was previously not considered as an important degradation pathway for pesticides in soil. Recently, however, evidence has emerged to suggest that photoinduced transformations can, in some instances, be significant. Although pesticide molecules are not directly transformed by solar radiation, because of low absorbance between 290 and 400 nm wavelengths, indirect photodegradation may still be an important factor. Zayed *et al.* (1994) reported that the degradation of DDT to DDE in soil was enhanced by exposure to sunlight. Over a 90-day exposure period, 65% of the initial DDT remained compared with 91% in the control. It is also known that photodegradation proceeds faster in moist soils than in dry soils (e.g. Klehr *et al.*, 1983).

Retention mechanisms

Sorption–desorption

Sorption encompasses the physico-chemical processes by which a pesticide molecule present in soil solution binds to the soil particles. It is the major process that attenuates the mobility and accessibility of chemicals to target organisms. Sorption processes may vary from complete reversibility to total irreversibility. The extent of sorption depends on the properties of soil and the compound, which include size, shape, configuration, molecular structure, chemical functions, solubility, polarity, polarizability and charge distribution of interacting species, and the acid–base nature of the pesticide molecule (Bailey and White, 1970; Senesi, 1992; Pignatello and Xing, 1996). Sorption may be purely physical, as with van der Waals forces, or chemical in nature, as with electrostatic interactions (Senesi, 1992, 1993). In soils, one or more of these mechanisms may occur simultaneously, depending on the nature of the functional group and the acidity of the system. Chemical reactions between unaltered pesticides or their metabolites often lead to the formation of stable chemical linkages, resulting in an increase in the persistence of the residue in soil, while causing it to lose its chemical identity (Berry and Boyd, 1985; Calderbank, 1989; Bollag et al., 1992; Dec and Bollag, 1997). From a toxicological perspective, binding of pesticides to soil constituents, notably soil organic matter, leads to: (i) a decrease of material available to interact with biota; (ii) a reduction in the toxicity of the compound; and (iii) immobilization of the compound, thereby reducing its leaching and transport properties (Berry and Boyd, 1985; Calderbank, 1989; Bollag et al., 1992; Dec and Bollag, 1997).

It is generally accepted that organic matter and clay minerals are the principal soil constituents involved in pesticide sorption. Therefore both the amount and nature of the OM, in terms of its composition and its stage of degradation, and the clays in terms of their charge density determine the extent of sorption and desorption. However, sorption and desorption of a pesticide molecule in soil can also be significantly influenced by other soil factors such as pH, soil solution composition, or the presence of competitive co-solvents and ions (Kookana *et al.*, 1990; Brusseau *et al.*, 1991).

The reversibility of sorption reactions determines whether the soil solid phase provides a temporary or long-term home for pesticides. When the process is reversible, the pesticide can be released back in solution, in response to a decrease in its solution concentration. Hysteresis, which represents differences between sorption–desorption isotherms is a frequently observed phenomenon and is still rather poorly understood (DiToro and Horzempa, 1982; Brusseau and Rao, 1989; Nicholls, 1991). Hysteresis could be due to: (i) experimental artefacts such as changes in sorption capacity arising from shaking and centrifugation; (ii) incomplete equilibration; and (iii) degradation or transformation of the pesticide during experimentation (Kookana *et al.*, 1998). Nevertheless, true hysteresis as a result of the entrapment of pesticide molecules in the organic matter matrix or within mineral particles has clearly been observed (Pignatello, 1990; Pignatello and Xing, 1996; Xing and Pignatello, 1997).

Sequestration

Non-polar and hydrophobic compounds can undergo sequestration during prolonged residence or 'ageing' in soils. 'Ageing' refers to the increased contact time between a chemical and soil, which may allow a compound to become more strongly associated with soil components over time. Ageing may result from the formation of stronger bonds with increased residence times in soil or physical entrapment/occlusion of the compound in the soil organic matter or mineral lattice. This may include the formation of covalent bonds of parent compounds or their degradation products.

Sorption and sequestration can be viewed as one process, which is initially fast and then shifts to a slower sustained rate (Huang *et al.*, 1996; Pignatello and Xing, 1996; Kan *et al.*, 1998). This notion does not imply that the slow phase is dependent on the initial fast sorption and will not occur unless the latter is completed. Although sequestration is a form of sorption, there is no chemical or physical barrier that will prevent the two occurring simultaneously and having independent time courses (Pignatello and Xing, 1996; Dec and Bollag, 1997). An important difference between the two processes is their kinetics. Sorption occurs within minutes following addition of the chemical to soil, whereas sequestration may take much longer to become important.

Adsorption processes, which are involved predominantly in the initial phase of overall sorption, are generally ascribed to H-bonding, van der Waals forces, electrostatic attraction, and co-ordination reactions; that is, to mechanisms that are expected to occur instantaneously on contact of xenobiotic molecules with the active surfaces of the soil matrix. Diffusion phenomena, on the other hand, combined with sorption in remote microsites within the soil matrix, appear to be the domain of ageing and sequestration (Pignatello and Xing, 1996; Xing and Pignatello, 1996).

Bound residue formation

Bound residues are those residues that are not extractable by organic solvents and, therefore, are not detected in normal residue analysis. They have been defined as 'chemical species originating from pesticides, used according to good agricultural practice, that are unextracted by methods which do not significantly change the chemical nature of these residues' (Roberts, 1984). A detailed review of the current state of knowledge regarding the formation and biological/environmental significance of bound pesticide residues in soils has been completed, with a detailed discussion of the factors that affect their formation and fate (Gevao *et al.*, 2000). The environmental significance of bound residues is still subject to debate, but in the context of this chapter the formation of bound residues represents an important process that will prevent/reduce pesticides and POPs from being subject to leaching, volatilization and uptake into biological systems, processes that are considered in the next section.

Dissipation and Transport

Pesticides that enter the soil environment are subject to a variety of degradative and transport processes. The overall dissipation of a pesticide from soil results from a combination of loss mechanisms such as microbial degradation, chemical hydrolysis, photolysis, volatility, leaching and runoff. The degree to which each mechanism will contribute to the overall loss of the pesticide is in turn dependent on the physicochemical properties of the soil and the chemicals themselves, environmental factors including temperature and precipitation, and management practices (e.g. application rate, formulation type) (Weber and Miller, 1989). These variables can operate concurrently, are interdependent and difficult to quantify separately in the field.

The contribution made by each of the loss mechanisms to the overall dissipation is generally assessed by conducting laboratory studies where greater control is possible (Gile and Gillett, 1979; Toole and Crosby, 1989; Wang and Jones, 1994a; Kruger *et al.*, 1996). These studies provide both quantitative and qualitative data, enabling the kinetics and mechanisms of the loss to be identified. Typical dissipation patterns have been conceptualized by the decay curves shown in Fig. 4.3. Pesticides with short half-lives exhibit relatively rapid dissipation; most of the residues are ultimately removed from the soil (curve A). Such decay curves are typical of compounds which are extremely volatile, water-soluble or easily degraded (or a combination of these) such as soil fumigants like ethylene dibromide (Pignatello and Cohen, 1990). Studies with such compounds are often misleading because many suggest that the pesticides are completely lost from the soil. In practice, a small residual component, of concentrations less than typical contemporary analytical detection limits, may persist in the longer term, as was the case for ethylene dibromide detected in the top soils of an agricultural soil 19 years after application (Steinberg *et al.*, 1987).

At the other extreme, pesticides and POPs which are non-volatile, relatively water-insoluble and recalcitrant are characterized by curve B. Such pesticides are highly sorbed and persist in the soil over long periods. They often have high soil:water partition coefficients, and are desorbed very slowly into the aqueous phase, limiting their rate of dissipation by volatilization, leaching or biodegradation. Examples of pesticides that fit into this category are chlorinated insecticides including DDT, aldrin, endrin and dieldrin (Kjeller *et al.*, 1991).

Most pesticides fall within these two extremes, exhibiting biphasic decay curves (curves B, C and D), whereby a preliminary short period of rapid dissipation is followed by a much longer period characterized by a slower loss rate. As a consequence, a portion of the applied pesticide persists in the soil for a long time. These residual fractions are slowly reversible and should not be confused with the bound (non-extractable) residues (Lichtenstein, 1980; Führ *et al.*, 1991; Dec and Bollag, 1997).

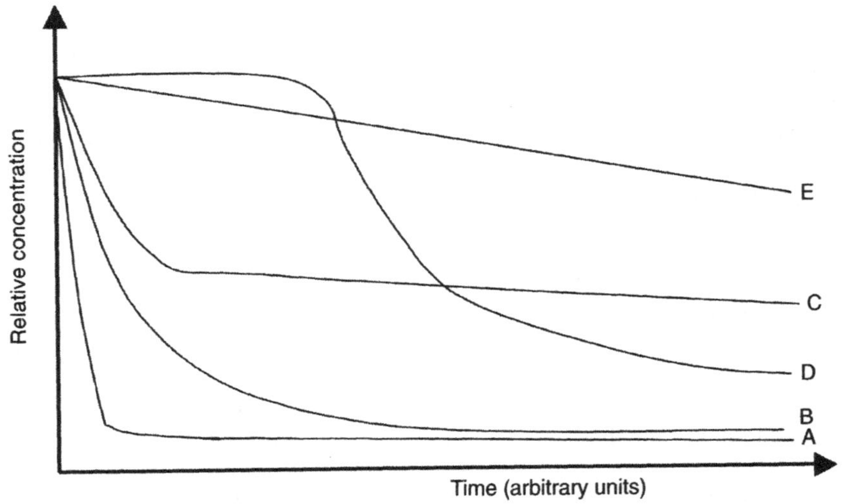

Fig. 4.3. Hypothetical loss profiles for organic chemicals in sewage sludge-amended soil. A–E are discussed in the text.

Pesticides applied to soil are highly beneficial to the crops being grown, but only as long as they remain in the root zone of the soil. If a portion of the compound leaves the target soil compartment, then the chemicals can no longer complete their intended purpose. The major pathways through which pesticides leave the root zone are volatilization to the atmosphere, leaching to subsoil and groundwater, runoff to surface water systems, and uptake by plants and animals. Each of these transport pathways is considered in the following sections.

Volatilization

During and after the application of pesticide to agricultural soils, a substantial fraction of the dosage may enter the atmosphere and be transported over varying distances downwind of the target. Volatilization of applied pesticides from soil is one of the important pathways for loss. The rate and extent of the emission during application depends primarily on the application technique and the type of formulation used, whereas the emission after application depends primarily on the properties of the pesticide, soil and environmental conditions (Spencer *et al.*, 1973; Nash, 1988).

Volatilization rates of pesticides from non-adsorbing surfaces are directly proportional to their vapour pressures. The actual rates of loss are almost entirely dependent on external conditions that affect movement away from the evaporating surface, such as surface roughness, wind speed and air turbulence (Nash, 1988; Nash and Gish, 1989; Spencer and Cliath, 1990).

Soil fumigants form a class of pesticides that are unique, since their vapour pressures are many orders of magnitude higher than those of other pesticides. Because of their high volatility, most of the applied dosage is lost to the atmosphere. Soil incorporated pesticides volatilize at a greatly reduced rate, dependent not only on the equilibrium distribution between the air, water and soil matrix as related to vapour pressure, solubility and adsorption coefficients, but also on their rate of movement to the soil surface (Nash, 1983). Volatilization from soil involves desorption of the pesticide from the soil, movement to the soil surface and vaporization into the atmosphere (Nash, 1988). Shallow soil incorporation and dry soil surfaces greatly restrict volatilization losses. Hill *et al.* (1989) found that 50% of trifluralin was lost in 3–7 h, and 90% in 2–5 days when applied to a moist soil surface. Much smaller losses were measured with soil incorporation: 22% loss in 120 days when incorporated to 2.5 cm and 3.5% loss in 90 days when incorporated to 7.5 cm. Heptachlor had similar volatilization rates from moist soil and foliage: *c.* 90% in 2–7 days. However, rates were much lower when the heptachlor was incorporated to a shallow depth of 7.5 cm: 7% in 167 days.

Temperature is an important factor that influences the rate of volatilization of pesticides from soils, mainly through its effect on vapour pressure. The vapour pressure for most pesticides increases 3–4 times for each 10°C increase in temperature (Spencer *et al.*, 1969, 1973; Nash, 1988). Temperature may also influence volatility of soil incorporated pesticides through its effect on movement of chemicals to the surface by diffusion or mass flow in the evaporating water, or its effect on the soil–water sorption–desorption equilibrium. In such cases, increases in temperature are usually associated with increases in volatilization rate. However, complicating factors may occur with unexpected results. There is, for example, a potential increase in the rate of drying of soil with increase in temperature, thereby decreasing the vapour density with a resultant decrease in volatilization rate at higher temperatures.

In agricultural soils, ploughing/tilling will increase volatilization of pesticides from soil by regularly replenishing the soil surface with compounds that would probably otherwise not have come to the surface. Soil biota, like earthworms, aid this process since it has been estimated that earthworms can turn over the topsoil every 5–20 years

(Muller-Lemans and van Dorp, 1996). Another emission pathway for pesticides into the atmosphere occurs when pesticides are sorbed on to soil particles and entrained into the atmosphere on wind blown particles (Glotfelty and Schomberg, 1989).

Leaching and downward transport mechanisms

Leaching is a fundamental soil process whereby constituents, dissolved or suspended in soil solution phase, are lost from the soil profile by the action of percolating liquid water. It has been identified as a major cause for the occurrence of pesticides in groundwater. The factors which appear to have the most significant influence on pesticide mobility in soils are: physical properties of soil, chemical properties of the pesticide, climatic conditions at the time of application, adsorption, pesticide formulation and method of application (Burnside *et al.*, 1963; Enfield *et al.*, 1982; Kan and Tomson, 1990; Kookana *et al.*, 1998).

Pesticides and POPs are transported through the soil profile in gas, solution and colloidal/particulate phases. Movement in the gas phase would be important for those pesticides with a high vapour pressure, such as soil fumigants. Soil porosity has been found to be one of the most important factors affecting the diffusion of soil fumigants (Hanson and Nex, 1953). This method of movement is probably more important than downward water flow for the transport of those pesticides which are highly volatile. Similarly, air diffusive transport will also play a dominant role in the eventual loss of pesticides from the soil by volatilization (Spencer and Cliath, 1973; Spencer and Cliath, 1990).

The prerequisites for leaching of pesticides are entrance of the compound into solution and sorption of the compound on to soil particles. Additionally, the moisture content of the soil at the time of pesticide application and the evapotranspiration ratio influence the leaching of pesticides into soil. Pore size and pore-size distribution affect the rate at which water enters and moves through the soil. The rate of water movement can, in turn, influence the nature of the equilibrium between the pesticide in solution and that sorbed on to soil particles and colloids (Vinten *et al.*, 1983; White, 1985; Brown *et al.*, 1995; Jarvis, 1998). As the clay content of a soil increases, there is a tendency for aggregation of soil particles. The presence of an impervious layer horizon influences the rate of water movement and would probably increase the extent of both desorption and pesticide movement (Brown *et al.*, 1995; Jarvis, 1998).

In moist soils, it has been calculated that several years would be required for as little as 1% of soil applied pesticides to migrate by diffusion to a depth of *c.* 50 cm. It appears that percolating water is the principal means of movement of the relatively non-volatile pesticides, and that diffusion in soil water is important only for transport over very small distances. Therefore, the pesticides of low aqueous solubility and relatively higher sorption are not expected to leach through the soil profile to any significant extent (Spencer and Cliath, 1973; Spencer *et al.*, 1973; Nash, 1988).

It has been demonstrated that earthworm burrows are effective flow pathways for water and solutes (Edwards *et al.*, 1989, 1993). Consequently, a large proportion of the soil matrix may not interact directly with flowing water and equilibrium concentrations of solute throughout the soil profile are not attained. When water moves along preferential pathways, less water is available for the leaching of the bulk mass of the chemical. In such a case, a small portion of chemicals moves faster, but a large portion moves slower than in the absence of preferential flow (Edwards *et al.*, 1993; Jarvis, 1998). A comprehensive review of preferential flow mechanisms and their effect on solute transport is provided by Jarvis (1998). In contrast to movement by rapid transport, pesticides that leach by slow processes are considered to be subject to microbial degradation in the root zone (Flury, 1996).

Another phenomenon which can contribute to the rapid leaching of even strongly sorbed pesticides through the soil profile is 'colloid-assisted transport'. Clay and organic matter colloids can act as vectors for pesticide transport through soil, just as they do for surface transport. Dissolved organic carbon in the form of macromolecules can increase the solubility and hence the movement of sparingly soluble organic compounds (Enfield *et al.*, 1982). The effect is more pronounced for chemicals with a high octanol–water partition coefficient (K_{ow}). Ballard (1971) observed enhanced mobility of DDT when urea was added to the soil to disperse the humic acids. Vinten *et al.* (1983) reported the vertical transport of DDT, adsorbed to suspended colloids, in soil columns. In the field, Jorgensen and Frederica (1992) observed transport of DDE adsorbed to clay minerals in a fractured clayey till.

Surface runoff

Runoff is defined as 'water and any dissolved or suspended matter it contains that leaves a plot, field, or small single-cover watershed in surface drainage' (Leonard, 1990). Offsite movement of pesticides via overland flow and their impact on surface and groundwater quality has been a major concern for the last three decades. Specifically, pesticide runoff includes dissolved, suspended particulate, and sediment-adsorbed pesticide that is transported by water from a treated land surface. Leonard (1990) has given a detailed account of the processes that govern the nature and extent of overland flow. The factors that affect runoff losses include: (i) physicochemical properties of the pesticide; (ii) application rate of the pesticide; (iii) the time of application in relation to a precipitation event; (iv) method of application; and (v) the formulation used (Wauchope, 1978; Cohen and Steinmetz, 1986; Leonard, 1990; Shipitalo *et al.*, 1990; Flury, 1996).

The proportion of pesticides lost through overland flow is often directly dependent on the concentration in the first few centimetres of soil, and total loss in each runoff event declines exponentially with time (Leonard, 1990). An examination of pesticide data from a number of watersheds revealed that runoff concentrations over a wide range of storm conditions were strongly correlated with pesticide concentrations in the surface 10 mm of watershed soil. It was therefore concluded that this 10 mm surface zone could be used as a descriptor or predictor of the amount of pesticide available for runoff (Leonard, 1990). Because pesticides deposited on soil surfaces are subjected to environmental extremes, their dissipation rates are often faster than those in bulk soil (Nash, 1983, 1988). Therefore, the pesticides that are either incorporated in soil or move into soil relatively rapidly are less prone to off-site movement via overland flow. In other words, highly sorbed pesticides, such as paraquat, are more likely to move due to the erosion of pesticide-enriched finer soil particles by surface runoff (Nash, 1983, 1988). This occurs by a phenomenon known as 'raindrop stripping', a process whereby outer soil layers are removed as a result of raindrop impact, which preferentially removes finer particles (clay and organic matter) from the parent soil. Since non-ionic pesticides are hydrophobic, organic compounds have a greater affinity for organic matter and clay; such surface erosion will lead to a higher proportion of transport of sorbed pesticides in runoff.

Urban runoff is an important contributor of non-point source pollution (Trichell *et al.*, 1968; Bucheli *et al.*, 1998). Urban runoff may contain pesticides used for non-agricultural purposes, for example railway tracks, industrial yards, parks, garden lawns and treatment plants. In such applications, pesticides are often used in excess of their application rates to agricultural soils. Another potential source of pesticides in urban runoff is washoff of atmospheric dusts containing pesticide residues that have settled on streets and other impervious urban surfaces (Chevreuil *et al.*, 1996; Bucheli *et al.*, 1998; van Dijk and Guicherit, 1999). Little information is available on specific pesticides in urban runoff.

Effective Management Practices that Reduce Runoff to Water Bodies

As mentioned previously, the most important variable determining the concentrations of pesticides in runoff is the amount of pesticide at the soil surface during a precipitation event. Obviously, the application rates, formulation types, application methods and type of chemical are all important factors that affect the ultimate concentrations in runoff.

A variety of management practices are known to reduce potential non-point source pollution from pesticides. These include: (i) the use of alternative pesticides; (ii) crop rotations; (iii) integrated pest management systems to soil conservation; (iv) substitution of crops; and (v) use of mechanical procedures (Novotny and Chesters, 1981). The most effective management practice for reducing concentrations in runoff depends on the mode of transport in runoff and whether or not a given management practice affects runoff volume, sediment yields, or both (Leonard, 1990).

The effects of soil erosion control practices on pesticide runoff depend on the sorption characteristics of the pesticide and degree of reduction of fine sediment transport (Leonard, 1990). Sediment yield and sorbed pesticide in runoff reduce concomitantly, although not in the same proportion because control practices tend to reduce transport of coarse particles more than fine particles; therefore, the capacity for sorbed chemical transport per unit sediment mass is increased (McDowell *et al.*, 1981). Ritter *et al.* (1974), for example, showed that conservation practices that reduce runoff volumes also reduced losses of atrazine and propachlor.

Conservation tillage systems are also known to reduce pesticide losses in runoff although they may also act to enhance losses under certain conditions. In one such situation where pesticide losses in runoff from non-till systems are enhanced soluble pesticides are applied to crop residues and runoff occurs soon after application, and quickly after rainfall begins. These conditions may occur under a combination of high initial soil water content, intense rainfall and low infiltration rates. Conversely, no-till or the presence of crop residues may reduce herbicide runoff under conditions where crop residues reduce surface sealing and maintain higher infiltration rates; therefore, time to runoff increases and the total runoff volumes are reduced. In this case, most of the soluble pesticide will be leached into the soil before runoff begins. Runoff of relatively insoluble pesticides sorbed to sediment will be reduced under no-till because sediment yields are reduced (Leonard, 1990).

Impact of Pesticides in the Aquatic Environment

When runoff from a treated area containing pesticide enters a watercourse, concentrations are rapidly diluted and are also partitioned among various compartments. The properties of pesticides of particular relevance to their overall environmental behaviour include sorption, leaching, vaporization, degradation and, in terms of biological effect, bioaccumulation. The solubility of pesticides is, however, the most important property that determines the effect of pesticides in the aquatic environment.

If the pesticide in question is ionizable, such as an acid, base or salt compound, the solubility is likely to be greater than that of the less polar materials such as the chlorinated hydrocarbons, organophosphorus compounds or the carbamates. Sedimentation and burial is often the most important removal pathway for hydrophobic pesticides in lakes and other water bodies. Chemicals introduced into the water column undergo interactions with biotic and abiotic particles. The particle reactive chemicals (those having log $K_{ow} > 4$) sorb to, or partition into, aqueous particles. The fate and residence time of these chemicals is linked closely to the fate of the particles. If the partition coefficient of a chemical is high and the degradation rate is low, the compound will accumulate in organisms of the food chain with successive increase at each step (Mackay and Clark, 1991).

Impact on aquatic organisms

Pesticides, by virtue of their nature as poisons, would be expected to have dose-related effects on both target and non-target organisms. The vast majority of pesticides are relatively non-selective, with a broad spectrum of effects. The effects of pesticides on target and non-target species of aquatic invertebrates have been discussed in many good reviews (e.g. Mulla *et al.*, 1981; Muirhead-Thompson, 1987). In the discussion that follows, a brief general overview is presented.

Aquatic organisms can be affected by pesticides either directly through toxicity, or indirectly by some changes in the environment. The direct effects may be either a rapid destruction of the tissue in contact with the pesticide or a slower disturbance of some physiological process. There is evidence to suggest that the impact of pesticides on aquatic organisms might be related to several factors including the age of the organism (Hudson *et al.*, 1972), its feeding habits and its habitat (Eisler and Jacknow, 1985). In lakes, for instance, toxicity of pesticides to fish was significantly related to depth, stratification and biological activity (Eisler and Jacknow, 1985). The classic examples of pesticide-related fish mortality were the massive fish kills in the Mississippi and Atchafalaya rivers in Louisiana between 1960 and 1963 where 3.5 million fish died in 1962 alone. The highly toxic insecticide, endrin, was singled out as the major cause of this tragedy that affected more than nine species of fish (Mount and Putnicki, 1966).

Acute toxicities of pesticides are dramatic; on the other hand, chronic toxicities resulting from the exposure to sublethal concentrations of pesticides are much more insidious and difficult to identify. Depending on the type of organism involved, exposure to sublethal concentrations may cause genetic, physiological, or behavioural changes in the target and non-target species (Madhun and Freed, 1990). Cases of increased tolerance or resistance, reproduction impairment, inhibition of brain enzyme activities, growth reduction or inhibition are among the more insidious effects documented. The two types of toxicities for a given pesticide vary, however, with water temperature, water chemistry and biological factors such as age, sex, size and health conditions, as well as species (Johnson, 1968).

The effects of pesticides discussed above are meant to illustrate the potential impact of pesticides when present in high concentrations in the aquatic environment. However, appropriate use of pesticides based on recommended application rates is generally expected to cause little adverse impact on the environment. Pesticide concentrations in streams and significant water bodies are much lower than those measured in edge-of-field runoff because of processes of dilution, sedimentation, vegetation trapping and degradation during transport. Where detectable, concentrations are commonly in the low $\mu g\ l^{-1}$ to $mg\ l^{-1}$ range. Other examples of pesticide transport and impacts are given in Cameron *et al.*, Chapter 17, and Harrod and Theurer, Chapter 7, this publication.

Case Studies

Specific examples regarding the fate of compounds are presented below for two types of delivery mechanisms to agricultural soils. The first case study discusses the fate of POPs, which are delivered to agricultural systems principally or exclusively by atmospheric deposition. The second case study discusses the fate of a pesticide that is deliberately added to agricultural farms as a herbicide. These two examples will highlight the important processes involved in determining the eventual fate of chemicals reaching agricultural systems through the two different delivery mechanisms.

Atmospheric delivery: the case of polychlorinated dibenzo-p-dioxins and furans

Polychlorinated dibenzo-*p*-dioxins and furans are examples of a POP chemicals that enter the agricultural ecosystem principally by atmospheric deposition. Welsch-Pausch and McLachlan (1998) measured the input–output fluxes of this group of chemicals on a dairy farm located 25 km away from any major city in Germany. The farm was 800 ha in size, of which about 60% comprised grassland and 40% tilled.

Using a series of assumptions, various flux terms were estimated and the information was used to elucidate the behaviour of these chemicals in the farm system. Figure 4.4 gives an illustration of the fate of selected congeners in this agricultural system. Typically, input was dominated by atmospheric deposition (> 99%), with a minute fraction coming from imported feed (typically < 1%). The major processes were retention by soil (*c.* 80–90%) and plant harvest (*c.* 10–20%). Metabolism in cows following ingestion of contaminated feed, soil or grass was important as a loss mechanism for the low chlorinated congeners (e.g. Cl_4DD/Fs, 17%). However, for higher chlorinated ones, absorption exceeded depuration and as such the compounds bioaccumulated.

The result of this mass balance exercise helped to identify the major processes that are important in determining the fate of

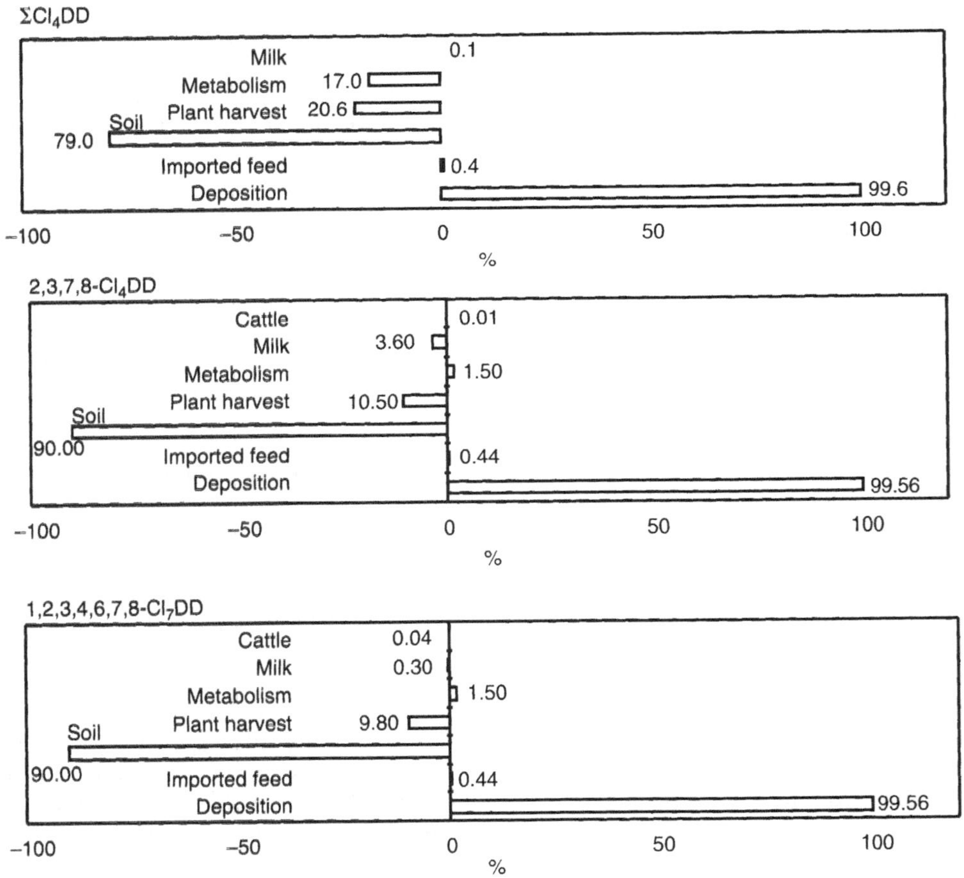

Fig. 4.4. Mass balance representation of the major fate processes for selected polychlorinated dibenzo-*p*-dioxins congeners on a farm. The graph shows that atmospheric deposition provides the major input of persistent organic pollutants to farms and retention in soil dominates their fate following deposition.

POP chemicals and any chemical that is delivered to agricultural systems exclusively via atmospheric deposition. It shows how insignificant other loss processes discussed in the text are for POP chemicals although they were not assessed in the study.

Pesticides deliberately added to soil

The major processes determining the fate of compounds that are added to soil are entirely different from those described above. Figge *et al.* (1983) used an experimental chamber to study the major loss processes for field applied compounds. The test system was designed so that complex physical, chemical and biological processes can occur in a similar way to the natural environment. Using a whole array of sensors, it was possible to maintain the same climatic conditions in the chamber as in the field over the 30-day experimental period. ^{14}C-tracer techniques were used to study the fate of seven compounds with a wide range of physico-chemical properties.

The amount of radioactivity found in the different compartments for all seven compounds is summarized in Fig. 4.5. Different loss mechanisms were important for different compounds as will be expected. The fate of toluene, for example, with a high vapour pressure, was dominated by volatilization with degradation, retention in soil and plant uptake, contributing *c.* 2% each. On the other hand, plant uptake and retention in soil were the dominant processes determining the fate of perylene. Because of low vapour pressure and water solubility, the proportions found in air and leacheate were extremely low. The phenols were degraded to a larger extent than the other compounds. Plant uptake, soil retention, volatilization and microbial degradation governed their fates. The extent of each of these processes was different for the different compounds reflecting the differences conferred on them by the substituting group or pattern.

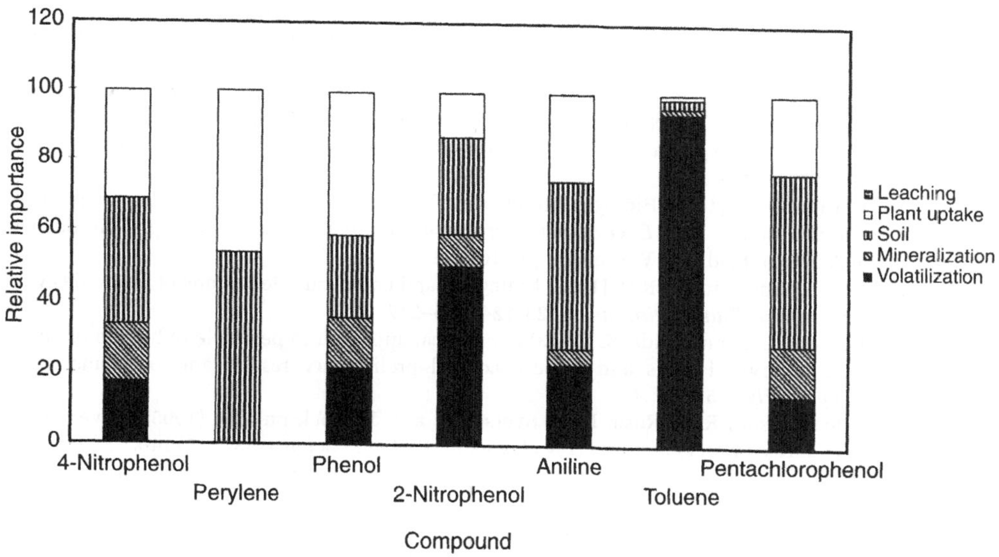

Fig. 4.5. Fate of various compounds with widely different physico-chemical properties in soil. The graph shows that the major loss processes that influence the eventual fate of a compound depend on its properties and differ from one compound to the next (adapted from Figge *et al.*, 1983).

Concluding Remarks

Soils acts as a repository for many pesticides and POPs that are deliberately applied or deposited from the atmosphere. While the chemical is in soil, it is subject to a host of loss (e.g. volatilization, biotic and abiotic degradation, leaching, runoff) and retention (e.g. sorption–desorption, sequestration, bound residue formation) processes. The loss pathways of importance in the context of this book are volatilization to air and subsequent deposition to surface waters, and off-site movement (leaching and runoff) from the farms into surface waters. The concentrations of pesticides in streams and other water bodies are significantly lower than those in edge-of-field runoffs because of dilution, sedimentation, vegetation trapping and degradation during transport. Where detectable, concentrations are often in the low μg l^{-1} mg l^{-1} range. The rate and amounts of runoff are highly variable and depend on the application rates to land, environmental conditions, soil properties, chemical properties, the form in which the compound is added and method and timing of application and the persistence of the pesticide or POP.

References

Akesson, N.B. and Yates, W.E. (1964) Problems relating to application of agricultural chemicals and resulting drift residues. *Annual Review of Entomology* 9, 285–318.

Alexander, M. (1994) *Biodegradation and Bioremediation*. Academic Press, San Diego.

Bailey, G.W. and White, J.L. (1970) Factors influencing the adsorption, desorption, and movement of pesticides in soils. *Residue Reviews* 32, 29–92.

Ballard, T.M. (1971) Role of humic carrier substances in DDT movement through forest soil. *Soil Science Society of America Proceedings* 28, 145–147.

Ballschmiter, K. and Wittlinger, R. (1991) Interhemisphere exchange of hexachlorocyclohexanes, hexachlorobenzenes, polychlorobiphenyls, and 1,1,1-trichloro-2,2-*bis*(p-chlorobiphenyl)ethane in the lower troposphere. *Environmental Science and Technology* 25, 1103–1110.

Beck, A.J. and Jones, K.C. (1995) Kinetic constraints on the *in-situ* remediation of soils contaminated with organic chemicals. *Environmental Science and Pollution Research* 2, 244–252.

Bergstrom, L. and Stenstrom, J. (1998) Environmental fate of chemicals in soil. *Ambio* 27, 16–23.

Berry, D.F. and Boyd, S.A. (1985) Decontamination of soil through enhanced formation of bound residues. *Environmental Science and Technology* 19, 1132–1133.

Bidleman, T.F. (1988) Atmospheric processes: wet and dry deposition of organic compounds are controlled by their vapour-particle partitioning. *Environmental Science and Technology* 22, 361–367.

Bidleman, T.F. (1999) Atmospheric transport and air-surface exchange of pesticides. *Water, Air and Soil Pollution* 115, 115–166.

Bollag, J.M. and Liu, S.-Y. (1990) Biological transformation processes of pesticides. In: Cheng, H.H. (ed.) *Pesticides in the Soil Environment: Processes, Impacts, and Modelling*. Soil Science Society of America, Madison, Wisconsin, pp. 169–211.

Bollag, J., Myers, C.J. and Minard, R.D. (1992) Biological and chemical interactions of pesticides with soils. *Science of the Total Environment* 123/124, 205–217.

Bower, C.C., Penrose, L.J. and Dodds, K. (1993) A practical approach to pesticide reduction on apple crops using supervised pests and disease control–preliminary results and problems. *Plant Protection Quarterly* 8, 57–62..

Brown, C.D., Hodgkinson, R.A., Rose, D.A., Syers, J.K. and Wilcockson, S.J. (1995) Movement of pesticides to surface waters from a heavy clay loam. *Pesticide Science* 43, 131–140.

Bruce, A.M. and Davis, R.A. (1989) Sewage sludge disposal: current and future options. *Water Science Technology* 21, 1113–1128.

Brusseau, M.L. and Rao, P.S.C. (1989) Sorption nonideality during organic contaminant transport in porous media. *Critical Reviews in Environmental Control* 19, 3–99.

Brusseau, M.L., Wood, A.L., Suresh, P. and Rao, C. (1991) Influence of organic co-solvents on the sorption kinetics of hydrophobic organic chemicals. *Environmental Science and Technology* 25, 903–910.

Bucheli, T.D., Muller, S.R., Heberle, S. and Schwarzenbach, R.P. (1998) Occurrence and behaviour of pesticides in rainwater, roof runoff, and artificial stormwater infiltration. *Environmental Science and Technology* 32, 3457–3464.

Burnside, O.C., Fenster, C.R. and Wicks, G.A. (1963) Dissipation and leaching of monoron, simazine, and atrazine in Nebraska soils. *Weeds* 11, 209–213.

Buser, H.R. (1990) Atrazine and other *s*-triazine herbicides in lakes and in rain in Switzerland. *Environmental Science and Technology* 24, 1049–1058.

Calderbank, A. (1989) The occurrence and significance of bound pesticide residues in soil. *Environmental Contamination and Toxicology* 108, 71–103.

Centner, T.J. (1998) Unwanted agricultural pesticides: state disposal programs. *Journal of Environmental Quality* 27, 736–742.

Centner, T.J. and Gunter, L.F. (1999) Financing the disposal of unwanted agricultural pesticides. *Environment International* 25, 635–646.

Chapman, R.A. and Cole, C.M. (1982) Observations on the influence of water and soil pH on the persistence of insecticides. *Journal of Environmental Science and Health* B17, 487–504.

Chevreuil, M., Garmouma, M., Teil, M.J. and Chesterikoff, A. (1996) Occurrence of organochlorines (PCBs, pesticides) and herbicides (triazines, phenylureas) in the atmosphere and in the fallout from urban and rural stations of the Paris area. *The Science of the Total Environment* 182, 25–37.

Cliath, M.M. and Spencer, W.F. (1972) Dissipation of pesticides from soils by volatilisation of degradation products. I. Lindane and DDT. *Environmental Science and Technology* 6, 910–914.

Cohen, M.L. and Steinmetz, W.D. (1986) Foliar washoff of pesticides by rainfall. *Environmental Science and Technology* 20, 521–523.

Dec, J. and Bollag, J. (1997) Determination of covalent and non-covalent binding interactions between xenobiotic chemicals and soil. *Soil Science* 162, 858–874.

DiToro, D.M. and Horzempa, L.M. (1982) Reversible and resistant components of PCB adsorption–desorption isotherms. *Environmental Science and Technology* 16, 594–602.

Edwards, W.M., Shipitalo, M.J., Owens, L.B. and Norton, L.D. (1989) Water and nitrate movement in earthworm burrows within long-term no-till cornfields. *Journal of Soil and Water Conservation* 44, 240–243.

Edwards, W.M., Shipitalo, M.J., Owens, L.B. and Dick, W.A. (1993) Factors affecting preferential flow of water and atrazine through earthworm burrows under continuous no-till corn. *Journal of Environmental Quality* 22, 453–457.

Eisler, R. and Jacknow, J. (1985) *Toxaphene Hazard to Fish, Wildlife and Invertebrates: a Synoptic Review.* United States Wildlife Services Biological Report 85. United States Government Print Office, Washington, DC.

Elliott, J.G. and Wilson, B.J. (1983) *The Influence of Weather on the Efficiency and Safety of Pesticide Application. The Drift of Pesticides.* Report No. 3. British Crop Protection Council, Surrey, UK.

Enfield, C.G., Carsel, R.F., Cohen, S.Z., Phan, T. and Walters, D.M. (1982) Approximating pollutant transport to groundwater. *Groundwater* 20, 711–722.

Falconer, K.E. (1998) Managing diffuse environmental contamination from agricultural pesticides: an economic perspective on issues, policy options, with particular reference to Europe. *Agriculture, Ecosystems and Environment* 69, 37–54.

Figge, K., Klahn, J. and Koch, J. (1983) Testing of chemicals by evaluation of their distribution and degradation patterns in an environmental standard system. *Regulatory Toxicology and Pharmacology* 3, 199–215.

Flury, M. (1996) Experimental evidence of transport of pesticides through field soil: a review. *Journal of Environmental Quality* 25, 25–45.

Fries, G.F. (1996) Ingestion of sludge applied organic chemicals by animals. *Science of the Total Environment* 185, 93–108.

Führ, F., Steffens, W., Mittelstaedt, W. and Brumhard, B. (1991) Lysimeter experiments with ^{14}C-labelled pesticides – an agroecosystem approach. *Pesticide Chemistry* 14, 37–47.

Garthwaite, D.G., Thomas, M.R. and Hart, M. (1995) *Pesticide Usage Survey Report 127: Arable Farm Crops in Great Britain 1994.* MAFF, London.

Getzin, L.W. (1981) Degradation of chlorpyrifos in soil – influence of autoclaving, soil moisture, and temperature. *Journal of Economic Entomology* 74, 707–713.

Gevao, B., Semple, K.T. and Jones, K.C. (2000) Bound pesticide residues in soil: a review. *Environmental Pollution* 108, 3–14.

Gile, J.D. and Gillett, J.W. (1979) Fate of selected fungicides in a terrestrial laboratory ecosystem. *Journal of Agricultural Food and Chemistry* 27, 1159–1164.

Glotfelty, D.E. and Schomberg, C.J. (1989) Volatilization of pesticides from soils. In: Sawney, B.L. (ed.) *Reactions and Movements of Organic Chemicals in Soils.* American Society of Agronomy, Madison,Wisconsin.

Gohre, K. and Miller, G.C. (1986) Photooxidation of thioether pesticides on soil surfaces. *Journal of Agricultural Food and Chemistry* 31, 1164–1168.

Grover, R., Maybank, J., Caldwell, B.C. and Wolf, T.M. (1997) Airborne off-target losses and deposition characteristics from a self-propelled, high speed and high clearance ground sprayer. *Canadian Journal of Soil Science* 77, 493–500.

Hanson, W.J. and Nex, R.W. (1953) Diffusion of ethylenedibromide in soils. *Soil Science* 76, 209–215.

Hill, B.D., Inaba, D.J. and Charnetski, W.A. (1989) Dissipation of deltamethrin applied to forage lucerne. *Journal of Agricultural Food and Chemistry* 37, 1150–1153.

Himel, C.M., Loats, H. and Bailey, G.W. (1990) Pesticide sources to the soil and principles of spray physics. In: Cheng, H.H. (ed.) *Pesticides in the Soil Environment: Processes, Impacts and Modelling.* Vol. 2. Soil Science Society of America, Madison, Wisconsin, pp. 7–50.

Hoff, R.M., Strachan, W.M.J., Sweet, C.W., Chan, C.H., Shackleton, M., Bidleman, T.F., Brice, K.A., Burniston, D.A., Cussion, S., Gatz, D.F., Harlin, K. and Schroeder, W.H. (1996) Atmospheric deposition of toxic chemicals to the Great Lakes: A review of data through 1994. *Atmospheric Environment* 30, 3505–3527.

Huang, W., Schlautman, M.A. and Weber, W.J. Jr (1996) A distributed reactivity model for sorption by soils and sediments 5: the influence of near-surface characteristics in mineral domains. *Environmental Science and Technology* 30, 2993–3000.

Hudson, R.H., Tucker, R.K. and Haegele, M.A. (1972) Effect of age on sensitivity: acute oral toxicity of 14 pesticides to mallard ducks of several ages. *Toxicology and Applied Pharmacology* 22, 556–561.

Jarvis, N.J. (1998) Modelling the impact of preferential flow on solute transport in soils. In: Selim, H.M. and Ma, L. (eds) *Physical Non-equilibrium in Soils: Modelling and Applications.* Ann Arbor Press, Ann Arbor, Michigan, pp. 195–222.

Johnson, D.L. (1968) Pesticides and fishes – a review of selected literature. *Transactions of American Fisheries Society* 97, 398–424.

Jones, K.C. and Sewart, A.P. (1997) Dioxins and furans in sewage sludges: a review of their occurrence and sources in sludge and their environmental fate, behaviour, and significance in sludge-amended agricultural systems. *Critical Reviews in Environmental Science and Technology* 27, 1–85.

Jones, M. (1993) Agricultural clean sweep: Waste pesticide removal 1988–1992. Chicago, IL. USEPA Region 5 USEPA.

Jorgensen, P.R. and Frederica, J. (1992) Migrations of nutrients, pesticides and heavy metals in a fractured clayey till. *Geotechnique* 42, 67–77.

Kan, A.T. and Tomson, M.B. (1990) Groundwater transport of hydrophobic organic compounds in the presence of dissolved organic matter. *Environmental Toxicology and Chemistry* 9, 253–263.

Kan, A.T., Fu, G., Hunter, M., Chen, W., Ward, C.H. and Tomson, M.B. (1998) Irreversible sorption of neutral hydrocarbons to sediments: experimental observations and model predictions. *Environmental Science and Technology* 32, 892–902.

Karnatz, A. (1991) *Obsolete Pesticide Disposal Project. Montpelier, VT.* Vermont Department of Agriculture, Food and Markets.

Kirkland, K. and Fryer, J.D. (1972) Degradation of several herbicides in a soil previously treated with MCPA. *Weed Research* 12, 90–95.

Kjeller, L.-O., Johnston, A.E. and Rappe, C. (1991) Increases in the polychlorinated dibenzo-*p*-dioxin and -furan content of soils and vegetation since the 1840s. *Environmental Science and Technology* 25, 1619–1627.

Klehr, M., Iwan, J. and Riemann, J. (1983) An experimental approach to the photolysis of pesticides adsorbed on soil – Thidiazuron. *Pesticide Science* 14, 359.

Kookana, R.S. and Aylmore, L.A.G. (1994) Estimating pollution potential of pesticides to groundwater. *Australian Journal of Soil Research* 32, 1141–1155.

Kookana, R.S., Gerritse, R.G. and Aylmore, L.A.G. (1990) Effect of organic co-solvent on adsorption and desorption of Linuron and Simazine in soil. *Australian Journal of Soil Research* 28, 717–725.

Kookana, R.S., Baskaran, S. and Naidu, R. (1998) Pesticide fate and behaviour in Australian soils in relation to contamination and management of soil and water: a review. *Australian Journal of Soil Research* 36, 715–764.

Korpraditskul, R., Katayama, A. and Kuwatsuka, S. (1993a) Chemical and microbial degradation of atrazine in Japanese and Thai soils. *Journal of Pesticide Science* 18, 77–83.

Korpraditskul, R., Katayama, A. and Kuwatsuka, S. (1993b) Degradation of atrazine by soil bacteria in the stationary-phase. *Journal of Pesticide Science* 18, 293–298.

Kruger, E.L., Rice, P.J., Anhalt, J.C., Anderson, T.A. and Coats, J.R. (1996) Use of undisturbed soil columns under controlled conditions to study the fate of [^{14}C]deethylatrazine. *Journal of Agricultural Food and Chemistry* 44, 1144–1149.

Laskowski, D.A., Swann, R.L., McCall, P.J. and Bidlack, H.D. (1983) Soil degradation studies. *Residue Reviews* 85, 139–147.

Lehmann, R.G. and Miller, J.R. (1989) Soil catalysed hydrolysis of fluoroxypyrmethylheptyl ester. *Weed Research* 29, 385–389.

Leonard, R.A. (1990) Movement of pesticides into surface waters. In: Cheng, H.H. (ed.) *Pesticides in the Soil Environment: Processes, Impacts and Modelling.* Soil Science Society of America, Madison, Wisconsin, pp. 309–349.

Lichtenstein, E.P. (1980) 'Bound' residues in soils and transfer of soil residues in crops. *Residue Reviews* 76, 147–153.

Ligocki, M.P. and Pankow, J.F. (1989) Measurements of the gas/particle distributions of atmospheric organic compounds. *Environmental Science and Technology* 23, 75–83.

Ligocki, M.P., Leuenberger, C. and Pankow, J.F. (1985) Trace organic compounds in Rain – II. Gas scavenging of neutral organic compounds. *Atmospheric Environment* 19, 1609–1617.

Mackay, D. and Clark, K.E. (1991) Predicting the environmental partitioning of organic contaminants and their transfer to biota. In: Jones, K.C. (ed.) Organic Contaminants in the Environment, Environmental Pathways and Effects. Environmental Management Series, Elsevier Applied Science, London, pp. 159–188.

Madhun, Y.A. and Freed, V.H. (1990) Impact of pesticides on the environment. In: Cheng, H.H. (ed.) *Pesticides in the Soil Environment: Processes, Impacts and Modelling*, SSSA Book Series, no. 2. Madison, Wisconsin, pp. 429–466.

Majewski, M. S. and Capel, P. D. (1995) *Pesticides in the Atmosphere*, Ann Arbor Press, Ann Arbor, Michigan.

Majewski, M., Desjardins, R., Rochette, P., Pattey, E., Seiber, J. and Glotfelty, D. (1993) Field comparison of an eddy accumulation and aerodynamic gradient system for measuring pesticide volatilization fluxes. *Environmental Science and Technology* 27, 121–128.

Maybank, J., Yoshida, K. and Grover, R. (1978) Spray drift from agricultural pesticide applications. *Journal of Air Pollution Control Association* 28, 1009–1014.

McConnell, L.L., LeNoir, J.S., Datta, S. and Sieber, J.N. (1998a) Wet deposition of current-use pesticides in the Sierra Nevada mountain range, California, USA. *Environmental Toxicology and Chemistry* 17, 2554.

McConnell, L.L., LeNoir, J.S., Datta, S. and Sieber, J.N. (1998b) Wet deposition of current-use pesticides in the Sierra Nevada mountain range, California, USA. *Environmental Toxicology and Chemistry* 17, 1908–1916.

McDowell, L.L., Willis, G.H., Murphree, C.E., Southwick, L.M. and Smith, S. (1981) Toxaphene and sediment yields in runoff from a Mississippi USA delta watershed. *Journal of Environmental Quality* 10, 120–125.

Moorman, T.B. (1993) Pesticide degradation by soil microorganisms: environmental, ecological and management effects. In: Hartfield, J.L. and Stewart, B.A. (eds) *Soil Biology: Effects on Soil Quality.* Lewis Publishers, London, pp. 121–165.

Mount, D.I. and Putnicki, G.J. (1966) Summmary report on the 1963 Mississippi fish kill. *Transactions of North American Wildlife Conference* 31, 177–184.

Muir, D.C.G., Omelchenko, A., Grift, N.P., Savoie, D.A., Lockhart, W.L., Wilkinson, P. and Brunskill, G.J. (1996) Spatial trends and historical deposition of polychlorinated biphenyls in Canadian mid-latitude and Arctic Lake sediments. *Environmental Science and Technology* 30, 3609–3617.

Muirhead-Thompson, R.C. (1987) *Pesticide Impact on Stream Fauna with Special Reference to Macroinvertebrates.* Cambridge University Press, Cambridge.

Mulla, M.S., Mian, L.S. and Kawecki, J.A. (1981) Distribution, transport, and fate of the insecticides malathion and parathion in the environment. *Residue Reviews* 81, 1–159.

Muller-Lemans, H. and van Dorp, F. (1996) Bioturbation as a mechanism for radionuclide transport in soil: relevance of earthworms. *Journal of Environmental Radioactivity* 31, 7–20.

Nash, R.G. (1983) Comparative volatilization and dissipation rates of several pesticides from soils. *Journal of Agricultural Food and Chemistry* 31, 210–217.

Nash, R.G. (1988) Dissipation from soil. In: Grover, R. (ed.) *Environmental Chemistry of Herbicides*, Vol. 1. CRC Press, Boca Raton, Florida, pp. 131–169.

Nash, R.G. and Gish, T.J. (1989) Halogenated pesticide volatilization and dissipation from soil under controlled conditions. *Chemosphere* 18, 2353–2362.

Nicholls, P.H. (1991) Organic contaminants in soils and groundwater. In: Jones, K.C. (ed.) *Organic Contaminants in the Environment*. Elsevier Applied Science, London, pp. 87–132.

Novotny, V. and Chesters, G. (1981) *Handbook of Nonpoint Pollution: Sources and Management*. Van Nostrand Reinhold, New York.

Pankow, J.F., Storey, J.M.E. and Yamasaki, H. (1993) Effects of relative humidity on gas/particle partitioning of semivolatile organic compounds to urban particulate matter. *Environmental Science and Technology* 27, 2220–2226.

Parkin, C.S. and Merritt, C.T. (1988) The measurement and prediction of spray drift. *Aspects of Applied Biology* 17, 351–361.

Pignatello, J.J. (1990) Slowly reversible sorption of aliphatic halocarbons in soils. II. Mechanistic aspects. *Environmental Toxicology and Chemistry* 9, 1117–1126.

Pignatello, J.J. and Cohen, S.Z. (1990) Environmental chemistry of ethylene dibromide in soil and groundwater. *Review of Environmental Contamination and Toxicology* 112, 1–47.

Pignatello, J.J. and Xing, B. (1996) Mechanisms of slow sorption of organic chemicals to natural particles. *Environmental Science and Technology* 30, 1–11.

Racke, K.D., Steele, K.P., Yoder, R.N., Dick, W.A. and Avidov, E. (1996) Factors affecting the hydrolytic degradation of chlorpyrifos in soil. *Journal of Agricultural Food and Chemistry* 44, 1582–1592.

Rawn, D.F.K., Halldorson, T.H.J., Lawson, B.D. and Muir, D.C.G. (1999a) A multi-year study of four herbicides in air and precipitation from a small prairie watershed. *Journal of Environmental Quality* 28, 898–906.

Rawn, D.F.K., Halldorson, T.H.J., Turner, W.N., Woychuk, R.N., Zakrevsk, J.G. and Muir, D.C.G. (1999b) A multi-year study of four herbicides in surface water of a small prairie watershed. *Journal of Environmental Quality* 28, 906–917.

Ritter, W.F., Johnson, H.P., Lovely, W.G. and Molnau, M. (1974) Atrazine, propachlor, and diazinon residues on small agricultural watersheds. *Environmental Science and Technology* 8, 37–42.

Roberts, T.R. (1984) Non-extractable pesticide residues in soils and plants. *Pure and Applied Chemistry* 56, 945–956.

Roeth, F.W. (1986) Enhanced herbicide degradation in soil with repeat application. *Weed Science* 2, 45–65.

Sanchez, P.A. (1976) *Properties and Management of Soils in the Tropics*. John Wiley & Sons, New York.

Scow, K. (1993) Effects of sorption-desorption and diffusion processes on the kinetics of biodegradation of organic chemicals in soil. In: *Sorption and Degradation of Pesticides and Organic Chemicals in Soil*. Soil Science Society of America, Madison, Wisconsin, pp. 73–114.

Senesi, N. (1992) Binding mechanisms of pesticides to soil humic substances. *Science of the Total Environment* 123/124, 63–76.

Senesi, N. (1993) Nature of interactions between organic chemicals and dissolved humic substances and the influence of environmental factors. In: Beck, A.J., Jones, K.C., Hayes, M.H.B. and Mingelgrin, U. (eds) *Organic Substances in Soil and Water: Natural Constituents and their Influences on Contaminant Behaviour*, Vol. 135. Royal Society of Chemistry, Cambridge, pp. 73–101.

Sethunathan, N., Adhya, T.K. and Raghu, K. (1982) Microbial degradation of pesticides in tropical soils. In: Matsumura, F. and Krishnamurty, C.R. (eds) *Biodegradation of Pesticides*. Plenum Press, New York, pp. 91–115.

Shipitalo, M.J., Edwards, W.M., Dick, W.A. and Owens, L.B. (1990) Initial storm effects on macropore transport of surface-applied chemicals in no-till soil. *Soil Science Society of America Journal* 54, 1530–1536.

Skipper, H.D. and Turco, R.F. (eds) (1995) *Bioremediation Science and Applications*. Soil Science Society of America, Madison, Wisconsin.

Smith, A.E. and Lafond, G.P. (1990) Effects of long-term phenoxyalkanoic acid field applications on the rate of microbial degradation. *ACS Symposium Series* 426, 14–22.

Spencer, W.F. and Cliath, M.M. (1973) Pesticide volatilization as related to water loss from soil. *Journal of Environmental Quality* 2, 284–289.

Spencer, W.F. and Cliath, M.M. (1990) Movement of pesticides from soil to the atmosphere. In: Kurtz, D.A. (ed.) *Long Range Transport of Pesticides.* Lewis Publishing Co., Chelsea, Michigan, pp. 1–26.

Spencer, W.F., Cliath, M.M. and Farmer, W.J. (1969) Vapour density of soil-applied dieldrin as related to soil-water content, temperature, and dieldrin concentration. *Soil Science Society of America Proceeding* 33, 509–512.

Spencer, W.F., Farmer, W.J. and Cliath, M.M. (1973) Pesticide volatilization. *Residue Review* 49, 1–47.

Steinberg, S.M., Pignatello, J.J. and Sawhney, B.J. (1987) Persistence of 1,2-dibromoethane in soils: entrapment in intraparticle micropores. *Environmental Science and Technology* 21, 1201–1207.

Stevenson, F.J. (1976) Organic matter reactions involving pesticides in soil. In: Kaufman, D.D., Still, G.G., Paulson, G.D. and Bandal, S.K. (eds) *Bound and Conjugated Pesticide Residues*, Vol. 29. ACS Symposium Series, Washington, DC, pp. 180–207.

Suett, D.L. and Jukes, A.A. (1990) Some factors influencing the accelerated degradation of mephosfolan in soils. *Crop Protection* 9, 44–51.

Tal, A., Rubin, B., Katan, J. and Aharonson, N. (1989) Involvement of microorganisms in accelerated degradation of EPTC in soil. *Journal of Agricultural Food and Chemistry* 38, 1100–1105.

Toole, A.P. and Crosby, D.G. (1989) Environmental persistence and fate of fenoxaprop-ethyl. *Environmental Toxicology and Chemistry* 8, 1171–1176.

Tortensson, N.T.L., Stark, J. and Goransson, B. (1975) The effect of repeated application of 2,4-D and MCPA on their breakdown in soil. *Weed Science* 15, 159–164.

Trichell, D.W., Morton, H.L. and Merkle, M.G. (1968) Loss of herbicides in runoff water. *Weed Science* 16, 447–449.

Van Dijk, H.F.G. and Guicherit, R. (1999) Atmospheric dispersion of current-use pesticides: a review of the evidence from monitoring studies. *Water, Air, and Soil Pollution* 115, 21–70.

Vinten, A.J.A., Yaron, B. and Nye, P.H. (1983) Vertical transport of pesticides into soil when adsorbed on suspended particles. *Journal of Agricultural Food and Chemistry* 28, 662–664.

Waite, D.T., Grover, R., Westcott, N.D., Irvine, D.G., Kerr, L.A. and Sommerstad, H. (1995) Atmospheric deposition of pesticides in a small southern Saskatchewan watershed. *Environmental Toxicology and Chemistry* 14, 1171–1175.

Wang, M. and Jones, K.C. (1994a) Behaviour and fate of chlorobenzenes (CBs) introduced into soil-plant systems by sewage sludge application: a review. *Chemosphere* 28, 1325–1360.

Wang, M. and Jones, K.C. (1994b) Uptake of chlorobenzenes by carrots from spiked and sewage sludge-amended soil. *Environmental Science and Technology* 28, 1260–1267.

Wania, F. and Mackay, D. (1996) Tracking the distribution of persistent organic pollutants: control strategies for these contaminants will require a better understanding of how they move around the globe. *Environmental Science and Technology* 30, 390A–396A.

Ward, N., Clark, J., Lowe, P. and Seymour, S. (1993) *Water Pollution from Agricultural Pesticides.* Centre for Rural Economy, University of Newcastle-upon-Tyne.

Wauchope, R.D. (1978) The pesticide content of surface water draining from agricultural fields – a review. *Journal of Environmental Quality* 7, 459–472.

Weber, J.B. (1971) Behaviour of organic pesticides in soils. *Proceedings of the Soil Science Society* 14, 74–118.

Weber, J.B. (1972) Interaction of organic pesticides with particulate matter in aquatic and soil systems. In: *Fate of Organic Pesticides in the Aquatic Environment*, Vol. 111. pp. 55–120.

Weber, J.B. and Miller, C.T. (1989) Movement of organic chemicals over and through soils. In: Sawhney, B.L. and Brown, K. (eds) *Reactions and Movement of Organic Chemicals in Soils.* Soil Science Society of America, Madison, Wisconsin, pp. 305–334.

Welsch-Pausch, K. and McLachlan, M.S. (1998) Fate of airborne polychlorinated dibenzo-*p*-dioxins and dibenzofurans in agricultural ecosystems. *Environmental Pollution* 102, 129–137.

White, J.L. (1976) Clay-particle interactions. In: Kaufman, D.D., Still, G.G., Paulson, G.D. and Bandal, S.K. (eds) *Bound and Conjugated Pesticide Residues*, Vol. 29. ACS Symposium Series, Washington, DC, pp. 208–218.

White, R.E. (1985) The influence of macropores on the transport of dissolved and suspended matter through soil. In: Stewart, B.A. (ed.) *Advances in Soil Science*, Vol. 3. Springer Verlag, New York, pp. 95–120.

Wild, S.R. and Jones, K.C. (1992) Organic chemicals entering agricultural soils in sewage sludges: screening for their potential transfer to crop plants and livestock. *Science of the Total Environment* 119, 85–119.

Willis, G.H., McDowell, L.L., Harper, L.A., Southwick, L.M. and Smith, S. (1983) Seasonal disappearance and volatilization of Toxaphene and DDT from a cotton field. *Journal of Environmental Quality* 12, 80–85.

Wilson, S.C., Alcock, R.E., Sewart, A.P. and Jones, K.C. (1997) Persistence of organic contaminants in sewage sludge-amended soil: a field experiment. *Journal of Environmental Quality* 26, 1467–1477.

Wolfe, N.L., Mingelgrin, U. and Miller, G.C. (1990) Abiotic transformations in water, sediments and soil. In: Cheng, H.H. (ed.) *Pesticides in the Soil Environment: Processes, Impacts and Modelling.* Soil Science Society of America, Madison, Wisconsin, pp. 103–168.

Xing, B. and Pignatello, J.J. (1996) Time-dependent isotherm shape of organic compounds in soil organic matter: Implications for sorption mechanism. *Environmental Toxicology and Chemistry* 15, 1282–1288.

Xing, B. and Pignatello, J.J. (1997) Dual-mode sorption of low-polarity compounds in glassy poly(vinyl chloride) and soil organic matter. *Environmental Science and Technology* 31, 792–799.

Yoshida, T. and Castro, T.F. (1970) Degradation of g-BHC in rice soils. *Soil Science Society of America Proceedings* 34, 440–442.

Zayed, S.M.A.D., Mostafa, I.Y. and El-Arab, A.E. (1994) Chemical and biological release of ^{14}C-bound residues from soil treated with ^{14}C-p,p′-DDT. *Journal of Environmental Science Health* B29, 169–175.

5 Heavy Metals

W. de Vries,[1] P.F.A.M. Römkens,[1] T. van Leeuwen[2] and J.J.B. Bronswijk[3]
[1]*Alterra Green World Research, PO Box 44, 6700 AA, Wageningen, The Netherlands;*
[2]*Agricultural Economic Research Institute (LEI), PO Box 29703, 2502 LS, The Hague, The Netherlands;* [3]*National Institute of Public Health and Environment (RIVM), PO Box 1, 3720 BA, Bilthoven, The Netherlands*

Key words: agroecosystems, balances, critical limits, critical loads, dynamic models, heavy metals, leaching, soil accumulation, target values, transfer functions

Introduction

Problems related to an excess of heavy metals

Heavy metals are elements that have a high density and belong largely to the transition group of the periodic table. Most of these metals are also called trace elements, stressing their relatively low abundance in soils. The elements that are generally considered include arsenic, Cd, Co, chromium, Cu, Hg, Mo, Ni, Pb, selenium, vanadium and Zn (De Haan and Zwerman, 1978). Some are referred to as micronutrients, indicating the small quantities needed by organisms. Established micronutrients for plants are Cu, Mo and Zn but there is evidence that Ni is a micronutrient as well and legumes require Co to fix N (McBride, 1994). Additional essential micronutrients for animals (including humans) are Co, Cr, Se and V. In soils, there can be a deficiency of micronutrients, especially in many forest soils. Conversely, many trace elements, including all the micronutrients, can reach levels in soils that are toxic to plants and soil life and can lead to toxicity to animals and humans. Some of the most toxic are Hg, Cd and Pb, for which no biological function is known, which are particularly toxic to soil organisms, higher animals and humans (McBride, 1994). Copper, Ni, Co and to a lesser extent Zn are particularly toxic to plants (phytotoxic). When leached to surface waters, the metals can cause toxification problems to aquatic organisms (e.g. Hare and Tessier, 1996; Crommentuijn *et al.*, 1997). The concern for aquatic ecosystems is specifically related to the adverse impact of Hg on fish, violating food quality criteria in many Scandinavian surface waters (Meili, 1997).

In this chapter, we focus on the problems of excess inputs of Cd, Pb, Cu and Zn in agroecosystems. In several countries, there is great concern about the excessive inputs of heavy metals, specifically Cd, Cu and Zn in agriculture (e.g. Moolenaar and Lexmond, 1998). An excess of these in agroecosystems may result in agricultural products with unacceptable levels, violating food quality criteria, and even causing reduced crop production (Alloway, 1990; Fergusson, 1990). Apart from adverse impacts on food quality and crop growth, elevated metal concentrations may affect soil organisms, including microorganisms (Bååth, 1989), nematodes (Bengtsson and Tranvik, 1989) and earthworms (Ma and van der Voet, 1993). Protection of soil microorganisms is essential to sustain so-called 'life support functions', such as decomposition processes, which control the nutrient cycle of elements. Finally, elevated inputs may cause an accelerated leaching of metals to ground and surface waters, thus affecting drinking water quality and aquatic organisms, respectively (Crommentuijn *et al.*, 1997). High fertilizer applications and acid atmospheric deposition, combined with insufficient liming, may also cause a decrease in pH and thus an increase in heavy metal availability, aggravating the problem of deteriorating food quality, metal leaching and impacts on soil organisms. This problem is specifically of concern in the eastern European countries where liming has been strongly diminished since the last decade (Várallyay, 1993).

A simplified overview of major pathways of heavy metals in agroecosystems, including the most relevant receptors, is given in Fig. 5.1. Major pathways are soil to solution transfer (mobilization) followed by plant uptake and leaching to groundwater and surface water. Soil erosion may also cause metal transfer to surface water (Haygarth and Jarvis, 1999).

Metal balances

Insight into present metal accumulation and leaching rates in agricultural systems can be derived from balances describing all inputs to the soil, from both fertilizers/animal manure (sometimes also sewage sludge) and atmospheric deposition, and all outputs in terms of plant uptake and leaching. Such metal balances can be derived for field, farm and national scales.

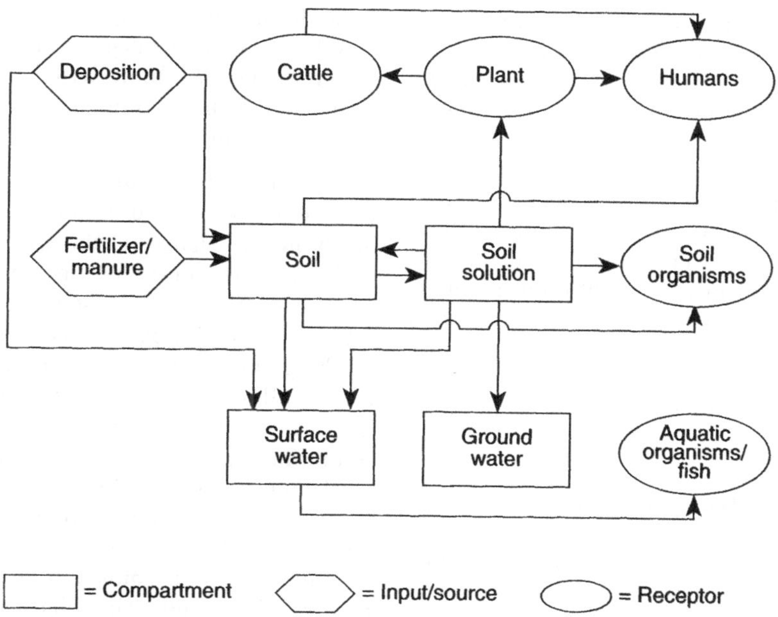

Fig. 5.1. Overview of major pathways of heavy metals in agroecosystems.

A field-scale balance refers to the inputs and outputs to and from the soil compartment (the plough layer) of individual fields, thus allowing the calculation of accumulation in those fields. Field-scale balances enable a direct link with criteria for the protection of soil and other relevant environmental receptors. A farm-scale balance refers to the inputs and outputs as determined at the farm gate, thus showing the characteristic metal flows on to the farm as a whole and allowing the fine-tuning of metal management at the farm level (Moolenaar, 1998). National balances, finally, give an overview of metal flows on a national scale, and are thus relevant for more general agricultural policies (De Boo, 1995). More insight into metal balances is given below (see also Beegle *et al.*, Chapter 8, this volume, for a general discussion on nutrient balances).

Present metal inputs can furthermore be compared with long-term acceptable inputs, or critical loads in view of the current concern over sustainability of agriculture (De Vries and Bakker, 1998). The concept of critical metal loads is based on a static concept. It refers to long-term acceptable metal inputs and aims to maintain the soil in steady-state with respect to inputs and outputs. Such a comparison gives insight into the necessary changes in the management practices of agricultural land. The aim of sustainable metal management in agroecosystems is to ensure that the soil continues to fulfil its functions in agricultural production, by not restricting nutrient cycling or limiting soil biodiversity. In this context, sustainability can be defined as the situation where: (i) no further net accumulation of heavy metals occurs; or (ii) accumulation of heavy metals is below critical limits in defined compartments (e.g. soil/soil solution, plants or animal organs). Critical limits for heavy metals are generally derived from acute toxicity data, such as concentrations at which 50% of a certain species is affected (EC 50) or even dies (LC 50), or from chronic toxicity data, such as no observed effect concentrations or NOECs (OECD, 1992). Test organisms in terrestrial systems are microbe-mediated processes, earthworms or arthropods and plants. In aquatic systems the test organisms are algae, crustaceans and fish.

In order to get an insight into future

metal accumulation and leaching rates, use has to be made of models that include the dynamics of uptake and leaching, by relating those outputs to concentrations in the soil. Such models allow the prediction of metal concentrations in soil, soil solution and plants in time at a given input. They also allow the calculation of time periods before a critical value metal concentration in soil, soil solution or plants is exceeded and time periods that are needed to arrive at steady-state by taking the present metal inputs and metal status of the soil into consideration.

Aim of the chapter

At the time of writing, several papers had been written describing present metal inputs and outputs at the farm or field scale (e.g. Reiner *et al.*, 1996) or the dynamics in metal fluxes at the field scale (e.g. Moolenaar and Beltrami, 1998). An integrated approach that: (i) includes metal balances at the farm scale while making use of field information; and (ii) illustrates the use of critical limits in calculating long-term acceptable metal inputs and the time period in which those limits are violated, is, however, still lacking. The major aim is to present and apply such an integrated approach for agro-ecosystems, thus allowing the evaluation of metal management over time. More specifically, the application aims to answer the following questions:

- What are the major sources of metal inputs in different farming systems?
- To what extent do the present inputs exceed the farm outputs and what is the relative contribution of leaching and soil accumulation in response to this excess?
- What is the change in accumulation and leaching over time and what are the steady-state metal concentrations in soil and soil solution that will ultimately be reached?
- What are critical metal loads in different farming systems in view of different environmental effects?
- What is the percentage of farms where a certain critical limit will be exceeded when the metal input continues at its present load?

In order to reach this objective, we first present an overview of literature results on heavy metal balances at the farm and field scale. We then describe the methods that were used to assess present, critical and future pools and fluxes, focusing on uptake, leaching and accumulation rates. This section is followed by a summary of calculated heavy metal balances (present, critical and future fluxes) at approximately 100 agricultural monitoring sites in The Netherlands, using the tools and methods described before. A review of management options to minimize metal accumulation and leaching is then described. The chapter finishes with a discussion and conclusions related to the research questions posed.

Heavy Metal Balances at the National Scale and Farm-field Scale: a Literature Review

At present, most metal balances in the literature refer to national scales. Examples are balances for Cd, Cu, Pb and Zn in Finland (Mäkela-Kurtto, R., 1996), Sweden (Andersson, 1992), Denmark (Hovmand, 1984) and The Netherlands (De Boo, 1995). These balances are based on a static concept, in which records are kept of the input and output flows without considering the relation between contents in the soil (stock) and outputs. Consequently, dynamics of metal fluxes cannot be predicted. Furthermore, site specific aspects with respect to hydrology and soil characteristics influencing metal leaching and crop uptake are averaged out. Use can be made of top-down and bottom-up approaches as discussed by Moolenaar (1999a,b). In the top-down approach, use is made of a national bookkeeping system of inputs of feedstuff, animal manure, mineral fertilizers, sewage sludge, compost, etc. and outputs in crops, milk, meat, etc. together with generic data on metal concentrations (e.g. De Boo, 1995). In the bottom-up approach, farm-scale results are aggregated by assuming that the farms studied are representative for specific farm types, which represent a certain percentage of the agricultural sector in the country (e.g. Poppe *et al.*, 1994). In principle, this approach offers a more detailed insight into specific flows at the national scale, but upscaling of a limited number of farm or field balances to the national scale is prone to large uncertainties.

The difference between farm- and field-scale balances is illustrated in Fig. 5.2. Farm-scale balances consider inputs to and outputs from the farm, thus neglecting the internal flows, for example through uptake by grass (in the case of animal husbandry) followed by the production of manure that is applied on the farm. Inputs to a farm include products that are bought, such as fertilizer, animal manure, feed (concentrates) medicines and roughage (see Fig. 5.2, which shows only the major inputs). The input to the field, however, includes only fertilizer and animal manure (both brought to the farm and produced internally) together with atmospheric deposition. On arable farms, there is no difference with respect to net uptake by the crop in a field-scale balance and output in a farm-scale balance. On grassland, however, the output (or net uptake) in a farm-scale balance includes the net removal in milk and meat, whereas it includes the above ground removal of grass in a field-scale balance (see Fig. 5.2).

In The Netherlands, field-scale metal balances have been derived for typical agroecosystems, such as grassland and arable land (e.g. van Driel and Smilde, 1990; Moolenaar and Lexmond, 1998). An example of a farm-scale balance is presented in Reiner *et al.* (1996). Major conclusions from both field- and farm gate balance studies conducted so far are that the inputs of Cd, Pb, Cu and, to a lesser extent, also Zn, exceed the net outputs (Ferdinandus, 1989; van Driel and Smilde, 1990; Christensen and Tell, 1991; Moolenaar, 1998). This implies that metal accumulation takes place in agricultural systems, with potential for

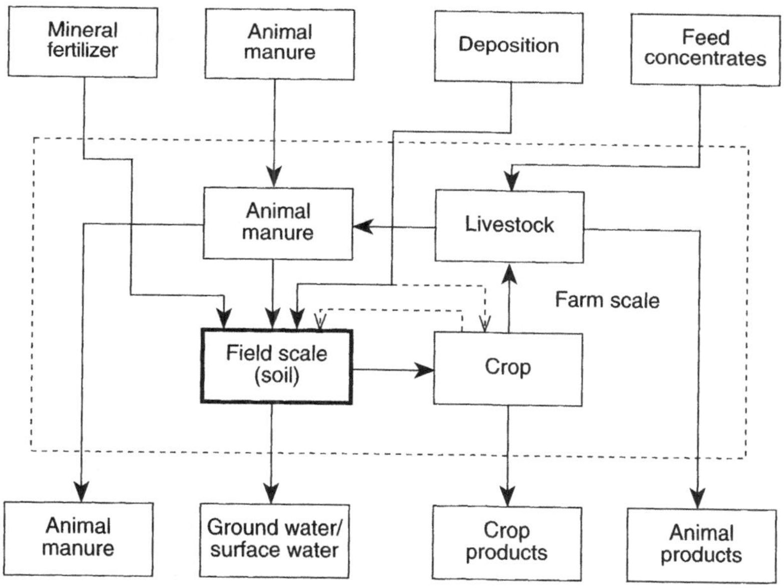

Fig. 5.2. Inputs, internal flows and outputs of heavy metals illustrating the difference between a farm-scale and a field-scale metal balance.

increased leaching. In forest systems, however, a significant depletion of all metals is observed in The Netherlands as a result of high leaching losses (Groot *et al.*, 1997).

Most examples of metal balances are based either on data from a few individual experimental farms or on average data. Conclusions drawn from most studies are therefore based on average values for input (e.g. fertilizer, manure, atmospheric deposition) and output (e.g. crop uptake, leaching) data on a large scale and do not distinguish between various types of management. Moolenaar and Lexmond (1998), however, showed that both management strategies and farm type (mixed farms versus dairy or arable crop production) have a large impact on the size of the surplus, with lower surpluses, or even a near balance, in mixed farms. An example showing the importance of the farming system with respect to the metal balance is given in Table 5.1 which includes recent data for farms with arable land on sandy soils and grassland with cattle production on peat in The Netherlands (Groot *et al.*, 1998). The data in Table 5.1 illustrate the very important contribution of animal manure to the net surplus load of Cu and Zn on arable land. For grassland systems, concentrates and other feeds contribute most to the net surplus load of Cu and Zn. Fertilizers are the dominant source of Cd, although atmospheric deposition is also important, which is comparable to results in other Western European countries (e.g. Hovmand, 1984). Atmospheric deposition is still the major source of Pb in these systems despite the significant reduction resulting from the shift to lead-free fuel. For all elements studied here, there is a significant imbalance between the net supply and removal of metals.

In farm gate balances, crop uptake and leaching (if included at all) are often considered to be constant. To overcome these limitations, dynamic balances have been developed, but until now, those balances refer to the field level. In this approach, not only current input–output data are included (manure, fertilizer, etc.) but also variable terms, such as soil specific leaching or crop uptake, to estimate long-term consequences of management strategies (e.g. van der Zee *et al.*, 1995; Moolenaar, 1998). Although this approach requires more information than the steady-state balance, it can be used on both field and farm scale to evaluate the effect of specific activities (fertilizer addition, choice of crop, etc.) on the net metal balance in time.

Table 5.1. Average farm-gate metal balances (g $ha^{-1}year^{-1}$) for arable land on sandy soils (17 sites) and grassland with cattle production on peat (17 sites) in The Netherlands (after Groot *et al.*, 1998).

	Arable land (sand); *N* = 17				Grassland (peat); *N* = 17			
Source	Cd	Pb	Cu	Zn	Cd	Pb	Cu	Zn
Fertilizer	2.1	9	22	76	1.5	3	7	30
Manure	0.7	8	323	566	0.1	1	44	66
Atmospheric deposition	0.7	32	6	23	0.8	41	9	40
Feedstuff	0	0	0	0	0.6	7	95	321
Other	0.1	1	23	88	0	0	4	10
Total supply	3.6	51	374	753	3	52	160	467
Total removal[a]	1	2	44	241	0	0	3	50
Leaching	1	9	56	163	0.3	22	44	97
Net surplus	1.6	40	274	349	2.7	30	113	320

[a] Total removal includes metals in animal products (milk, meat, etc.) as well as export of manure.

Locations and Methodological Approaches: a Case Study from The Netherlands

This section describes the methods that were used to calculate balances for Cd, Cu, Pb and Zn at the farm-field scale using data from locations that are part of The Netherlands' National Soil Monitoring Network (NSMN). With respect to the balances, a distinction is made between methods calculating present fluxes, long-term acceptable or critical fluxes and future fluxes assuming a continuing metal input at present day rates. The scale considered in the NSMN is the farm scale, since information on the input from manure, fertilizer, etc. and the output by crop uptake is only available at this level. Average information on soil properties at the field scale was, however, used to make an adequate estimate of leaching from the fields and accumulation in the soils belonging to the farm. Since the farms were located on comparable soil types, the danger of averaging out soil information is limited. In this approach, the advantages of field-scale information are combined with those of the farm scale, this being the scale at which all problems are investigated in this chapter.

The monitoring sites

The monitoring programme

The monitoring sites were part of the NSMN and have been operational since 1993 (Groot *et al.*, 1997, 1998). The primary objective of NSMN is to determine the changes in soil quality in The Netherlands over time. A secondary objective is to assess the quality of soil and upper groundwater. Within the network, the topsoil (0–10 cm depth), subsoil (30–50 cm) and uppermost groundwater are (planned to be) repeatedly sampled at 200 locations. Ten categories, of different combinations of soil type and land use, are distinguished. Each category has 20 sampling locations; two categories, with 40 locations in total, are sampled each year taking 5 years to complete the sampling and analyses of all locations.

Description of included locations

The locations included in this chapter are related to the first 3 years of study, and limit the sites to agricultural systems. In 1993, 35 dairy-cattle farms were sampled in the sandy regions of The Netherlands. In 1994, 20 intensive cattle farms with a high input of animal manure were sampled and 20 forested sites (not included in our calculations). In 1995, finally, sampling was carried out at 18 cattle farms on peat soils and at 19 arable farms on sandy soils. For some of those farms relevant data were missing. A total of 84 locations were thus included in our calculations. A summary of the major features of the locations included is given in Table 5.2 and an overview of the locations in Fig. 5.3. The approximate metal balances (Table 5.1) refer to the last two farm types in Table 5.2, that is, dairy cattle on peat and arable crops on sand.

Sampling and analyses

Soil sampling was carried out in such a way that it represents the whole farm, to allow a comparison of the observed soil quality with the metal inputs and outputs to and from the farm. Consequently, one average value for soil quality was derived at each location

Table 5.2. Characteristics of the locations included in present analysis in terms of major land use and soil type.

Type of farm	Major land use	Major soil type	Number of sites
Intensive livestock	Grass, maize	Sand	20
Dairy cattle (cows)	Grass	Sand	35
Dairy cattle (cows)	Grass	Peat	18
Arable	Wheat	Sand	19

Fig. 5.3. Geographical distribution of the farm locations in The Netherlands, distinguished by farm type.

(farm) by taking 320 soil samples from the topsoil, evenly distributed over all fields belonging to the farm. These 320 samples were randomly pooled in four buckets (approximately 1 kg of soil), each bucket thus containing 80 soil subsamples, which were mixed thoroughly. The four pooled samples were taken to the laboratory and analysed by standard methods. In this way, the spatial variation within the farm could be estimated.

The average metal concentrations and their ranges are comparable, with the exception of the dairy cattle farms on peat (Table 5.3). The relatively high concentrations in those farms are due to the much higher organic matter and clay contents, compared with all other farm types. Ranges in those contents, which were also measured at the sites, together with information on pH (pH-KCl and pH-H_2O) are

Table 5.3. Average concentrations and ranges of Cd, Pb, Cu and Zn in the four farm types. Values in brackets give the range between 5% and 95%.

	Metal concentration (mg kg^{-1})							
Type of farm	Cd		Pb		Cu		Zn	
Intensive cattle	0.43	(0.23–0.90)	15	(10–21)	17	(12–27)	31	(21–49)
Cattle (sand)	0.29	(0.15–0.52)	12	(8.2–17)	16	(9.1–27)	31	(18–46)
Cattle (peat)	0.72	(0.40–0.99)	42	(11–98)	86	(22–287)	124	(43–226)
Arable	0.26	(0.13–0.41)	19	(9.3–34)	31	(10–60)	31	(18–61)

presented in Table 5.4. The soil properties shown in Table 5.4 all influence the metal availability, and thus have impacts on concentrations and behaviour of metals in soil, soil solution and crops.

Calculation of present metal fluxes

The metal fluxes included in the calculation were all the major metal inputs (fertilizer, animal manure, food concentrates and atmospheric deposition) and outputs (crop and animal products, leaching) as presented in Fig. 5.2. The possible impact of soil erosion was neglected since all sites are located in flat areas. A description of the calculation procedures is given below.

Metal input

All farms that were sampled in the NSMN were selected from farms participating in the Farm Information Network (BIN) of the Agricultural Economics Research Institute (LEI). LEI-BIN is a representative test sample of 1500 farms where farm management and economics are monitored for 5–6 years.

Using the BIN data-set, the heavy metal inputs at the farm level were determined by multiplying the farm specific data for inputs with generic fixed metal concentrations for each type of input, which were obtained from the literature. Metal concentrations available for approximately 50 different types of fertilizer and animal manure were used (Groot *et al.*, 1997). The ranges in those concentrations are shown in Table 5.5. Heavy metal concentrations in fertilizers and organic manure were derived from Smilde (1986) and Hoogervorst (1991), respectively. Metal concentrations in feed concentrates were calculated from the use of raw materials used to produce the feed concentrates per specific animal type, using data on metal concentrations in those raw materials (Hanegraaf *et al.*, 1991). Atmospheric deposition data were based on model calculations including emission estimates for the various metals and the emission–deposition model TREND (van Jaarsveld and Onderdelinden, 1993), describing the transport and chemical transformations in the atmosphere (van Jaarsveld, 1994).

Table 5.4. Average values and ranges of organic matter and clay contents and pH in the four farm types. Values in brackets give the range between 5% and 95%.

Type of farm	Organic matter (%)		Clay (%)		pH-H_2O		pH-KCl	
Intensive Livestock	5.5	(3.9–7.2)	4.5	(3.0–5.9)	5.9	(5.4–6.2)	5.4	(4.8–5.8)
Dairy sand	7.5	(3.6–17)	3.8	(1.9–9.3)	5.9	(5.4–6.3)	5.0	(4.6–5.5)
Dairy peat	250	(14–34)	26	(4.4–46)	5.9	(5.3–6.8)	5.0	(4.4–6.3)
Arable	12	(5.3–21)	3.3	(2.2–7.0)	6.0	(5.5–6.5)	5.0	(4.5–5.6)

Table 5.5. Ranges in heavy metal concentrations in fertilizers, animal manure and feed concentrates in The Netherlands used in the calculations of metal balances.

	Metal concentration (mg kg^{-1})			
Product	Cd	Pb	Cu	Zn
N-fertilizer	0.10–0.40	0.70–2.7	2.6–14	7.8–87
P-fertilizer	7.4–75	2.0–17	24–114	160–904
K-fertilizer	0.04–0.35	3.9–12	14	0.70–20
NPK-fertilizer	7.5–24	0.90–7.5	14–50	74–186
Organic manure	0.03–0.30	0.34–3.6	4.1–60	9.8–220
Feed concentrates	0.05–0.14	0.80–1.7	16–160	78–87

Metal output

The outputs of heavy metals were calculated by multiplication of farm-specific quantities of products (in arable crops, milk, livestock, roughage and animal manure) from LEI-BIN with generic metal concentrations in those products. Table 5.6 presents the metal concentrations in milk, cattle, hay, fodder and most crops grown in arable farming; from Stoop and Rennen (1990, 1991) and Stoop *et al.* (1992, 1993).

Metal leaching

A distinction was made between the leaching rate from the plough layer and from the unsaturated zone. The annual leaching rate was calculated by multiplying estimated annual precipitation excesses by estimated annual average dissolved metal concentrations. For the plough layer (topsoil of 0–10 cm or 0–30 cm), the annual average dissolved concentration was estimated from the measured total metal concentrations and soil properties, using so-called transfer functions. For the unsaturated zone, use was made of measured metal concentrations in phreatic groundwater. The precipitation excess was calculated as:

$$PE = (1 - fr_{int}) \cdot P - E_s - E_t \quad (5.1)$$

with

$$E_t = E_{t,ref} + fr_{tr} \cdot (P - 780) \quad (5.2)$$

where:

PE = precipitation excess (mm year^{-1})
P = precipitation (mm year^{-1})
E_s = soil evaporation (mm year^{-1})
E_t = transpiration (mm year^{-1})
$E_{t,ref}$ = reference transpiration at a precipitation of 780 mm year^{-1} (mm year^{-1})
fr_{int} = interception fraction
fr_{tr} = transpiration fraction

Table 5.6. Concentrations of heavy metal in farm output products used to calculate the total metal output from the farm.

	Metal concentration (mg kg^{-1})			
Product	Cd	Pb	Cu	Zn
Milk	0.002	0.003	0.12	3.7
Cattle	0.02	0.148	1.0	4
Hay	0.011	0.22	1.1	40
Fodder	0.013	0.22	9	41
Wheat	0.06	0.14	3.2	28
Potatoes	0.027	0.03	1.2	0.37

According to Equation 5.1, interception evaporation is implicitly described as a fraction of the precipitation. Precipitation data for the various sites were based on an overlay of interpolated precipitation normals from 280 stations over the period 1950–1980 at 10 km × 10 km grid cells. Interception fractions were set at 0.05 for grassland and 0.1 for maize and arable land (De Visser and De Vries, 1989). The sum of evaporation and transpiration was estimated as a function of land use and soil type. Average annual values were 420 mm for grassland on sand, 330 mm for maize and arable land on sand and 480 mm for grassland on peat.

There are various approaches to derive total dissolved metal concentrations from total soil metal concentrations as illustrated in Fig. 5.4. The simplest is a direct empirical approach relating both concentrations, while accounting for the impact of major soil properties influencing the sorption relationship (e.g. Janssen *et al.*, 1996). The most mechanistic approach is to relate the free metal ion activity to the reactive metal concentration (metals that can be extracted by a weak acid or a complexing agent such as EDTA), accounting for the impact of soil properties and major ions in soil solution competing with the metals. Such an approach (e.g. Bril, 1995), however, requires additional relationships relating: (i) the reactive metal concentration to the total soil concentration; and (ii) the total dissolved concentration to the free metal ion activity, using a (simple) complexation model (De Vries and Bakker, 1998).

In the present study, we used an intermediate approach, relating the total dissolved metal concentration to the reactive soil metal concentration, according to:

$$\log[M]_{ss} = \alpha_0 + \alpha_1 \log M_{re} + \alpha_2 \log(\%OM) + \alpha_3 \log(\%clay) + \alpha_4 \log DOC + \alpha_5 \, pH - H_2O \quad (5.3)$$

where:

$[M]_{ss}$ = concentration of heavy metal M in the soil solution (mol l^{-1})
M_{re} = reactive concentration of heavy metal M in the soil (mol kg^{-1})
DOC = dissolved organic carbon concentration (mg l^{-1})

Using this approach, the impact of complexation is empirically accounted for by including DOC as an explanatory variable. Values for the various regression coefficients were derived from batch experiments with 49 soil samples from Dutch locations (Römkens *et al.*, 2001), as shown in Table 5.7.

The reactive metal concentration, which was estimated with mild (0.43 M) HNO_3 extraction, was related to the total concentration (aqua regia digestion), according to:

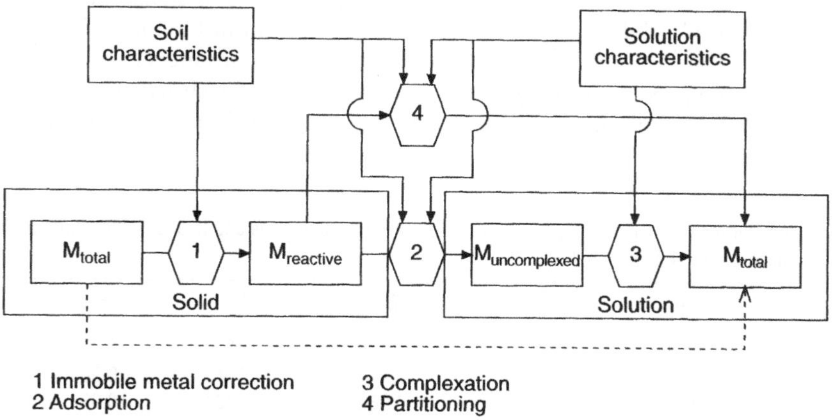

Fig. 5.4. Diagram illustrating different possibilities for calculating total dissolved metal concentrations from total concentrations in the soil solid phase.

Table 5.7. Values for the coefficients α_0–α_5 in the relationship relating dissolved total concentrations and reactive soil concentrations of Cd, Cu, Pb and Zn, according to Equation 5.3.

Metal	α_0	α_1	α_2	α_3	α_4	α_5	R^2	se-y_{est}[a]
Cd	1.288	1.142	−0.735	−0.276	0.318	−0.453	0.74	0.46
Cu	−5.217	0.517	−0.517	—	0.731	—	0.56	0.38
Pb	−2.165	0.860	−0.860	—	0.542	−0.323	0.50	0.66
Zn	1.274	1.023	−0.611	−0.403	0.410	−0.568	0.56	0.38

[a] Standard error of the estimated metal concentration (y estimate) expressed, on a logarithmic basis.

$$\log M_{re} = \beta_0 + \beta_1 \log M_{soil} + \beta_2 \log(\%OM) + \beta_3 \log(\%clay) \quad (5.4)$$

where M_{soil} = total concentration of heavy metal M in soil (mol kg^{-1}).

Values for the various coefficients were again derived from the same 49 soils as shown in Table 5.8 (Römkens *et al.*, 2001).

The two relationships (Equations 5.3 and 5.4) were applied at each site, using the measured total metal concentrations and soil pH, organic matter and clay content. Since DOC data for the soil solution were not available, those concentrations were estimated from the equation:

$$\log DOC = 1.99 + 0.94 \log(\%OM) - 0.34 \log(Al + Fe) - 0.07 \text{ pH-KCl} \quad R^2 = 0.52 \quad (5.5)$$

where Al + Fe is the sum of oxalate extractable Al and Fe concentrations (mmol kg^{-1}), which was derived from the other soil properties according to:

$$\log(Al + Fe) = 1.36 + 0.41 \log(\%OM) - 0.35 \log(\%clay) \quad R^2 = 0.80 \quad (5.6)$$

The very indirect approximation of DOC, based on the batch experiments with 49 soils (Römkens *et al.*, 2001) causes a relatively large uncertainty in the estimated dissolved metal concentration in the plough layer. Nevertheless, it was considered better to include this variable, since neglecting DOC in Equation 5.3 caused a very strong decrease in accounting for the variability (value of R^2) of the metal concentration in soil solution.

Metal accumulation

The metal accumulation in the mineral topsoil was calculated from the net input to the farm (this being assumed to be equal to the net soil input by neglecting pool changes on the farm level) minus the leaching from the soil according to:

$$M_{ac} = M_{in} - M_{out} - M_{le} \quad (5.7)$$

where:

M_{in} = the total metal input to the farm (g ha^{-1} year^{-1})

M_{out} = the total metal output from the farm in the form of crop and animal products (g ha^{-1} year^{-1})

M_{le} = the leaching from the mineral topsoil (g ha^{-1} year^{-1}).

Table 5.8. Values for the coefficients β_0–β_3 in the relationship relating reactive and total soil concentrations of Cd, Cu, Pb and Zn, according to Equation 5.4.

Metal	β_0	β_1	β_2	β_3	R^2	se-y_{est}[a]
Cd	0.0107	1.258	−0.179	−0.183	0.97	0.11
Cu	−0.307	0.9845	0.111	−0.004	0.96	0.18
Pb	−0.099	1.011	0.064	−0.113	0.93	0.17
Zn	−0.850	1.221	0.32	−0.236	0.96	0.11

[a] Standard error of the estimated metal concentration (y estimate) expressed, on a logarithmic basis.

The accumulation in the mineral subsoil was calculated from the difference between the metal leaching from the plough layer and the unsaturated zone.

Prediction of future metal fluxes and of time periods to reach a steady-state situation

Because of accumulation or release, the soil metal concentrations change in time, thus influencing the leaching and uptake. Changes in soil metal concentrations were calculated according to:

$$M_{soil}(t) = M_{soil}(t-1) + \frac{M_{ac}(t-1)}{\rho T \times 10} \quad (5.8)$$

where: ρ = bulk density of the soil (kg m^{-3}); T = soil thickness (m).

Bulk density was derived from relationships with the organic matter and clay content for mineral (Hoekstra and Poelman, 1982) and peat (van Wallenburg, 1988) soils. Changes in metal leaching due to a change in the soil metal concentration were derived from Equations 5.3 and 5.4. The metal output was also recalculated, since the change in soil metal concentrations affects the plant metal concentrations. Metal concentrations in the plant were related to soil concentrations, while accounting for the impact of soil properties influencing metal availability, according to:

$$\log M_{plant} = \alpha + \beta \log(\%OM) + \gamma \log(\%clay) + \delta \, pH\text{-}KCl + \log(M_{soil}) \quad (5.9)$$

Values for the various coefficients were derived for Cd, Cu, Pb and Zn in grass, wheat and maize, these being the main crops on the farms, based on different Dutch data-sets (Römkens and de Vries, 2001). In general, relationships were better for Cd and Zn than for Cu and Pb. Furthermore, in some cases, different relationships were found for polluted and unpolluted soils. As an example, the results for Cd and Zn in unpolluted soils are presented in Table 5.9.

The total output at the farm gate (meat and milk in the case of cattle farms) was scaled against the changes in metal concentrations in the crops, according to:

$$M_{outl}(t) = M_{outl}(t0) \times \left(frM_{gr} \frac{M_{gr}(t)}{M_{gr}(t0)} + frM_{m} \frac{M_{m}(t)}{M_{m}(t0)} + frM_{w} \frac{M_{w}(t)}{M_{w}(t0)} \right) \quad (5.10)$$

where:

frM_{gr}, frM_{m}, frM_{w} = fractions of the farm covered by grass, maize and wheat

M_{gr}, M_{m}, M_{w} = metal concentrations in grass, maize and wheat (mg kg^{-1}).

Because of changes in leaching and outflow, the estimates of metal accumulation (Equation 5.7) changed over time. Time periods to reach steady-state were calculated iteratively by requiring that the change in metal concentration was less than 0.01%

Table 5.9. Values for the coefficients α–ε in the relationship relating total concentration of Cd and Zn in different plants and in soil according to Equation 5.9.

Metal	Crop	α	β	γ	δ	ε	R^2	se-y_{est}[a]
Cd	Grass	—	−0.24	—	−0.10	0.47	0.53	0.22
	Maize	0.17	−0.21	−0.05	−0.07	0.44	0.58	0.12
	Wheat	0.43	−0.44	—	−0.17	0.36	0.44	0.20
Zn	Grass	2.06	1.09	−1.05	−0.09	0.41	0.49	0.11
	Maize	0.91	—	−0.57	−0.10	0.93	0.54	0.11
	Wheat	1.32	—	−0.24	−0.06	0.45	0.56	0.09

[a] Standard error of the estimated metal concentration (y estimate) expressed, on a logarithmic basis.

in 1 year. Practically, metal accumulation was negligible when using this criterion. The procedure to calculate those time periods is further illustrated in Fig. 5.5.

Calculation of critical metal loads

The concept of critical metal loads has been described extensively in De Vries and Bakker (1998). It is defined as the load that will not lead to either: (i) accumulation of heavy metals; or (ii) concentrations of heavy metals above critical limits in defined compartments (e.g. soil/soil solution, groundwater, surface water, plants or animal organs) in a steady-state situation. The first so-called precautionary or 'stand still' principle implies that the present metal concentration in the soil is considered the critical limit above which no further increase is accepted. In the second risk-based approach, the critical limits are based on adverse effects on (parts of) the ecosystem. Examples are food quality criteria for metals in crops, animal food or animal organs such as the kidney and the liver (Huinink, 1999) or soil quality criteria related to adverse impacts on soil organisms (Bååth, 1989; Witter, 1992). In this approach, some accumulation may thus be accepted in unpolluted areas, whereas a decline in metal concentrations (implying a lower metal input than metal output) may be required in polluted areas. Using this approach, critical metal loads were simply calculated according to De Vries and Bakker, (1998):

$$M_{in}(\text{crit}) = M_{out} + M_{le}(\text{crit}) \tag{5.11}$$

where:

$M_{in}(\text{crit})$ = critical total metal input to the farm (g ha^{-1} year^{-1})
$M_{le}(\text{crit})$ = critical metal leaching rate (g ha^{-1} year^{-1}).

The critical metal leaching rate was derived by multiplying the precipitation excess by a critical dissolved concentration, which was related to: (i) the present soil metal concentration (critical load based on the stand still principle); (ii) a target value of soil metal concentrations used in Dutch environmental policy; (iii) a critical dissolved concentration in view of impacts on plants and surface water organisms; or (iv) a critical metal concentration in plants based on food quality criteria (Cd and Pb) and phytotoxicity levels (Cu and Zn).

When using soil concentrations, the soil–soil solution relationships described above (Equations 5.3 and 5.4) were used to calculate related critical dissolved concentrations. Target values of total soil concentrations in The Netherlands (hot aqua regia extraction of soils sieved to < 2 mm) are related to the organic matter and clay content according to:

$$M_{tv} = A + B\ \%\text{Clay} + C\ \%\text{OM} \tag{5.12}$$

Values used are given in Table 5.10.

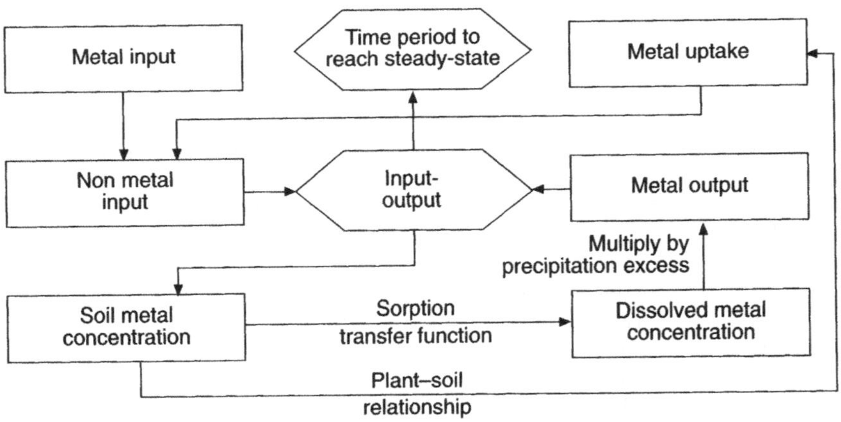

Fig. 5.5. Diagram illustrating the calculation of time periods to reach steady-state.

Critical dissolved metal concentrations for plants were directly derived from available literature (Table 5.11). For soil organisms, data are sparse and only related to total soil metal concentrations (e.g. Witter, 1992). Consequently, available data for aquatic organisms were used as a surrogate (Table 5.11). Those limits are also of relevance in view of the runoff of metals to ditches. Use of critical metal concentrations in plants required the assessment of soil solution–plant relationships. This was done indirectly by: (i) first assessing soil–plant relationships to calculate critical concentrations in the soil (Equation 5.9), followed by; (ii) application of the soil–soil solution relationships described above (Equations 5.3 and 5.4). Critical limits used for metal concentrations in soil solution and in plants grown at the study sites are shown in Table 5.11.

Present, Critical and Future Metal Concentrations and Metal Fluxes at Agricultural Sites in The Netherlands

Present metal balances

An overview of the average metal fluxes and their most common ranges for all the study farms (Table 5.12) shows that on average there is an accumulation in both the plough layer and the unsaturated zone. In the plough layer, calculated leaching is sometimes, however, higher than the net input causing a negative accumulation (release) of heavy metals (see the 5% values in Table 5.12). The accumulation of

Table 5.10. Constants used to calculate target total soil metal concentrations in The Netherlands (Equation 5.12), including the value for a standard soil (10% organic matter and 25% clay).

Metal	A (mg kg^{-1})	B	C	M_{tv} (mg kg^{-1})
Cd	0.4	0.007	0.021	0.8
Pb	50	1	1	85
Cu	15	0.6	0.6	36
Zn	50	3	1.5	140

Table 5.11. Critical limits for Cd, Cu, Pb and Zn in soil water, surface water and plants related to impacts on soil organisms, aquatic organisms and plants (food quality criteria or phytotoxicity levels).

	Dissolved metal concentrations (mg m^{-3})		Plant metal concentrations (mg m^{-3})		
Metal	Plants[a]	Aquatic organisms[b]	Grass	Maize	Wheat
Cd	2.0	0.34	0.83[c]	0.10[d]	0.15[e]
Cu	2.5	1.1	25	25[f]	10[f]
Pb	15	11	11.6[c]	10[d]	0.15[e]
Zn	25	6.6	—	460[f]	560[f]

[a] Based on Tyler (1992), using the second lowest LOEC data from laboratory studies with culture solutions reported by Bahlsberg Påhlsson (1989), divided by a safety factor of 10.

[b] Based on NOEC data for aquatic organisms, using a statistical extrapolation method while assuming a log-logistic distribution (after Crommentuijn *et al.*, 1997).

[c] Based on critical Cd and Pb concentrations in the kidney of cows of 2.5 and 1.0, respectively, divided by a plant–kidney transfer factor of 2.99 for Cd and 0.086 for Pb (after Huinink, 1999).

[d] Based on critical metal concentrations in animal feed (after Huinink, 1999).

[e] Based on food quality criteria for wheat.

[f] Based on lowest leaf contents in leaf tissue showing toxicity in plants (after Smilde, 1976; Kabata Pendias and Pendias, 1984).

Table 5.12. Average fluxes of Cd, Pb, Cu and Zn on all 84 study farms. Values in brackets give the range between 5% and 95% (90 percentile range).

	Metal flux (g ha^{-1} year^{-1})							
Metal source	Cd		Pb		Cu		Zn	
Input	3.8	(0.62–8.7)	98	(37–222)	269	(63–608)	742	(190–1511)
Output	0.24	(0.03–1.1)	0.52	(0.09–2.1)	8.6	(1.5–37)	82	(31–207)
Leaching plough layer	2.4	(0.63–5.8)	52	(22–111)	131	(84–224)	461	(163–1003)
Accumulation plough layer	1.2	(−3.0–5.7)	46	(−67–189)	129	(−157–460)	199	(−780–1081)
Leaching unsaturated zone	1.3	(0.06–4.3)	9.1	(1.4–30)	37	(7.1–81)	234	(34–741)
Accumulation unsaturated zone	2.3	(−0.76–6.2)	89	(24–214)	223	(1.3–546)	426	(−221–1279)

Cd and Zn in the subsoil (the soil layer between the bottom of the plough layer and the phreatic groundwater level) is much less than for Pb and Cu. Considering the whole unsaturated zone, there is always accumulation of Pb and Cu but not of Cd and Zn (see the 5% values in Table 5.12), illustrating the much higher mobility of Cd and Zn.

The contribution of various sources of farm inputs to the total input of metals depends on the element (Table 5.13). In general, feed concentrates contribute most to the input of Cu and Zn to livestock systems, whereas fertilizers are the dominant source of Cd, and atmospheric deposition is the major source of Pb (compare with Table 5.1). Negative inputs in manure are due to a larger removal from the farm than external input to the farm. This only takes place at intensive cattle farms.

A comparison of the average metal balances for the four farm types shows that the input of Cu and Zn is highest on the intensive cattle farms, whereas Cd and Pb input is highest on the dairy cattle farms on sand. For all the metals, the average input is lowest on the dairy cattle farms on peat, with the exception of Pb, where the input is comparable to arable land (Table 5.14). The high Pb inputs to the dairy cattle farms on sand reflect the location of those farms in high Pb deposition areas. On average, there is a net loss (negative accumulation) of all metals from the plough layer in the dairy cattle farms on peat, where input is lowest with the exception of Cd. In all other farm types there is, on average, an accumulation in the plough layer, with the exception of Cd in intensive cattle farms and Pb in arable land (Table 5.14).

The most distinct differences occur for Cu and Zn balances, with highest (mainly agricultural) inputs in intensive cattle farms and lowest inputs at cattle farms on peat. The ranges in net input and in metal leaching from both the plough layer and the unsaturated zone (Fig. 5.6) show that net metal mobilization (leaching being higher

Table 5.13. Average inputs of Cd, Pb, Cu and Zn in fertilizer, manure, deposition and feed concentrates to all 84 study farms.

	Metal flux (g ha^{-1} year^{-1})				
Metal	Fertilizer	Manure	Deposition	Feed concentrates	Total
Cd	2.1	0.023	0.83	0.82	3.8
Pb	9.5	−1.3	79	11	98
Cu	17	44	9.1	199	269
Zn	85	67	40	550	742

Table 5.14. Average fluxes of Cd, Pb, Cu and Zn for the four farm types. Both leaching and accumulation refer to the plough layer (0–30 cm).

		Metal flux (g ha^{-1} year^{-1})			
Metal	Type of farm	Input	Output	Leaching	Accumulation
Cd	Intensive cattle	3.3	0.093	4.2	−1.0
	Cattle sand	4.6	0.046	2.4	2.1
	Cattle peat	3.0	0.039	1.2	1.7
	Arable	3.6	0.97	1.5	1.1
Pb	Intensive cattle	61	0.72	41	19
	Cattle sand	170	0.15	42	127
	Cattle peat	52	0.22	70	−18
	Arable	50	1.3	63	−15
Cu	Intensive cattle	398	4.1	104	289
	Cattle sand	206	2.4	110	94
	Cattle peat	160	2.2	178	−21
	Arable	362	31	155	176
Zn	Intensive cattle	1170	63	340	767
	Cattle sand	663	50	435	177
	Cattle peat	467	46	633	−212
	Arable	710	199	464	47

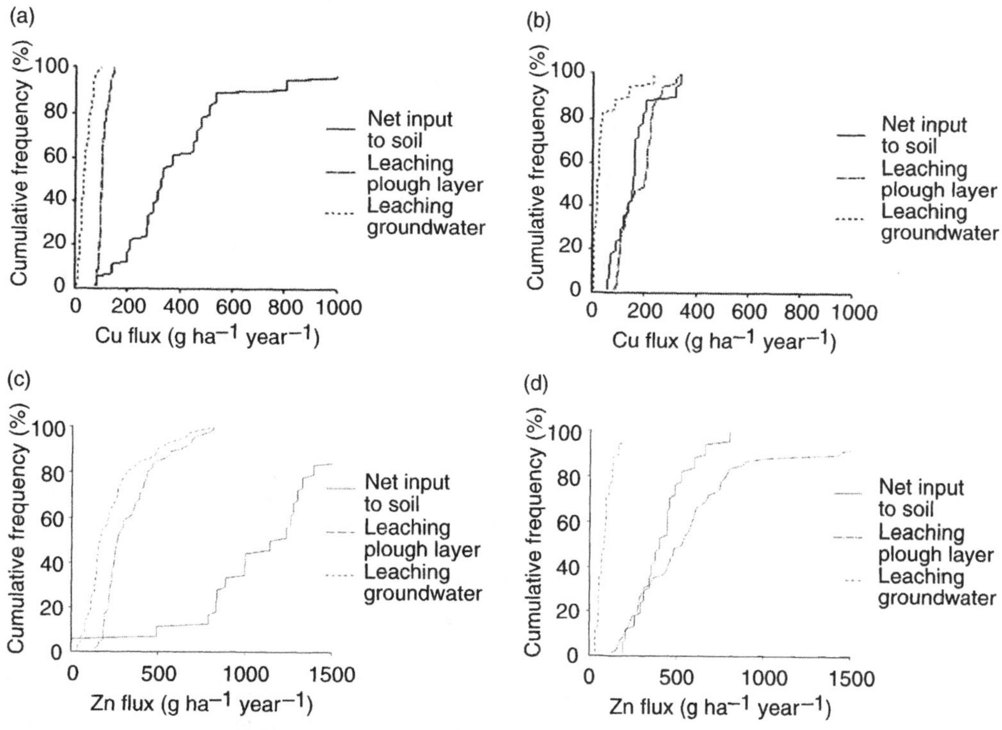

Fig. 5.6. Ranges in the leaching flux of Cu and Zn from intensive cattle farms on sand (a, c) and cattle farms on peat (b, d).

than the net input) never occurs in intensive cattle farms. In the cattle farms on peat, leaching from the plough layer is generally higher than the net input, but leaching to the unsaturated zone is lower, indicating that metal accumulation takes place even in these farming systems, but that this occurs deeper down in the soil profile.

Future uptake, leaching and accumulation rates of metals

Soil metal concentrations change over time, thus influencing the leaching, uptake and output from the farm. Ultimately, a steady-state will be reached where soil metal accumulation is negligible and the excess farm input (farm input minus farm output) is completely leached from the soil profile. These changes are illustrated in Table 5.15. The results indicate that the total metal output from the farm is hardly affected by the change in soil concentrations. For Pb and to a lesser extent Cu, this is mainly due to a lack of relationship between plant and soil metal concentrations. For Cd and Zn, the relative changes are larger, but in absolute terms the change is small, specifically in cattle farms. The differences between leaching at steady-state and at the beginning of the simulation were larger for Pb and Cu (an average increase of approximately 100%) than for Cd and Zn (an average increase of approximately 50%). Within a 500 year period, however, the relative changes were comparable (all approximately 40–50%). This is because at most of the sites, the metals in the plough layer are in steady-state within 500 years for Cd and Zn (nearly negligible accumulation), whereas it takes a much longer time for Pb and Zn. More information on the time periods to reach steady-state is given in Table 5.16. Results show that those time periods not only differ

Table 5.15. Average output, leaching and accumulation fluxes of Cd, Pb, Cu and Zn in the plough layer at different time periods for all 84 study farms.

		Metal flux (g ha^{-1} $year^{-1}$)			
Farm type	Situation	Cd	Pb	Cu	Zn
Output	Present	0.24	0.52	8.6	82
	500 year	0.27	0.52	8.2	94
	Steady-state	0.27	0.51	8.3	95
Leaching	Present	2.4	52	131	461
	500 year	3.5	73	193	639
	Steady-state	3.5	96	260	647
Accumulation	Present	1.2	46	129	199
	500 year	0.023	25	74	8.3
	Steady-state	0.0015	1.9	7.5	0.37

Table 5.16. Averages of time periods to reach steady-state for Cd, Pb, Cu and Zn on all considered 84 farms. Values in brackets give the range between 5% and 95% (90 percentile ranges).

	Time period to reach steady-state (years)							
Farm type	Cd		Pb		Cu		Zn	
Intensive cattle	410	(92–643)	1331	(642–2239)	2416	(611–4696)	444	(208–989)
Cattle sand	352	(147–551)	1794	(775–2959)	1272	(172–3006)	302	(83–640)
Cattle peat	271	(99–443)	776	(217–1765)	364	(125–849)	176	(48–328)
Arable	493	(109–712)	1527	(0–2649)	1747	(209–4037)	382	(65–160)

quite strongly between the different metals but also between different farming types. In general, steady-state is reached within 100–1000 years for Cd and Zn, but it can last up to 3000 years for Pb and even 4000 years for Cu (Table 5.16).

Estimated changes in leaching and accumulation rates for Cd and Pb were largest on the dairy cattle farms on sand, where the net input is highest. The estimated Cu and Zn accumulation was largest in intensive cattle farms, with the highest (mainly agricultural) inputs, but the average change in Cu leaching rates was comparable with cattle farms.

The changes in both soil accumulation and soil leaching were reflected in changes in metal concentrations in both the soil solid phase and the soil solution. Differences in steady-state and present soil metal concentrations were highest for Cu (an average fivefold increase) followed by Pb and Cd (an increase of approximately 55–60%) and Zn (an increase of approximately 30%). The differences in steady-state and present dissolved metal concentrations are approximately 100% for Pb and Cu and 50% for Cd and Zn, comparable to the leaching fluxes. During the period to steady-state, the increase in soil metal concentrations at several sites was such that it exceeded the target values used in Dutch pollution policy, specifically for Cu, as illustrated in Table 5.17. Similarly, the increase in dissolved metal concentrations at several sites was such that it exceeded critical limits related to adverse effects on plants (Table 5.18). These results indicate that the present loads do exceed the long-term critical loads at those sites. This aspect is discussed further in the following section.

Critical metal loads

A comparison of critical metal loads related to various adverse effects is presented (Table 5.19) for comparison with the range in present loads (Table 5.12). This indicates that present loads are generally higher than critical loads, especially for Cu. Approximately 65–75% of the sites present loads which caused an accumulation in the soil (violation of the stand still principle) for all the metals included (Table 5.20). At the other sites, metal leaching was higher than the net input,

Table 5.17. Average present, critical and steady-state soil metal concentrations. Values in brackets give the range between 5% and 95% (90 percentile ranges).

	Metal concentration (mg kg^{-1})							
Metal status	Cd		Pb		Cu		Zn	
Present	0.40	(0.16–0.88)	34	(11–122)	20	(8.9–54)	50	(19–166)
Critical	0.70	(0.51–1.3)	70	(58–114)	27	(20–53)	247	(122–594)
Steady-state	0.62	(0.18–1.7)	55	(13–122)	105	(4.9–381)	64	(13–147)

Table 5.18. Average present, critical and steady-state dissolved metal concentrations. Values in brackets give the range between 5% and 95% (90 percentile ranges).

	Metal concentration (mg mg^{-3})							
Metal status	Cd		Pb		Cu		Zn	
Present	0.8	(0.22–2.0)	18	(8.5–38)	44	(31–80)	160	(54–350)
Critical	2.0		15		25		250	
Steady-state	1.2	(0.31–2.5)	33	(12–73)	88	(25–210)	220	(50–520)

Table 5.19. Average critical metal loads related to the stand still principle, targets values for soil used in Dutch pollution policy and critical limits for soil solution related to adverse effects on plants. Values in brackets give the range between 5% and 95% (90 percentile ranges).

	Critical metal loads (g ha^{-1} $year^{-1}$)							
Effect	Cd		Pb		Cu		Zn	
Stand still	2.6	(0.77–5.8)	52	(23–111)	140	(86–231)	543	(225–1076)
Soil target values	5.7	(1.7–11)	121	(38–188)	168	(119–228)	4,336	(1180–11,458)
Plants	6.1	(5.1–7.8)	45	(38–53)	82	(65–116)	817	(672–1054)
Minimum	2.4	(0.77–5.1)	39	(23–49)	82	(65–116)	496	(225–878)

Table 5.20. Percentage of sites exceeding soil target values and critical limits in the soil solution related to plant impacts.

		Exceedance (%)			
Effect	Situation	Cd	Pb	Cu	Zn
Accumulation	Present	74	73	74	67
Exceedance soil target	Present	3.0	5.7	11	0
value	Steady-state	25	27	61	2.7
Exceedance critical limit	Present	5.1	45	100	14
soil solution	Steady-state	17	85	94	32

thus leading to decreased metal concentrations. The present loads caused a large increase in the number of sites exceeding the target values for Cu used in Dutch pollution policy (from 11 to 61%), the largest changes taking place on intensive cattle farms and arable land. For Cd and Pb, the increase was lower (from approximately 5 to 25%), the largest changes taking place on cattle farms on peat for Cd and on sand for Pb, and for Zn it was nearly negligible. The increase in sites exceeding critical limits for impacts on plants was highest for Pb. The percentage of sites exceeding critical Cu limits hardly changed since the excess was already 100% at the start of the simulation (Table 5.20). This result indicates that the critical limit of 25 mg m^{-3} for Cu might be too low. In general, the number of sites violating food quality criteria for Cd and Pb or phytotoxicity levels for Cu and Zn were much lower for all metals, both at the start of the simulation and in the steady-state situation.

Time periods to reach soil target values at sites were nearly always less than 1000 years for Cd and Zn (on average between 200–300 years), but could go up to more than 3000 years for Pb and Cu (Table 5.21). Time periods to reach critical soil solution concentrations were generally higher for Cd and Zn and lower for Pb and Cu.

Management Options to Minimize Metal Accumulation and Metal Leaching

The metal balances of Cd, Cu and Zn show a net surplus that is directly related to activities at the farm scale. For Pb, however, the net surplus is related mainly to atmospheric deposition (see Table 5.1), and farm-level management options to reduce the surplus in the soil are unavailable. Only measures taken on a national or international scale (reduction of Pb emission by industry and traffic for example) will have a significant effect on the metal balance for Pb.

Table 5.21. Average time periods to reach critical soil and soil solution concentrations at sites where these are ultimately reached. Values in brackets give the range between 5% and 95% (90 percentile ranges).

	Time period to reach steady state (years)							
Effect	Cd		Pb		Cu		Zn	
Soil	197	(12–1008)	1178	(307–3120)	465	(0–4815)	297	(193–599)
Soil solution	355	(44–1084)	343	(0–2411)	280	(0–4560)	470	(8–1331)

Management options that are available for the other metals to reduce the net surplus or prevent metals from leaching from the soil (the latter, however, always leads to an increase in the net surplus) can be divided into three categories: (i) farming system management; (ii) crop management; or (iii) soil management.

Farming system management options to reduce metal inputs

The choice of amendment (e.g. manure, compost, sludge or mineral fertilizers) has a large impact on the metal balance. Reduction of the use of manure will especially reduce the input of Cu and Zn in arable land and, to a lesser extent, also Cu in grassland systems. A reduction of the Zn content in feedstuff could result in a significant lowering of the Zn imbalance. Reiner *et al.* (1996) showed that the average net surplus in different farms decreased from 700 to 40 g ha^{-1} $year^{-1}$ for Cu and from 2800 to 300 g ha^{-1} $year^{-1}$ for Zn when mineral fertilizers were used instead of manure. Similarly, Moolenaar (1998) showed that the use of mineral fertilizer resulted in a close to zero net surplus for both Cu and Zn in a dairy cattle farm, instead of a net Cu surplus of 103 g ha^{-1} $year^{-1}$ in an ecological treatment with goat and cattle manure and a net Zn surplus of 800 g ha^{-1} $year^{-1}$ in a conventional management scheme. However, the shift from manure to, for example, inorganic fertilizer will lead to an increase in the net surplus of Cd in soils (Velthof *et al.*, 1996; Moolenaar, 1998). Moolenaar showed an increase in the Cd surplus from 1 g ha^{-1} $year^{-1}$ in the conventional management scheme to 5 g ha^{-1} $year^{-1}$, when using mineral fertilizers.

Important management options also include the type of farming system. The net surplus of Cd, Pb, Cu and Zn in an integrated mixed farming system in The Netherlands was much lower than those shown in Table 5.1. A net removal was observed (−26 g ha^{-1} $year^{-1}$) for Cu and the net surplus for Cd and Zn was much lower as well (0.4 and 33 g ha^{-1} $year^{-1}$ for Cd and Zn, respectively) because of the low amount of manure used, in combination with fertilizers with a low Cd content (Moolenaar, 1998).

Crop management options to increase metal uptake

In an integrated mixed farming system, crop uptake of Zn by grass can be as high as 96% of the total output (Moolenaar, 1998), and ranges from 47% for Cd, 65% for Pb to 79% for Cu. In a mixed farm (with both arable land and grassland), the uptake by grass is quantitatively more important than that by arable crops. This is due to both a higher crop uptake rate and a higher net dry matter production of grass compared with arable crops. The total uptake of metals will thus increase at higher ratios of grassland to arable land. Also, choice of crop (in the case of arable land) can alter the total uptake significantly. Net metal removal from the soil by crops will, for example, be much higher for sugarbeet, with both a high biomass production and a higher metal uptake rate than with potatoes or wheat, with a low to very low uptake rate. Other factors, such as rooting depth, should also be considered, since the uptake from a limited rooting zone (e.g. 10 cm in the case of grass) will result in a more pronounced change in the metal content of the topsoil, compared with deep rooting crops with a similar yield and metal uptake rate.

Soil management options to influence the ratio between metal leaching and accumulation

In most non-calcareous agricultural soils, soil pH is maintained between 6.0 and 7 by regular application of lime. However, especially on grassland soils or soils used for foodcrops on sandy soils (e.g. maize), pH-KCl can be as low as 4 to 4.5 (see Table 5.4). At these pH levels, metal leaching can be significant, especially for metals such as Zn and Cd. Increased metal availability may have impacts on soil life but it may also lead to an exceedance of quality standards or target values for groundwater quality. Maintenance of soil pH at near neutral levels (between 6 and 7) is thus an effective tool to minimize leaching losses. Reduced leaching losses, of course, lead to higher accumulation rates in the soil. The impact of liming all soils to a pH-H_2O of 6.0 on soil leaching losses is illustrated in Fig. 5.7 for Cd and Zn, the two metals most affected by a change in soil pH. The change in metal leaching is substantial, causing a decrease in the percentage of sites where critical limits for dissolved metal concentrations are exceeded. Inversely, the lower leaching losses cause a larger accumulation, thus causing ultimately higher percentages of sites exceeding soil target values. This impact, however, only occurs after very long periods.

Discussion and Conclusions

The results of this study are focused on Dutch agroecosystems. Nevertheless, the results are likely to be applicable to agroecosystems in most industrialized countries, considering the comparable inputs in other countries as shown in the literature review. In view of the questions related to metal balances, posed at the beginning of this chapter, the following general conclusions can be drawn:

- The major sources of metal inputs vary between different farming systems. For grassland systems feedstuff contributes most to the inputs of Cu and Zn, whereas animal manure is the most important contributor to the input of those metals on arable land. Fertilizers are the dominant source of Cd, whereas atmospheric deposition is generally the major source of Pb.

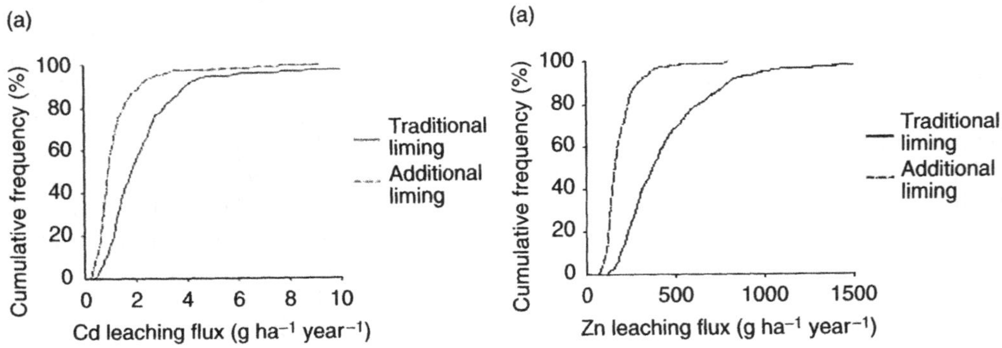

Fig. 5.7. Present leaching rates of Cd (a) and Zn (b) from the plough layer before and after liming of all sites to a pH-H_2O of 6.0.

- For the Dutch agroecosystems, the present inputs of metals exceed the farm outputs at approximately 70–75% of the sites for all the considered metals. This is an indication that present loads to agroecosystems in industrialized countries are generally higher than critical loads based on the stand still principle. In other words, the present loads generally cause an increase in soil metal concentrations. The highest accumulation is likely for Cu and Zn in intensive cattle farms, with highest (mainly agricultural) inputs. Net metal mobilization did not occur at any of the farms in the Dutch case study.
- Present metal inputs can cause changes in metal accumulation and metal leaching over a period of several hundreds or even thousands of years, depending on metal and farming type considered. In general, steady-state is reached within 100–1000 years for Cd and Zn, but it can last up to 3000 years for Pb and even 4000 years for Cu. Those timescales are an indication of the transition times in fertilized agroecosystems.
- The steady-state soil metal concentrations that will ultimately be reached can differ strongly from the present metal concentrations. In the Dutch case study, the steady-state Cu concentration was approximately fivefold greater than the present concentration. For the other metals, the changes were much less (an increase of approximately 60% for Pb and Cd and of approximately 30% for Zn). The average increase in dissolved metal concentrations (and in leaching fluxes) was approximately 100% for Pb and Cu and 50% for Cd and Zn, an indication of likely changes in comparable systems.
- Critical metal loads vary in view of different environmental effects and for different farming systems. The predicted increase in the percentage of farms where soil target values used in Dutch pollution policy will be exceeded when the metal input continues at its present load was large for Cu (from 11 to 61%), intermediate for Cd and Pb (from approximately 5 to 25%), and nearly negligible for Zn. Similarly, the present loads caused an increase in the percentage of sites exceeding critical limits for dissolved metal concentrations for Cd, Pb and Zn related to adverse effects on plants. This indicates that present metal loads in agroecosystems in most industrialized countries are likely to increase nonpoint source pollution problems.

In other countries, the impact of soil erosion can be considerable with respect to the transfer to surface water, but this mechanism is not important in flat countries such as The Netherlands. The conclusions that are drawn with respect to critical loads are strongly dependent on the reliability of the critical limits used. There is, for example, a clear inconsistency in critical limits related to adverse effects on plants. For Cu, the critical limits in plants related to phytotoxicity are hardly ever exceeded, whereas the percentage of sites exceeding critical limits for dissolved Cu concentrations is nearly 100%. This is an indication that more research is needed into reliable critical limits for soil, soil solution and plants. The results are also influenced by the reliability of soil–soil solution relationships and soil–plant relationships. More insight into this reliability can be obtained from an uncertainty analysis, studying the impact of the uncertainty in those relationships on the dynamics in leaching and accumulation of metals. Nevertheless, the major conclusions related to the slow dynamics and the change in leaching and accumulation behaviour of those metals in different farming systems are relevant both in The Netherlands and in comparable systems elsewhere.

References

Alloway, B.J. (1990) *Heavy Metals in Soils.* Blackie and John Wiley & Sons, Glasgow.

Andersson, A. (1992) *Trace Elements in Agricultural Soils: Fluxes, Balances and Background Values,* Report 4077. Swedish Environmental Protection Agency, Solna.

Bååth, E. (1989) Effects of heavy metals in soil on microbial processes and populations. A literature review. *Water, Air, and Soil Pollution* 47, 335–379.

Bahlsberg-Påhlsson, A.M. (1989) Toxicity of heavy metals (Zn, Cu, Cd, Pb) to vascular plants. A literature review. *Water, Air, and Soil Pollution* 47, 287–319.

Bengtsson, G. and Tranvik, L. (1989) Critical metal concentrations for forest soil invertebrates. A review of the limitations. *Water, Air, and Soil Pollution* 47, 381–417.

Bril, J. (1995) Transfer functions between adsorption constants and soil characteristics. In: Reinds, G.J., Bril, J., de Vries, W., Groenenberg, J.E. and Breeuwsma, A. (eds) *Critical and Present Loads for Cadmium, Copper and Lead for European Forest Soils,* Report 96. DLO Winand Staring Centre, Wageningen, The Netherlands, pp. 31–50.

Christensen, T.H. and Tell, J. Chr. (1991) *Sustainable Management of Heavy Metals in Agriculture; Example: Cadmium. Proceedings of International Conference on Heavy Metals in the Environment,* Vol. 1. CEP Consultants, Edinburgh, pp. 40–49.

Crommentuijn, T., Polder, M.D. and van de Plassche, E.J. (1997) *Maximum Permissible Concentrations and Negligible Concentrations for Metals, Taking Background Concentrations into Account.* RIVM report no 601501 001, National Institute of Public Health and the Environment, Bilthoven, The Netherlands.

De Boo, W. (1995) *Heavy Metal Balance for Dutch Agricultural Soils in 1992 and 2000: Consequences of the Waste Materials Policy* (in Dutch). Bleiswijk, The Netherlands.

De Haan, F.A.M. and Zwerman, P.J. (1978) Pollution of soil. In: Bolt, G.H. and Bruggenwert, M.G.M. (eds) *Development in Soil Science 5A: Soil Chemistry A. Basic Element.* Elsevier Science, pp. 192–271.

De Visser, P.H.B. and De Vries, W. (1989) *De Gemiddelde Waterbalans van Bos-, Heide- en Graslandvegetaties,* Rapport 2085. Stichting voor Bodemkartering, Wageningen.

De Vries, W. and Bakker, D.J. (1998) *Manual for Calculating Critical Loads of Heavy Metals for Terrestrial Ecosystems. Guidelines for Critical Limits, Calculation Methods and Input Data,* Report 166. DLO Winand Staring Centre, Wageningen, The Netherlands.

Ferdinandus, G. (1989) *Calculation of Heavy Metal Balances for Soils* (In Dutch). Report Technical Committee Soil Protection (TCB). The Hague, The Netherlands.

Fergusson, J.E. (1990) *The Heavy Elements. Chemistry, Environmental Impact and Health Effects.* Pergamon, Oxford.

Groot, M.S.M., Bronswijk, J.J.B., Willems, W.J., de Haan, T. and del Castilho, P. (1997) *National Soil Quality Monitoring Network: Results of 1994,* Report no. 714801017. National Institute for Public Health and the Environment (RIVM) Bilthoven, The Netherlands.

Groot, M.S.M., Bronswijk, J.J.B., Willems, W.J., de Haan, T. and del Castilho, P. (1998) *National Soil Quality Monitoring Network: Results of 1995,* Report no. 714801024. National Institute for Public Health and the Environment (RIVM) Bilthoven, The Netherlands.

Hanegraaf, M., van der Mey, T. and de Graaf, H. (1991) *Natuur en Milieu in Landbouwmodellen.* Milieubiologie, Leiden, R.U.

Hare, L. and Tessier, A. (1996) Predicting animal cadmium concentrations in lakes. *Nature* 380, 430–432.

Haygarth, P.M. and Jarvis, S.C. (1999) Transfer of phosphorus from agricultural soils. *Advances in Agronomy* 66, 195–249.

Hoekstra, C. and Poelman, J.N.B. (1982) *Dichtheid van Gronden Gemeten aan de Meest Voorkomende Bodemeenheden in Nederland,* Rapport nr. 1582. Stichting voor Bodemkartering, Wageningen.

Hoogervorst, N.J.P. (1991) *Het Landbouw-scenario in de Nationale Milieuverkenning 2; Uitgangspunten en Berekeningen.* RIVM-rapportnr. 251701005. National Institute for Public Health and the Environment (RIVM) Bilthoven, The Netherlands.

Hovmand, M.F. (1984) Cycling of Pb, Cu, Zn, and Ni in Danish agriculture. In: Berglund, S., Davis, R.D. and Hermite, P.L. (eds) Commission of the European Communities. *Utilization of Sewage Sludge on Land: Rates of Application and Long-term Effects of Metals.* Reidel Publishers, Dordrecht, The Netherlands, pp. 166–185.

Huinink, J. (1999) *Functiegerichte Bodemkwaliteitssystematiek. 2 Functiegerichte Bodemkwaliteits Waarden.* Informatie en Kennis Centrum Landbouw, Ede, The Netherlands.

Janssen, R.P.T., Pretorius, P.J., Peijnenburg, W.J.G.M. and van den Hoop, M.A.G.T. (1996) *Determination of Field-based Partition Coefficients for Heavy Metals in Dutch Soils and the Relationships of these Coefficients with Soil Characteristics,* Report 719101023. National Institute of Public Health and the Environment, Bilthoven, The Netherlands.

Kabata Pendias, A. and Pendias, H. (1984) *Trace Elements in Soils and Plants.* CRC Press, Boca Raton, Florida.

Ma, W. and van der Voet, H. (1993) A risk-assessment model for toxic exposure of small mamalian carnivores to cadmium in contaminated natural environments. *The Science of the Total Environment* (Supplement 1993), 1701–1714.

Mäkela-Kurtto, R. (1996) Heavy Metal Balances in Finnish Cultivated soils. Unpublished report on The Agricultural Research Centre of Finland.

McBride, M.B. (1994) *Environmental Chemistry of Soils.* Oxford University Press, Oxford.

Meili, M. (1997) Mercury in lakes and rivers. In: Sigel, A. and Sigel, H. (eds) *Mercury and its Effects on Environment and Biology,* Metal Ions in Biological Systems, Volume 34, Marcel Dekker Inc, New York, pp. 21–51.

Moolenaar, S.W. (1998) Sustainable management of heavy metals in agro-ecosystems. PhD thesis, Wageningen Agricultural University.

Moolenaar, S.W. (1999a) Heavy metal balances, part I. General aspects of cadmium, copper, zinc, and lead balance studies in agroecosystems. *Journal of Industrial Ecology* 2, 45–60.

Moolenaar, S.W. (1999b) Heavy metal balances, part II. Management of cadmium, copper, lead, and zinc in European agroecosystems. *Journal of Industrial Ecology* 3, 41–53.

Moolenaar, S.W. and Beltrami, P. (1998) Heavy-metal balances of an Italian soil as affected by sewage sludge and Bordeaux mixture applications. *Journal of Environmental Quality* 27, 828–835.

Moolenaar, S.W., and Lexmond, Th.M. (1998) Heavy metal balances of agroecosystems in The Netherlands. *Netherlands Journal of Agricultural Science* 46, 171–192.

OECD (1992) Report of the OECD workshop on the extrapolation of laboratory aquatic toxicity data to the real environment. *OECD Environmental Monographs* vol. 59, Paris.

Poppe, K.J., Brouwer, F.M., Mulder, M., and Welten, J.P.P.J. (1994) *Agriculture, Environment and Economics; data for 1990 and 1991* (in Dutch). LEI-DLO Periodieke rapportage, Den Haag, pp. 68–91.

Reiner, I., Lampert, C., Piterkova, M. and Brunner, P.H. (1996) *Stoffbilanzen Landwirtschaft-licher Böden von Ausgewählten Betriebstypen bei Verwendung von Klärschlamm und Kompost.* BKKB 2 Endbericht, Technical University, Vienna.

Römkens, P.F.A.M. and de Vries, W. (2001) Derivation of soil-specific quality standards for Cd and Zn. *Journal of Environmental Quality* (in press).

Römkens, P.F.A.M., Groenenberg, J.E., Tiktak, A., Bril, J. and de Vries, W. (2001) Derivation of partition relationships to calculate Cd, Cu, Pb, Ni and Zn solubility and activity in soil solutions. *Alterra Green World Research Report* No. 305 (in press).

Smilde, K.W. (1976) *Toxische Gehalten aan Zware Metalen (Zn, Cu, Cr, Ni, Pb en Cd) in Grond en Gewas; een Literatuuroverzicht.* Haren, IB-nota 25.

Smilde, K.W. (1986) *Zware-metaalgehalten van Kunstmeststoffen en Aanvoer van Zware Metalen via deze Meststoffen op Landbouwgronden,* Nota 154. Instituut voor Bodemvruchtbaarheid (IB), Haren, The Netherlands

Stoop, J.M. and Rennen, A.J.M. (1990) *Schadelijke Stoffen voor Land- en Tuinbouw: Lood.* Centrum voor landbouw en Milieu, Utrecht.

Stoop, J.M. and Rennen, A.J.M. (1991) *Schadelijke Stoffen voor Land- en Tuinbouw: Cadmium.* Centrum voor landbouw en Milieu, Utrecht.

Stoop, J.M., Leemans, R.J.D. and Rennen, A.J.M. (1992) *Schadelijke Stoffen voor Land- en Tuinbouw: Zink.* Centrum voor Landbouw en Milieu, Utrecht.

Stoop, J.M., Leemans, R.J.D. and Rennen, A.J.M. (1993) *Schadelijke Stoffen voor Land- en Tuinbouw: Koper.* Centrum voor Landbouw en Milieu, Utrecht.

Tyler, G. (1992) *Critical Concentrations of Heavy Metals in the Mor Horizon of Swedish Forests,* Report 4078. Swedish Environmental Protection Agency, Solna, Sweden.

Van Driel, W. and Smilde, K.W. (1990) Micronutrients and heavy metals in Dutch agriculture. *Fertilizer Research* 25, 115–126.

Van Jaarsveld, J.A. (1994) *Atmospheric Deposition of Cadmium, Copper, Lead, Benzo (a) pyrene and Lindane over Europe and its Surrounding Marine Areas*, Report 222401002. National Institute of Public Health and the Environment, Bilthoven, The Netherlands.

Van Jaarsveld, J.A. and Onderdelinden, D. (1993) *TREND; an Analytical Long-term Deposition Model for Multi-scale Purposes*, Report 228603009. National Institute of Public Health and the Environment, Bilthoven, The Netherlands.

Van Wallenburg, C. (1988) *De Dichtheid van Moerige Lagen*, Nota project 799. Stichting voor Bodemkartering, Wageningen, The Netherlands.

Van der Zee, S.E.A.T.M., Moolenaar, S.W. and Lexmond, Th.M. (1995) Heavy metal immision and fate according to a dynamic balance approach. In: van den Brink, W.J., Bosman, R., and Arendt, F. (eds) *Contaminated Soil '95*. Kluwer Academic Publishers, Dordrecht, The Netherlands, pp. 283–292.

Várallyay, G. (1993) *Soil Data Bases, Soil Mapping Soil Information – and Soil Monitoring Systems in Hungary*. National Report for the FAO/ECE International Workshop of Soil Conservation Monitoring Systems, Budapest, 14–17 September 1993. FAO.

Velthof, G.L., van Erp, P.J. and Moolenaar, S.W. (1996) Optimising fertilizer plans of arable farming systems. II. Effects of fertilizer choice on inputs of heavy metals. *Meststoffen 1996*, 74–80.

Witter, E. (1992) *Heavy Metal Concentrations in Agricultural Soils Critical to Microorganisms*, Report 4079. Swedish Environmental Protection Agency, Solna, Sweden.

6 Human Enteric Pathogens

D.L. Jones,[1] G. Campbell[2] and C.W. Kaspar[3]
[1] *School of Agricultural and Forest Sciences, University of Wales, Bangor, Gwynedd LL57 2UW, UK;* [2] *Department of Plant and Soil Science, University of Aberdeen, Aberdeen AB24 3UU, UK;* [3] *Department of Food Microbiology and Toxicology, Food Research Institute, University of Wisconsin, Madison, WI 53706-1187, USA*

Key words: *Campylobacter*, cattle, *Cryptosporidium*, drinking water, *Escherichia coli*, food poisoning, *Listeria*, pathogen, *Salmonella*, slurry

Introduction

Food-borne illnesses represent a significant public health problem with major economic and social consequences. Although difficult to fully assess in terms of health care and loss of working hours, estimates of the annual economic cost of food and water poisoning exceed £1 billion in the UK, US$23 billion in the USA and AUS$2 billion in Australia and New Zealand (Sockett and Roberts, 1991; CDNANZ, 1997; Jones, 1999). The emergence of food-borne pathogens such as strains of *Salmonella*, *Campylobacter* and *Escherichia coli* present major health concerns. Trends in food-borne illness in the developed world, including North America, Europe and Australia, clearly indicate that the incidence of food-borne infections is increasing and that it is likely to remain a public health threat well into the foreseeable future (Fig. 6.1). The increase in the number of food-borne infections has been attributed to a number of causal agents including the emergence of new pathogen strains (e.g. *E. coli* O157), changing patterns in food process-

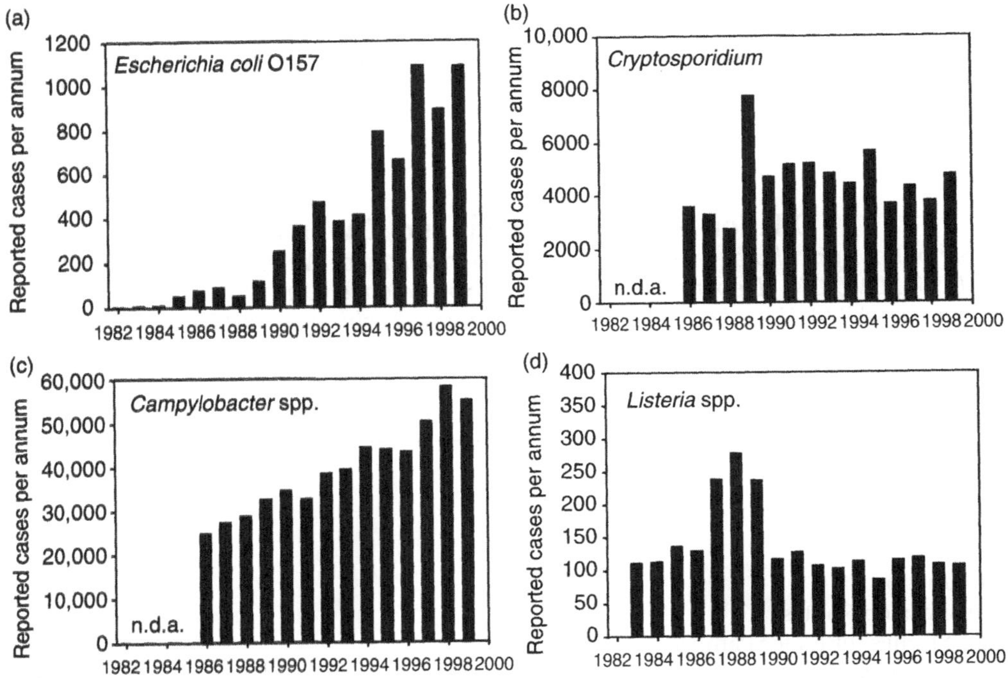

Fig. 6.1. Number of cases of food poisoning by enteric pathogens in England and Wales since 1982 (n.d.a. indicates no data available; data source: Public Health Laboratory Services, London).

ing, retailing and consumption, reduced immune systems and changes in farming practices (CDNANZ, 1997).

Food-borne pathogens are intrinsically linked to farming environments and can enter the food production chain or water supply network at many stages from farm gate to consumer plate. Animals and vegetation carry a large proportion of microbes at the time of harvest. Some of these organisms causing disease in humans do not cause disease in the gastrointestinal tract of animals and, therefore, detection often goes unnoticed until outbreaks occur. Particularly for foods of animal origin and watercourses close to farms, a low baseline of contamination is often impossible to avoid. However, this does not mean that control is not possible. It has been clearly demonstrated that pathogen populations are highly dependent on farm management strategy (e.g. organic versus conventional) and can vary widely over time and space (Jones, 1999). While the consumption of contaminated food remains the primary route of food-borne illness, the ingestion of contaminated water and the use of contaminated water for crop irrigation are commonly implicated pathogen transmission routes in rural areas. A trend towards eating fresh, unprocessed foods and processed foods in which preservatives have been omitted also appears to permit the growth of food-borne pathogens and has been linked with numerous outbreaks worldwide (Altekruse and Swerdlow, 1996). When this is twinned with an increasing proportion of the population possessing a heightened susceptibility to food-borne infection, control measures need to be set in place to deal with this issue. In particular, the young and old, pregnant women and immuno-compromised people are most at risk, while those people exposed to low levels of pathogens remain at low risk (i.e. those in farming or food processing environments; Ragletti and Untermann, 2000).

Transmission Routes

While undoubtedly the major transmission route of human pathogens is via the consumption of contaminated food, water also provides a common infection pathway. This may include the direct drinking of contaminated surface or groundwater, but also through subconscious consumption of water during recreational activity. The major routes by which water becomes contaminated with enteric pathogens are summarized in Fig. 6.1 and are described as follows:

- Contamination from horizontal overland flow of water from agricultural fields.
- Contamination of groundwater from the vertical movement of water which has previously percolated through agricultural fields.
- Contamination of recreational water by wild animals and birds which have previously visited contaminated areas.
- Contamination of surface and groundwater as a result of land spreading of organic wastes including:
  - animal slurry and manure
  - raw sewage or processed sewage sludges
  - food processing waste
  - abattoir waste
  - wastewater from dairy parlours, etc.

Although food and water can transmit a wide range of human pathogens, this review aims to address four of the common food- and water-borne pathogens in the developed world: *Campylobacter*, *Listeria monocytogenes*, *E. coli* O157 and *Cryptosporidium* (Table 6.1). Within this chapter, we will address the general ecology of these microorganisms in farming and aquatic environments, highlighting their environmental prevalence. We will also present evidence on their persistence, potential transmission routes and human impact.

Case Studies for Major Pathogens

Campylobacter

The Gram-negative motile bacterium *Campylobacter* was first isolated in 1913; however, its recognition as a human enteric pathogen was not fully realized until 1972 (Dekeyser *et al.*, 1972). Of the many recognized species of *Campylobacter*, *Campylobacter jejuni* is the most predominant cause of enteric disorders (90%) while *Campylobacter coli*, *Campylobacter lari*, *Campylobacter hyointestinalis* and *Campylobacter upsaliensis* have also been identified as causing disease in humans (USFDA, 1992; HMSO, 1993). *Campylobacter* is typically present in most environmental samples as a slender, spirally curved rod; however, under adverse environmental conditions it has the potential to become coccoid in shape and enter a viable but non-cultureable (VBNC) state (HMSO, 1993). Although some *Campylobacter* species are able to grow in ambient air, most of those responsible for enteric diseases require less O_2 than is present in air (termed microaerobic); optimal growth occurs at 3–10% O_2 and 2–0% CO_2 (Bolton and Coates, 1983; Jones *et al.*, 1993). The organism can also undergo limited anaerobic growth in some media (Takata *et al.*, 1995). *Campylobacter* is described as a thermophilic organism where growth occurs at temperatures ranging from 30 to 45°C reflecting the temperature of its primary animal hosts (HMSO, 1993). From this perspective, numbers of the pathogen typically decline

Table 6.1. Annual incidence of pathogens per 100,000 of the human population.

Pathogen	UK	USA	Australia
E. coli O157	2.0	2.4	< 1
Listeria spp.	0.2	0.5	0.4
Campylobacter spp.	98	20	100
Cryptosporidium	9	2.3	—

rapidly after release from the host into the environment. However, under certain conditions, prolonged survival can occur in some environmental niches.

Human pathophysiology of Campylobacter

In many countries, *Campylobacter* has become recognized as the major cause of food-borne illness (Table 6.1; HMSO, 1993). The annual number of reported *Campylobacter* cases has steadily risen over the last 20 years with *c.* 50,000 annual cases reported in the UK and 3,000,000 in the USA. The typical worldwide incidence rate is approximately 100:100,000; mortality occurs in 0.01–0.1% of cases (USFDA, 1992; HMSO, 1993). Human *Campylobacter* infections are typically bimodal with peaks in late spring and autumn (Skirrow, 1991; HMSO, 1993). After ingestion, *Campylobacter* multiplies in the lower gastrointestinal tract causing damage to gut epithelial cells and producing typical disease symptoms such as diarrhoea, intestinal cramps, vomiting and fever. The incubation period is 2–5 days with symptoms lasting from 1 to 7 days. Unlike *Salmonella,* where the infective dose is typically greater than 10^5 cells, ingestion of *c.* 500 cells is sufficient to cause symptom development (USFDA, 1992; HMSO, 1993). Typically, the severity of the symptoms is not correlated with infectious dose, although the initial likelihood of developing symptoms does appear to be (Robinson, 1981). Whether a particular individual develops clinical symptoms is determined by a number of factors. However, greatest outbreaks occur in immuno-compromised patients, the elderly, the newborn and pregnant women (HMSO, 1993). In developed countries, non-symptomatic carriage of the pathogen is rare. This contrasts with developing countries where carriage of the pathogen can persist for more than 1 month after infection (Kapperud *et al.*, 1992).

Reservoirs of Campylobacter *in farm animals*

The primary reservoir of *Campylobacter* in the environment is within the gastrointestinal tract of a wide variety of wild and domestic animals (cattle, pigs, sheep, cats, dogs and birds). In most adult animals, the presence of *Campylobacter* results in no detectable illness or pathology. However, in young animals both *C. jejuni* and *C. coli* are often associated with enteric disease (Stanley *et al.*, 1998a). As domestic and wild animals carry and excrete large numbers of *Campylobacter* there is a continuous flow of the pathogen into the environment.

Cattle are thought to be the largest reservoir of *Campylobacter* in agricultural environments and the main source of water contamination occurs by the transmission routes shown in Fig. 6.2. As cattle and pigs

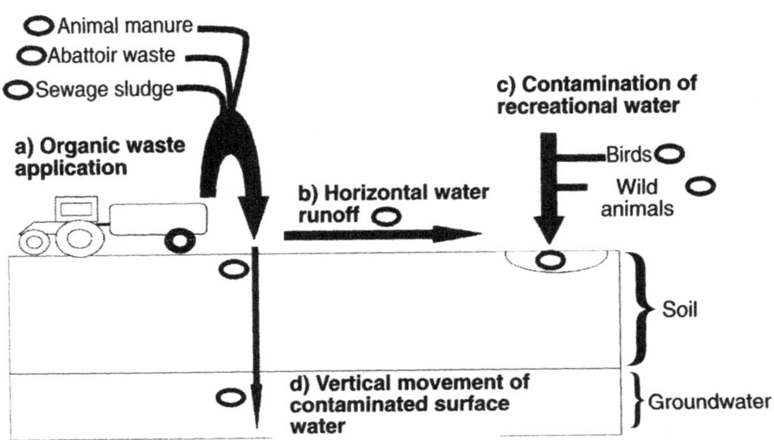

Fig. 6.2. Major routes by which water becomes contaminated with enteric pathogens (open ovals).

are born without the pathogen, infection of livestock is thought to occur primarily via the consumption of faecally contaminated water and vegetation (Weijtens *et al.*, 1997, 1999). Grazing herds with access to rivers or streams have been shown to have a greater incidence of infection than those drinking chlorinated water supplies (Humphrey and Beckett, 1987). Cattle herd infection rates typically vary from 70 to 90%; infection exhibits a distinct seasonality (bimodal), similar to the peaks reported for human infections (Koenraad *et al.*, 1997; Stanley *et al.*, 1998b). Sheep have also been shown to be widespread carriers of *Campylobacter*. One study reported that between 50 and 90% of UK sheep were infected with *Campylobacter* and that carriage was again seasonal with peaks in late spring/early summer (Stanley *et al.*, 1998a). *Campylobacter* can also be found in the intestinal tracts of a wide variety of birds and domestic poultry. Within the majority of these cases, *Campylobacter* is carried in a non-pathogenic, commensural association within the intestine (Palmgren *et al.*, 1997). Cross-infection can easily occur in domestic flocks as the infectious dose may be as low as 50 colony forming units (cfu) for birds (van de Giessen *et al.*, 1992). Typical rates of infection within the intestines and faeces of domestic and wild animals range from 10^3 to 10^9 cfu g^{-1} (Wallace *et al.*, 1997a,b).

Presence and survival of Campylobacter *in soil, vegetation and organic wastes*

As *Campylobacter* typically only multiplies at elevated temperatures, in most circumstances growth in soil and vegetation can be expected to be very low. However, in terms of survival, *Campylobacter* is able to remain viable in a range of environments at 4°C for up to 7 days (HMSO, 1993). Other environmental constraints include its inability to tolerate desiccation, low pH (< pH 5), exposure to O_2 and UV exposure (HMSO, 1993).

As contamination of pig and cattle faeces is common at rates of 10^2–10^6 cfu g^{-1}, it is not surprising that bio-solids can carry a heavy burden of pathogens when applied to land (Koenraad *et al.*, 1997; Weijtens *et al.*, 1997; Stanley *et al.*, 1998b). Despite this, limited pathogen survival (< 24 h) after application of contaminated organic waste to soil has been reported. Human faeces are also known to contain more than 10^6 cfu g^{-1} in infected cases and consequently *Campylobacter* can be commonly found in sewage and untreated water (Fricker and Park, 1989; Jones *et al.*, 1990). Land disposal of sewage sludge should, therefore, be considered as a potential transmission route. Untreated abattoir waste, which is also applied to land in many countries, has also been shown to be heavily contaminated with *Campylobacter* (> 10^3 ml^{-1}), and has been implicated in watercourse contamination (Jones, unpublished; Jones *et al.*, 1990).

Survival of Campylobacter *in water*

As a consequence of faecal contamination, *Campylobacter* has regularly been isolated from both surface waters and seawater and shown to be responsible for many human disease outbreaks (HMSO, 1993; Koenraad *et al.*, 1997). *Campylobacter* survival rates in freshwaters can be up to 4 months with survival greatest at 4°C (Rollins and Colwell, 1986; Thomas *et al.*, 1999). Survival time, however, is highly dependent on strain type, previous growth conditions, water quality and environmental conditions (Fig. 6.3; Buswell *et al.*, 1998, 1999). Seasonality in the occurrence of *Campylobacter* has also been demonstrated in surface water samples. Interestingly, however, peaks do not correspond with those of human infection (Jones *et al.*, 1990). Aeration also influences survival time, with greatest persistence in non-aerated water (Rollins and Colwell, 1986; Chynoweth *et al.*, 1998). In freshwater, *Campylobacter* can enter a VBNC state, a response thought to be linked to starvation, sub-optimal temperatures and unfavourable environmental conditions. While in the VBNC state, cells maintain a basal metabolic activity but are incapable of undergoing growth on selective growth

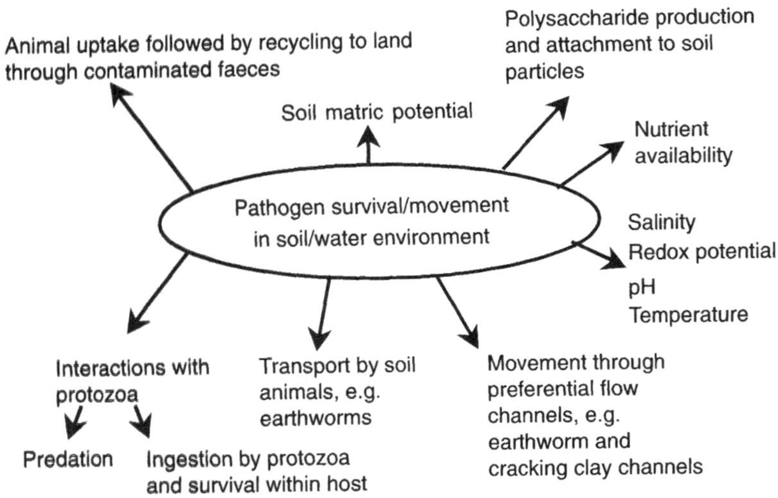

Fig. 6.3. Schematic representation of the major factors affecting enteric pathogen persistence in the environment.

media. There is also some evidence to suggest that *Campylobacter* can remain infectious in the VBNC state although this remains controversial (Rollins and Colwell, 1986; Jones *et al.*, 1991; Koenraad *et al.*, 1997). In a number of European studies measurement of *Campylobacter* contamination of wastewaters indicated colony numbers of between 0.5 and 500 cfu ml^{-1} (Koenraad *et al.*, 1994).

As well as surface water, groundwater contamination has been implicated in the longitudinal spread of *Campylobacter*. In some cases, contamination has been linked directly to nearby dairy farms (Stanley *et al.*, 1998c). Environmental conditions in subsurface aquifers may also favour the survival of *Campylobacter* because of the low redox potential, low O_2 concentrations, low temperature and protection from desiccation and UV irradiation (Fig. 6.3; Stanley *et al.*, 1998c).

Listeria monocytogenes

Listeria can be found worldwide, occupying many environmental niches including the intestines of most domestic and wild animals within agricultural ecosystems (Welshimer, 1968). *Listeria* is also common in both freshwater and marine environments. Although most outbreaks of listeriosis are associated with the consumption of contaminated milk and dairy products, a causal link between agriculture and water contamination has been implicated on many occasions (Beuchat and Brackett, 1990; Bibb *et al.*, 1990; van Renterghem *et al.*, 1991; Garcia *et al.*, 1996). *L. monocytogenes* is the dominant cause of human infection; however, other species including *Listeria innocua, Listeria ivanovi* and *Listeria seeligeri* may also cause illness. *Listeria* is described as an aerobic to facultatively anaerobic non-spore forming Gram-positive rod shaped bacterium that is psychotrophic, giving the potential to multiply at low temperatures (2–5°C).

Human pathophysiology of Listeria

Although *L. monocytogenes* was first identified over 100 years ago, the first human outbreak of listeriosis was not documented until 1981. This outbreak was associated with the consumption of sheep manure-contaminated salad vegetables (Donnelly and Baigent, 1986). *L. monocytogenes* causes the disease listeriosis in humans, which clinically manifests itself in a range of symptoms including septicaemia, meningitis, abortion, encephalitis and 'flu-like' illness. Typically, severe listeriosis is fatal in about 20–30% of reported cases. Severe listeriosis infections normally fall into at least three categories: (i) pregnant women, with infection frequently undetected until abortion; (ii) the newborn; and (iii), the older generation or immuno-suppressed patients. For this reason *L. monocytogenes* is often seen as an opportunist as symptoms rarely develop in healthy individuals. It has been proposed that healthy humans regularly harbour *L. monocytogenes* at least transiently in the gut and an estimated 10 to 20% of the population carries the pathogen at any one time (Welshimer, 1968; Watkins and Sleath, 1981). The incidence of listeriosis in the developed world is approximately 1:200,000 (Table 6.1); levels have remained relatively constant over the last 20 years. In the USA, where there are approximately 1800 cases reported annually, it has been estimated that the disease is responsible for a loss of US$200–300 million to the US economy.

Reservoirs of Listeria *in farm animals*

Typically, 50–80% of farm environments are contaminated with *L. monocytogenes*; many of the isolated strains are pathogenic. However, strains that are pathogenic to humans are commonly non-lethal to animals (Welshimer, 1968). As *L. monocytogenes* is present in the intestines of some farm animals, it is not surprising that it is also present in the faeces of these animals at rates > 10^2 cfu g^{-1}. *L. monocytogenes* is frequently described as a facultative intracellular parasite that can be shed in the milk of cows as well as in faeces. Studies looking at the incidence of *L. monocytogenes* in faeces of farm animals have typically shown that 2–50% of pig, cattle and sheep faeces are contaminated with the pathogen (Skovgaard and Morgen, 1988; Skovgaard and Nørrung, 1989; van Renterghem *et al.*, 1991). As with other human food-borne pathogens such as *E. coli* O157 and *Campylobacter* spp., *L. monocytogenes* appears to show seasonal shedding from animals; higher numbers are present in faecally contaminated vegetation between July and September (MacGowan *et al.*, 1994). The degree of pathogen contamination in faecal material also appears to be highly correlated with the type of feedstuff. Inferior quality feed most likely to cause clinical outbreaks in animals has been associated with wet hay, poorly stored silage and vegetables (Grønstal, 1980). In contrast, dry feed in which *L. monocytogenes* does not proliferate under normal storage is less likely to cause outbreaks (Skovgaard and Nørrung, 1989).

Presence and survival of Listeria *in soil, vegetation and organic wastes*

In agreement with the information presented above (i.e. a widespread incidence of *L. monocytogenes* within farming environments), surveys have shown a similar widespread degree of soil contamination. *L. monocytogenes* can be frequently detected in soil, associated litter and vegetation, indicating that the pathogen can exist in the wider environment as a saprophyte on both soil organic matter and plants (Weiss and Seelinger, 1975). Typical concentrations in soil range from 1 to 500 cfu g^{-1} (MacGowan *et al.*, 1994; Dowe *et al.*, 1997). The greatest numbers of *L. monocytogenes* isolated from soil have also been shown to be correlated with the highest numbers in contaminated cattle faeces (MacGowan *et al.*, 1994). Based on these observations, it has been suggested that *L. monocytogenes* is as common in soil as the almost ubiquitous *Bacillus* spp. Studies on *L. monocytogenes* survival in soil have revealed that the pathogen can persist for up to 10 months and grow in soil when applied in contaminated waste samples (Dowe *et al.*, 1997). Generally, greatest survival is observed in non-sandy, fertile soils held close to field capacity (Welshimer, 1960; Watkins and Sleath, 1981; Dowe *et al.*, 1997). Persistence of *L. monocytogenes* for up to 10 years after incorporation of animal residues in soil has been reported (Welshimer, 1968). In contrast, however, other studies have indicated that *L. monocytogenes* is incapable of surviving in soil for longer than 2 months and that for certain strains, soil is not a natural reservoir (van Renterghem *et al.*, 1991).

Surveys of farm vegetation indicate that pathogen contamination is common: up to 10–30% of crops can test positively for *L. monocytogenes* (Weiss and Seelinger, 1975; Heisick *et al.*, 1989). Although the optimum temperature for pathogen growth is between 30 and 37°C, *L. monocytogenes* can grow and survive over a wide temperature range (2–45°C). One feature of *L. monocytogenes* is its ability to grow quickly when stored at low temperatures (2–5°C) on vegetation. Cellular doubling has been reported to take place within 4 days and survival and growth at low temperatures exceeds 2 months (Beuchat *et al.*, 1986; Sizmur and Walker, 1988; Steinbruegge *et al.*, 1988; Jones *et al.*, 1997). Pathogen survival, however, is plant dependent; crops such as carrots are less susceptible to *L. monocytogenes* than other root crops (e.g. potatoes), possibly because of the release of anti-microbial compounds such as phenolics (Beuchat and Brackett, 1990). Further, crops not in contact with soil are also less likely to become contaminated with *L. monocytogenes* (Heisick *et al.*, 1989). Survival on vegetation is also dependent on initial contamination load and the indigenous microflora (Francis and O'Beirne, 1998).

Many organic wastes applied to land, including sewage sludge, farm and industrial wastewater, food processing waste, abattoir waste, animal manures and slurries, contain *L. monocytogenes* (Watkins and Sleath, 1981; Al-Ghazali and Al-Azawi, 1986; MacGowan *et al.*, 1994). Where contamination has been detected, *L. monocytogenes* numbers typically range from 10^2 to 10^5 cfu ml^{-1}, with numbers often exceeding those of other pathogens such as *Salmonella* (Watkins and Sleath, 1981; Fenlon, 1985). Al-Ghazali and Al-Azawi (1988) showed that *L. monocytogenes* can survive in stored sewage cake for up to 1 year and that, after land application of *L. monocytogenes*, pathogen levels remain constant for up to 2 months (Watkins and Sleath, 1981).

L. monocytogenes has also been detected frequently in grass silage, and a causal link between listeriosis in animals and the production of badly fermented silage is well established (Fenlon, 1985). In poorly prepared silage, where the pH rises above 4.2, *L. monocytogenes* can proliferate, whereas in properly prepared silage the production of lactic acid and subsequent pH drop (< 4.2) suppresses pathogen growth (Grønstal, 1979; Fenlon, 1985; Beuchat *et al.*, 1986). In contrast, *L. monocytogenes* survival in cattle and pig manure indicates a rapid decline until the pathogen is undetectable within 8 weeks (van Renterghem *et al.*, 1991).

Birds within farm environments have also been implicated as a potential transmission route for *L. monocytogenes* within agroecosystems and transfer from farm to surrounding water bodies (Fig. 6.2; Garcia *et al.*, 1996). In particular, seagulls have been implicated in lateral pathogen transfer. In one study, 5–30% of birds were carriers of *L. monocytogenes* with higher incidence on those at sewage disposal sites (Fenlon, 1985). It is well known that birds will forage in grassland, possibly picking up or depositing the pathogen. If grass is used for silage production, cattle may become infected.

Survival of Listeria *in water*

L. monocytogenes has been detected in a range of waters including freshwater, estuarine and marine environments with typical concentrations ranging from 1 to 10^2 cfu ml^{-1} (Frances *et al.*, 1991; Monfort *et al.*, 2000). Water has also been implicated as a transmission route in human and animal outbreaks of listeriosis (Jemmi and Keusch, 1994; Garcia *et al.*, 1996). *L. monocytogenes* can survive for 10–30 days in tap water at 30°C and for 7–110 days at 5–10°C (Varma and Lyer, 1993). In pond water, *L. monocytogenes* has been reported to survive for between 7 and 60 days while similar values have been obtained for seawater. Survival is, however, highly dependent on strain, temperature and environmental conditions (Fig. 6.3; Botzler *et al.*, 1975; Franco *et al.*,

1990; Bremer *et al.*, 1998). Survival is greatest at low temperatures where there is little competition from other microorganisms and predation (Monfort *et al.*, 2000). *L. monocytogenes* can also enter a VBNC state in seawater (Bremer *et al.*, 1998).

Escherichia coli *O157:H7*

Most strains of *E. coli* are harmless commensals found in the intestinal tracts of warm-blooded animals. However, of the 250+ serotypes identified, approximately one-third are known to cause illness in animals or man. Diarrhoeagenic *E. coli* are divided into five classes based on their virulence attributes as well as clinical symptoms and pathogenic features, and are described in detail in Nataro and Kaper (1998). *E. coli* O157:H7 and other serotypes (i.e. O111:NM and O26:H11) belong to the group of pathogenic *E. coli* referred to as enterohaemorrhagic *E. coli*. Most outbreaks involving enterohaemorrhagic *E. coli* in North America and Europe involve serotype O157:H7 strains (Griffin, 1995; Thomas *et al.*, 1996), although other serotypes have been associated with outbreaks (Griffin, 1995). Serotype O157:H7 strains have several distinguishing characteristics from other *E. coli* that assist in their isolation and identification.

In addition to serological differences from other *E. coli*, serotype O157:H7 strains grow slowly or not at all at >42°C in selective media, do not ferment sorbitol, or produce functional β-glucuronidase (Doyle and Schoeni, 1984). *E. coli* O157:H7 is not heat resistant (D_{145} = 24 seconds) compared with *Salmonella* (D_{145} = 42 seconds) but survives in frozen ground beef (−20°C) for 12 months with little or no decline in numbers (Doyle and Schoeni, 1984; Ansay *et al.*, 1999). Although primarily used in epidemiological investigations, molecular typing methods such as insertion fingerprinting, genomic fingerprinting and ribotyping can also discriminate serotype O157:H7 strains from other *E. coli* as well as genomic variation between strains within this serotype (Arbeit *et al.*, 1990). Virulence determinants also separate *E. coli* O157:H7 from other *E. coli* (Kaper and O'Brien, 1998). Collectively, these characteristics of *E. coli* O157:H7 have facilitated biochemical identification and the development of DNA probes and primers for its detection (Samadpour *et al.*, 1990).

Clinical manifestations and epidemiology of E. coli *O157:H7*

E. coli O157:H7 causes a spectrum of illnesses in humans ranging from mild diarrhoea to haemorrhagic colitis, and life-threatening haemolytic uraemic syndrome (HUS) and thrombotic thrombocytopenic purpura (Griffin and Tauxe, 1991; Griffin, 1995). In diagnosed cases, patients usually experience non-bloody diarrhoea 1–2 days following infection which progresses to bloody diarrhoea that lasts between 4 and 10 days (Tarr, 1995). Approximately 10% of children under the age of 10 years that are culture-positive for *E. coli* O157:H7 develop HUS, a leading cause of acute kidney failure in children in the United States (Griffin and Tauxe, 1991; Griffin, 1995). Children with HUS often require dialysis (50% of HUS patients) and there is 3–5% mortality. Thrombotic thrombocytopenic purpura more commonly occurs in older adults and involves the clinical manifestations of HUS plus neurological symptoms (Griffin, 1995). The complications associated with *E. coli* O157:H7 infection contribute to the substantial economic impact associated with this pathogen. Annual cost estimates are US$400–900 million in the USA (Roberts and Buzby, 1998) and £30 million in the UK (Jones, 1999).

Ground beef and beef remain the foods most frequently associated with outbreak of disease. A variety of other foods have also been confirmed or are suspected to transmit this pathogen, including: apple cider, lucerne sprouts, cantaloupe, coleslaw, cheese curds, fermented sausage, lettuce, mayonnaise, milk, salad and yoghurt (Besser *et al.*, 1993; Morgan *et al.*, 1993; Weagant *et al.*, 1994; CDC, 1995). Surveillance data from 1998 in the USA identified 42 confirmed outbreaks of *E. coli* O157:H7 infection, which is an increase from the average of 31 outbreaks year^{-1} occurring from 1994 to 1997 (CDC, 1999a,b). Ground beef was the suspected or confirmed vehicle in 11 outbreaks and other foods accounted for 10 outbreaks. Six outbreaks occurred in day care centres and four water-borne outbreaks were reported (CDC, 1999a,b). The outbreaks in day care centres and those involving swimming pools demonstrate the importance of person-to-person transmission and humans as a source of this pathogen.

Person-to-person and water-borne transmission of *E. coli* O157:H7 indicate that the infectious dose for this pathogen is low, perhaps as low as 10–100 cells. This is supported by data from epidemiological investigations which have found approximately 10–1000 cells of *E. coli* O157:H7 per gram of raw ground beef linked with an outbreak (CDC, 1993). Gorden and Small (1993) proposed that acid tolerance is a contributing factor to the infectious dose of a pathogen. Acid-tolerant pathogens such as *Shigella* and *E. coli* O157:H7 have a low infectious dose because of their ability to survive the acidic conditions of the stomach. Acid-sensitive pathogens, such as some serotypes of *Salmonella*, require 10^5 to 10^{10} cells to cause infection while < 500 shigellae can result in illness. Acid tolerance in *E. coli* O157:H7 has an important role during passage through both cattle and mice (Price *et al.*, 2000). In addition to gastric passage, acid tolerance contributes to the ability of this pathogen to survive in low-pH foods such as apple cider, fermented sausage, fruits and mayonnaise (Besser *et al.*, 1993; Miller and Kaspar, 1993; Weagant *et al.*, 1994). The ability of *E. coli* O157:H7 and other pathogens to tolerate acid conditions is likely to be a major factor contributing to survival or proliferation in the intestinal tract of animals or low pH environments (i.e. silage).

Sources and distribution of E. coli *O157:H7*

A number of animals can carry *E. coli* O157:H7 including birds, cattle, deer, flies, raccoon, sheep and humans (Faith *et al.*, 1996; Wallace *et al.*, 1997b; Shere *et al.*, 1998). Cattle are thought to be the primary reservoir of *E. coli* O157:H7 strains because of outbreaks and cases associated with the consumption of ground beef (CDC, 1999a,b). However, the contribution of asymptomatic human carriers should not be overlooked and would be consistent with humans as a primary host for the other groups of pathogenic *E. coli* (Tarr, 1995). In cattle, the prevalence of *E. coli* O157:H7 is approximately 2%; 7% of the herds contain one or more animals shedding the organism (Hancock *et al.*, 1994; Faith *et al.*, 1996). A recent study by Elder *et al.* (2000) found *E. coli* O157 or O157:H7 in at least one animal in 72% of feed lots sampled in the Midwestern USA, and the overall prevalence in faeces was 28%. These results are difficult to compare with those obtained solely for serotype O157:H7 and to determine their public health significance because *E. coli* O157:H7 is infrequently associated with human illness. Likewise, surveys for Shiga-like toxin-producing *E. coli* other than serotype O157:H7 strains have found that they are ubiquitous and prevalent in foods, particularly ground beef (Read *et al.*, 1990; Griffin, 1995). The significance of Shiga-toxin-producing *E. coli* other than serotype O157:H7 strains in animals and foods requires further investigation.

There have been a number of studies that have evaluated management practices for a correlation with the presence of *E. coli* O157:H7 in cattle (Orskov *et al.*, 1987; Hancock *et al.*, 1994; Faith *et al.*, 1996; Hancock *et al.*, 1997; Herriott *et al.*, 1998; Shere *et al.*, 1998). The grouping of weaned calves (Shere *et al.*, 1998) and, in one study, manure spreading, have been positively correlated (Hancock *et al.*, 1994). However, a more recent study did not find manure-handling practices associated with the presence of *E. coli* O157:H7 (Hancock *et al.*, 1997). Animal drinking water contaminated with *E. coli* O157:H7 is one mode of transmission of this organism between cattle (Faith *et al.*, 1996; Hancock *et al.*, 1998; Shere *et al.*, 1998). As a possible intervention strategy, frequent cleaning of water tanks and the use of high-turnover water tanks are being evaluated as potential methods for decreasing the transmission and prevalence of O157:H7 in cattle. A complicating factor in on-farm studies is the nature of *E. coli* O157:H7 shedding in cattle, which is typically sporadic, at low numbers and of limited duration (Besser *et al.*, 1997; Hancock *et al.*, 1997; Shere *et al.*, 1998). The numbers of *E. coli* O157:H7 in cattle faeces ranged from 2.0×10^2 to 8.7×10^4 cfu g^{-1} and continued for 1–16 weeks (Besser *et al.*, 1997; Shere *et al.*, 1998). These data indicate that *E. coli* O157:H7 is a transient member of the microbial flora of cattle and similar observations have been made for other strains of *E. coli* (Jarvis *et al.*, 2000). In the study by Jarvis *et al.* (2000), cattle harboured five genotypically different *E. coli*, determined using pulsed-field gel electrophoresis, which varied between animals and diet. Thus, sporadic shedding of *E. coli* O157:H7 by cattle for a limited duration may not be unique. Despite the comparatively low numbers of *E. coli* O157:H7 in cattle faeces, the low infectious dose of this human pathogen is problematic and creates multiple challenges to the development of effective waste management practices.

Survival of E. coli *O157:H7 in manure, soil and on vegetation*

The application of cattle faeces containing *E. coli* O157:H7 to fields has the potential to contribute to the propagation and environmental spread of this bacterium. Human health can be impacted if human food crops come in contact with contaminated animal faeces. Two studies (Wang *et al.*, 1996; Kudva *et al.*, 1998) on the survival of *E. coli* O157:H7 in animal faeces found that the pathogen can persist for > 30 days at 22–23°C and survival is extended at lower temperatures. In addition, *E. coli* O157:H7 survived for 21 months in a manure pile stored outside under fluctuating environmental conditions (Kudva *et al.*, 1998). Survival of the pathogen was lower in faecal slurries and was undetectable after 5 days of incubation at 23°C (Kudva *et al.*, 1998). While storage of manure slurries reduced the number of *E. coli* O157:H7, storage during winter months may be less effective because lower temperatures promote greater pathogen survival.

Information related to the survival or migration of *E. coli* O157:H7 in soils is limited. In a study by Gagliardi and Karns (2000), *E. coli* O157:H7 was able to replicate in, and migrate through, cores of various soil types. Numbers of the pathogen in leachate correlated with NH_4^+ and NO_3^- concentrations, and the numbers exceeded inoculum levels in all treatments (i.e. soil types, tilled and no till, and rainfall amounts) except in intact clay loam cores (Gagliardi and Karns, 2000). Survival of non-pathogenic *E. coli* exceeded 60 days at 25°C and 100 days at 4°C (Bogosian *et al.*, 1996) and may be extended beyond that by residing within soil protozoa (Barker *et al.*, 1999).

In comparison with the paucity of information on the survival of *E. coli* O157:H7 in soils, there is much known about its survival in foods, including fruits and vegetables (Abdul-Raouf *et al.*, 1993; Besser *et al.*, 1993; Beuchat, 1999; Janisiewicz *et al.*, 1999). Outbreaks of this pathogen involving apple cider (Besser *et al.*, 1993; Miller and Kaspar, 1993) focused attention on the acid tolerance properties of this organism, as noted above, and its ability to survive in acidic foods (i.e. fruits). Numbers of *E. coli* O157:H7 declined only by approximately 1 $\log_{10}$ cfu ml^{-1} in refrigerated apple cider (pH 3.8) after 14 days (Besser *et al.*, 1993; Miller and Kaspar, 1993). There are reports that *E. coli* O157:H7 can replicate in wounds of Golden Delicious apples (Janisiewicz *et al.*, 1999). Contamination of apples may occur by flies (Janisiewicz *et al.*, 1999) or immersion of apples in contaminated water (Buchanan *et al.*, 1999). *E. coli* O157:H7 also replicates on contaminated cantaloupe and watermelon stored at 25°C, and numbers remain constant on fruit stored at 4°C. Similarly, *E. coli* O157:H7 numbers increase on shredded lettuce and sliced cucumber stored at 12°C and 21°C although decline in culturable numbers is observed at 5°C (Abdul-Raouf *et al.*, 1993). *E. coli* O157:H7 survived on inoculated lettuce for 15 days of storage at 4°C, even when the initial inoculum was 100 cfu g^{-1} (Beuchat, 1999). In addition, *E. coli* O157:H7 is able to replicate to greater than 10^6 cfu g^{-1} during sprouting over a 48 h period, thereafter numbers remained unchanged during cold storage (Taormina and Beuchat, 1999). In summary, these studies demonstrate that contamination of fruit and vegetable crops with *E. coli* O157:H7 contaminated faeces should be avoided.

Survival of E. coli *O157:H7 in water*

Two recent outbreaks of water-borne *E. coli* O157:H7 have raised concerns about the contamination of well water, particularly after heavy rain. Preliminary findings have implicated contaminated well water as the source of *E. coli* O157:H7 in outbreaks in New York State in 1999 and Ontario, Canada, in 2000 (CDC, 1999a,b; Associated Press, 2000). The New York State outbreak occurred at a county fair where 921 people reported diarrhoea after attending the fair. Eleven children developed HUS and two people died. Drinking water and beverages made with water from a suspect well were linked with illness (CDC, 1999a,b). Although the outbreak and statistics associated with the Ontario outbreak are still being compiled, at least seven people have died and 2000 others have been infected from contaminated drinking water.

The survival of *E. coli* O157:H7 in water is temperature dependent and analogous to persistence in faeces, soil and foods (Fig. 6.3). At low temperature (8°C), numbers are reported to decrease by approximately 1–100 cfu ml^{-1} after 91 days, whereas, O157:H7 cells are detectable after 49–84 days at 25°C (initial number 10^3 cfu ml^{-1}) (Wang and Doyle, 1998). Survival in water is also dependent on the source of the water and the biotic community. In bottled water, for example, *E. coli* O157:H7 survives at 22°C for > 300 days depending on the water source (Warburton *et al.*, 1998). The ability of *E. coli* O157:H7 to persist in water, combined with its low infectious dose, contributes to the public health concerns associated with this pathogen and the need to manage faecal contamination of natural waters by point and non-point sources.

Cryptosporidium

Cryptosporidium parvum is a parasitic protozoan that inhabits a broad host range of animals (Dubey *et al.*, 1990; Graczyk *et al.*, 1998). It is disseminated to new hosts or humans by the faecal–oral route, and commonly involves contaminated water and in some cases food (Goodgame, 1996). Oocysts are the infectious stage of the *Cryptosporidium* life cycle and are shed in the faeces of infected animals and humans. Following ingestion by a suitable host, the oocysts pass to the small intestine where they excyst. Sporozoites emerge that invade the epithelial cells of the small intestine and form spheroidal trophozoites and then a schizont. There are several nuclear divisions followed by merozoites budding from the schizont surface which emerge from the host cell. The merozoites then infect other enterocytes and complete several cycles of reproduction (schizogony). Eventually, merozoites enter a sexual reproduction cycle whereby male and female gametocytes are formed. Fertilization results in the formation of zygotes that develop into oocycsts which are passed in the faeces. Oocysts are shed in faeces at concentrations as high as 10^6 g^{-1}. Oocysts are destroyed by temperature extremes (Dubey *et al.*, 1990) and freezing is detrimental to survival (Fayer and Nerad, 1996). However, oocysts tolerate chlorine, as demonstrated in laboratory studies, leading to outbreaks involving chlorinated drinking water (Fayer, 1995).

Several methodologies have been devised for the detection of oocysts in faeces, food, soil and water (Kuczynska and Shelton, 1999) including several PCR assays capable of detecting 1–10 oocycsts (Laberge *et al.*, 1996; Di-Giovanni *et al.*, 1999). In contrast to bacteriological methods that use selective enrichment to increase numbers of bacterial pathogen, concentration methods are employed for oocysts because they do not replicate in culture media (Musial *et al.*, 1987; Di-Giovanni *et al.*, 1999; Kuczynska and Shelton, 1999).

Clinical manifestations and epidemiology

As with most human pathogens, the symptoms presented by *C. parvum* infection are host dependent (i.e. age, immune status, etc.) and range from asymptomatic infection to severe and persistent diarrhoea that lasts for months. Symptoms sometimes include fever, nausea, and vomiting in addition to diarrhoea. Infection is primarily limited to the intestinal tract, particularly the small intestine, but the colon and liver may be affected in some individuals, especially the immuno-compromised. In a majority of cases, diarrhoea is self-limiting and resolved within 2–3 weeks. *C. parvum* oocysts are not host specific and are capable of infecting a variety of vertebrates including birds, cats, dogs, ferrets, opossums, pigs, raccoons, reptiles, rodents, ruminants and humans (De Graaf *et al.*, 1999; Sreter and Varga, 2000).

C. parvum is estimated to cause 300,000 illnesses and 66 deaths each year in the USA (Mead *et al.*, 2000). Water-borne *Cryptosporidium* is responsible for the greatest percentage of such outbreaks (MacKenzie *et al.*, 1994; PHLS Communicable Disease Surveillance Centre, 1996; Regan *et al.*, 1996). In 1993, over 400,000 people were involved in a water-borne outbreak of cryptosporidiosis in the state of Wisconsin (MacKenzie *et al.*, 1994). This massive outbreak was caused by *Cryptosporidium* oocysts that passed through the filtration system of one water treatment plant. Food has only recently been established as a vehicle of transmission for *Cryptosporidium* (Millard *et al.*, 1994; Besser-Wiek *et al.*, 1996). Person-to-person transmission may also occur, particularly in day-care facilities. However, water contaminated with oocysts from agricultural overland flow or sewage is responsible for the majority of the cases of cryptosporidiosis.

Sources and distribution of Cryptosporidium

Cryptosporidium is carried by a wide variety of animals (Graczyk *et al.*, 1998; De Graaf *et al.*, 1999), which contributes to the wide distribution of oocysts found in the environment. Infectious oocysts are frequently present in lakes, rivers and streams (Rose, 1988; Lechevallier *et al.*, 1991; Rose *et al.*, 1991; Lisle and Rose, 1995) and this prevalence is primarily responsible for water-borne cryptosporidiosis.

Food-borne cryptosporidiosis was only recently documented and involved unpasteurized apple cider (Millard *et al.*, 1994). It is thought that the cider became contaminated from apples that had fallen on the ground in a cow pasture. Another outbreak in 1995 involved chicken salad that was prepared in a day-care home by a food handler who had recently changed an infant's nappy that was asymptomatic (Besser-Wiek *et al.*, 1996). *C. parvum* oocysts are inactivated by pasteurization and are sensitive to desiccation and freezing (Robertson *et al.*, 1992). Thus, oocysts are inactivated in foods with appropriate heating or freezing but can persist on fresh foods such as fruits and vegetables. For example, a survey of fresh vegetables in Costa Rica (Monge and Chinchilla, 1996) found *Cryptosporidium* oocysts in samples of cilantro (5.0%), lettuce (2.5%) and other vegetables (carrot, cucumber, radish and tomato, 1.2%).

Survival of Cryptosporidium *in manure and soil*

The survival of *C. parvum* oocysts in manure and soil is influenced by the biota, freeze–thaw cycles, temperature and water potential (Fig. 6.3; Jenkins *et al.*, 1999). Exposure of oocysts to soil or animal waste materials (piles), temperatures between 35 and 50°C, and decreased water activity decreases viability of oocysts. Additionally, freeze–thaw cycles reduce numbers of infectious oocysts to < 1% while control suspensions maintain 50% infectious oocysts (Jenkins *et al.*, 1999). Oocysts (applied at 10^8 oocysts per core) are capable of migrating through irrigated soil cores of clay loam and silty loam soils but not loamy sand soil (Mawdsley *et al.*, 1996a). The majority of the oocysts are found in the top 2 cm of soil and, in the study by Mawdsley *et al.* (1996b) that employed low-permeability soil blocks supported at an angle of 7.5° and irrigated at regular intervals, oocysts were detected in overland flow (runoff) for 21–70 days from the time of inoculation. The number of oocysts decreased with distance from the inoculation point. At 70 cm or greater from the point of inoculation, oocysts could not be detected in soil at any depth. These data demonstrate that oocysts remain primarily suspended in water and that overland flow must be controlled to prevent contamination of waterways.

Survival of Cryptosporidium *in water*

Cryptosporidium oocysts survive for several months in surface waters (Medema *et al.*, 1997; Fayer *et al.*, 1998a), but survival is influenced by several environmental factors including predators, salinity, sunlight and temperature as noted above. Temperature is particularly important to the viability of oocysts in water; optimal survival occurs between 0 and 20°C (Fayer *et al.*, 1998b). The time necessary for a 1-log reduction in numbers of viable oocysts in river water was 40–160 days at 15°C and 100 days at 5°C. Similarly, *C. parvum* oocysts remained infectious for 4–12 weeks in artificial seawater microcosms of varying salinities that were incubated at 10 or 20°C (Fayer *et al.*, 1998b). Higher salinity and temperature decreased the length of time oocysts remained viable. In another study involving artificial seawater, oocysts were infectious for 1 year when incubated at 6–8°C (Tamburrini and Pozio, 1999). Oocysts are removed from water by filter feeders, such as oysters and mussels, and remain infectious (Fayer et al., 1998a; Tamburrini and Pozio, 1999).

One challenge to the control or inactivation of *Cryptosporidium* oocysts in water is their resistance to chlorine. Fayer *et al.* (1998a) demonstrated that oocysts remained infective following exposure to 5.25% sodium hypochlorite for 2 h. Exposure of oocysts to environmental conditions encountered in water does not increase the effectiveness of chlorine inactivation (Chauret *et al.*, 1998). Ozone treatment of water is more effective: 1.1 mg l^{-1} ozone for 6 minutes eliminates infectivity of 10^4 oocysts, and chlorine dioxide (0.4 mg l^{-1}) significantly reduces infectivity after 15 min (Peeters *et al.*, 1989). The ability of *C. parvum* oocysts to tolerate chlorine and persist in water indicate that the control or reduction of faecal contamination of natural waters by point and non-point sources is key to the control of cryptosporidiosis.

Conclusions

Human enteropathogenic organisms such as *Campylobacter, Cryptosporidium* and *E. coli* O157 are virulent human pathogens whose global incidence is increasing. Typically, farm animals such as cattle are the primary reservoir of these pathogens within the environment, and the main sources of human infection are consequently associated with direct consumption of contaminated food products (e.g. beef) or contact with animal-contaminated environments (e.g. animal faeces-contaminated drinking water). The presence of enteric pathogens in resources that were previously presumed to be safe (e.g. cattle manure, cattle slurry, water) has serious implications for the land-based disposal of organic wastes and irrigation of food crops. Further, it also has serious ramifications for the protection of surface and groundwater drinking supplies and public access to pastureland and recreational resources (e.g. swimming in lakes/rivers). However, while soil and vegetation can be expected to directly influence the survival of enteric pathogens, there is a paucity of knowledge concerning the behaviour and survival of pathogens in agricultural environments to currently allow a full risk assessment. From the information available, it is apparent that enteric pathogen survival in the wider environment away from the animal host is controlled by many environmental and strain-specific factors and that dispersal occurs by many vectors. Formulation of policy and management strategies for the protection of land and water resources around farming environments is therefore paramount to the preservation of human health.

Acknowledgements

This work is supported by funding provided by the Biotechnology and Biological Sciences Research Council (BBSRC-Agri-Food). Thanks are also due to staff at the Public Health Laboratory Service, UK, for provision of data.

References

Abdul-Raouf, U.M., Beuchat, L.R. and Ammar, M.S. (1993) Survival and growth of *Escherichia coli* O157:H7 in ground, roasted beef as affected by pH, acidulants, and temperature. *Applied and Environmental Microbiology* 59, 2364–2368.

Al-Ghazali, M.R. and Al-Azawi, S.K. (1986) Detection and enumeration of *Listeria monocytogenes* in a sewage-treatment plant in Iraq. *Journal of Applied Bacteriology* 60, 251–254.

Al-Ghazali, M.R. and Al-Azawi, S.K. (1988) Storage effects of sewage-sludge cake on the survival of *Listeria monocytogenes*. *Journal of Applied Bacteriology* 65, 209–213.

Altekruse, S.F. and Swerdlow, D.L. (1996) The changing epidemiology of food-borne diseases. *The American Journal of the Medical Sciences* 311, 23–29.

Ansay, S.E., Darling, K.A. and Kaspar, C.W. (1999) Survival of *Escherichia coli* O157:H7 in ground beef patties during storage at 2, −2, 15 and then −2°C, and −20°C. *Journal of Food Protection* 62, 1243–1247.

Arbeit, R.D., Arthus, M., Dunn, R., Kim, C., Selander, R.K. and Goldstein, R. (1990) Resolution of recent evolutionary divergence among *Escherichia coli* from related lineages: the application of pulsed-field gel electrophoresis to molecular epidemiology. *Journal of Infectious Diseases* 161, 230–235.

Associated Press (2000) *E. coli* eyed in three more Ontario deaths. 16 June 2000. http://www.seattlep-i.com:80/local/bact13.shtml

Barker, J., Humphrey, T.J. and Brown, M.W. (1999) Survival of *Escherichia coli* O157 in a soil protozoan: implications for disease. *FEMS Microbiology Letters* 173, 291–295.

Besser, R.E., Lett, S.M., Weber, J.T., Doyle, M.P., Barrett, T.J., Wells, J.G. and Griffin, P.M. (1993) An outbreak of diarrhoea and hemolytic uremic syndrome from *Escherichia coli* O157:H7 in fresh-pressed apple cider. *Journal of the American Medical Association* 269, 2217–2220.

Besser, T.E., Hancock, D.D., Pritchett, L.C., McRae, E.M., Rice, D.H. and Tarr, P.I. (1997) Duration of detection of faecal excretion of *Escherichia coli* O157:H7 in cattle. *Journal of Infectious Diseases* 175, 726–729.

Besser-Wiek, J.W., Forfang, J., Hedberg, C.W., Korlath, J.A., Osterholm, M.T., Sterling, C.R. and Garcia, L. (1996) Foodborne outbreak of diarrhoeal illness associated with *Cryptosporidium parvum* – Minnesota, 1995. *Morbidity and Mortality Weekly Report* 45, 783–784.

Beuchat, L.R. (1999) Survival of enterohaemorrhagic *Escherichia coli* O157:H7 in bovine faeces applied to lettuce and the effectiveness of chlorinated water as a disinfectant. *Journal of Food Protection* 62, 845–849.

Beuchat, L.R. and Brackett, R.E. (1990) Inhibitory effects of raw carrots on *Listeria monocytogenes*. *Applied and Environmental Microbiology* 56, 1734–1742.

Beuchat, L.R., Brackett, R.E., Hao, D.Y.Y. and Conner, D.E. (1986) Growth and thermal inactivation of *Listeria monocytogenes* in cabbage and cabbage juice. *Canadian Journal of Microbiology* 32, 791–795.

Bibb, W.F., Gellin, B.G., Weaver, R., Schwartz, B., Plikaytis, B.D., Reeves, M.W., Pinner, R.W. and Broome, C.V. (1990) Analysis of clinical and food-borne isolates of *Listeria monocytogenes* in the United States by multilocus enzyme electrophoresis and application of the method to epidemiological investigations. *Applied and Environmental Microbiology* 56, 2133–2141.

Bogosian, G., Sammons, L.E., Morris, P.J., O'Neil, J.P., Heitkamp, M.A. and Weber, D.B. (1996) Death of the *Escherichia coli* K-12 strain W3110 in soil and water. *Applied and Environmental Microbiology* 62, 4114–4120.

Bolton, F.J. and Coates, D. (1983) A study of the oxygen and carbon dioxide requirements of thermophilic Campylobacters. *Journal of Clinical Pathology* 36, 829–834.

Botzler, R.G., Cowan, A.B. and Wetzler, T.F. (1975) Survival of *Listeria monocytogenes* in soil and water. *Journal of Wildlife Disease* 10, 204–212.

Bremer, P.J., Osborne, C.M., Kemp, R.A. and Smith, J.J. (1998) Survival of *Listeria monocytogenes* in sea water and effect of exposure on thermal resistance. *Journal of Applied Microbiology* 85, 545–553.

Buchanan, R.L., Edelson, S.G., Miller, R.L. and Sapers, G.M. (1999) Contamination of intact apples after immersion in an aqueous environment containing *Escherichia coli* O157:H7. *Journal of Food Protection* 62, 444–450.

Buswell, C.M., Herlihy, Y.M., Lawrence, L.M., McGuiggan, J.T.M., Marsh, P.D., Keevil, C.W. and Leach, S.A. (1998) Extended survival and persistence of *Campylobacter* spp. in water and aquatic biofilms and their detection by immunofluorescent-antibody and rRNA-staining. *Applied and Environmental Microbiology* 64, 733–741.

Buswell, C.M., Herlihy, Y.M., Keevil, C.W., Marsh, P.D. and Leach, S.A. (1999) Carbon load in aquatic ecosystems affects the diversity and biomass of water biofilm consortia and the persistence of the pathogen *Campylobacter jejuni* within them. *Journal of Applied Microbiology* 85, 161–167.

CDC (Centers for Disease Control and Prevention) (1993) Update: multistate outbreak of *Escherichia coli* O157:H7 infections from hamburgers – Western United States, 1992–1993. *Morbidity and Mortality Weekly Report* 42, 258–263.

CDC (Centers for Disease Control and Prevention) (1995) *Escherichia coli* O157:H7 outbreak linked to commercially distributed dry-cured salami – Washington and California, 1994. *Morbidity and Mortality Weekly Report* 44, 157–159.

CDC (Centers for Disease Control and Prevention) (1999a) *1998 FoodNet Surveillance Report.* http://www.cdc.gov/ncidod/dbmd/foodnet/annual/98_surv.htm

CDC (Centers for Disease Control and Prevention) (1999b) Surveillance for outbreaks of *Escherichia coli* O157:H7 infection summary of 1998 data. http://www.cdc.gov/ncidod/dbmd/diseaseinfo/csteec98.pdf

CDNANZ (1997) *Foodborne Disease – Towards Reducing Foodborne Illness in Australia.* Technical Report Series No. 2, Foodborne Disease Working Party for the Communicable Diseases Network of Australia and New Zealand, Commonwealth Department of Health and Family Services, Commonwealth of Australia, Canberra.

Chauret, C., Nolan, K., Chen, P., Springthorpe, S. and Sattar, S. (1998) Ageing of *Cryptosporidium parvum* oocysts in river water and their susceptibility to disinfection by chlorine and monochloramine. *Canadian Journal of Microbiology* 44, 1154–1160.

Chynoweth, R.W., Hudson, J.A. and Thom, K. (1998) Aerobic growth and survival of *Campylobacter jejuni* in food and stream water. *Letters in Applied Microbiology* 27, 341–344.

De Graaf, D.C., Vanopdenbosch, E., OrtegaMora, L.M., Abbassi, H. and Peeters, J.E. (1999) A review of the importance of cryptosporidiosis in farm animals. *International Journal for Parasitology* 29, 1269–1275.

Dekeyser, P., Gossuin-Detrain, M., Butzler, J.P. and Sternon, J. (1972) Acute enteritis due to related *Vibrio*: first positive stool culture. *Journal of Infectious Diseases* 15, 390–392.

Di-Giovanni, G.D., Hashemi, F.H., Shaw, N.J., Abrams, F.A., LeChevallier, M.W. and Abbaszadegan, M. (1999) Detection of infectious *Cryptosporidium parvum* oocysts in surface and filter backwash water samples by immunomagnetic separation and integrated cell culture-PCR. *Applied and Environmental Microbiology* 65, 3427–3432.

Donnelly, C.W. and Baigent, G.J. (1986) Method for flow cytometric detection of *Listeria monocytogenes* in milk. *Applied and Environmental Microbiology* 52, 689–695.

Dowe, M.J., Jackson, E.D., Mori, J.G. and Bell, C.R. (1997) *Listeria monocytogenes* survival in soil and incidence in agricultural soils. *Journal of Food Protection* 60, 1201–1207.

Doyle, M.P. and Schoeni, J.L. (1984) Survival and growth characteristics of *Escherichia coli* associated with haemorrhagic colitis. *Applied and Environmental Microbiology* 48, 855–856.

Dubey, J.P., Speer, C.A. and Fayer, R. (1990) *Cryptosporidiosis of Man and Animals.* CRC Press, Boca Raton, Florida.

Elder, R.O., Keen, J.E., Siragusa, G.R., Barkocy-Gallagher, G.A., Koohmaraie, M. and Laegreid, W.W. (2000) Correlation of enterohaemorrhagic *Escherichia coli* O157 prevalence in faeces, hides, and carcasses of beef cattle during processing. *Proceedings of the National Academy of Sciences, USA* 97, 2999–3003.

Faith, N.G., Shere, J.A., Brosch, R., Arnold, K.W., Ansay, S.E., Lee, M.S., Luchansky, J.B. and Kaspar, C.W. (1996) Prevalence and clonal nature of *Escherichia coli* O157:H7 on dairy farms in Wisconsin. *Applied and Environmental Microbiology* 62, 1519–1525.

Fayer, R. (1995) Effect of sodium hypochlorite exposure on infectivity of *Cryptosporidium parvum* oocysts for neonatal BALB/c mice. *Applied and Environmental Microbiology* 61, 844–846.

Fayer, R. and Nerad, T. (1996) Effects of low temperatures on viability of *Cryptosporidium parvum* oocysts. *Applied and Environmental Microbiology* 62, 1431–1433.

Fayer, R., Graczyk, T.K., Lewis, E.J., Trout, J.M. and Farley, C.A. (1998a) Survival of infectious *Cryptosporidium parvum* oocysts in seawater and eastern oysters (*Crassostrea virginica*) in the Chesapeake Bay. *Applied and Environmental Microbiology* 64, 1070–1074.

Fayer, R., Trout, J.M. and Jenkins, M.C. (1998b) Infectivity of *Cryptosporidium parvum* oocysts stored in water at environmental temperatures. *Journal of Parasitology* 84, 1165–1169.

Fenlon, D.R. (1985) Wild birds and silage as reservoirs of *Listeria* in the agricultural environment. *Journal of Applied Bacteriology* 59, 537–543.

Frances, N., Hornby, H. and Hunter, P.R. (1991) The isolation of *Listeria* species from freshwater sites in Cheshire and North Wales. *Epidemiology and Infection* 107, 235–238.

Francis, G.A. and O'Beirne, D. (1998) Effects of the indigenous microflora of minimally processed lettuce on the survival and growth of *Listeria innocua*. *International Journal of Food Science and Technology* 33, 477–488.

Franco, M.A., Santini, P.I., Korol, S., Moretton, J., Dominguez, N. and Aquino, M.D. (1990) In vitro survival of pathogenic microorganisms in recreational water. *Review of Latin American Microbiology* 32, 149–152.

Fricker, C.R. and Park, R.W.A. (1989) A 2-year study of the distribution of thermophilic Campylobacters in human, environmental and food samples from the reading area with particular reference to toxin production and heat-stable serotype. *Journal of Applied Bacteriology* 66, 477–490.

Gagliardi, J.V. and Karns, J.S. (2000) Leaching of *Escherichia coli* O157:H7 in diverse soils under various agricultural management practices. *Applied and Environmental Microbiology* 66, 877–883.

Garcia, E., de Paz, M., Rodriguez, J.L., Gaya, P., Medina, M. and Nunez, M. (1996) Exogenous sources of *Listeria* contamination in raw ewe's milk. *Journal of Food Protection* 59, 950–954.

Goodgame, R.W. (1996) Understanding intestinal spore-forming protozoa: cryptosporidia, microsporidia, Isospora. and Cyclospora. *Annual of Internal Medicine* 124, 429–441.

Gorden, J. and Small, P.L. (1993) Acid resistance in enteric bacteria. *Infection and Immunity* 61, 364–367.

Graczyk, T.K., Fayer, R., Trout, J.M., Lewis, E.J., Farley, C.A., Sulaiman, I. and Lal, A.A. (1998) *Giardia* sp. cysts and infectious *Cryptosporidium parvum* oocysts in the faeces of migratory Canada geese (*Branta canadensis*). *Applied and Environmental Microbiology* 64, 2736–2738.

Griffin, P.M. (1995) *Escherichia coli* O157:H7 and other enterohaemorrhagic *Escherichia coli*. In: Blaser, M.J., Smith, P.D., Ravdin, J.I., Greenberg, H.B. and Guerrant, R.L. (eds) *Infections of the Gastrointestinal Tract*. Raven Press, New York, pp. 739–761.

Griffin, P.M. and Tauxe, R.V. (1991) The epidemiology of infections caused by *Escherichia coli* O157:H7, other enterohaemorrhagic *E. coli*, and the associated hemolytic uremic syndrome. *Epidemiological Reviews* 13, 60–98.

Grønstal, H. (1979) Listeriosis in sheep – isolation of *Listeria monocytogenes* from grass silage. *Acta Veterinary Scandanavica* 20, 492–497.

Grønstal, H. (1980) Listeriosis in sheep – *Listeria monocytogenes* in sheep fed hay or grass silage during pregnancy – immunological state, white blood cell, total serum protein and serum iron. *Acta Veterinary Scandanavica* 21, 1–10.

Hancock, D.D., Besser, T.E., Kinsel, M.L., Tarr, P.I., Rice, D.H. and Paros, M.G. (1994) The prevalence of *Escherichia coli* O157:H7 in dairy and beef- cattle in Washington state. *Epidemiology and Infection* 113, 199–207.

Hancock, D.D., Rice, D.H., Herriott, D.E., Besser, T.E., Ebel, E.D. and Carpenter, L.V. (1997) Effects of farm manure-handling practices on *Escherichia coli* O157 prevalence in cattle. *Journal of Food Protection* 60, 363–366.

Hancock, D.D., Besser, T.E., Rice, D.H., Ebel, E.D., Herriott, D.E. and Carpenter, L.V. (1998) Multiple sources of *Escherichia coli* O157 in feedlots and dairy farms in the Northwestern USA. *Preventative Veterinary Medicine* 38, 11–19.

Heisick, J.E., Wagner, D.E., Nierman, M.L. and Peeler, J.T. (1989) *Listeria* spp. found on fresh-market produce. *Applied and Environmental Microbiology* 55, 1925–1927.

Herriott, D.E., Hancock, D.D., Ebel, E.D., Carpenter, L.V., Rice, D.H. and Besser, T.E. (1998) Association of herd management factors with colonization of dairy cattle by Shiga toxin-positive *Escherichia coli* O157. *Journal of Food Protection* 61, 802–807.

HMSO (1993) Interim report on *Campylobacter*. Advisory Committee on the Microbiological Safety of Food, HMSO, London.

Humphrey, T.J. and Beckett, P. (1987) *Campylobacter jejuni* in dairy cows and raw milk. *Epidemiology and Infection* 98, 263–269.

Janisiewicz, W.J., Conway, W.S., Brown, M.W., Sapers, G.M., Fratamico, P. and Buchanan, R.L. (1999) Fate of *Escherichia coli* O157:H7 on fresh-cut apple tissue and its potential for transmission by fruit flies. *Applied and Environmental Microbiology* 65, 1–5.

Jarvis, G.N., Kizoulis, M.G., Diez-Gonzalez, F. and Russell, J.B. (2000) The genetic diversity of predominant *Escherichia coli* strains isolated from cattle fed various amounts of hay and grain. *FEMS Microbiology Ecology* 32, 225–233.

Jemmi, T. and Keusch, A. (1994) Occurrence of *Listeria monocytogenes* in fresh-water fish farms and fish-smoking plants. *Food Microbiology* 11, 309–312.

Jenkins, M.B., Walker, M.J., Bowman, D.D., Anthony, L.C. and Ghiorse, W.C. (1999) Use of sentinel system for field measurements of *Cryptosporidium parvum* oocyst inactivation in soil and animal waste. *Applied and Environmental Microbiology* 65, 1998–2005.

Jones, C.E., Shama, G., Jones, D., Roberts, I.S. and Andrew, P.W. (1997) Physiological and biochemical studies on psychrotolerance in *Listeria monocytogenes*. *Journal of Applied Microbiology* 83, 31–35.

Jones, D.L. (1999) Potential health risks associated with the persistence of *Escherichia coli* O157 in agricultural environments. *Soil Use and Management* 15, 76–83.

Jones, D.M., Sutcliffe, E.M. and Curry, A. (1991) Recovery of viable but non-culturable *Campylobacter jejuni*. *Journal of General Microbiology* 137, 2477–2482.

Jones, D.M., Sutcliffe, E.M., Rios, R., Fox, A.J. and Curry, A. (1993) *Campylobacter jejuni* adapts to aerobic metabolism in the environment. *Journal of Medical Microbiology* 38, 145–150.

Jones, K., Betaieb, M. and Telford, D.R. (1990) Correlation between environmental monitoring of thermophilic *Campylobacters* in sewage effluent and the incidence of *Campylobacter* infection in the community. *Journal of Applied Bacteriology* 69, 235–240.

Kaper, J.B. and O'Brien, A.D. (1998) *Escherichia coli O157:H7 and Other Shiga Toxin-producing E. coli Strains.* ASM Press, Washington, DC.

Kapperud, G., Lassen, J., Ostroff, S.M. and Aasen, S. (1992) Epidemiology of *Campylobacter jejuni* infections. *Epidemiological Revue* 5, 157–176.

Koenraad, P.M.F.J., Hazeleger, W.C., Vanderlaan, T., Beumer, R.R. and Rombouts, F.M. (1994) Survey of *Campylobacter* spp in sewage plants in The Netherlands. *Food Microbiology* 11, 65–73.

Koenraad, P.M.F.J., Rombouts, F.M. and Notermans, S.H.W. (1997) Epidemiological aspects of thermophilic *Campylobacter* in water-related environments: A review. *Water Environment Research* 69, 52–63.

Kuczynska, E. and Shelton, D.R. (1999) Method for detection and enumeration of *Cryptosporidium parvum* oocysts in faeces, manures, and soils. *Applied and Environmental Microbiology* 65, 2820–2826.

Kudva, I.T., Blanch, K. and Hovde, C.J. (1998) Analysis of *Escherichia coli* O157:H7 survival in ovine or bovine manure and manure slurry. *Applied and Environmental Microbiology* 64, 3166–3174.

Laberge, I., Ibrahim, A., Barta, J.R. and Griffiths, M.W. (1996) Detection of *Cryptosporidium parvum* in raw milk by PCR and oligonucleotide probe hybridization. *Applied and Environmental Microbiology* 62, 3259–3264.

Lechevallier, M.W., Norton, W.D. and Lee, R.G. (1991) Occurrence of *Giardia* spp. and *Cryptosporidium* spp. in surface water supplies. *Applied and Environmental Microbiology* 57, 2610–2616.

Lisle, J.T. and Rose, J.B. (1995) *Cryptosporidium* contamination of water in the USA and UK: a mini-review. *Aqua-Oxford* 44, 103–117.

MacGowan, A.P., Bowker, K., Mclauchlin, J., Bennett, P.M. and Reeves, D.S. (1994) The occurrence and seasonal-changes in the isolation of *Listeria* spp. in shop bought food stuffs, human faeces, sewage and soil from urban sources. *International Journal of Food Microbiology* 21, 325–334.

MacKenzie, W.R., Hoxie, N.J., Proctor, M.E., Gradus, M.S., Blair, K.A., Peterson, D.E., Kazmierczak, J.J., Addiss, D.G., Fox, K.R., Rose, J.B. and Davis, J.P. (1994) A massive outbreak in Milwaukee of *Cryptosporidium* infection transmitted through the public water supply. *New England Journal of Medicine* 331, 161–167.

Mawdsley, J.L., Brooks, A.E. and Merry, R.J. (1996a) Movement of the protozoan pathogen *Cryptosporidium parvum* through three contrasting soil types. *Biology and Fertility of Soils* 21, 30–36.

Mawdsley, J.L., Brooks, A.E., Merry, R.J. and Pain, B.F. (1996b) Use of a novel soil tilting table apparatus to demonstrate the horizontal and vertical movement of the protozoan pathogen *Cryptosporidium parvum* in soil. *Biology and Fertility of Soils* 23, 215–220.

Mead, P.S., Slutsker, L., Dietz, V., McCaig, L.F., Bresee, J.S., Shapiro, C., Griffin, P.M. and Tauxe, R.V. (2000) *Food-related Illness and Death in the United States.* Emerging Infectious Diseases 5, no. 5. Available from http://www.cdc.gov/ncidod/eid/vol5no5/mead.htm

Medema, G.J., Bahar, M. and Schets, F.M. (1997) Survival of *Cryptosporidium parvum, Escherichia coli*, faecal enterococci and *Clostridium perfringens* in river water: influence of temperature and autochthonous microorganisms. *Water Science and Technology* 35, 249–252.

Millard, P.S., Genscheimer, K.F., Addiss, D.G., Sosin, D.M., Beckett, G.A., Houck-Jankoski, A. and Hudson, A. (1994) An outbreak of cryptosporidiosis from fresh-pressed apple cider. *Journal of American Medical Association* 272, 1592–1596.

Miller, L.G. and Kaspar, C.W. (1993) *Escherichia coli* O157:H7 acid tolerance and survival in apple cider. *Journal of Food Protection* 57, 460–464 [errata 57, 645].

Monfort, P., Piclet, G. and Plusquellec, A. (2000) *Listeria innocua* and *Salmonella panama* in estuarine water and seawater: a comparative study. *Water Research* 34, 983–989.

Monge, R. and Chinchilla, M. (1996) Presence of *Cryptosporidium* oocysts in fresh vegetables. *Journal of Food Protection* 59, 202–203.

Morgan, D., Newman, C.P., Hutchinson, D.N., Walker, A.M., Rowe, B. and Majid, F. (1993) Verotoxin producing *Escherichia coli* O157 infections associated with the consumption of yoghurt. *Epidemiology and Infection* 111, 181–187.

Musial, C.E., Arrowood, M.J., Sterling, C.R. and Gerba, C.P. (1987) Detection of *Cryptosporidium* in water by using polypropylene cartridge filters. *Applied and Environmental Microbiology* 53, 687–692.

Nataro, J.P. and Kaper, J.B. (1998) Diarrheagenic *Escherichia coli. Clinical Microbiology Reviews* 11, 142–201 [erratum 11, 403].

Orskov, F., Orskov, I. and Villar, J.A. (1987) Cattle as reservoir of verotoxin-producing *Escherichia coli* O157:H7. *Lancet* 2, 276.

Palmgren, H., Sellin, M., Bergstrom, S. and Olsen, B. (1997) Enteropathogenic bacteria in migrating birds arriving in Sweden. *Scandinavian Journal of Infectious Diseases* 29, 565–568.

Peeters, J.E., Mazas, E.A., Masschelein, W.J., Villacorta-Martinez-De-Maturana, I. and Debacker, E. (1989) Effect of disinfection of drinking water with ozone or chlorine dioxide on survival of *Cryptosporidium parvum* oocysts. *Applied and Environmental Microbiology* 55, 1519–1522.

PHLS Communicable Disease Surveillance Centre (1996) Surveillance of water-borne disease and water quality: January to June 1996. *Communicable Disease Report* 6, 301–302.

Price, S.B., Cheng, C.M., Kaspar, C.W., Wright, J.C., DeGraves, F.J., Penfound, T.A., CastanieCornet, M.P. and Foster, J.W. (2000) Role of rpoS in acid resistance and faecal shedding of *Escherichia coli* O157:H7. *Applied and Environmental Microbiology* 66, 632–637.

Ragletti, S. and Untermann, F. (2000) Prevalence and characteristics of verotoxin-producing *Escherichia coli* (VTEC) in stool samples from asymptomatic human carriers working in the meat processing industry in Switzerland. *Journal of Applied Microbiology* 88, 335–341.

Read, S.C., Gyles, C.L., Clarke, R.C., Lior, H. and McEwen, S. (1990) Prevalence of verocytotoxigenic *Escherichia coli* in ground beef, pork, and chicken in southwestern Ontario. *Epidemiology and Infection* 105, 11–20.

Regan, J., McVay, R., McEvoy, M., Gilbert, J., Hughes, R., Tougaw, T., Parker, E., Crawford, W., Johnson, J., Rose, J., Boutros, S., Roush, S., Belcuore, T., Rains, C., Munden, J., Stark, L., Hartwig, E., Pawlowicz, M., Hammond, R. and Hopkins, R. (1996) Outbreak of cryptospridiosis at a day camp – Florida, July–August 1995. *Morbidity and Mortality Weekly Report* 45, 442–444.

Roberts, T. and Buzby, J.C. (1998) Estimated costs of *E. coli* O157:H7 disease in the United States. In: *VTEC '97. Conference Proceedings of the 3rd International Symposium and Workshop on Shiga*

Toxin (Verocytotoxin) – producing Escherichia coli *Infections, 22–26 June 1997, Baltimore, Maryland.*

Robertson, L.J., Campbell, A.T. and Smith, H.V. (1992) Survival of *Cryptosporidium parvum* oocysts under various environmental pressures. *Applied and Environmental Microbiology* 58, 3494–3500.

Robinson, D.A. (1981) Infective dose of *Campylobacter jejuni* in milk. *British Medical Journal* 282, 1584.

Rollins, D.M. and Colwell, R.R. (1986) Viable but nonculturable stage of *Campylobacter jejuni* and its role in survival in the natural aquatic environment. *Applied and Environmental Microbiology* 52, 531–538.

Rose, J.B. (1988) Occurrence and significance of *Cryptosporidium* in water. *American Water Works Association Journal* 80, 53–58.

Rose, J.B., Gerba, C.P. and Jakubowski, W. (1991) Survey of potable water supplies for *Cryptosporidium* and Giardia. *Environmental Science and Technology* 25, 1393–1400.

Samadpour, M., Liston, J., Ongerth, J.E. and Tarr, P.I. (1990) Evaluation of DNA probes for detection of Shiga-like toxin-producing *Escherichia coli* in food and calf faecal samples. *Applied and Environmental Microbiology* 56, 1212–1215.

Shere, J.A., Bartlett, K.J. and Kaspar, C.W. (1998) Longitudinal study of *Escherichia coli* O157:H7 dissemination on four dairy farms in Wisconsin. *Applied and Environmental Microbiology* 64, 1390–1399.

Sizmur, K. and Walker, C.W. (1988) *Listeria* in prepacked salads. *Lancet* 1, 1167.

Skirrow, M.B. (1991) Epidemiology of *Campylobacter enteritis. International Journal of Food Microbiology* 12, 9–16.

Skovgaard, N. and Morgen, C.A. (1988) Detection of *Listeria* spp. in faeces from animals, in feeds, and in raw foods of animal origin. *International Journal of Food Microbiology* 6, 229–242.

Skovgaard, N. and Nørrung, B. (1989) The incidence of *Listeria* spp. in faeces of Danish pigs and in minced pork meat. *International Journal of Food Microbiology* 8, 59–63.

Sockett, P.N. and Roberts, J.A. (1991) The social and economic impacts of salmonellosis. *Epidemiology and Infection* 107, 335–347.

Sreter, T. and Varga, I. (2000) Cryptosporidiosis in birds – a review. *Veterinary Parasitology* 87, 261–279.

Stanley, K.N., Wallace, J.S., Currie, J.E., Diggle, P.J. and Jones, K. (1998a) Seasonal variation of thermophilic Campylobacters in lambs at slaughter. *Journal of Applied Microbiology* 84, 1111- 1116.

Stanley, K.N., Wallace, J.S., Currie, J.E., Diggle, P.J. and Jones, K. (1998b) The seasonal variation of thermophilic Campylobacters in beef cattle, dairy cattle and calves. *Journal of Applied Microbiology* 85, 472–480.

Stanley, K., Cunningham, R. and Jones, K. (1998c) Isolation of *Campylobacter jejuni* from groundwater. *Journal of Applied Microbiology* 85, 187–191.

Steinbruegge, E.G., Maxcy, R.B. and Liewen, M.B. (1988) Fate of *Listeria monocytogenes* on ready to serve lettuce. *Journal of Food Protection* 51, 596–599.

Takata, T., Wai, S.N., Takade, A., Sawae, Y., Ono, J. and Amako, K. (1995) The purification of a groel-like stress protein from aerobically adapted *Campylobacter jejuni. Microbiology and Immunology* 39, 639–645.

Tamburrini, A. and Pozio, E. (1999) Long-term survival of *Cryptosporidium parvum* oocysts in seawater and in experimentally infected mussels (*Mytilus galloprovincialis*). *International Journal for Parasitology* 29, 711–715.

Taormina, P.J. and Beuchat, L.R. (1999) Behaviour of enterohaemorrhagic *Escherichia coli* O157:H7 on lucerne sprouts during the sprouting process as influenced by treatments with various chemicals. *Journal of Food Protection* 62, 850–856.

Tarr, P.I. (1995) *Escherichia coli* O157:H7: clinical, diagnostic, and epidemiological aspects of human infection. *Clinical Infectious Diseases* 20, 1–10.

Thomas, A., Cheasty, T., Frost, J.A., Chart, H., Smith, H.R. and Rowe, B. (1996) Vero cytotoxin-producing *Escherichia coli*, particularly serogroup O157, associated with human infections in England and Wales: 1992–4. *Epidemiology and Infection* 117, 1–10.

Thomas, C., Hill, D.J. and Mabey, M. (1999) Evaluation of the effect of temperature and nutrients on the survival of *Campylobacter* spp. in water microcosms. *Journal of Applied Microbiology* 86, 1024–1032.

USFDA (1992) Campylobacter jejuni. *Bad Bug Book – Foodborne Pathogenic Microorganisms and Natural Toxins Handbook.* US Food and Drug Administration, Washington, DC.

Van de Giessen, A., Mazurier, S.I., Jacobs-Reitsma, W., Janson, W., Berkens, P., Ritmeester, W. and Werneds, K. (1992) Study on the epidemiology and control of *Campylobacter jejuni* in poultry broiler flocks. *Applied and Environmental Microbiology* 58, 1913–1917.

Van Renterghem, B., Huysman, F., Rygole, R. and Verstraete, W. (1991) Detection and prevalence of *Listeria monocytogenes* in the agricultural ecosystem. *Journal of Applied Bacteriology* 71, 211–217.

Varma, P.R.G. and Lyer, T.S.G. (1993) Viability of *Listeria monocytogenes* in water. *Fish Technology* 30, 164–165.

Wallace, J.S., Stanley, K.N., Currie, J.E., Diggle, P.J. and Jones, K. (1997a) Seasonality of thermophilic *Campylobacter* populations in chickens. *Journal of Applied Microbiology* 82, 219–224.

Wallace, J.S., Cheasty, T. and Jones, K. (1997b) Isolation of Vero cytotoxin-producing *Escherichia coli* O157 from wild birds. *Journal of Applied Microbiology* 82, 399–404.

Wang, G.D. and Doyle, M.P. (1998) Survival of enterohaemorrhagic *Escherichia coli* O157:H7 in water. *Journal of Food Protection* 61, 662–667.

Wang, G., Zhao, T. and Doyle, M.P. (1996) Fate of enterohaemorrhagic *Escherichia coli* O157:H7 in bovine faeces. *Applied and Environmental Microbiology* 62, 2567–2570.

Warburton, D.W., Austin, J.W., Harrison, B.H. and Sanders, G. (1998) Survival and recovery of *Escherichia coli* O157:H7 in inoculated bottled water. *Journal of Food Protection* 61, 948–952.

Watkins, J. and Sleath, K.P. (1981) Isolation and enumeration of *Listeria monocytogenes* from sewage, sewage-sludge and river water. *Journal of Applied Bacteriology* 50, 1–9.

Weagant, S.D., Bryant, J.L. and Bark, D.H. (1994) Survival of *Escherichia coli* O157:H7 in mayonnaise and mayonnaise-based sauces at room and refrigerated temperatures. *Journal of Food Protection* 57, 629–631.

Weijtens, M.J.B.M., van der Plas, J., Bijker, P.G.H., Urlings, H.A.P., Koster, D., van Logtestijn, J.G. and Veld, J.H. (1997) The transmission of *Campylobacter* in piggeries; an epidemiological study. *Journal of Applied Microbiology* 83, 693–698.

Weijtens, M.J.B.M., Reinders, R.D., Urlings, H.A.P. and van der Plas, J. (1999) *Campylobacter* infections in fattening pigs; excretion pattern and genetic diversity. *Journal of Applied Microbiology* 86, 63–70.

Weiss, J. and Seelinger, H.P.R. (1975) Incidence of *Listeria monocytogenes* in nature. *Applied Microbiology* 30, 29–32.

Welshimer, H.J. (1960) Survival of *Listeria monocytogenes* in soil. *Journal of Bacteriology* 80, 316–320.

Welshimer, H.J. (1968) Isolation of *Listeria monocytogenes* from vegetation. *Journal of Bacteriology* 95, 300–303.

7 Sediment

T.R. Harrod[1] and F.D. Theurer[2]

[1]*National Soil Resources Institute, North Wyke Research Station, Okehampton, Devon EX20 2SB, UK;* [2]*National Water and Climate Center, Natural Resources Conservation Service, USDA, 7413 Cinnabar Terrace, Gaithersburg, MD 20879-4575, USA*

Key words: agriculture, compaction, erosion, extended drainage network, overland flow, sediment, soil degradation, timeliness of land use, turbidity

Introduction

Any traffic on, or disturbance of, wet soil risks suspension of that soil as mud or sediment in turbid water and its transfer towards watercourses. This process can be triggered by movements of farm machinery or implements, animals and even people. On some degraded soils the impact of raindrops alone can be sufficient. Whereas in the USA sediment is seen as the single greatest polluter of water (Waters, 1995), it has not been regarded as important in north-west Europe, where pollution is seen primarily as chemical in character. In addition to its primary physical impacts, sediment is important as a vector for P, a gamut of residues of agrochemicals and microbes, each being moved across and from the land in turbid water. Soil degradation primarily mobilizes soil particles, but livestock manures spread to land are further sources of suspended sediment that can enter watercourses via overland flow and other pathways.

The viewpoint presented here is from north-west Europe, tempered by insights from the rest of the world. The hydrological and textural range of soils in this region of the world is considerable. The climate is oceanic and highly seasonal; wet but mild in the winter half year, relatively dry in the summer with modest soil moisture deficits, particularly in the lowlands. Limited frost and snowfall mean that overland flow and associated sediment are generated and transferred throughout the winter half year, rather than in short-lived spring floods typical of continental, cool temperate zones.

European agriculture underwent unprecedented changes in the second half of the 20th century (RCEP, 1996). The combination of robust financial support with many technical and commercial advances, encouraged specialization along with an industrialized approach to farming. As well as conventional farming with 'broad-acre' arable cropping and livestock grazing in fields, our considerations include both outdoor horticulture and field vegetable growing, in addition to the upland moors and mountains. Many horticultural holdings produce three crops in 2 years, placing the land under great pressure. The sheep grazing the uplands, thanks to agricultural support structures, are owned by farmers. At the turn of the 21st century, European agriculture is still experiencing many changes. Support payments remain part of the farming economy, but they are being reduced and directed away from production and towards environmental protection and benefit. Consequently, farmers are seeking to diversify and bring in new income, broadening the range of sediment sources from farming.

This chapter discusses the various means by which soil becomes sediment and turbid discharges and their transportation towards watercourses. Some examples of the impacts of sediment pollution are considered, followed by discussion of mitigation.

Agriculture and Sources of Fine Sediment

Background

Generation of sediment by agriculture originates from damage to the soil by ill-timed or ill-sited use of the land. Mostly it happens under wet conditions, by animals' hooves, cultivations, excavations and vehicles travelling on the land, and by people walking (Theurer *et al.*, 1998). All these take place both on stream banks and more widely throughout the catchment. Mud and slurried soil is produced, and the soil's infiltration capacity reduced, while pathways for enhanced transport of turbid flow to watercourses are created.

Although all land is subject to geomorphological processes, notably weathering and erosion, the interest here is on the acceleration of sediment generation and delivery above natural background rates, especially the suspended wash load of finer particles and aggregates. Erosion of river banks during spates is undoubtedly a natural geomorphological process, occurring in non-agricultural and farmed catchments. It is widely accepted that agriculture has modified hydrological regimes from those of the original post-glacial temperate forest, increasing stream discharges and subsequent bank erosion. The evidence is present at the landscape scale, within some soil profiles and from archaeological sources. In some landscapes, colluvial soils are widely developed (Colborne and Staines, 1987), although downslope displacement during cultivation can be as powerful a mechanism as overland flow. Sediment transfer can take place within soils. Lessivage, the downward translocation of fine clay (conventionally, < 0.02 μm) to form argillans or clay coatings, in subsoils, is widely recognized across the world (FitzPatrick, 1986), particularly in regions with marked contrasts between dry and wet seasons. Coarser siltans, composed of particles between 2 and 60 μm, present in some subsoils, have been attributed to phases of accelerated dispersion and erosion of soils, as following forest clearance.

Less widely recognized is the impact of hundreds if not thousands of years of settlement and agriculture on surface hydrology. The man-made, terraced and bunded rice paddies of South East Asia are extreme examples. However, much of north-west Europe, with hedges, banks, lanes, sunken roads, drains, ditches, etc., can be perplexing when trying to decide just where surface water will flow, contrasting sharply with the more pristine landscapes of parts of North America.

Generation and transport of sediment is strongly influenced by soil hydrology, particle size distribution (texture), slope and the way the land is used or abused. Although traditional soil classification may appear esoteric to the non-specialist, its structured interpretation aids the understanding of how farming can transform soil into sediment. Intrinsic soil hydrology, described in the UK with the Hydrology of Soil Types (HOST) methodology by Boorman *et al.* (1995), ranges from soils that are seldom wet, through to permanently wet peat. The assessment for soil workability of Thomasson (1982) characterizes the ease of cultivation and the relative susceptibility to structural damage, and by that hydrological degradation and risk of sediment generation through enhanced overland flow. The similarly structured scheme for rating poaching risk/trafficability for grassland (Harrod, 1979), can be applied to sediment generation from that land use. Susceptibility to erosion by water when land is under arable management varies with soil texture and slope (it appears to be relatively independent of soil hydrology), and has been assessed by various schemes (Morgan, 1985; Evans, 1990; Palmer, 1993). Lee (1979) provides a means for assessing risk of delivery of sediment from farm slurry spreading.

Damage to stream banks is commonplace wherever animals are allowed unrestricted access to the watercourse. Banks are trampled and poached, the sward is broken or destroyed. Sometimes termed 'bank erosion', this is clearly different from bank erosion by fluvial processes, such as meander migration. The former is irregular and scalloped, whereas natural erosion forms a linear or gently articulated vertical face. Most of the work by water managers (e.g. NRA, 1995; Walker, 1997) regarding fine sediment delivery is at present focused on streambed and banks. Clearly, bank erosion must be taken seriously, since any sediment is present quickly in the watercourse; however, catchments as a whole have received little attention.

Arable farming

The soil is most vulnerable to accelerated erosion during preparation for sowing and planting, before the establishment of the crop's canopy and after harvest of some crops. Overland flow is encouraged by the production of fine, smooth seedbeds, by cultivation up and down slope, by bed working and damage to soils during late harvests, as with maize, sugarbeet and potatoes. In north-west Europe, techniques used elsewhere to conserve soil and water, such as contour cultivation or conservation tillage (Unger, 1984), are rarely implemented.

Soils with large fine sand and silt contents are vulnerable, both because they are easily worked and favoured for tillage, and because they readily form caps (when rain impact detaches silt and sand grains, blocking pores) and lose their surface infiltration capacity. This encourages overland flow, which can lead to rill and gully erosion, generating sediment that may reach watercourses. Less obvious, but possibly more widespread, is diffuse, turbid surface flow which fails to cut channels (sheet erosion) and flows overland to watercourses or reaches them after infiltration into field drains.

Compaction is a further form of hydrological degradation of soil, caused when the land is used while too moist. It comes about through the weight of machinery or animals, shearing by rotating powered wheels or walking hooves and vibration from machinery. These primarily damage topsoils, although in severe cases they can affect the subsoil. Subsoils are vulnerable to compaction from an annual farming practice, ploughing, when one set of tractor wheels is run directly on the subsoil in the furrow, creating a plough pan. Marshall (1787) commented on plough pan formation long ago saying 'the soil is turned, and returned, in the same manner it would be, if spread on a floor of stone'. Compaction degrades natural soil aggregates, reducing their porosity and the soil's permeability, increasing the likelihood of overland flow. In soils with impeded internal drainage, inherent soil wetness is compounded.

Organic farming may offer some protection against physical degradation of the soil by raising topsoil organic matter content. However, beyond that, triggers of erosion and overland flow, such as ill-timed or inappropriate cultivation or other use, are just as likely under organic regimes.

Livestock farming

Livestock farming in north-west Europe involves a number of practices, which, through ill-timed use, may reduce soil infiltration rates and generate sediment, particularly on impermeable soils. Included are poaching (grazing stock treading the surface of moist or wet soil) and traffic by both animals and machinery. For dairy cattle, high stocking rates on pastures, coupled with the twice-daily journey back and forth to the milking parlour, results in much disturbance of gateways, tracks, verges, hedges, banks and fords through watercourses. Many beef cattle and nearly all sheep are outwintered, resulting in poaching and producing muddy water. In recent winters, lowland pastures have seen increased sheep numbers, with the inevitable turbid surface water discharges. This is a knock-on effect from preventing overgrazing of upland heather moorland in environmentally sensitive areas. In recent years 'extended grazing' has been adopted by some British farmers. Managed sensitively on freely draining soils, damage to the soil and sward can be minimal. On heavy land, or without careful management, extended grazing provides risks of degradation of the pastures and sediment generation. Remedial activity following stream bank erosion by livestock has been mentioned above. It must be commented that access to ditches and minor watercourses by cattle and sheep is commonplace in livestock farming areas, with comparable impacts on banks to those along rivers.

Vehicular traffic by farmers taking out forage or checking outwintered stock ruts and churns the soil, particularly at gateways, feeding racks and other bottlenecks, providing sediment and channels for its transport. Feeding racks are often near to streams to give stock access to drinking water. Similar slurrying and rutting takes place around big bale stores when they are drawn on during the winter.

Outdoor pig keeping has expanded in recent years. Unlike other outdoor livestock, pigs can remain in the same paddocks for *c.* 2 years. These piggeries can generate substantial sediment. The pigs root, puddle and poach the soil, accelerating overland flow. Additionally, there is daily vehicular traffic along the access races, encouraging turbid discharges and transferring it downslope along ruts.

Manures from housed cattle, pigs and poultry are commonly spread to land. Much ill-timed spreading takes place in late autumn and winter when the crops (grass included) are unable to use the applied nutrients, and the soil is wet or frozen, posing a high risk of overland or subsoil flow. Such application breaches the spirit of the *Code of Good Agricultural Practice* (MAFF, 1998a). There is then the risk of sediment being generated and transported both from damaged soil and from the manures. Slurry, a common form of livestock manure, is more likely to flow into water than solid wastes, such as straw-based farmyard manure. Low-rate irrigation is common for disposal of dilute farm wastes (dirty water) from dairy washings, yard runoff, etc., which rarely has storage. Therefore, they are used all year round, even in very wet weather and are also used frequently for spreading liquors separated from slurries, making more potential organic sediment available, as well as other pollutants.

Land drainage

Land drainage is a common practice on wet or waterlogged soils (Trafford, 1970). It was started by the Romans and has involved the excavation of ditches and surface grips, and burial of field drains, in addition to canalization of watercourses. Early field drains were made of stones and brushwood. Since the 19th century, baked clay tiles and pipes have been used, but from the 1960s continuous slotted plastic pipe has become the standard. During and since the Second World War, British agriculture saw unprecedented investment in land drainage. Encouraged by agricultural mechanization, it was supported by generous grants from government, in an era when production from home resources was a politically desirable contribution to self-sufficiency, and at times a stark matter of survival. In the last 10 years, grant aid has been discontinued, resulting in a large reduction in land drainage.

The strategies were to lower water tables in groundwater-affected soils and improve permeability of surface wet soils (Trafford, 1970). When draining permeable soils, ditch and pipe drainage, often with pumped outfalls, produce a rapid and dramatic response. On impermeable soil, benefits are less dramatic, and ditches and field drains need supplementation with secondary treatments. These comprise mole ploughing (semi-permanent, continuous, cylindrical channels drawn through stable, clayey subsoils) and subsoiling (the disruption of dense, impermeable subsoils to increase their porosity). Such drains require gravel backfill to connect them to the surface or to the secondary drainage work, although this is not always done, particularly on small grassland farms. The post-war grant aid required high installation standards, to maximize the removal of water. Consequently, on clay soils, as across much of lowland England, arable fields have pipe drains and ditches, ensuring immediate delivery to watercourses.

Agricultural drainage systems, designed to remove water from the soil surface and profile, have proved equally effective in transferring pollutants (Haygarth *et al.*, 1998), including soil-derived sediment. While this aspect of soil erosion is little studied, there are several pertinent points. In its pristine state, soil can filter out sediment as water percolates through it. However, where there is preferential flow, the soil matrix can be bypassed. Bypass flow can be a natural soil property or can be created by the farmer. Clay dominated soils in all but the wettest parts have subsoil fissures that open by shrinkage in summer. In the autumn, rainwater goes down these fissures until the soil matrix re-wets and swells. Worm and other animal burrows and pores left by roots create other pathways for preferential movement of water. Drainage works, including secondary treatments, similarly provide bypass routes.

Delivery of sediment from field drains can be substantial. There is evidence that, in at least some circumstances, it is topsoil mobilized by rain impact, overland flow and sheet erosion, which remains in suspension in water and reaches the drains via bypass routes in the soil. Relationships are likely between amounts moved in this way and soil textures and free-lime contents. Silty and fine sandy soil particles are most readily detached (Morgan, 1986). Patterns dictated by particle size are modified in soils rich in lime, its presence increasing soil stability. Field drainage using clay tiles required careful butting of each (30 cm long) pipe, to minimize intrusion of soil from the trench sides. In unstable soils, sediment intrusion is prevented only by protection with fibre or coarse mineral filter covers, a standard not always achieved in agricultural drainage.

Within the field, land drainage can reduce hydrological responses by removing water received from previous rainfall or by increasing storage capacity through secondary treatments. However, at a larger scale associated improvements to ditches and minor watercourses are likely to counter that effect (Robinson, 1990) by speeding water movement.

Ditch erosion

Ditches can be sources of sediment. Flowing water scours the ditch floor and where the banks are incompletely vegetated, soil can spill into the ditch. This risk increases when grass is shaded-out by ditch-side shrubs. Access by livestock or burrowing by wild animals encourages further sediment delivery (Harrod and Fraser, 1998). As on stream banks, these processes can operate on both dry and moist soils, unlike most other sediment generation. In general, ditch erosion is greater in livestock farming areas, which tend to be in wetter districts and have more drain flow. Direct defecation into the water may add suspended organic sediment. Where medieval land enclosure gives fields boundaries several hundred years old, continued erosion of ditches can produce gullies metres deep. These can be sufficiently common to have dialect names; e.g. 'goyle' in south-west England. In arable areas, hedges are fewer and ditch banks are routinely mown, maintaining a grassy sward, minimizing ditch erosion. In the uplands shallow grips dug as surface drains are known (Stewart, 1963) to have eroded into deep gullies within a decade or so.

Upland erosion

Sheep and cattle graze most of the upland moors and mountains (Evans and Felton, 1987). In places, overgrazing, foot and vehicle traffic, air pollution and peat cutting have degraded the vegetation, exposing the soil to erosion. This can be on spectacular scales, with marked sedimentation in lakes and reservoirs. While part of this erosion is of historical origin, post-war upland sheep numbers have increased substantially in response to headage payments. Overgrazing can destroy the protective cover of vegetation, whereas exclusion of stock can re-establish it.

Novel and non-agricultural sources of sediment

An important response by farmers to late 20th century pressures has been diversification, with the introduction of novel crops and livestock, plus changes in the use of land. The circumstances likely to generate sediment from novel crops (such as *Miscanthus* or flax) and livestock (for example water buffalo or ostriches) are little different from those already considered. However, where land use alters fundamentally there is potential for accelerated generation of fine sediment. Important changes include the development of farm woodlands and forestry, novel productive and leisure use of land and adjacent water, conversion of redundant barns and construction of new buildings. Also non-agricultural activities sharing or annexing farmland can generate sediment, so that the overall balance of rural sediment sources needs to be appreciated. Although rarely on the scale of conventional farming practices, these can be important through their nature, location or timing.

Farm woodlands and forestry

Much of the land used for trees is on ground poorly suited to agriculture, for example in and around the uplands. Siting is commonly on steep, often acid but freely draining soils, or on wet land, which may be impermeable, affected by high groundwater or on blanket peat. As with other land uses, sediment generation comes from the working or disturbance of the soil under wet conditions. Inherently wet soils in or near the uplands with their long wet seasons are particularly vulnerable, and are safely workable only during very dry summer conditions. Work quoted by Nesbit (1996) suggested that a two- to fivefold increase in sediment generation occurred after afforestation of moorland. Cultivation or draining prior to planting presents initial risks, further disturbance coming with the construction and use of roadways and in harvesting.

New construction

The areal extent of soil disturbance from new construction, either on the farm for agricultural buildings, conversions of redundant barns, facilities for visitors, or for housing, transport, industry, etc., is limited, compared with field-scale farming. Nevertheless, there is substantial potential for both sediment generation and transfer (CIRIA, 2000) as the work can involve heavy machinery and continues almost regardless of ground conditions. Much new construction goes on after stormwater drainage has been installed, providing ready access for turbid emissions to watercourses. Awareness of construction as a source of particulate pollution remains low.

Mineral extraction

Farmers have traditionally quarried stone and gravel for roads, buildings, drain backfill and other purposes. This continues, with the risk of associated discharges of turbid water. Some extraction of gravel from stream beds occurs, both casually by riparian land users and commercially. The risk of immediate release of fine sediment is obvious. In the long term, removal of only a few centimetres of gravel can destabilize the bed and banks for up to a kilometre upstream, causing bed degradation, accelerated bank erosion and meandering, and increased sediment yield downstream. At the industrial scale, quarrying and open-cast working for a range of minerals includes aggregate/gravel, sand, peat, coal, china, ball and brick clays, along with other materials. In many circumstances, these workings have the potential to generate turbid water. In some cases, use of water is integral to the process and the undertakings are controlled over discharges to watercourses.

Water-based activities

Many farms contain riparian land, natural or long-established lakes and ponds. Others have constructed water features as part of environmental or economic diversification. Where these connect with watercourses there is the potential for sediment transfers if they are used. Any use is better suited to lakes and ponds well isolated from sensitive watercourses. A few examples from Theurer *et al.* (1998) serve as illustration.

Duck rearing, preparatory to sporting shooting, can concentrate birds and traffic at feeding sites, causing localized, severe damage to banks, releasing sediment into the water. Many farms have opened up lakes and ponds for fishing, while riparian fishing rights are prized assets. Sediment generation from these is usually minimal, although some favoured spots suffer damage to banks. Fish farms, established commercially, or as attractions at farm visitor centres can, within the terms of their consents, discharge low-density organic matter of faecal and other origins, much of which is relatively coarse (up to a few mm across). Watercress growing, known to be a source of sediment, has a long history in various groundwater-fed catchments, especially in the chalk country of southern England.

Other land uses

Soil disturbance by non-agricultural activity can cause compaction with associated loss of infiltration and enhanced overland flow. Hooves, feet, wheels and tracks have the same effect on wet ground, making a slurry of the soil, damaging, wearing and breaching swards and producing channels and ruts. The damage to soil along some popular footpaths, bridleways and byways, has been recognized for decades (Agate, 1983). However, the scale and potential for sediment transfer has grown with the expansion of public enthusiasm for outdoor activities and with the availability of all-terrain vehicles. Military manoeuvres, vehicle testing and off-road driving courses, utilize some land shared with farmers. The need to operate regardless of ground conditions, often with large numbers of large and powerful vehicles, inevitably produces turbid water and wheelings for it to run downslope, threatening watercourses.

Connections between land-derived sediment and watercourses

Any sediment mobilized by land use becomes significant to the aquatic environment if there is a serious prospect of it reaching watercourses. Its delivery there depends on interactions of soil and site hydrology, weather and season, and land use activities (Harrod and Fraser, 1998). In landscapes with impermeable soils, that risk is high. The natural drainage network is dense, with numerous rivers and streams. Potential for water storage within the soil is small and water transfers rapidly to watercourses. In the winter, responses are to all but the lightest rainfall. Farmers have always sought to improve drainage by ditching, installing pipes and by secondary drainage treatments. The vulnerability of impermeable soils to structural damage by ill-timed traffic, cultivation or poaching adds to the susceptibility.

Watercourses are much less frequent in landscapes dominated by freely draining soils. However, physiographic relief may be greater and relict dry valley networks remain. Aside from restricted areas on alluvium, artificial drainage is rare. Sediment can generate when land use degrades the surface soil structure or leaves it unprotected by vegetation. In some instances overland flow will not reach a watercourse, but percolates into the soil beforehand. Where it does, it is via an extended, ephemeral, drainage network (Harrod and Fraser, 1998), which includes natural features such as dry valley floors and concavities as well as others with man-made origins. These comprise permanent elements, such as hedge lines, gateways, footpaths, tracks, lanes and roadways; the latter usually have drainage systems connected to watercourses, as well as temporary features including livestock tracks, furrows, cultivation marks, tramlines, wheelings and ruts. When activated by sufficient rainfall and overland flow, the extended drainage network can be effective in carrying fine sediment from the land to water bodies, often over distances in excess of a kilometre.

Impacts of Sediment Derived from Farmland

The generation and transport of soil-derived sediment from farmland results in physical changes. Detachment of eroding soil may entrain whole aggregates as clasts along flowing rills, or selectively remove individual fine sand or silt-sized grains by rain impact. Dispersion of aggregated soil may occur in transport. Poaching by grazing livestock mobilizes sediment somewhat differently. Once moving the sediment will be sorted by size and mass; the saltation load of stones and sand is easily deposited while the wash load of silt and clay-sized particles/aggregates and organic matter travels further, sometimes depositing as colluvium, sometimes reaching watercourses. The association of chemically active materials with the clay and organic matter means that sorting can leave pollution hotspots in colluvial sites or concentrate pollutants reaching water. Once it enters water, sediment has numerous impacts. Extreme quantities may block channels and cause flooding, blanket stream beds altering their ecology, or may insidiously infiltrate the gravel bed. In addition, any attached nutrients or pollutants have their own impacts on rivers and streams. A few examples follow.

Sediment and salmonids

Historical background

Salmon and trout in north-west Europe have declined from great abundance in pre-industrial times to the present, where they are absent from former habitats or where the threat of extinction to sparse, residual populations is real. Causes of their reduction may be numerous and complex; sediment intrusion into spawning gravels is one of them and has potential for serious impacts.

Netboy (1968) gives a history of the atlantic salmon. In the early 8th century the Venerable Bede commented that Britain 'had the greatest plenty of salmon'. At the time of the Domesday survey in 1086 (Ordnance Survey, 1862), fishery owners paid 'renderings' to the crown. Often the value was expressed in eels (Easton, 1978), and the common statement that renderings were for 1000 salmon or 100 eels gives an idea of the abundance of salmon. Netboy's (1968) accounts show rivers in north-west Europe and New England to be well stocked with salmon before the Industrial Revolution. Martin, as late as 1785, noted that the Mersey 'greatly abounds with salmon'. By the late 18th century, industrialization was under way in Britain. Untreated discharges from factories and slums went into the rivers as a matter of course. As well as killing aquatic life in rivers, pollution presented barriers to migrating fish. Water abstraction and the construction of weirs and dams compounded the degradation.

Stream beds, sediment and salmonids

The view of Theurer *et al.* (1998), from these historical sources, was that, prior to human interference with rivers and the land, and particularly industrialization, all gravel-bedded streams in England and Wales would have supported salmonids. Rivers on hard rocks with little or no superficial geological deposits are least likely to produce fine sediment during spates. The hard igneous rocks of the Scottish Highlands, the slates of the Lake District, Wales and Cornwall, are examples of this geology. At the other extreme are rivers in soft beds and banks with little or no gravel, as in the English clay vales. Some rocks, such as the Carboniferous shales, easily weather to clay in some soil-forming environments.

Catchment hydrology plays its part too. Groundwater-fed rivers have smaller suspended solid loads than surface water catchments. The chalk of northern France and southern and eastern England provides extensive examples of groundwater streams. In some of the chalk valleys, upwelling groundwater saturated with calcium carbonate complicates the picture. Thick peat and soft algal marl cover some chalkland floodplains, providing soft, readily eroded material that may add suspended sediment.

The present

The present atmosphere in the UK is one of crisis for anadromous salmonids. Twentieth century intensification of farming and other rural land use has contributed chemical pollutants and increased sources of sediment. Other possible causes of their decline may be environmental changes and over-fishing, both in rivers and at sea. Industrialized fishing in the North Atlantic, where salmon spend much of their lives, will have had direct effects on them and their source of food.

Over a century of British legislation and improved practices has greatly reduced pollution in rivers from the Thames and Trent downwards (Mills, 1989), at great expense to the nation as a whole. Limited salmon restoration has been achieved, as in the Thames. There is an opportunity to refill restored rivers with their indigenous fauna rather than allowing them to be colonized by inappropriate species, or to remain as clean but sterile rivers and streams, as in the acidified upper Wye.

Fine sediment intrusion and salmonid redds

Spawning is the beginning of the life cycle of salmonids. A gravid female spawns after digging a redd in bed gravels in a riffle, where water flow through the substrate is optimal for survival of her eggs. She sweeps the smaller gravel and any surrounding interstitial fine sediment up into the flowing column of water. The gravel falls immediately downstream over any previously deposited eggs, but the re-suspended fine sediment is swept away. The completed redd contains clean interstices and less fine sediment than the surrounding undisturbed substrate. The bottom of a single redd, the egg zone, can hold upwards of 5000 embryos. Water flowing through the egg zone carries dissolved O_2 throughout the incubation period, unless subsequent significant intrusion of fine sediment into the redd reduces its permeability. Depending on the species and water temperature, the embryo/fry spend 60–180 days in the redd. More complete description of redd building is given in Alonso *et al.* (1996), including their field measurements in the Tucannon River, a tributary of the Columbia River in northwestern USA.

That study demonstrated that, during base flow, suspended sediment content in the water column of streams is minimal. However, during spates, or events on the catchment surface generating sediment, the wash load of suspended silt (0.2–60 μm), clay (< 0.2 μm), fine sand (60 μm–0.2 mm)-sized particles and aggregates, plus organic matter, increases dramatically. Turbid water intrudes into the interstices of bed gravels and any redds that are present. There is insufficient energy to remove the intruded wash load, which obstructs subsequent interstitial flow, and therefore delivery of dissolved O_2. Additionally, the intruded material may include organic matter with an oxygen demand competing with that of the eggs or fry. Further events compound the intrusion of fine material and the threat to the salmonids. This process is insidious.

To illustrate this process, Fig. 7.1, developed by Theurer *et al.* (1998), shows the Sediment Intrusion and Dissolved Oxygen (SIDO) computer model of Alonso *et al.* (1996). Since data for UK stream geometry, hydraulics and salmonids (atlantic salmon or sea trout) were not readily available, the input data for Fig. 7.1 were taken from published data in the SIDO documentation, based on chinook salmon and conditions in the Tucannon River.

Some examples

Case studies quoted by Theurer *et al.* (1998) from south-west England indicate recent suspended sediment loads in streams in agricultural catchments which are known to have supported salmonids. At base flow, amounts were between 2 and 7 mg l^{-1}. During winter spates, triggered by light rain of between 1.4 and 3.5 mm h^{-1}, these rose, in the Mount's Bay streams (Harrod, 1992), to between 12 and 1720 mg l^{-1}, with median values of 59 mg l^{-1} and mean 298 mg l^{-1}. Overland flow from agricultural land entering these watercourses at the same time had a range of 58 to 8000

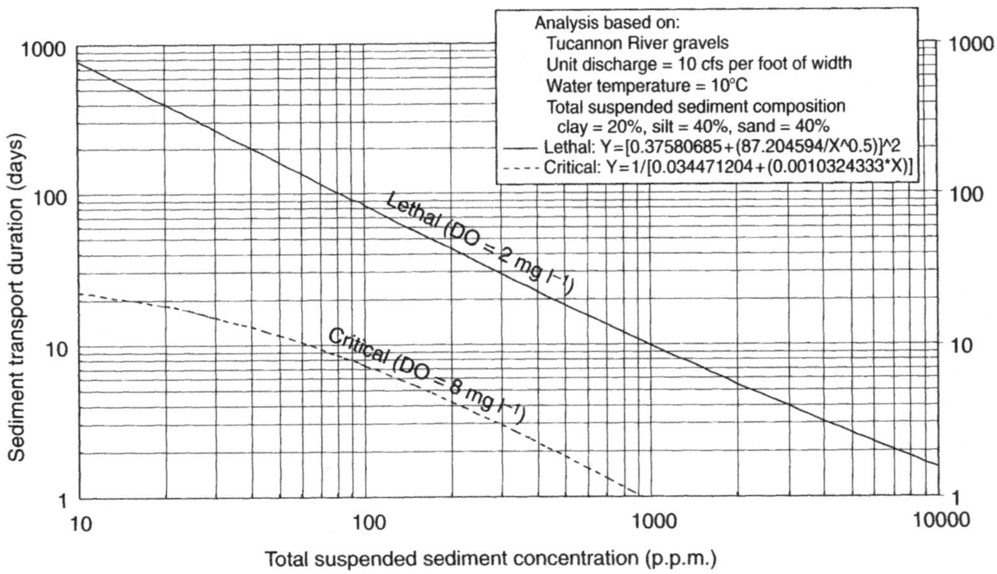

Fig. 7.1. Dissolved oxygen (DO) in a salmonid redd: simulation of duration of suspended sediment transport for lethal and critical DO in redd egg zone (after Alonso *et al.*, 1996).

mg l^{-1} suspended solids (median 520, mean 1102 mg l^{-1}). In the studied sites this resulted in mean movements of 4.98 (range 0.068–31.0) kg suspended solids ha^{-1} h^{-1} from slopes of < 7°. Table 7.2 gives analyses, based on long-term rainfall data, showing that light and moderate winter rainfall, which triggered these discharges, accounts for about 70% of overland flow from agriculturally degraded soils. That is not to deny that occasional heavy rainfall causes major erosion/sediment delivery events, but shows that low intensity background amounts are also relevant.

Although most hydrological activity is in the winter half year, summer rain can cause sediment discharges. Purvis (1997) reports on a highly turbid discharge, associated with heavy rain on to potatoes and field vegetables near Ross on Wye, on 5 August 1997. Routine water quality samples that day contained suspended solids at 1555 mg l^{-1}. Background amounts during dry weather were 8 mg l^{-1} on 14 July and 18 mg l^{-1} on 29 August.

The atlantic salmon, once so abundant in many British rivers, uses the autumn spates to reach its spawning grounds. Clearly, its eggs and fry are at mortal risk from sediment intrusion throughout the winter, just when the hazard is greatest.

Applying the values in Fig. 7.1 (with the reservation that it is based on North American data) to Mount's Bay and Ross on Wye is sobering. A median suspended sediment load in runoff events of 59 mg l^{-1}, as at Mount's Bay, would reduce the dissolved O_2 to critical levels in little over 10 days. Lethal levels would be reached in 140 days. Concentrations of the order observed by Purvis (1997) near Ross on Wye, during the incubation period, would be lethal within 6 or 7 days.

Transfer of pollutants: pesticides and phosphorus

Pollutants can enter watercourses from the land on soil-derived sediment. Many pesticides and their metabolites are readily sorbed by the soil, as are forms of the major plant nutrient, P.

The suspended sediment data quoted above from the Mount's Bay catchment were gathered in studies of pesticide pollution in west Cornwall (Harrod, 1992). Winter use of the land, exploiting the area's mild climate, compacts the soil, causing overland flow and sediment generation. For many years the insecticide aldrin was applied to daffodil and potato fields. Aldrin, along with its metabolite dieldrin (biological half-life 8 years), was carried on the eroded soil, some getting into streams via the extended drainage network. Aldrin and dieldrin enter aquatic organisms via the respiratory system as well as by direct ingestion. They are lipophilic, and bioaccumulate in higher members of the food chain, such as eels and trout.

Table 7.1. Dieldrin in soil, colluvium (μg kg^{-1}) and water (μg kg^{-1} suspended solids), Mount's Bay (after Harrod, 1994).

Dieldrin in:	Mean	Range
Source soils	264	10–1100
Colluvium	358	20–710
Concentration: *colluvium* / source soil	3.6	0.1–15.8
Source soils	249	72–932
Overland flow entering streams	821	300–1632
Concentration: *overland flow* / source soil	5.8	1.0–22.1
Receiving streams	116	27–272

The overland flow and soil erosion in west Cornwall mostly affected soils with inherently free drainage. However, after winter use causing compaction and capping, overland flow was consistently observed to begin in rain intensities of less than 1.5 mm h^{-1}. Comparison of dieldrin amounts in colluvium and sediment entering streams with those in the source soils (Table 7.1) shows enhancement during erosion and transport averaging 3.6- and 5.8-fold, respectively. During the studied events in rainfall of between 1.4 and 3.5 mm h^{-1}, the mean dieldrin load leaving polluting fields around Mount's Bay was 4.0 mg ha^{-1} h^{-1}.

Sharpley and Rekolainen (1997) and Leinweber *et al.* (Chapter 2, this volume) summarize the environmental implications of agricultural P for the aquatic environment. Phosphorus is necessary for the farmer to maintain crop productivity, but biological activity in receiving waters will be stimulated following agriculturally trivial transfers. Existing both in solution and as particles, P is subject to transformations of its form and bioavailability during its transfer across the land and in water bodies. According to Sharpley and Rekolainen (1997), P associated with soil mineral and organic particles accounts for 60–90% of P transferred from cultivated land. They state that discharges from grassland and forest contain little sediment, and P is largely in solution. However, this is perhaps an oversimplification: for example, agricultural intensification in the Torridge catchment, reported by Naismith *et al.* (1996), accelerated sediment delivery, largely involving grassland improvement. This apparent contradiction may be resolved by remembering that many soils are impermeable and readily poached during the grazing season, not an uncommon circumstance in north-west Europe.

Table 7.2 shows the importance of low intensity/high frequency rainfall for sediment and associated particulate P losses from capped, freely draining, arable soils in the UK. The efficiency of sediment delivery to water following such erosion requires further investigation. As already noted, on impermeable soils the hydrological connection between the field and watercourse is very effective. In parallel research (Harrod and Fraser, 1999) it was established that at least part of the turbid discharge from 70% of eroding arable fields in freely draining sites reached watercourses, some of it without attenuation, using extended, ephemeral drainage networks.

Table 7.2. Mean rainfall intensity (mm h^{-1}) October to March (1974–1995) at Abbotsinch, Glasgow, Scotland, and Lyneham, Wiltshire, England, with estimated overland flow (mm per winter) from soil with infiltration of 1 mm h^{-1}, plus flow-weighted concentrations of total P (mg l^{-1}) and suspended sediment (g l^{-1}). (After Fraser, 2000.)

	Rainfall intensity (mm h^{-1})	Duration (hours per winter)	Overland flow (mm per winter)	Total phosphorus (mg l^{-1})	Suspended sediment (g l^{-1})
Abbotsinch					
(Annual mean rain 1035 mm)	1–2	150	75	7.7	2.85
	2–4	75	151	7.3	6.36
	4–10	12	70	16.0	6.51
	> 10	0.2	2.3	5.7	0.70
	Total	238	299		
Lyneham					
(Annual mean rain 709 mm)	1–2	85	42	7.7	2.85
	2–4	37	73	7.3	6.36
	4–10	6	36	16.0	6.51
	> 10	0.2	2.3	5.7	0.70
	Total	128	154		

Mitigation of the Impacts of Sediment

Ideally, mobilization of sediment should be prevented at source. Techniques for this, or for intercepting sediment during its transfer, have been developed for soil and water conservation across the world over the millennia (Unger, 1984). In humid, cool, temperate agriculture the solutions can be summarized in terms of timeliness and appropriateness of land use and cultivations.

Compared with the dramatic scale of soil erosion in some countries, soil erosion and sediment generation on agricultural land in north-west Europe may appear minor. In the UK, the Royal Commission on Environmental Pollution (RCEP, 1996) did not view erosion of arable land as a major concern, although some irreversible off-farm effects were acknowledged. A further step forward will be for agriculturists to see soil erosion primarily as the transfer of particles as much via turbid overland flow or drain discharges as from easily identified rills and gullies. That is, shifting the emphasis from sole concern with the degradation of the agricultural resource by channelled erosion to include soil detachment and sediment transfers.

There has not been a culture of soil and water conservation in UK agriculture, either on farms or administratively. In the past, some estates required tenants to redistribute eroded soil, or pay dilapidations, but these have largely lapsed by default. Similarly, many tenancies specify 'good husbandry'. When enforced this understandably tends to be directed at agricultural aspects, although at least one major landowner is considering its re-interpretation to encourage environmental protection. At the advisory level, charging and commercialization arrived in the 1980s and 1990s. At the same time, environmentally driven needs for re-education of the industry became apparent. So far, resources for that re-education have not been available on an appropriate scale, to the detriment of both the environment and farmers. While the *Codes of Good Agricultural Practice* (MAFF, 1998a,b) promote protection of water, soil and air, they are expressed in broad terms. They stop short of making recommendations for specific cropping and livestock management activities. In the USA, Australia and New Zealand, detailed best management practices (BMPs) are encouraged as a means of protecting water and conserving soil.

Implementation of BMPs, as suggested by the Environment Agency (1997), possibly through whole-farm management planning, should reduce agricultural emissions of sediment. Inevitably, there will be costs to farming and inertia in uptake, although changes in approach, such as waste minimization or reduced tillage, may provide long-term savings in fertilizer, fuel and other costs. Recommendation 26 by the RCEP (1996), attaching environmental 'cross compliance' to support payments, could spur adoption of BMPs. In some catchments there are attempts to encourage BMP uptake, as in the Environment Agency's 'Landcare' project situated in the Hampshire Avon basin, and the Westcountry Rivers Trust's 'Tamar 2000'. Recent Ministry of Agriculture publications on controlling soil erosion (MAFF, 1999) have acknowledged the wider definition of soil erosion as the transfer of soil particles, as well as the risk of consequent degradation of the water environment.

The impacts of farm woodland and forestry operations can be minimized by fully considering site conditions. The Forestry Commission's guidelines (1993) detail good practice for forestry. When fully implemented this will raise standards of practice, encouraging planning of operations in ways sensitive to specific sites. Ploughing and draining are particularly hazardous wherever they increase water flow, soil erosion and sediment delivery to watercourses. Good practice avoids any unnecessary cultivation or other soil disturbance. Where furrows or drains are needed, they should follow the contour with minimal gradients along them. Sensitively constructed buffer strips should reduce the sediment reaching streams but, as in other situations, the finer sediment is least likely to be removed. Harvesting of trees involves heavy machinery on the land. Poorly timed work compacts and ruts wet ground, accelerating overland flow, soil erosion and sediment transfer. Damage to ditches and watercourses is a further risk when felling and clearing. Forest roadways are both sources and means of transport for sediment. Soil exposed as cuttings, embankments and ditches during road construction is easily eroded. Vehicle movements in wet weather on unmetalled roads generate mud and turbid water, which flow along them or their drainage systems, often adding further erosion. Crossing points of ditches, streams and rivers are especially vulnerable.

In the past, there has been little attempt to control sediment from construction. Unlike the situation in the USA this continues for small developments. Recently, the Environment Agency for England and Wales has attempted to reduce turbid discharges through management of major projects from their design onwards. This has included the 1997 'Building a cleaner future' national campaign and the production of Pollution Prevention Guidance leaflets (Environment Agency, 2001).

Where, despite all efforts, unacceptable amounts of sediment continue to affect spawning beds, water jetting to clean the gravel and raking to loosen it, may help under limited conditions. However, these are not effective unless they penetrate sufficiently below the egg zone of redds to provide a sump with at least 1 year's storage for intruded fine sediment. Constriction of streams to increase velocity is another approach. While it may prevent deposition of fine sediment on the stream bed, it will be less effective in preventing intrusion during spates. These procedures are costly, temporary, and can be counter-productive; the real solution remains in keeping fine sediment out of watercourses in the first instance.

Conclusion

Sediment entering water represents a dilemma common to all issues of agricultural pollution: the reconciliation of farmers' priorities with the needs of the environment. Agriculture in Europe remains largely a response to public policies. Some aspects of these policies were established when food shortages were fresh in people's memory. Now the public expects full shelves in the supermarket all year round. With a few exceptions, farmers grow crops and rear livestock in ways strongly influenced by support payments and quotas. Often these, and the resultant overproduction and overstocking, have encouraged practices detrimental to the environment, in the ways described above.

This might be seen as environmental degradation by government and European Union policies, funded by the taxpayer. At the same time, substantial amounts of money are spent redressing matters, in particular for protection of water. Funding of this is again from public sources, as well as by farmers who have contributed to agriculturally non-productive investment to avoid pollution and the risk of fines that might ensue. So the public pays both ways. Despite the high value of water once it reaches the water service companies, for most farmers money changes hands only when they get things wrong. For everything else of wholesome quality and market value that leaves the farm there is payment. Why should this not be done for water as well? If this were achieved, there would be real ownership of the problem by those able to prevent or mitigate it.

Farming has evolved to maximize agricultural benefits, not to meet environmental concerns, which have come forward only in recent years. Turbid overland flow, as an example, can be a serious pollutant, but so far the message has not got across to many environmentalists, let alone farmers. There is powerful long-term pressure, notably via the World Trade Organization, for support payments to be moved from agricultural production to environmental benefits. Such fundamental change will present an opportunity for reconciliation of the needs of the environment and agriculture. A priority, which BMPs will begin to address, will be for education on the links between land use and the environment, particularly regarding sources, causes, pathways and mitigation of water pollution. This will require an effective extension service, supported by sound science, with robust credentials in both environmental and agricultural matters, readily available to all farmers and land managers.

Supposing that, one day, agriculturally derived pollution, including sediment, were solved and became part of history, the question would arise as to what end has all the effort been put? Wholesome water must be a benefit to society, particularly for water supply. But has it all just created, at great expense, clean, near-sterile rivers or niches for inappropriate or alien aquatic communities, or is there a real chance of restoring rivers to their pre-industrial state when many of them teemed with salmon?

References

Agate, M. (1983) *Footpaths.* British Trust for Conservation Volunteers, Wallingford, UK.

Alonso, C.V., Theurer, F.D. and Zachmann, D.W. (1996) *Sediment Intrusion and Dissolved Oxygen Transport Model-SIDO.* Technical Report No. 5, USDA-ARS National Sedimentation Laboratory, Oxford, Mississippi,

Boorman, D.B., Hollis, J.M. and Lilley, A. (1995) *Hydrology Of Soil Types: a Hydrologically-Based Classification of the Soils of the UK,* Report No. 126. Institute of Hydrology, Wallingford UK.

CIRIA (Construction Industry Research and Information Association) (2000) *Control of Water Pollution from Construction – Guidance for Consultants and Contractors.* CIRIA, London.

Colborne, G.J.N. and Staines, S.J. (1987) *Soils in Somerset I (Sheet ST41/51),* Soil Survey Record No. 111. Lawes Agricultural Trust, Harpenden, UK.

Easton, K.W. (1978) *The history of the fisheries of the River Trent.* Dissertation for Licentiate of Fisheries Managers. Institute of Fisheries Management, West Bridgeford, Nottingham, UK.

Environment Agency (1997) *Best Management Practices to Reduce Pollution from Agriculture,* R & D Technical Report P40. Environment Agency, Bristol, UK.

Environment Agency (2001) *General Guide to the Prevention of Pollution of Controlled Waters.* Pollution Prevention Guidance Leaflet (PPG01) http://www.environment-agency.gov.uk/business/ppg/?version=1

Evans, R. (1990) Soils at risk of accelerated erosion in England and Wales. *Soil Use and Management,* 6, 125–131.

Evans, R. and Felton, M. (1987) Hill livestock compensatory allowances and upland management. In: Bell, M.D. and Bunce, R.G.H. (eds) *Agriculture and Conservation in the Hills and Uplands.* Institute of Terrestrial Ecology, Merlewood, UK, pp. 66–72.

FitzPatrick, E.A. (1986) *An Introduction to Soil Science.* Longman, Harlow, UK.

Forestry Commission (1993) *The Forests and Water Guidelines.* HMSO, London.

Fraser, A.I. (2000) Towards a systematic framework for quantifying and extrapolating diffuse phosphorus transfer from agricultural fields to the national scale. PhD thesis, Cranfield University, Silsoe, UK.

Harrod, T.R. (1979) Soil suitability for grassland. In: Jarvis, M.G. and Mackney, D. (eds) *Soil Survey Applications.* Soil Survey Technical Monograph No. 13, Harpenden, UK, pp. 51–70.

Harrod, T.R. (1992) *Land Use Practices and Pathways of Pesticide Residues in the Mount's Bay Catchment, Cornwall.* Project. No. 82/3982. Soil Survey and Land Research Centre, Okehampton, UK.

Harrod, T.R. (1994) Runoff, soil erosion and pesticide pollution in Cornwall. In: Rickson, R.J. (ed.) *Conserving Soil Resources.* European Perspectives. CAB International, Wallingford, UK, pp. 105–115.

Harrod, T.R. and Fraser, A.I. (1998) *Phosphorus Loss from Agriculture,* Project NT1003. Soil Survey and Land Research Centre Report to MAFF, Okehampton, UK.

Harrod, T.R. and Fraser, A.I. (1999) *A Systematic Approach to National Budgets of Phosphorus Loss through Soil Erosion and Surface Runoff at National Soil Inventory Nodes,* Project NT1014. Soil Survey and Land Research Centre Report to MAFF, Okehampton, UK.

Haygarth, P.M., Hepworth, L. and Jarvis, S.C. (1998) Forms of phosphorus transfer in hydrological pathways from soil under grazed grassland. *European Journal of Soil Science* 49, 65–72.

Lee, J.W. (1979) Slurry acceptance. In: Jarvis, M.G. and Mackney, D. (eds) *Soil Survey Applications.* Soil Survey Technical Monograph No. 13, Harpenden, UK, pp. 83–100.
MAFF (1998a) *Code of Good Agricultural Practice for the Protection of Water,* PB 0585. MAFF, London.
MAFF (1998b) *Code of Good Agricultural Practice for the Protection of Soil,* PB 0617. MAFF, London.
MAFF (1999) *Controlling Soil Erosion. Advice for Preventing Erosion by Water in Lowland England and Wales,* PB 4262. MAFF, London.
Marshall, W. (1787) *The Rural Economy of Norfolk.* Cadell, London.
Martin, B. (1785) *The Natural History of England.* Owen, London.
Mills, D. (1989) *Ecology and Management of Atlantic Salmon.* Chapman and Hall, London.
Morgan, R.P.C. (1985) Assessment of soil erosion risk in England and Wales. *Soil Use and Management* 1, 127–131.
Morgan, R.P.C. (1986) *Soil Erosion and Conservation.* Longman, Harlow, UK.
Naismith, I.A., Wyatt, R., Gulson, J. and Mainstone, C.P. (1996) *The Impact of Land Use on Salmonids, a Study of the River Torridge Catchment.* R & D Report No. 30 to the National Rivers Authority, Bristol, UK.
Nesbit, T.R. (1996) The efficacy of the 'Forests and water guidelines' in controlling pollution. In: Petchy, A.M., D'Arcy, B.J. and Frost, C.A. (eds) *Diffuse Pollution and Agriculture.* Scottish Agricultural College, Aberdeen.
Netboy, A. (1968) *The Atlantic Salmon. A Vanishing Species?* Faber and Faber, London.
NRA (1995) *Understanding Riverbank Erosion from a Conservation Perspective. A Strategy for the Management of Salmon in England and Wales,* NY-4/95–2.5k-E-ANFQNRA. National Rivers Authority, Bristol, UK.
Ordnance Survey (1862) *Domesday Book; the Great Survey of William the Conqueror, A.D. MLXXXVI.* Ordnance Survey Office, Southampton, UK.
Palmer, R.C. (1993) *Risk of Soil Erosion in England and Wales by Water on Land Under Winter Cereals,* 1:625,000 scale map. Soil Survey and Land Research Centre, Silsoe, UK.
Purvis, W.K. (1997) *Environment Agency Technical Memorandum* TM/EAE/97/50. Environment Agency, Welsh Region, South Eastern Area, Environmental Appraisal Unit, Cardiff.
RCEP (Royal Commission on Environmental Pollution) (1996) *Nineteenth Report. Sustainable Use of Soil.* HMSO, London.
Robinson, M. (1990) *Impact of Improved Land Drainage on River Flows,* Report No. 113. Institute of Hydrology, Wallingford, UK.
Sharpley, A.N. and Rekolainen, S. (1997) Phosphorus in agriculture and its environmental implications. In: Tunney, O.H., Carton, O.T., Brookes, P.C. and Johnston, A.E. (eds) *Phosphorus Loss From Soil to Water.* CAB International, Wallingford, UK, pp. 1–53.
Stewart, L. (1963) *Investigations into Migratory Fish Propagation in the Area of the Lancashire River Board.* Barber, Lancaster, UK.
Theurer, F.D., Harrod, T.R. and Theurer, M., (1998) *Sedimentation and Salmonids in England and Wales,* R & D project P2–103, Environment Agency. Institute of Hydrology, Wallingford, UK.
Thomasson, A.J. (1982) Soil and climatic aspects of workability and trafficability. In: *Proceedings of the 9th Conference of the International Soil Tillage Research Organisation.* ISTRO, Osijek, Croatia, pp. 551–557.
Trafford, B.D. (1970) Field drainage. *Journal of the Royal Agricultural Society* 131, 129–152.
Unger, P.W. (1984) *Tillage Systems for Soil and Water Conservation.* FAO Soils Bulletin 54, Rome.
Walker, J. (1997) *Fine Sediment in the River Keer. Geomorphological Assessment,* Report T86/12/1. Environment Agency, Warrington, UK.
Waters, T.F. (1995) *Sediment in Streams; Sources, Biological Effects and Controls.* American Fisheries Society, Bethesda, Maryland. 251pp.

8 Nutrient Balances

Douglas B. Beegle,[1] Les E. Lanyon[1] and J. Thomas Sims[2]
[1]Department of Crop and Soil Sciences, The Pennsylvania State University, 116 ASI Building, University Park, PA 16802, USA; [2]Department of Plant and Soil Science, 163 Townsend Hall, University of Delaware, Newark, DE 19717-1303, USA

Key words: cropping, Delaware, farm, nitrate, nitrogen, nutrient balance, phosphorus, poultry

Introduction

The issues surrounding nutrient balance vary considerably around the world. Table 8.1 provides a comparison of the N balance for arable land in Rwanda, arable land in The Netherlands, grassland in the UK, and a high intensity poultry farm in Delaware, USA. These N budgets show both extremes of the balances. The first is a negative balance, where there are concerns with soil nutrient depletion. Here, providing inputs to achieve adequate crop production are of paramount importance. At the other extreme is the intensive poultry farm in Delaware where there is a significant

Table 8.1. Comparison of nitrogen balance (kg ha^{-1} $year^{-1}$) for A, arable land in Rwanda; B, arable land in The Netherlands; C, grassland in the UK; and D, intensive poultry production in Delaware, USA.

	Rwanda[a]	Netherlands[a]	UK[a]	Delaware
Inputs	19	250	450	671
Output	72	250	370	426
Balance	−53	0	+80	+245

[a] Janssen (1999).

excess N balance. In this case the greatest concern is with the potential environmental impact from the excess nutrients. As these data illustrate there is a wide range in N balance; however, there is a general trend towards greater specialization and intensification in agriculture and, especially in the developed world, a trend towards greater excess nutrient balances.

Nutrient balance can be looked at from many perspectives. A common focus is to look at nutrient balance at the soil and plant level. The emphasis of soil fertility management has been to provide balanced nutrition for optimum production of crops. The concern has been with optimizing the economic return from nutrients used to produce a crop based on the expected crop response from adding nutrients. This emphasis grew out of a chronic negative nutrient balance in agriculture. Barnes, in 1892 in a Pennsylvania State Board of Agriculture report (Barnes, 1892), emphasized Leibig's Law: 'The manure produced in the course of husbandry is not sufficient to permanently maintain the fertility of a farm; it lacks the constituents which are annually exported in the shape of grain, hay, milk, and livestock.' Overcoming this imbalance has been the driving force behind the development of the science of soil fertility management. This continues to be an issue in areas with a negative nutrient balance such as the Rwanda example in Table 8.1.

Today the main focus of nutrient balance is on excess nutrients, as illustrated by the intensive poultry example in Table 8.1, and the potential threat that this represents to water quality. To many, this imbalance is perceived to be the result of mismanagement by farmers. While it is true that mismanagement has been, and can still be, a source of nutrient imbalance, the factors that influence the overall balance are much more complex. To understand this issue it will help to understand how agriculture has changed from the 1800s to today.

History of Nutrient Use

Traditionally, crop and livestock farms produced outputs that resulted from the almost exclusive use of on-farm resources (Fig. 8.1a). A large proportion of the plant nutrients that were in the crops produced as feed for the animals were returned to the farm fields in manure. The output of this traditional farm is animals or animal products resulting in a relatively small proportion of the crop nutrients leaving the farm. On a traditional, self-sufficient farm, fewer nutrients will be returned to the fields in the manure than were harvested in the crop through the sale of the animal products and nutrient losses in manure production, storage and handling. Consequently, the nutrient balance will be negative. The efficient return of nutrients to the fields was essential to maintain the level of crop production on the farm. Unless there is gross mismanagement, a farm such as this is less likely to lose nutrients to the environment; instead the farm will need to import nutrients to overcome its negative nutrient balance.

A variety of actions have created a policy path promoting specialization in agriculture and the development of a feed-grain/animal production system (Lanyon,

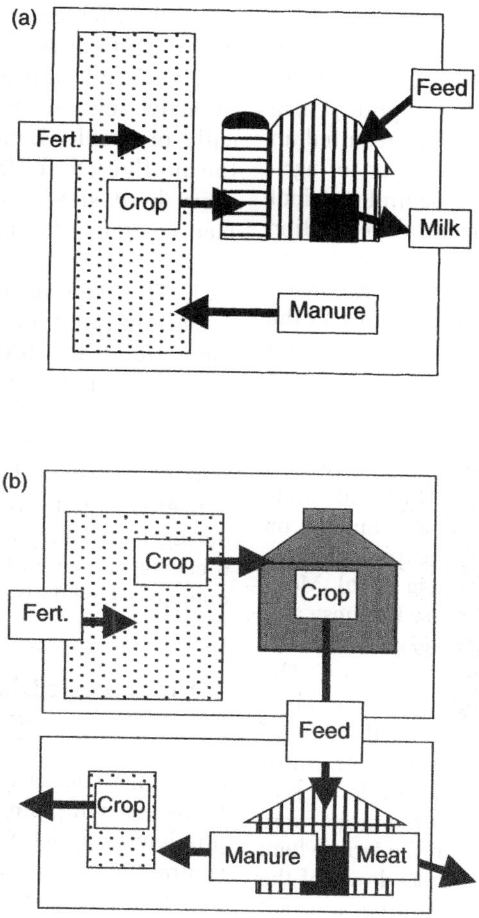

Fig. 8.1. Nutrient cycling for (a) a traditional crop–livestock farm with nutrients recycling within the farm, and (b) specialized crop farm linked to an intensive livestock or poultry farm with no recycling of nutrients. Nutrients usually accumulate in the limited cropland on the livestock farm.

1994). Probably the most important was the development of the fertilizer industry. Because of a shortage of N for use in manufacturing munitions to fight the Second World War, industrial N_2 fixation capacity was greatly expanded. At the end of the war, these plants were not mothballed but were used for N fertilizer production. Mass production of other fertilizer (e.g. P and K) soon followed.

The ready availability of N fertilizer helped to overcome the chronic negative N balance that limited crop production on many farms. This has made it possible to offset the losses of plant nutrients in animal production and manure handling from a traditional, self-sufficient farm. If the nutrients previously available on the farm were inadequate to meet potential crop productivity, fertilizer made it possible to build up soil fertility to achieve higher potential crop production. This is the familiar nutrient response to fertilization that was so important to the widespread adoption of fertilizer by farmers. This also enabled some farms to specialize in grain crop production because they no longer had to depend on forage legumes and animal manures to help supply the N, P and K requirements of the crops.

At the same time, other factors such as improved transport systems, new crop and animal production technologies and, in some countries, government policy also promoted specialization. This resulted in major changes in the organization of agriculture. The driving force behind nutrient balance on farms changed from dealing with the on-farm scarcity of nutrients and the need to efficiently recycle them for on-farm feed production, to responding to the greater efficiencies of production and resulting economic incentives that arose from specialization.

The success of crop production on cash-crop farms in specialized geographic regions has also made it possible to concentrate large numbers of non-ruminant animals on small land areas, usually completely separate from crop production (Fig. 8.1b). Most, if not all, of the feed necessary for these animals can be economically transported to the farm where the animals are housed. The cash-crop farm and the intensive, modern livestock farm are connected by the flow of feed to the livestock farm. However, nutrients in this flow often do not cycle back to the original locations where the crops were produced but accumulate on the farm where the livestock are located. Breakdown of the nutrient cycle and the resulting nutrient imbalances are significant features of the nutrient management situations encountered on these farms. Table 8.2 shows some examples from Pennsylvania (USA) farms of the P imbalances that can occur. The cash-crop farm is very close to balance as would be expected, but as the off-farm feed inputs increase the excess P balance increases also.

While this greater specialization and use of off-farm sources of feed can result in significant nutrient imbalances, these intensive animal operations are economically profitable. Data from dairy farms in the north-eastern USA illustrate this. As dairy herd size increased the purchased feed costs for the farm increased and the cropland area per cow decreased (Fig. 8.2). This can result in nutrient imbalances and pollution. However, the driving force for this trend is the dramatic increase in net earnings per farm that do not fully include the environmental costs of animal production. Managing nutrients to address these additional environmental concerns is not simply a matter of eliminating bad management, because most of these farmers would be considered good managers based on economic criteria.

The Strategic Dimension to Nutrient Balance

Non-point source nutrient pollution from farms is a symptom of a more fundamental problem in agriculture. The localized concentration of nutrients in animal agriculture has less to do with the need of local crops for nutrients than with the consequences of the structure of modern animal agriculture. Solving these problems will require strategic changes in our agricultural systems well beyond on-farm management changes. To meet this challenge will require innovative broad-scale policy and management approaches.

Table 8.2. Phosphorus balance (kg P ha^{-1} $year^{-1}$) on some representative farms in Pennsylvania (Bacon *et al.*, 1990; Lanyon, 1990; and unpublished).

	Cash crop 30 ha	Dairy 65 Holsteins 41 ha	Pig 1280 pigs 30 ha	Poultry 75,000 layers 30 ha
Fertilizer	18	11	—	—
Feed	—	29	101	1647
Output	16	12	19	502
Balance	+2	+28	+82	+1145

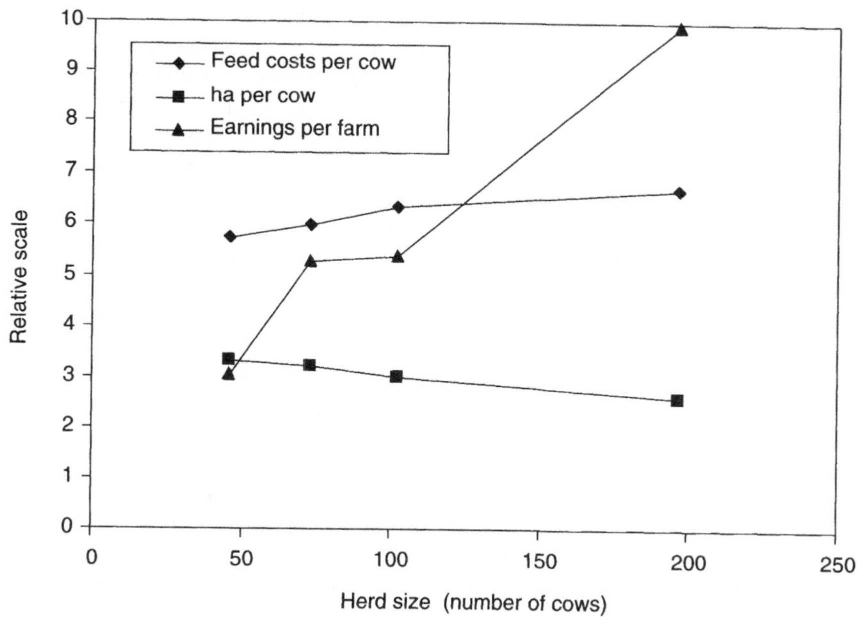

Fig. 8.2. Trends in dairy farm benchmarks for the north-eastern USA for 1991 indicating that increasing herd size results in more purchased feed and less land per cow which increases the nutrient imbalances; however, farm earnings increase (Snow, 1992).

It is strategic management that determines the goals of the farm and what resources will be incorporated into the farm operation to meet them. In most parts of the developed world, nutrients are no longer scarce resources that can limit farm activities. Therefore, many recent strategic plans have focused on a variety of business, personal and social goals; the fate of nutrients is only a secondary consequence of the management process. Problems of excess nutrient imbalances are the result of strategic decisions, not tactical mismanagement.

Tactical management, development and implementation of nutrient management plans, is where most of the emphasis has been placed in programmes related to nutrient management for improving water quality. However, unless tactical management is guided by appropriate strategic goals and due consideration of the other strategic messages farmers and agribusinesses receive, the desired outcome in terms of nutrient balance and thus water quality will not be achieved. We must resist the urge to go directly to the tactical management before we have addressed the strategic issues of the problem. At the same time we must make sure that the tactical nutrient management plan can be operationally implemented. Finally, it is important to remember that on many farms all of these distinct levels of management may reside within one person, the farmer. Thus, not only must people who work with farmers recognize the importance of these different levels of management, but the farmers themselves must also realize these differences.

Controlling nutrient flows and balances at the appropriate scale

Algal blooms in surface water resulting from high P concentrations, or high NO_3^- concentrations, in groundwater above safe drinking water limits are examples of symptoms that have led to increased dissatisfaction with existing nutrient management in agriculture. The typical remediation response is to begin where symptoms are observed and trace backwards to identify the source of the contamination. Once the source is identified, then one or more recommendations are made for a tactical management change. This trace back and stop approach has been effective for identification and remediation of point sources of pollution. This approach has also been relied on as the primary means to reduce agricultural non-point source pollution, yet the root of the non-point source problem with nutrients in animal agriculture is not tactical mismanagement, it is the excessive accumulation of nutrients on the farm. This is an outcome of strategic management that goes well beyond the farm scale. The management of nutrient flows and balances is really related to the business and community development programmes of the public sector in an area and global decisions made by governments and corporate enterprises. The decisions depend on a wide range of biological, economic, and socio-technical factors at many levels of agriculture and society.

Biological controls

Characteristics of plant and animal biology have a strong but not always consistent influence on nutrient flows and balances. On a cash-crop farm the nutrient flow is relatively simple: nutrients leave the farm in the crop produced and fertilizer or other sources of nutrients are brought on to the farm to replace this removal. On this type of farm, there is a direct linkage between biology, field nutrient balance and the expected economic returns to the farming activities. Those concerned with water quality protection have rarely advocated supplying less than the nutrients required for the economic optimum availability. They have instead emphasized maximizing the efficiency of nutrient use (rate, timing and method of application).

On a livestock farm that is primarily feed-self-sufficient, the most important consideration in managing nutrients is managing the biological efficiency of the whole farm to minimize the need for external inputs of nutrients. Here again, biological controls can be effective for encouraging nutrient balance. However, biological inefficiency in crop production does not always translate directly into economic inefficiency of the farm. In this recent era of readily available, inexpensive nutrients, animal-based agriculture has prospered based on the flow of nutrients to the animal operations in crops prepared as animal feeds. Thus, nutrients usually do not cycle back to their original locations. This often results in more manure produced and applied to cropland in limited areas near where animal production is concentrated.

Even though biological efficiency would dictate nutrient recycling, there are no economic incentives to limit the nutrient flows and imbalances in specialized animal production operations. Thus, strategic management for crop or animal producers does not include a strong biological constraint on nutrient accumulation at the animal feeding locations. Instead, the impacts of the accumulations are viewed as externalities or consequences of production that are experienced outside the production system. To expect a manager to voluntarily make a decision to invest in the reduction of externalities may not be realistic, especially after the animal facility is in place and the operating budgets established.

There are some opportunities to enhance the efficiency of nutrient utilization by animals that can reduce the magnitude of the nutrient flows and balances. For instance, non-ruminants are not able to synthesize all the necessary amino acids or fully utilize the P in the maize grain and soybean oil meal-based diets that are commonly used. Overfeeding of total protein to compensate for the low availability of one or a few amino acids or supplementation of the diets with readily available P to compensate for low P availability are among the techniques used to overcome the liabilities of these raw materials. However, these approaches generally increase the excess nutrient levels. New technologies, such as amino acid supplementation, which reduces the total protein content of pig rations, and adding phytase to feeds to increase the availability of P to pigs and poultry, can reduce the total N and P in the diets of these animals and in the resulting manure. When production per animal is the most critical guiding criterion of successful strategic farm management, convincing farmers and other business decision makers to try these ration manipulations will be difficult. Also, this will require a whole new research base in animal nutrition.

Increasingly in modern agriculture, there is little biological reason for nutrients not to accumulate and there is no crop production reason for them to be returned to where the crops were grown. The function of crops or animals within this organization is significant, but is not a meaningful focus for remedial action to address issues of nutrient imbalance and the resulting threat to water quality. As long as fertilizers are available to meet crop needs, this form of agricultural organization can be biologically and economically successful.

Economic controls

Specialized animal production does not occur in isolation from the host of economic controls over agricultural production. Market place rewards for large-scale marketing of animal products, policy programmes and financial advantages that encourage specialized crop production are significant contributors to the strategic management environment of both farmers and businesses that makes concentrated animal production economically practical. Economic controls of nutrient flows and balances have tended to encourage nutrient accumulations in excess of crop production because of the positive economic feedback to concentrated animal feeding operations. Some control programmes have used economic tools such as taxes on inputs or emissions, subsidies for pollution reduction, purchase or sale of emission rights, and funding research and education in attempts to rectify the consequences of nutrient accumulation. To date, these tools have had limited application to nutrient flow and balance problems. The success of programmes based on these economic controls has been limited because the control efforts designed to balance nutrients are often in direct conflict with the positive incentives created by the wide range of other economic influences which are causing the imbalances. There is also the question of how much effect these local, state or even national economic programmes can have in the context of the global economy.

At the farm level on a cash-crop farm, anything that improves the efficiency of nutrient utilization and thus minimizes losses should increase the economic returns to the farm. Thus, there is a strong economic incentive for good nutrient management consistent with water quality protection. On an animal production farm, neither crop production nor fertilizer use is directly connected to the output of such farms. For example, on a dairy farm the main consideration in crop management is producing adequate, high-quality feed for the cows because the money is made from the dairy. Farm performance depends on the animal husbandry skills of the farmer, not just success in crop production. On such a farm, the decisions about plant nutrient use in the fields are not as sensitive to the economic or agronomic criteria in crop production as on the cash-crop farm. Often agronomic and economic principles of crop production are compromised in the effort to ensure that there will be adequate, high-quality feed for the livestock and that management resources are not diverted from the animal enterprise.

As the intensity of animal production increases, the application of nutrients to fields becomes less closely related to the major production activity of the farm; that is, selling animals or animal products. This will often result in an excess of nutrients on the farm and a high potential for environmental problems. The field-based economic and agronomic incentives that can be employed effectively to manage nutrients on a cash-crop farm, and that will also minimize negative environmental impacts, are not as significant on the intensive livestock production-oriented farm. Further, field-based agronomic practices may be of limited effectiveness in treating the total quantity of nutrients on the farm because of the small land area on the farm relative to the number of animals. It is very likely that plant nutrient management to protect environmental quality cannot be accomplished solely on the farm where the animals are housed. Successful management of nutrients to protect the environment will depend on support from off-farm people and organizations. Neighbours with land sufficiently low in nutrients to need manure application could provide land for manure distribution. Off-farm organizations may deal with manure hauling to locations where the manure can be used directly or transformed into another product such as compost. However, improved nutrient management for environmental protection on these farms will almost always come with a significant economic cost, such as expenses associated with hauling, storage and application of manure away from the farm.

Social controls

While successful marketing of animal products creates positive economic feedback to the farm (Fig. 8.3), perceived or actual consequences, such as degraded water quality, of other outputs from agricultural practices can generate negative feedback for intensive agricultural production. The negative feedback mechanisms can be direct, but informal, complaints to the operators, or more formal requests for curative action by enforcement agencies if the discharges violate existing codes. Either of these approaches is most effective when an acute discharge occurs that can be readily perceived by those who feel threatened by the consequences. This acute situation is not typical of non-point source pollution problems. The impacts of chronic nutrient discharges from agricultural production are more difficult to measure and social responses to unacceptable discharges may only come after a threshold in water quality or in perceived performance of a water resource has been crossed or a public catastrophe occurs. Responses in the past have emphasized voluntary actions by farmers that were promoted through educational

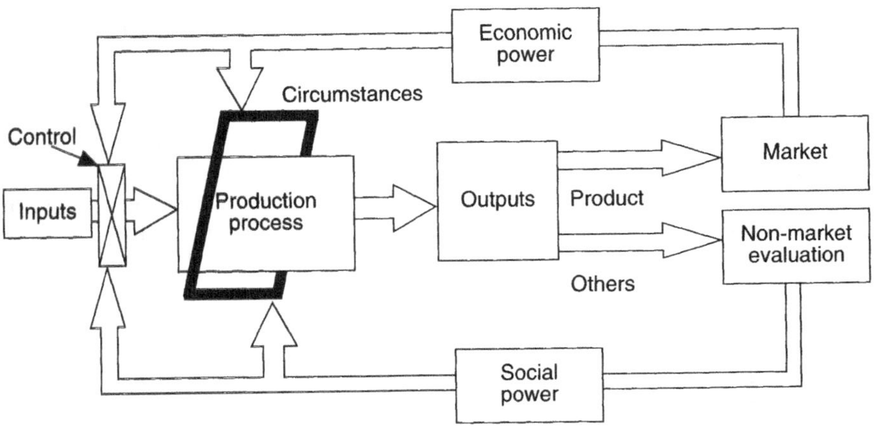

Fig. 8.3. Farmers are faced with dual and often contradictory feedback mechanisms when making management decisions about their inputs and production process. They must balance the positive feedback from the market and the often negative feedback from the social consequences for animal agriculture (Lanyon and Thompson, 1996). Reproduced with permission from *Animal Agriculture and the Environment: Nutrients, Pathogens, and Community Relations*, NRAES-96, by NRAES, the Natural Resource, Agriculture, and Engineering Service, Cooperative Extension.

efforts, cost-sharing (or subsidy) programmes, and the provision of technical assistance through government-sponsored programmes. These approaches have been most successful in situations in which the farmer also gains some direct benefit when a recommended practice, often referred to as a best management practice (BMP), is adopted or implemented.

In the more difficult situations, especially when a substantial excess nutrient balance is present on an animal enterprise, voluntary participation by farmers may not achieve adequate nutrient loss reductions. Because voluntary participation is being viewed as achieving insufficient response in many situations, government programmes are beginning to be formulated and implemented that require participation.

Mandatory approaches to nutrient loss reduction often include one or more of the following: taxes, quotas, nutrient management plans, practice standards and permits. Experience with most of these programmes is limited to a small number of countries, states or watersheds so their impacts are not yet clear. However, the concept of 'polluter pays' is becoming a bigger part of social controls on nutrient flows and balances.

The dilemma associated with these new mandatory programmes is that they do not offer individual farmers or agribusinesses operating in the targeted areas much flexibility. Farmers and agribusiness still receive contradictory messages coming from the economic drivers promoting concentrated animal production and the negative feedback from society that requires reduction in the flow of nutrients to environmental air and water systems. The self-interest of the adopter at the core of the traditional practice-based methods of intervention is no longer an appropriate basis for prompting change. The symptoms of nutrient imbalances that are observed in air and water systems cannot be readily treated by simply increasing the biological efficiency of the production process itself. Approaches will have to recognize symptoms, and not mistake them as the problem. This is likely to contribute to the reorganization of agricultural production. A process to address the implications at the necessary scale and a course of action for informed decision making will have to emerge from the interactions of all stakeholders if the framework is to be sufficiently broad and flexible. The possible consequences are difficult to foresee.

Controlling nutrient flows and balances is challenging. The lessons learned from point source pollution remediation are not likely to be very equitable, even if effective, in dealing with non-point source pollution. The strict concept of polluter pays may not fit the current organization of agriculture. It will weigh too heavily on individual farms where the symptoms occur, rather than on the system that is at the root of the problem. Effective alternative approaches may promote the value of pollution prevention to a broad range of the stakeholders and be compatible with a dynamic, evolving and equitable organization of agriculture. Pollution prevention as a marketable output of agriculture may reconcile the positive feedback from the market for traditional agricultural products with the negative feedback in reaction to the social concerns about the non-market consequences of production.

On-farm Nutrient Management Based on Nutrient Balance

Nutrient management plans will vary according to the nutrient balance for the farm. The intensity of the animal production system is a key factor in the nutrient balance on a farm. The intensity of the animal production system can be estimated using either the animal density on the farm, as indicated by the animal liveweight per area available for manure application, or the proportion of feed imported on to the farm. Using this criterion, nutrient management efforts can be appropriately targeted. The following nutrient balance categories are helpful in selecting appropriate nutrient management strategies as general guidelines to help farmers make management decisions. The categories discussed here are based on N balance. However, P is becoming a greater concern and a similar categorization can be made: in this case the animal densities for the categories will be one-half to one-quarter of those used for N.

Intensity of animal production systems

Farms characterized by animal densities < 1400 kg liveweight ha^{-1} available for manure application, or $< 60\%$ of the feed for the animals imported on to the farm will usually have a negative nutrient balance. There is usually more cropland than is needed to meet the feed requirements of the animals and consequently there is insufficient manure to meet total crop nutrient needs. In this category, the strategy will be to maximize efficiency of nutrient use by utilizing soil tests and manure analysis to assure distribution and timing of manure applications to maximize crop response to the nutrients from the manure and minimize purchase of commercial fertilizer. Many of the standard BMPs for manure management will be applicable. Practices such as storing manure, applying it in the spring to non-legume crops, or incorporating it immediately, will be recommended. The environmental impact of these operations should be minimal except where there is gross mismanagement. Any management changes in this category of operation will usually have only a small beneficial effect on the environment. Ultimately, there may be substantial economic benefits to the farmer for developing and implementing an improved manure management plan on this category of farm.

Medium intensity farms have animal densities between 1400 and 2500 kg liveweight ha^{-1} available for manure application, and 60–90% of the feed for the animals comes from elsewhere. These farms generally do not produce enough feed for the animals and there is likely to be enough manure to meet total crop nutrient needs. Most farms in this group should be close to being in balance for N and the strategy will be to utilize soil tests and manure analysis in conjunction with appropriate management practices to balance, as closely as possible, nutrients available in manure with crop needs over the entire rotation. In most cases, the emphasis will be on maximizing the safe utilization of the manure nutrients and strategic management to maintain whole farm balance into the future. Where there is a slight nutrient excess, efficiency may be intentionally compromised to increase utilization of the manure as long as it does not create an environmental threat. An example of this is applying manure to legumes. Even though the legumes may not need the N in the manure, they will utilize some of it. While the farm as a whole may be close to balance, there can be serious imbalances within the farm and careful management will be needed to provide the most favourable economic situation while protecting the environment. There is good potential for environmental benefits from improved management on these farms, but generally the economic impact of these management changes will be small. A detailed manure management plan will probably be necessary and other changes in the overall farm management, such as altering the cropping system, to increase nutrient utilization, may be necessary to achieve balance. Most farmers in this group will probably want to take advantage of technical assistance from public agencies and/or private consultants in developing and implementing a manure management plan.

High intensity farms have animal densities > 2500 kg liveweight ha^{-1} available for manure application and over 90% of the feed for the animals comes from off the farm. On these farms, the feed demands greatly exceed the capacity of the cropland and thus livestock manure production significantly exceeds total crop nutrient needs resulting in an excess nutrient balance. The objective will be to utilize every available means to remove all excess manure not needed for crop production. Alternative off-farm uses for the manure will need to be explored, which for many will mean locating a market for the manure and arranging the logistics of transport and appropriate application. Determining the maximum amount of manure that can be safely applied to the farm will be the main emphasis for nutrient management planning. However, in most cases the available land and the high residual nutrient levels in the soil may severely restrict on-farm use of the manure. Detailed nutrient management plans will also be important for those farms where the manure is ultimately utilized. This group of farms has the highest potential to have negative impact on the environment. In many cases, unless a favourable marketing arrangement can be developed, implementing improved nutrient management on this group of farms will have a negative economic impact on the farm. Assistance from public agencies and private consultants, manure brokers and manure haulers will be required. Unfortunately, these forms of assistance are not currently well developed in many regions of the world.

Nitrogen vs. phosphorus balance

A major issue in manure management policy today is whether nutrient management plans should be balanced for N or P. Both N and P can have a negative impact on water quality. A major concern with N is NO_3^- contamination of groundwater by leaching from the soil (see Hatch *et al.*, Chapter 1 and Chadwick and Chen, Chapter 3, this volume). High NO_3^--N in groundwater is thought to have adverse health effects. Controlling the source of N is important because it is very difficult to control its subsequent transfer. If runoff and erosion losses of N are reduced by applying conservation practices, other losses such as NH_3 volatilization, denitrification and leaching usually increase. Data from Sharpley and Smith (1994) show that when conservation tillage was used in wheat production, overland flow decreased but infiltration and N loss by leaching increased (Fig. 8.4). The most practical methods of minimizing N loss from cropland are to balance rates as close to crop removal as possible and to time N applications to avoid periods of the year when loss potentials are high.

The major concern with P is eutrophication of surface waters (Correll, 1998). Biological activity of many surface inland water bodies is P limited and when P inputs are increased, algal and other plant growth are stimulated. The resulting eutrophication restricts the use of surface waters for aesthetics, fisheries, recreation, industry and drinking water. While P may move through the soil into groundwater in some situations, such as in sandy soils with shallow water tables or where there is artificial drainage in soils with macropore structure (Sims *et al.*, 1998), the major pathway for P loss is by overland flow carrying P with detached soil particles. Management practices to limit P loss focus on reducing soil erosion and overland flow. Until recently, it was generally assumed that P losses could be controlled solely by reducing erosion. However, as noted above, P leaching can be significant in some situations and overland flow has been found to contribute significantly to losses of highly bioavailable soluble P in certain situations. This is especially true where conservation

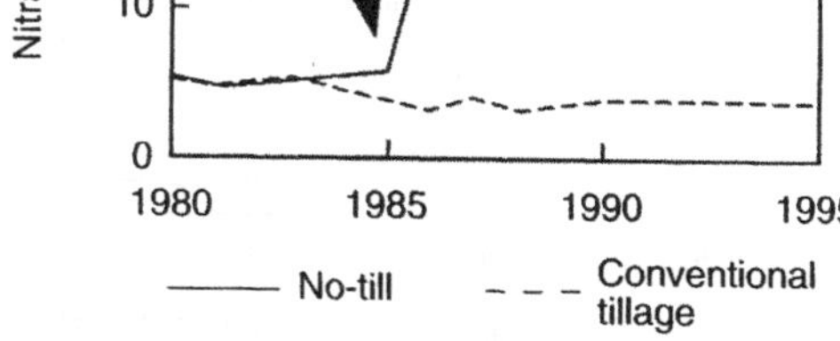

Fig. 8.4. Using no-till in a wheat production system to reduce erosion and runoff losses of nutrients results in increased infiltration and thus greater nitrate loss by leaching (Sharpley and Smith, 1994; A.N. Sharpley, University Park, Pennsylvania, 2000, personal communication).

tillage systems are used for erosion control. Sharpley and Smith (1994) showed that converting to a no-till wheat production system dramatically reduced total P loss (Fig. 8.5). However, conservation tillage systems preclude deep incorporation of manure and thus high nutrient contents develop near the soil surface. This means that runoff water is in intimate contact with highly nutrient enriched soil which may result in high soluble P concentrations in the runoff water.

While it is desirable that both N and P be balanced within farms, the controversy over N *vs.* P nutrient management plans results from the difference between their ratio in manure and their requirement by crops. Animal manures typically have an N:P ratio of *c.* 4 while crops require a N:P ratio of *c.* 8 (Fig. 8.6), so that if manure application rates are based on achieving a balance between the N applied in the manure and the N requirement of the crops, excess P will be applied. While this imbalance may not be an agronomic concern, it raises an environmental concern with the fate of this excess P. If manure application rates are based on achieving a balance between the P applied in manure and P requirement of the crop, at least twice as much area will be required for manure application and purchase of N fertilizer will be necessary. Many intensive animal production operations have less and less land area per animal and this presents a practical problem in properly balancing manure nutrients. Basing manure application rates on P will often carry a significant cost to the operation for excess manure disposal off-farm and the expense of purchasing supplemental N.

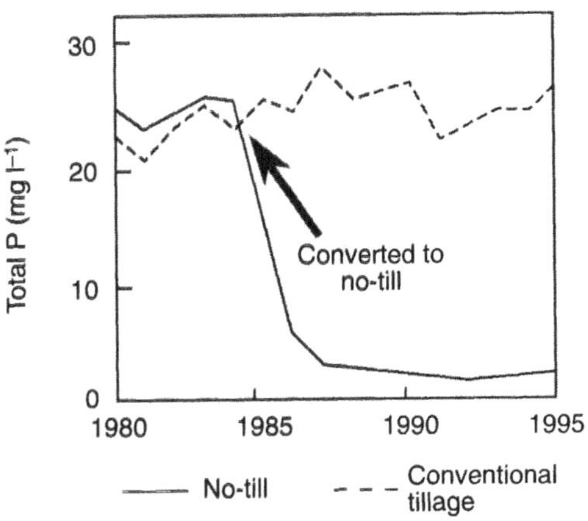

Fig. 8.5. Using no-till in a wheat production system to reduce erosion and runoff losses results in a significant reduction in total P loss (Sharpley and Smith, 1994; A.N. Sharpley, University Park, Pennsylvania, 2000, personal communication).

Most manure application guidelines and regulations are based on balancing N. This is driven by practical and economic considerations and on the assumption that best management practices can be developed and implemented which will reduce P losses to surface water even when excess is applied. However, because of high accumulations of P near the soil surface conservation practices such as no-till may not reduce P loss as much as was thought. The ultimate goal of nutrient management is to balance P so that no excess is applied. Because of the major changes required and the potential cost of a P-based nutrient management system, some compromise will be necessary on this issue.

Integrating nitrogen and phosphorus balance on farms

Fortunately, there are alternative management options that can contribute to achievement of nutrient balance. Standard BMPs such as proper timing and application of manure and soil conservation practices are important in all nutrient management plans designed to minimize nutrient impact on water quality. Manure treatment such as adding Al or Fe sulphate or waste materials such as fly ash to lower the solubility of P in manure can be effective in reducing P loss to water (Moore and Miller, 1994). While these BMPs do not correct the nutrient imbalance, they can help to minimize immediate environmental impacts of the imbalance. Cropping systems can sometimes be modified, particularly on livestock and dairy farms where forage crops are important, to better

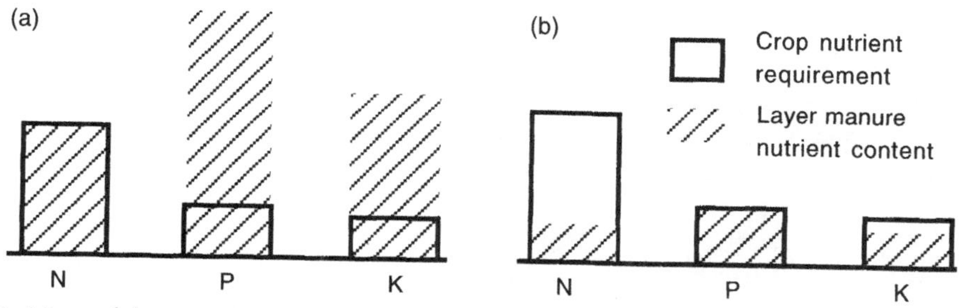

Fig. 8.6. Mismatch between manure nutrients and crop requirements for maize and layer manure. (a) When layer manure is applied to balance N, excess P and K are usually applied. (b) When layer manure is applied to balance P, a much lower rate of manure is applied and thus significantly more land and additional N will usually be required. (Example uses book values for manure nutrients and crop removal values.)

balance the crop requirements with the manure nutrients (Ford, 1994). This option is less feasible on pig and poultry farms. Improved ration formulation in animal feeds can have a significant impact on the nutrient levels in manure (Lynch and Caffrey, 1997). The use of phytase and low phytate maize in non-ruminant animal feeds to improve the efficiency of P utilization by the animals has also been successful (Ertl *et al.*, 1998). Greater utilization of grain P by animals results in less supplemental P coming on to the farm and thus a better P balance.

For nutrient loss to occur, there must be a source and a mechanism for transport to water. A key concept to effectively managing nutrient pollution is to focus on where these two factors overlap, that is, a high source coupled with high potential for transport. Current thinking recognizes that not all areas on a landscape contribute equally to nutrient losses. Recent research in the USA has shown that the majority of the P lost from agricultural fields to water may come from a limited area in many watersheds and from a few storm events (Gburek and Sharpley, 1998). The emphasis is on identifying such critical source areas for P loss to surface waters and then imposing application limitations and/or BMPs on those critical areas while allowing flexibility to apply manure based on N in non-critical areas. This targeting approach will also maximize the positive impact from programmes designed to help farmers to reduce water quality problems from nutrient imbalances. The P Index as proposed by Lemunyon and Gilbert (1993) is an example of a tool that is being modified to help managers to delineate these critical source areas. In the P Index, soil tests and nutrient applications are indicators of sources of P, and overland flow, leaching and proximity to water indicate areas where there is a high potential for transport of P. Together, these factors are used to delineate the critical source areas for P loss (Fig. 8.7). Sharpley (1995) has shown that there is a relationship between the P Index and total P loss from a watershed. This approach appears to be a much more rational way to address the environmental concerns with P in an economically practical way than proposals to immediately achieve balance by simply limiting manure application on a broad scale solely on the basis of soil test P. Similarly, critical source areas for N could be delineated and used to guide N management in a watershed.

It is unlikely that any of the above approaches, or others that may come out of this active research, will individually solve the problem. However, a set of tools will be provided that can be integrated into a management plan, based on the site-specific situation, to effectively address concerns that have been expressed about imbalances in agricultural nutrients without taking the extreme approach of a strict limit on P applications.

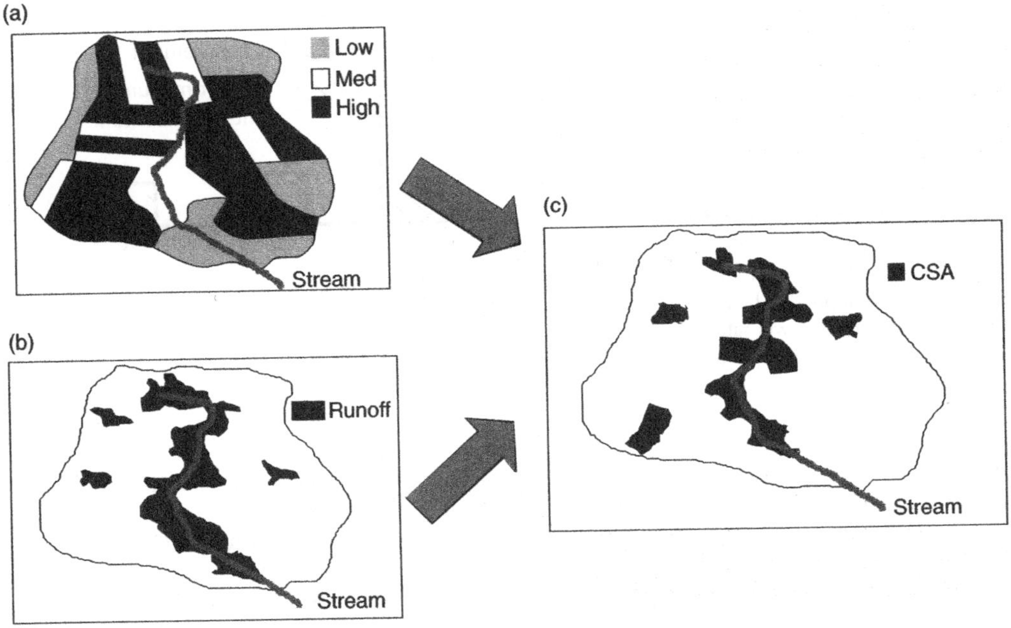

Fig. 8.7. Critical source area management, as illustrated here with an agricultural watershed in Pennsylvania, involves identifying the critical source areas for nutrient loss which can be targeted for site-specific management. Areas where there is a high source of P (a), such as high soil test levels or high applications of P, and high transport potential (b) are the critical source areas (c). These areas may be targeted for best management practices (BMPs) or restrictions on nutrient applications. (Adapted from A.N. Sharpley, University Park, Pennsylvania, 2000, personal communication: see also McDowell *et al.*, Chapter 9, this volume).

Case Study of Nutrient Balance: Delaware's Animal Based Agriculture

Many of the concepts in this chapter can be illustrated by consideration of the nutrient management and environmental quality issues faced today by the state of Delaware, located on the Delmarva (Delaware–Maryland–Virginia) Peninsula of the USA Atlantic coastal plain (Fig. 8.8). Delaware has a number of serious water quality problems related to non-point source pollution by agricultural nutrients, including groundwater contamination by NO_3^- and eutrophication of surface fresh waters and coastal estuaries by N and P (Hamilton and Shedlock, 1992; Sims *et al.*, 1999; Sims, 2000). Delaware agriculture is dominated by one of the most geographically concentrated poultry production industries in the USA (Sims, 1997; Cabrera and Sims, 2000). Past research has shown that NO_3^--N from animal manures and fertilizers is leaching into groundwaters (Ritter and Chirnside, 1987; Andres, 1995) and that P in many Delaware soils has accumulated to levels well in excess of the values needed for economically optimum crop yields, raising concerns about soluble and particulate P losses in runoff and leaching (Sims *et al.*, 2000). In recent years, public and governmental concerns about Delaware agriculture's role in non-point source pollution of ground and surface waters, and particularly the impacts of animal agriculture on water quality, have intensified markedly. In 1999, as a result of these concerns, the state of Delaware passed a Nutrient Management Act that established the Delaware Nutrient Management Commission (DNMC) with the following charge: 'To regulate those activities involving the generation and application of nutrients in order to help improve and maintain the quality of Delaware's

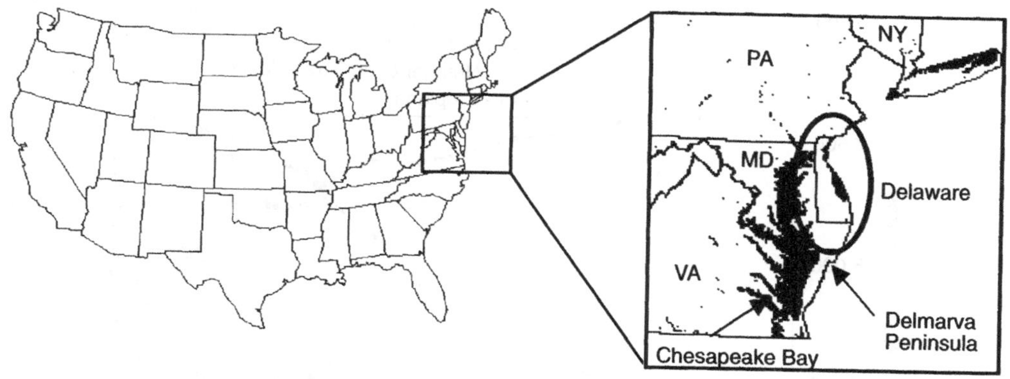

Fig. 8.8. Location of Delaware on the Delmarva Peninsula along the Atlantic Coastal Plain of the eastern USA.

ground and surface waters and to meet or exceed federally mandated water quality standards, in the interest of overall public welfare' (DDA, 1999). This case study provides an overview of the nutrient balance situation in Delaware and outlines some of the options now being considered to sustain agriculture and improve the quality of Delaware's environment.

Overview of Delaware agriculture

Poultry production dominates Delaware agriculture, accounting for 75–80% of the total cash receipts from farm marketing (DDA, 1998). About 260 million broiler chickens are produced each year, generating an estimated 275,000 Mg of broiler litter (a mixture of manure, urine and the wood shavings/sawdust used as bedding), which is almost exclusively used as a fertilizer material for the production of 235,000 ha of grain and vegetable crops. Other livestock production is limited, with a total inventory of 29,000 cattle and calves (dairy and beef) and 31,000 pigs (DDA, 1998).

The poultry industry in Delaware is vertically integrated, with poultry growers working under contracts from four large companies. These companies provide the young chicks, the feed, the medication and veterinary advice, remove the chickens at the end of each growth period (*c.* 5–6 weeks per flock), and handle the processing, marketing and retailing of the poultry meat. The growers' responsibilities include owning, building and maintaining the poultry houses, caring for the chickens during the growout period, and using or disposing of all wastes and by-products generated during poultry production (manures, litters, dead animals). Not all growers are crop farmers; a recent survey estimated that 60% of the poultry growers on the Delmarva Peninsula did not raise crops (Michel *et al.*, 1996) and of those, only 14% had more than 100 ha available for land application of manure.

The poultry industry grew dramatically in Delaware during the 1970s and 1980s, while at the same time the number of farms and the amount of cropland declined as the state has urbanized (Fig. 8.9). Most (> 90%) broilers are grown in Sussex County, Delaware, which has about 100,000 ha of cropland, primarily used for the production of maize, soybeans and small grains (wheat and barley which are mainly used as feed for the poultry). However, the size of the industry requires the import of large amounts of grain from other regions. Based on recent industry and state estimates 50% of the maize and 60% of the soybeans are imported from other states (Fig. 8.10; DDA, 1998; DPI, 1998). These imports represent significant nutrient imports to the state, comparable in magnitude to commercial fertilizer sales. For example, in 1998, N and P imported to Delaware in maize and soybeans amounted to 18,600 and 2000 Mg

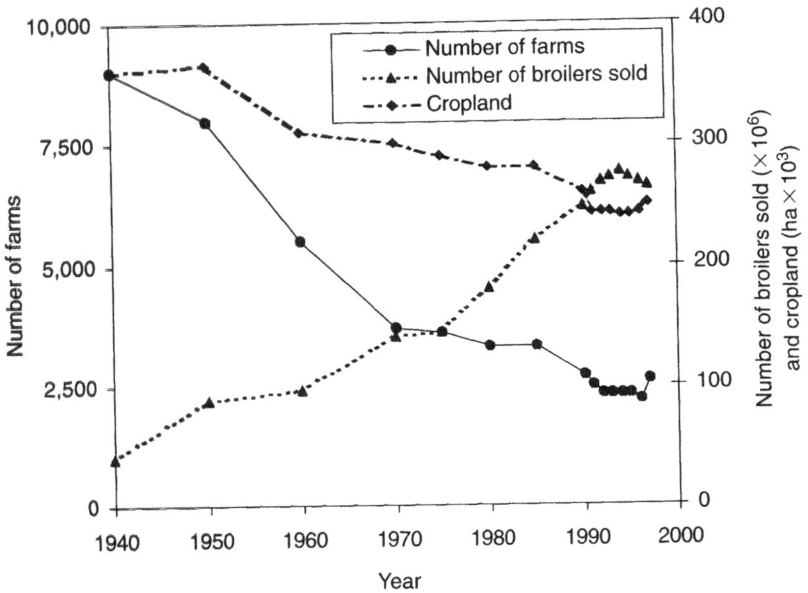

Fig. 8.9. Trends for poultry production, cropland and number of farms in Delaware, 1940–2000.

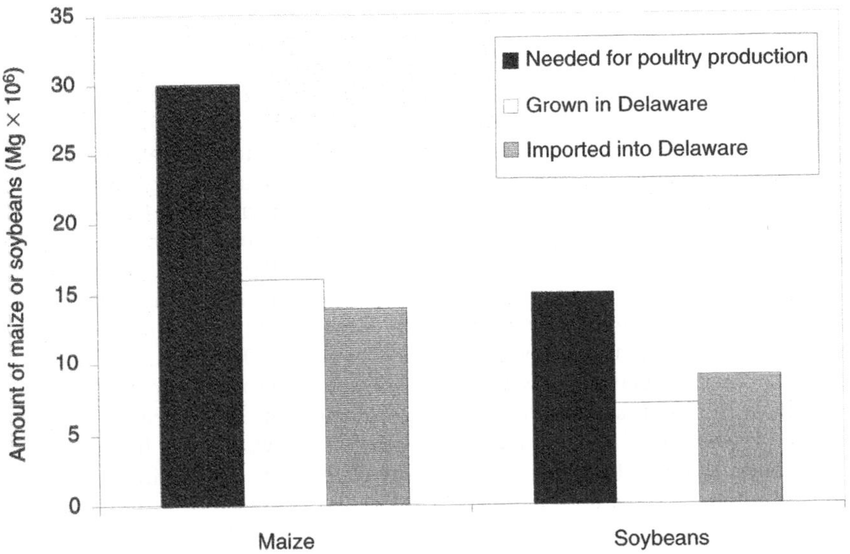

Fig. 8.10. Comparison of maize and soybean required for poultry production in Delaware with the amount of each grain grown and imported into the state.

per state year^{-1} (*c.* 80 kg N ha^{-1} year^{-1} and 9 kg P ha^{-1} year^{-1}), respectively, compared with fertilizer sales of 20,700 (N) and 3500 (P) Mg per state year^{-1} (*c.* 87 kg N ha^{-1} year^{-1} and 15 kg P ha^{-1} year^{-1}).

Nutrient balance issues in Delaware

The primary nutrient management issue in Delaware is one of surplus. Nutrient mass balance calculations clearly show that the importation of feed nutrients in combination with sales of commercial fertilizers, results in large annual surpluses of N and P in excess of the needs of economically optimum crop yields. There are no alternatives to land application of animal manures that are currently viable. The continual presence of nutrient surpluses in any agricultural system can usually be expected to exacerbate nutrient losses at all scales, from individual fields to watersheds. While it is well understood that some nutrient losses are inevitable because agriculture is not 100% efficient, sustainable agricultural systems must develop nutrient management practices, including properly balanced nutrient inputs and outputs, that minimize these losses to prevent the degradation of air, soil and water quality. In the following discussion we provide nutrient balance analyses for Delaware's predominantly animal-based agriculture to illustrate the strategic, tactical and operational aspects of this issue, including potential options to improve agricultural nutrient use.

Strategic dimension of Delaware's nutrient balance

The strategic dimension of the nutrient balance issue focuses on the overall goals of production, the external factors that affect these goals and the top-level, long-term policy and management changes needed to improve the efficiency of nutrient use. This dimension usually focuses on larger scales, such as the whole farm, or in the case of Delaware, on politically defined areas such as the state as a whole or a particular county, or on geographically identified areas, such as a critical watershed. The DNMC operates at the strategic scale as it attempts to develop equitable, long-term, statewide solutions.

Sound strategic nutrient management planning begins with a reasonable approximation of the current nutrient balance on the farm or other defined area to provide an estimate of the imbalance and thus the scale of the solutions that will be required. In Delaware, the estimated surplus is 83 kg total N ha^{-1} $year^{-1}$ and 30 kg total P ha^{-1} $year^{-1}$ (calculated on a state-wide basis) (Table 8.3). The surplus is even greater when calculated for Sussex County, Delaware, the site of most of the poultry production and location of almost half the cropland in the state (Table 8.4; 121 kg total N ha^{-1} $year^{-1}$ and 51 kg total P ha^{-1} $year^{-1}$). Note, however, that nutrient surpluses are much lower in New Castle and Kent Counties where the predominant agricultural activities are cash grain production and vegetable farming, not intensive animal agriculture (Table 8.4). In fact, New Castle County has a small P deficit, primarily because many of the soils in that county are considered 'low' or 'medium' in soil test P and thus in need of P fertilization (Sims, 1998).

Obviously, the nature and magnitude of the nutrient balance not only will vary between geographic areas (i.e. counties) but also will vary between different types of farming/animal production operations. This is clearly shown in Table 8.5, which provides N and P mass balances for three poultry farms with varying production intensities. These nutrient mass balances were calculated by comparing the inputs to the farm in feed, animals, fertilizers and biological N fixation with the outputs from the farm in harvested crops and animal products. In all cases there is a significant surplus of N and P, ranging from 28 to 56% of the N inputs and from 45 to 73% of the P inputs. After removal of the animals for processing and marketing, most of the surplus N and all of the surplus P will initially be found in the by-products of poultry production (litters, manures, dead animals). Some of the N (*c.* 20–30%) will be lost by volatilization of NH_3-N directly from the poultry houses. For the two farming operations ('typical' and 'high' intensity) the surplus N and P will then be divided between crop removal, storage in soils on the farm and losses by leaching, volatilization, denitrification and runoff. The large amounts of surplus N and P present on these two farms after animal and crop nutrient removal are accounted for (109–243 kg N ha^{-1} $year^{-1}$ and 30–72 kg P

Table 8.3. Nutrient mass balance (Mg year^{-1} × 10^3) for the state of Delaware, determined by comparison of nutrients produced or sold in Delaware annually with crop nutrient requirements at realistic yield goals.[a]

Nutrients produced or sold in Delaware vs. crop nutrient requirements	Nutrient available or required	
	Nitrogen	Phosphorus
Nutrients produced or sold:		
Commercial fertilizers	20.7	3.5
Poultry litter/manure	14.3	5.2
Dairy manure	0.86	0.13
Beef manure	0.15	0.04
Swine manure	0.17	0.06
Municipal biosolids	0.83	0.41
Total	37.0	9.4
Nutrients required by crops:		
Maize	8.5	0.53
Soybeans	0	1.12
Wheat	3.2	0.35
Barley	1.4	0.15
Sorghum	0.58	0.08
Processing vegetables	2.8	0.20
Fresh market vegetables	0.88	0.06
Hay	0.68	0.07
Total required by all crops	18.0	2.6
State-wide annual nutrient balance:		
Per state year^{-1}	19.0	6.8
kg ha^{-1} year^{-1}	83	30

[a] Assumptions:
1. Nutrients produced or sold: Information on commercial fertilizers was obtained from the Delaware Department of Agriculture (DDA, 1998). Nutrients available from animal manures are based on number of animals grown in Delaware (DDA, 1998) and standard values of the amount of manure produced per animal and the total and plant available nutrient contents of the manures (USDA-SCS, 1992).
2. Crop nutrient requirements are based on nutrient recommendations of the University of Delaware for realistic yield goals of each crop (Sims and Gartley, 1996) and the amount of each crop grown in Delaware (DDA, 1998). Phosphorus recommendations were also based on soil test P results, using state-wide soil test P summary information from the University of Delaware.

ha^{-1} year^{-1}), are clearly of concern from any perspective, whether economic efficiency of animal and crop production or environmental quality.

Developing nutrient management plans (animal and crop) that minimize nutrient losses to water and air and exporting surplus nutrients to other farming operations or other beneficial uses are primary goals in these settings. Economically, exporting these animal by-products (and nutrients) is the chief concern for the non-crop farming systems.

Given the state, county and farm-scale nutrient surpluses associated with poultry production in Delaware, what should be the strategic goals for nutrient management today? While a number of questions still remain, several options are emerging, all of which recognize that large-scale, long-term solutions are required. To achieve perfect nutrient balance for Delaware would require a strategic plan with enormous implications for agriculture and the state's economy, one that would need to reduce agricultural inputs of N by about 19,000 Mg year^{-1} and P by about 6800 Mg year^{-1} (Table 8.3). To fully accomplish a plan such as this, it would be necessary to eliminate commercial fertilizer sales and replace commercial fertil-

Table 8.4. Nutrient mass balance (Mg year^{-1} × 10^3) for Delaware's three counties, determined by comparison of nutrients produced or sold in each county annually with crop nutrient requirements at realistic yield goals.[a]

Nutrients produced or sold vs. crop nutrient requirements	New Castle N	New Castle P	Kent N	Kent P	Sussex N	Sussex P
Nutrients produced or sold:						
Commercial fertilizers	3.2	0.6	6.9	1.2	10.5	1.8
Poultry litter/manure	n/a	n/a	1.6	0.6	12.6	4.6
Other manures and biosolids	0.7	0.3	0.7	0.2	0.6	0.2
Total	3.9	0.9	9.2	1.9	23.8	6.6
Nutrients required by crops	2.7	0.9	6.3	0.9	9.7	0.7
County annual nutrient balance:						
Per county year^{-1}	1.2	–0.1	3.0	1.1	14.1	5.9
kg ha^{-1} year^{-1}	35	–2	39	30	121	51

[a] Assumptions:
1. Assumptions used for nutrients produced or sold and crop requirements are identical to those in Table 8.1.
2. n/a: information not available due to confidentiality of reporting.

Table 8.5. Estimated nitrogen (N) and phosphorus (P) mass balance (Mg per farm year^{-1}) for three Delaware poultry (broiler) operations with varying production intensities.

Inputs or outputs	Intensity of poultry production[a]					
	Typical		High		Non-crop farming	
	N	P	N	P	N	P
Inputs						
Feed	30.6	6.2	61.2	12.4	30.6	6.2
Animals	0.3	< 0.1	0.6	< 0.1	0.3	< 0.1
Fertilizers	4.4	0.5	0.8	0.3	n/a	n/a
Biological N fixation	4.5	0	4.5	0	n/a	n/a
Total inputs	39.8	6.7	67.1	12.7	30.9	6.2
Outputs						
Animals	13.7	1.7	27.5	3.4	13.7	1.7
Crops	15.1	2	15.1	2	n/a	n/a
Total outputs	28.8	3.7	42.6	5.4	13.7	1.7
Nutrient surplus						
Per farm year^{-1}	11.0	3.0	24.5	7.3	17.2	4.5
kg ha^{-1} year^{-1}	109	30	243	72	n/a	n/a
% of inputs	28	45	37	57	56	73

[a] Assumptions:
1. A typical Delaware poultry farm has two broiler houses, each with a production capacity of 22,000 birds per flock, and produces five flocks year^{-1}. The farm has 100 ha of cropland used for the production of maize (35 ha), soybeans (45 ha) and small grains (20 ha). A high-intensity poultry farm has four broiler houses and same amount of cropland and types of crops. A non-farming poultry operation has two broiler houses and no cropland.
2. Feed and animal nutrient inputs and outputs were calculated using standard information on poultry feeds and the composition of young chicks and mature broiler chickens (Sims and Vadas, 1997).
3. Fertilizer inputs are based on crop needs at realistic yield goals using the recommendations of the University of Delaware (Sims and Gartley, 1996). Note that (i) fertilizer inputs were reduced based on the amount of N and P contained in the broiler litter produced on the farm; and (ii) inputs of fertilizer P were based on soil test P, using state-wide soil test P summary information from the University of Delaware, and assuming that starter P fertilizer was used only for the maize.
4. Crop outputs are based on USDA Natural Resources Conservation Services estimates of N and P concentrations in the harvested maize, soybean and wheat grain.
5. Nitrogen surplus includes any NH_3-N volatilized from broiler houses during poultry production and from broiler litters during storage and application to cropland.

izers with animal manures. This would require state and/or industry subsidies to transport manures and also acceptance of the economic impacts that are associated with major reductions in fertilizer sales. Another option is to find alternative uses to land application for virtually all animal manures produced in the state, particularly poultry litter (*c.* 275,000 Mg year^{-1}), and to reduce fertilizer sales of P.

As strategic plans have begun to be developed for Delaware agriculture, several key questions have emerged, most of which remain unanswered. First, is perfect nutrient balance the appropriate strategic goal for Delaware or is some degree of nutrient surplus acceptable today and in the long term, given the state's dual goals of sustaining agriculture and improving water quality? Second, are nutrient surpluses more acceptable in some areas than others owing to differences in risk of nutrient loss to ground and surface waters because of variations in topography, water body sensitivity and nutrient management practices? Third, if, because of environmental concerns, no or only a minimal surplus is the strategic goal, what alternatives to land application of animal manures are most suitable and economically viable? This question assumes that developing large-scale alternative uses for animal manures is more desirable than near-elimination of commercial fertilizer sales. Two options already supported by the state and the poultry industry are a pelletization plant that will eventually export *c.* 75,000 Mg of broiler litter year^{-1} from the state for use as fertilizer in other regions and a state-subsidized animal waste relocation programme that will move raw poultry litters locally (distances > 10 km) from farms with nutrient surpluses to those with nutrient deficits. Other options under consideration include large-scale incineration of broiler litter for energy or steam production, exports of manures to land reclamation projects in nearby states (e.g. coal mine spoils), and industry-wide modifications of poultry diets to reduce the amount of N and P excreted as manures.

Tactical and operational dimensions of Delaware's nutrient balance

The tactical dimension of nutrient balance emphasizes the implementation of practices that will achieve strategic goals, while the operational dimension focuses on the details of the specific tasks that must be implemented. Because the strategic dimensions of nutrient management for Delaware are still in flux, the tactical and operational dimensions have changed very little. The emphasis today remains on traditional BMPs such as applying nutrients at rates that are consistent with soil test results and realistic crop yield goals, using efficient storage, handling and application practices for manures and fertilizers, and educational programmes on nutrient management. However, some significant changes will occur in the next few years as a result of the passage of Delaware's 1999 Nutrient Management Act, including: (i) nutrient management plans required for all farms with more than eight animal units (1 animal unit = 454 kg liveweight) or that apply nutrients to more than 4 ha of crop land; (ii) N applications to be based on realistic yield goals, defined as the best yields from 4 of the past 7 years; (iii) P applications to 'high P' (as yet undefined) soils must not exceed the amount of P removed by the harvested portions of crops grown in the following 3 years; (iv) a nutrient management certification programme will be established to certify nutrient generators, private nutrient applicators, commercial nutrient applicators, and nutrient consultants. These requirements are considerably more stringent than the traditional BMPs that have been used in Delaware, particularly the limitation of manure and fertilizer P applications based on soil P.

Nutrient mass balance analyses for Delaware show substantial state-wide surpluses of N and P, the greatest associated with intensive poultry production in Sussex County. Strategic planning to develop long-term solutions to the state's nutrient imbalances is being undertaken by the newly created DNMC and addresses not only agricultural management practices but also the social, political and economic aspects of nutrient balance needed to effectively implement the tactical and operational dimensions of nutrient management at all scales. The effectiveness of this approach will be determined by actions taken in the next decade, but the fact that strategic planning is under way gives reason for optimism today.

Conclusion

The ultimate solutions to environmental problems related to nutrients will be to balance inputs with outputs. This will be a major challenge because the trends towards greater specialization and intensification of agriculture, especially in the developed world, are generally resulting in increasing excess nutrient balances. This may require major changes in agricultural systems, such as changing the structure and management of animal and crop production systems or creating new infrastructure to redistribute manure nutrients economically. In the meantime there are a number of options to minimize the immediate impact of nutrient imbalances on the environment. Using a combination of approaches will be required in order to make progress in protecting the environment. In other areas with chronic negative nutrient balance the challenge will continue to be providing adequate nutrient inputs and efficient nutrient recycling to overcome the limitations of this imbalance on crop production.

References

Andres, A.S. (1995) Nitrate loss via groundwater flow, coastal Sussex County, Delaware. In: Steele, K. (ed.) *Animal Waste and the Land Water Interface*. Lewis Publishers-CRC, New York, pp. 69–76.

Bacon, S.C., Lanyon, L.E. and Schlauder, R.M. Jr (1990) Plant nutrient flow in the managed pathways of an intensive dairy farm. *Agronomy Journal* 82, 755–761.

Barnes, J.P. (1892) Soil and Agriculture, Official Doc. No. 7, Agriculture of Pennsylvania, Commonwealth of Pennsylvania, Harrisburg, pp. 222–226.

Cabrera, M.L. and Sims, J.T. (2000) Beneficial uses of poultry by-products: challenges and opportunities. In: Power, J. (ed.) *Beneficial Uses of Agricultural, Municipal, and Industrial By-products.* Soil Science Society America, Madison, Wisconsin.

Correll, D.L. (1998) The role of phosphorus in the eutrophication of receiving waters: a review. *Journal of Environmental Quality* 27, 261–266.

DDA (Delaware Department of Agriculture) (1998) *Delaware Agricultural Statistics Summary for 1998.* Delaware Agricultural Statistics Service, Dover.

DDA (Delaware Department of Agriculture) (1999) *Delaware's Nutrient Management Act of 1999* (House Substitute 1 for House Bill 250). Delaware Department of Agriculture, Dover, Delaware.

DPI (Delmarva Poultry Industry, Incorporated) (1998) *Look What the Poultry Industry is Doing for Delmarva: 1998 Facts about Delmarva's Broiler Industry.* Delmarva Poultry Industry Incorporated, Georgetown, Delaware.

Ertl, D.S., Young, K.A. and Raboy, V. (1998) Plant genetic approaches to phosphorus management in agricultural production. *Journal of Environmental Quality* 27, 299–304.

Ford, S.A. (1994) *Economics of Pasture Systems: Dairy Economics,* Fact Sheet 1. The Pennsylvania State University, College of Agricultural Sciences, Cooperative Extension Service, University Park, Pennsylvania.

Gburek, W.J. and Sharpley, A.N. (1998) Hydrologic controls on phosphorus loss from upland agricultural watersheds. *Journal of Environmental Quality* 27, 267–277.

Hamilton, P.A. and Shedlock, R.J. (1992) *Are Fertilizers and Pesticides in the Ground Water? A Case Study of the Delmarva Peninsula.* US Geological Survey Circular 1080, Denver, Colorado.

Janssen, B.H. (1999) Basics of budgets, buffers, and balances of nutrients in relation to sustainability of agroecosystems. In: Smaling, E.M.A., Oenema, O. and Fresco, L.O. (eds) *Nutrient Disequilibria in Agroecosystems: Concepts and Case Studies.* CAB International, Wallingford, UK, pp. 27–56.

Lanyon, L.E. (1990) Developing and implementing nutrient management systems for Pennsylvania farms. Final Project Report, Pennsylvania Department of Environmental Resources, Pennsylvania State University, College of Agricultural Sciences, Department of Agronomy, University Park, Pennsylvania.

Lanyon, L.E. (1994) Dairy manure and plant nutrient management issues affecting water quality and the dairy industry. *Journal of Dairy Science* 77, 1999–2007.

Lanyon, L.E. (1995) *A Nutrient Management Approach for Pennsylvania: Exploring Performance Criteria*, Agronomy Facts # 38D. College of Agricultural Sciences, Penn State University, University Park, Pennsylvania.

Lanyon, L.E. and Thompson, P.B. (1996) Changes in farm production. In: *Proceedings of Animal Agriculture and the Environment Conference.* Northeast Regional Agricultural Engineering Service, Ithaca, New York, pp. 15–23.

Lemunyon, J.L. and Gilbert, R.G. (1993) The concept and need for a phosphorus assessment tool. *Journal Production Agriculture* 6, 483–486.

Lynch, P.B. and Caffrey, P.J. (1997) Phosphorus requirements for animal production. In: Tunney, H., Carton, O.T., Brooks, P.C. and Johnston, A.E. (eds) *Phosphorus Loss From Soil to Water.* CAB International, Wallingford, UK.

Michel, K., Bacon, J.R., Gempesaw, C.M. and Martin, J.H. Jr (1996) *Nutrient Management by Delmarva Poultry Growers: a Survey of Attitudes and Practices*, FREC RR96–01. College of Agricultural Science, University of Delaware, Newark, Delaware.

Moore, P.A. and Miller, D.M. (1994) Decreasing phosphorus solubility in poultry litter with aluminum, calcium, and iron amendments. *Journal of Environmental Quality* 23, 325–330.

Ritter, W.F. and Chirnside, A.E.M. (1987) Influence of agricultural management practices on nitrates in the water table aquifer. *Biological Wastes* 19, 165–178.

Sharpley, A.N. (1995) Identifying sites vulnerable to phosphorus loss in agricultural runoff. *Journal of Environmental Quality* 24, 947–951.

Sharpley, A.N. and Smith, S.J. (1994) Wheat tillage and water quality in the southern plains. *Soil Tillage Research* 30, 33–38.

Sims, J.T. (1997) Agricultural and environmental issues in the management of poultry wastes: recent innovations and long-term challenges. In: Rechcigl, J. (ed.) *Uses of By-Products and Wastes in Agriculture.* American Chemical Society, Washington, DC, pp. 72–90.

Sims, J.T. (1998) Phosphorus soil testing: Innovations for water quality protection. *Communications in Soil Science and Plant Analysis* 29, 1471–1489.

Sims, J.T. (2000) Advances in animal waste management for water quality protection: a case study of the Delmarva peninsula. In: *Proceedings of Managing Nutrients and Pathogens from Animal Agriculture,* Harrisburg, Pennsylvania, 30 March 2000. Natural Resource, Agriculture and Engineering Service, Cooperative Extension, Ithaca, New York.

Sims, J.T. and Gartley, K.L. (1996) *Nutrient Management Handbook for Delaware.* Cooperative Bulletin No. 59. University of Delaware, Newark, Delaware.

Sims, J.T. and Vadas, P.A. (1997) *Nutrient Management Planning for Poultry-grain Agriculture*, Fact Sheet ST-11. College of Agricultural Science, University of Delaware, Newark, Delaware.

Sims, J.T., Simard, R.R. and Joern, B.C. (1998) Phosphorus loss in agricultural drainage: Historical perspective and current research. *Journal of Environmental Quality* 27, 277–293.

Sims, J.T., Goggin, N. and McDermott, J. (1999) Nutrient management for water quality protection: Integrating research into environmental policy. *Water Science Technology* 39, 291–298.

Sims, J.T., Edwards, A.C., Schoumans, O.F. and Simard, R.R. (2000) Integrating environmental soil phosphorus testing into environmentally-based agricultural management practices. *Journal of Environmental Quality* 29, 60–72.

Snow, C.T. (1992) *The Northeast Dairy Farm Summary.* Farm Credit Banks for Springfield, Springfield, Massachusetts.

USDA-Soil Conservation Service (1992) *Agricultural Waste Management Field Handbook*, Part 651. United States Government Printing Office, Washington, DC.

Section 2

Hydrology: the Carrier and Transport of Water Pollution

Introduction: Modelling Hydrological and Nutrient Transport Processes

Keith Beven
Institute of Environmental and Natural Sciences, Lancaster University, Lancaster LA1 4YQ, UK

Water is the primary transport mechanism for solutes and sediments to reach stream channels. A proper understanding of the hydrology of a site is therefore crucial to understanding, predicting and managing nutrient and contaminant exports from agricultural catchment areas. It is probably true to say that we now understand quite well the principles that are governing hydrological processes in agricultural areas (see the perceptual model outlined in Beven, 2001a). We do not expect any great surprises in the future in terms of our qualitative understanding of hydrological processes and already in the perceptual model we can recognize all the potential complexity of those processes. Such complexity will include, for example, the redistribution of rainfall intensities by the vegetation canopy, the heterogeneity of soil properties, preferential flows in structured soils, local generation and channelling of overland flows, reinfiltration of surface runoff downslope, the local generation of perched water tables and subsurface flows through fractured bedrock. Even a cursory examination of the literature (and of the chapters that follow in this section), however, will reveal that this range of possible processes does not commonly feature in either analysis or modelling of catchment responses. Recognition of the principles underlying the processes clearly does not easily translate into practical applications of those principles.

The reasons for this are discussed in detail in Beven (2001b), in the context of distributed hydrological modelling. Difficulties exist in terms of problems of nonlinearity, scale, uniqueness of place, equifinality (i.e. the concept that many different parameter sets, or model structures, might provide simulations that are acceptable in reproducing the available observations) of models and estimation of uncertainty. These problems are not due to lack of understanding, but to the difficulty of applying that understanding at particular application sites in the face of heterogeneity, complex local flow pathways and limitations in measurement techniques. Essentially, the uniqueness of a site is impossible to characterize in detail, particularly for subsurface flow pathways (Beven, 2000). The implication is that any predictive models will be inherently uncertain. This has led to an 'alternative blueprint' for distributed hydrological modelling (Beven, 2002) that has important implications for the prediction of other variables that depend on water as the carrier. If there is uncertainty in predicting water flows, there must be additional uncertainties in predicting solute and sediment fluxes that should be taken into account in predicting exports from catchment areas.

This 'alternative blueprint' is framed in terms of mapping the landscape into a model space. The model space will have the dimensions of the parameters necessary to characterize a particular place or catchment area. It might also include multiple model structures or sets of boundary conditions, where there is some uncertainty about what is an appropriate structure to use. The common search for a single 'optimal' parameter set to represent a catchment of interest is then equivalent to mapping to a single point in the model space. There is now enough evidence from wider searches of the model space using Monte Carlo simulation (i.e. using a large number of model runs with parameter sets and sometimes initial and boundary conditions randomly selected from some chosen prior distributions) to suggest that there may be many models that will provide acceptable simulations of the data available (the problem of equifinality), particularly if we recognize the limitations of the data available in representing the real response of the system.

It is important that the model space should include a sufficient range of functional responses to be able to predict the observed behaviour at the site. If this is not the case then the model will fail, even if uncertainty is taken into account. An interesting case in point here is the modelling of the export of agricultural P from catchment areas. For a long time it was assumed that the flow through soils could be described by Darcy's law, and transport by an advection–dispersion equation (ADE). Both assume good mixing in the soil solution and for the case of reactive solutes such as P, good contact with the mineral and organic materials of the soil. Batch experiments suggested that P would sorb strongly on to the soil particles and therefore should not move far into the soil, so that it would be gradually available to the plants for which it was intended to fertilize. This was reinforced by profiles of P content of the soil measured in the field. These typically showed high total P content in the topsoil, rapidly dropping with increasing depth from the surface. The conclusion was drawn that agricultural fertilizers would not result in significant concentrations in drainage waters and stream channels. Thus far, therefore, the models and observations were consistent.

It was gradually realized, however, that the models were not then consistent with observations made of P concentrations in drainage waters, stream channels and lakes that were being subjected to increasing incidences of eutrophication. A common initial reaction to this was that, since this could not be due to agricultural fertilizer sources, it must be due to the other anthropogenic sources. This was not totally consistent with other observations, for example of pesticide and herbicide concentrations in drainage waters, stream channels and even groundwaters. Most biocides also sorb very strongly on to particulate matter and should not move far in the soil. Eventually, it was realized that the models had to be revised (see for example Gish and Shirmohammadi, 1991) to include a preferential flow component. In this case, therefore, a description of the processes based on Darcy's law and the ADE, while consistent with some of the observations, did not have sufficient functionality to remain consistent with all the observations (although these concepts still underlie models in use today). It is in this way, by rejection of models that are inconsistent with observations, that our science will progress.

This type of methodology then focuses attention not only on the nature of model structures but also on the value of information and observations in predicting water flows and the associated fluxes of solutes and sediments. This information, as seen in the following chapters, can come from a variety of sources. There is general information about sites that comes from databases, particularly distributed GIS databases on topography, soils, vegetation, nutrient inputs, etc. There is information extrapolated from other sites, such as the use of pedo-transfer functions to estimate the hydraulic characteristics of soils from textural information. Finally, there may be some information derived from observations at the site itself that can be used either to refine parameter estimates for that site or to check model predictions for consistency.

All of these sources of information have limited utility in setting up models at a particular site. Thus any model predictions will be associated with some uncertainty. If we start to recognize that uncertainty, then we may be able to start to focus on the value of data in constraining that uncertainty. At the current time, we are not able to judge properly the value of different types of data or model structures in constraining predictive uncertainty and hence in assessing different management strategies. This is a very important issue that will need to be addressed in future. The future of distributed modelling lies not so much in the development of new model structures (except in so far as we are forced to by inadequate predictions), but in the application of models and their use over a period of time in specific places or catchments.

This is very important because long-term application in specific places implies an increasing potential for model evaluation, post-simulation audits and definition of model deficiencies. This suggests that, over time, there should be a greater possibility for learning about the uniqueness of different places, building up both qualitative and quantitative knowledge in a form that can be used to refine the representation of the processes at that site. This will be one way of making use of the increased computer power that will be available in the future, to: (i) build a system that will store or re-run the results of past simulations in a form that can be compared with a current situation; (ii) identify where there is drift or error in the simulations or where the model functionality seems inadequate; and (iii) act as a repository for information, knowledge and understanding about specific catchment areas such that local model representations of those areas can be improved. It is now possible to model the hydrology of the globe, albeit crudely and with some uncertainty. More modestly and more importantly it should also now be possible to model at a much smaller scale and in detail. This would still have uncertainty, but provides a means of gradually learning about the particular characteristics and idiosyncrasies in the response of particular places.

In this respect, therefore, the chapters in this section represent a starting point in this process. They provide the techniques to start characterizing and modelling the hydrology and nutrient fluxes at specific sites. In the future, we will need to go a step further into more detailed evaluations of different potential model representations of a site to learn something about what observations will be most valuable in site assessment within an uncertainty estimation framework.

References

Beven, K.J. (2000) Uniqueness of place and the representation of hydrological processes. *Hydrology and Earth System Science* 4(2) 203–213.

Beven, K.J. (2001a) *Rainfall-Runoff Modelling – the Primer.* John Wiley & Sons, Chichester, UK.

Beven, K.J. (2001b) How far can we go in distributed hydrological modelling. *Hydrology and Earth System Sciences* 15, 1–12.

Beven, K.J. (2002) Towards an alternative blueprint for a physically-based digitally simulated hydrologic response modelling system. *Hydrological Processes* 16(2), 189–206.

Gish, T.J. and Shirmohammadi, A. (eds) (1991) *Preferential Flow.* American Society of Agricultural Engineers, St Joseph, Michigan.

9 Hydrological Source Management of Pollutants at the Soil Profile Scale

R.W. McDowell,* A.N. Sharpley, P.J.A. Kleinman and W.J. Gburek
USDA–ARS, Pasture Systems and Watershed Management Research Unit, Curtin Road, University Park, PA 16802-3702, USA

Key words: fertilizer, hydrology, management, manure, nitrate, overland flow, phosphorus, soil, subsurface flow

Introduction

The successful control of non-point source pollution must target the availability of pollutants within the soil (source), as well as the potential hydrological controls on their movement (transport). Depending on the pollutant, either source or transport management may need to be emphasized. This chapter focuses on concerns over the non-point transport of N and P to ground and surface waters from agricultural land at the soil profile scale, which provides an excellent example of the importance of source and transport management to prevent nutrients with widely differing behaviour in the soil from becoming pollutants in the water environment.

*Present address: AgResearch Invermay, Private Bag 50034, Mosgiel, New Zealand.

The movement of N and P from soil to water is influenced by many factors, such as the type of inputs (fertilizer and manure applied), outputs dependent on soil type and management, and transport processes dependent on environmental conditions (Fig. 9.1). Whereas NO_3^--N is highly mobile and easily transported via multiple pathways, the affinity of P to soil constituents imposes conditions that restrict its potential for transport. In the past, this affinity supported a perspective that detachment and subsequent erosion (i.e. movement of nutrients with soil particles) was required for P transport from land to water. Although erosion certainly represents the most dramatic means of transporting P from land to water, transport of leached dissolved P in flow via percolation through the soil profile is now seen as a major concern (Heathwaite *et al.*, 1997; Sims *et al.*, 1998).

Both N and P are essential inputs for optimum crop production, but an excess of either produces a nutrient build-up: a source that is potentially transportable. The risk of nutrient transport from land to water is dependent on hydrological conditions. Consequently, understanding and combining both source and transport factors enables us to quantify and manage the potential for nutrient movement from land to water and is key to site risk-assessment tools, such as the P Index (Lemunyon and Gilbert, 1993) and the combined N and P Indices (Heathwaite *et al.*, 2000; Sharpley and Gburek, 2000). With this in mind, this chapter examines source and transport factors influencing the loss of pollutants from soil to water at the soil profile scale as defined in Haygarth *et al.* (2000). Emphasis is placed on N and P, and in particular P, which is currently receiving considerable scrutiny in research and legislative arenas.

Source Factors

Source factors determine the availability of pollutants to transport mechanisms, such as erosion and leaching, and represent the interaction of soil biological, chemical and physical properties with the pollutant of concern. For highly soluble pollutants, source factors often provide the best management opportunities for pollution control. For pollutants with high affinities to soil solids, management of transport factors is often as important as source factor management.

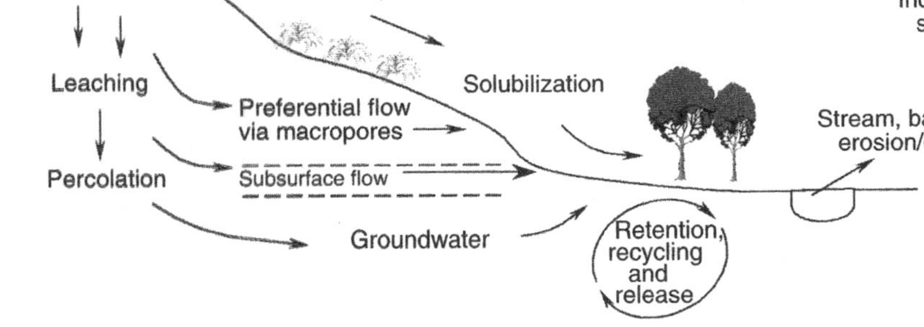

Fig. 9.1. The inputs, outputs and processes during transfer of pollutants to surface waters in agricultural soils.

Nitrogen

Source factors are a key control of N losses from agricultural lands, as altering N transport is difficult to manage (Beegle, 2000). Nitrate in the soil profile during periods when precipitation exceeds evapotranspiration is at risk of loss to water. This NO_3^- is derived from fertilizer, manure, the decomposition of plant residues and the mineralization of soil organic matter (Sharpley and Lord, 1997). If a crop is correctly fertilized, NO_3^- levels at all times will be low and little fertilizer N will contribute to losses (Davies and Sylvester-Bradley, 1995). However, excessive N inputs as fertilizer or manure or detrimental environmental conditions leading to poor crop growth will increase losses (Schepers *et al.*, 1991).

The risk of NO_3^- availability for leaching is fundamentally linked to crop management. For instance, losses from unfertilized grassland, where plant cover may be present all year, may be close to zero. Conversely, losses from intensively grazed pastures receiving large quantities of fertilizer and/or manure may be more than 100 kg N ha^{-1} (Korevaar and Den Boer, 1990). Lord and Mitchell (1998) showed that even in unfertilized cereal crops, NO_3^- losses were not much less than the European Community limit of 50 mg NO_3^--N l^{-1} at the economic optimum N requirement (Fig. 9.2). This is caused by NO_3^- leached in the winter derived from N that is mineralized during late summer and autumn, when arable crop uptake is low. Mineralization of N is affected little by fertilizer inputs, and it is only when N inputs exceed crop requirements and fertilizer N contributes directly to nitrate losses under leaching conditions (Fig. 9.2). Therefore, management of NO_3^- leaching from soils must concentrate on using a balance approach incorporating consideration of form, rate, timing and method of N application (Sharpley *et al.*, 1998a). In addition, crop selection and rotation can encourage retention. For example, legumes require no supplemental N inputs and can scavenge residual N left over in the soil from the previous crop (Mathers *et al.*, 1975; Muir *et al.*, 1976).

The importance of cropping system management to source risk is further illustrated by Fig. 9.3, which overlays hypothetical root development for several crops with predicted NO_3^--N leaching patterns (adapted from Sharpley *et al.*, 1998b). Continuous cropping with maize (*Zea mays*) results in root development that

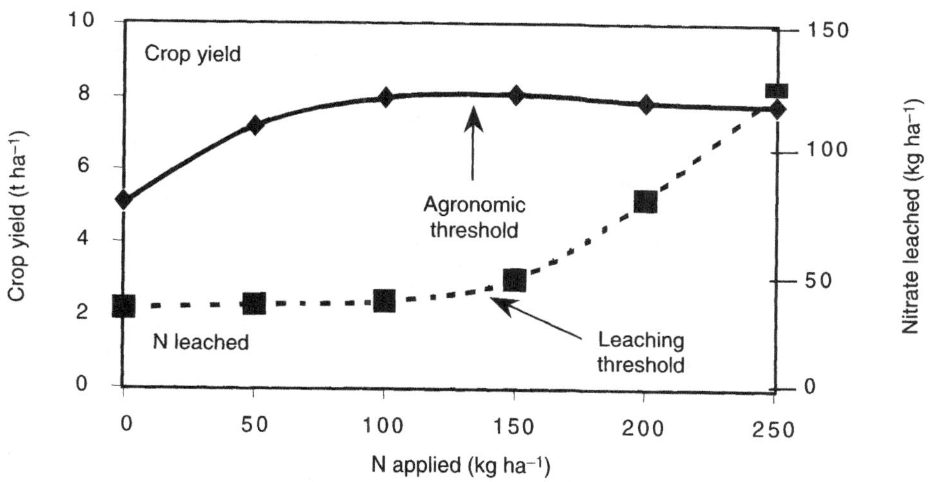

Fig. 9.2. The effect of fertilizer N on arable crop yield and nitrate leaching (adapted from Sharpley and Lord, 1997).

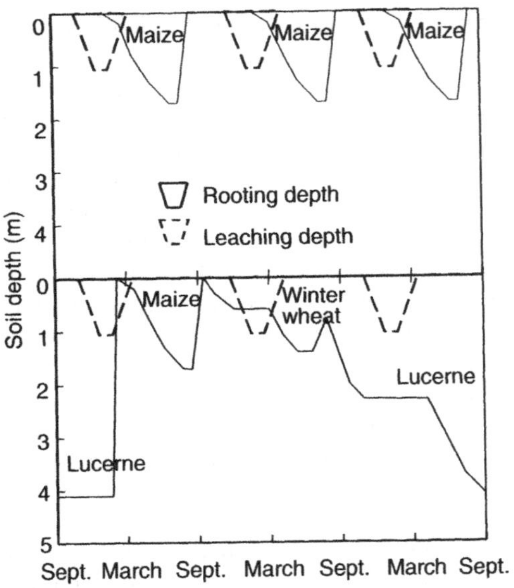

Fig. 9.3. Root development of continuous maize and a maize–winter wheat–lucerne rotation in relation to soil drainage over a 3-year period (adapted from Sharpley *et al.*, 1998b).

does not overlap with periods of high transport potential (leaching), resulting in increased availability of soil NO_3^--N to subsurface flow. However, rotational cropping of maize–winter wheat–lucerne increases NO_3^- retention by improving efficiency derived from deeper root distribution at times when leaching risk is high.

Phosphorus

Sources of P in overland and subsurface flow from agricultural soils include release from soil and plant material, added fertilizer and manure, as well as that attached to eroded soil material. Increases in P loss in overland flow have been measured after the application of fertilizer P and manure (Table 9.1). Rate, time and method of application, form of fertilizer or manure, amount and time of rainfall and vegetative cover influence these losses after application. The portion of applied P transported in overland flow is greater from conventional than conservation-tilled catchments (Table 9.1). However, McDowell and McGregor (1984) found that fertilizer P application to no-till maize reduced P transport, probably through an increased vegetative cover afforded by addition of fertilizer. As expected, the transfer of applied P in subsurface tile drainage is appreciably lower than in overland flow (Table 9.1). Although it is difficult to distinguish between losses of fertilizer or manure and native soil P without the use of expensive and hazardous radio tracers, transfers of applied P in overland flow are generally less than 10% of that applied, unless rainfall immediately follows application or where overland flow has occurred on steeply sloping, poorly drained and/or frozen soils. The high proportion of manurial P in overland flow may result from high rates of manure application and often less flexibility in application timing than for fertilizer (Table 9.1).

Phosphorus application method and timing is important in relation to rainfall, and influences the transfer of P in overland flow. For example, several studies have shown a decrease in P transfer with an increase in the length of time between P application and overland flow (Westerman *et al.*, 1983; Edwards and Daniel, 1993b; Sharpley, 1997). This decrease can be attributed to the reaction of added P with soil and dilution of applied P by infiltrating water from before overland flow.

Incorporation of manure into the soil profile, either by tillage or subsurface placement, reduces the potential for P transfer in overland flow (Fig. 9.4). For example, the dissolved P concentration of overland flow 2 days after 100 kg P ha^{-1} dairy manure was applied to the surface of a Berks silt loam was 2.75 mg l^{-1}. When the same amount of manure was incorporated by ploughing to a depth of 10 cm, overland flow dissolved P (here referred to as P soluble at < 0.45 μm) was 1.70 mg l^{-1} and when placed 5 cm below the soil surface, dissolved P in overland flow was only 0.15 mg l^{-1} (Fig. 9.4).

The transfer of P in overland flow after fertilizer application is dependent on the solubility of the fertilizer applied and its reaction with the soil. For example,

Table 9.1. Effect of fertilizer and manure application on P loss in overland flow and fertilizer application on P loss in tile drainage (kg ha^{-1} $year^{-1}$).

Land use	P added	Phosphorus loss 'Dissolved'	Phosphorus loss Total	Percentage applied[a]	Reference and location
Overland flow					
Fertilizer					
Grass	0	0.02	0.22		McColl *et al.*, 1977;
	75	0.04	0.33	0.1	New Zealand
No-till maize	0	0.70	2.00		McDowell and McGregor, 1984;
	30	0.80	1.80		Mississippi
Conventional till maize	0	0.10	13.89		
	30	0.20	17.70	12.7	
Wheat	0	0.20	1.60		Nicholaichuk and Read, 1978;
	54	1.20	4.10	4.6	Saskatchewan, Canada
Grass	0	0.50	1.17		Sharpley and Syers, 1976;
	50	2.80	5.54	8.7	New Zealand
Grass	0	0.17	0.23		Uhlen, 1988;
	24	0.25	0.31	1.2	Norway
	48	0.42	0.49	1.0	
Dairy manure[b]					
Lucerne	0	0.10	0.10		Young and Mutchler, 1976;
–spring	21	1.90	3.70	17.1	Minnesota
–autumn	55	4.80	7.40	13.3	
Maize	0	0.20	0.10		
–spring	21	0.20	0.60	2.4	
–autumn	55	1.00	1.60	4.7	
Poultry manure					
Grass	0	0.10	0.40		Westerman *et al.*, 1983;
	95	1.40	12.40	12.6	North Carolina
Swine manure					
Fescue	0	0.10	0.10		Edwards and Daniel, 1993a;
	19	1.50	1.50	7.4	Arkansas
	38	4.80	3.30	8.4	
Artificial drainage					
Maize	0	0.13	0.42		Culley *et al.*, 1983;
	30	0.20	0.62	0.7	Ontario, Canada
Oats	0	0.10	0.29		
	30	0.20	0.50	0.7	
Potatoes + wheat + barley					Catt *et al.*, 1997; Woburn, UK
Minimal till	102	0.26	8.97	8.8	
Conventional till	102	0.35	14.38	14.1	
Lucerne	0	0.12	0.32		
	30	0.20	0.51	0.6	
Grass 0–30 cm	16	0.25	0.38	2.3	Haygarth *et al.*, 1998;
0–85 cm	16	0.89	1.77	11.1	Devon, UK
Grass 0–30 cm	32	0.12	0.38	1.1	Heathwaite *et al.*, 1997;
30–80 cm	32	0.76	1.77	5.5	Devon, UK
Grass	0	0.08	0.17		Sharpley and Syers, 1979;
	50	0.44	0.81	1.3	New Zealand

[a] Percentage P applied lost in runoff.

[b] Manure applied in either spring or autumn.

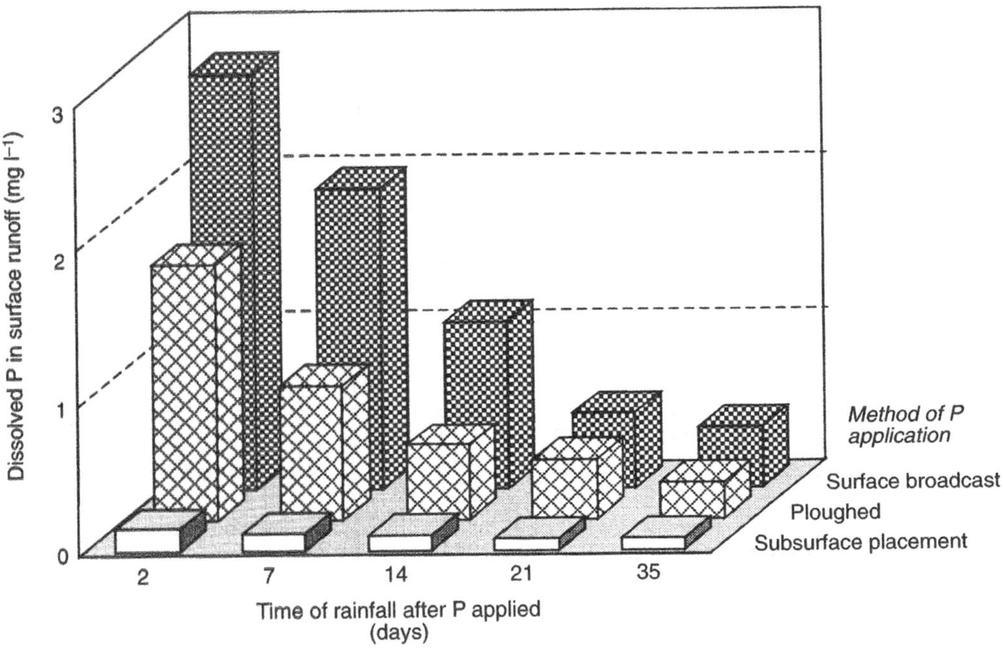

Fig. 9.4. The effect of application method and timing of rainfall after application of dairy manure (100 kg P ha^{-1}) on the concentration of P in surface runoff from a grassed Berks silt loam.

Sharpley *et al.* (1978) observed a slightly greater dissolved P loss in overland flow following the application of monocalcium P (the main P component of superphosphate) to a permanent pasture in New Zealand (2.80 kg ha^{-1}), compared with that from dicalcium P (2.17 kg ha^{-1}), a slow-release fertilizer. This difference was attributed to a more rapid dissolution of mono than dicalcium P at the soil surface. However, an appreciably greater transfer of sediment-bound P with dicalcium (4.92 kg ha^{-1}) than monocalcium P (2.63 kg ha^{-1}) results from an increase by transport of the less soluble dicalcium P particles in overland flow.

Because of the variable paths and time of water flow through a soil with subsurface drainage, the sources of P controlling transfers in subsurface pathways are more complex than for overland flow. Subsurface flow includes tile drainage and natural subsurface flow, where tile drainage is percolating water intercepted by artificial systems, such as mole and tile drains (Fig. 9.1). Generally, the P concentration in water percolating through the soil profile is small because of sorption of P by P-deficient subsoils (Haygarth *et al.*, 1998). Exceptions occur in acid organic or peaty soils, where the sorption affinity and capacity for P is low because of the predominantly negatively charged surfaces and the complexing of Al and Fe by organic matter (Duxbury and Peverly, 1978; Miller, 1979; White and Thomas, 1981). Similarly, P is more susceptible to movement through sandy soils with low P sorption capacities; in soils which have become waterlogged, leading to conversion of Fe^{3+} to Fe^{2+} and the mineralization of organic P; and with preferential flow through macropores and earthworm holes (Bengston *et al.*, 1992; Sharpley *et al.*, 1998a).

Studies have suggested that leaching of P from plant material in different stages of growth and decay may account for seasonal fluctuations and differences from catchment to catchment in the transfer of P in overland flow (Gburek and Heald, 1974). The increased amounts of dissolved P in overland flow from lucerne plots (33 g

ha^{-1}) compared with forested (4 g ha^{-1}), maize (11 g ha^{-1}) and oat plots (16 g ha^{-1}) during several simulated rainfall events (7.4–12.2 cm^{-1}) over a 2-year period were attributed to larger amounts of P leached from the lucerne (Wendt and Corey, 1980).

Other contaminants

Compared with N and P, less information is available on the transport of other chemicals from inputs to the soil. Many trace elements, including all of the micronutrients, can reach concentrations in the soil that are toxic to plants and microorganisms. Some of the most toxic are Hg, Pb, Cd, Cu, Ni and Co. The first three are especially toxic to higher animals, while the last are more phytotoxic. Each element interacts with the soil in different ways, such as forming complexes with organic matter, chemisorption and co-precipitation with minerals and precipitating as insoluble carbonates, sulphides, oxides or phosphates. Although many of the mechanisms are similar to those by which N and P are transferred from the soil, discussion of each individual element is beyond the scope of this chapter and is covered in texts such as Allen *et al.* (1998).

Organic pollutants in the soil, such as pesticides and hydrocarbons, interact with the soil in two ways, by sorption and by biological or chemical degradation. Hence, the basis for predicting the behaviour of organic materials in soil and, in turn, their ability to leach from the soil, requires knowledge of the extent to which these sorption or degradation processes occur. The water solubility and sorption of an organic pollutant are dependent on criteria such as: the functional groups on the organic pollutant and their acidity or basicity; its molecular size and shape; and its polarity and polarizability. The chemical degradation of an organic pollutant is dependent on factors such as: soil physicochemical environment (E_h, pH, temperature, etc.), degree of accessibility if sorbed, and the susceptibility of the structure to catalytic enzymatic reactions. The characteristics of organic pollutants vary widely; for example McDowell *et al.* (1997) showed that the acetolactate synthase-inhibiting herbicides flumetsulam and imazapyr exhibited degradation half-lives of 24 and 155 days, respectively. Flumetsulam was leached beyond a depth of 25 cm in significant concentrations after 14 days (12.5% of that applied to the soil surface), while the highly sorbed imazapyr was largely confined to the top 5 cm. As part of the requirement for registering pesticides for use, most of the criteria for interaction with the soil are established and can be found in texts such as Tomlin (1991).

Transport Factors

Non-point source pollutants are transported via a variety of hydrological pathways, including overland and subsurface flow. The importance of these pathways regarding environmental impact depends on the source factors described above. Even when considering hydrological factors at a soil profile scale, it is necessary to account for landscape variables, as these processes often control overland flow and leaching potential.

Nitrogen

Because of the solubility of NO_3^-, factors affecting N transport can be equated with those governing water movement through soils and bedrock. Soil physical properties affecting infiltration as well as the amount of water draining through the soil profile (determined by rainfall and irrigation intensity) regulate N transport potential. An implication of NO_3^- solubility is that concentrations in overland flow are little affected by the route taken or the distance to the stream. Thus, most areas within a catchment have an equal chance of contributing to long-term NO_3^- loss. As a result, source factors tend to govern the magnitude of N loss, while transport factors dictate the delay caused by percolation through various soil horizons.

The concentration of NO_3^--N is greatly affected by the volume of water draining the soil profile. Fig. 9.5 illustrates the coincidence of NO_3^- concentrations in tile drain effluent and tile discharge from a recently manured field at Willsboro Farm, Cornell University, New York. Declines in effluent NO_3^- concentrations coincide with peaks in tile discharge and may be attributed to effluent dilution caused by increased flow through the soil profile (Goerhing *et al.*, 1999).

Hydrological factors also play a role in the volatile loss of N to the atmosphere from denitrification. Denitrification occurs when water remains in anaerobic zones. As a result, soil drainage class is a key indicator of denitrification potential. For instance, Hanson *et al.* (1994) found a strong correlation between soil drainage class and mean denitrification rates.

The management of N losses can be difficult. For example, irrigation timed to reduce water movement below the root zone and, therefore, NO_3^- leaching potential can have a significant effect on N leached (Sharpley *et al.*, 1998a). However, control of N loss via one pathway often aggravates N loss by another pathway. For example, Sharpley and Smith (1994) showed that the adoption of conservation tillage practices in wheat production reduced N losses in overland flow but increased leaching losses.

Phosphorus

The transport of dissolved P is initiated by the desorption, dissolution and extraction of P from soil and plant material (Fig. 9.1). These processes occur when rainfall interacts with a thin layer of surface soil (1–5 cm) before leaving the field as overland flow (Sharpley, 1985a). Although the proportion of rainfall and depth of soil involved are difficult to quantify in the field, they will be highly dynamic due to variations in rainfall intensity, soil tilth and vegetative cover.

During the initial phases of detachment and transport of sediment, the finer-sized fractions of source material are eroded preferentially; thus, the P content and reactivity of transported material is greater than source soil on a mass basis. This has led to the determination of enrichment ratios (ER) for P, calculated as the ratio of the concentration of P in the sediment (eroded soil) to that in source soil. Enrichment ratio values of 1.3 and 3.3 for total and extractable P

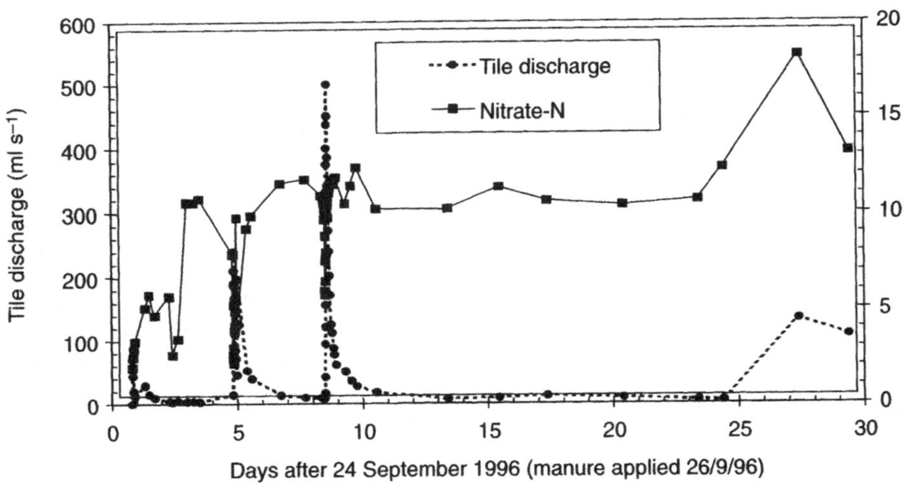

Fig. 9.5. Relationship of effluent NO_3^--N concentration to discharge from a tile drain servicing a manure-applied field (adapted from Geohring *et al.*, 1999). Increases in saturated flow through the soil profile are responsible for the lowered effluent nitrate concentrations.

(0.001 M H_2SO_4 extractable P), respectively, were observed by Rogers (1941), whereas Massey and Jackson (1952) observed values between 1.9 and 2.2 for 'water soluble plus pH 3 extractable' P. More recently, Sharpley (1985b) observed that the enrichment of plant available P (Bray-1; 1.2–6.0) was greater than for other forms of P (total, inorganic and organic) (1.2–2.5) in overland flow from several soils under simulated rainfall. Phosphorus sorption–desorption characteristics, buffer capacity, sorption index and equilibrium P concentration (EPC) were also enriched in overland flow sediment compared with source soil (Sharpley, 1985b).

The effect of erosion on particulate P transport is illustrated by a 15-year study of overland flow from several grassed and cropped catchments in the Southern Plains (Sharpley *et al.*, 1991; Smith *et al.*, 1991). As erosion from native grass and no-till and conventional-till wheat at El Reno, Oklahoma, increased, particulate P was a greater portion of P transported in overland flow, although amounts transported varied (0.08–10 Mg ha^{-1} year^{-1}) with management and associated fertilizer P application (Fig. 9.6). Accompanying the increase in particulate P movement is a relative decrease in dissolved P movement (Fig. 9.6).

Subsurface drainage of a soil can reduce P transport via enhanced infiltration and drainage and thereby decreased overland flow volumes. For example, Bengston *et al.* (1988) found that annual total P transfer in overland flow from a clay loam soil under maize in the lower Mississippi Valley was reduced by an average 36% in 6 years in fields with 1-m deep tile drains (5.0 kg ha^{-1} year^{-1}), compared with undrained soil (7.8 kg ha^{-1} year^{-1}). Similarly, Haygarth *et al.* (1998) found that draining land reduced the transfer of total P by *c.* 30% and attributed this to sorption as P flowed through the soil. The reduction in P loss must be considered in terms of a potential increase in NO_3^--N loss in tile drained fields. However, the transport of P in tile drainage can be responsive to rainfall and land management, when preferential flow through macropores, old root channels and earthworm holes exists.

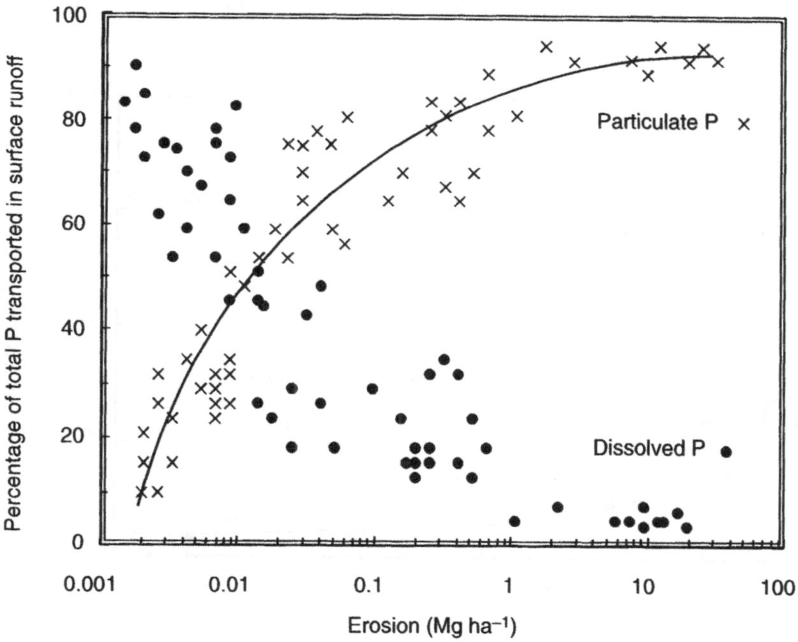

Fig. 9.6. Percentage of total P as dissolved and particulate P as a function of erosion in surface runoff from catchments at El Reno, Oklahoma.

Re-evaluating several studies of P and N transport in tile drainage highlights the role of preferential pathways (Sharpley *et al.*, 1976; Sharpley and Syers, 1979, 1981). Flow, P and N transport from two tiles draining 1.1 ha (Area A) and 0.8 ha (Area B) of a Tokomaru silt loam under permanent pasture in New Zealand, were monitored for 4 years (1974–1977). The tile drains were situated at 75-cm soil depth. During typical flow events, the concentration of dissolved P increased prior to peak flow and then gradually decreased (Fig. 9.7). The initial increase in dissolved P indicated a dynamic source or pool of P in the soil, which can be released to infiltrating water, bypassing P-deficient subsoil material as it travels to the 75-cm deep tile drain. An opposite concentration pattern was observed for NO_3^--N because of the rapid dilution with infiltrating surface water (Fig. 9.7).

Over a 4-year period, a rapid response of P transport in tile drainage to surface management was observed (Sharpley and Syers, 1979). In late June 1975, 30 kg P ha^{-1} of fertilizer were applied to one of the drained areas (area A). The other drained area (area B) was not fertilized but was grazed in three sections by 100 dairy cattle at a stocking rate of 300 cattle ha^{-1} 2 h^{-1} over 36 h at the beginning of August, as part of the normal grazing plan for the farm. During this period (April to September) the drains flowed continuously. The concentration of dissolved P in drainage from both areas increased 1 day after fertilizer application and grazing, with maximum values attained after only 1 week (Fig. 9.8). Clearly, the transport of P in preferential flow through the soil profile can be rapid.

If continuity of the preferential flow pathways is disrupted, however, P transport through the profile can be slowed. For example, before grazing, the flow hydrograph for the discharge from both drained fields was similar (Fig. 9.9a). In the first storm event after grazing, overland flow was observed on field B but not on field A. As a result, discharge from the grazed field increased more gradually than that from the ungrazed field and peak flow from the grazed field was less (Fig. 9.9b). The hydrograph for discharge from the grazed field was still less well defined than that from the ungrazed field 3 weeks after grazing (Fig. 9.9c). The reduction in tile drainage

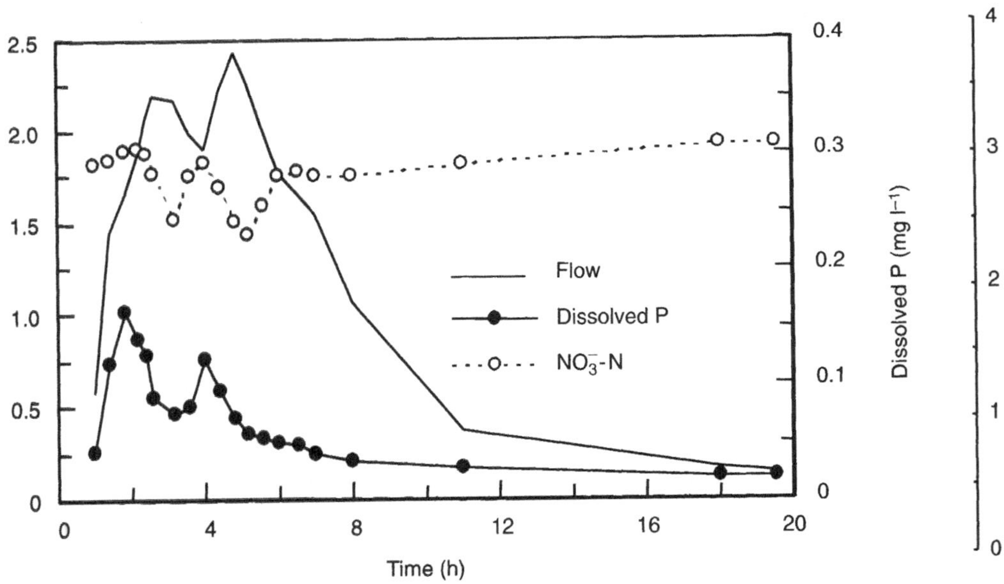

Fig. 9.7. Flow, dissolved P, and NO_3^--N during a typical accelerated tile drainage flow event for a 1.1 ha field on a dairy farm, Palmerston North, New Zealand.

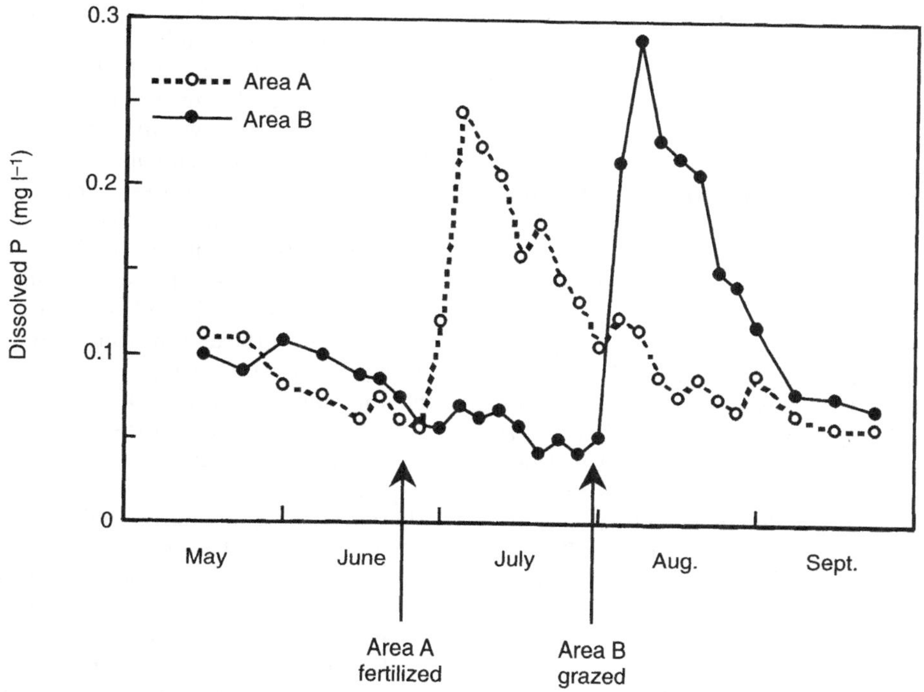

Fig. 9.8. Mean dissolved P concentration in tile drainage before and after fertilizer P applied (area A) and grazing (area B) on a dairy farm, Palmerston North, New Zealand.

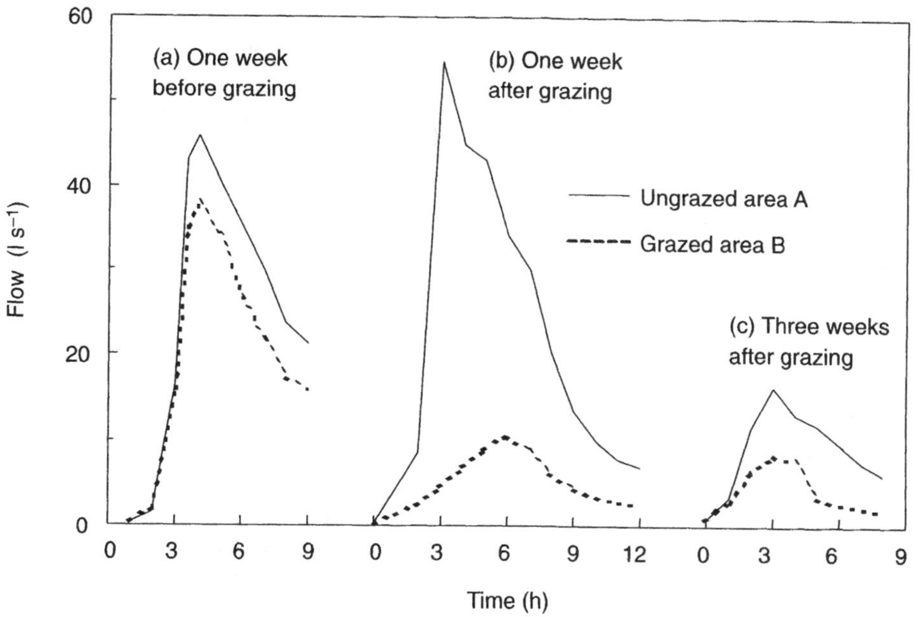

Fig. 9.9. Water discharge rate from ungrazed (area A) and grazed (area B) fields on a dairy farm, Palmerston North, New Zealand, (a) 1 week before grazing, (b) 1 week after grazing, and (c) 3 weeks after grazing.

volume can be attributed to the disruption of macropores at the top of the soil profile and to surface 'pugging', which reduced infiltration capacity and preferential flow.

Management of Source and Transport Factors

Environmental soil phosphorus testing

Environmental concerns about high soil P concentrations in excess of plant needs, means that new methods of soil P testing are being used to indicate when P enrichment of overland flow and subsurface flow may become unacceptable. As these methods are developed, much attention has focused on linking soil-specific testing with the potential for P loss (Sharpley *et al.*, 1999).

A common approach has been to adopt agronomic soil P standards, following the rationale that soil P in excess of crop requirements is vulnerable to removal by overland flow or leaching. Since agronomic standards already exist for soil test P, this approach requires little investment in research and can be readily implemented. However, we must be careful of how we interpret such soil test results for environmental purposes (Fig. 9.10). Interpretations from soil test reports (i.e. low, medium, optimum, high, etc.) are based on the expected response of a crop to P and not on soil P release to overland flow or subsurface flow (Sharpley *et al.*, 1996). Some would simply extend the levels that are used for interpretation for crop response and say a soil test that is above the level at which a crop response is expected is in excess of crop needs and therefore is potentially polluting (Fig. 9.10). However, it cannot be assumed that there is a direct relationship between the soil test calibration for crop response to P and overland flow enrichment potential.

Much research has shown that P concentrations in overland flow and drainage water are related to soil test P concentra-

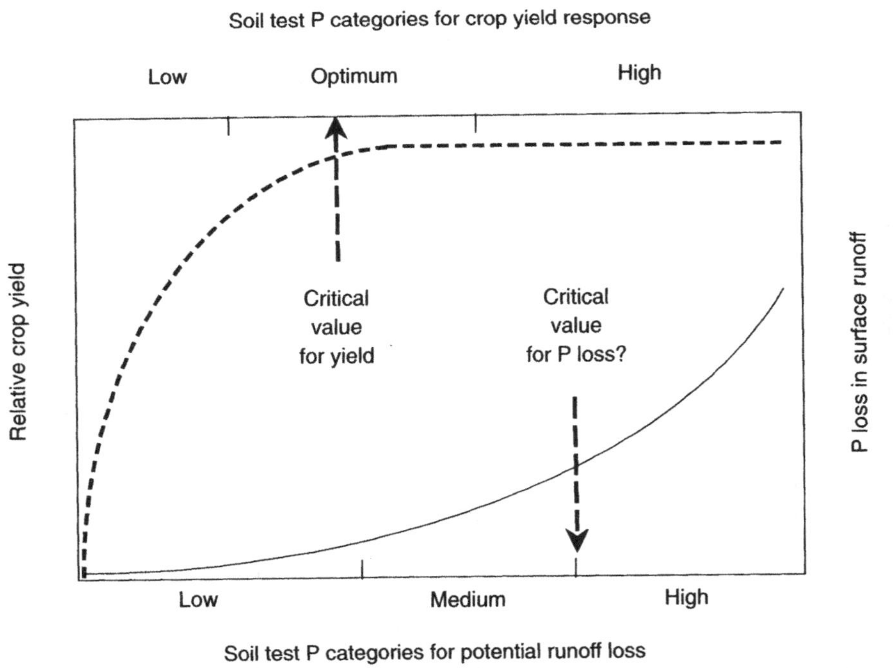

Fig. 9.10. Relative crop yield and P loss in surface runoff in comparison with soil test P categories for potential runoff loss.

tion in topsoil. For example, Pote *et al.* (1996) found that DRP concentration in overland flow was linearly related to Mehlich-3 P (r^2 = 0.72), Bray-I P (r^2 = 0.75), Olsen P (r^2 = 0.72), P extracted by distilled water (r^2 = 0.82), P sorbed by iron oxide paper (r^2 = 0.82), P extracted by acidified ammonium oxalate (r^2 = 0.85) and P saturation (r^2 = 0.77). Smith *et al.* (1998) noted a curvilinear relationship between DRP concentration in drainage water from lysimeters of six soils from lowland England and topsoil (0–15 cm depth) Olsen P concentrations, at Olsen P concentrations > 70 mg l^{-1}. Sibbeson and Sharpley (1997) gave examples of linear and curvilinear relationships between DRP concentrations in overland flow and soil test P concentrations and noted that linear relationships may become curvilinear if a wider range of soil test P concentrations were examined.

One environmental measure of soil P, which was developed in The Netherlands by Breeuwsma and Silva (1992) to assess P leaching potential, is soil P saturation (percentage saturation = available P/P fixation maximum). The role of soil P saturation as an indicator of P loss potential is assumed from the observation that soil P saturation is strongly correlated to P desorption, such that P desorption increases at higher degrees of soil P saturation (Sibbeson and Sharpley, 1997). Indeed, many studies have correlated soil P saturation with P in overland flow (Sharpley, 1995; Pote *et al.*, 1996, 1999), as well as with P in leachate (McDowell and Condron, 1999). In The Netherlands, a threshold soil P saturation of 25% has been established above which the potential for P movement in surface and groundwaters becomes unacceptable (Breeuwsma and Silva, 1992).

A variety of soil extractants have been evaluated as indicators of potential P transfer by relating extracted P to P in overland flow or subsurface leachate. Beauchemin *et al.* (1996) note that 'a measure of soil capacity for P accumulation must be accompanied by a measure of desorbability to fully assess the risk of contamination of drainage waters by P leaching'. Furthermore, Ryden and Syers (1975) stated that, to establish the relationship between P additions, fertilizer additions and P in particulate or aqueous phases of overland flow, a desorption or 'support medium should reflect the cation status as well as the ionic strength of the aqueous phase of the system'. With the possible exception of macropore or bypass flow, the longer residence time of subsurface leachate as it flows through the soil implies that a soil extractant of higher ionic strength is required than would be used for overland flow. Soil extractions with water and 0.01 M $CaCl_2$ have shown promise in estimating the dissolved P concentrations of overland flow and subsurface leachate, respectively (Fig. 9.11) (McDowell and Sharpley, 1999; McDowell and Condron, 1999). Using these desorption approaches in conjunction with Fe-oxide strips and gels, we can determine the quantity of desorbable P in overland flow or soils over the medium and long term (Sharpley, 1993; Freese *et al.*, 1995).

Analysis of soil P data for environmental purposes represents an important step in the development of recommendations for farmers and policy makers. One innovative environmental approach to data analysis uses a split-line model to determine a soil P threshold or 'change point', which separates the relationship between soil P and dissolved P in overland flow into two sections, one with greater P loss per unit increase in soil P concentration or percentage saturation than the other (Heckrath *et al.*, 1995; McDowell and Condron, 1999; Hesketh and Brookes, 2000). McDowell and Sharpley (1999) showed that a similar change point occurred between soil test P (Mehlich-3 P) concentrations of the 0–5 cm soil layer and dissolved P in either overland flow or subsurface drainage water (Fig. 9.12). The change point identifies a critical soil P concentration above which the potential for P loss is substantially greater.

The change point has potential for use in site evaluation of P loss potential (e.g. the P Index described below). McDowell and Trudgill (2000) showed in a small catchment in Devon (UK) that the extrac-

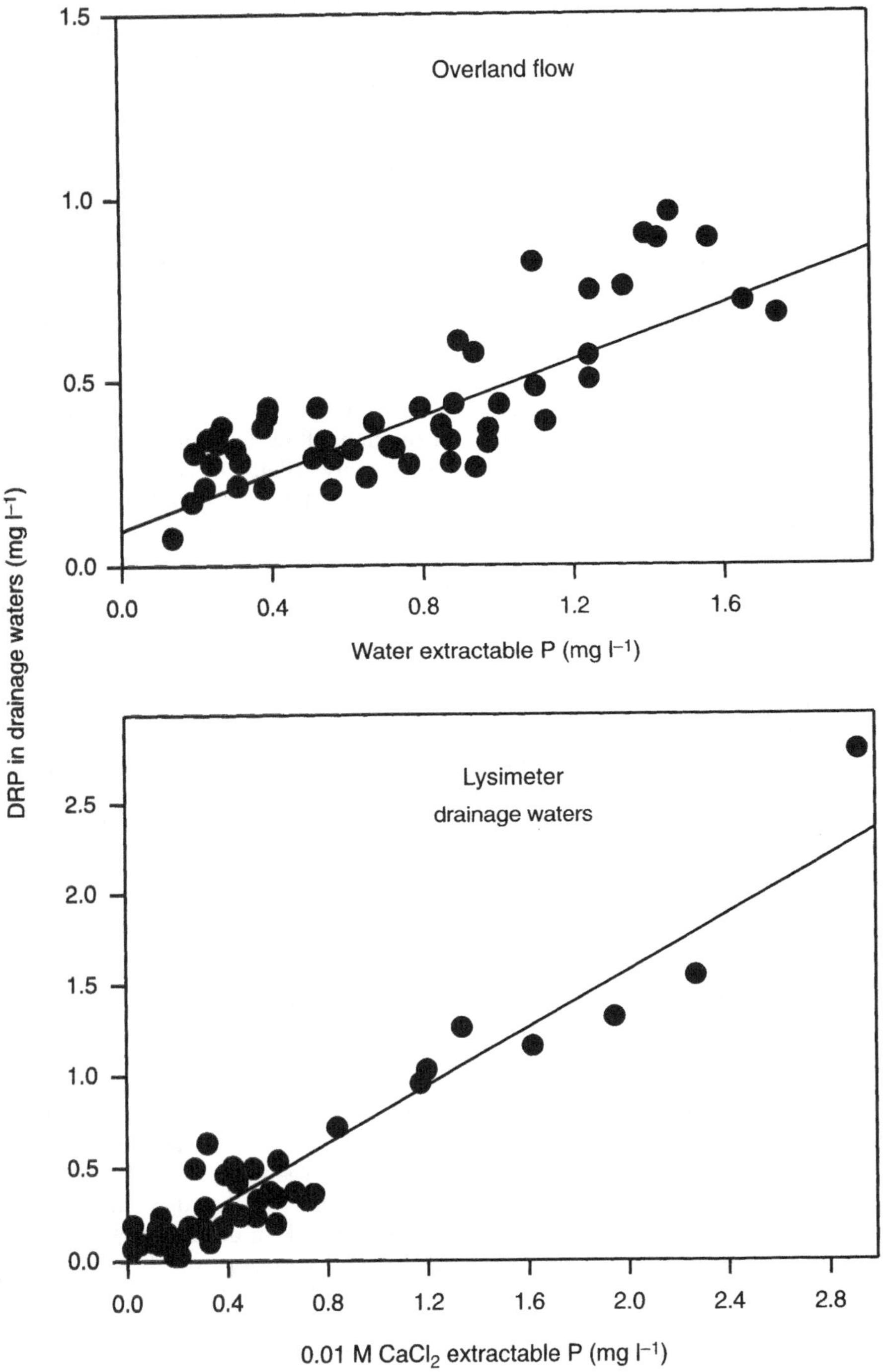

Fig. 9.11. The relationship between the concentration of dissolved P in overland flow and subsurface drainage from 30 cm deep lysimeters and the water and 0.01 M $CaCl_2$ extractable soil P concentration, respectively, of surface soil (0–5 cm) from a central Pennsylvania catchment (adapted from McDowell and Sharpley, 1999).

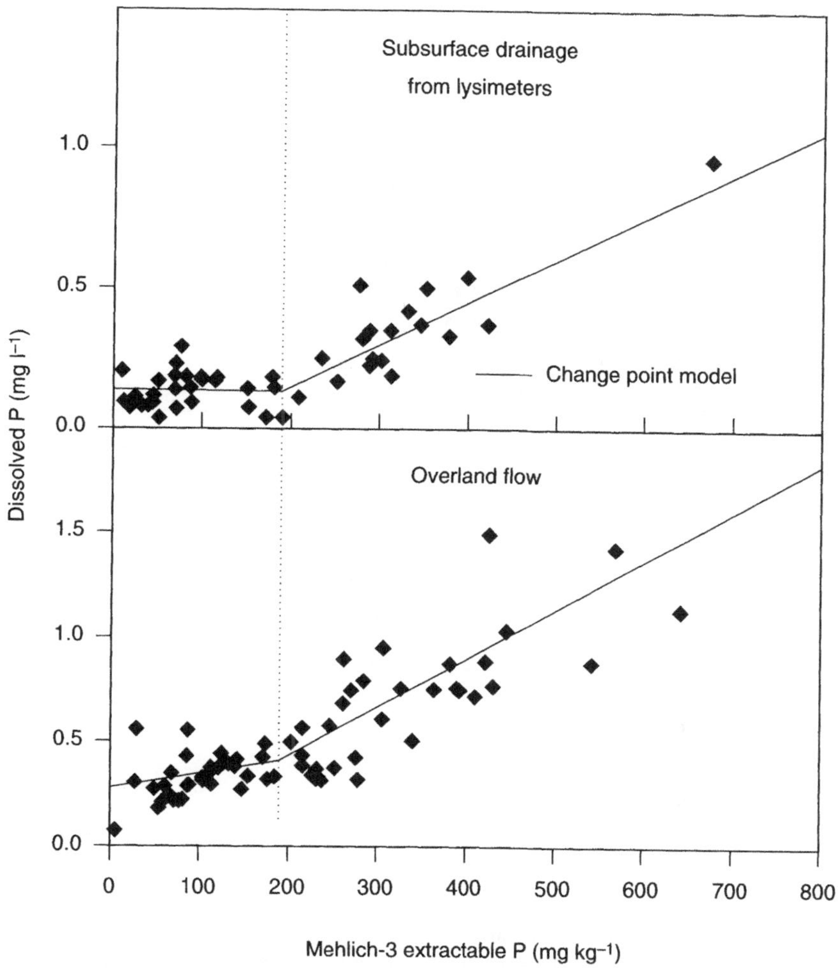

Fig. 9.12. The relationship between the concentration of dissolved P in overland flow and subsurface drainage from 30 cm deep lysimeters and the Mehlich-3 extractable P concentration of surface soil (0–5 cm) from a central Pennsylvania catchment (adapted from McDowell and Sharpley, 1999). Dotted lines indicate 'change points'.

tion of soil with $CaCl_2$ was related to the weekly export of soluble P from the catchment over a year. These authors showed that despite seasonal variation in both $CaCl_2$-P and Olsen P, the change point remained virtually constant at a mean value of 31 mg kg^{-1} Olsen P. The effect of seasonal variation was greatest between the months of February and June. However, it was found that this caused only the number and spread of data points around the change point to vary (Fig. 9.13). McDowell and Condron (1999, 2000) showed that the change point was an intrinsic property of the soil, effectively representing an approximation in the changing behaviour of an inverse desorption isotherm (quantity–intensity relationship). In one soil type with a wide range of pH, more than one change point can be detected, one for soils at pH > 5.8 and another for soils at less than this pH. Control of $CaCl_2$-P was attributed to Al and Fe oxides in soils with pH below 5.8, and the influence of Ca in soils with pH above 5.8.

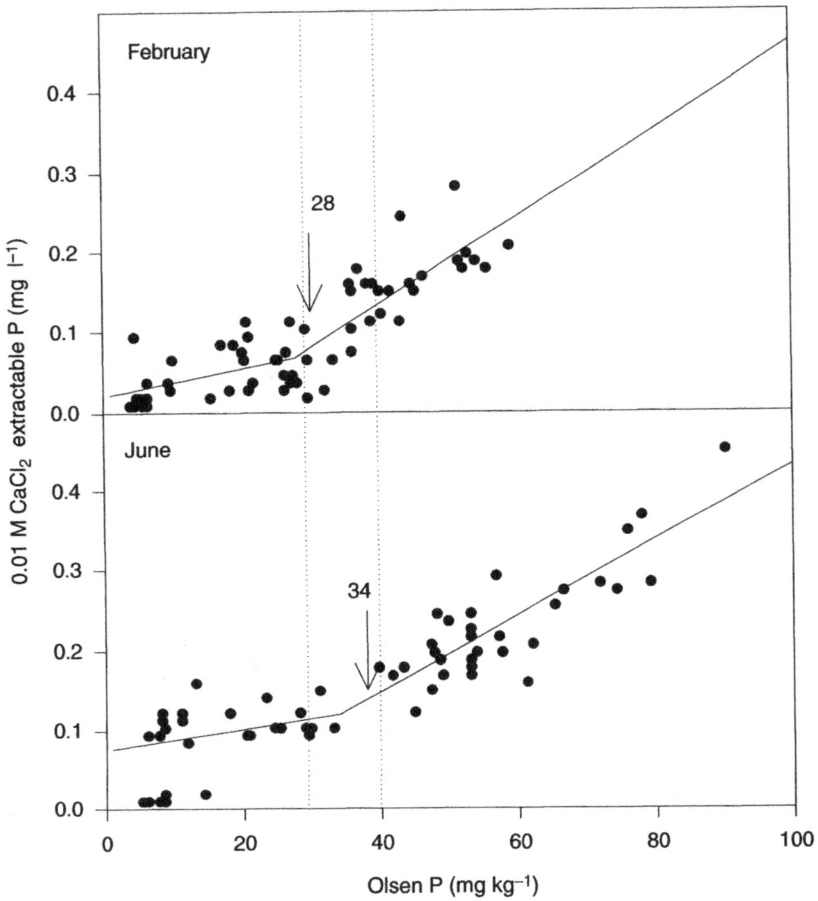

Fig. 9.13. The relationship between $CaCl_2$-P and Olsen P for soils in a small catchment in Devon (UK), showing the range of the change point (between the dotted vertical lines) and the number and variation of data points around the change point in February and June (adapted from McDowell and Trudgill, 2000).

Integrated source and transport management

On-farm N inputs can usually be more easily balanced with plant uptake than can P inputs. In the past, strategies for the remediation of N and P loss have been developed separately and implemented at the farm or catchment scale. Because the sources, pathways and sinks controlling N and P loss are different, remedial efforts directed to either N or P individually can be detrimental towards the other nutrient. In confined livestock operations, for example, large amounts of manure can be produced which, if applied to the land on a N basis to minimize NO_3^- leaching, can increase the soil P concentration and enhance the potential for P loss (Sims, 1997). Conversely, reducing the loss of P in overland flow by using conservation tillage may enhance the NO_3^- leaching (Sharpley, 2000).

Incorporation of added P increases interactions between P and the soil, reducing the availability of that P to transport mechanisms. Surface application of P (e.g. top-dressed manure) greatly increases source risk relative to methods that incorporate added P (e.g. injection of manure). This applies both to overland transport via overland flow and to subsurface transport by preferential or macropore flow: the higher the concentration of P at the soil surface, the higher the concentration of P in overland flow and leachate.

In addition to application methods, a number of management practices are now being explored to control source risk. Stout *et al.* (1999) determined that gypsum produced as a coal combustion by-product reduces P solubility in soil without significantly reducing plant-available P. Elsewhere, Sharpley (1999) showed that deep tillage can reduce soil test P by as much as 65%. While such options do address the proximate concern of excessive soil P levels, they should not be seen as solutions to the greater problem of continuing the over-application of P to soils. Furthermore, in the case of deep tillage, the trade-off between reduced soil test P levels and increased susceptibility to erosion must be considered.

Such positive and negative impacts of conservation practices on N and P loss potential need to be considered in the development of practical and sound remedial measures. One approach is the use of an indexing procedure to identify agricultural soils and management practices that might affect water quality. Such an index would be helpful for field personnel with limited resources working with land managers, to identify areas vulnerable to N and P loss and to suggest management alternatives to decrease risk. Strategies to minimize N and P loss are most effective if vulnerable areas within a catchment are targeted, rather than implementation of general strategies over a broad area.

Case study: use of nitrogen and phosphorus indices on an experimental catchment in central Pennsylvania, USA

In July 1996, soils sample (0–5 cm depth) were collected on a 30 m grid within a 39.5 ha sub-catchment of Mahantango Creek, a tributary of the Susquehanna River and ultimately the Chesapeake Bay. Soils were air-dried, sieved (< 2 mm), and Mehlich-3 extractable P determined (Mehlich, 1984). The dominant soils are loamy skeletal to fine loamy, mixed, mesic families of Typic Dystudepts (Alvira, Berks and Calvin series), Typic Hapludults (Hartleton series) and Typic Fragiudults (Watson series). Slopes range from 1 to 20%. The catchment is characterized by mixed land use (50% soybean, wheat or maize, 20% pasture, 30% woodland). The catchment receives an average rainfall of 1100 mm $year^{-1}$. Stream flow is 450 mm $year^{-1}$.

Two site assessment indices, an N index and a P index, were applied at a 5 m cell scale over the entire catchment. USDA-NRCS soil surveys provided information for transport factors (e.g. erosion and overland flow classes, permeability and texture). Source factors were derived from soil test P (as Mehlich-3 P) and management information on an individual field basis (rate and timing) provided by the farmers. Fertilizer application averaged 30 kg P and 112 kg N ha^{-1} $year^{-1}$ to soybeans. Manured fields received differing rates, ranging from 60 m^3 ha^{-1} $year^{-1}$ pig slurry (approximately 75 kg P ha^{-1} $year^{-1}$ and 300 kg N ha^{-1} $year^{-1}$) to 5 Mg ha^{-1} $year^{-1}$ poultry manure (approximately 225 kg P ha^{-1} $year^{-1}$ and 480 kg N ha^{-1} $year^{-1}$) (Eck and Stewart, 1995; Sharpley *et al.*, 1998a).

The P index employed (Table 9.2) was based on the original P Index described by Lemunyon and Gilbert (1993) and the modified P index of Gburek *et al.* (2000), which rank transport (overland flow and erosion) and source (soil test P, applied P type, rate and method) factors controlling P loss from agricultural fields. Within the index, the product of transport factors is multiplied by the sum of source characteristics for a site (Table 9.2). This requires the coincidence of source and transport factors in order to be able to identify sites as highly vulnerable to P loss. In the original P Index of Lemunyon and Gilbert (1993), a site could be ranked as highly vulnerable to P loss if source factors alone were high, even though no overland flow or erosion occurred. A change was made to the P index of Gburek *et al.* (2000) by incorporating a sliding scale for soil test P and a change point, whereby at soil test P concentrations above the change point

Table 9.2. The modified P indexing system to rate the potential P loss in surface runoff from source and transport soil characteristics (modified from Sharpley *et al.*, 2000).

Part A	Phosphorus transport characteristics				
Transport factors	None (0.6)	Low (0.7)	Medium (0.8)	High (0.9)	Very high (1.0)
Soil erosion (Mg ha^{-1})	Negligible	< 10	10–20	20–30	> 30
Runoff Class	Negligible	Infrequent irrigation on well-drained soils	Moderate irrigation on soils with slopes < 5%	Frequent irrigation on soils with slopes of 2–5%	Frequent irrigation on soil with slopes > 5%
Contributing distance (m)	None (0.2) > 170	Low (0.4) 170–130	Medium (0.6) 130–80	High (0.8) 80–30	Very High (1.0) < 30
Part B	**Phosphorus soil characteristics**				
Source factors	Very low	Low	Medium	High	Very high
Soil test P			Soil test P (mg P kg^{-1})		
Loss rating value			Soil test P × 0.21[a] × CP[b]		
Fertilizer P rate			Fertilizer rate (kg P ha^{-1})		
P fertilizer application method and timing	Placed with planter or injected more than 5 cm deep	Incorporated < 1 week after application	Incorporated >1 week or not incorporated following application in May–October	Incorporated > 1 week or not incorporated following application in Nov–April	Surface applied on frozen or snow covered soil
	0.2	0.4	0.6	0.8	1.0
Loss rating value	Fertilizer P application rate × loss rating for fertilizer P application method and timing				
Manure P rate			Manure Application (kg ha−1)		
P manure application method and timing	Placed with planter or injected more than 5 cm deep	Incorporated < 1 week after application	Incorporated > 1 week or not incorporated following application in May–October	Incorporated > 1 week or not incorporated following application in Nov–April	Surface applied on frozen or snow covered soil
	0.2	0.4	0.6	0.8	1.0
Loss rating value	Fertilizer P application rate × loss rating for fertilizer P application method and timing				

[a] A correction factor of 0.2 for soil test P based on the field data which showed a fivefold greater runoff P concentration with a unit increase in manure or fertilizer P addition than soil test P increase (as Mehlich-3 P) (Sharpley and Tunney, 2000).

[b] A factor is included to account for the change point in soil test P, where the ratio of the slope of the split-line model after the change point to the slope of the split-line model before the change point is multiplied by soil test P. For this case study this relates to the slopes in Fig. 9.12.

P index rating = Part A (erosion runoff x return period) x Part B (soil + fertilizer + manure P loss rating values)

soil test P was multiplied by the ratio of the two slopes to give the potential increase in P loss by overland flow. The slopes used relate to those seen in Fig. 9.12 and better represent the relationship of overland flow to soil P.

An N index to assess site vulnerability to N loss was developed, based on the principles set out by Kissel *et al.* (1982) and derived from the structure of the P Index (Table 9.3). The amount of NO_3^- leached from agricultural lands depends on soil properties, volume of water draining through soil and the amount escaping plant uptake that moves below the root zone. Source factors for N (fertilizer and manure

Table 9.3. The N index to rate the potential loss in leaching from site characteristics determining source and transport factors (modified from Heathwaite *et al.*, 2000).

Part A	Nitrogen transport characteristics				
Transport factors	None (0)	Low (1)	Medium (2)	High (4)	Very high (8)
Texture	Clay	Clay loam to silty clay loam	Loam to silt loam	Loamy fine sand to coarse sandy loam	Sand
Permeability ($cm\ h^{-1}$)	< 1.5	1.6–4	4.1–14.9	15–50	> 50
Part B	**Nitrogen source characteristics**				
Source factors	None (0)	Low (1)	Medium (2)	High (4)	Very high (8)
Fertilizer rate ($kg\ N\ ha^{-1}$)	None applied	1–50	51–150	151–300	> 300
Application method for fertilizer	None	Placed with planting	Incorporated immediately before crop	Surface applied > 3 months before crop or incorporated < 3 months before crop	Incorporated > 3 months before crop
Manure N rate ($kg\ N\ ha^{-1}$)	None applied	1–50	51–150	151–300	> 300
Application method for manure	None	Placed with planting	Incorporated immediately before crop	Surface applied > 3 months before crop or incorporated < 3 months before crop	Incorporated > 3 months before crop

N index rating = Part A (texture rating × permeability rating) × Part B (sum of source characteristic ratings)

N, application method) are weighted and multiplied by transport factors (texture, permeability) to achieve a final weighting. Index rankings were used to estimate impacts of manure and fertilizer applications on water quality (Table 9.3).

Within the catchment, Mehlich-3 P ranged from 7 to 788 mg kg^{-1} (Fig. 9.14) and was generally distributed as a function of land use and field boundaries. Soils in wooded areas had low Mehlich-3 P values (< 30 mg kg^{-1}), grazed pastures had from 100 to 200 mg kg^{-1}, and cropped fields receiving manure and fertilizer applications were, in most cases, above 200 mg kg^{-1}. Near-stream areas that were wet for much of the year had lower Mehlich-3 P values (< 100 mg kg^{-1}), reflecting their limited productive value and low levels of P application. In total, 52% of the soils in the catchment had Mehlich-3 P concentrations in excess of those required for optimum crop yield (> 100 mg kg^{-1}), with 33% above 200 mg kg^{-1}. Of the remaining 48% of soils at or below optimum soil test P concentrations, P application would be recommended on only 14% (30–100 mg kg^{-1}), as the remaining 34% of the soils were wooded.

These findings present important insight into the implications of P-based management of land-applied manure. If source-based agronomic criteria were imposed on farmers, many soils would be excluded from further manure application since nearly 80% of cultivated and pasture soils were sufficiently high (> 100 mg kg^{-1}) for no crop response to be expected. An alternative scenario could involve the use of the change point, giving the land user a simple point in soil test P concentration not to exceed. This may be less environmentally and economically restrictive because the change point is often above optimum soil test P concentration for plant growth. For example, the Broadbalk soil (FAO Tax: Chromic Luvisol, USDA Tax: Aquic Paleudalf) at Rothamsted (Harpenden, UK)

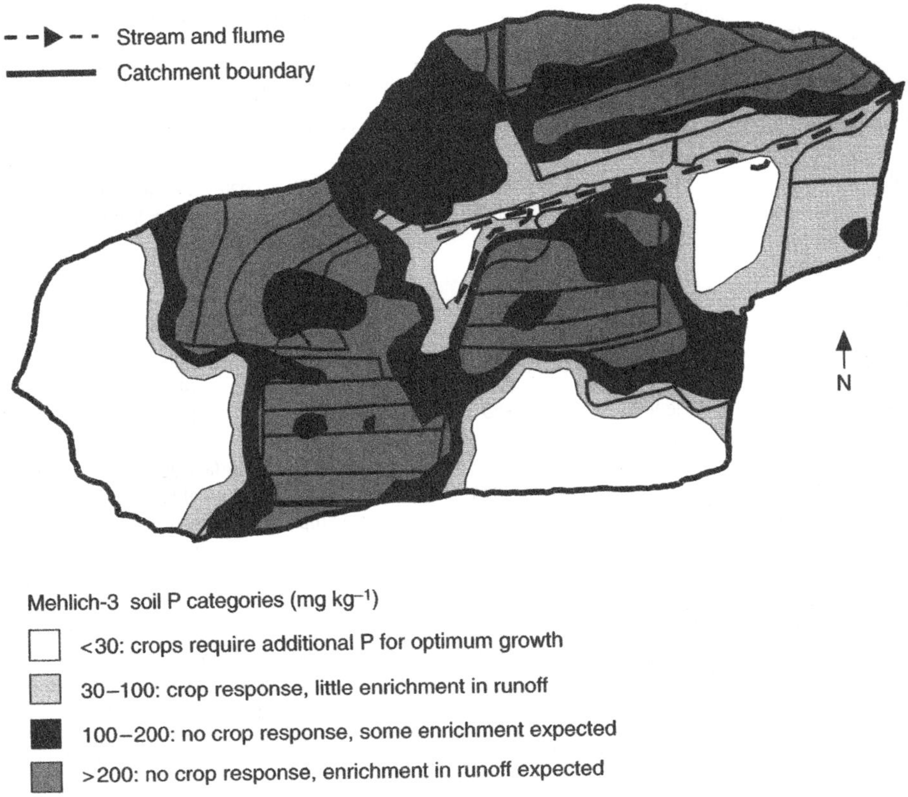

Fig. 9.14. Distribution of Mehlich-3 soil P within an experimental catchment in central Pennsylvania, (0–5 cm soil depth).

has a change point at 55 mg Olsen P kg^{-1} (Heckrath *et al.*, 1995), well above the required optimum growth of potatoes, sugarbeet, winter wheat and spring barley (Johnston and Poulton, 1997). Within the experimental catchment, a change point was established for most of the soils in the catchment at 185 mg Mehlich-3 P kg^{-1} for overland flow. This value is similar to that proposed by several US states of 200 mg Mehlich-3 P kg^{-1} (Sharpley *et al.*, 1996), which if imposed would render 33% of the land (representing 63% of the cultivated area) ineligible for further manure application. However, such source control management ignores the role of transport factors in P loss that govern P movement to the stream channel, restricting management in areas that are of no actual risk. Such 'blanket' approaches to P management are unnecessarily restrictive.

Source and transport factors derived from Tables 9.2, 9.3 and 9.4 for P and N are illustrated in Fig. 9.15. Spatial trends in the relative risk of P and N loss as determined by the combination of source and transport factors in the N and P site assessment indices are shown in Fig. 9.16. Critical source areas are defined as regions of high or very high source and transport loss potential (Table 9.4). Such zones, with 'high' risk of P loss, represent < 4% of the catchment area, while only 0.5% of the catchment was identified as having a 'very high' risk of P loss, and 'medium' and 'low' risk areas accounted for 21% and 75% of the remaining catchment area, respectively. Medium and high P loss areas were largely, but not wholly, in near-stream zones. Some areas with elevated risk of P loss can be found at distance from the stream, occurring in areas with

Table 9.4. Interpretation of the N and P index (modified from Heathwaite *et al.*, 2000).

Index rating		
P	N	Generalized interpretation of the N and P index
< 25	< 3	*Low* potential for N and P loss. If current farming practices are maintained, there is a low probability of adverse impacts on surface waters
25–75	3–8	*Medium* potential for N and P loss. The chance for adverse impacts on surface waters exists, and some remediation should be taken to minimize the probability of P loss
75–150	9–18	*High* potential for N and P loss and adverse impacts on surface waters. Soil and water conservation measures and a P management plan are needed to minimize the probability of P loss
> 150	> 18	*Very high* potential for N and P loss and adverse impacts on surface waters. All necessary soil and water conservation measures and a P management plan must be implemented to minimize P loss

very high source risk and moderate transport risk. Not all near-stream soils present high risk of P loss because poor drainage limits their production potential and few P inputs are directed at them.

Much larger catchment areas are susceptible to leaching; 80% of the catchment was identified as having medium to high risk of leaching (Fig. 9.15). The large number of sites vulnerable to N loss reflects the greater mobility of NO_3^-; areas prone to leaching are far more extensive than those prone to overland flow. Source areas at risk are both less defined and less important compared with P, although some 'hotspots' are apparent.

The delineation of catchment areas contributing P relative to N suggests that a prudent approach to manure management should begin with assessing site vulnerability to N and P loss. For P, remedial strategies and management of both source and transport factors include modifying fertilizer and manure applications and implementation of measures to restrict overland flow and erosion. In many cases, slight changes in field boundaries to align with P management boundaries could greatly reduce risk of P loss (e.g. Fig. 9.16a, sites 1 and 2). Given the very high, but localized, risk of P loss from site 3 (Fig. 9.16a), remedial management options might be better focused on the portion of the field that is very highly vulnerable to P than on the entire field.

Management of source factors, such as rate, timing and method of N input, as well as crop selection and rotation, best address N loss. Transport factors affecting N loss are more difficult to control, as leaching is pervasive compared with overland flow (Fig. 9.15). For fields with high risk of N loss, N applications should be carefully managed to maximize crop uptake. In order to minimize levels of residual NO_3^-, management must take into account crop N needs, timing of greatest N uptake and mineralization potential of organic N. Thus, management interpretations derived from use of site assessment indices in the experimental catchment support a 'precision agriculture' approach to farming, even in the relatively small fields associated with north-eastern USA.

Conclusions

Source factors for P and N play an important role in the risk of diffuse nutrient losses from agricultural lands. Because of the differing mobility of P and N in the environment, source factors are much more important to N loss potential than to P loss potential, as P loss is restricted primarily to source areas where high P availability and high transport potential overlap. The measurement and integration of source and transport risk require an understanding of nutrient sources, soil properties, hydrological conditions and cropping system management.

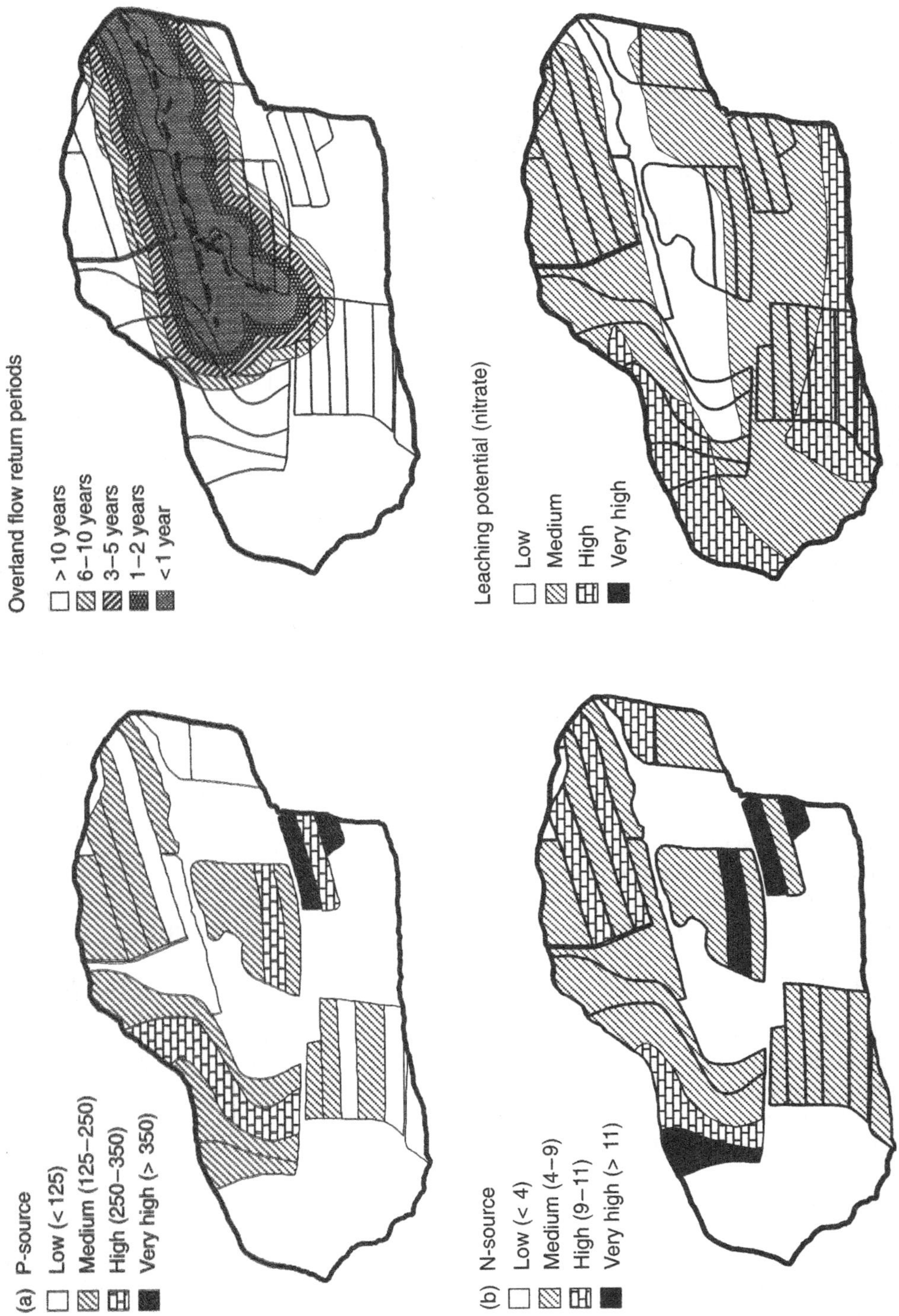

Fig. 9.15. Source (left) and transport (right) factors for P (a) and N (b) applied to an experimental catchment in central Pennsylvania.

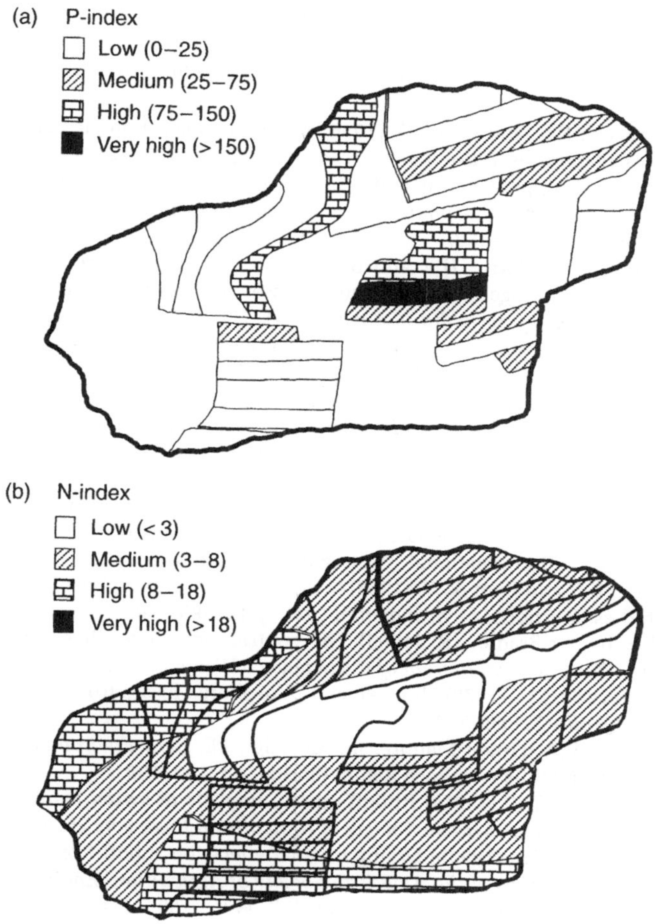

Fig. 9.16. Combined source and transport P (a) and N (b) indices applied to an experimental catchment in central Pennsylvania; sites marked are discussed in the text.

In adopting an integrated approach to land-applied manure management, a degree of bias towards either N or P could be introduced. This bias would depend on particular management objectives for each catchment, as well as objectives related to nutrient control. For instance, if a site's risk of P loss is low, but NO_3^- leaching potential is high, priority should be given to N-based management rather than to P-based management. Conversely, in catchment areas with high overland flow or erosion potential, management should be P based.

Legislation has been introduced in some eastern US states and The Netherlands that would base manure inputs to land on P rather than N. Such legislation overrides landscape and hydrological variables that affect nutrient transport, and therefore ignores the differential susceptibility of certain areas to nutrient loss. Prevention of P and N loss from catchments should focus on defining, targeting and remediating source areas of P that combine high soil P levels with high erosion and overland flow potential, and source areas of N with high permeability. However, while both source and transport factors can be combined to yield an estimate of overall risk of P loss, true control of P losses ultimately requires balance of nutrient inputs and outputs at the farm gate.

References

Allen, H.E., Garrison, A.W. and Luther, G.W. (eds) (1998) *Metals in Surface Waters.* Ann Arbor Press, Chelsea, Michigan.

Beauchemin, S., Simard, R.R. and Cluis, D. (1996) Phosphorus sorption–desorption kinetics of soil under contrasting land uses. *Journal of Environmental Quality* 25, 1317–1325.

Beegle, D. (2000) Integrating phosphorus and nitrogen management at the farm level. In: Sharpley, A.N. (ed.) *Agriculture and Phosphorus Management: the Chesapeake Bay.* Lewis Publishers, Boca Raton, Florida, pp. 159–168.

Bengston, R.L., Carter, C.E., Morris, H.F. and Bartkiewics, S.A. (1988) The influence of subsurface drainage practices on nitrogen and phosphorus losses in a warm, humid climate. *Transactions of the American Society of Agricultural Engineers* 31, 729–733.

Bengston, L., Seuna, P., Lepisto, A. and Saxena, R.K. (1992) Particle movement of meltwater in a subdrained agricultural basin. *Journal of Hydrology* 135, 383–398.

Breeuwsma, A. and Silva, S. (1992) *Phosphorus Fertilization and Environmental Effects in The Netherlands and the Po Region (Italy),* Report 57. Agricultural Research Department, The Winand Staring Centre for Integrated Land, Soil and Water Research, Wageningen, The Netherlands.

Catt, J.A., Johnston, A.E. and Quinton, J.N. (1997) Phosphate losses in the Woburn erosion reference experiment. In: Tunney, H., Carton, O.T., Brookes, P.C. and Johnston, A.E. (eds) *Phosphorus Loss from Soil to Water.* CAB International, Wallingford, UK, pp. 374–377.

Culley, J.L.B., Bolton, E.F. and Bernyk, V. (1983) Suspended solids and phosphorus loads from a clay soil: I. Plot studies. *Journal of Environmental Quality* 12, 493–498.

Davies, D.B. and Sylvester-Bradley, R. (1995) The contribution of fertilizer nitrogen to leachable nitrogen in the UK. A Review. *Journal of the Science of Food and Agriculture* 68, 399–406.

Duxbury, J.M. and Peverly, J.H. (1978) Nitrogen and phosphorus losses from organic soils. *Journal of Environmental Quality* 7, 566–570.

Eck, H.V. and Stewart, B.A. (1995) Manure. In: Rechcigl, J.E. (ed.) *Environmental Aspects of Soil Amendments.* Lewis Publishers, Boca Raton, Florida, pp. 169–198.

Edwards, D.R. and Daniel, T.C. (1993a) Runoff quality impacts of swine manure applied to fescue plots. *Transactions of the American Society of Agricultural Engineers* 36, 81–80.

Edwards, D.R. and Daniel, T.C. (1993b) Drying interval effects on runoff from fescue plots receiving swine manure. *Transactions of the American Society of Agricultural Engineers* 36, 1673–1678.

Freese, D., Lookman, R., Merckx, R. and van Riemsdijk, W.H. (1995) New method for assessment of long-term phosphate desorption from soils. *Soil Science Society of America Journal* 59, 1295–1300.

Gburek, W.J. and Heald, W.R. (1974) Soluble phosphate output of an agricultural watershed in Pennsylvania. *Water Resources Research* 10, 113–118.

Gburek, W.J., Sharpley, A.N., Heathwaite, L. and Folmar, G.J. (2000) Phosphorus management at the watershed scale: a modification of the phosphorus index. *Journal of Environmental Quality* 29, 130–144.

Goerhing, L.D., Wright, P.E., Steenhuis, T.S. and Walter, M.F. (1999) *Subsurface Drain Water Quality Impacts from Manure Applications,* NABEC Paper No. 99-9905. Presented at Northeast Agricultural/Biological Engineering Conference, Lancaster, Pennsylvania.

Hanson, G.C., Groffman, P.R. and Gold, A.J. (1994) Denitrification in riparian wetlands receiving high and low groundwater nitrate inputs. *Journal of Environmental Quality* 23, 917–922.

Haygarth, P.M., Hepworth, L. and Jarvis, S.C. (1998) Forms of phosphorus transfer in hydrological pathways from soil under grazed grassland. *European Journal of Soil Science* 49, 65–72.

Haygarth, P.M., Heathwaite, A.L., Jarvis, S.C. and Harrod, T.R. (2000) Hydrological factors for phosphorus transfer from agricultural soils. *Advances in Agronomy* 69, 153–178.

Heathwaite, A.L., Griffiths, P., Haygarth, P.M., Jarvis, S.C. and Parkinson, R.J. (1997) Phosphorus loss from grassland soils: implications of land management for the quality of receiving waters. In: *Freshwater Contamination,* Proceedings Rabat Symposium, April-May 1997, IHAS 243, pp. 177–186.

Heathwaite, A.L., Sharpley, A.N. and Gburek, W.J. (2000) A conceptual approach for integrating phosphorus and nitrogen management at watershed scales. *Journal of Environmental Quality* 29, 158–167.

Heckrath, G., Brookes, P.C., Poulton, P.R. and Goulding, K.W.T. (1995) Phosphorus leaching from soils containing different phosphorus concentrations in the Broadbalk experiment. *Journal of Environmental Quality* 24, 904–910.

Hesketh, N. and Brookes, P.C. (2000) Development of an indicator for risk of phosphorus leaching. *Journal of Environmental Quality* 29, 105–110.

Johnston, A.E. and Poulton, P.R. (1997) Defining critical levels of available soil phosphorus for agricultural crops. In: Tunney, H., Carton, O.T., Brookes, P.C. and Johnston, A.E. (eds) *Phosphorus Loss from Soil to Water*, CAB International, Wallingford, UK, pp. 441–445.

Kissel, D.E., Bidwell, O.W. and Kientz, J.F. (1982) *Leaching Classes of Kansas Soils*. Kansas Experimental Station Bulletin 6641, Manhattan, Kansas.

Korevaar, H. and Den Boer, D.J. (1990) *Practical Measures to Reduce Nutrient Losses from Grassland Systems*, Proceedings No. 301. The Fertiliser Society, London.

Lemunyon, J.L. and Gilbert, R.G. (1993) The concept and need for a phosphorus assessment tool. *Journal of Production Agriculture* 6, 483–486.

Lord, E.I. and Mitchell, R.D. (1998) Effect of nitrogen inputs to cereals on nitrate leaching from sandy soils. *Soil Use and Management* 14, 78–83.

Massey, H.F. and Jackson, M.L. (1952) Selective erosion of soil fertility constituents. *Soil Science Society of America Proceedings* 16, 353–356.

Mathers, A.C., Stewart, B.A. and Blair, B. (1975) Nitrate-nitrogen from soil profiles by lucerne. *Journal of Environmental Quality* 4, 403–405.

McColl, R.H.S., White, E. and Gibson, A.R. (1977) Phosphorus and nitrate runoff in hill pasture and forest catchments, Taita, New Zealand. *New Zealand Journal of Marine and Freshwater Research* 11, 729–744.

McDowell, R.W. and Condron, L.M. (1999) Developing a predictor for phosphorus loss from soil. In: Currie, L.D. (ed.) *Best Soil Management Practices for Production* (Fertilizer and Lime Research Centre 12th Annual Workshop). Massey University, Palmerston North, New Zealand, pp. 153–164.

McDowell, R.W. and Condron, L.M. (2000) Chemical nature and potential mobility of phosphorus in fertilized grassland soils. *Nutrient Cycling in Agroecosystems* 57, 225–233.

McDowell, L.L. and McGregor, K.C. (1984) Plant nutrient losses in runoff from conservation tillage corn. *Soil Tillage Research* 4, 79–91.

McDowell, R.W. and Sharpley, A.N. (1999) Relating soil phosphorus release to the potential for phosphorus movement in surface and subsurface runoff. In: *Agronomy Abstracts 1999*. American Society of Agronomy, Salt Lake City, Utah.

McDowell, R.W. and Trudgill, S.T. (2000) Variation in phosphorus loss from a small catchment in South Devon, UK. *Agriculture Ecosystems and Environment* 79, 143–157.

McDowell, R.W., Condron, L.M., Main, B.E. and Dastgheib, F. (1997) Dissipation of imazapyr, flumetsulam and thifensulfuron in soil. *Weed Research (Oxford)* 37, 381–389.

Mehlich, A. (1984) Mehlich-3 soil test extractant: a modification of Mehlich-2 extractant. *Communications in Soil Science and Plant Analysis* 15, 1409–1416.

Miller, M.H. (1979) Contribution of nitrogen and phosphorus to subsurface drainage water from intensively cropped mineral and organic soils in Ontario. *Journal of Environmental Quality* 8, 42–48.

Muir, J., Boyce, J.S., Seim, E.C., Mosher, P.N., Deibert, E.J. and Olson, R.A. (1976) Influence of crop management practices on nutrient movement below the root zone in Nebraska soils. *Journal of Environmental Quality* 5, 255–259.

Nicholaichuk, W. and Read, D.W.L. (1978) Nutrient runoff from fertilized and unfertilized fields in western Canada. *Journal of Environmental Quality* 7, 542–544.

Pote, D.H., Daniel, T.C., Sharpley, A.N., Moore, P.A. Jr, Edwards, D.R. and Nichols, D.J. (1996) Relating extractable soil phosphorus to phosphorus losses in runoff. *Soil Science Society of America Journal* 60, 855–859.

Pote, D.H., Daniel, T.C., Nichols, D.J., Sharpley, A.N., Moore, P.A. Jr, Miller, D.M. and Edwards, D.R. (1999) Relationship between phosphorus levels in three ultisols and phosphorus concentrations in runoff. *Journal of Environmental Quality* 28, 170–175.

Rogers, M.T. (1941) Plant nutrient losses by erosion from a corn, wheat, clover rotation on Dunmore silt loam. *Soil Science Society of America Proceedings* 6, 263–271.

Ryden, J.C. and Syers, J.K. (1975) Rationalization of ionic strength and cation effects on phosphate sorption by soils. *Journal of Soil Science* 26, 395–406.

Schepers, J.S., Francis, D.D., Alberts, E.E. and Frank, K.D. (1991) Maize production impacts on groundwater quality. *Journal of Environmental Quality* 20, 12–16.

Sharpley, A.N. (1985a) Depth of surface soil-runoff interaction as affected by rainfall, soil slope, and management. *Soil Science Society of America Journal* 49, 1010–1015.

Sharpley, A.N. (1985b) The selective erosion of plant nutrients in runoff. *Soil Science Society of America Journal* 49, 1527–1534.

Sharpley, A.N. (1993) An innovative approach to estimate bioavailable phosphorus in agricultural runoff using iron oxide-impregnated paper. *Journal of Environmental Quality* 22, 597–601.

Sharpley, A.N. (1995) Dependence of runoff phosphorus on extractable soil phosphorus. *Journal of Environmental Quality* 24, 920–926.

Sharpley, A.N. (1997) Rainfall frequency and nitrogen and phosphorus in runoff from soil amended with poultry litter. *Journal of Environmental Quality* 26, 1127–1132.

Sharpley, A.N. (1999) Soil inversion by plowing decreases surface soil phosphorus content. In: *Agronomy Abstracts 1999*. American Society of Agronomy, Madison, Wisconsin.

Sharpley, A.N. (2000) Fate and transport of phosphorus in agriculture and the environment. In: Lander, C.H. (ed.) *Natural Resource Conservation Service National Agronomy Manual*, Part 503.29. US Govt Printing Press, Washington, DC.

Sharpley, A.N. and Gburek, W.J. (2000) Integrated nutrient management for agricultural landscapes. In: McFarland, M.J. and Sanderson, M.A. (eds) *Integrated Land Management of Land Application of Animal Wastes*. American Society of Agricultural Engineers.

Sharpley, A.N. and Lord, E.J. (1997) The loss of nitrogen and phosphorus in agricultural runoff: processes and management. In: van Cleemput, O., Haneklaus, S., Hofman, G., Schnug, E. and Vermoesen, A. (eds) *Fertilization for Sustainable Plant Production and Soil Fertility*. 11th International World Fertilizer Congress, Ghent, Belgium, pp. 548–563.

Sharpley, A.N. and Smith, S.J. (1994) Effects of cover crops on surface water quality. In: Hargrove, W.L. (ed.) *Cover Crops for Clean Water Conference Proceedings*. Soil and Water Conservation Society, Ankeny, Iowa, pp. 41–49.

Sharpley, A.N. and Syers, J.K. (1976) Phosphorus transport in surface runoff as influenced by fertilizer and grazing cattle. *New Zealand Journal of Science* 19, 277–282.

Sharpley, A.N. and Syers, J.K. (1979) Loss of nitrogen and phosphorus in tile drainage as influenced by urea application and grazing animals. *New Zealand Journal of Agricultural Research* 22, 127–131.

Sharpley, A.N. and Syers, J.K. (1981) Amounts and relative significance of runoff types in the transport of nitrogen into a stream draining an agricultural watershed. *Water, Air and Soil Pollution* 10, 299–308.

Sharpley, A.N. and Tunney, H. (2000) Phosphorus strategies to meet agricultural and environmental challenges of the 21st century. *Journal of Environmental Quality* 29, 176–181.

Sharpley, A.N., Syers, J.K. and O'Connor, P.W. (1976) Phosphorus inputs into a stream draining an agricultural watershed. *Water, Air and Soil Pollution* 6, 39–52.

Sharpley, A.N., Syers, J.K. and Gregg, P.E.H. (1978) Transport of phosphorus from superphosphate and dicalcium phosphate in surface runoff from pasture. *New Zealand Journal of Science* 21, 307–310.

Sharpley, A.N., Smith, S.J., Williams, J.R., Jones, O.R. and Coleman, G.A. (1991) Water quality impacts associated with sorghum culture in the Southern Plains. *Journal of Environmental Quality* 20, 239–244.

Sharpley, A.N., Daniel, T.C., Sims, J.T. and Pote, D.H. (1996) Determining environmentally sound soil phosphorus levels. *Journal of Soil Water Conservation* 51, 160–166.

Sharpley, A.N., Meisinger, J.J., Breeuwsma, A., Sims, J.T., Daniel, T.C. and Schepers, J.S. (1998a) Impacts of animal manure management on ground and surface water quality. In: Hatfield, J.L. and Stewart, B.A. (eds) *Animal Waste Utilization: Effective Use of Manure as a Soil Resource*. Sleeping Bear Press, Inc., Ann Arbor, Michigan, pp. 173–242.

Sharpley, A.N., Gburek, W.J. and Folmar, G.J. (1998b) Integrated phosphorus and nitrogen management in animal feeding operations for water quality protection. In: *Animal Feeding Operations and Ground Water: Issues, Impacts and Solutions*. National Ground Water Association, St Louis, Missouri, pp. 72–95.

Sharpley, A.N., Daniel, T.C., Wright, B., Kleinman, P.J.A., Sobecki, T., Parry, R. and Joern, B. (1999) National research project to identify sources of agricultural phosphorus loss. *Better Crops* 83, 12–15.

Sharpley, A.N., Kleinman, T.C. and McDowell, R.W. (2000) Innovative management of agricultural phosphorus to protect soil and water quality. *Communications in Soil Science and Plant Analysis* (in press).

Sibbeson, E. and Sharpley, A.N. (1997) Setting and justifying upper critical limits for phosphorus in soils. In: Tunney, H., Carton, O.T., Brookes, P.C. and Johnston, A.E. (eds) *Phosphorus Loss from Soil to Water*. CAB International, Wallingford, UK, pp. 151–176.

Sims, J.T. (1997) Agricultural and environmental issues in the management of poultry wastes: recent innovations and long-term challenges. In: Rechcigl, J. and MacKinnin, H.C. (eds) *Uses of By-products and Wastes in Agriculture*. American Chemical Society, Washington, DC, pp. 72–90.

Sims, J.T., Simard, R.R. and Joern, B.C. (1998) Phosphorus losses on agricultural drainage: Historical perspectives and current research. *Journal of Environmental Quality* 27, 277–293.

Smith, K.A., Chalmers, A.G., Chalmers, B.J. and Christie, P. (1998) Organic manure phosphorus accumulation, mobility and management. *Soil Use and Management* 14, 154–159.

Smith, S.J., Sharpley, A.N., Naney, J.W., Berg, W.A. and Jones, O.R. (1991) Water quality impacts associated with wheat culture in the Southern Plains. *Journal of Environmental Quality* 20, 244–249.

Stout, W.L., Sharpley, A.N., Gburek, W.J. and Pionke, H.B. (1999) Reducing phosphorus export from croplands with FBC fly ash and FGD gypsum. *Fuel* 78, 175–178.

Tomlin, C. (1991) *The Pesticide Manual*. British Crop Protection Council, Royal Society of Chemistry, Farnham, UK.

Uhlen, G. (1988) Surface runoff losses of phosphorus and other nutrient elements from fertilized grassland. *Norwegian Journal of Agricultural Science* 3, 47–55.

Wendt, R.C. and Corey, R.B. (1980) Phosphorus variations in surface runoff from agricultural lands as a function of land use. *Journal of Environmental Quality* 9, 130–136.

Westerman, P.W., Donnely, T.L. and Overcash, M.R. (1983) Erosion of soil and poultry manure – a laboratory study. *Transactions of the American Society of Agricultural Engineers* 26, 1070–1078.

White, R.E. and Thomas, G.W. (1981) Hydrolysis of aluminum on weakly acidic organic exchangers: implications for phosphorus adsorption. *Fertiliser Research* 2, 159–167.

Young, R.A. and Mutchler, C.K. (1976) Pollution potential of manure spread on frozen ground. *Journal of Environmental Quality* 5, 174–179.

10 Hydrological Mobilization of Pollutants at the Field/Slope Scale

David Nash,[1] David Halliwell[1] and Jim Cox[2]
[1]*Agriculture Victoria Ellinbank, RMB 2460, Ellinbank, VIC 3821, Australia;*
[2]*CSIRO Land and Water, PMB 2, Glen Osmond, SA 5064, Australia*

Key words: detachment, dissolution, interflow, mobilization, overland flow, phosphorus, pollutants, scale, sedimentation, subsurface flow

Introduction

Agricultural production can drastically alter the hydrological balance of landscapes, thereby adversely affecting water quality. Conceptually, most agricultural catchments contain countless 'point sources' of potential pollutants. These pollutants become problems when they are mobilized into aquifers, tributaries and streams (Pigram, 1986; Hodgkin and Hamilton, 1993; Sharpley *et al.*, 1994; Gerriste, 1996; Baldwin, 1998).

The concentrations and loads of pollutants originating from agricultural catchments reflect the characteristics of the agricultural landscapes (e.g. soils, slope) and the pathways that the water has taken. The hydrology of the catchment determines how water moves, its physical characteristics such as flow and volume, and area and time of contact with soil.

This chapter investigates the effects of field/slope hydrology on the mobilization of pollutants. To facilitate discussion, our functional definition of field/slope will be an area of land without significant channelled flow, which removes the additional complexity of in-stream processes that alter pollutant concentrations and forms. Field-scale hydrology can be divided into surface and subsurface pathways comprising infiltration and saturation excess overland flow and interflow, and vertical matrix and macropore flow (Fig. 10.1). Infiltration or saturation excess overland flow and near-surface interflow are important for the generation of pollutants near the soil surface. However, the transport mechanisms and pathways of pollutants through the landscape often determine the concentrations and loads exported.

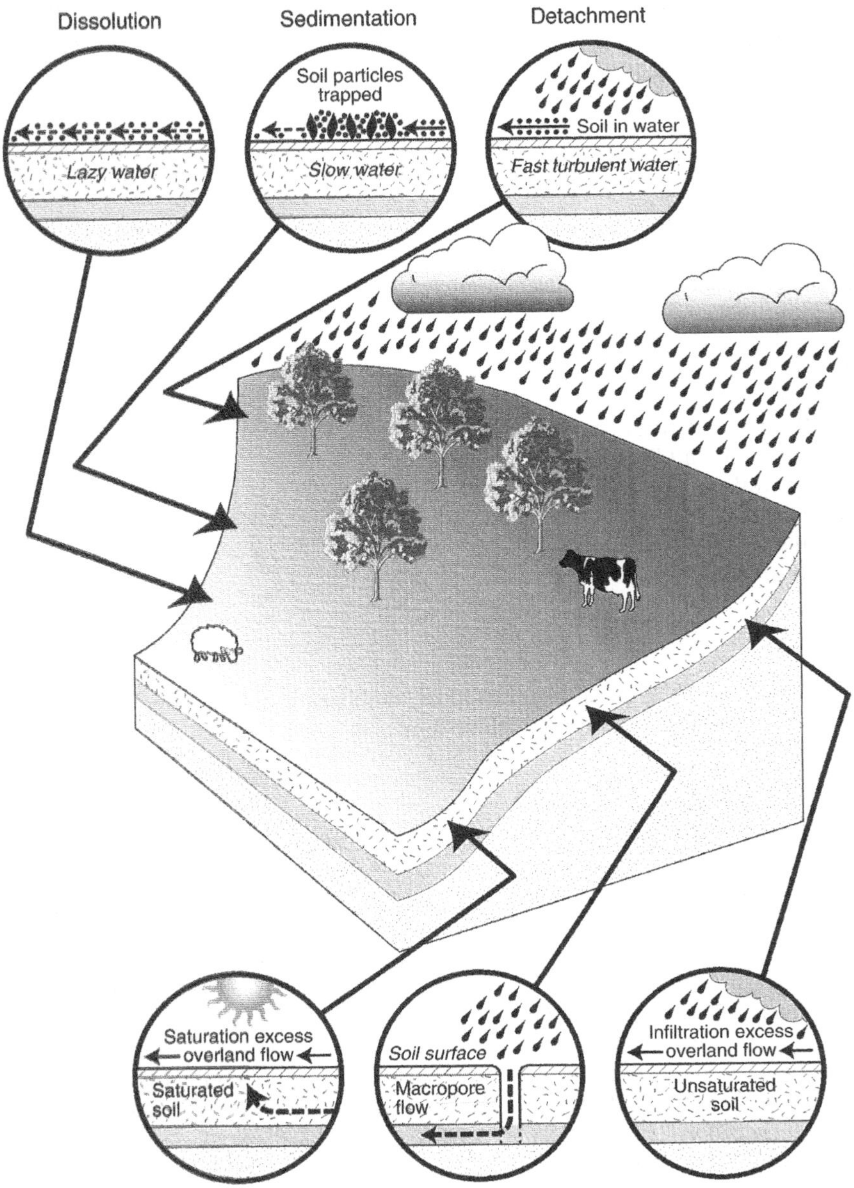

Fig. 10.1. Hydrology and pollutant mobilization processes at the field/slope scale.

Pollutant Mobilization Processes: Detachment and Dissolution

Detachment and dissolution are primarily responsible for pollutant mobilization in agricultural systems (White, 1987, 1988; Shainberg *et al.*, 1994; Nash and Halliwell, 1999a,b). While detachment is a physical process, dissolution is affected by the chemistry of the pollutants and surrounding materials, and the available reaction time. Distinguishing between detachment and dissolution processes is generally difficult. Most analytical procedures use particle size, typically the ability to pass through a 0.45 μm filter, to delineate between the two (Nash *et al.*, 2000). For the purposes of this chapter, we will use the term 'dissolved' to define pollutants that pass through a 0.45 μm filter, recognizing that this may include some pollutants attached to colloids (Beckett and Hart, 1993; Haygarth *et al.*, 1997).

Detachment commences with the mobilization of fine particles (sediments) and associated pollutants from aggregates and other soil materials by physical processes including raindrop impact, cultivation and flowing water, and physico-chemical processes such as slaking and dispersion (Leeper and Uren, 1997; Davies and Payne, 1988). These materials are transported by overland flow at a rate that is related to the kinetic energy of the water (Shainberg *et al.*, 1994). Factors that increase water velocity and turbulence generally increase detachment and transport rates. The process of surface soil detachment has been well reviewed elsewhere (Smith and Wischmeier, 1962; Anon, 1965; Kelley, 1983).

Factors affecting dissolution include the solubility of the pollutant in water, its sorption characteristics and the presence or absence of chelating agents or substances that block sorption sites. However, the most important factor is the 'residence time' or the time available for soil and water interaction (Sharpley *et al.*, 1994; Kirkby *et al.*, 1997; Kookana *et al.*, 1998; Nash and Halliwell, 1999a).

Pollutant Mobilization in Surface Pathways

Surface pathways

The pathways for water and pollutant transport in surface soil are shown in Fig. 10.1. Unfortunately, these pathways are rarely as well defined as the diagram suggests and in most landscapes infiltration excess overland flow, saturation excess overland flow and interflow are difficult to distinguish.

Infiltration excess overland flow

Infiltration excess overland flow occurs where the infiltration rate for the soil profile is exceeded. The volume and rate of infiltration excess overland flow depend on the rate of water addition, surface soil infiltration rate and soil hydraulic conductivity (Emmett, 1978). Consequently, factors such as slaking and dispersion that alter infiltration behaviour, and soil compaction and structural deterioration that affect hydraulic conductivity (Hillel, 1980; Cresswell *et al.*, 1992), also affect infiltration excess overland flow. At a field scale, infiltration excess overland flow tends to increase down slopes as both surface and subsurface run-on from higher areas increases the hydraulic load, and hence the probability of overland flow in receiving areas (Fig. 10.2).

Saturation excess overland flow

Saturation excess overland flow is characterized by saturation of the soil over which it is moving. For many soil profiles, saturated excess overland flow is a special case of infiltration excess overland flow. Generally, infiltration is occurring, albeit at a negligible rate, because of the low hydraulic conductivity of the underlying strata. Soils with a permeable A-horizon overlying a heavy clay, less permeable B-horizon (Chittleborough, 1992) are prone to such conditions (Cox and McFarlane, 1995). However, saturation excess overland flow also occurs where groundwater or interflow rises to the surface in discharge zones, often

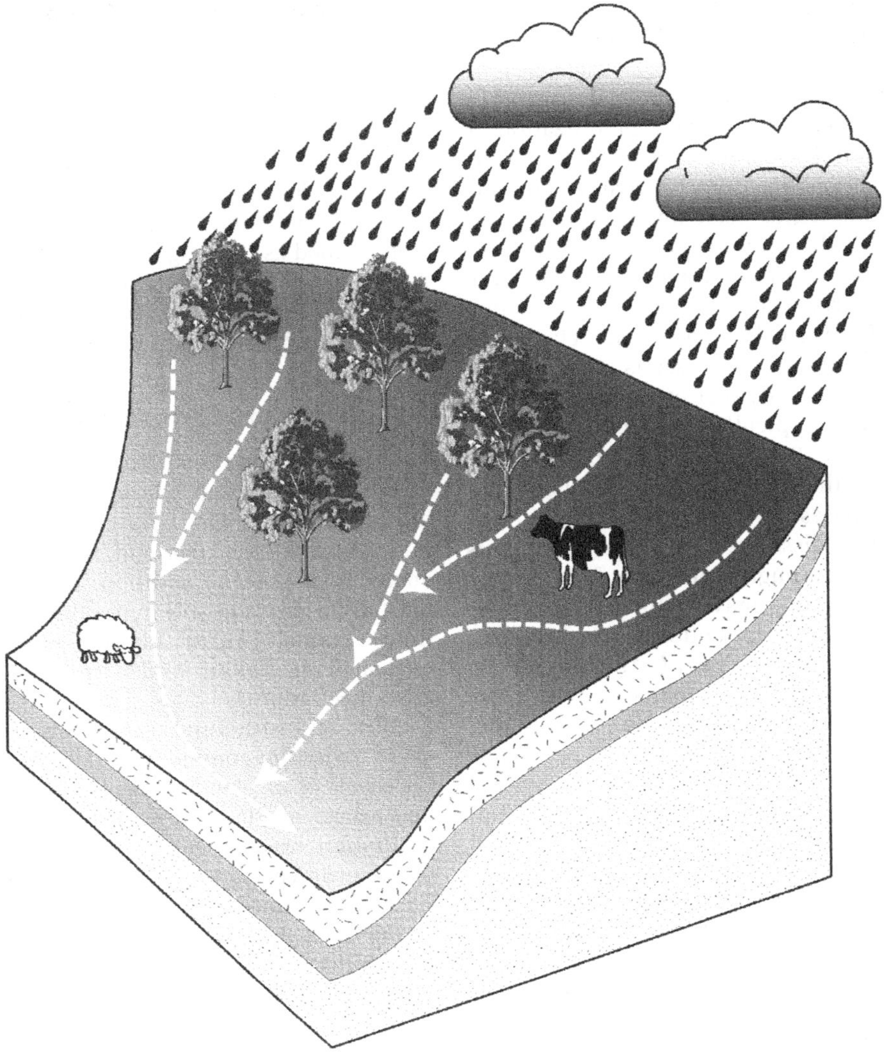

Fig. 10.2. Hydrology changes where flows converge downslope.

at the break of slope (Rulon *et al.*, 1985; Gerits *et al.*, 1990; Moore and Foster, 1990).

Interflow

Interflow describes water moving vertically into the soil without reaching groundwater and then laterally until discharged as overland flow (Amerman, 1965) at the change of slope or where the profile has been disrupted, for example, by clearing or ploughing. Often used interchangeably with the term 'throughflow' (Kirkby and Chorley, 1967), interflow can occur in the pollutant rich layer immediately below the soil surface or at the interface of the A/B or B/C horizons (Cox and McFarlane, 1995; Fleming and Cox, 1998; Stevens *et al.*, 1999). Waters moving at these interfaces are best referred to as A/B or B/C interflows and treated as components of subsoil hydrology (Stevens *et al.*, 1999).

As in the case of overland flow, interflow accumulates down a slope and when the incident water and interflow exceed the transmissivity (i.e. hydraulic conductivity × soil thickness), infiltration/saturation excess overland flow is likely. Interestingly, interflow and infiltration excess overland flow can occur simultaneously, especially in naturally anisotropic soils or soils with a compaction layer near to the surface. As infiltration/saturation excess overland flow and interflow immediately below the soil surface (0–20 mm) are difficult to functionally separate, the term overland flow will be used to describe them collectively.

Overland flow: theory

Overland flow is one of the more difficult hydrological concepts to mathematically describe as water addition (rainfall) and infiltration vary temporally and spatially (Nielsen *et al.*, 1973). In addition, flow may be laminar (subcritical), turbulent (supercritical) or a combination of both (Daugherty *et al.*, 1989).

One-dimensional overland flow down a plane land surface is described in Equations 10.1 and 10.2 (Hairsine *et al.*, 1992). Incorporating run-on and groundwater additions into the infiltration term and assuming no change in soil profile storage, these equations describe the combined effects of saturated excess and infiltration excess overland flows. The equations imply, as we might expect, that the changing water flux is inversely proportional to changes in water depth with time, infiltration rate and the rate of water addition, which is usually rain intensity.

$$\delta q/\delta x + \delta D/\delta t = R \qquad (\mathrm{m\ s^{-1}}) \qquad (10.1)$$

where: q is the water flux; x is the distance down the plane; D is the depth of water; t is time; and R is the excess rainfall rate.

$$R = P - I - \mathrm{d}E/\mathrm{d}t \qquad (\mathrm{m\ s^{-1}}) \qquad (10.2)$$

Where: P is the rate of water addition (usually rainfall rate); I is the infiltration rate; dE/dt is the rate of change of depressional storage.

Using the kinematic flow approximation, the relationship between water flux and flow depth can also be described by the resistance equation (Equations 10.3 and 10.4) (Emmett, 1978; Moore and Foster, 1990). For turbulent flow, the K term includes the effects of hydraulic roughness and slope while for laminar flow, K reflects the slope and kinematic viscosity. The value of the exponent (M) is 1.67 for turbulent flow and 3.00 for laminar flow (Emmett, 1978; Gerits *et al.*, 1990), implying that depth increases more rapidly for turbulent than for laminar flow.

$$q = KD^M \quad (\mathrm{m^3\ s^{-1}}\ \text{per metre width of slope}) \qquad (10.3)$$

where D is the depth of water; K is a constant; M is a constant.

$$D = (q/K)^{(1/M)} \qquad (\mathrm{m}) \qquad (10.4)$$

In a field, vegetation and soil materials impede water flow creating an overall 'disturbed flow' pattern, which is distinct from the irregularity of turbulent flow near the obstacles (Daugherty *et al.*, 1989). As a result, the depth of water increases with the highest down-slope velocities experienced at the water surface (Fig. 10.3).

Detachment processes

Mobilization of particulate pollutants requires sediment detachment and transport. The complexity of real systems makes it difficult to generalize about the impacts of hydrology on the level of detachment especially since laminar flow does not have the same detachment potential as disturbed or turbulent flow. As a general rule, increased flow rates lead to increased potential for detachment and this is likely at higher rain intensities, on steeper slopes, on lower infiltration soils with fewer impediments to flow and on slopes with significant run-on.

An important part of the detachment process is the sorting of pollutants during transport based on their surface area and density. The disturbed flow created by perturbations at the soil surface enriches

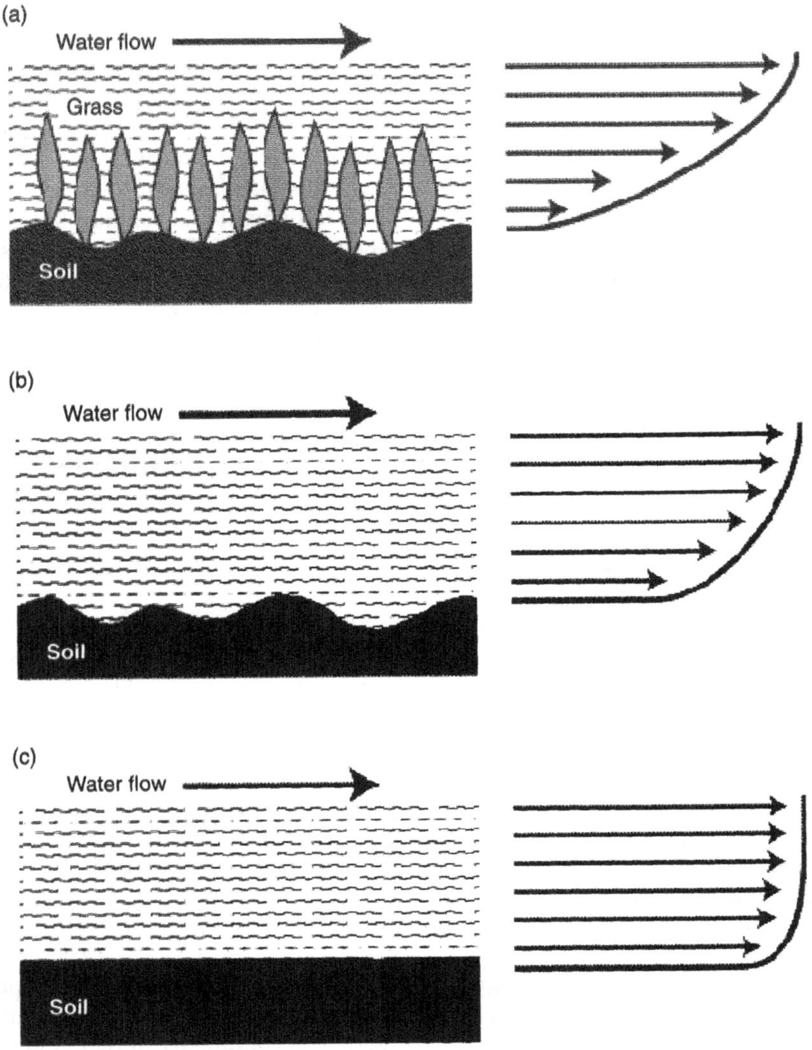

Fig. 10.3. Velocity profile of water in (a) pasture, (b) a ploughed field, and (c) a ploughed and previously eroded field.

runoff water with pollutants contained in low-density detrital material and those already sorbed to colloids, compared with the source materials (Sharpley, 1980, 1985). Enrichment ratios vary depending on the pollutant in question and soil treatment (Sharpley, 1980, 1985). A number of authors have shown that as the concentration of detached soil increases, the enrichment ratio decreases and approaches unity (Massey and Jackson, 1952; Menzel, 1980). This suggests that pollutant loads are higher, but concentrations lower, in more intense flows and has implications for water sampling strategies at this scale.

Dissolution processes

With the exception of a few pollutants, such as P which can be extracted directly from plants, the source of dissolution for most pollutants is close to the soil surface (Nash and Halliwell, 1999a). Here, pollutants, including agrochemicals and nutrients, accumulate because plants draw them upwards. Unlike detachment, dissolution occurs in a region where the velocity of the water is low (Fig. 10.3) and the migration of material into streamlines is determined by desorption, dissolution and diffusion rates rather than physical processes. A notable exception would be where raindrop impact on overland flow or detachment increases the surface area of a pollutant source such as faecal materials (Leonard and Wauchope, 1980; Ahuja *et al.*, 1981; Ahuja and Lehman, 1983).

The most important factors affecting dissolution are the volume of soil explored by the water, often only the first 20 mm layer (Sharpley *et al.*, 1988), and the contact time (Addiscott and Thomas, 2000). Where the depth of interaction between soil and water is essentially fixed, for example on stable soils, and source materials are not limiting, the load of pollutants exported depends almost exclusively on the contact time. It follows that factors affecting flow rates such as the rain intensity, slope and depth of flow, have less effect on pollutant loads than factors affecting contact time between soil and water such as storm length and duration of flow (Haygarth and Jarvis, 1999; Haygarth *et al.*, 2000). Using similar logic and compared with detachment under the same conditions, dissolution is equally likely on upper and lower slopes.

Once mobilized, species such as NO_3^--N, which do not strongly sorb to most soils, are generally transported without modification. In most cases the concentrations in overland flow are well below the solubility product (K_{sp}) ensuring that the pollutant remains in solution. Consequently, the load is determined by the quantities of the pollutant initially mobilized while the pollutant concentration is primarily determined by the flow.

The transport of dissolved pollutants that sorb to soil is more complex (Table 10.1). The mechanisms responsible for sorption include ion exchange, cation bridging, charge transfer, H^+ bonding and van der Waals interactions (Kookana *et al.*, 1998). While some pollutants are predominantly sorbed to clay colloids, for example paraquat and diquat (Kookana and Aylmore, 1993), others such as atrazine (Bowmer, 1971) appear to sorb primarily to organic matter. Further, the sorption of pollutants to sediments in transit (Sharpley *et al.*, 1981b) reduces solution concentrations, especially where both detachment and dissolution processes are significant and sediment source and deposition areas are located within a field (Fig. 10.1).

Pollutant Mobilization in Subsurface Pathways

Historically, it has been assumed that most pollutants do not move through soils containing substantial amounts of clay. The time-dependent sorption processes occurring mainly at the surfaces of iron and aluminium oxides and calcium and magnesium carbonates (Rajan *et al.*, 1974; Rolston *et al.*, 1975; Bolt, 1976; Bonneau and Souchier, 1982; Trudgill, 1990; Cresser *et al.*, 1993) were thought to remove most species from subsurface waters. Exceptions were believed to include ions with a low affinity for soil, such as NO_3^--N (Troeh *et al.*, 1980), very sandy soils (Ozanne *et al.*, 1961; Mansell *et al.*, 1977; Peverill *et al.*, 1977; Sharpley *et al.*, 1993) and highly organic soils where soluble organic matter facilitated transport either directly or by coating the active sites for sorption (Pierzynski *et al.*, 1994).

Recent work has demonstrated that relatively high fluxes of contaminants can move through some soils as well as in overland flow (Smetten *et al.*, 1994; Fleming and Cox, 1998; Stevens *et al.*, 1999). The relative importance of surface and subsurface pathways for pollutant transport has been studied at plot scale (Greenhill *et al.*, 1983a,b; Kirkby *et al.*, 1996) and sub-catchment scale (Fleming *et al.*, 1996; Stevens *et al.*, 1999) and found to be partly dependent on the macroporosity of the subsoil clays (Cox *et al.*, 2000).

Table 10.1. Sorption coefficients of some pesticides from reports on Australian soils (Kookana *et al.*, 1998).

	Important soil properties		
Pesticide	pH	Org. C (%)	K or K_d (ml g^{-1})
Atrazine	4.3–8.1[a]	0.3–6.0	0.24–8.4
	4.7–6.5[b]	0.46–1.0	0.39–0.55
	5.3–8.4[c]	0.7–1.8	0.6–4.4
	3.3–5.5[d]	0.1–3.0	0.35–24.9
Desethylatrazine	3.3–5.5[e]	0.1–3.0	0.30–90.0
Hydroxyatrazine			0.30–7.5
Carbaryl	4.8–8.4[c]	0.11–3.8	1.4–2.45
Chlorbromuron			0.25–10.40
Chlortoluron			0.13–3.54
Diuron			0.21–4.99
DMU			0.16–8.85
Fenuron			0.08–1.09
Monuron			0.08–2.63
Monolinuron			0.08–2.45
Metoxuron			0.06–1.99
Metobromuron			0.16–3.40
Permethrin			26.4–1074.6
Linuron			0.16–6.04
Linuron	5.0–6.9[d]	0.3–2.7	4.68–38.17
			0.56–19.66
Chlorsulfuron	5.8[b]	0.46	0.44
Triasulfuron			0.11
Simazine	5.0–6.9[d]	0.3–2.7	0.53–13.87
	4.7–6.5[b]	0.46–1.0	0.24–0.37
	3.3–5.5[e]	0.1–3.0	0.38–25.2
Glyphosate	3.3–5.5[e]	0.1–3.0	0.26–500
Aminomethyl phosphonic acid			2.1–100
Strychnine	4.7–7.3[d]	1.0–2.4	20.82–173.19

[a] 1:2.5 soil : 0.02 M $CaCl_2$
[b] 1:2 soil : 0.01 M $CaCl_2$
[c] 1:5 soil : water
[d] 1:5 soil : 0.01 M $CaCl_2$
[e] 1:1 soil : 0.005 M $CaCl_2$

On hillslopes, water first infiltrates vertically and then moves laterally as interflow, often above a restricting layer such as a clay subsoil (Fig. 10.1). As water moves through the soil, exchange takes place between water-borne constituents and the soil matrix. Soil material may be deposited or mobilized. These sorption processes are complex, and depend largely on the soil and water chemistry. As the accumulated interflow moves down slope it can express itself as interflow in another horizon or in an extreme case as saturated excess overland flow. Thus, water and pollutants transported on the C horizon of 'leaky' texture-contrast soils in the upper parts of landscapes (Kirkby *et al.*, 1996) may leave the catchment as overland flow or flow on the B horizon (Stevens *et al.*, 1999). The residence time in these pathways depends on a few parameters that are easily measured such as the flow gradient, path length and hydraulic conductivity and can therefore be readily calculated.

Where water moves slowly through the soil matrix (matrix flow) many pollutants, especially particulate materials and those that sorb to soil, are removed. However, where water moves quickly into and through soils via stable macropores (fissures or biopores), large quantities of water and contaminants can be rapidly transported downslope. Many studies have demonstrated the ability of macropore flow to transport pollutants rapidly through soil (Bottcher *et al.*, 1981; Richard and Steenhuis, 1988; van Ommen *et al.*, 1989; Kladivko *et al.*, 1991), especially those having artificial drainage (Nash and Halliwell, 1999a), when compared with matrix flow (Peverill and Douglas, 1976; Anderson and Bourma, 1977a,b; Kanchanasut *et al.*, 1978; Seyfried and Rao, 1987; Singh and Kanwar, 1991; Small *et al.*, 1994).

Originally, it was thought that to initiate macropore flow, the soil must be saturated, with water ponded over the entire soil surface (Beven and Germann, 1982). Later studies have shown that macropore flow could occur under non-ponded conditions provided that saturated layers formed very near the soil surface (Adreini and Steenhuis, 1990) or at distinct horizon boundaries or structural discontinuities (Steenhuis *et al.*, 1990). However, there is now growing evidence that macropore flow may occur without any such saturated zones. Methylene blue has been used to trace the pathways of water flow in unsaturated, undisturbed cores of a well structured, dry clay soil and this demonstrated that over half of the applied water left the cores via bypass or macropore flow (Booltink and Bourma, 1991). It has also been demonstrated that water from an unsaturated soil matrix can enter an artificially created macropore if the macropore walls are already wet (Philip *et al.*, 1989). It would appear that, particularly in well aggregated soils, individual peds may become saturated or develop a film of water on their surfaces that can be channelled into nearby macropores without entering the unsaturated soil matrix (Bouma *et al.*, 1979; Russell and Ewel, 1985; Seyfried and Rao, 1987; Sollins and Radulovich, 1988; Booltink and Bourma, 1991; Trojan and Linden, 1992).

The Effects of Scale and Pollutant Mobilization

Many studies have investigated pollutant mobilization processes at a field scale. Some studies have involved modelling aspects of the hydrological system and investigated specific processes that might be expected to occur at a field scale (Fuhr *et al.*, 1997). In other studies, plots have been used to measure hydrological attributes and pollutant mobilization under both simulated and natural rainfall (Ahuja *et al.*, 1982; Greenhill *et al.*, 1983a,b,c; Armstrong, 1990). At an even larger scale, fields, sometimes encompassing entire slopes, have been monitored (Haygarth and Jarvis, 1996a,b; Nash and Murdoch, 1997; Fleming and Cox, 1998; Nash *et al.*, 2000). While each approach is valid, each has its limitations in representing field-scale hydrological pathways.

Laboratory studies

Repacked soil cores are often used to investigate the vertical mobility of pollutants (Nelson *et al.*, 1998; Shaw and Burns, 1998), including microorganisms (Abu-Ashour *et al.*, 1998), in soil. Similarly, boxes of repacked soil have been used to investigate downslope processes associated with overland flow (Ahuja, 1982; Ahuja *et al.*, 1982; Ahuja and Lehman, 1983). The uniformity of the medium minimizes between-replicate variation, thereby simplifying experiments. Although the preparation of the medium alters the physical and biological attributes compared with field soils, especially macro-porosity, repacked cores have been usefully applied in laboratory comparisons of pollutant, soil and environmental treatments (Topp *et al.*, 1994; Starrett *et al.*, 1995, 1996).

Lysimeter studies

Lysimeters containing intact soil monoliths have been used to study hydrology and pollutant mobilization for over 150 years (Howell et al., 1991). These 'undisturbed' soils retain most of the natural structure and pore size distribution, and some of the biological components found in the field. Unfortunately, the preferential flow of water and solutes in the cavity between the monolith and lysimeter wall (McNeal and Reeve, 1964; Tokenuga, 1988) and the natural variation in soil structure limit their use, particularly those with a smaller surface area. In recent years, improvements in the design of lysimeters (Cameron et al., 1990, 1992; Fuhr et al., 1997) have increased their value in studying one-dimensional vertical flow and solute transport in both saturated and unsaturated soils.

An important application of lysimeters has been to compare soil hydrology, especially macropore flow, between sites. For example, a study in the Adelaide Hills of South Australia compared infiltration characteristics of P in lysimeters (undisturbed cores) from different sites and positions in the landscape, using a range of artificial rain intensities and drying treatments (Cox et al., 2000). Macropores, macropore flow and applied P in drainage were shown to be greater in dry soils and higher in the toposequence. These findings suggest that fertilizer applied to the toeslope, an area of low macroporosity and likely saturated excess overland flow, may have a disproportionately large off-site impact, compared with other landscape positions, and that fertilizer application to dry soils may be agronomically inefficient.

Plot studies

Plots are another important method used to investigate the mobilization of pollutants. Ranging from c. 1 m^2 up to 100 m^2 or more, many plots are distinguished from lysimeters and repacked soils by not being hydrologically isolated from surrounding soil. At their simplest, plots are small areas of soil divided by surface barriers that provide a relatively inexpensive way of studying field-scale processes, especially surface soil hydrology, under controlled conditions. Our primary understanding of detachment processes, for example, has been largely developed through the use of field plots and repacked soil subjected to artificial rainfall (Loch and Donnollan, 1982; Loch and Foley, 1992).

The 'hydrologic response time', 'time to equilibrium' or 'time of concentration' for plots (Lettenmaier and Wood, 1992; Dingman, 1994); that is, the time taken for water from the furthest point of the plot (as determined by path length rather than physical length) to reach the monitoring point, is relatively short. Consequently, under natural rainfall, the hydrographs for plots have a short time-span compared with their larger field-scale counterparts (Dingman, 1994). In addition, the intensity of natural rainfall generally declines logarithmically with time (Smith, 1992). As an approximation, the peak flow is proportional to the rain intensity for the period over which the entire catchment, in this case the plot, is contributing to flow. Consequently, peak flows per unit area from a plot are disproportionately greater than from a field. The short, intense flows from replicated plot-scale experiments, especially when coupled with the turbulence created by raindrop impact at higher rain intensities, are useful for comparing detachment treatments (Ahuja, 1986). Unfortunately, while the principles developed at the plot scale generally apply at the field scale, quantitative extrapolation from plot to field scale is difficult (Beven, 1995; Bloschl and Sivalapan, 1995). In part, this reflects the spatial and temporal variation in rainfall, soil and consequent flow characteristics, which cause both detachment and sedimentation to occur simultaneously at different positions in the same landscape.

Plots are less often used to study dissolution processes. The short contact time between water and soil, and high kinetic energy of the overland flow that enhances detachment minimizes dissolution, and both in-stream (Sharpley et al., 1981b) and in-sample (Haygarth et al., 1995) processes tend to mask dissolution products that sorb to soil.

Field-scale studies

Field-scale monitoring has been used to investigate dissolution and detachment of pollutants and their transport in many parts of the world. Unlike plots, where the hydrological properties of replicates are essentially uniform, field-scale monitoring sites have variable hydrology often necessitating the use of different statistical techniques (Cooke *et al.*, 1995; Clausen *et al.*, 1996; Nash *et al.*, 2000). Field-scale monitoring sites vary in area up to many hectares and the greater period of contact and possible interaction at the larger scale allows greater expression of dissolution processes. For natural rainfall, the hydrographs from the sites are characteristically flatter and longer than from plot-scale experiments (Dingman, 1994).

Unfortunately, most field-scale monitoring studies involve some modification of the system being studied. Where load determinations are important, surface and subsurface drains are often used to define the area from which pollutants are mobilized. By draining the subsoil and preventing run-on, these drains would be expected to increase infiltration and decrease overland flow. Similarly, monitoring systems that use drains to measure pollutant flows in subsoil will increase lateral flow and infiltration at the expense of overland flow. This effect is exacerbated where weirs or flumes that collect water are used for overland flow measurement.

It is possible to make a number of generalizations about export pathways. The physical size of most detached products prevents them from moving through soil, so pollutants that are mobilized by detachment processes are generally exported in overland flow. Notable exceptions are the pollutants attached to fine colloids that move through macropores (Frenkel *et al.*, 1978; Jacobsen *et al.*, 1997; Cox *et al.*, 2000). Consequently, factors that increase the kinetic energy of overland flow near the soil surface such as high intensity rainfall, a lack of vegetation, high slopes and run-on, also increase the export of detached pollutants.

Pollutants mobilized by dissolution can be sorbed on to soil during transit, making it more difficult to generalize about these materials. Pollutants that do not sorb to soil or organic matter are exported in both surface and subsurface pathways. Depending on the site at which mobilization occurs, either at the soil surface or within the soil profile, and the relative amounts of water moving over or through the soil, either pathway may predominate. Similarly, pollutants that are sorbed to soil materials can be exported through surface or subsurface pathways. As an approximation, the level of soil and water interaction determines the relative importance of the pathways. Consequently, the highest concentrations of pollutants that sorb on to soil materials are generally found in overland flow. However, significant concentrations of these types of pollutants have been found in macropore flow where the level of soil and water interaction is also reduced compared with vertical matrix flow. The concentrations of these materials in water is further complicated by the fact that other soil components, especially dissolved organic matter, can reduce the sorptive capacity of the soil either directly or through chelation of the pollutant itself (Nash and Halliwell, 1999a).

Unfortunately, while we can generalize about the concentrations of pollutants exported by various pathways, less is known about the overall loads. Being the product of concentration and flow volume, the loads of pollutants exported through the different pathways depend on the hydrology of the field. For example, in many environments, overland flow represents a small percentage ($< 10\%$) of the total water applied. While the concentration of pollutants in overland flow may be high compared with that in subsurface water, the opportunity for the export of greater loads through subsurface pathways is considerable (Stevens *et al.*, 1999).

Mathematical modelling is a useful way of investigating and in some cases quantifying pollutant mobilization at the field scale (Trudgill, 1995). Predictive tools that are used include regression equations, empirical relationships and simulation models (Sharpley *et al.*, 1981a, 1982, 1985; Sharpley and Smith, 1989; Cooke *et al.*, 1995; Hutson and Wagenet, 1995; Wagenet and Hutson, 1996; Nash and Murdoch, 1997; Nash *et al.*, 2000). For investigating pollutant mobilization, comprehensive models that simulate soil, water and chemical processes are by far the most appealing. Unfortunately, as the processes they describe vary both spatially and temporally these simulation models are often extremely complex requiring vast data-sets for calibration (Cambardella *et al.*, 1994; Wagenet and Hutson, 1995). None the less, simulation models provide our best opportunity for generalizing the results from empirical studies of pollutant mobilization to the wider landscape.

Conclusions

Hydrology drives the transfer of waterborne pollutants regardless of scale. However, the scale chosen to study the phenomenon determines, to a large extent, the dominant mobilization process (i.e. detachment or dissolution) and transport pathways (i.e. surface or subsurface) that are measured. This implies that the dominant process and pathway may change with scale, which is true, but it is more important to understand the limitations in measurements made at every scale. Unless these limitations are considered, our interpretation of the results may be flawed. For example, many field-scale monitoring systems enhance water movement in subsurface pathways. The installation of surface and subsurface drains prevents run-on, promoting infiltration, with a subsequent reduction in overland flow. Consequently, load determinations of pollutant exports may be biased.

In all discussions of pollutant mobilization at a field scale, an important question is which pathway, surface or subsurface, dominates in respect to pollutant loads? For systems where the mobilization process is dissolution, there is in fact no generally correct answer because it depends on the properties of the pollutant in question and the relative volumes of water exported by each pathway. For example, it is generally agreed that the export of NO_3^- to groundwater is of concern, predominantly through subsurface pathways. This contrasts to P export where often the concentrations measured in overland flow are much higher than those found in subsurface waters because of the sorption of P to soil. The water volumes exported in surface pathways are often much smaller than those from subsurface export. Clearly, there is potential for greater loads of P to be exported through larger volumes of less concentrated solutions in subsurface pathways than in lower volumes of more concentrated solutions in overland flow. In systems where detachment is the dominant mobilization process, subsurface pathways appear to be less important than surface pathways as the detached particles are generally filtered from the water as it moves through soil.

The many studies of pollutant mobilization have emphasised the importance of understanding hydrology and scale in the context of a landscape. Rarely do processes, such as detachment and dissolution occur in isolation. Factors that increase the volume, velocity and turbulence of overland flow, such as impaired infiltration, high intensity storms, run-on, reduced soil cover, cultivation and high slopes, increase detachment compared with dissolution. Conversely, stable soils with good ground cover that reduce the kinetic energy of overland flow favour dissolution. It follows that, in studies using small plots, the comparatively high kinetic energy of overland flow is likely to suggest that detachment is the most important mobilization mechanism. Assuming all else is equal, in the same study conducted at a larger scale, the hydrograph would be less peaked and considerably longer, conditions that favour dissolution. Moving to a larger scale again, run-on would increase water depth (Equation 10.3) and consequently the velocity of overland flow (Fig. 10.1), again favouring detachment.

Clearly, it is difficult, if not impossible, to increase scale as a single factor. However, there would appear to be an optimum scale for studying particular pollutant mobilization processes in any environment, and understanding the implications of that scale dependence is critical if experimental results are to form the basis of remedial strategies. Perhaps in the future, models that simulate the fundamental processes involved in pollutant mobilization will allow us to integrate across scales and generalize across environments.

References

Abu-Ashour, J., Joy, D.M., Lee, H., Whiteley, H.R. and Zelin, S. (1998) Movement of bacteria in unsaturated soil columns with macropores. *Transactions of the American Society of Agricultural Engineers* 41, 1043–1050.

Addiscott, T.M. and Thomas, D. (2000) Tillage, mineralization, and leaching: phosphate. *Soil and Tillage Research* 53, 255–273.

Adreini, M.S. and Steenhuis, T.S. (1990) Preferential flow paths under conventional and conservational tillage. *Geoderma* 46, 85–102.

Ahuja, L.R. (1982) Release of a soluble chemical from soil to runoff. *Transactions of the American Society of Agricultural Engineers* 25, 948–960.

Ahuja, L.R. (1986) Characterization and modeling of chemical transfer to runoff. *Advances in Soil Science* 4, 149–188.

Ahuja, L.R. and Lehman, O.R. (1983) The extent and nature of rainfall–soil interaction in the release of soluble chemicals to runoff. *Journal of Environmental Quality* 12, 34–40.

Ahuja, L.R., Sharpley, A.N., Yamamoto, M. and Menzel, R.G. (1981) The depth of rainfall–runoff–soil interaction as determined by ^{32}P. *Water Resources Research* 17, 969–974.

Ahuja, L.R., Sharpley, A.N. and Lehman, O.R. (1982) Effect of soil slope and rainfall characteristics on phosphorus in runoff. *Journal of Environmental Quality* 11, 9–13.

Amerman, C.R. (1965) The use of unit-source watershed data for runoff prediction. *Water Resources Research* 1, 499–508.

Anderson, J.L. and Bourma, J. (1977a) Water movement through pedal soils I. Saturated flow. *Soil Science Society of America Journal* 41, 413–418.

Anderson, J.L. and Bourma, J. (1977b) Water movement through pedal soils II. Unsaturated flow. *Soil Science Society of America Journal* 41, 419–423.

Anon (1965) *Soil Erosion by Water*, Paper 81. Food and Agricultural Organisation of the United Nations, Rome.

Armstrong, J.L. (1990) Runoff and soil loss from bare fallow plots at Inverell, New South Wales. *Australian Journal of Soil Research* 28, 659–675.

Baldwin, D.S. (1998) Reactive organic phosphorus revisited. *Water Research* 32, 2265–2270.

Beckett, R. and Hart, B.T. (1993) Use of field flow fractionation techniques to characterize aquatic particles, colloids, and macromolecules. In: Buffle, J. and van Leewen, H.P. (eds) *Environmental Particles*. Lewis Publishers, Boca Raton, Florida, pp. 165–205.

Beven, K. (1995) Linking parameters across scales: subgrid parameterizations and scale dependent hydrological models. *Hydrological Processes* 9, 507–525.

Beven, K. and Germann, P. (1982) Macropores and water flow in soils. *Water Research* 32, 2265–2270.

Bloschl, G. and Sivalapan, M. (1995) Scale issues in hydrological modelling: a review. *Hydrological Processes* 9, 251–290.

Bolt, G.H. (1976) Adsorption of anions by soil. In: Bolt, G.H. and Bruggenwert, M.G.M. (eds) *Soil Chemistry: A. Basic Elements*. Elsevier Scientific Publishing, New York, pp. 91–95.

Bonneau, M. and Souchier, B. (1982) *Constituents and Properties of Soil*. Academic Press, New York.

Booltink, H.W.G. and Bourma, J. (1991) Physical and morphological characterization of bypass flow in a well structured clay soil. *Soil Science Society of America Journal* 55, 1249–1254.

Bottcher, A.B., Monke, E.J. and Huggins, L.F. (1981) Nutrient and sediment loadings from a subsurface drainage system. *Transactions of the American Society of Agricultural Engineers* 24, 1221–1226.

Bouma, J., Jongerius, J.A. and Schoonderbeck, D. (1979) Calculation of saturated hydraulic conductivity of some pedal clay soils using morphometric data. *Soil Science Society of America Journal* 43, 261–264.
Bowmer, K.H. (1971) Atrazine adsorption and control of *Echinochloa* in Australian soils. *Proceedings of the Weed Society of NSW* 4, 12–20.
Cambardella, C.A., Moorman, T.B., Novak, J.M., Parkein, T.B., Karlen, D.L., Turco, R.F. and Konopka, A.E. (1994) Field scale variability of soil properties in Central Iowa soils. *Soil Science Society of America Journal* 58, 1501–1511.
Cameron, K.C., Harrison, D.F., Smith, N.P. and Mclay, C.D.A. (1990) A method to prevent edge-flow in undisturbed soil cores and lysimeters. *Australian Journal of Soil Research* 28, 879–886.
Cameron, K.C., Smith, N.P., Mclay, C.D.A., Fraser, P.M., McPherson, R.J., Harrison, D.F. and Harbottle, P. (1992) Lysimeters without edge flow: an improved design and sampling procedure. *Soil Science Society of America Journal* 56, 1625–1628.
Chittleborough, D.J. (1992) Formation and pedology of duplex soils. *Australian Journal of Experimental Agriculture* 32, 815–825.
Clausen, J.C., Jokela, W.E., Potter, F.I. and Williams, J.W. (1996) Paired watershed comparison of tillage effects on runoff, sediment, and pesticide losses. *Journal of Environmental Quality* 25, 1000–1007.
Cooke, R.A., Mostaghimi, S. and McClellan, P.W. (1995) Application of robust regression to the analysis of BMP effects in paired watersheds. *Transactions of the American Society of Agricultural Engineers* 38, 93–101.
Cox, J.W. and McFarlane, D.J. (1995) The causes of waterlogging in shallow soils and their drainage in south western Australia. *Journal of Hydrology* 167, 175–194.
Cox, J.W., Kirkby, C.A., Chittleborough, D.J., Fleming, N.K. and Smythe, L.J. (2000) Mobility of phosphorus through intact soil cores collected from the Adelaide Hills South Australia. *Australian Journal of Soil Research* 38, 973–990.
Cresser, M., Killiam, K. and Edwards, T. (1993) *Soil Chemistry and its Applications*. Cambridge University Press, Cambridge.
Cresswell, H.P., Smiles, D.E. and Williams, J. (1992) Soil structure, soil hydraulic properties and the soil water balance. *Australian Journal of Soil Research* 30, 265–283.
Daugherty, R.L., Franzini, J.B. and Finnemore, E.J. (1989) *Fluid Mechanics with Engineering Applications*. McGraw-Hill, Singapore.
Davies, D.B. and Payne, D. (1988) Management of soil physical properties. In: Wild, A. (ed.) *Russell's Soil Conditions and Plant Growth*. John Wiley & Sons, New York, pp. 412–448.
Dingman, S.L. (1994) *Physical Hydrology*. Prentice Hall International, New Jersey.
Emmett, W.W. (1978) Overland flow. In: Kirkby, M.J. (ed.) *Hillslope Hydrology*. John Wiley & Sons, Chichester, UK, pp. 145–176.
Fleming, N.K. and Cox, J.W. (1998) Chemical losses off dairy catchments located on a texture-contrast soil: carbon, phosphorus, sulfur, and other chemicals. *Australian Journal of Soil Research* 36, 979–995.
Fleming, N.K., Cox, J.W., Chittleborough, D.J. and Nash, D. (1996) *Optimising Phosphorus Fertiliser Use on Dairy Farms (Catchment Study) South Australia*. CRC for Soil and Land Management, Adelaide, South Australia.
Frenkel, H., Goertzen, J.O. and Rhoades, J.D. (1978) Effects of clay type and content, exchangeable sodium percentage, and electrolyte concentration on clay dispersion and soil hydraulic conductivity. *Soil Science Society of America Journal* 42, 32–39.
Fuhr, F., Hance, R.J., Plimmer, J.R. and Nelson, J.O. (1997) *The Lysimeter Concept – Environmental Behaviour of Pesticides*. American Chemical Society, Washington, DC.
Gerits, J.J.P., De Lima, J.L.M.P. and van Den Brock, T.M.W. (1990) Overland flow and erosion. In: Anderson, M.G. and Burt, T.P. (eds) *Process Studies in Hillslope Hydrology*. John Wiley & Sons, Chichester, UK, pp. 173–214.
Gerriste, R.G. (1996) Transport times of dissolved inorganic phosphate in soils. *Journal of Environmental Quality* 25, 107–110.
Greenhill, N.B., Peverill, K.I. and Douglas, L.A. (1983a) Nutrient concentrations in runoff from pasture in Westernport, Victoria. *Australian Journal of Soil Research* 21, 139–145.
Greenhill, N.B., Peverill, K.I. and Douglas, L.A. (1983b) Nutrient loads in surface runoff from sloping perennial pastures in Victoria, Australia. *Australian Journal of Soil Research* 26, 503–506.

Greenhill, N.B., Peverill, K.I. and Douglas, L.A. (1983c) Surface runoff from sloping, fertilised perennial pastures in Victoria, Australia. *New Zealand Journal of Agricultural Research* 26, 227–231.

Hairsine, P.B., Moran, C.J. and Rose, C.W. (1992) Recent developments regarding the influence of soil surface characteriztics on overland flow and erosion. *Australian Journal of Soil Research* 30, 249–264.

Haygarth, P.M. and Jarvis, S.C. (1996a) Pathways and forms of phosphorus losses from grazed grassland hillslopes. In: Anderson, M.G. and Brooks, S.M. (eds) *Advances in Hillslope Processes*. John Wiley & Sons, Chichester, UK, pp. 283–294.

Haygarth, P.M. and Jarvis, S.C. (1996b) Soil derived phosphorus in surface runoff from grazed grassland lysimeters. *Water Research* 31, 140–148.

Haygarth, P.M. and Jarvis, S.C. (1999) Transfer of phosphorus from agricultural soils. *Advances in Agronomy* 66, 195–249.

Haygarth, P.M., Ashby, C.D. and Jarvis, S.C. (1995) Short-term changes in the molybdate reactive phosphorus of stored soil waters. *Journal of Environmental Quality* 24, 1133–1140.

Haygarth, P.N., Warwick, M.S. and House, W.A. (1997) Size distribution of colloidal molybdate reactive phosphorus in river waters and soil solution. *Water Research* 31, 439–448.

Haygarth, P.M., Heathwaite, A.L., Jarvis, S.C. and Harrod, T.R. (2000) Hydrological factors for phosphorus transfer from agricultural soils. *Advances in Agromomy* 69, 153–178.

Hillel, D. (1980) *Fundamentals of Soil Physics*. Academic Press, New York.

Hodgkin, E.P. and Hamilton, B.H. (1993) Fertilisers and eutrophication in southwestern Australia: setting the scene. *Fertiliser Research* 36, 95–103.

Howell, T.A., Schneider, A.D. and Jensen, M.E. (1991) History of lysimeter design and use for evapotranspiration measurements. In: Allen, R.G. (ed.) *Lysimeters for Evapotranspiration and Environmental Measurements*. American Society for Civil Engineering, New York, pp. 1–9.

Hutson, J.L. and Wagenet, R.J. (1995) The application of chemical equilibrium in solute transport models. *Chemical Equilibrium and Reaction Models* 42, 97–112.

Jacobsen, O.H., Moldrup, P., Larsen, C., Konnerup, L. and Petersen, L.W. (1997) Particle transport in macropores of undisturbed soil columns. *Journal of Hydrology* 196, 185–203.

Kanchanasut, P., Scotter, D.R. and Tillman, R.W. (1978) Preferential solute movement through larger soil voids. II. Experiments with saturated soil. *Australian Journal of Soil Research* 16, 269–276.

Kelley, H.W. (1983) *Keeping the Land Alive. Soil Erosion – its Causes and Cures*. FAO, Rome.

Kirkby, C.A., Chittleborough, D.J., Smetten, K.R.J. and Cox, J.W. (1996) Water, phosphate, clay and DOC movement through a texture contrast soil. In: *Australian and New Zealand National Soils Conference, Melbourne, Australia*. Australian Society for Soil Science Inc., pp. 141–142.

Kirkby, C.A., Smythe, L.J., Cox, J.W. and Chittleborough, D.J. (1997) Phosphorus movement down a toposequence from a landscape with texture contrast soils. *Australian Journal of Soil Research* 35, 399–417.

Kirkby, M.J. and Chorley, R.J. (1967) Throughflow, overland flow and erosion. *Bulletin of the International Association for Science Hydrology* 12, 5–21.

Kladivko, E.J., van Scoyoc, G.E., Monke, E.J., Oates, K.M. and Pask, W. (1991) Pesticide and nutrient movement into subsurface tile drains on a silt loam soil in Indiana. *Journal of Environmental Quality* 20, 264–270.

Kookana, R.S. and Aylmore, L.A.G. (1993) Retention and release of diquat and paraquat herbicides in soil. *Australian Journal of Soil Research* 31, 97–109.

Kookana, R.S., Baskaran, S. and Naidu, R. (1998) Pesticide fate and behaviour in Australian soils in relation to contamination and management of soil and water: a review. *Australian Journal of Soil Research* 36, 715–764.

Leeper, G.W. and Uren, N.C. (1997) *Soil Science – an Introduction*, 5th edn. Melbourne University Press, Melbourne, Australia.

Leonard, R.A. and Wauchope, R.D. (1980) The pesticide sub-model. In: Knisel, W.G. (ed.) *CREAMS: a Field Scale Model for Chemicals, Runoff and Erosion from Agricultural Management Systems*. USDA, Washington, DC, pp. 88–112.

Lettenmaier, D.P. and Wood, E.F. (1992) Hydrologic forecasting. In: Maidement, D.R. (ed.) *Handbook of Hydrology*. McGraw-Hill, New York, pp. 26.1–26.30.

Loch, R.J. and Donnollan, T.E. (1982) Field rainfall simulator studies on two clay soils of the Darling Downs, Queensland. I. The effects of plot length and tillage orientation on erosion processes and runoff and erosion rates. *Australian Journal of Soil Research* 21, 33–46.

Loch, R.J. and Foley, J.L. (1992) Effects of plot size on size distributions of water-stable material at the soil surface under simulated rain. *Australian Journal of Soil Research* 30, 113–118.

Mansell, R.S., Selim, H.M., Kanchanasut, P., Davidson, J.M. and Fiskell, J.G.A. (1977) Experimental and simulated transport of P through sandy soils. *Water Resources Research* 13, 189–194.

Massey, H.F. and Jackson, M.L. (1952) Selective erosion of soil fertility constituents. *Soil Science Society of America Proceedings* 16, 353–356.

McNeal, B.L. and Reeve, R.C. (1964) Estimations of boundary flow errors in laboratory hydraulic conductivity measurements. *Soil Science Society of America Proceedings* 28, 713–714.

Menzel, R.G. (1980) Enrichment ratios for water quality modelling. In: Knisel, W.G. (ed.) *CREAMS: a Field Scale Model for Chemicals, Runoff, and Erosion from Agricultural Management Systems.* USDA, Washington, DC, pp. 486–492.

Moore, I.D. and Foster, G.R. (1990) Hydraulics and overland flow. In: Anderson, M.G. and Burt, T.P. (eds) *Process Studies in Hillslope Hydrology.* John Wiley & Sons, Chichester, UK, pp. 215–254.

Nash, D. and Halliwell, D. (1999a) Fertilisers and phosphorus loss from productive grazing systems. *Australian Journal of Soil Research* 37, 403–429.

Nash, D. and Halliwell, D. (1999b) Understanding phosphorus transfers from land to water. *Agricultural Science* 12, 37–42.

Nash, D. and Murdoch, C. (1997) Phosphorus in runoff from a fertile dairy pasture. *Australian Journal of Soil Research* 35, 419–429.

Nash, D., Hannah, M., Halliwell, D. and Murdoch, C. (2000) Factors affecting phosphorus transfer from a pasture based grazing system. *Journal of Environmental Quality* 29, 1160–1165.

Nelson, S.D., Letey, J., Farmer, W.J., Williams, C.F. and Ben-Hur, M. (1998) Facilitated transport of naporpmide by dissolved organic matter in sewage sludge-amended soil. *Journal of Environmental Quality* 27, 1194–1200.

Nielsen, D.R., Biggar, J.W. and Erh, K.T. (1973) Spatial variability of field-measured soil-water properties. *Hilgardia* 42, 215–259.

Ozanne, P.G., Kirkton, D.J. and Shaw, T.C. (1961) The loss of P from sandy soils. *Australian Journal of Agricultural Research* 12, 409–423.

Peverill, K.I. and Douglas, L.A. (1976) Use of undisturbed cores of surface soil for investigating leaching losses of sulphur and phosphorus. *Geoderma* 16, 193–199.

Peverill, K.I., Douglas, L.A. and Greenhill, N.B. (1977) Leaching losses of applied P and S (phosphorus and sulphur) from undisturbed cores of some Australian surface soils. *Geoderma* 19, 91–96.

Philip, J.R., Knight, J.H. and Waechter, R.T. (1989) Unsaturated seepage and subterranean holes: Conspectus and exclusion problem for circular cylindrical cavities. *Water Resources Research* 25, 16–28.

Pierzynski, G.M., Sims, J.T. and Vance, G.F. (1994) *Soils and Environmental Quality.* Lewis Publishers, Boca Raton, Florida.

Pigram, J.J. (1986) *Issues in the Management of Australia's Water Resources.* Cheshire, Australia.

Rajan, S.S.S., Perrott, K.W. and Saunders, W.M.H. (1974) Identification of phosphate reactive sites of hydrous alumina from proton consumption during phosphate adsorption at constant pH values. *Journal of Soil Science* 25, 438–447.

Richard, T.L. and Steenhuis, T.S. (1988) Tile drain sampling of preferential flow on a field scale. *Journal of Contaminant Hydrology* 3, 307–325.

Rolston, D.E., Rauschkolb, R.S. and Hoffman, D.L. (1975) Infiltration of organic phosphate compounds in soil. *Soil Science Society of America Proceedings* 39, 1089–1094.

Rulon, J., Rodway, R. and Freeze, A. (1985) The development of multiple seepage faces on layered slopes. *Water Resources Research* 21, 1625–1636.

Russell, A.E. and Ewel, J.J. (1985) Leaching from a tropical andept during big storms: a comparison of three methods. *Soil Science* 139, 181–189.

Seyfried, M.S. and Rao, P.S.C. (1987) Solute transport in undisturbed columns of an aggregated tropical soil: preferential flow effects. *Soil Science Society of America Journal* 51, 1434–1444.

Shainberg, I., Laflen, J.M., Bradford, J.M. and Norton, L.D. (1994) Hydraulic flow and water characteriztics in rill erosion. *Soil Science Society of America Journal* 158, 1007–1012.

Sharpley, A.N. (1980) The enrichment of soil phosphorus in runoff sediments. *Journal of Environmental Quality* 9, 521–526.

Sharpley, A.N. (1985) The selective erosion of plant nutrients in runoff. *Soil Science Society of America Journal* 49, 1527–1534.

Sharpley, A.N. and Smith, S.J. (1989) Prediction of soluble phosphorus transport in agricultural runoff. *Journal of Environmental Quality* 18, 313–316.

Sharpley, A.N., Ahuja, L.R. and Menzel, R.G. (1981a) The release of soil phosphorus to runoff in relation to the kinetics of desorption. *Journal of Environmental Quality* 10, 386–391.

Sharpley, A., Menzel, R., Smith, S., Rhoades, E. and Olness, A. (1981b) The sorption of soluble phosphorus by soil material during transport in runoff from cropped and grassed watersheds. *Journal of Environmental Quality* 10, 211–215.

Sharpley, A.N., Smith, S.J. and Menzel, R.G. (1982) Prediction of phosphorus losses in runoff from southern plains watersheds. *Journal of Environmental Quality* 11, 247–251.

Sharpley, A.N., Smith, S.J., Berg, W.A. and Williams, J.R. (1985) Nutrient runoff losses as predicted by annual and monthly soil sampling. *Journal of Environmental Quality* 14, 354–360.

Sharpley, A.N., Ahuja, L.R. and Smith, S.J. (1988) Chemical transport of agricultural runoff: model improvement and application. In: *Modelling Agricultural, Forest, and Rangeland Hydrology.* American Society of Agricultural Engineers, Chicago, Illinois, pp. 142–155.

Sharpley, A.N., Daniel, T.C. and Edwards, D.R. (1993) Phosphorus movement in the landscape. *Journal of Productive Agriculture* 6, 492–499.

Sharpley, A.N., Chapra, S.C., Wedepohl, R., Sims, J.T., Daniel, T.C. and Reddy, K.R. (1994) Managing agricultural phosphorus for protection of surface waters: Issues and options. *Journal of Environmental Quality* 23, 437–451.

Shaw, L.J. and Burns, R.G. (1998) Biodegradation-transport interactions of pulse applied 2,4-D in repacked soil column microcosms. *Journal of Environmental Quality* 27, 1472–1478.

Singh, P. and Kanwar, R.S. (1991) Preferential solute transport through macropores in large undisturbed saturated soil columns. *Journal of Environmental Quality* 20, 295–300.

Small, D., McDonald, J. and Wales, B. (1994) *Alternative Farming Practices Applicable to the Dairy Industry.* Department of Agriculture, Melbourne, Victoria, Australia.

Smetten, K.R.J., Kirkby, C.A. and Chittleborough, D.J. (1994) Hydrologic response of undisturbed soil cores to simulated rainfall. *Australian Journal of Soil Research* 32, 1175–1187.

Smith, D.D. and Wischmeier, W.H. (1962) Rainfall erosion. *Advances in Agronomy* 14, 109–148.

Smith, J.A. (1992) Precipitation. In: Maidment, D.R. (ed.) *Handbook of Hydrology.* McGraw-Hill, New York, pp. 3.1–3.47.

Sollins, P. and Radulovich, R. (1988) Effects of soil physical structure on solute transport in a weathered tropical soil. *Soil Science Society of America Journal* 52, 1168–1173.

Starrett, S.K., Luke, S.E., Christians, N.E. and Austin, T.A. (1995) Comparing chloride transport in undisturbed and disturbed soil columns under turfgrass conditions. *Communications in Soil Science and Plant Analysis* 26, 1283–1290.

Starrett, S.K., Christians, N.E. and Austin, T.A. (1996) Comparing dispersivities and soil chloride concentrations of turfgrass covered undisturbed and disturbed soil columns. *Journal of Hydrology* 180, 21–29.

Steenhuis, T.S., van Es, H.M., Parlange, J.Y., Baveye, P.C., Walter, M.F., Geohring, L.D., Hutson, J.L., Richard, T.L., Bell, J.L., Boll, J., Gannon, J., Liu, Y., Sanford, W.E., Alexander, M., Bryant, R.B., Selker, J.S., Tan, Y., Vandevivere, P., Verheyden, S.M.L. and Vermeulen, J. (1990) Hydrology and the environment. *New York's Food and Life Sciences Quarterly* 20, 15–19.

Stevens, D.P., Cox, J.W. and Chittleborough, D.J. (1999) Pathways of phosphorus, nitrogen and carbon movement over and through texturally differential soils, South Australia. *Australian Journal of Soil Research* 37, 679–693.

Tokenuga, T.K. (1988) Laboratory permeability errors from annular wall flow. *Soil Science Society of America Journal* 52, 24–27.

Topp, E., Smith, W.N., Reynolds, W.D. and Khan, S.U. (1994) Atrazine and metolachlor in soils incubated in undisturbed cores, repacked cores and flasks. *Journal of Environmental Quality* 23, 693–700.

Troeh, F.R., Hobbs, J.A. and Donahue, R.L. (1980) *Soil and Water Conservation for Productivity and Environmental Protection.* Prentice-Hall Inc., Englewood Cliffs, New Jersey.

Trojan, M.D. and Linden, D.R. (1992) Microrelief and rainfall affects on water and solute movement in earthworm burrows. *Soil Science Society of America Journal* 56, 727–733.

Trudgill, S.T. (1990) *Barriers to a Better Environment.* Belhaven Press, London.

Trudgill, S.T. (1995) *Solute Modelling in Catchment Systems.* John Wiley & Sons, Chichester, UK.

Van Ommen, H.C., van Genuchten, M.T., van der Molen, W.H., Kijksma, R. and Hulshoj, J. (1989)

Experimental and theoretical analysis of solute transport from a diffuse source of pollution. *Journal of Hydrology* 105, 225–251.

Wagenet, R.J. and Hutson, J.L. (1995) Consequences of scale-dependency on application of chemical leaching models: a review of approaches. In: El-Kadi, A.I. (ed.) *Groundwater Models for Resources Analysis and Management*. CRC Press, Boca Raton, Florida, pp.169–184.

Wagenet, R.J. and Hutson, J.L. (1996) Scale-dependency of solute transport modeling/GIS applications. *Journal of Environmental Quality* 25, 499–510.

White, I. (1988) Measurement of soil physical properties in the field. In: Steffan, W.L. and Denmead, O.T. (eds) *Flow and Transport in the Natural Environment: Advances and Applications*. Springer-Verlag, Berlin, pp. 59–85.

White, R.E. (1987) *Introduction to the Principles and Practice of Soil Science*, 2nd edn. Blackwell Scientific Publications, London.

11 Modelling Hydrological Mobilization of Nutrient Pollutants at the Catchment Scale

Yves J.P. Van Herpe,[1] Peter A. Troch,[2] Paul F. Quinn[3] and Stephen Anthony[4]
[1]*Laboratory of Hydrology and Water Management, Ghent University, Ghent, Belgium;* [2]*Department of Water Management, Wageningen Agricultural University, Wageningen, The Netherlands;* [3]*Water Resources Systems Research Laboratory, Newcastle University, Newcastle upon Tyne, UK;* [4]*ADAS Research and Development, Woodthorne, Wergs Road, Wolverhampton W16 8TQ, UK*

Key words: hydrology, land use, leaching, model, nitrate, overland flow, uncertainty, vadoze zone

Introduction

Nitrate loading in groundwater and surface waters demonstrates a general trend of increase, most notably in the past three decades (Heathwaite *et al.*, 1996). Although NO_3^- pollution results from a range of sources, evidence suggests that non-point source pollution from agriculture is the major contributor to the increase in NO_3^- loss to the aquatic environment (Heathwaite *et al.*, 1993). While further research is still needed to obtain a better understanding of NO_3^- transport in rural catchments, there is an urgent need to build our existing knowledge into modelling tools for use at the catchment scale. Such tools can then aid the development of sound policy and best management practices. In order to accurately evaluate these management strategies, there is a need for easy to use, efficient, transparent modelling approaches. Modelling an integrated water management system requires specialized operational models that can operate at the catchment scale, are appropriate to the specific goals of the policy maker and can be driven by available data. The use of models for policy making also requires an understanding of the model uncertainty. The difficulties of tracing the effects of model parameter uncertainty indicate that a model dependent on fewer parameters is more justifiable at the catchment scale than a complex modelling approach (Beven, 1989; Anthony *et al.*, 1996).

Model complexity and predictive uncertainty remain the key issues when simulating nutrient mobilization at the catchment scale. It is difficult to argue whether or not complex physical models or simple empirical models are a better approach to catchment simulation; however, the authors believe that uncertainty in input information is high and that the impacts of this uncertainty must be propagated through to the final model output. Thus, a model with a simple structure allows the error in inputs to be transferred to the model output in a transparent and unambiguous way. Hence, a land use manager can set policy in the light of this final uncertainty estimate. The reduction of uncertainty remains a long-term goal of the nutrient pollution community and uncertainty analysis should not be avoided in our catchment models. This chapter carries out a full quantitative uncertainty analysis of the catchment hydrological model. The purpose of the analysis is to show in a clear way the significant uncertainty that exists in any model (even a simple model). The GLUE (Generalized Likelihood Uncertainty Estimation) procedure (Beven, 1993) allows us to analyse each model parameter in turn and evaluate its model sensitivity. While the analysis concludes that uncertainty in our models is high, it also points to some simple steps that can be taken to reduce the model uncertainty.

The choice of a catchment model structure and the parameters used must always be justified. Traditionally, there is a split between physically based distributed modellers and the empirical or quasi-physical modellers (Kalma and Sivapalan, 1995; Singh, 1995; Sorooshian and Gupta, 1995). This philosophical debate of model choice often detracts from the main issue of problem solving at the catchment scale. A modeller that has used both complex physical models and simple operational models tends to be more useful to the overall project. Ultimately, confidence in our model approaches will be dependent on the confidence in our input data estimation and the physical justification of those parameters. Thus an experienced modeller should be able to recommend an appropriate model structure and make a prudent choice of the data-sets needed to parametrize a model. Here we will use an approach first described as the Minimum Information Requirement (MIR) approach (Anthony *et al.*, 1996; Quinn *et al.*, 1999). A MIR can be defined as the simplest model structure that satisfies the modelling needs of the policy maker while ensuring that the model parameters retain physical significance. The approach uses a simple but justifiable model structure for simulating flow and nutrients at the catchment scale. However, the basis of the parameters and the model functions used are approximations to the output of complex physically based models or detailed field experiments. Physically based models (and the experiments they attempt to simulate) are most suited to small plot or field scale experiments. At a local 'research' scale, the effort needed to parametrize physically based models can be achieved and thus the model output is justified. Thus, basic cause and effect relationships can be studied and great practical knowledge and expertise can be gained. The simple, low parameter MIR models then mimic the output of these more complex models (i.e. MIR models simulate the output time series of flow and nutrients). A careful choice is made to make sure that the key driving parameters used in the MIR relate closely to those used in the more complex model. The MIR parameters, although approximations, do have some physical meaning and can be interpreted in a meaningful way. The goal is to use physical models at a local research scale and simple lumped, quasi-physical models at the catchment scale.

The chapter illustrates two MIR models, the first is a lumped hydrological model that uses a statistical representation of land units. The model structure is justified, in that each land unit is directly correlated with the total input of nutrients: in essence a nutrient input index. Thus, land use data and its associated husbandry can be derived from most national GIS databases. A combined leaching function and flow transfer component then act as a transport index which is correlated with soil properties and the catchment response time. Full uncertainty analysis is carried out on this model. The model shows both spatial and temporal fluxes in nitrate input and loss for the Zwalm catchment in Belgium.

The second model is a full implementation of TOPMODEL (Quinn and Beven, 1993), which is a quasi-physical model that reflects the role of topography in flow generation. The model simulates the operation of shallow water table activity and variable source areas (Dunne and Black, 1970). The choice to complicate the original model with an extra spatial component may seem to contradict the initial MIR premise; however, the model can be justified because:

1. The Zwalm catchment has a clear set of topographically controlled flow processes that are similar in nature to variable source area activity (Gyasi-Agyei *et al.*, 1995; Troch *et al.*, 1995; Verhoest *et al.*, 1998);
2. The needs of the policy makers require a spatial component to test certain land use scenarios, such as the establishment and operation of buffer strips and wetlands (Nieswand *et al.*, 1990; Pionke *et al.*, 1996; Johnes and Heathwaite, 1997).

The goal is to keep the number of extra variables and data-sets low and to show clearly all steps in the analysis. It can also be shown that, given existing topographic data-sets and our understanding of flow processes, the net benefit of adding a topographic component to the model is justified. The ability to improve the pattern of the nutrient input index and transport loss index can be of great use to policy makers. The argument is that one should only complicate the model complexity when needed.

Model Concept

The NO_3^- mobilization and transport models reported in this chapter are based on the conceptual runoff generation model TOPMODEL proposed by Beven and Kirkby (1979). The first model is a lumped version of TOPMODEL (Van Herpe *et al.*, 1998) and the second model is close to the original TOPMODEL, with an added NO_3^- component.

TOPMODEL assumes a simple relationship between catchment storage deficit and local water table position, as controlled by the catchment topography. In essence, the model correlates the degree of wetness of the catchment with the location of the water table (and saturated areas) by combining a series of soil moisture stores and a function for routing water from the soil to the outlet of the catchment. The formulation of the TOPMODEL stores used in this chapter has been adopted from Quinn (1991), Beven (1991) and Quinn and Beven (1993).

Lumped Approach

Hydrological module

A schematic formulation of the stores used in the lumped version of the model is illustrated in Fig. 11.1. The lumped model operates at a daily time step and adopts a linear root zone store, with a maximum storage capacity *SRMAX* (m). *SRMAX* approximates to the field capacity of the soil and no flow occurs in the unsaturated zone until the field capacity is exceeded.

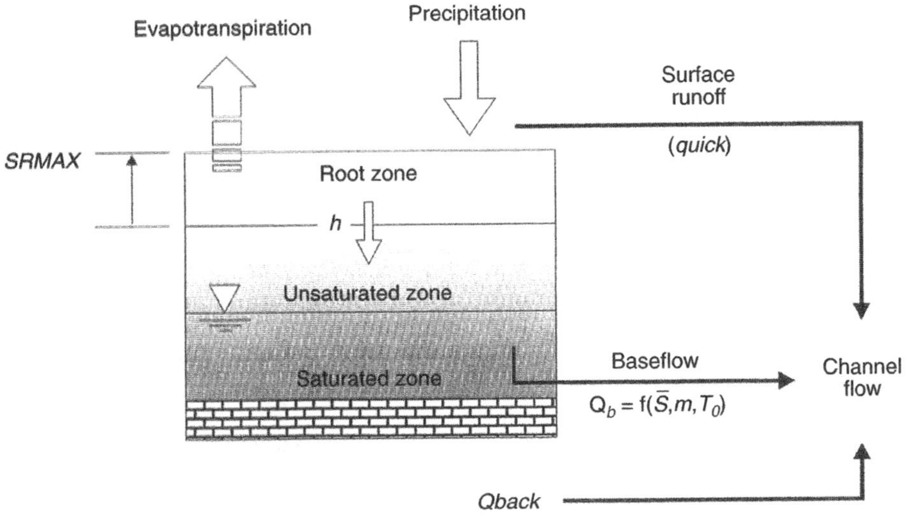

Fig. 11.1. Schematic of the storage elements and parameters of the lumped hydrological model formulation.

An exponential saturated zone store controls the rate at which the subsurface saturated flow reaches the channel. Evaporation from the root zone store is allowed at the estimated potential rate until complete depletion of the store. A constant overland flow coefficient *quick* (%) generates instantaneous overland flow estimated as a proportion of the rainfall. Vertical flux h (m) from the root zone to the water table occurs whenever the root zone storage capacity is exceeded. Daily output from the saturated zone store is given by the baseflow term Q_b (m):

$$Q_b = Q_0 \cdot e^{-\frac{S}{m}} \qquad (11.1)$$

$$Q_0 = A \cdot e^{(\ln T_0 - \lambda)} \qquad (11.2)$$

where Q_0 (m) is the baseflow discharge, per unit catchment area, when the average catchment storage deficit S (m) equals zero, m is a model parameter (m) that defines the variation of saturated hydraulic conductivity with depth, A is the total catchment area (m^2), λ is an approximation of the catchment mean topographic index (see the section describing the NO_3^- leaching module), and T_0 is the saturated lateral transmissivity (m^2 day^{-1}). The catchment average storage deficit S for every time step t is updated by subtracting the unsaturated zone recharge and adding the baseflow from the previous time step:

$$S_t = S_{t-1} - h_t \times q_{b,t-1} \qquad (11.3)$$

where h (m) is the vertical water flux from the root zone to the water table. q_b is the baseflow term Q_b as calculated in Equation 11.1. Initial flow and soil moisture conditions (Troch *et al.*, 1993) and values for the model parameters m and Q_0 (Beven *et al.*, 1995) can be assessed by recession flow analysis. *Qback* can be used to cancel out any systematic water balance error, for example any artificial flow loss or gain in the catchment. The model is driven by daily catchment mean precipitation and potential evapotranspiration time series and discharge data at the catchment outlet.

Because of the high solubility of the NO_3^-, its transport is intimately linked with the hydrological pathways controlling nutrient transport from the land to the stream. In this respect, a lumped NO_3^- transport catchment model can simulate the loss of diffuse sources of near surface using TOPMODEL equations in a lumped form. Thus, TOPMODEL has been extended to incorporate a NO_3^- transport module (see Fig. 11.2).

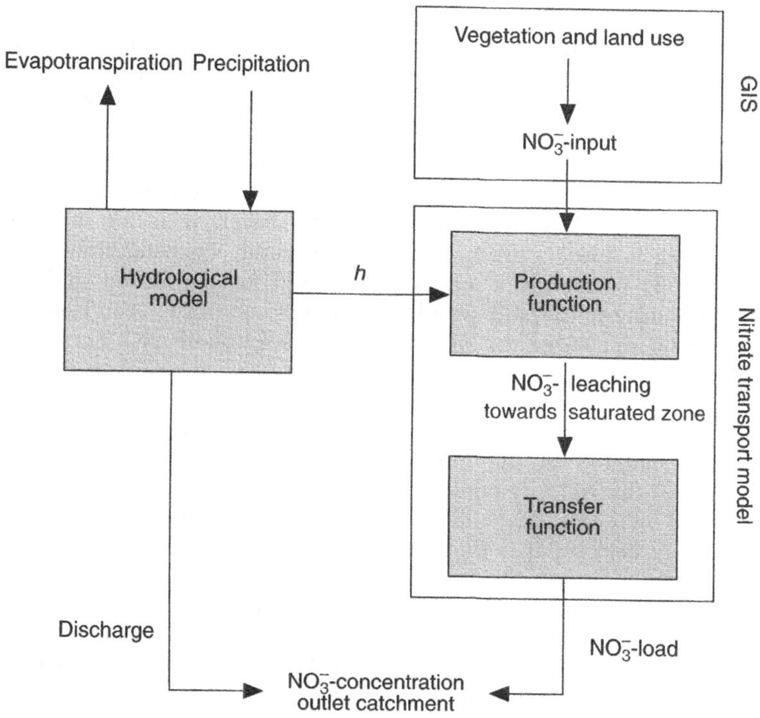

Fig. 11.2. Scheme of the lumped nitrate transport model concept and forcing data.

The results of the hydrological module are used to control the various functions of the NO_3^- transport module. The transport module can be considered as a transfer function, transforming input from a NO_3^- production function to an output signal representing an estimate of the river NO_3^- concentration. A simple mass balance mixing of the *Qback* flow and *Quick* flow with the event leachate flow gives the final flow and NO_3^- concentration. A detailed formulation of the model components can be found in Van Herpe *et al.* (1998).

Nitrate leaching module

The NO_3^- transport model considers two transport mechanisms: (i) the production function: a vertical export mechanism controlling the release of NO_3^- from the soil root zone; and (ii) the transfer function: a lateral export mechanism controlling the appearance of NO_3^- in the stream channel network.

Production function

During rainfall events, part of the rainwater percolates and displaces a portion of the current NO_3^- within the soil profile. The NO_3^- concentration of the leachate depends on the amounts within the soil profile, which is estimated from an empirical model. A simple MIR model function is used to simulate the concentration in the leachate. This leaching function can emulate the output of physically based models (see Anthony *et al.*, 1996). The leaching model expresses the proportion of soil NO_3^- leached as a function of increasing soil drainage efficiency:

$$P = 1.111\varepsilon - 0.203\varepsilon^3 \quad \text{when } \varepsilon \le 1.35 \qquad (11.4)$$

$$P = 1.0 \quad \text{when } \varepsilon > 1.35 \qquad (11.5)$$

$$\varepsilon = \frac{H}{\phi} \qquad (11.6)$$

where P is the proportion of the available NO_3^- leached, ε is the drainage efficiency (m), H is the cumulative soil drainage (m) (i.e. the integrated amount of water draining through the soil during a finite time interval) and ϕ is the soil moisture content at field capacity ($m^3 m^{-3}$). Hence, in combination with the amount of NO_3^- present in the soil profile, the NO_3^- volume that leached during a rainfall event of known intensity can be predicted. The model is applied to leaching seasons starting after the harvest period, assuming a certain residual NO_3^- amount in the soil profile, which continuously decreases during the leaching season and is reset at the onset of the next leaching season. In this way, the effects of catchment memory on potential leaching are incorporated in the model, since the history of the NO_3^- dynamics in the soil is reflected by the residual soil contents. Thus, the key data input is the total amount of NO_3^- applied by farmers for each land unit and how much residual NO_3^- remains from previous years. While models do exist to estimate this amount, the basis of this value is largely derived from field measurements and empirical knowledge. Any map of land cover is thus a map of NO_3^- available to leaching before the onset of the winter rainfall period.

Transfer function

The transfer function is based on a flushing hypothesis, that postulates a relationship between river NO_3^- concentrations and the extent to which the catchment is saturated (Hornberger *et al.*, 1994). The flushing hypothesis states that a catchment will release the available sources of N to adjacent water as a function of the soil saturation deficit. Accordingly, the transfer function used here states that NO_3^- concentration of the subsurface flow increases proportionally as the catchment saturation deficit declines (that is, the corresponding rise of the water table). The NO_3^- transfer function consists of two components defining the relationship between the saturation deficit (calculated in the hydrological part) and concentration in the river. The first component of the transfer function defines a lower boundary condition for the catchment saturation deficit, S_0, and for the baseflow concentration, c_0, the so-called background concentration. The lower boundary condition is reached during dry periods, when river NO_3^- concentrations are not influenced by recent rainfall events, but are merely a function of those present in baseflow. This baseflow originates from older subsurface flow and the NO_3^- content of this baseflow is an indication for the background concentration c_0, which is assumed to be constant under unchanging land use conditions (Van Herpe *et al.*, 1999). The background NO_3^- concentration can also be used to simulate dry weather flow from urban areas, which can be relatively high (Quinn *et al.*, 1999). The background value S_0 is set through the value of the *Qback* parameter (see above).

The second component of the transfer function represents the dynamic part. Initially, a first measurement of the river NO_3^- concentration serves as the initial condition for the transport module, while the hydrological model component calculates the saturation deficit for the corresponding time step based on the initial flow (see Quinn and Beven, 1993). For the case of no rainfall, the transfer function hypothesis will produce a linear drop in NO_3^- between the initial condition and the lower boundary condition. Thus, the saturation deficit will proceed towards S_0, while the baseflow concentration will decrease towards c_0. For the case of a rainfall event that generates leachate, the saturation deficit will fall as the water table shallows. Assuming the volume of the water table recharge per unit area is h (as calculated in the hydrological module), the concentration c_L of this leachate can be calculated from Equation 11.7. The decline in saturation deficit produces new values for S and c, obtained through Equation 11.4. If the baseflow concentration at the start of the rainfall event was c_1, together with the calculated saturation deficit S_1, the new values c_2 and S_2 are:

$$S_2 = S_1 - h \tag{11.7a}$$

$$c_2 = \alpha c_1 + \beta c_L \tag{11.7b}$$

$$\alpha = \frac{S_0 - S_1}{h + S_0 - S_1} \quad \beta = \frac{h}{h + S_0 - S_1} \tag{11.8}$$

Equation 11.7 is based on the concept that as the saturation deficit increases, relatively more deep groundwater will feed the baseflow. As the subsurface flow to the river continues, the saturation deficit increases and relatively more of the deeper groundwater adds to the total river flow. Note that the transfer function implicitly accounts for a NO_3^- reduction occurring in the aquifer. However, it is difficult to justify exactly the mechanism of NO_3^- reduction as either a mixing of differing sources, or a true denitrification process.

Application to the Zwalm catchment

In order to illustrate the applicability of the model, it was used to simulate NO_3^- concentrations in the surface water at the outlet of the rural Zwalm catchment. The Zwalm catchment is situated about 20 km south of Ghent in East-Flanders, Belgium (50°45′48″ N to 50° 54′16″ N and 3°40′17″ E to 3°50′15″ E), and is a tributary of the Scheldt river (Fig. 11.3). It is a 5th Strahler-order basin with a total drainage area of 114.3 km², and a drainage density

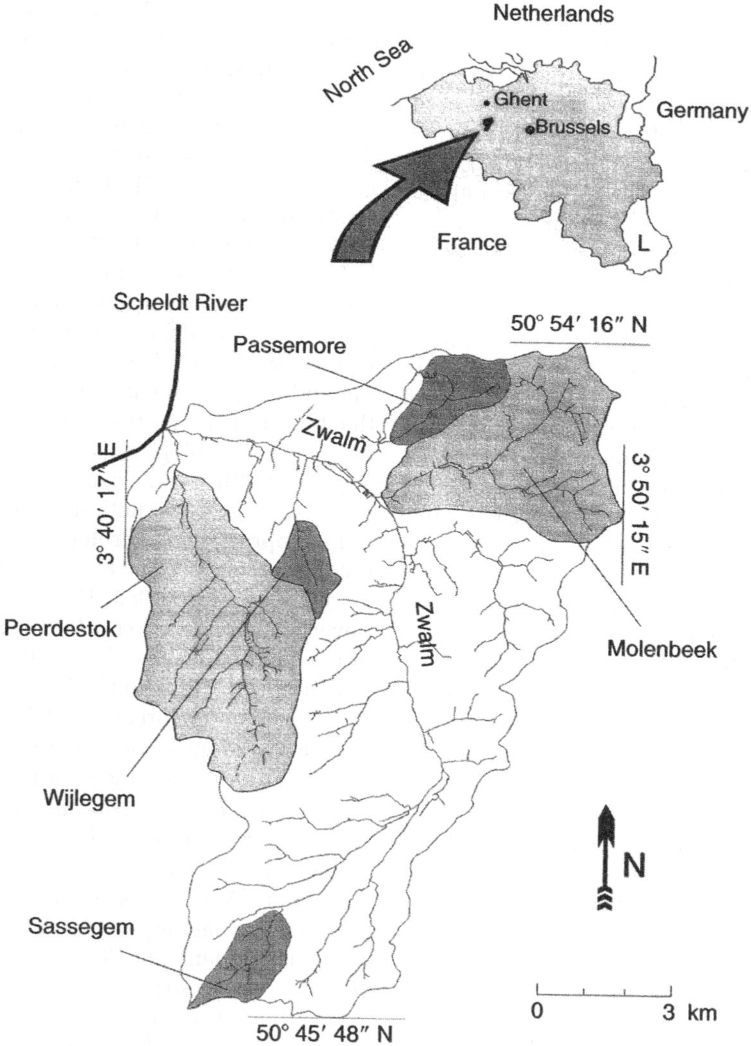

Fig. 11.3. Location of the Zwalm catchment and some subcatchments.

of 1.55 km km^{-2}. Rolling hills and mild slopes, with a maximum elevation difference of about 150 m, characterize the topography. The main surficial geological formations are Tertiary and Quaternary deposits. During the last glacial period (Wurm), the Tertiary layers were covered by aeolian sandy-loam and loam deposits. The depth of this aeolian cover strongly depends on the geomorphological characteristics of the region, and is estimated to range between 0 m on steep slopes (> 7%) oriented to the south and 15 m on mild slopes (< 2%) oriented to the east. The valley bottoms were partly covered with alluvium (originating from soil erosion on the steep slopes) and peat during the Holocene. According to the Belgian Soil Map, minor isolated patches of sand and clay cover < 5% of the total catchment area. Land use is mainly agricultural (arable crop farming and permanent pasture), while the south of the catchment is partly forested (9%). The degree of urbanization is about 11%, mainly clustered in three small towns in the eastern part of the catchment. The climatic regime can be considered as humid temperate with a mean annual rainfall of 780 mm distributed uniformly over the year and a mean annual pan evapotranspiration of 450 mm. The yearly mean air temperature is 9.8°C, the average of the coldest month (January) is 2.6°C and the average of the warmest month (July) 17.1°C. The Zwalm catchment is described in detail in De Troch (1977).

The Zwalm data-set contains daily hydrometeorological data and daily river NO_3^- concentrations. A 1-year simulation (1991–1992) for the Zwalm catchment was performed with a daily time step, a leaching season starting on 25 September and an initial estimated soil NO_3^- content of 40 kg NO_3^--N ha^{-1}, which is an estimate derived from local experimental and agronomic expertise. As such, this input is prone to high uncertainty. The sensitivity and impact of the input parameter uncertainty estimate on the final model is high (see Quinn *et al.*, 1999). Quinn *et al.* (1999) concluded that it is difficult to reduce the error in the input N data and that it is best derived from survey data. The uncertainty analysis shown here is for the hydrological model only.

In order to avoid traditional calibration approaches such as interactive calibration or use of a best fit hillclimbing (Rosenbrock, 1960), a full Monte Carlo procedure was used. Monte Carlo analysis forms the basis of the Generalized Sensitivity Analysis proposed by Spear and Hornberger (1980). All model inputs are randomly perturbed within a physically realistic range, as drawn from the literature. A set of 10,000 data-sets was generated using a uniform sampling strategy that is, by allowing all variables to vary independently. The model is run for each randomly generated set (i.e. 10,000 times), each run giving a different model output. In that respect, it was possible to identify a single 'optimal' set from the 10,000 simulations by comparing simulated and observed discharge values, using the Nash and Sutcliffe simulation efficiency (Nash and Sutcliffe, 1970) as the goodness of fit criterion. The resulting simulations for flow and NO_3^- concentrations, showing good correspondence with the observed values, are depicted in Fig. 11.4. This simple optimization of parameter values to give the 'best fit' is by far the most common way in which hydrological models have been applied in the past. Several other studies, however, suggest limitations to this conventional approach to hydrological model calibration (Sorooshian and Gupta, 1983; Hornberger *et al.*, 1985; Duan *et al.*, 1992; Beven, 1993; Freer *et al.*, 1996). Hence, other methodologies for calibration and uncertainty estimation of models were introduced or are already widely applied. A commonly adopted methodology for assessing the predictive uncertainty associated with model parametrizations is the GLUE methodology (Beven and Binley, 1992). In essence an extension of the Generalized Sensitivity Analysis (see above), the GLUE methodology attempts to explicitly recognize the fundamental limitations of representing hydrological processes with contemporary models. An application of the GLUE approach can be found in the following section.

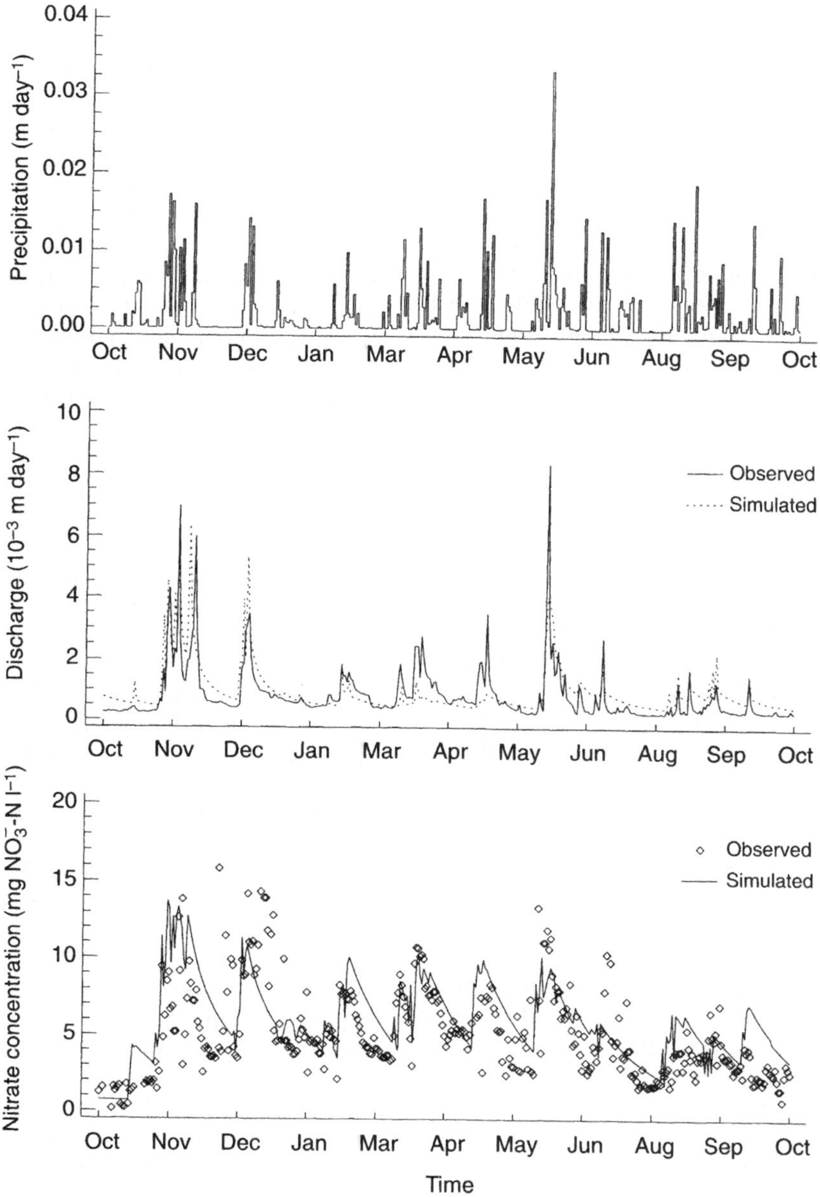

Fig. 11.4. Rainfall, runoff hydrographs and nitrate concentrations at the outlet of the Zwalm catchment for one leaching season (October 1991 – October 1992). Simulated values are obtained using the lumped model version.

Uncertainty analysis

Before applying the GLUE procedure it is worth reviewing the purpose of uncertainty analysis. There are basically three sources of uncertainty:

1. Measurement uncertainty, where instrument error and sampling strategies must be considered.

2. Process uncertainty, where the physics of the processes is not fully understood and some form of approximation is needed.

3. Model uncertainty, where the complexity and architecture of the model control how input errors are propagated to the final output.

Model uncertainty is also related closely to problems of parameter estimation because of the calibration and validation procedures. In almost every model application, an attempt is made to resolve the problem of measurement and process uncertainty, but all too frequently we 'hope' that our physical understanding of processes and our deterministic sampling regimes are adequate and that the model operates correctly. However, when we upscale to the catchment, our limited ability to sample heterogeneity and understand the complex interactions of processes inevitably gives rise to error and increase the model uncertainty. Model complexity and the resulting predictive uncertainty thus becomes a critical issue (Beven, 1989). Equally, the calibration process of any model can be brought into question. If modellers just uses a best-fit procedure, as seen above, they can get the impression that they have a good working model. However, GLUE analysis always demonstrates the occurrence of a wide range of model parameter sets that can give acceptable simulations. This phenomenon of multiple behavioural parameter sets with a similar likelihood of generating almost identical outputs, has been called equifinality (Beven, 1993; Franks *et al.*, 1998).

The GLUE procedure clearly shows that equifinality and error propagation are a fundamental problem to modellers. However, GLUE analysis can also point us towards some ways of resolving uncertainty or at least help us to choose:

1. A sensible parsimonious model structure;
2. Which data-sets to measure in the field to validate our models;
3. Which parameters are the most important to the final model output and may be worthy of more investment in time and money.

Using the Monte Carlo approach, 10,000 simulations are made for which the output of total flux is calculated and analysed. The model outputs used here are the cumulative saturation deficit, cumulative baseflow and cumulative discharge for the hydrological module, and cumulative NO_3^- fluxes for the transport module. The cumulative outputs for the hydrological model are plotted against each of the input variables, resulting in a fuzzy plot (Fig. 11.5). Figure 11.5 shows that the choice of any sensible input value can give virtually any output flux. Take, for example, the *m* parameter, where a wide range of output fluxes can be generated no matter which initial input value is chosen. The root cause of this problem is variable interdependence, that is, other factors and equations compensate for the change in flux caused by changing the *m* data alone.

The ultimate problem of model parameter interdependence is equifinality, that is, more than one combination can give the same resultant flux. It is very difficult to say which unique set is the best. There are many combinations that can give a good fit to the observed data. Equally, a change in the calibration data-set could give a differing set of 'best' fit. A failure to acknowledge this problem will result in a high predictive uncertainty when simulating future scenarios (Beven, 1989; Beven and Binley, 1992; Freer *et al.*, 1996; Franks *et al.*, 1998). Hence, there is a need to understand the uncertainty of input data (measurement and process uncertainty) and how input error is represented in the model output (model uncertainty). The authors thus recommend the use of complex models only at a local, 'data rich' scale and simple MIR models at the catchment scale. Simple models still contain high uncertainty but the uncertainty in the inputs and the final output is quite clear and allows policy decisions to be set on a more realistic probability/risk basis. At this stage of model development, confidence in parameters is based on an *a priori* likelihood weighting, which is conditioned by physical process knowledge, good validation data-sets and experience.

If model simplicity and field observations help us to understand the model uncertainty better, how can the GLUE procedure help to determine a better model? It

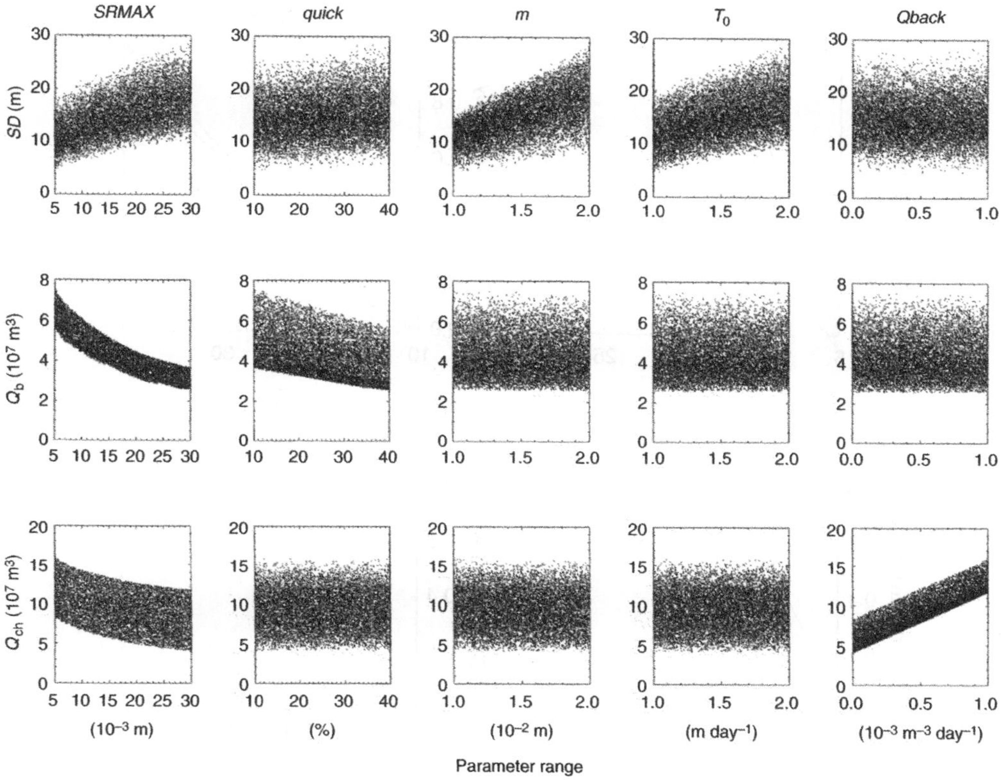

Fig. 11.5. Cumulative fluxes of saturation deficit (*SD*), baseflow (Q_b) and discharge (Q_{ch}) plotted against the hydrological parameters *SRMAX, quick, m,* T_0 and *Qback.*

is possible to study the behaviour of each parameter by analysing the model output further. The parameter sets are subdivided into discrete classes of the model output by ranking the 10,000 model realizations, for each set. The rank position is based on the resulting cumulative fluxes, according to cumulative flux totals. The 10,000 simulations are divided into ten groups of 1000 realisations, thus forming ten performance classes. For each parameter, the distribution of each defined class can be visualized against the three considered cumulative fluxes, yielding cumulative frequency plots (Fig. 11.6). Figure 11.6 shows the cumulative frequency plot for the cumulative saturation deficit and the five hydrological characteristics. Together with the fuzzy plots (Fig. 11.5), the cumulative frequency plots (Fig. 11.6) can be interpreted to determine the sensitivity range of the model outputs for each of the hydrological characteristics. A straight line on Fig. 11.6 indicates that the parameter is insensitive across the ranges and if all ten lines lie on the same straight line then the parameter has no effect on the model, it is in fact redundant. If the line of one of the subsets deviates from a straight line, the parameter is significantly affecting the flux within that subset. If all the lines deviate from the straight line, then the parameter is sensitive across its complete range. As could be expected, the maximum storage capacity of the root zone, *SRMAX*, is strongly related to all three of the considered model outputs. Higher values for *SRMAX* correspond with higher volumes of water that need to be stored in the root zone before gravity drainage towards the saturated zone can occur. This results in higher values for the cumulative saturation deficit, and, inherently, in lower values for

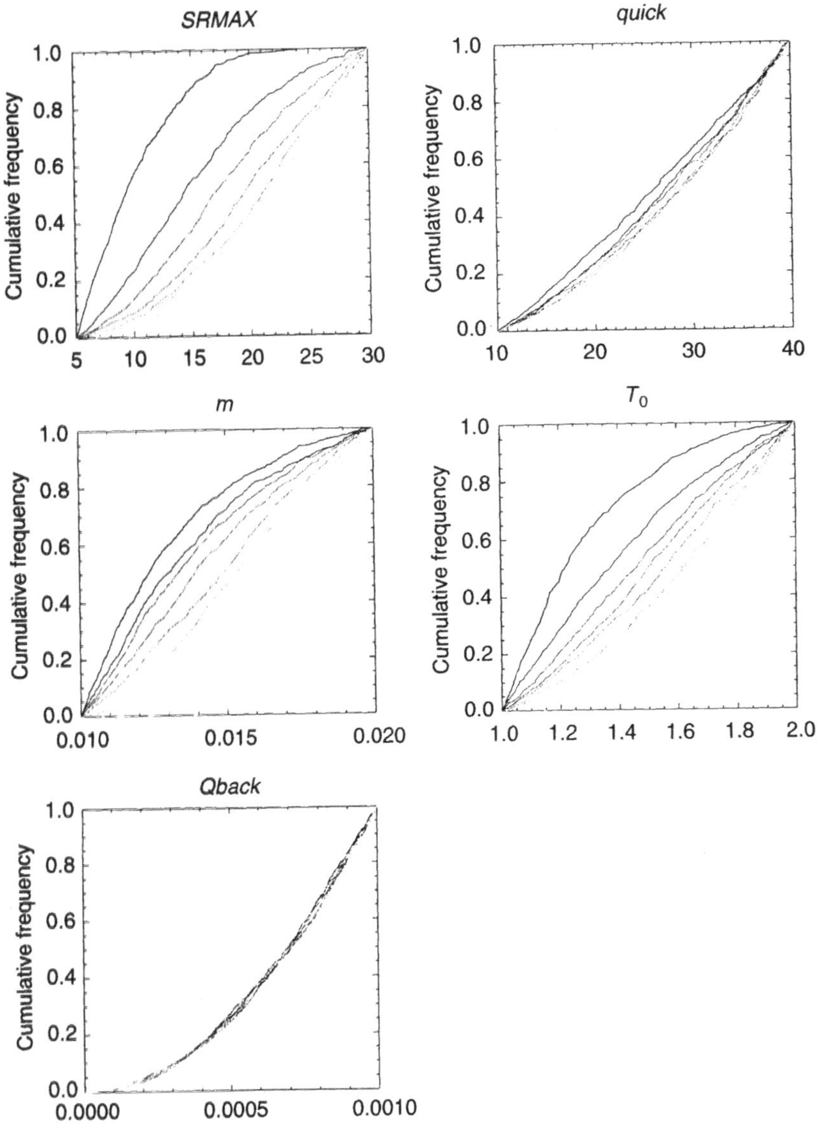

Fig. 11.6. Cumulative frequency plot of each performance class for the saturation deficit (*SD*) versus the different parameter ranges.

the generated baseflow and discharge. The overland flow coefficient, *quick*, appears to have a small influence (see Figs 11.5 and 11.6) on the cumulative saturation deficit, but is strongly associated with the cumulative baseflow. Higher values for *quick* generate more overland flow and less infiltration, resulting in lower baseflow volumes, while the generated discharge is not influenced. The decay parameter *m* and the lateral transmissivity T_0 have a considerable influence on the cumulative saturation deficit, because higher values for *m* and T_0 correspond with higher values for the saturated hydraulic conductivity, K_s (m day^{-1}]. This results in faster drainage or increased baseflow and discharge, which consequently leads to increased values for the saturation deficit. Since *Qback* is only used for eliminating systematic errors from the simulated discharge (see Fig. 11.1), this parameter only affects the cumulative discharge, and has no effect on the cumulative baseflow and cumulative saturation deficit.

The analysis shows clearly the behaviour of each parameter across its sensible range. Also, the behaviour can be interpreted in a physically reasonable way. The analysis allows us to evaluate the sensitivity of each model parameter and also deduce which field measurements would improve the model output. It is clear that *SRMAX* is dominating the model and hence quantifying this term a priori would be beneficial. T_0 is also very sensitive and again may be worthy of detailed analysis. However, a crucial point is that T_0 is behaving in a similar manner to *SRMAX*. Both are soil physical parameters; therefore it may be better to integrate the two together, creating a single soil type characteristic. The benefit to the model is a reduction in the total number of parameters and that the modeller can identify an effective value (and its associated error) based on a soil map and the greater part of the model uncertainty can be grouped together. The parameter still has some physical meaning and a skilled modeller will know whether certain land use change effects tend to increase or decrease its value.

The parameters *quick* and *Qback* are relatively insensitive to the model. Both could be dropped from the model without seriously affecting the model performance. However, the user may still need a background flow value. It may be relatively simple to set these values to a constant. Better still would be to evaluate *Qback* from observed flow data. In fact, given a time series of flow, *Qback* can be taken as the summer dry weather flow. Thus, this parameter falls out of the model analysis and becomes a measured input. Having a good estimate of summer flows for a catchment is, therefore, beneficial to the model.

The remaining parameter *m* is also very sensitive; however, it is relatively easy to estimate the value of *m* if observed flow data are available. The recession limb of a number of observed storms can be used to estimate *m* (Beven *et al.*, 1995). Not all catchments have flow records but it could be argued that it is cost effective to measure the flow in channels if you are working at the catchment scale. As we also use catchment flow data to validate the model, then the benefits of spending time and money on flow measurements should not be underestimated. It is equally reasonable to argue that we should measure nutrient at the same time in the same place. The future ideal MIR modelling approach would be a simple hydrological model based on a soil parameter and number of flow control parameters derived directly from flow measurements.

The impact of the input uncertainty can also be shown based on the 10,000 parameter sets generated from independent uniform distributions. If the top 1000 simulations are treated as 'good' model simulations, then the predictive uncertainty in the output can be estimated (see Freer *et al.*, 1996). This allows the 90% uncertainty bounds to be derived for both the hydrograph and NO_3^- concentration simulations (see Fig. 11.7). The full implication of even a modest uncertainty bound of 90% is quite clear when looking at the NO_3^- simulation in Fig. 11.7. As modellers, it is important to understand that our ability to represent processes must be tempered by a careful understanding of the model uncertainty.

Simulating land use impacts and anthropogenic influences

Watershed models for integrated water management offer the opportunity for systematic analysis of management policies. Therefore, the model needs to produce reliable simulation results, both for existing situations and for future catchment management strategies. The environmental parameters considered here were land use and climate. Land use changes like afforestation and deforestation, changing from undeveloped lands to residential or agricultural lands, and shifts towards more (or less) intensively fertilized crops on agricultural lands, will affect the residual soil NO_3^- content after harvest; that is, at the beginning of winter drainage. This change is reflected in the amount of NO_3^- leached

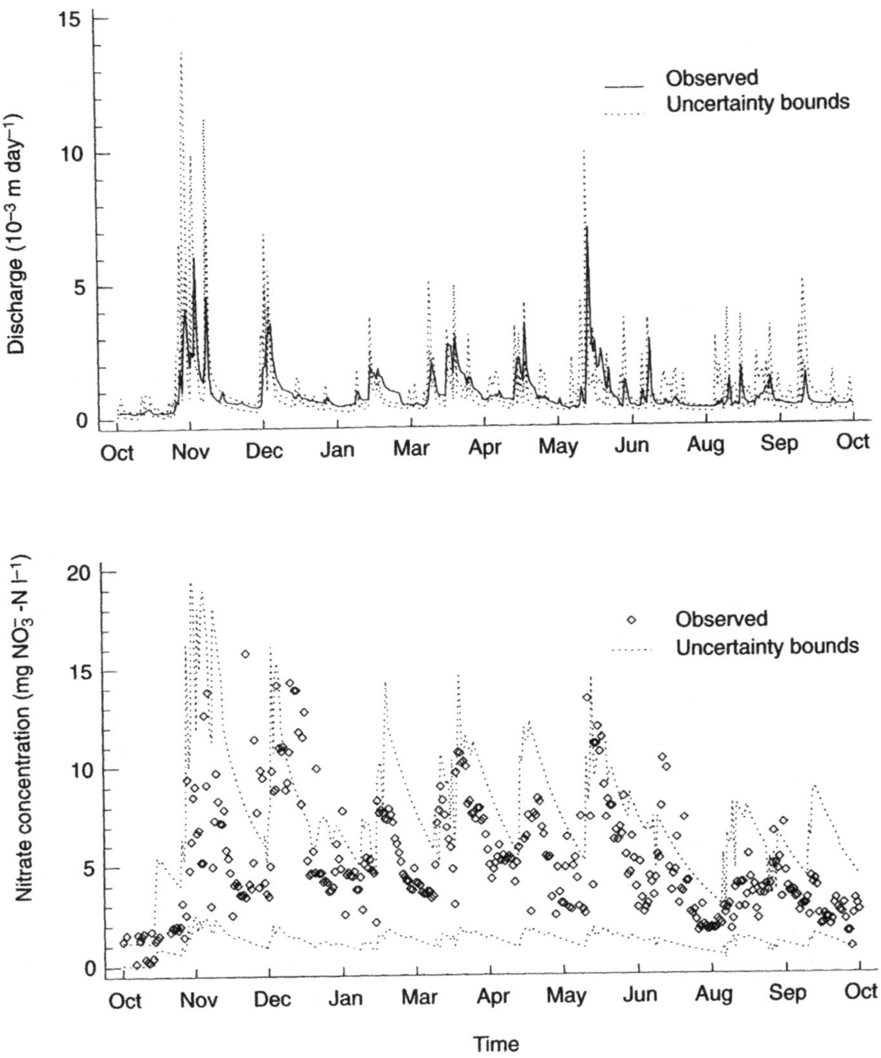

Fig. 11.7. Zwalm catchment simulations of hydrographs and nitrate concentrations. 90% uncertainty bounds for the October 1991 – October 1992 period are calculated from 1000 behavioural parameter sets.

towards the aquatic environment. Hence, the impact of land use changes on the NO_3^- concentrations of the surface water at the outlet of the Zwalm was simulated by increasing and decreasing the initial soil NO_3^- content by 25%. The impact of climate changes on the quantitative and qualitative characteristics of the surface water at the outlet of the Zwalm catchment was simulated by increasing and decreasing rainfall intensity by 25%. This type of climate or land use change scenario still has high uncertainty levels but it is possible to show the likely impacts of land use and climate change. Despite the uncertainty in the final prediction, policy makers can still deduce the likely impact of their management strategies. The results for the produced NO_3^- concentrations at the catchment outlet, under changing land use and climatic conditions are depicted in Fig. 11.8, together with an indication of the EC Drinking Water Directive 80/788 limit (11.3 mg NO_3^--N l^{-1}).

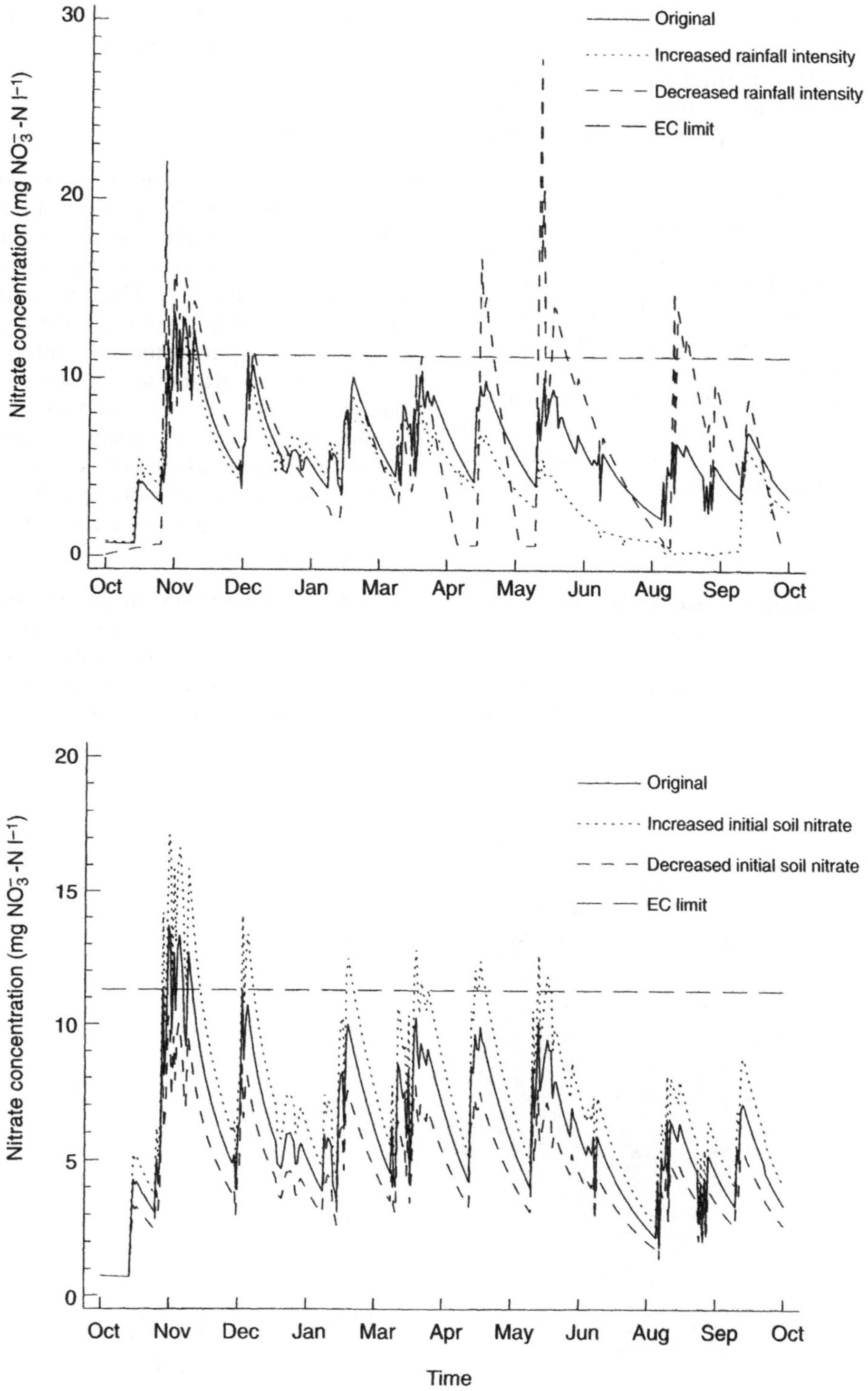

Fig. 11.8. Impact of climate and land use changes on the nitrate concentrations at the outlet of the Zwalm catchment. The initial model simulations are compared with the different environmental impact scenarios.

Distributed Approach

A spatially distributed version of the model was also implemented. It is argued that the introduction of a topographic flow component is justified, as the Zwalm catchment exhibits variable source area activity and many policy decisions relate closely to a specific topographic location, namely the riparian buffer zone area. In order to disaggregate the lumped NO_3^- export, the model is applied using a semi-distributed approach, based on an index of topographic similarity. By introducing a topographic distribution, the model can also be spatially validated against the observed soil NO_3^- content under different crop types.

Soil moisture redistribution and runoff production

Soil moisture redistribution through saturated throughflow and variable source areas are represented in TOPMODEL via an index of hydrologic similarity (λ_i):

$$\lambda_i = \ln\left(\frac{a}{\tan\beta}\right)_i \tag{11.9}$$

where λ_i is the topographic index for location i, $\tan\beta$ is the local slope, used to approximate the hydraulic gradient of the saturated zone, and a is the upslope drained area per unit contour width (m). The topographic index was calculated from the Zwalm digital terrain map using an automated, multiple-flow direction algorithm (Jenson and Domingue, 1988). A representation of the distribution of the topographic indices within the Zwalm catchment is shown in Fig. 11.9. Daily local saturation deficit S_i (m) due to gravity drainage is calculated as:

$$S_i = S + m\,(\lambda - \lambda_i) \tag{11.10}$$

where S (m) is the mean catchment saturation deficit, m (m) defines the variation of saturated hydraulic conductivity with depth (see above), and λ is the catchment's areal averaged topographic index:

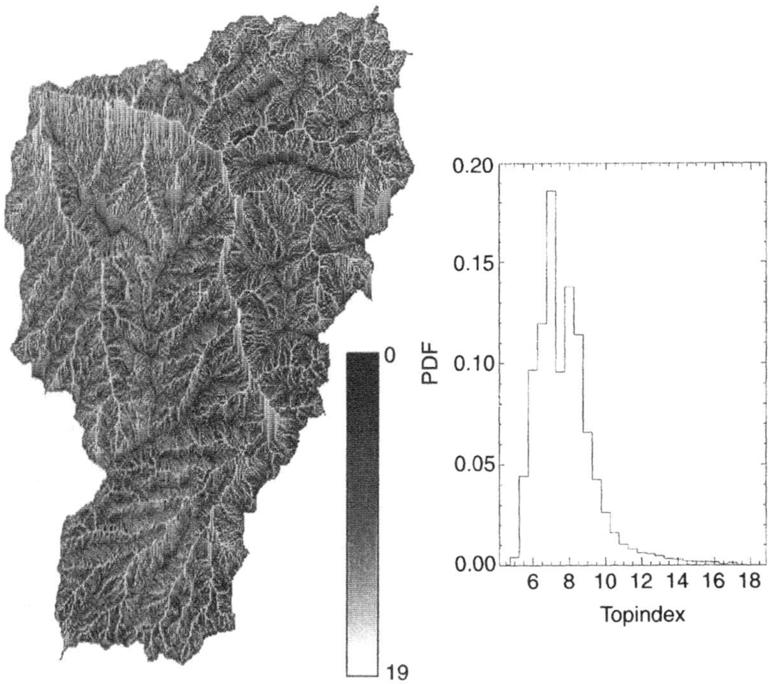

Fig. 11.9. Topographic index pattern and distribution function for the Zwalm catchment.

$$\lambda = \frac{1}{A}\sum_i \ln\frac{a}{\tan\beta} \quad (11.11)$$

with A (m^2) the entire surface area of the catchment. Saturation excess overland flow is produced for all locations where $S_i < 0$. This type of overland flow is produced when the soil becomes saturated to the surface from below (i.e. a rising water table), owing to both rainfall inputs and downslope subsurface flow. The area of saturation (the contributing area) will expand and contract during rainfall events. The overland flow component Q_{se} due to the saturation excess mechanism is obtained from:

$$Q_{se}(t) = \frac{A_c}{A}P \quad (11.12)$$

where A_c (m^2) is the contributing area, and P (m) is the rainfall intensity per unit time t.

Daily output from the saturated zone store is given by Equation 11.1, while the catchment average storage deficit S (m) can be obtained again by subtracting the unsaturated zone recharge and adding the baseflow from the previous time step (see Equation 11.3).

Vertical soil fluxes in the vadoze zone

TOPMODEL uses a root zone detention store and a basic soil column with two variable depth layers, an unsaturated zone and a saturated zone. The boundary between the saturated and unsaturated zone is defined by the soil moisture deficit S (m). Within the soil column, downslope transmissivity T is assumed to decay exponentially with this deficit:

$$T(S) = T_0 e^{-\frac{S}{m}} \quad (11.13)$$

where T_0 (m^2 day^{-1}) is the lateral transmissivity when the soil is just saturated. Daily storage deficit in the unsaturated zone $S_{rz,i}$ (m) is modelled as:

$$\Delta S_{rz,i} = -h_i + PA_i \quad (11.14)$$

where h_i (m) is the vertical drainage from the unsaturated zone, P (m) is rainfall intensity, and A_i is the relative proportion of the catchment characterized by its topographic increment λ_i. Vertical flux from the root zone to the water table occurs by gravity drainage whenever field capacity is exceeded. Expressed in terms of storage deficit, the vertical flux h at any point i is:

$$h_i = \frac{S_{uz,i}}{S_i t_d} \quad (11.15)$$

where $S_{uz,i}$ (m) is storage in the unsaturated zone, S_i is the local saturated zone deficit due to gravity drainage, and t_d (m^{-1}) is a time delay constant. Evaporation is allowed at the full potential rate for water draining freely in the unsaturated zone and for predicted areas of surface saturation.

Nitrate leaching module

The semi-distributed hydrological model calculates the amount of vertical drainage h_i for every soil column i characterized by its topographic index λ_i. Accordingly, the vertical NO_3^- flux calculated by the production function (Equation 11.4), will be higher under low topographic index values compared with high topographic index values. As the transfer function is not a spatial function, the effect of the lateral export cannot be validated against field experiments, but sites with high topographic index values will appear biased towards high soil NO_3^- concentrations. This is an implicit spatial effect that implies a flushing mechanism. The flushing effect is reflected in the simulated river NO_3^- concentrations, as the modelled lateral NO_3^- export is related to the saturation deficit of the catchment (see Equation 11.7). Hence, zones characterized by a frequent shallowing of the water table, that is zones with high topographic index values, exert a higher lateral NO_3^- export compared with zones where the water table is always situated at a greater depth. The zones dominant in the NO_3^- export mechanism of leaching (vertical export) and zones dominant in flushing (lateral export) are visualized in Fig. 11.10.

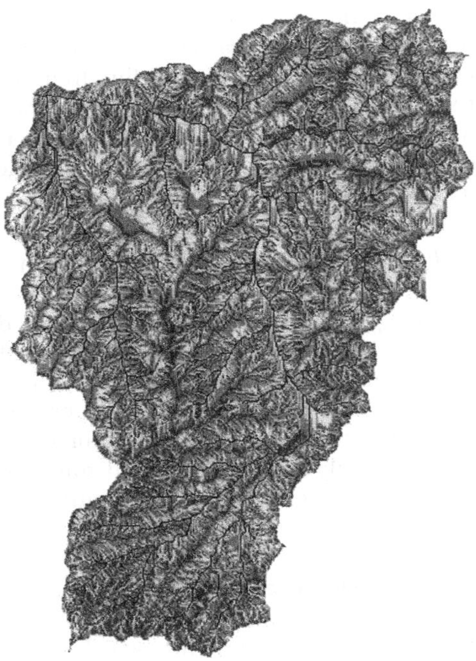

Fig. 11.10. Transport potential index map for the Zwalm catchment, indicating leaching dominated zones ($\lambda_i < 7.0$), intermittent zones ($7.0 < \lambda_i < 8.5$), flushing dominated zones ($8.5 < \lambda_i < 11.5$) and the channel network ($\lambda_i > 11.5$).

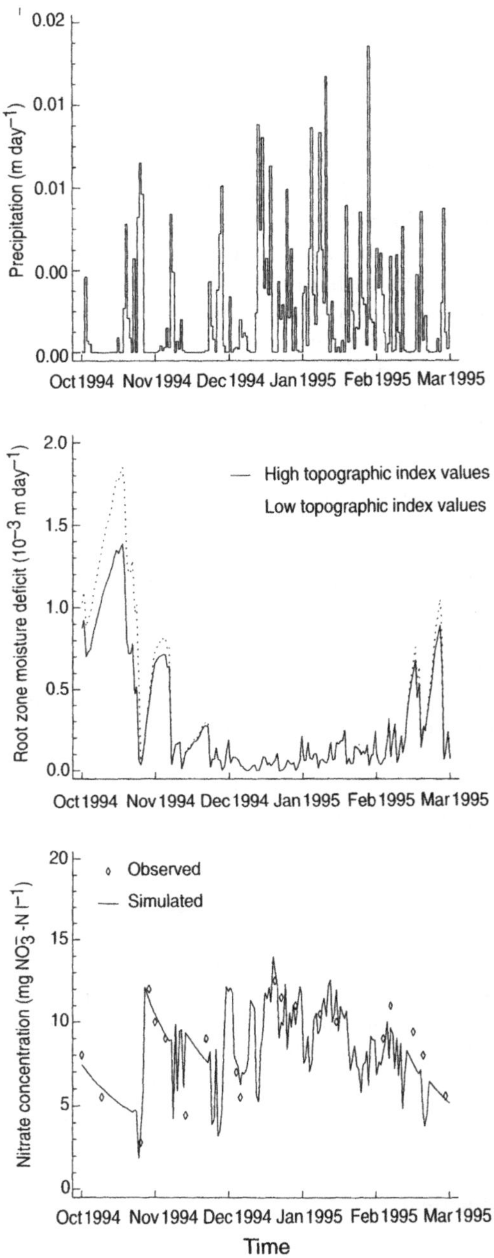

Fig. 11.11. Rainfall, soil moisture deficit and nitrate concentrations at the outlet of the Zwalm catchment for the winter drainage period October 1994 – March 1995. Simulated values are obtained using the semi-distributed model version.

Model results

The semi-distributed model was applied to the Zwalm catchment for the winter drainage condition. The simulation results are shown for the leaching period from 1 October 1994 until 31 March 1995 in Fig. 11.11. The model is able to simulate the temporal variability of the NO_3^- concentrations at the outlet of the catchment, as well as the spatial variability of the soil moisture and the soil NO_3^- concentration (see Fig. 11.11). In that respect, the spatially distributed model can be validated against local NO_3^- measurements obtained via field surveys. Land use within the Zwalm catchment at the onset of winter drainage was obtained from a supervised classification on a Landsat TM image of 14 October 1994 (Verhoest *et al.*, 1998). The corresponding residual soil NO_3^- content is depicted in Fig. 11.12. The catchment average soil NO_3^-

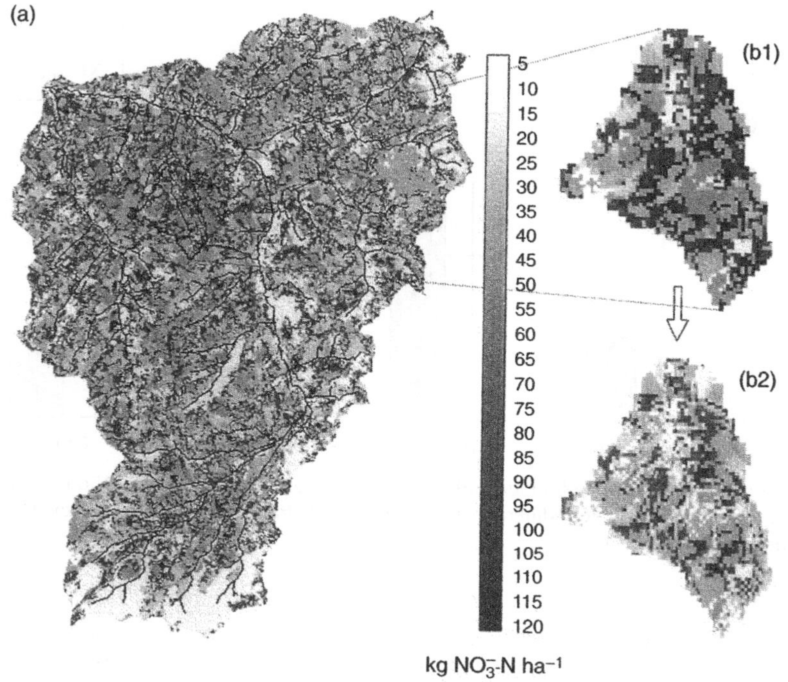

Fig. 11.12. Observed and simulated soil NO_3^--N contents within the Zwalm catchment: (a) observed soil nitrate contents for the entire catchment before the winter drainage period from October 1994 – March 1995, (b1) observed soil nitrate contents for the Wijlegem subcatchment before the winter drainage period (October 1994), (b2) simulated residual soil nitrate contents for the Wijlegem subcatchment after the winter drainage period (March 1995).

content is based on the land use classification and field observations of the upper 90 cm, amounting to 38.2 kg NO_3^--N ha^{-1} for the entire Zwalm catchment and 51.1 kg NO_3^--N ha^{-1} for the Wijlegem subcatchment (see inserts a and b1 in Fig. 11.12). Field observations for four different land use classes (sugarbeet, potatoes, maize and grassland) were carried out before and after the winter drainage period. The simulated average soil NO_3^- content for the Zwalm catchment after winter drainage was 20.1 kg NO_3^--N ha^{-1}, corresponding to an export of 18.1 kg NO_3^--N ha^{-1} during this 6-month period. Comparable export values for the Zwalm catchment were found by De Troch (1985) for the 1970–1973 period and by Lauryssen *et al.* (1994) for the 1991–1992 period. Simulated average soil NO_3^- content for the Wijlegem subcatchment after winter drainage amounted to 31.2 kg NO_3^--N ha^{-1}, corresponding to a NO_3^- export of 19.9 kg NO_3^--N ha^{-1} during this winter period (see insert b2 in Fig. 11.12). Differences between field observations and modelled soil NO_3^- contents, in the Wijlegem subcatchment, after winter drainage and for the land use types under consideration are depicted in Fig. 11.13. Model predictions for residual soil NO_3^- under sugarbeet and grassland are significantly lower than the observed values. This may be attributed to the fact that sugarbeet and grassland fields are more predominant on locations characterized by high topographic index values, that is, in the flushing zone. As the model only considers the vertical leaching mechanism at the field scale, this results in an underestimation of NO_3^- export. No plausible explanation could be provided for why the model predictions for the residual soil NO_3^- under potatoes and maize which were significantly higher than the observed values.

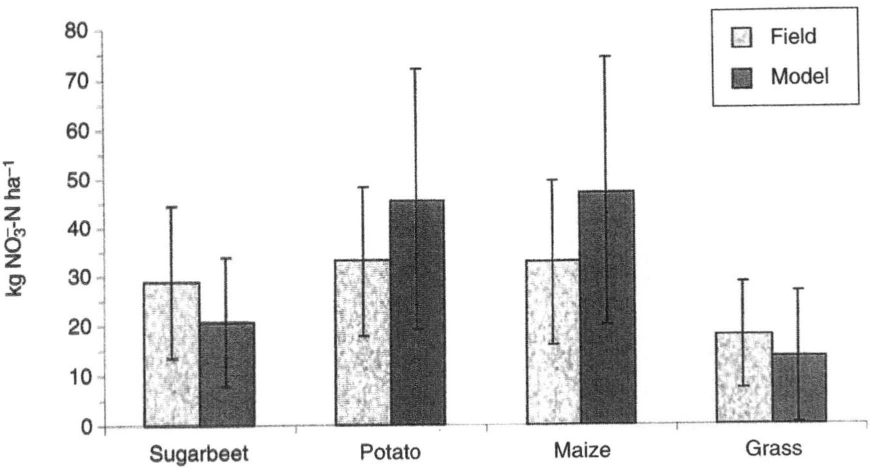

Fig. 11.13. Comparison between observed and predicted average residual soil NO_3^--N contents within the Wijlegem subcatchment, under four different land use classes. Nitrate concentrations are expressed as average values per land use class.

Conclusions

This chapter has presented two models for studying the hydrological mobilization of nutrients at the catchment scale. The first is a low parameter lumped version of TOPMODEL that uses land use units to approximate to a nutrient input index and a combined nutrient production and transfer function. The second model distributes the soil moisture and nutrient activity, based on topography, to reflect zones of vertical leaching and zones of flushing. Both models have been applied to the Zwalm catchment.

The complexity of the models is kept as low as possible, but the models are still able to satisfy the goals of land use planners. The simple structure allows a full and transparent uncertainty analysis to be carried out and thus an estimate of the uncertainty can be passed on to the end user. Uncertainty analysis demonstrated that models (even simple models) have a high uncertainty. At this time there is a need for nutrient models to communicate a realistic uncertainty estimate to end users, rather than to attempt to resolve or ignore the uncertainty problem. GLUE analysis allows us to evaluate how the model is operating and can point us towards a better model structure by determining which parameters are redundant or which can be combined. The analysis indicates which measurements are cost effective in catchment scale studies; as such the authors strongly recommend the acquisition of flow and nutrient time series within a local river system.

The introduction of a topographic model component is justified because of the evidence of variable source area activity in the Zwalm catchment, and the need for policy makers to implement topographically related policy on buffer zones and wetlands. The model is still relatively simple and the new component adds greatly to our ability to disaggregate the sources of nutrient mobilization. The introduction of a topographic component enables the end user to evaluate management strategies at both the catchment and the field scale.

As all model parameters have some physical meaning, the model can still represent any change in land use or climatic conditions. Thus, the probable impacts of any land use or policy change on catchment water quality can be estimated at an appropriate scale in a quasi-physical manner, while future policy can be set in the full knowledge of the model uncertainty.

References

Anthony, S., Quinn, P. and Lord, E. (1996) Catchment scale modelling of nitrate leaching. *Aspects of Applied Biology. Modelling in Applied Biology: Spatial Aspects* 46, 23–32.

Beven, K. (1989) Changing ideas in hydrology: the case of physically-based models. *Journal of Hydrology* 105(1–2), 157–172.

Beven, K. (1991) Spatially distributed modelling: conceptual approach to runoff prediction. In: Bowles, D.S. and O'Connel, P.E. (eds) *Recent Advances in the Modelling of Hydrologic Systems.* Kluwer, Dordrecht, pp. 373–387.

Beven, K.J. (1993) Prophecy, reality and uncertainty in distributed hydrological modelling. *Advances in Water Resources* 16, 41–51.

Beven, K.J. and Binley, A.M. (1992) The future of distributed models: model calibration and uncertainty prediction. *Hydrological Processes* 6, 279–298.

Beven, K.J. and Kirkby, M.J. (1979) A physically based, variable contributing area model of basin hydrology. *Hydrological Sciences Bulletin* 24(1), 43–69.

Beven, K., Lamb, R., Quinn, P., Romanowicz, R. and Freer, J. (1995) Topmodel. In: Singh, V. (ed.) *Computer Models of Watershed Hydrology.* Water Resources Publications, Colorado, pp. 627–668.

De Troch, F.P. (1977) Study of the surface water hydrology of an East Flemish river catchment: the Zwalm river at Nederzwalm. PhD dissertation, Faculty of Applied Sciences, University of Gent, Belgium (in Dutch).

De Troch, F. (1985) Surface water quantity-quality relationships and dissolved loads in a rural catchment in Flanders, Belgium. In: *IWRA Vth World Congress on Water Resources. Water Resources for Rural Areas and their Communities,* 9–15 June, Brussels, Belgium, pp. 423–430.

Duan, Q.S., Sorooshian, S. and Gupta, V.K. (1992) Effective and global optimisation for conceptual rainfall-runoff models. *Water Resources Research* 28, 1015–1031.

Dunne, T. and Black, R.D. (1970) An experimental investigation of runoff production in permanent soils. *Water Resources Research* 6, 478–490.

Franks, S.W., Gineste, P., Beven, K. and Merot, M. (1998) On constraining the predictions of a distributed model: the incorporation of fuzzy estimates of saturated areas into the calibration process. *Water Resources Research* 34(4), 787–797.

Freer, J., Beven, K. and Ambroise, B. (1996) Bayesian estimation of uncertainty in runoff prediction and the value of data: an application of the glue approach. *Water Resources Research* 32(7), 2161–2173.

Gyasi-Agyei, Y., De Troch, F.P. and Troch, P.A. (1995) A dynamic hillslope response model in a geomorphology based rainfall-runoff model. *Hydrological Processes* 9, 575–587.

Heathwaite, A., Burt, T. and Trudgill, S. (1993) The nitrate issue. In: Burt, T., Heathwaite, A. and Trudgill, S. (eds) *Nitrate. Processes, Patterns and Management.* John Wiley & Sons, Chichester, UK, pp. 3–21.

Heathwaite, A.L., Johnes, P. and Peters, N. (1996) Trends in nutrients. *Hydrological Processes* 10, 263–293.

Hornberger, G., Beven, K., Cosby, B. and Sappington, D.E. (1985) Shenandoah watershed study: Calibration of a topography-based, variable contributing area hydrological model to a small forested catchment. *Water Resources Research* 21(12), 1841–1850.

Hornberger, G., Bencala, K. and McKnight, D. (1994) Hydrological controls on dissolved organic carbon during snow melt in the Snake River near Montezuma, Colorado. *Biogeochemistry* 25, 147–165.

Jenson, S.K. and Domingue, J.O. (1988) Extracting topographic structure from digital elevation data for geographic information system analysis. *Photogrammatic Engineering and Remote Sensing* 54(11), 1593–1600.

Johnes, P.J. and Heathwaite, A.L. (1997) Modelling the impact of land use changes on water quality in agricultural catchments. *Hydrological Processes* 11(3), 269–286.

Kalma, J.D. and Sivapalan, M. (1995) *Scale Issues in Hydrological Modelling,* Advances in Hydrological Processes Series. John Wiley & Sons, Chichester, UK.

Lauryssen, F., Tack, F. and Verloo, M. (1994) Nitrogen load in the Zwalm river catchment. *Water* 75, 46–49 (in Dutch).

Nash, J.E. and Sutcliffe, J. (1970) River flow forecasting through conceptual models: discussion of principles. *Journal of Hydrology* 10, 282–290.

Nieswand, G.H., Hordon, R.M., Shelton, T.B., Chavooshian, B.B. and Blair, S. (1990) Buffer strips to protect water supply reservoirs. A model and recommendations. *Water Resources Bulletin* 26(6), 959–966.

Pionke, H.B., Gburek, W.J., Sharpley, A.N. and Schnabel, R.R. (1996) Flow and nutrient export patterns for an agricultural hill-land watershed. *Water Resources Research* 32(6), 1795–1804.

Quinn, P.F. (1991) The role of digital terrain analysis in hydrological modelling. PhD dissertation, Lancaster University, UK.

Quinn, P.F. and Beven, K.J. (1993) Spatial and temporal predictions of soil moisture dynamics, runoff, variable source areas and evapotranspiration for Plynlimon, Mid-Wales. *Hydrological Processes* 7, 425–448.

Quinn, P.F., Anthony, S. and Lord, E. (1999) Basin-scale nitrate simulation using minimum information requirement approach. In: Trudgill, S., Walling, D. and Webb, B. (eds) *Water Quality Processes and Policy*. John Wiley & Sons, Chichester, UK.

Rosenbrock, H.H. (1960) An automatic method of finding greatest or least value of a function. *Computer Journal* 3, 175–184.

Singh, V.P. (1995) Watershed modeling. In: Singh, V. (ed.) *Computer Models of Watershed Hydrology*. Water Resources Publications, Colorado, pp. 1–22.

Sorooshian, S. and Gupta, V.K. (1983) Automatic calibration of conceptual rainfall-runoff models: the question of parameter observability and uniqueness. *Water Resources Research* 19, 260–268.

Sorooshian, S. and Gupta, V.K. (1995) Model calibration. In: Singh, V. (ed.) *Computer Models of Watershed Hydrology*. Water Resources Publications, Colorado, pp. 23–68.

Spear, R. and Hornberger, G. (1980) Eutrophication in Peel Inlet: 2. Identification of critical uncertainties via Generalised Sensitivity Analysis. *Water Research* 14, 43–49.

Troch, P.A., De Troch, F.P. and Brutsaert, W. (1993) Effective water table depth to describe initial conditions prior to storm rainfall in humid regions. *Water Resources Research* 29, 427–434.

Troch, P.A., De Troch, F.P., Mancini, M. and Wood, E.F. (1995) Stream network morphology and storm response in humid catchments. *Hydrological Processes* 9, 575–587.

Van Herpe, Y., Troch, P.A., Callewier, L. and Quinn, P.F. (1998) Application of a conceptual catchment scale nitrate transport model on two rural river basins. *Environmental Pollution* 102 (S1), 569–577.

Van Herpe, Y., Troch, P.A., Callewier, L. and De Troch, F.P. (1999) A conceptual model for simulation of catchment scale nitrate transport. In: Heathwaite, L. (ed.) *Impact of Land-Use Change on Nutrient Loads from Diffuse Sources*, IAHS Publication no. 257. Institute of Hydrology, Wallingford, UK, pp. 181–191.

Verhoest, N.E., Troch, P., Paniconi, C. and De Troch, F.P. (1998) Mapping basin scale variable source areas from multitemporal remotely sensed observations of soil moisture behavior. *Water Resources Research* 34(12), 3235–3244.

12 Pollutant–Sediment Interactions: Sorption, Reactivity and Transport of Phosphorus

D.S. Baldwin,[1] and A.M. Mitchell[1] and J.M. Olley[2]
[1]CRC for Freshwater Ecology and Murray–Darling Freshwater Research Centre, PO Box 921, Albury, NSW 2640, Australia; [2]CSIRO Land and Water, PO Box 1666, Canberra, ACT 2601, Australia

Key words: adsorption, phosphorus, release, sediments, transport

Introduction

Sediments are complex mixtures of minerals, organic material and biota found suspended in the water column and at the bottom of water bodies. They can vary in size from coarse sands and gravels through to very fine clay-rich muds. Although ostensibly derived from soils and rocks, sediments often behave very differently from their terrestrial counterparts. These differences can be attributed to extensive microbial activity reworking organic and inorganic constituents of the sediments, and oxidation and reduction reactions resulting in changes to the surface coatings of the sediment grains.

Sediments play a number of key roles in the movement and transformation of pollutants derived from agriculture. Pollutants often form strong associations with sediment particles so that their fates are often intimately linked. Sediment deposition may be an important pathway for the removal of a pollutant from the water column. Conversely, sediment re-suspension and transport can be an important mechanism for the movement of the pollutant through an aquatic ecosystem. However, sediment–pollutant interactions often go beyond simple sorption and transport processes. Biotic and abiotic processes associated with sediments may transform the pollutant. Clearly, the interplay between the sediment and pollutants can be quite complex and in this chapter we explore some of this complexity by examining the interaction between sediments and P. While we focus on P–sediment interactions, many other pollutants such as heavy metals and pesticides are transported in association with sediment particles. Consequently, mechanisms analogous to those underlying P–sediment interactions will influence the transport and transformations of these other pollutants associated with sediments (e.g. Allen, 1995; Stumm and Morgan, 1996).

In attempting to understand any system it is useful to start with a conceptual framework around which ideas and discussion can be structured. Phosphorus in the aquatic environment can be viewed as residing in one of four pools: (i) dissolved in the water column; (ii) associated with suspended sediments; (iii) deposited in bed sediments; or (iv) incorporated into the biota (Fig. 12.1). The arrows in the diagram indicate the exchanges between each of these 'pools'. These exchange processes are discussed briefly in the following paragraph; some are then covered in more detail in the subsequent text. Upstream inputs (arrow A) into a water body (river reach or lake) can add P into any of the four defined pools. Water and sediment chemistries and biological activity then control the exchange between each of the pools in the water body. Adsorption and desorption exchanges occur between the dissolved pool and the sediment bound pools (arrows B and C). Phosphorus is also released from the sediment pools to the dissolved pool by the mineralization of organic matter present both in the bed and in suspended sediments. Phosphorus is

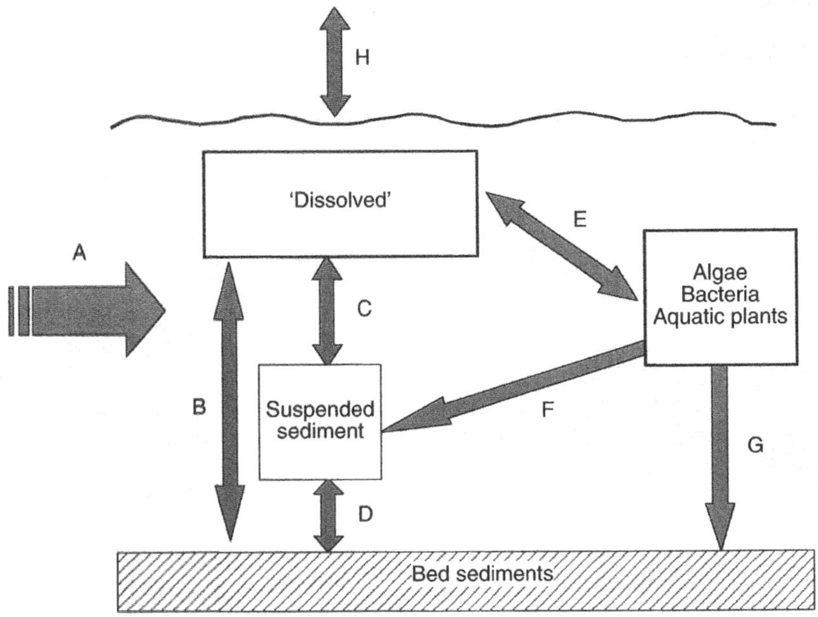

Fig. 12.1. A stylized diagram of the phosphorus cycle in aquatic systems (see text for definitions).

incorporated into the biological pool from the dissolved pool through the growth of algae, bacteria and aquatic plants (arrow E). For simplicity, organisms further along the food chain have been omitted from the diagram. The release of P from the biological pool back to the dissolved pool occurs either through direct excretion (arrow E) or mineralization during decomposition following the death of the organism (arrows F to C and G to B). Exchange between the bed sediment and the suspended sediment pools occurs through the processes of sedimentation and re-suspension (arrow D).

The Sources and Transport of Sediment Phosphorus

Phosphorus can be delivered to water bodies in either dissolved form or in association with organic or mineral particles. The fraction of P delivered in each of these forms depends on the geology, land use, land cover, hydrology and the nature of the drainage network (Sharpley *et al.*, 1995; Jordan *et al.*, 1997; Carpenter *et al.*, 1998; Harris, 1999). Water and sediment chemistries and biological activity in the receiving waters then control the form in which the P is transported through the aquatic system.

Phosphorus and other chemical inputs into the aquatic environment are classified as point or non-point sources. Point sources, such as municipal sewage treatment plants and effluent release from animal feedlots, discharge into the aquatic environment at defined points, usually through drains or pipes. In contrast, non-point sources such as runoff from agricultural land enter the aquatic environment by numerous pathways and over wide areas. In heavily populated or industrialized areas point sources can dominate the pollutant delivery. However, in much of the developed world significant effort has been expended in controlling pollutant delivered from point sources so that in countries such as the USA non-point sources now dominate (Carpenter *et al.*, 1998). Delivery of P from non-point sources is a function of land use, population densities, agricultural practices, urban development (Harris, 1999) and the drainage density (Sharpley *et al.*, 1995).

Sediment particles present within the aquatic system can be delivered by erosion of surface soils and stream banks and beds. From the perspective of contaminant source and management, it is erosion of surface soil from agricultural areas that poses the greatest hazard to aquatic systems. Surface soils are likely to have enhanced concentrations of P from biological cycling (Richardson, 1994) and from fertilizer addition (Norrish and Rosser, 1983). Sediments derived from the erosion of stream banks and beds will consist mainly of subsoils, which only contain native P that is derived from rock weathering. The relative amount of surface soil and subsoil-derived sediment in streams depends on the age of the channel network. In systems which have experienced recent channel rejuvenation (gully formation), subsoil-derived sediments dominate (see for example Olley *et al.*, 1993; Wallbrink *et al.*, 1998).

Phosphorus loads in rejuvenated channel systems tend to be dominated by suspended particulate transport. Most of the P in these systems is derived from the subsoil particles (for example, Olley, 1996). In clear water systems, typically draining sandy catchments, most of the P is transported in dissolved form (e.g. Peel-Harvey River systems in Western Australia: Birch, 1982).

Incorporation of Dissolved Phosphorus into Sediments

Suspended and bottom sediments have long been seen as important sinks for the removal of dissolved P from aquatic systems (downward arrows C and B in Fig. 12.1). Abiotic pathways for the incorporation of dissolved PO_4^{3-} into sediments include adsorption and precipitation/co-precipitation reactions (Holtan *et al.*, 1988).

Sorption reactions

A sorption reaction involves the removal of PO_4^{3-} from solution by concentration in a solid phase that may be physically or chemically mediated. Physically mediated sorption results in a small decrease in energy and is reversible; that is, the adsorbed phosphate may desorb. In contrast, chemisorption is partly or completely irreversible and results in a large decrease in energy.

The chemical composition of the sediments dictates number and type of P sorption sites. The chemical composition of sediment is determined by the type of rock it is derived from, its weathering history and the biological activity that it has been exposed to (Stone and English, 1993).

As a consequence of the complexity of the different sediment components it is rather difficult to determine the reactions governing the sorption of phosphate in natural sediments. However, a number of studies have been performed on individual clay minerals, metal oxides and hydroxides (Holtan *et al.*, 1988). Brinkman (1993) suggests that the most important sites for anion adsorption are surface coatings of Fe^{3+} and Al oxides and oxyhydroxides. Hydroxyl groups (OH^-) are formed at the solid–water interface of the oxides for stability and charge neutrality. The basis for adsorption of PO_4^{3-} is the exchange with the surface hydroxyl groups (Fig. 12.2). The PO_4^{3-} is bound to the metal centre on the mineral surface by covalent bond(s) through ligand exchange with mineral surface hydroxyl groups (Goldberg and Sposito, 1985); that is, P forms an inner-sphere complex with the metal ion (Stumm and Morgan, 1996; Fig. 12.2).

The extent of binding will depend in part on the pH of solution, the number of exchangeable hydroxyl groups per unit area of the sediment and the specific surface area of the sediment. The pH of the reaction is important. It is easier for an anion to approach a mineral surface if the net charge on that surface is neutral or positive. The charge at a metal(oxy)hydroxide surface will depend on the degree of protonation of the surface hydroxyl groups (Fig. 12.3). The pH at which the bulk surface has no net charge is known as the point of zero charge (pH_{pzc}). At pH levels less than the pH_{pzc} the net surface charge is positive. The pH_{pzc} of common Fe minerals is typically in the range of *c.* pH 6.5–9, while the pH_{pzc} typically ranges from *c.* 5 for α–Al(OH_3) to about 9 for more crystalline Al (oxy)hydroxides (Baldwin *et al.*, 1995; Stumm and Morgan, 1996).

Manganese minerals may also adsorb PO_4^{3-} in the same manner because, like Fe and Al, Mn oxides also form hydroxides (Boström *et al.*, 1988). However, Coleman (1996) reports that the pH_{pzc} for manganese dioxide is *c.* pH 2, much lower than would be expected in ambient sediment conditions. Nevertheless, Yao and Millero (1996) report significant adsorption of P to manganese dioxide in seawater at neutral pH. However, these authors conclude that Mn oxides will only become important sinks for P at high P concentrations, that is, once Fe (and presumably Al) surface sites have become saturated.

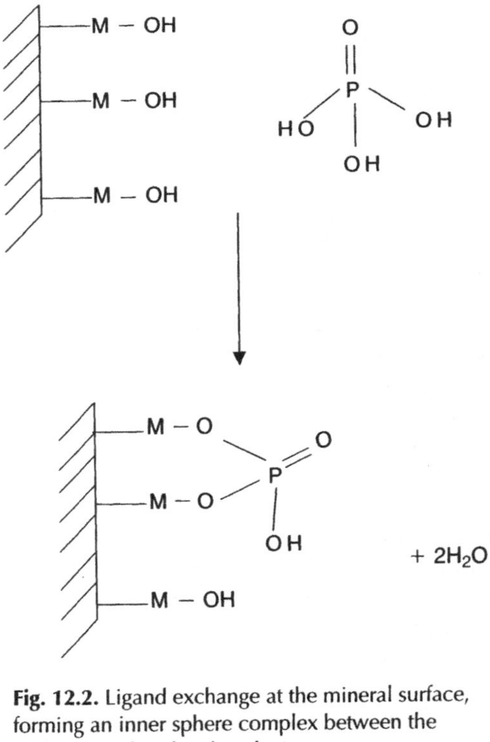

Fig. 12.2. Ligand exchange at the mineral surface, forming an inner sphere complex between the surface metal and a phosphate group.

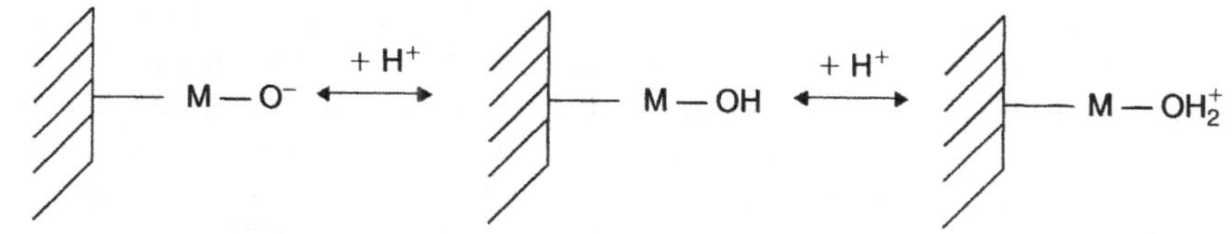

Fig. 12.3. Successive addition of protons to a mineral surface.

It is noteworthy that kinetic studies of P adsorption to sediments often show an initial, very rapid uptake (in the order of minutes) followed by a significantly slower rate (periods of days to months; e.g. Detenbeck and Brezonik, 1991). This biphasic behaviour has been attributed to a rapid uptake of P on the surface of the mineral followed by a significantly slower diffusion from the surface P to the interior of the particle (Barrow, 1983; Detenbeck and Brezonik, 1991). The extent of diffusion of P into the mineral matrix is dependent on the crystallinity of the mineral phase (Barrow, 1999). For example, Strauss *et al.* (1997) have shown that P does not diffuse at all into the matrix of highly crystalline goethite but does diffuse slowly into the crystal matrix of less crystalline goethite phases.

Other sorption reactions are possible with Al and Fe oxides as they have the general stoichiometric formula M_2O_3 and often a hexagonal, close-packed crystal structure with metal atoms on two-thirds of the octahedral positions (Brinkman, 1993). Small cations may sorb on vacant octahedral positions thereby increasing the positive charge of the colloids. In this instance, the P does not necessarily form a chemical bond with surface minerals but rather is held in close proximity to the mineral surface by ion-paring (i.e. an outer-sphere complex is formed). Alternatively, cations may substitute with the metal ions resulting in a more negative surface charge with resultant decrease in P affinity. These sorption and substitution reactions are controlled in part by the ionic composition of the surrounding water (Brinkman, 1993). For example, Yao and Millero (1996) report a higher affinity of P for MnO_2 surfaces when magnesium (Mg^{2+}) and Ca^{2+} ions were added to solution.

Changing sediment–phosphorus affinity

From the preceding discussion it is clear that the affinity of sediments for P will depend to a large part on the nature of the sediment particle's surface. We have already discussed how changes in solution pH can affect sediment–P affinity by changing the charge on the mineral surface. Other factors can also influence sediment–P affinities, including sorption of other substances on mineral surfaces (particularly dissolved organic matter) and ageing (i.e. increasing crystallinity) of the mineral phase.

Dissolved organic matter (DOM) in natural waters consists of a large suite of compounds derived from both aquatic and terrestrial origins (Volk *et al.*, 1997). While it is not possible to assign unequivocally a given chemical structure to DOM, numerous studies have shown that DOM often contains easily ionizable functional groups including phenols, hydroxyls and carboxylic acids (Volk *et al.*, 1997). As such, DOM tends to be anionic at pH levels normally found in aquatic systems (e.g. Miles *et al.*, 1983). The negatively charged DOM can readily sorb on to typically positively charged mineral surfaces, forming a sphere of organic complexes surrounding the mineral particle and thereby competing with P for binding sites (Tipping, 1981; Gu *et al.*, 1994). Sibanda and Young (1986) showed that both humic and fulvic acids could compete strongly with PO_4^{3-} for binding sites on synthetic Fe (goethite) and Al (gibbsite) oxides, particularly at lower pH values. Similarly, van Berkel and Beckett (1997) have proposed a model of suspended sediments in which a mineral or clay core is almost entirely coated by organic matter. They suggest that this organic coating inhibits P adsorption. Using field-flow fractionation combined with ^{33}P uptake experiments, they showed that there was a fourfold increase in P affinity when organic matter was stripped from natural particles (in the range of 0.1–0.5 μm) with hydrogen peroxide. The effect was not as evident for larger particles suggesting an incomplete coverage of the larger particle by organic matter (van Berkel and Beckett, 1997).

The degree of crystallinity of the adsorbing mineral phases (particularly Fe and Al) will also change a sediment's affinity for P (Lijklema, 1980). Amorphous oxyhydroxides tend to have a higher surface area and more binding sites than more crystalline oxides (e.g. Baldwin *et al.*, 1995). As minerals age they can become more crystalline. For example, amorphous $Fe(OH)_3$ can dehydrate to poorly crystalline ferrihydrite ($Fe_5HO_8.4H_2O$) and ultimately to crystalline haematite (α-Fe_2O_3) (Schwertmann and Cornell, 1991).

One determinand of mineral ageing in shallow freshwater ecosystems is sediment exposure to air and subsequent drying (Sah *et al.*, 1989), as a consequence of either drought or water drawdown. However, the effects of sediment exposure/desiccation on subsequent P affinity are far from straightforward. Qiu and McComb (1994) showed through adsorption isotherm measurements that there was a decrease in the PO_4^{3-} affinity of wetland sediment that had been allowed to dry out in the laboratory. Their study is consistent with a report from Twinch (1987) on sediments taken from a large water-storage reservoir in South Africa. Twinch showed that there was a measurable decrease in the PO_4^{3-} buffering capacity and an increase in the PO_4^{3-} equilibrium concentration when sediments were air-dried in the laboratory. Similarly, Baldwin (1996a) showed that the P affinity along a sediment transect that covered submerged, wet-littoral, and exposed heavily desiccated sediments decreased by about three orders of magnitude as a consequence of desiccation. It was suggested that this increase in affinity was due almost entirely to the degree of crystallinity of Fe minerals in the sediments. Submerged sediments contained reduced iron (Fe^{2+}) species that presumably were oxidized rapidly to amorphous Fe^{3+} oxyhydroxides under the conditions of the experiment and hence had a high affinity for P. As the sediments were allowed to dry out, it was suggested that desiccation led to ageing of the minerals and hence a decrease in affinity.

In contrast to these studies, Jacoby *et al.* (1982) reported a net reduction in dissolved P levels in the water column above sediments that had been exposed to the air and subsequently re-submerged, suggesting an increase in P affinity. Similarly, DeGroot and van Wijck (1993) showed that affinity for P in sediments from a wetland in southern France increased after drying. They ascribed the increased affinity to oxidation of ferrous sulphides to amorphous Fe oxyhydroxides on exposure of the sediments to air. Taken together, these results suggest that the change in P affinity of sediments after drying will depend on both the degree of reduction of the sediments prior to exposure and the length of exposure. Reduced iron phases will oxidize rapidly to amorphous oxyhydroxides on exposure and hence the affinity for P should increase (De Groot and van Wijck, 1993; Baldwin, 1996a). Extended periods of drying will result in the dehydration of amorphous oxyhydroxides, with concurrent increase in crystallinity (Sah *et al.*, 1989; Baldwin, 1996a).

Precipitation reactions

The removal of two or more components from solution by their mutual combination forming a new solid-phase compound is termed precipitation (Holtan *et al.*, 1988). Authigenic (within the sediment) phosphatic mineral formation occurs when pore waters become supersaturated with respect to an authigenic phosphatic phase. The authigenic minerals most commonly cited as forming in sediments of natural water bodies are Ca^{2+}- and Fe^{3+} phosphates (Ruttenberg, 1992), but may also include Fe^{2+} phosphates such as vivianite (Emerson, 1976; Emerson and Widmer, 1978; Postma, 1981). However, precipitation of PO_4^{3-} with mineral phases as an early diagenic, authigenic process is thought to be a small contributor to the mineral-PO_4^{3-} compounds within sediments, the majority being allogenically derived (i.e. from adjacent geological deposits and surface soils). Furthermore, the kinetics for precipitation reactions of authigenic PO_4^{3-} minerals are often slow. Other physical, chemical and biological processes may be significantly more rapid (Enell and Lofgren, 1988). Therefore, PO_4^{3-} mineral precipitation as an early diagenic process is considered negligible within biologically active sediments where environmental conditions may change in hours (Enell and Lofgren, 1988).

Calcium minerals, particularly carbonates, have also been implicated in P cycling in sediments. Calcium apatite may be formed biogenically within the water body, sequestering PO_4^{3-} in sediments. This apatite is different from the insoluble allogenic apatite of igneous or metamorphic origin and is thought more likely to be a source of bioavailable P (Ruttenberg, 1992). Additionally, calcium and carbonate combine to form calcite which is also able to sorb/co-precipitate PO_4^{3-}, particularly in highly calcareous environments (Pettersson *et al.*, 1988; House, 1999). For example, Olila and Reddy (1997) found that formation of Ca-P compounds and/or calcite-P co-precipitation were the dominant mechanisms for P removal from a shallow eutrophic lake at pH levels above about pH 8.5. In contrast, Sherwood *et al.* (1987) suggested that most calcium carbonate-bound P is actually associated with the Fe^{3+} oxyhydroxide coatings rather than with the mineral calcium carbonate.

Remobilization of Phosphorus from Sediments

While sediments may represent the largest sink in aquatic systems they may also represent the major source of P (e.g. Pettersson, 1998). Remobilization of P from sediments to the overlying water involves two processes, release from sediment particles to the interstitial waters within the sediment and transfer from the interstitial water to the overlying water column.

Four broad pathways can be described for the release of P from sediment particles (upward arrows B and C in Fig. 12.1):

- changes in solution chemistry (particularly PO_4^{3-} concentration and pH) which result in shifts in surface P–solution P equilibrium;
- release of P from sedimentary biomass through either active excretion or cell death and subsequent mineralization;
- hydrolysis of sedimentary organic P compounds;
- reduction of Fe oxides and (oxyhydroxides) in anaerobic sediments.

The mechanism of release of P will obviously depend on the way that the P is bound to the sediment and will be influenced by the physico-chemical environment of the sediment.

For the P to remobilize from the pore water to the overlying water column, it must move from the sediment either by molecular diffusion or by mass transfer of the pore water. Molecular diffusion is a slow process and consequently is probably not important in the exchange of P from sediment pore water to the overlying water column (Scarlatos, 1997). Mass transfer occurs when the sediment–water interface is physically disturbed. The disturbance may be caused by high flow events which result in sediment bed scouring, anthropogenic causes (e.g. dredging), ebullition of gases (such as methane and CO_2) from the sediment entraining pore water, disturbance of sediments by bottom feeding organisms (e.g. carp), or by bio-irrigation caused by sediment dwelling invertebrates (e.g. tube worms).

Shifts in surface phosphorus – solution phosphorus equilibria

Desorbable-phosphorus

As noted in the first part of this chapter, P associates with sediment surfaces as either inner or outer sphere complexes, or it (co)precipitates from solution. In all cases the system is in some form of equilibrium:

$$X - P_{solid} \longleftrightarrow X_{solid\ or\ soln} + P_{soln}$$

where X represents the mineral surface. For desorption reactions, X will remain in the solid phase, in dissolution reactions X will go into solution. Therefore as the concentration of P in solution (P_{soln}) decreases (e.g. by biotic uptake) then the equilibrium reaction will shift to the right (Le Chatelier's principle: Chang, 1977) resulting in a net release of P from the sediment. The position of the equilibrium will depend on the thermodynamics of the reaction. For example, for a dissolution reaction of the type

$$XY_{solid} \longleftrightarrow X_{soln} + Y_{soln}$$

One can define a thermodynamic solubility product K_{sp} such that

$$K_{sp} = a_x * a_y$$

(Stumm and Morgan, 1996); that is, K_{sp} is equal to the product of the activity of X and Y in solution. In solutions of low ionic strength, activity is equivalent to concentration (Chang, 1977) so that

$$K_{sp} \approx [X_{soln}]\,[Y_{soln}]$$

For highly insoluble mineral phases such as apatite ($K_{sp} = 10^{-114}$: Stumm and Morgan, 1996), little or no dissolution will occur. However, for minerals with a higher solubility constant (e.g. $CaHPO_4$ $K_{sp} \approx 10^{-7}$: Stumm and Morgan, 1996) or where the P is bound as an outer sphere complex, the surface PO_4^{3-} may be readily desorbable. This readily desorbable P represents an important source of P to the overlying water column and hence is of ecological significance. In bulk sediments the size of the 'easily desorbable' or 'biologically active' pool of P has been estimated in a number of ways, including by chemical extraction or by using Fe-impregnated filter-papers ('iron strips'). In chemical extraction schemes, the easily desorbable P fraction is determined from the amount of P displaced from a sediment sample by a relatively concentrated salt solution such as 1 M, NH_4Cl (Hieltjes and Lijklema, 1980) or 1M $MgCl_2$ (Ruttenberg, 1992). In the iron-strip method, filter-papers impregnated with Fe oxides are equilibrated with slurry of the soil or sediment of interest (van der Zee *et al.*, 1987; House *et al.*, 1995). The supported Fe oxides act as an 'infinite' sink for any P desorbed from the sediment or soil (van der Zee *et al.*, 1987; Oliver, 1993). The bioavailable P is determined by first recovering the P from the iron-strip by extraction with sulphuric acid and then determining the concentration of soluble reactive P in the extractant.

Estimates of the fraction of P that is readily desorbable from sediments vary markedly. Baldwin (1996b) showed that repeated extractions of sediments from a number of wetlands and other water bodies with 1 M $MgCl_2$ removed < 2% of the total P found in the sediment. Similarly, Ruttenberg (1992) found that < 10% of the total P from a marine sediment could be extracted with $MgCl_2$. In contrast, Oliver (1993) showed that up to 80% of the total P in some suspended sediments from rivers in the Murray–Darling basin in Australia was readily desorbable (assessed with the 'Fe strip' method). Typical estimates for suspended sediments from this area ranged from 10 to 30% (R. Oliver, CRC for Freshwater Ecology, 1997, personal communication).

pH

Changes in pH can also affect the distribution of P between a sediment particle and solution. pH strongly influences the chemical reactivity of the sediment constituents causing a positive correlation between PO_4^{3-} release and pH (Boström *et al.*, 1982; Drake and Heaney, 1987). This is most likely due to the effect of pH on surface charge and the sorption capacity of metal oxides (Moore *et al.*, 1991). Andersen (1975) found that the P release from sediments increased as pH increased up to *c.* 9.5, and then decreased above this. This decrease was attributed to both desorption of P from Fe oxides and hydroxides by exchange with OH^- groups and the precipitation of P as hydroxyapatite. In support of this hydroxyapatite hypothesis, Driscoll *et al.* (1993) reported an increase in P release with decreasing pH in the highly calcareous Onondaga Lake.

Drake and Heaney (1987) report that whole water column values of pH cannot be utilized to determine pH mediated release of P from sediments, as the buffering capacity of the sediments generates a steep pH gradient near the sediment–water interface. This buffering capacity is a result of the acidic products of bacterial metabolism, resisting any increase in pH in the water column (Marsden, 1989). This phenomenon may also limit the effect of high pH on the desorption of P from Fe or the precipitation of P with carbonate to the sediment surface. However, the steep pH gradients near the sediment–water interface may be destroyed by turbulence (Drake and Heaney, 1987) or bioturbation, that is, the reworking/re-suspension of bottom sediments by organisms (Marinelli and Boudreau, 1996). This may significantly affect the pH-mediated P exchange with the water column by re-suspending P-rich sediment into high pH water with a subsequent release of P. However, turbulence will also result in a lowering of pH within the epilimnion due to CO_2 diffusion across the air–water interface, which may then reduce the rate of P release from the sediments (Drake and Heaney, 1987).

Release of phosphorus from sedimentary biomass

Sediments are often thought of as inert mixtures of minerals and recalcitrant organic matter. However, sediments contain a rich mixture of living organisms which may include macrofauna such as tube worms and other burrowing invertebrates, macro and micro algae and, most importantly, microorganisms, particularly bacteria. Indeed, using simple calculations it is possible to estimate that 1 m^3 of fine organic sediment would contain *c.* 500 g of living bacteria. Therefore, a substantial proportion of available sediment P may be bound in bacterial biomass (Boström *et al.*, 1988; Gächter and Meyer, 1993; Waara *et al.*, 1993). Gächter and Meyer (1993) report that in oligotrophic lakes, the P incorporated in benthic bacterial biomass may exceed the annual bioavailable P deposition several times, while in eutrophic lakes it may be equivalent to 1 year's deposition. This pool of P can be mobilized either through direct excretion of P by some microorganisms or after cell death.

Facultative anaerobic bacteria may take up and store P as inorganic polyphosphates under aerobic conditions and release this luxury store of P as PO_4^{3-} under anaerobic conditions (Wentzel *et al.*, 1986; Gächter *et al.*, 1988; Davelaar, 1993; Gächter and Meyer, 1993). These polyphosphate bacteria are reported to comprise up to 20% of their dry weight as P under oxic conditions. It has been postulated that under anaerobic conditions these bacteria take up low molecular weight fatty acids, particularly acetate, which is first protonated outside the bacterial cell (Wentzel *et al.*, 1986). This acetic acid is taken up by the cell; the acetate is metabolized in the synthesis of poly-β-hydroxybutyrate; the excess proton associated with acetic acid is then excluded from the cell as a protonated PO_4^{3-} ion.

While poly-P granules have been identified in a number of natural sediments (Hupfer *et al.*, 1995; Baldwin, 1996b) the importance of this mechanism to remobilization of P from sediments is still unclear. For example, Waara *et al.* (1993) tested the P release hypothesis on *Pseudomonas* species and found no dramatic flux of P from the bacteria under anoxic conditions. They further stated, however, that they did not have enough evidence to reject the hypothesis. The explanation that they gave for this was that bacteria present in the natural environment might behave differently from their test species.

Decomposition of dead organic material (mineralization) can result in the mobilization of P from sediments (Driscoll *et al.*, 1993). Mobilization is principally a result of the liberation of mineral nutrients during the decomposition processes. Mobilization may occur simply as a result of cell lysis, for example following sediment desiccation (Schoenberg and Oliver 1988; Qiu and

McComb, 1995), as a result of bacterial activity (Gächter and Meyer, 1993) or through excretion by detritivores (Schaus *et al.*, 1997). Indeed Hudson *et al.* (1999) suggest that P regeneration through the ingestion/excretion of suspended particles (including algae and organic detritus) is the main source of P to lake planktonic communities.

In addition to direct release, mineralization has a number of other effects that can impact on P mobilization. Mineralization of dead organic matter by sedimentary bacteria and other detritivores utilizes oxygen, consequently depleting O_2 concentrations in the sediments and creating a sediment O_2 demand (Gächter and Meyer, 1993; Waara *et al.*, 1993). When combined with the subsequent respiration of anaerobic bacteria and their requirement for electron acceptors, the ultimate consequence is a strongly reducing environment. Furthermore, the release of acidic fermentation products also lowers the pH of sediments (Marsden, 1989). The resulting CO_2-rich hypolimnetic layers and sediments will, to some extent, dissolve precipitated calcite during sedimentation, solubilizing any co-precipitated P (Boström *et al.*, 1988).

Hydrolysis of organic-phosphorus compounds in the sediment

Organic-P compounds (compounds which contain P covalently bound to organic molecules, usually through PO_4^{3-} ester bonds) are ubiquitous in aquatic environments (Herbes *et al.*, 1975; De Groot, 1990). Although the exact nature of many of these compounds is largely unknown (see e.g. Baldwin, 1998) compounds such as DNA and RNA (Karl and Bailiff, 1989; and references therein), c-AMP (Franko and Wetzel, 1982), inositol hexaphosphate (Minear *et al.*, 1988; Golterman, 1993) and polyphosphates (Hupfer *et al.*, 1995; Baldwin, 1996b) have been isolated from aquatic environments. Ortho-phosphate has also been shown to be incorporated into high molecular weight humic acids (e.g. Jones *et al.*, 1988). Organic-P compounds can be hydrolysed through either biotically or abiotically mediated processes.

Extracellular biotically mediated hydrolysis of organic-P substrates occurs through the action of exo-enzymes such as phosphatase (Jansson *et al.*, 1988) and phytase (Herbes *et al.*, 1975). These enzymes are produced by aquatic microbiota and released into the immediate environment surrounding the organism. The enzymes then catalyse the hydrolysis of the substrate to produce free PO_4^{3-}.

Other bacterial exudates may also play a role in P release. Some sediment and soil bacteria are known to produce compounds that have an extremely high affinity for Fe, for example siderophores (Fett *et al.*, 1998). The siderophores can mobilize Fe from mineral phases (e.g. Hersman *et al.*, 1995) with concurrent release of any associated P.

The hydrolysis of organic-P substrates can also be facilitated by mineral phases. It has been shown that Fe (amorphous Fe oxyhydroxide, goethite and haematite), Mn (α- and β-MnO_2 and akhtetenskite) Al (α- and β-Al_2O_3) and titanium (anatase and rutile) minerals can all facilitate the hydrolysis of model organic-P compounds (Baldwin *et al.*, 1995). It is assumed that the chemical mechanisms responsible for facilitated hydrolysis of organic-P compounds by mineral phases is the same as seen for exoenzymes (Jones *et al.*, 1984). The organic-P molecule binds to the surface of the mineral through a metal centre. On coordination, electron density is drawn away from the P atom towards the metal centre. The electron deficient P atom is then susceptible to nucleophilic attack by an hydroxyl group. It is still unclear whether or not the hydroxyl group comes from the mineral surface or from solution (Coleman, 1996). As an aside, it has also been demonstrated that the hydrolysis of pesticides can also be facilitated by mineral phases (Torrents and Stone, 1991).

It has been suggested that the hydrolysis of organic-P compounds, particularly inositol hexaphosphates in sediments, may be an important mechanism for the release of P from sediments (Golterman, 1995). However, the importance of hydrolytic reactions relative to other sedimentary processes is still open to question.

Reduction of iron oxides and (oxyhydroxides) in anaerobic sediments

As a consequence of oxygen consumption through microbial activity, sediments can become depleted of O_2. It has long been recognized that significant release of P from the sediments can occur under these anoxic aquatic conditions (Einsele, 1936; Mortimer, 1941). The P release was often accompanied by the co-release of dissolved Fe^{2+}. There is still some conjecture about the mechanisms responsible for P release from the anaerobic part of the sediments; however, it is becoming increasingly obvious that the microbiota are responsible for a significant proportion of this release (e.g. Roden and Edmonds, 1997; Mitchell and Baldwin, 1998). A number of mechanisms have been postulated for this P release which directly involve anaerobic bacteria.

Iron reduction

Iron-reducing bacteria use Fe^{3+} oxides and oxyhydroxides as the terminal electron acceptor for anaerobic respiration. In other words these bacteria catalyse the reduction of solid ferric minerals to dissolved ferrous ions. Any PO_4^{3-} that is associated with the solid mineral surface will be released when the surface is reduced. It has been shown by Lovley *et al.* (1991) that Fe^{3+} reduction in sediments is almost entirely mediated by bacteria. However, recent work by Roden and Edmonds (1997) suggests that much of the P released by the reduction of Fe^{3+} phases may form insoluble complexes with Fe^{2+} (e.g. vivianite). Roden and Edmonds (1997) contend that P can be displaced from these insoluble ferrous phosphates by sulphide (S^{2-}) produced as a consequence of SO_4^{2-} reduction (see below).

Sulphate reduction

Sulphate has been linked with the release of P from many aquatic systems (e.g. Boström *et al.*, 1988; Caraco *et al.*, 1989). Sulphate reducing bacteria use SO_4^{2-} as the terminal electron acceptor for anaerobic respiration. The respiratory end product of this reaction is hydrogen sulphide. Sulphide is a strong enough reducing agent to facilitate the reduction of solid ferric minerals to dissolved ferrous ions with concurrent P release (Boström *et al.*, 1988). This reaction is favoured by the insolubility of one of the reaction products, iron sulphide. As noted above, it has been suggested that S^{2-} can also displace P from insoluble Fe^{2+} phases (Roden and Edmonds, 1997); again, the reaction is favoured by the insolubility of the reaction product, FeS. Golterman (1995) has questioned the validity of this hypothesis. Using P adsorbed to a large excess of synthetic iron oxyhydroxide, he showed that a large excess of H_2S was required before P was released. Presumably PO_4^{3-} liberated from the sediment surface by reduction of Fe by sulphide was re-adsorbing to non-reduced ferric sites. A similar process may occur in anaerobic sediments if not all the Fe binding sites have been reduced or if other binding sites (e.g. Al or Ca centres) are available. Significant adsorption of P to anoxic sediments has been found in both marine (Krom and Berner, 1980) and freshwater (e.g. Baldwin *et al.*, 2000) systems.

Transfer of phosphorus from interstitial to surface water

The movement of P from the interstitial to the overlying water can occur either through molecular diffusion or by the mass transfer of pore water. Molecular diffusion is a slow process and consequently probably is not important in the exchange of P from sediment pore water to the overlying water column (Scarlatos, 1997). Mass transfer occurs when the sediment–water interface is physically disturbed.

Sediment–Pollutant Interactions

Pollutants often can form strong associations with aquatic sediments, so that in order to explore the fate of a pollutant in an aquatic ecosystem it is often necessary to understand pollutant–sediment interactions. Sediment deposition can as act as an important sink for pollutants in aquatic environments, but sorption of a pollutant to sediment particles can also result in the concentration of the pollutant. Therefore, bottom sediments often contain significantly higher concentrations of a given pollutant than is dissolved in the overlying water column. For example, the concentration of P in natural waters is often in the range of 10–50 μg kg^{-1}, P concentrations in depositional sediment can be in the range of 1–10 g kg^{-1} (Baldwin, 1996b). As seen for P, biotic and abiotic processes occurring within sediment deposits may result in the re-release of the pollutant. Often the rate of release back into the aquatic ecosystem is slow relative to episodic inputs of the pollutant from point and non-point sources, prompting Jannsen and Carpenter (1999) to characterize pollutant input into ecosystems from the sediment as a 'slow variable'. Therefore, sorption on to sediments followed by sediment deposition and then slow re-release of the pollutant back into the water column can be seen as buffering the potential impact of episodic pollution events. However, this buffering also serves to prolong the impact of a given pollutant. For example, removal of P inputs into lakes often does not result in a noticeable decrease in P concentrations in lake water, because of re-release of P from the sediments (Sas, 1989; Cooke *et al.*, 1993). Therefore, sediments can be seen as temporally shifting pollutant impacts. Interaction with sediments can also shift the impact of a pollutant spatially. A pollutant can enter a waterway, become attached to sediment particles and be transferred a long distance downstream before the sediments are deposited and re-release of the pollutant occurs.

Conclusion

In this chapter we have attempted to show that the interactions between sediment and a given pollutant (in this case P) can be complex. More often than not a number of competing pathways are available for the sorption, transport, transformation and re-release of pollutants. The importance of any given process will depend on a number of interacting factors including the nature of the sediment (composition, mineralogy, particle coatings, drying history, etc.), the microbial community structure and activity within the sediment, and the chemical composition of the water both overlying and within the sediment. Nevertheless, regardless of specific mechanisms it can be seen that both suspended and bottom sediments play an important role in influencing the behaviour of pollutants in the aquatic environment. The importance of the individual pathways to the fate of a given pollutant in any aquatic environment will no doubt vary both spatially and temporally and is, as yet, largely unresolved. This is not surprising given the complexity of the processes involved. The information in this chapter will hopefully contribute to some clarity at both the local and the global level.

References

Allen, H.E. (1995) *Metal Contaminated Aquatic Sediments.* Ann Arbor Press, Ann Arbor, Michigan.

Andersen, J.M. (1975) Influence of pH on release of phosphorus from lake sediments. *Archiv für Hydrobiologie* 76, 411–419.

Baldwin, D.S. (1996a) Effects of exposure to air and subsequent drying on the phosphate sorption characteristics of sediments from a eutrophic reservoir. *Limnology and Oceanography* 41, 1725–1732.

Baldwin, D.S. (1996b) The phosphorus composition of a diverse series of Australian sediments. *Hydrobiologia* 335, 63–73.

Baldwin, D.S. (1998) Reactive 'organic' phosphorus revisited. *Water Research* 32, 2265–2270.

Baldwin, D.S., Beattie, J., Coleman, L. and Jones, D. (1995) Phosphate ester hydrolysis facilitated by mineral phases. *Environment Science and Technology* 29, 1706–1709.

Baldwin, D.S., Mitchell, A.M. and Rees, G. (2000) The effects of *in situ* drying on sediment–phosphate interactions in sediments from an old wetland. *Hydrobiologia* 431, 3–12.

Barrow, N.J. (1983) A mechanistic model for describing the sorption and desorption of phosphate by soil. *Journal of Soil Science* 34, 733–750.

Barrow, N.J. (1999) The four laws of soil chemistry: the Leeper lecture 1998. *Australian Journal of Soil Science* 37, 787–829.

Birch, P.B. (1982) Phosphorus export from coastal plain drainage into the Peel-Harvey estuarine system of Western Australia. *Australian Journal of Marine and Freshwater Research* 33, 23–32.

Boström, B., Jansson, M. and Forsberg, C. (1982) Phosphorus release from lake sediments. *Archiv für Hydrobiologie Beiheft Ergebnisse Limnologie* 18, 5–59.

Boström, B., Andersen, J.M., Fleischer, S. and Jansson, M. (1988) Exchange of phosphorus across the sediment-water interface. *Hydrobiologia* 170, 229–244.

Brinkman, A.G. (1993) A double-layer model for ion adsorption onto metal oxides, applied to experimental data and to natural sediments of Lake Veluwe, The Netherlands. *Hydrobiologia* 253, 31–45.

Caraco, N.F., Cole, J.J. and Likens, G.E. (1989) Evidence for sulfate-controlled phosphorus release from sediments from aquatic systems. *Nature* 341, 316–317.

Carpenter, S.R., Caraco, N.F., Correll, D.L., Howarth, R.W., Sharpley, A.N. and Smith, V.H. (1998) Non point pollution of surface waters with phosphorus and nitrogen. *Ecological Applications* 8, 559–568.

Chang, R. (1977) *Physical Chemistry with Application to Biological Systems*. Macmillan Publishing, New York.

Coleman, L. (1996) The facilitated hydrolysis of phosphate esters by mineral phases. PhD thesis, University of Sydney, Sydney, Australia.

Cooke, G.B., Welch, E.B., Peterson, S.A. and Newroth, P.R. (1993) *Restoration and Management of Lakes and Reservoirs*. Lewis Publishing, Boca Raton, Florida.

Davelaar, D. (1993) Ecological significance of bacterial polyphosphate metabolism in sediments. *Hydrobiologia* 253, 179–192.

De Groot, C. (1990) Some remarks on the presence of organic phosphates in sediments. *Hydrobiologia* 207, 303–309.

De Groot, C. and van Wijck, C. (1993) The impact of desiccation of a freshwater marsh (Garcines Nord, Camargue, France) on sediment–water–vegetation interactions. Part one: the sediment chemistry. *Hydrobiologia* 252, 83–94.

Detenbeck, N.E. and Brezonik, P.L. (1991) Phosphorus sorption by sediments from a soft-water seepage lake. 1. An evaluation of kinetic and equilibrium models. *Environmental Science and Technology* 25, 396–403.

Drake, J.C. and Heaney, S.I. (1987) Occurrence of phosphorus and its potential remobilization in the littoral sediments of a productive English lake. *Freshwater Biology* 17, 513–523.

Driscoll, C.T., Effler, S.W., Auer, M.T., Doerr, S.M. and Penn, M.R. (1993) Supply of phosphorus to the water column of a productive hardwater lake: controlling mechanisms and management considerations. *Hydrobiologia* 253, 61–72.

Einsele, W. (1936) Uber die beziehungen des eisenkreislaufs zum phosphatkreislauf im eutrophen See. *Archiv für Hydrobiologie* 29, 664–686.

Emerson, S. (1976) Early diagenesis in anaerobic lake sediments: chemical equilibria in interstitial waters. *Geochimica et Cosmochimica Acta* 40, 925–934.

Emerson, S. and Widmer, G. (1978) Early diagenesis in anaerobic sediments – II. Thermodynamic and kinetic factors controlling the formation of iron phosphate. *Geochimica et Cosmochimica Acta* 42, 1307–1316.

Enell, M. and Lofgren, S. (1988) Phosphorus in interstitial water: methods and dynamics. *Hydrobiologia* 170, 103–132.

Fett, J.P., LeVier, K. and Guerinot, M.L. (1998) Soil microorganisms and iron uptake by higher plants. *Metal Ions in Biological Systems* 35, 187–214.

Franko, D.A. and Wetzel, R.G. (1982) The isolation of cyclic adenosine 3':5' monophosphate (c-AMP) from lakes of different trophic status: correlation with planktonic variables. *Limnology and Oceanography* 27, 27–38.

Gächter, R. and Meyer, J.S. (1993) The role of microorganisms in mobilization and fixation of phosphorus in sediments. *Hydrobiologia* 253, 103–121.

Gächter, R., Meyer, J. and Mares, A. (1988) Contribution of bacteria to the release and fixation of phosphorus in lake sediments. *Limnology and Oceanography* 33, 1542–1558.
Goldberg, S. and Sposito, G. (1985) On the mechanism of specific phosphate adsorption by hydroxylated mineral surfaces: a review. *Communications in Soil Science and Plant Analysis* 16, 801–821.
Golterman, H. (1993) On the presence of organic phosphate in some Camargue sediments: evidence for the importance of phytate. *Hydrobiologia* 252, 117–126.
Golterman, H. (1995) The role of the iron hydroxide-phosphate-sulfide system in the phosphate exchange between sediments and overlying water. *Hydrobiologia* 297, 43–54.
Gu, B., Schmitt, J., Chen, Z. and McCarthy, J.F. (1994) Adsorption and desorption of natural organic matter on iron oxide: Mechanism and models. *Environmental Science and Technology* 28, 38–46.
Harris, G. (1999) Comparison of the biogeochemistry of lakes and estuaries: ecosystem processes, functional groups, hysteresis effects and interactions between macro- and microbiology. *Marine and Freshwater Research* 50, 791–811.
Herbes, S.E., Allen, H.E. and Mancy, K.H. (1975) Enzymatic characterization of soluble organic phosphorus in lake water. *Science* 187, 432–434.
Hersman, L., Lloyd, T. and Sposito, G. (1995) Siderophore-promoted dissolution of haematite. *Geochimica et Cosmochimica Acta* 59, 3327–3330.
Hieltjes, A.H.M. and Lijklema, L. (1980) Fractionation of inorganic phosphates in calcareous sediments. *Journal of Environmental Quality* 9, 405–407.
Holtan, H., Kamp-Nielsen, L. and Stuanes, A.O. (1988) Phosphorus in soil, water and sediment: an overview. *Hydrobiologia* 170, 19–34.
House, W.A. (1999) The physico-chemical conditions for the precipitation of phosphate with calcium. *Environmental Technology* 20, 727–733.
House, W.A., Dennison, F.H. and Armitage, P.D. (1995) Comparison of the uptake of inorganic phosphorus to a suspended and stream bed sediment. *Water Research* 29, 767–779.
Hudson, J.J., Taylor, W.D. and Schindler, D.W. (1999) Planktonic nutrient generation and cycling efficiency in temperate lakes. *Nature* 400, 659–661.
Hupfer, M., Gächter, R. and Rüegger, H. (1995) Polyphosphates in lake sediments: ^{31}P NMR spectroscopy as a tool for its identification. *Limnology and Oceanography* 40, 610–617.
Jacoby, J.M., Lynch, D.D., Welch, E.B. and Perkins, M.A. (1982) Internal loading of a shallow eutrophic lake. *Water Research* 16, 911–919.
Jannsen, W. and Carpenter, S.R. (1999) Managing the resilience of lakes: a multi-agent modelling approach. *Conservation Ecology* 3, 15 (online) http://www.consecol.org/vol3/iss2/art15
Jansson, M., Olsson, H. and Pettersson, K. (1988) Phosphatases; origin, characteristics and function in lakes. *Hydrobiologia* 170, 157–175.
Jones, D.R., Lindoy, L.F. and Sargeson, A.M. (1984) Enhanced base hydrolysis of coordinated phosphate esters: The reactivity of an unusual cobalt (III) amine dimer. *Journal of the American Chemical Society* 106, 7807
Jones, R.I., Salonene, K. and DeHaan, H. (1988) Phosphorus transformations in the epilimnion of humic lakes: abiotic interactions between dissolved humic materials and phosphate. *Freshwater Biology* 19, 357–369.
Jordan, T.E., Correl, D.L. and Weller, D.E. (1997) Effects of agriculture on discharge of nutrient from coastal watersheds of Cheseapeake Bay. *Journal of Environmental Quality* 26/3, 836–848.
Karl, D.M. and Bailiff, M. (1989) The measurement and distribution of dissolved nucleic acids in aquatic environments. *Limnology and Oceanography* 34, 543–558.
Krom, M.D. and Berner, R.A. (1980) Adsorption of phosphate in anoxic marine sediments. *Limnology and Oceanography* 25, 797–806.
Lijklema, L. (1980) Interaction of orthophosphate with iron(III) and aluminium hydroxides. *Environmental Science and Technology* 14, 537–541.
Lovley, D.R., Phillips, E.J.P. and Lonergan, D.J. (1991) Enzymatic versus non-enzymatic mechanisms for Fe(III) reduction in aquatic sediments. *Environmental Science and Technology* 25, 1062–1067.
Marinelli, R.L. and Boudreau, B.P. (1996) An experimental and modelling study of pH and related solutes in an irrigated anoxic coastal sediment. *Journal of Marine Research* 54, 939–966.
Marsden, M.W. (1989) Lake restoration by reducing external phosphorus loading: the influence of sediment phosphorus release. *Freshwater Biology* 21, 139–162.

Miles, C.J., Tuschall, J.R. and Brezonik, P.L. (1983) Isolation of aquatic humus with diethylaminoethylcellulose. *Analytical Chemistry* 55, 410–411.

Minear, R.A., Seagers, J.E., Elwood, J.W. and Mulholland, P.J. (1988) Separation of inositol phosphates by high performance ion-exchange chromatography. *Analyst* 113, 645–649.

Mitchell, A. and Baldwin, D.S. (1998) Effects of desiccation/oxidation on the potential for bacterially mediated P release from sediments. *Limnology and Oceanography* 43, 481–487.

Moore, P.A., Reddy, K.R. and Graetz, D.A. (1991) Phosphorus geochemistry in the sediment-water column of a hypereutrophic lake. *Journal of Environmental Quality* 20, 869–875.

Mortimer, C. (1941) The exchange of dissolved substances between mud and water in lakes. 1 and 2. *Journal of Ecology* 29, 280–329.

Norrish, K. and Rosser, H. (1983) Mineral phosphate. In: *Soils: an Australian Viewpoint.* Division of Soils CSIRO, Melbourne, pp. 335–361.

Olila, O.G. and Reddy, K.R. (1997) Influence of redox potential on phosphate-uptake by sediments in two sub-tropical eutrophic lakes. *Hydrobiologia* 345, 45–47.

Oliver, R.L. (1993) Phosphorus dynamics and bioavailability in turbid waters. In: Robards, K., Helliwell, S. and Antolovich, M. (eds) *Proceedings of the National Workshop on Phosphorus in Australian Freshwaters.* Land and Water Resources Research and Development Corporation Occasional Paper No 03/93. LWRDC, Canberra, pp. 93–103.

Olley, J.M. (1996) Sources of suspended sediment and phosphorus to the Murrumbidgee River. In: Caitcheon, G.G., Donnelly, T., Olive, L.J., Olley, J.M., Murray, A.S., Short, D. and Wallbrink, P.J. (eds) Consultancy report No. 95–32

Olley, J.M., Murray, A.S., Mackenzie, D.H. and Edwards, K. (1993) Identifying sediment sources in a gullied catchment using natural and anthropogenic radioactivity. *Water Resources Research* 29/4, 1037–1043.

Pettersson, K. (1998) Mechanisms for internal loading of phosphorus in lakes. *Hydrobiologia* 374, 21–25.

Pettersson, K., Boström, B. and Jacobsen, O. (1988) Phosphorus in sediments – speciation and analysis. *Hydrobiologia* 170, 91–101.

Postma, D. (1981) Formation of siderite and vivianite and the porewater composition of a recent bog sediment in Denmark. *Chemical Geology* 31, 225–244.

Qiu, S. and McComb, A.J. (1994) Effects of oxygen concentration on phosphorus release from reflooded air-dried wetland sediments. *Australian Journal of Marine and Freshwater Research* 45, 1319–1328.

Qiu, S. and McComb, A.J. (1995) Planktonic and microbial contributions to phosphorus release from fresh and air-dried sediments. *Freshwater Research* 46, 1039–1045.

Richardson, A.E. (1994) Soil microorganisms and phosphorus availability. In: Pankhurst, C.F., Doube, B.M., Gupta, V.V.S.R. and Grace, P.R. (eds) *Soil Biota: Management in Sustainable Farming Systems.* CSIRO, Melbourne, Australia.

Roden, E.E. and Edmonds, J.W. (1997) Phosphate mobilization in iron-rich anaerobic sediments: Microbial Fe(III) oxide reduction versus iron-sulfide formation. *Archiv für Hydrobiologie* 139, 347–378.

Ruttenberg, K. (1992) Development of a sequential extraction method for different forms of phosphorus in marine sediments. *Limnology and Oceanography* 37, 1460–1482.

Sah, R.N., Mikkelsen, D.S. and Hafez, A.A. (1989) Phosphorus behaviour in flooded-drained soils. 2. Iron transformations and phosphorus sorption. *Soil Science Society of America Journal* 53, 1723–1729.

Sas, H. (1989) *Lake Restoration by Reduction of Nutrient Loading: Expectations, Experiences, Extrapolations.* Academia Verlag, Richarz, St Augustin, Germany.

Scarlatos, P.D. (1997) Experiments on water-sediment nutrient partitioning under turbulent, shear and diffusive conditions. *Water Air and Soil Pollution* 99, 411–425.

Schaus, M.H., Vanni, M.J., Wissing, T.E., Bremigan, M.T., Garvey, J.E. and Stein, R.A. (1997) Nitrogen and phosphorus excretion by detritivorous gizzard shad in a reservoir ecosystem. *Limnology and Oceanography* 42, 1386–1397.

Schoenberg, S.A. and Oliver, J.D. (1988) Temporal dynamics and spatial variation of algae in relation to hydrology and sediment characteriztics in the Okefenokee Swamp, Georgia. *Hydrobiologia* 162, 123–133.

Schwertmann, U. and Cornell, R.M. (1991) *Iron Oxides in the Laboratory.* VCH, Weinheim, Germany.

Sharpley, A.N., Hedley, M.N.J., Sibbesen, E., Hillbricht-Ilkowska, A., House, W.A. and Ryszkowski, L. (1995) Phosphorus transfers from terrestrial to aquatic ecosystems. In: Tiessen, H. (ed.) *Phosphorus in the Global Environment, Transfers, Cycles and Management.* John Wiley & Sons, Chichester, UK.

Sherwood, B.A., Sager, S.L. and Holland, H.D. (1987) Phosphorus in foraminiferal sediments from North Atlantic Ridge cores and in pure limestones. *Geochimica et Cosmochimica Acta* 51, 1861–1866.

Sibanda, H.M. and Young, S.D. (1986) Competitive adsorption of humus acids and phosphate on goethite, gibbsite and two tropical soils. *Journal of Soil Science* 37, 197–204.

Stone, M. and English, M.C. (1993) Geochemical composition, phosphorus speciation and mass transport of fine-grained sediment in two Lake Erie tributaries. *Hydrobiologia* 253, 17–29.

Strauss, R., Brümmer, G.W. and Barrow, N.J. (1997) Effects of crystallinity of goethite: II Rates of sorption and desorption of phosphate. *European Journal of Soil Science* 48, 101–114.

Stumm, W. and Morgan, J.J. (1996) *Aquatic Chemistry*, 3rd edn. John Wiley & Sons, New York.

Tipping, E. (1981) The adsorption of aquatic humic substances by iron oxides. *Geochimica et Cosmochimica Acta* 45, 191–199.

Torrents, A. and Stone, A.T. (1991) Hydrolysis of phenyl picolinate at the mineral/water interface. *Environmental Science and Technology* 25, 143–149.

Twinch, A.J. (1987) Phosphate exchange characteristics of wet and dried sediment samples from a hypertrophic reservoir: implications for the measurement of sediment phosphorus status. *Water Research* 21, 1225–1230.

Van Berkel, J. and Beckett, R. (1997) Estimating the effect of particle surface coatings on the adsorption of orthophosphate using sedimentation field-flow fractionation. *Journal of Liquid Chromatography and Related Technology* 20, 2647–2667.

Van der Zee, S.E.A.T.M., Fokkink, L.G.J. and van Riemsdjik, W.H. (1987) A new technique for assessment of reversibly adsorbed phosphate. *Soil Science Society of America Journal* 51, 599–604.

Volk, C.J., Volk, C.B. and Kaplan, L.A. (1997) Chemical composition of biodegradable dissolved organic matter in streamwater. *Limnology and Oceanography* 42, 39–44.

Waara, T., Jansson, M. and Petterson, K. (1993) Phosphorus composition and release in sediment bacteria of the genus *Psuedomonas* during aerobic an anaerobic conditions. *Hydrobiologia* 253, 131–140.

Wallbrink, P.J., Murray, A.S., Olley, J.M. and Olive, L. (1998) Determining sources and transit times of suspended sediment within the Murrumbidgee tributaries, NSW, Australia, using fallout ^{137}Cs and ^{210}Pb. *Water Resources Research* 34/4, 879–887.

Wentzel, M.C., Lotter, L.H., Loewenthal, R.F. and Marais, G.V.R. (1986) Metabolic behaviour of *Acinetobacter* spp. in enhanced biological phosphorus removal – a biochemical model. *Water South Africa* 124, 209–224.

Yao, W. and Millero, F.J. (1996) Adsorption of phosphate on manganese dioxide in seawater. *Environmental Science and Technology* 30, 536–541.

13 Quantifying Sediment and Nutrient Pathways within Danish Agricultural Catchments

Brian Kronvang, Ruth Grant, Anker R. Laubel and Morten L. Pedersen
National Environmental Research Institute, Department of Streams and Riparian Areas, Vejlsøvej 25, DK-8600 Silkeborg, Denmark

Key words: arable land, bank erosion, catchment budgets, groundwater, leaching, nitrogen, overland flow, pathways, phosphorus, soil erosion, suspended sediment, tile drainage water

Introduction

The current intensive agricultural production systems are net consumers of N and P of which part is left in arable soils with a risk for later losses to groundwater and surface water via different hydrological pathways. Moreover, agricultural production has in many areas altered the hydrological cycle because of intensive soil preparation, artificial draining and irrigation schemes. The alteration of the hydrological cycle in catchments has changed the strength and importance of different hydrological pathways. As an example, the artificial tile drainage of sloping fields has reduced the occurrence of overland flow but has at the same time increased the flow component from the unsaturated zone. This has increased the risk of loss of sediment and sediment-associated nutrients owing to the impact of preferential flow reaching tile drains (Grant *et al.*, 1996; Haygarth *et al.*, 1998; Laubel *et al.*, 1999a). Draining of wetlands and lakes, and straightening and channelization of watercourses have also decreased the hydraulic residence time in arable catchments and hence altered the natural buffer capacity for sediment and nutrient storage and removal in catchments (e.g. Kronvang *et al.*, 1999).

Sediment and nutrients are important ecological regulators in rivers and lakes (Kristensen and Hansen, 1994). Riverine suspended sediments (SS), which mainly originate from diffuse sources, have long been monitored for engineering purposes in connection with sedimentation and siltation problems in river channels, reservoirs and harbours (e.g. Walling, 1988). Thus, excess sediment transport in river channels can seriously affect the success of trout spawning because of siltation of spawning grounds and can cause damage to the macro-invertebrate community (Platts *et al.*, 1989). A high loss of dissolved and particulate P from arable fields to surface waters can seriously affect the ecology in rivers and lakes through an increased growth of benthic and pelagic algae resulting in eutrophication (Kristensen and Hansen, 1994; Jeppesen, 1998). In the case of N, leaching to groundwater may pose a threat to human health when extracted for drinking water. It is recommended that the concentration of NO_3^- does not exceed a threshold of 50 mg NO_3^- l^{-1}. Moreover, N is also the major cause of excess algal growth and hence eutrophication in coastal and open marine waters (Kronvang *et al.*, 1993).

The changes in land use and agricultural production that have occurred during the 20th century have undoubtedly increased the loss of sediment and nutrients to surface waters (Dearing *et al.*, 1981; Smith *et al.*, 1982; Foster *et al.*, 1990; Turner and Rabalais, 1991; Larsen *et al.*, 1999). Documentation of such changes have been based on either time series from river monitoring stations or sediment core dating in the case of sediment, thus reflecting the total response from land use changes in a catchment.

Quantification of the loss of sediment and nutrients along the different active hydrological pathways at the catchment scale is seldom made and, if so, mostly for short periods which limit the possibility of investigating the influence of any changes in hydrology, land use, etc. Such investigations, with the setting up of catchment budgets for sediment and nutrients including knowledge on sediment and nutrient storage, are urgently needed.

In this chapter, we quantify the N, P and SS loss via different hydrological pathways within small agricultural catchments in Denmark. We compare this information with the total sediment and nutrient export from the catchments in order to discuss the relative importance of each pathway by setting up catchment budgets. Linking the field to the stream will make it possible for us to investigate gaps in our present knowledge on sediment and nutrient fluxes in catchments concerning pathways and sinks for sediment and nutrients.

Problem Identification

Nitrogen cycling in Danish agriculture

An important indicator for the pollution potential from agricultural production is the consumption of chemical fertilizer and production of animal manure on arable land. Thus, several authors have documented a positive relationship between the consumption of chemical fertilizer in agriculture and the riverine N concentration or transport (e.g. Kronvang, 1996). In comparison with other countries, the consumption of N in chemical fertilizer in Danish agriculture is among the highest in Europe (European Environment Agency, 1995). The total N input to Danish arable land reached a maximum in the early 1980s following a strong increase in the consumption of N in chemical fertilizer during the 1960s and 1970s (Fig. 13.1a). The consumption of N in chemical fertilizer has decreased in Denmark since the early 1990s owing to a major national environmental plan from 1987 (Action Plan on the Aquatic Environment) which was reiterated in 1991 and 1997 (Jeppesen *et al.*, 1999). The Action Plans posed several restrictions on Danish farmers, especially on their storage facilities for liquid manure, application times for liquid manure and demands for utilization of N in animal manure (Jeppesen *et al.*, 1999).

Nitrogen cycling in Danish agriculture has been monitored since 1989 as part of a national monitoring programme in six small agricultural catchments, by questionnaire surveys at the field level and measurements of N in soil water, drainage water, groundwater and stream water (Kronvang *et al.*, 1995; Kronvang *et al.*, 1996). The six catchments represent typical Danish farming systems from crop production on the predominantly loamy soils on the islands and animal production on the sandy soils in Jutland. The data collected enabled us to establish an overall N-budget for small catchments which revealed the major inputs and outputs of N from arable soils (Fig. 13.2a). The total input of N to arable soils was higher in the sandy catchments with animal production (271 kg N ha^{-1}), than in the loamy catchments with predominantly plant production (215 kg N ha^{-1}). Nitrogen removal with harvested crops was nearly equal in the two types of catchments, whereas the measured leaching was nearly twice as high in the sandy catchments (118 kg N ha^{-1}) as in the loamy catchments (69 kg N ha^{-1}) (Fig. 13.2a).

Phosphorus cycling in Danish agriculture

Several authors have shown that P accumulates in the surfaces of agricultural soils because of the net input of P from agricultural production (European Environment Agency, 1995). An increased risk of P leaching and loss with subsurface drainage water has also been documented with increased soil P status (Heckrath *et al.*, 1995; Smith *et al.*, 1998). Moreover, P application with manure can give rise to

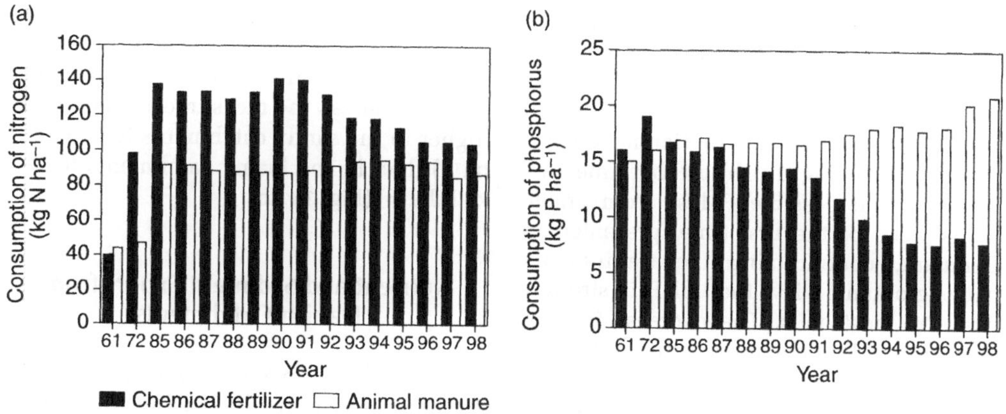

Fig. 13.1. Consumption of nitrogen (a) and phosphorus (b) in chemical fertilizer and animal manure on Danish agricultural land during the period 1961–1998.

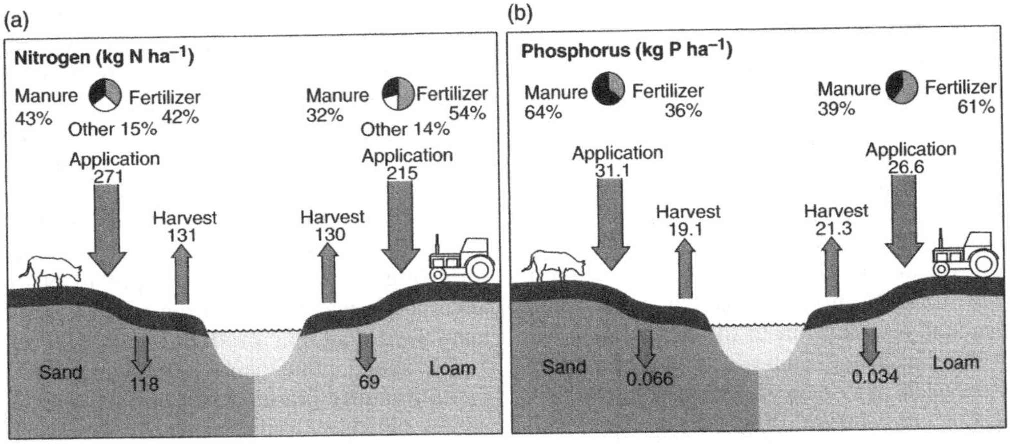

Fig. 13.2. The average input/output balance for nitrogen (a) and phosphorus (b) in three Danish sandy catchments and three Danish loamy catchments during the period 1989–1998.

excess P loss in surface runoff as documented by Smith *et al.* (1998). In contrast to N, no simple relationship exists between the consumption of P in fertilizer and animal manure and the riverine concentration or transport of P (e.g. Kronvang, 1996). The reason for this is presumably the many sources and pathways of importance for P loss to surface waters (urban wastewater, scattered dwellings, erosional sources and leaching from arable land), whereas, in most cases, N leaching from agricultural land is the predominant source.

The consumption of P in chemical and animal fertilizer in Danish agriculture is at a medium to low level in relation to the rest of Europe (European Environment Agency, 1995). The TP input to Danish arable land reached a maximum in the 1970s (Fig. 13.1b). The consumption of P in chemical fertilizer has decreased in Denmark since the late 1980s (Fig. 13.1b). At that time, evidence of the possible P loss from arable fields was, for the first time, presented in Denmark, contradicting the traditionally held agronomic view that all P is strongly fixed by most soils. The decrease in consumption of P in chemical fertilizer follows the decline found for N, but an increase was seen in the consumption of P in animal manure due to increased pig production during the 1990s in Denmark (Fig. 13.1b).

Phosphorus cycling in Danish agriculture is, as with N, monitored annually in six small agricultural catchments using the questionnaire surveys at the field level and measurements of P in soil water, drainage water, groundwater and stream water (Andersen *et al.*, 1999). The agricultural monitoring data collected enable us to establish an overall P budget for small catchments which reveals the major inputs and outputs of P from arable soils (Fig. 13.2b). The total input of P to arable soils is higher in the sandy catchments with animal production and hence production of animal manure (31.1 kg P ha^{-1}), than in the loamy catchments where crop production is predominant (26.6 kg P ha^{-1}). Phosphorus removal with harvested crops is nearly equal in the two types of catchments, whereas the measured P leaching is higher in the sandy catchments (0.063 kg P ha^{-1}) than in the loamy catchments (0.034 kg P ha^{-1}) (Fig. 13.2b).

Nutrient and Sediment Export from Danish Catchments

In Denmark, the export of N, P and SS from catchments is normally an integrated part of national or regional monitoring programmes (Iversen *et al.*, 1997). A

Nationwide Monitoring Programme of the Aquatic Environment was launched in Denmark in 1989 (Kronvang *et al.*, 1993). To illustrate the importance of hydrology and geology for nutrient losses from agricultural catchments and their impact on temporal variations, we focus on the extreme conditions for nutrient loss in Denmark by looking more closely at two classes of catchments (Table 13.1).

Nitrogen export

The N export from small agricultural catchments consists mostly of dissolved inorganic constituents and smaller amounts of organic N. Thus, NO_3^--N contributes on average 85% in loamy catchments and 78% in sandy catchments of the annual total N export. Ammonium N contributes only a small part of the total export, but is more than twice as important in the sandy catchments (5%), compared with the loamy catchments (2%). Organic N is therefore of minor importance (13–18%) when dealing with the total export from small agricultural catchments. In comparison, organic N is of much higher importance for the total N export from small undisturbed catchments where it contributes an average of 33%. The speciation of total N export from small agricultural catchments thus implies that the pathways are highly linked to the movement of water in the soil column and only to a minor degree to erosional processes.

The average annual N loss from agricultural land within small catchments was more than twice as high from loamy catchments (29.3 kg N ha^{-1}) as from sandy catchments (13.6 kg N ha^{-1}) during the 10 year monitoring period 1989–1998 (Fig. 13.3). In comparison, the average annual total N export from undisturbed catchments was 2.6 kg N ha^{-1} during the same period. The loamy catchments also experience higher inter-annual variations in export than is the case for sandy catchments (Fig. 13.3). The difference in export between the drought year of 1996 and the very wet year of 1994 amounted to 31.2 kg N ha^{-1} for the loamy catchments and 15.7 kg N ha^{-1} for the sandy catchments. The total N export during the drought year of 1996, which was a year with hardly any net precipitation and generation of percolating water through the root zone, amounted to 13.3 and 6.2 kg N ha^{-1} in the loamy and sandy catchments, respectively. The marked differences measured in the N export response to extreme hydrological conditions reveal important information on the nature and importance of the two main hydrological pathways: the unsaturated and saturated zone. In this case, the saturated zone seems to be of equal relative importance if the runoff in the drought year can be ascribed solely to baseflow.

Log-log relationships ($P < 0.001$) were developed between annual average runoff and annual average total N export from the loamy and sandy catchments (Fig. 13.4). The relationship which fitted best was a power function, revealing that export from the catchment increases with increasing runoff but to a lesser degree within extreme runoff years (Fig. 13.4). This pattern is typical for limited systems, in this case a catchment limited in production/mineralization of N in the root zone.

Table 13.1. Description of the two classes of catchments (loamy: $n = 10$ and sandy: $n = 5$) for average catchment area, average proportion of sandy soils and average land use characteristics.

	Catchment size (Km^2)	Sandy soil (%)	Agriculture (%)	Forest (%)	Natural (%)	Build-up (%)	Wetland (%)
Loamy catchments	7.4	10	74	13	2	8	3
Sandy catchments	13.8	96	73	10	7	7	3

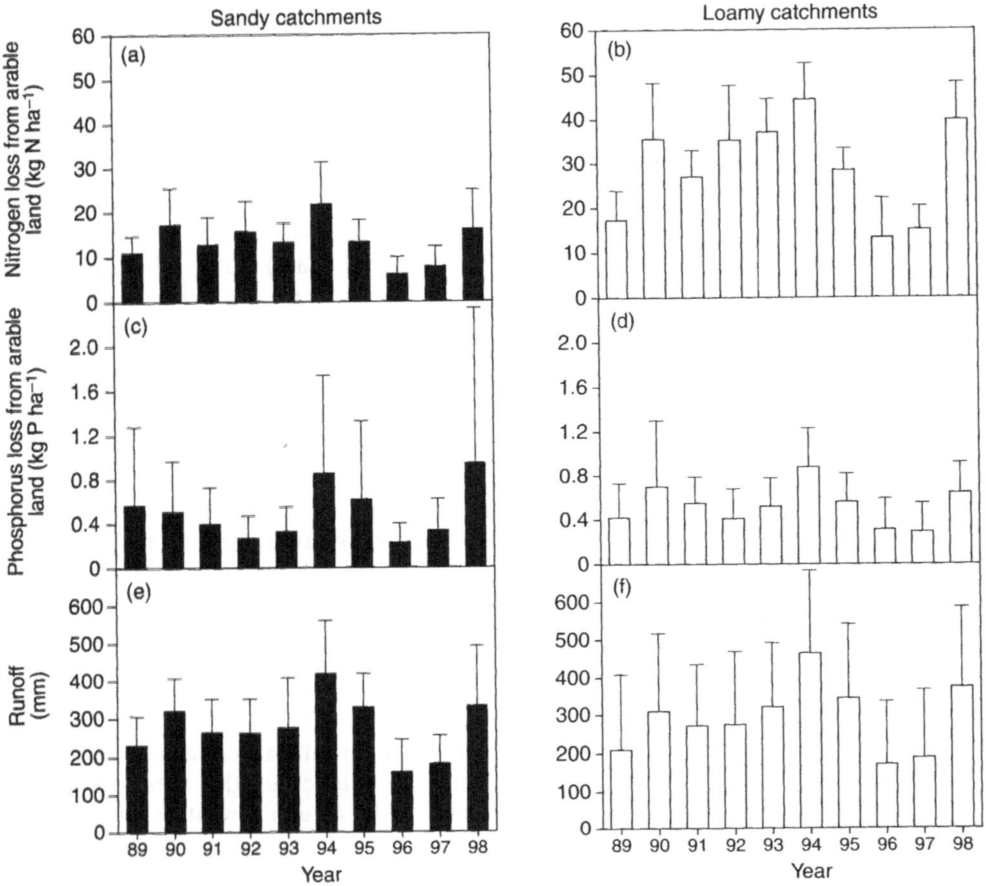

Fig. 13.3. Annual average total nitrogen export from five sandy (a) and ten loamy catchments (b), annual average total phosphorus export (c and d) and annual average runoff (e and f) calculated during the period 1989–1998.

Phosphorus export

Dissolved reactive P contributes an average of 38% of the TP export in both the loamy and the sandy catchments. Particulate P associated with inorganic and organic matter is therefore of great importance when dealing with the TP export from small agricultural catchments. In comparison, PP is of even greater importance for the TP export from small undisturbed catchments, contributing an average of 68%. The speciation of the TP export from small agricultural catchments thus implies that pathways are linked partly to the movement of water in the soil column and partly to erosional processes.

The average annual TP export from agricultural land within small catchments was similar from loamy (0.53 kg P ha^{-1}) and sandy catchments (0.51 kg P ha^{-1}) during the 10 year monitoring period 1989–1998 (Fig. 13.3). In comparison, the average annual TP export from undisturbed catchments was 0.079 kg P ha^{-1} during the same period. The TP export shown in Fig. 13.3 is based on fortnightly sampling and is therefore probably an underestimate of the actual loss rates (Kronvang and Bruhn, 1996). We investigate this underestimation each year by comparing the transport of TP from intensive time (hourly) or flow proportional sampling with the transport of TP from fortnightly sampling in a number of

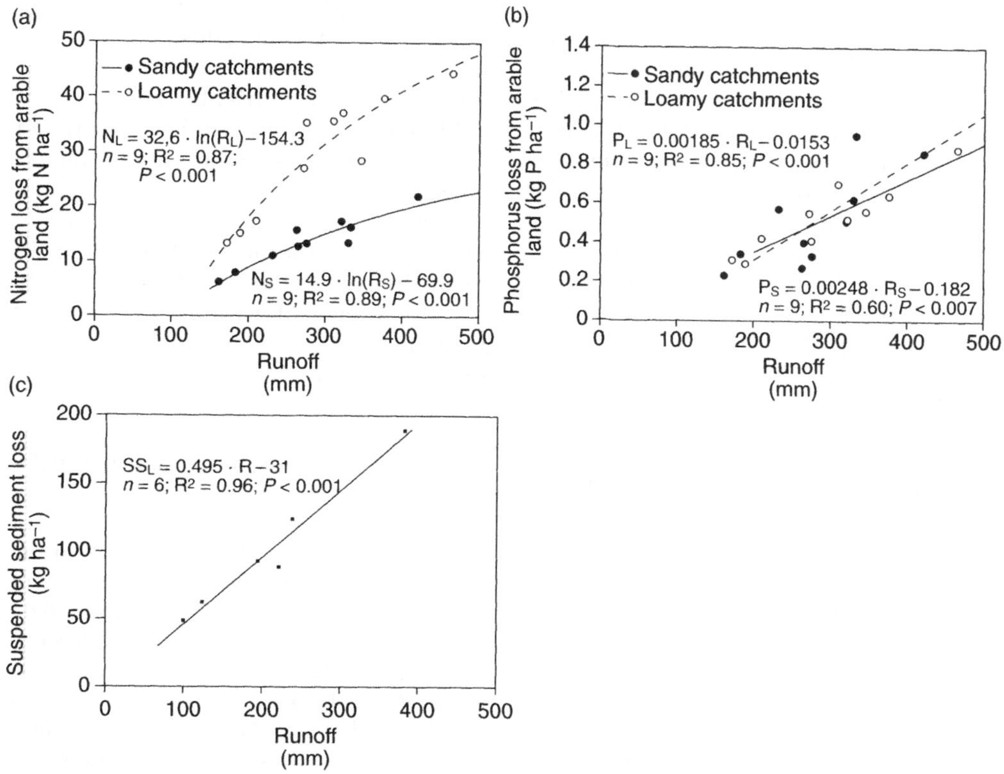

Fig. 13.4. Relationships between annual average runoff and annual average total nitrogen export (a), annual average total phosphorus export (b) and annual average suspended sediment export (c) from Danish catchments.

small agricultural catchments (Table 13.2). On average, fortnightly sampling underestimated the TP export by 0.116 kg P ha^{-1} in these small agricultural catchments during the 6-year period 1993–1998. We lack, however, information on the accuracy of annual TP export from undisturbed catchments.

Both loamy and sandy catchments experience a TP loss high in inter-annual variations (Fig. 13.3). The difference in TP loss between the drought year of 1996 and the wet year of 1994 amounted to 0.57 and 0.63 kg P ha^{-1} for the loamy and sandy catchments, respectively. The marked differences

Table 13.2. Annual average underestimation of the total P export from small agricultural catchments calculated as the difference between fortnightly sampling and intensive (hourly) time-proportional sampling during the period 1993–1998.

Year	Number of catchments	Underestimation of total P loss (kg P ha^{-1})	Coefficient of variation (%)	Annual runoff (mm)
1993	7	0.105	102	195
1994	9	0.228	118	382
1995	9	0.065	98	222
1996	9	0.067	76	100
1997	9	0.060	80	124
1998	9	0.170	59	239
Average		0.116	89	210

measured in the P loss response to extreme hydrological conditions reveal important information on the nature and importance of the main hydrological pathways. The delivery of particulate and dissolved P from erosional pathways (water erosion and overland flow, subsurface tile drainage water and bank erosion) can be anticipated to be at a minimum level in the drought year of 1996. On the other hand, these source areas and pathways must have been of great importance during the wet year of 1994.

Significant linear relationships were developed between annual average runoff and annual average TP loss from the loamy and sandy catchments (Fig. 13.4). The linear relationship indicates that TP in the form of both PP and DRP is available in the different source areas and can be mobilized and transported via the different hydrological pathways to the streams.

Suspended sediment export

The SS loss from eight small agricultural catchments was measured using intensive time or flow proportional sampling during 1993–1998. A significant relationship has been established between annual mean SS loss and annual mean runoff (Fig. 13.4). The SS loss increased linearly with increasing runoff which, as for the TP loss, indicates that new active source areas and hydrological pathways become important with increasing precipitation in the catchment. To a great extent, the importance of the new source areas and pathways delivering SS and associated substances to streams will be controlled by the selective transport of fine particles often enriched in, for example, P (Walling, 1988).

Nutrient Leaching from Danish Arable Fields

The interest in nutrient leaching from arable fields has formerly been restricted to agricultural research in Denmark. However, monitoring of nutrient leaching was introduced as an integrated part of the Danish Monitoring Programme in 1989 (Kronvang *et al.*, 1993). To illustrate the importance of various variables for nutrient leaching from arable fields we look at two soil classes (Table 13.3).

Nitrogen leaching from loamy and sandy fields

Nitrate constitutes by far the largest part of leached N, more than 90% of the TN in the percolating water (Table 13.4). The content of NH_4^+-N is very low, often between 0.01 and 0.1 mg N l^{-1}. Therefore, the difference between total N and NO_3^--N in percolating water is most likely to be dissolved/colloidal bound organic N. The observed difference is more pronounced at the beginning of the autumn than later in the winter. This indicates that a pulse of hydrolysed organic matter is leached through the root zone after the summer drying and the first re-wetting in the autumn (Birch, 1958).

The average annual NO_3^--N leaching from arable fields on loamy and sandy soils was 69 and 118 kg N ha^{-1}, respectively, over the 9-year monitoring period, 1990–1998. This difference is a combined effect of a high livestock density (i.e. larger N consumption) and greater percolation rate in the sandy soil types than in loamy soils. The greater percolation is an effect of both higher precipitation and lower water capacity and hence lower evaporation on sandy soils.

Table 13.3. Average precipitation, percolation of water, livestock density, crop types and N fertilization on two main Danish soil types, loamy soils and sandy soils during the period 1990–1997.

	Precipitation (mm)	Percolation (mm)	Livestock (LU ha^{-1})	Cereals (%)	Grass, maize, beet (%)	Inorganic N (kg N ha^{-1})	Organic N (kg N ha^{-1})
Loamy soils	780	322	0.9	76	24	122	35
Sandy soils	839	456	1.4	52	48	122	108

Table 13.4. Average annual concentration of total and NO_3^--N in percolating water from the root zone during the period 1993–1998.

	Total N ($mg\ N\ l^{-1}$)	NO_3^--N ($mg\ N\ l^{-1}$)
Loamy soils	16.3	15.3
Sandy soils	25.2	23.0

There is a very large inter-annual variation in N leaching from both loamy and sandy soils (Fig. 13.5). There was almost no leaching in the drought year of 1995/96 on the loamy soils, whereas N leaching amounted to 98 kg N ha^{-1} in the year with highest percolation (1993/94) (Fig. 13.5). Similarly, N leaching in the drought year of 1995/96 was relatively low (50 kg N ha^{-1}) on the sandy soils (Fig. 13.5). The highest leaching occurred in 1991/92 and 1992/93 on the sandy soils, 2 years with poor yield and hence a large pool of both inorganic and organic fertilizer N left in the soil at the end of the growing season.

We established a positive relationship ($P < 0.01$) between percolation and N leaching for the loamy soils, but that for the sandy soils was not significant. For both soil types the best relationship fitted a log-function revealing that the N leaching from arable fields increases with increasing percolation but to a lesser degree within extreme percolation years. This pattern indicates that there is a limited pool of nutrients in the root zone remaining after the growing season and that mineralization in the root zone is also limited.

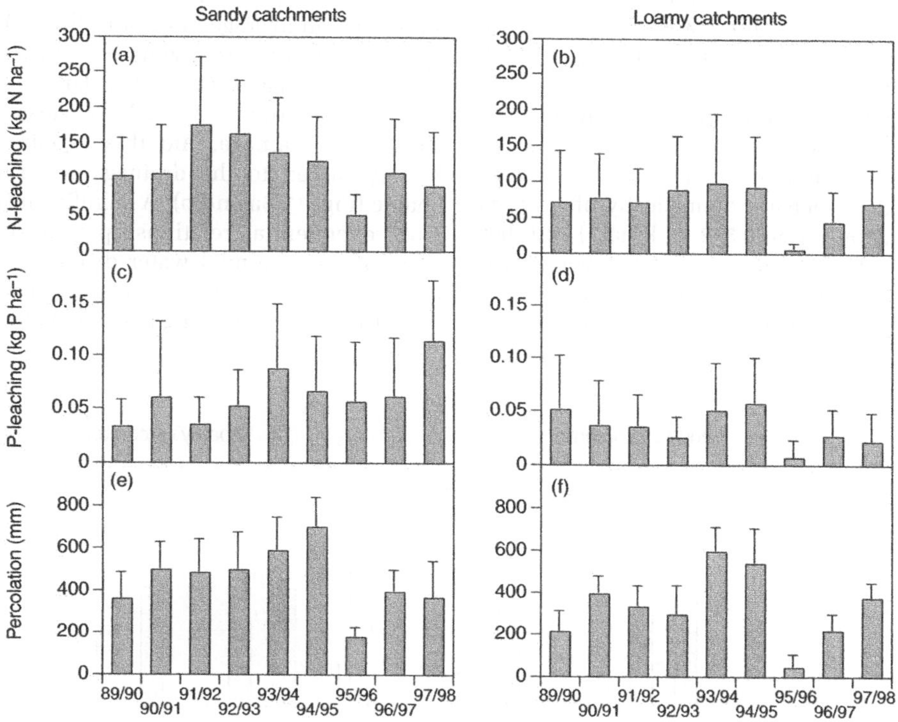

Fig. 13.5. Annual average nitrogen leaching from the root zone on sandy (a) and loamy soils (b), annual average phosphorus leaching (c and d) and annual average percolation of water (e and f) calculated during the period 1989/90 to 1997/98.

Phosphorus leaching from loamy and sandy fields low in soil phosphorus status

Phosphorus leaching from arable fields in Denmark is measured as DRP loss and is divided into two data-sets, fields with low or high soil P-contents. For 26 fields with low P content, the average annual leaching of DRP amounted to 0.034 kg P ha^{-1} for loamy soils and 0.066 kg P ha^{-1} for sandy soils (Fig. 13.5c and d). Phosphorus leaching is higher from the sandy soils than from the loamy soils and this may be attributed to higher percolation and a lower P adsorption capacity of these soils. For both soil classes there was a general tendency for P leaching to increase with increasing percolation; these relationships were, however, not significant.

Phosphorus leaching from fields high in soil phosphorus status

For five fields with high soil P-content, leaching was much greater (Fig. 13.6). For one loamy soil, P-leaching was constantly high (average annual P leaching 0.78 kg P ha^{-1}) although increasing with increasing percolation. This soil was characterized by having a high Olsen P content (107 mg P kg^{-1}). For one loamy soil and three sandy soils, leaching seemed to increase dramatically after slurry application (up to 0.5–2 kg P ha^{-1}) and then slowly decrease during the following years (Fig. 13.6). These soils also had a relatively high Olsen P content (> 47–68 mg P kg^{-1}).

Subsurface Drainage Water as a Nutrient and Sediment Pathway

Information on nutrient losses by subsurface tile drainage water was obtained from eight tile-drained sites on loamy areas (field size between 1 and 4 ha) in the Danish Monitoring Programme during 1989–1998. The data collected can be used to illustrate the importance of subsurface tile drainage water as a pathway for sediment and nutrients to surface water.

Nitrogen loss in subsurface tile drainage water

Average annual total N loss by subsurface tile drainage water amounted to 20.3 kg N ha^{-1} during 1990–1997. Nitrate N constituted the main part (95%) of the total loss, whereas organic N only constituted 5% (Fig. 13.7a). Monitoring of NO_3^--N leaching from the root zone and drainage water loss was carried out on the same sites. Results show that an average of 42% of the NO_3^--N leached out of the root zone is lost by subsurface drainage. There is a large inter-annual variation in NO_3^--N loss by subsurface drainage, and this variation is closely linked to the drainage water discharge (Fig. 13.8a and b). A significant ($P < 0.01$) exponential relationship was developed between annual water discharge and annual N loss, showing that total N loss increases with increasing discharge but to a

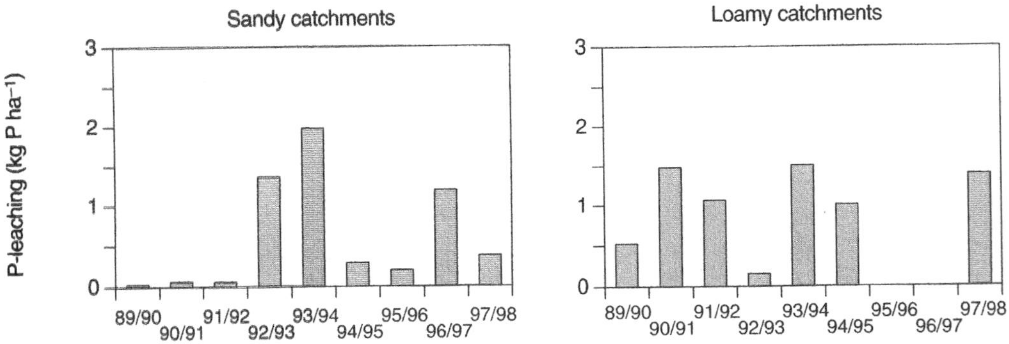

Fig. 13.6. Leaching of P from soils with a high soil P content and/or soils with a high slurry application.

lesser extent with extreme discharge years. This is the same pattern observed both for N-leaching from the root zone and for export from catchments.

Dissolved and particulate phosphorus loss in subsurface tile drainage water

In the case of P-leaching from the root zone there were two different groups of soils, those with low or high P contents. This is also the case in terms of drainage water loss. In the Monitoring Programme, seven drainage sites have a low soil P content and one site has a high content. The average annual TP loss from the sites with low P content amounted to 0.046 kg P ha^{-1} (Fig. 13.7b). On average, PP made up *c.* 50% and DRP the remainder. The temporal variation in P forms was, however, large during the monitoring period. Thus the PP loss varied from 0.002 to 0.062 kg P ha^{-1} with the largest PP loss measured in the 2 wet years 1993/94 and 1994/95. That PP is likely to be mobilized from the topsoil and delivered with preferential flow to tile drains during wet conditions was also documented by Laubel *et al.* (1999a), studying small experimental plots and arable fields.

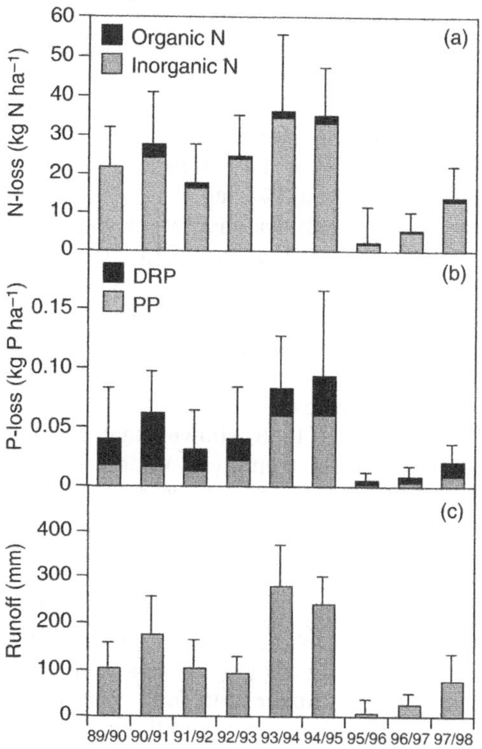

Fig. 13.7. Annual average total N loss via subsurface tile drainage water on loamy soils (a), annual average total P loss (b) and annual average runoff (c) calculated during the period 1989/90 to 1997/98.

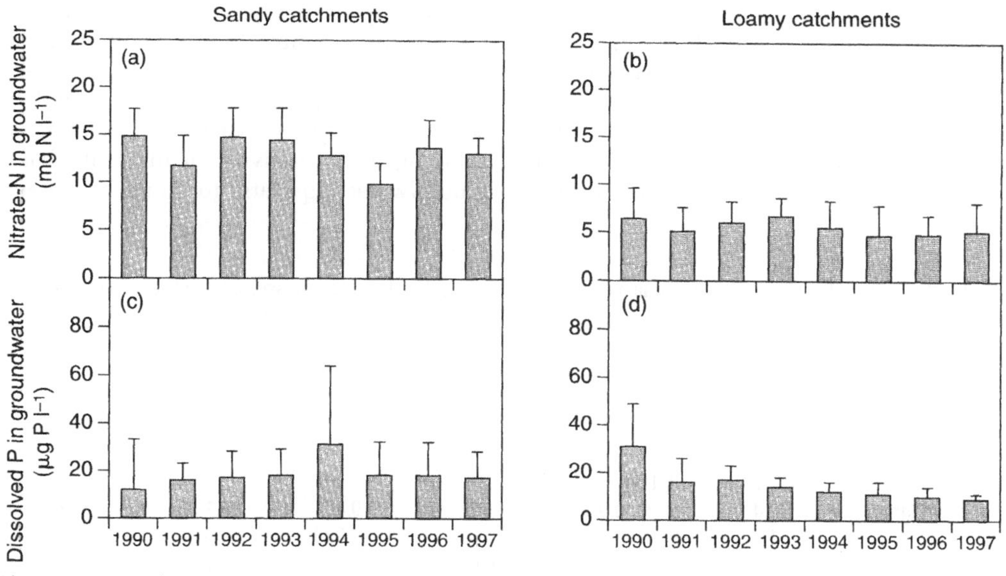

Fig. 13.8. Annual average concentrations of N (a and b) and phosphorus (P) (b) in upper groundwater (1.5–5.0 m) during 1989–1997. The standard deviation is shown on top of each bar.

For the seven soils with low P content, DRP and PP increased, in general, with increasing percolation, although no statistically significant relationships could be established. As mentioned above, one of the drained sites has a high P content (Olsen-P content 107 mg P kg^{-1}). At this site TP loss with drainage water was high, on average 0.155 kg P ha^{-1}, DRP making up 90% of TP. A significant linear relationship between percolation and DRP loss was documented. This indicates that there is no limitation to the source of dissolved P in this soil because of its high P content.

Grant *et al.* (1996) showed that the discrete sampling technique underestimates TP loss and SS loss by subsurface drainage. We have compared an intensive sampling (hourly) technique with discrete sampling, which was carried out on four tile-drained sites on loamy soils during 1993–1996. Three sites (2–4 ha) have crop rotations, whereas the fourth (13 ha) has crop rotations in the upper part and permanent grassland used for grazing in the lower part. For three of the drained sites with low P content the PP loss was 2–4 times higher with the intensive sampling compared with discrete sampling (Table 13.5). The intensive sampling technique had almost no effect on the estimate of DRP. One of the soils had a high P content: the soil in the grazed permanent grassland had an Olsen P content of *c.* 110 mg P kg^{-1}. Dissolved reactive and PP in drainage water from this site was much higher than from the other sites (Table 13.5). Apart from the P-rich topsoil, one reason for this could be that macropores are better developed on permanent grassland and that grazing and trampling cattle both disturb the soil surface and give inputs of P, thereby increasing the availability of fine particles enriched in P content.

Suspended sediment loss in subsurface tile drainage water

The SS loss via subsurface tile drainage water can contribute a high proportion of the total export from a catchment during single storm events (Kronvang *et al.*, 1997). Of the total SS export from a small Danish agricultural catchment (50% of which is artificially tile drained), drainage water accounted for an average of 25% (range: 3–69) of SS losses during 17 single storm events (Kronvang *et al.*, 1997). On an annual basis, subsurface drainage water accounted for 11–15% of the total catchment export of SS which ranged between 71 and 88 kg ha^{-1}.

Nutrient Concentrations in Danish Groundwater

Part of the leached dissolved N and P from the root zone on arable fields reaches the upper and deeper groundwater aquifers; the amount reaching the different aquifers is strongly dependent on the bio-geophysical processes governing N removal and P sorption/desorption reactions. There is a time lag in the delivery of nutrients from groundwater aquifers to surface waters

Table 13.5. Characteristics of the four drainage catchments, and estimated loss of dissolved reactive phosphorus (DRP) and particulate phosphorus (PP) under discreet or intensive sampling by subsurface drainage.

			DRP		PP	
Drainage site	Livestock (LU ha^{-1})	Crops	Discrete (kg P ha^{-1})	Intensive	Discrete (kg P ha^{-1})	Intensive
G1-rotation	0.7	Rotation				
G1-perm. grass	2.4	Perm. grazing	0.277	0.361	0.127	0.190
G2	1.5	Rotation	0.016	0.039	0.013	0.044
L1	0	Rotation	0.044	0.047	0.027	0.085
L2	1.5	Rotation	0.033	0.039	0.027	0.054

because of the travel time for water, which is strongly dependent on the hydraulic head and the hydraulic conductivity of the geological formations in the aquifers. However, the nutrient concentration in upper and deeper groundwater reveals important information both on groundwater quality and, combined with calculated groundwater flows, on the amount of nutrients delivered to surface waters via this hydrological pathway.

Upper groundwater

Monitoring of the annual average concentration of NO_3^--N and DRP in upper groundwater (1.5 m, 3 m and 5 m) in Denmark has been conducted in a total of 53 wells in three loamy catchments and 76 wells in three sandy catchments during the period 1990–1997. The concentration of NO_3^--N and DRP in upper groundwater reveals only small inter-annual variations on both sandy and loamy soils (Fig. 13.8). The concentration of NO_3^--N in upper groundwater was, however, consistently higher (twofold) beneath arable sandy soils than beneath loamy soils (13.1 mg N l^{-1} and 5.6 mg N l^{-1}, respectively) during the period 1990–1997. In the case of DRP, the average annual concentration was 1.4 times higher on sandy soils (20.6 μg P l^{-1}) than on loamy soils (15.0 μg P l^{-1}). A sudden increase in the DRP concentration on sandy soils in 1994 can possibly be explained by the very high precipitation and hence percolation of water from the root zone during 1994 (Fig. 13.8).

Deeper groundwater

The concentration of NO_3^--N and DRP decreases in general with depth as shown in Table 13.6. This is particularly true in the case of DRP in upper groundwater beneath loamy soils and can be explained by sorption reactions between DRP and iron/aluminium in deeper soils. The same decrease with depth is not seen on sandy soils which are also much lower in DRP concentration in the upper well at a depth of 1.5 m. A general decrease in the concentration of NO_3^--N with depth is observed under both sandy and loamy soils (Table 13.6). This can be explained by denitrification of NO_3^--N to gaseous N at depth.

Soil and Bank Erosion as Sediment and Nutrient Sources

Rill erosion and overland flow

Rill erosion has been measured over a 6-year period (1993/94 to 1998/99) on 140 field slope units covering all landscape types, climate gradients and dominant soil types in Denmark (Kronvang *et al.*, 2000). The average slope of the 140 slope units is 7%, which is greater than the typical slope of areas bordering Danish streams. A large number of the slope units exhibited no sign of rill erosion (Fig. 13.9a). Rill erosion was significantly higher (P = 0.05) on winter cereal fields and ploughed fields, compared with grass fields and untreated stubble (Fig. 13.9b).

Table 13.6. Average annual concentration of NO_3^--N and dissolved reactive phosphorus (DRP) in upper and deeper groundwater beneath arable sandy and loamy soils during the period 1990–1997.

	Sandy soils		Loamy soils	
Depth beneath terrain (m)	NO_3^--N (mg N l^{-1})	DRP (mg P l^{-1})	NO_3^--N (mg N l^{-1})	DRP (mg P l^{-1})
1.5	19.6	0.02	13.7	0.27
3.0	12.9	0.02	6.8	0.03
5.0	–	–	5.5	0.01
8–30	7.9	–	1.6	–

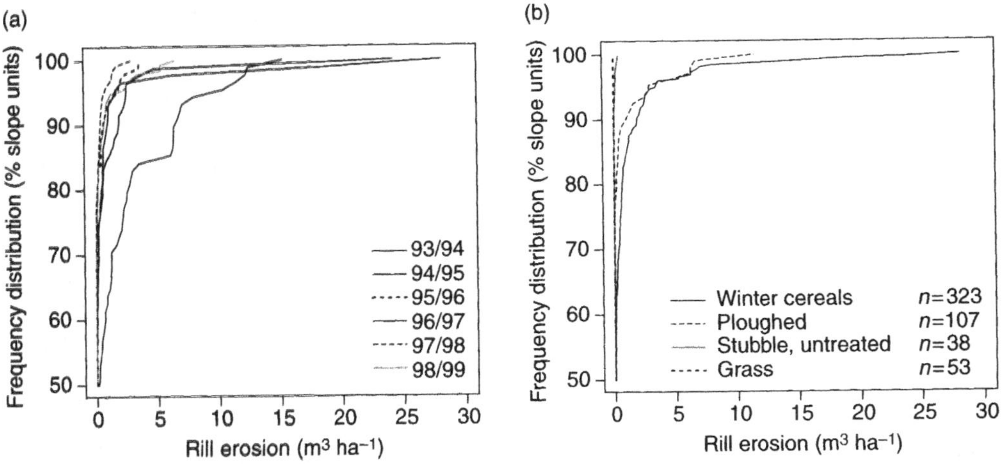

Fig. 13.9. Frequency distribution of rill erosion during 6 study years, 1993/94–1998/99 (a) (1993/94–1996/97: $n = 88$; 1997/98–1998/99: $n = 140$) and rill erosion in four types of slope units differing in management practice (b).

Rill erosion was significantly higher ($P < 0.05$) in 1993/94 than in the other years studied, as the winter was relatively cold (4.0°C) and humid (597 mm) with several thaw events (i.e. rain or melting of snow on frozen soil). Average annual rill erosion amounted to 0.68 m^3 ha^{-1} during the 6-year period, equivalent to 1.02 t sediment ha^{-1} when estimated based on annual simple means. If based on annual geometric mean values, however, the average annual rill erosion amounts to 0.31 m^3 ha^{-1} (Kronvang *et al.*, 2000). This is lower than the values documented by Uhlen and Lundekvam (1988) for different regions of Norway (700–3000 kg ha^{-1}), but similar to amounts of soil erosion found in the UK (Boardman, 1990) and Sweden (Alström and Bergman, 1990).

Annual total rill erosion and downslope sedimentation varied markedly over the 6 years. The soil volume lost over the field edge was highest in 1993/94 (1.02 m^3 ha^{-1}) meaning that 55% of the mobilized sediment from rills entered the buffer zone or stream. The soil loss was lowest in 1997/98 when only 0.02 m^3 ha^{-1} (14% of the mobilized sediment from rills) left the fields. The average annual soil loss from the fields was 0.30 m^3 ha^{-1} (44% of sediment from rill erosion) during the 6-year measuring period.

The potential loss of P associated with the soil loss over the field edge in the 140 slope units can be estimated roughly from the P content of the parent rill material (mean: 0.63 g total-P kg^{-1} per dry weight (DW). Based on the assumption that 30–80% of the P mobilized by rill erosion was lost from the fields, the average annual P loss from the fields was 0.17–0.45 kg P ha^{-1}. The N content of the topsoil was not measured in the study. Average annual N loss from the fields by rill erosion can, however, be roughly estimated at 0.24–10 kg N ha^{-1} based on the parent N content in Danish arable soils (1.0–1.5 g N kg^{-1} DW^{-1}). The estimated N and P loss from arable fields are only valid for medium to high soil erosion risk areas bordering small Danish lowland streams. Moreover, it is important to be aware that part of the N, P and sediment transported across the field edge via surface runoff may be trapped in existing buffer zones before entering the stream (Dillaha and Inamdar, 1997). In addition, dissolved P may also enter the stream via overland flow even without leaving signs of rills on the fields. Thus, the DRP concentration in surface runoff crossing the stream edge has been shown to vary considerably. The average DRP concentration was 0.181 mg P l^{-1}, but extremely high values were measured in surface runoff from slope units on which manure had been applied in the preceding period (> 0.700 mg P l^{-1}).

Bank erosion

Bank erosion was measured along 26 small, 1st order stream reaches in Denmark during 1 year (Laubel *et al.*, 2000). The median delivery of sediment by bank erosion to the streams was 0.010 m^3 m^{-1} of stream reach (one streamside only). In comparison, the mean bank erosion equalled 0.023 m^3 m^{-1} stream reach. The bank erosion rate was significantly higher (Student *t*-test: $P < 0.05$) from sandy than from loamy sites. The average TP content of the buffer zone topsoil alongside the 26 stream reaches was 0.71 mg P g^{-1} DW^1. If the P content of the bank material is equally high the delivery of PP to the stream via bank erosion would correspond to approximately 10–20 g P per m^{-1} stream reach (one streamside only).

The small agricultural catchments described in detail elsewhere in this chapter with respect to N and P cycling have similar small, 1st order streams to those situated along the 26 study streams discussed above. If we assume similar rates of bank erosion and similar P contents of the buffer zone, the potential PP loss by bank erosion in such agricultural catchments corresponds to 0.3–0.5 kg P ha^{-1}. The delivery of PP to streams via bank erosion is perhaps more realistically 0.1–0.4 kg P ha^{-1}, as bank erosion rates and possibly also the P content of bank material may normally be less than that measured in our study. By applying the typical N content of arable soils in Denmark we can estimate a potential PN loss via bank erosion in small agricultural catchments at 0.4–1.1 kg N ha^{-1}. Precipitation was higher during the study year of 1998/99 (871 mm) than normal precipitation during the period 1961–1990 (712 mm). Bank erosion may consequently have been somewhat higher in the study year than a long-term average. Earlier investigations have, however, also highlighted the importance of bank erosion as a sediment source in river systems (Walling *et al.*, 1999). In a Danish catchment bank erosion contributed 50–75% of the total SS export during the year 1994/95 (Laubel *et al.*, 1999b).

Nutrient Budgets for Danish Catchments: the Emission Approach

The information obtained on average annual N and P concentrations in the root zone, subsurface tile drainage water, upper groundwater and deeper groundwater from the intensively monitored three sandy and three loamy catchments was used to establish a catchment budget for N and P emissions via different hydrological pathways. The aim is twofold: (i) to assess the importance of each hydrological pathway at the catchment level; and (ii) to assess any discrepancies between the emission approach (sum of all emissions) and the actual measured export of N and P from agricultural land within the catchments.

Separation of stream discharge into two flow components

A deterministic precipitation/runoff model (NAM-model) has been set up for the five sandy and ten loamy catchments described in Table 13.1. The NAM-model is calibrated to the measured daily discharge from each catchment as described by Pedersen and Kronvang (1998). The NAM-model splits daily discharge into two flow components that can be assumed to mimic runoff from the unsaturated or saturated zones, respectively. The average annual runoff from the unsaturated and saturated zone has been calculated for the five sandy and ten loamy catchments during the 10-year period 1989–1998 (Table 13.7).

A nitrogen budget

The average annual measured export of total N from agricultural land within the five sandy catchments seemed to be most adequately described by applying the mea-

Table 13.7. Calculated annual N emissions from the unsaturated zone and saturated zone on agricultural land established as an average of five sandy and ten loamy catchments during the period 1989–1998, the wet period of 1993–1994 and the dry period of 1996–1997.

Nitrogen budget	R_{UZ} (10^3 m^3)	E_{RZ} (kg N ha^{-1}) (a)	E_{UG} (kg N ha^{-1}) (b)	R_{SZ} (10^3 m^3)	E_{DG} (kg N ha^{-1}) (c)	a+c (kg N ha^{-1})	b+c (kg N ha^{-1})	I_M (kg N ha^{-1})
Sandy catchments								
Average (1989–1998)	1.113	11.1	5.2	2.476	7.0	18.1	12.2	13.6
Wet period (1993/94)	1.524	11.3	7.4	2.959	8.3	19.6	15.7	17.6
Dry period (1996/97)	588	5.5	2.8	1.631	4.6	10.1	7.4	7.1
Loamy catchments								
Average (1989–1998)	1.815	32.5	9.4	1.370	2.0	34.5	11.4	29.3
Wet period (1993/94)	2.646	39.8	22.5	1.622	2.4	42.2	24.9	40.8
Dry period (1996/97)	1.017	17.6	4.7	939	1.4	19.0	6.1	14.2

R is runoff; E_{RZ} is emission from the root zone; E_{UG} is emission from upper groundwater; E_{DG} is emission from deeper groundwater; I_M is the emission of N measured as the average export from agricultural land within the catchments. The emission from the different hydrological pathways is calculated by applying average measured concentrations in each medium in three sandy and three loamy catchments.

sured N concentration in upper groundwater to the unsaturated flow component and N concentrations from deeper groundwater to the saturated flow component (Table 13.7). The sum of the emissions via the two hydrological pathways is 10% lower than the measured average N export from the catchments for the entire 9-year monitoring period and 11% in the wet period of 1993–1994 (Table 13.7). In contrast, the N emissions are higher than the measured N export from the catchments (4%) during the dry period (1996–1997) (Table 13.7). The fact that the measured N export from the sandy catchments is higher than the N emissions can be ascribed to the calculation of emissions for NO_3^--N and the N export for total N. Dissolved inorganic N constitutes an average of 82% of total N. In contrast, the higher N emission than N export during the dry period could be explained by the reduction of NO_3^- by denitrification in stream waters, which is of most significance in such dry years (Kronvang *et al.*, 1999).

The average annual measured export of N from agricultural land within the ten loamy catchments is most adequately described by applying the measured N concentration in the root zone to the unsaturated flow component and N concentrations from deeper groundwater to the saturated flow component (Table 13.7). The sum of the emissions via the two hydrological pathways is higher than the measured average N export from the catchments: 18%, 3% and 34% higher during the three investigated periods (Table 13.7). The reason for this is possibly that part of the unsaturated flow is derived from upper groundwater, which is lower in N concentrations than in the root zone.

The ratio between emissions from the unsaturated and the saturated zone is consistently below unity in the sandy catchments; on average 0.74 for the 10-year monitoring period and lowest in dry years (0.61) and highest in wet years (0.89). In contrast, the unsaturated/saturated N emission ratio is above unity (12–17) in the case of the loamy catchments. The low unsaturated/saturated N emission ratio for the sandy catchments corroborates the extremely large difference in average annual N leaching from the root zone (124 kg N ha^{-1}) and average annual N export (13.6 kg N ha^{-1}) during the 9-year monitoring period. Thus, 89% of the N leached from the root zone did not reach surface water during the monitoring period; this is probably due either to denitrification of NO_3^--N in groundwater or long lag times for the passsage of water through groundwater to surface water. In the case of the loamy catchments, 58% of the average annual N leached from the root zone was not recovered as an N export from the catchment during the 9-year monitoring period.

A phosphorus budget

The calculation of P emissions with the unsaturated and saturated flow components in the sandy and loamy catchments is shown in Tables 13.8 and 13.9. In the case of the loamy catchments, half of the unsaturated flow component is assumed to be subsurface tile drainage water. The calculated P emissions from the hydrological pathways contribute only 5–7% of the measured TP export in the case of the sandy catchments and 12–13% of the TP export in the loamy catchments (Tables 13.8 and 13.9). The total DRP emissions contributed nearly equal amounts of the measured DRP export from the two catchment types (sandy catchments: 13–18%; loamy catchments: 18–24%). The emission approach reveals that we cannot account for a measured DRP export of 0.168 kg P ha^{-1} in the sandy catchments and 0.152 kg P ha^{-1} in the loamy catchments. Only part of this can be explained by a maximum potential delivery of DRP from scattered dwellings in the catchments (0.060–0.120 kg P ha^{-1}). Substantial DRP emissions must therefore be derived either via surface runoff or from critical source areas (CSAs) in the catchment leaching high amounts of DRP. The CSAs will have to account for 8–10% of the agricultural land within the sandy and loamy catchments in order to explain the missing DRP export if we use a conservative estimate of P discharge from scattered dwellings of 0.090 kg P ha^{-1} and a 0.780 kg P ha^{-1} loss from CSAs.

All of the exported PP from the sandy catchments and 90–94% of the exported PP from the loamy catchments cannot be explained by the calculated PP emissions via the unsaturated and saturated flow components. Intensive sampling of tile drainage water would surely increase the PP loss but the same would be the case for the PP export from the catchments. On

Table 13.8. Calculated annual P emissions from the unsaturated zone and saturated zone on agricultural land established as an average of five sandy catchments during the period 1989–1998, the wet period of 1993–1994 and the dry period of 1996–1997.

Phosphorus budget	R_{UZ} (10^3 m^3)	E_{RZ} (kg P ha^{-1}) (a)	E_{UG} (kg P ha^{-1}) (b)	R_{SZ} (10^3 m^3)	E_{DG} (kg P ha^{-1}) (c)	a+c (kg P ha^{-1})	b+c (kg P ha^{-1})	I_M (kg P ha^{-1})
Average (1989–1998)	1.113	0.006	0.008	2.476	0.018	0.024	0.026	0.510
Wet period (1993/94)	1.524	0.006	0.014	2.959	0.026	0.032	0.040	0.595
Dry period (1996/97)	588	0.003	0.004	1.631	0.011	0.014	0.015	0.285

R is runoff; E_{RZ} is emission from the root zone; E_{UG} is emission from upper groundwater; E_{DG} is emission from deeper groundwater; I_M is the emission of total P measured as the average export from agricultural land within the catchments. The emission from the different hydrological pathways is calculated by applying average measured concentrations in each medium within three sandy catchments.

Table 13.9. Calculated annual P emissions from the unsaturated zone and saturated zone on agricultural land established as an average of ten loamy catchments during the period 1989–1998, the wet period of 1993–1994 and the dry period of 1996–1997.

Phosphorus budget	Average (1989–1998)	Wet period (1993/94)	Dry period (1996/97)
R_{UZ} (10^3 m^3)	1.815	2.646	1.017
E_{SD-DRP} (kg P ha^{-1}) (a)	0.017	0.013	0.007
E_{SD-PP} (kg P ha^{-1}) (b)	0.021	0.033	0.019
E_{UG} (kg P ha^{-1}) (c)	0.013	0.020	0.005
R_{SZ} (10^3 m^3)	1.370	1.622	939
E_{DG} (kg P ha^{-1}) (d)	0.019	0.020	0.008
a+c+d (kg P ha^{-1})	0.049	0.053	0.020
I_{M-DRP} (kg P ha^{-1})	0.201	0.287	0.108
b (kg P ha^{-1})	0.021	0.033	0.019
I_{M-PP} (kg P ha^{-1})	0.329	0.413	0.192

R is runoff; E_{SD} is emission from subsurface tile drainage water; E_{UG} is emission from upper groundwater; E_{DG} is emission from deeper groundwater; I_M is the emission of total P measured as the average export from agricultural land within the catchments. The emission from the different hydrological pathways is calculated by applying average measured concentrations in each medium within three loamy catchments.

average, the 'missing' PP amounted to 0.316 kg P ha^{-1} in the sandy catchments and 0.308 kg P ha^{-1} in the loamy catchments during the 9-year investigation period (Tables 13.8 and 13.9). These figures are, however, well within the above-mentioned potential loss of PP via water erosion and overland flow (0.17–0.45 kg P ha^{-1}) and via bank erosion (0.1–0.4 kg P ha^{-1}). On average, only 3% of the agricultural land in the catchments can be described as medium to high erosion risk areas. In general, the delivery of PP via surface runoff therefore seems of minor importance when compared with bank erosion.

Conclusions and Perspectives

In this Chapter we have demonstrated and quantified the importance of major hydrological pathways for N and P losses to surface water in Danish agricultural catchments. The N export from agricultural catchments is connected to a Normal Source Area (NSA) loss mainly influenced by relatively easily obtainable information on hydrology, soil type and agricultural practice. We are currently able to describe the N cycling on field and catchment levels adequately with more or less sophisticated model tools. We have demonstrated that combining results from a simple hydrological model with monitored N concentrations in the root zone, upper groundwater and deeper groundwater results in a reasonably well-established N export for two extreme types of small Danish catchments. The main uncertainty in describing the N cycling lies, however, in the inertia of N in the soil column relating to both the removal of NO_3^--N and the lag time with which water and N are transported to surface water from their point of origin. This can be shown with the large differences calculated between average annual N leaching from the root zone (118 kg N ha^{-1}) and the average annual N export (13.6 kg N ha^{-1}) for five Danish sandy catchments during a 9-year monitoring period. Thus, 89% of the N leached from the root zone is not recovered in the stream during this relatively long monitoring period either due to denitrification of NO_3^--N in groundwater or long lag times for the passage of water through groundwater to surface water. In comparison, 58% of the average annual N leached from the root zone is not recovered as an N export from ten loamy catchments during a similar long monitoring period.

In contrast to N, the combination of a hydrological model with measured P concentrations from NSAs in the root zone, subsurface tile drainage water, upper groundwater and deeper groundwater did not adequately describe the P export from agricultural land within the sandy and loamy catchments. The calculated NSA emissions contributed only 13–18% of average annual DRP export in the sandy catchments and 18–24% of the annual average TP export in the loamy catchments during the 9-year monitoring period. A maximum discharge of P from scattered dwellings can only explain part of the missing DRP in the budget. Thus, CSAs for DRP loss or DRP loss via overland flow must contribute substantial amounts of DRP to the streams. Utilizing a measured DRP loss of 0.78 kg P ha^{-1} from one CSA reveals that, on average, 8–10% of the agricultural land in the monitored catchments is CSAs. Soils with high P content and/or a large application of slurry may, however, contribute substantially to the P loss from catchments. In Denmark the annual soil testing in agriculture shows that about 50% of all soils have an Olsen P content above the level needed for optimal plant growth ($>$ 40 mg P kg^{-1}), and 10–15% of arable fields have an Olsen P content $>$ 60 mg P kg^{-1}. These soils may pose a potential threat in terms of P losses to water.

On average, PP constituted 62% of the TP export from both the sandy (total export: 0.51 kg P ha^{-1}) and the loamy catchments (total export: 0.53 kg P ha^{-1}) during the 9-year monitoring period. The majority of the PP export seems to be derived from bank erosion although water erosion and surface runoff can also contribute significant amounts in wet years with snow melting on frozen soils. We calculated that PP emission from subsurface tile drainage water on average contributes 6% of the PP export from the loamy catchments. As the PP loss from subsurface tile drainage water, measured by means of a discrete sampling strategy, is shown to underestimate actual PP losses by 50–215%, the value above is certainly a minimum estimate.

Assessments of source areas, hydrological pathways and sinks for sediment, N and P are a prerequisite for proper management strategies to be implemented at the catchment level. Catchment managers need operational tools that enable them to identify both NSAs and CSAs for the loss of sediment, N and P within a catchment. Moreover, catchment managers need tools that enable them to quantify the importance of active hydrological pathways and sinks within a catchment. The development of such tools that we based on our empirical knowledge demands, however, that we are able to establish relationships between the loss of sediment and nutrient forms and the factors governing sediment and nutrient fluxes at the catchment scale.

Acknowledgements

This study was partly supported by the Danish Ministry of Food, Agriculture and Fisheries under the research programme 'Land use – the farmer as land manager'. We also thank the Danish counties responsible for conducting the catchment monitoring in Denmark.

References

Alström, K. and Bergman, A. (1990) Water erosion on arable land in southern Sweden. In: Boardman, J., Foster, I.D.L. and Dearing, J.A. (eds) *Soil Erosion on Agricultural Land.* John Wiley & Sons, Chichester, UK, pp. 107–118.

Andersen, H.E., Kronvang, B. and Larsen, S.E. (1999) Agricultural practices and diffuse nitrogen pollution in Denmark: empirical leaching and catchment models. *Water Science and Technology* 39(12), 257–264.

Birch, H.E. (1958) The effect of soil drying on humus decomposition and nitrogen availability. *Plant and Soil* 10, 9–31.

Boardman, J. (1990) Water erosion on arable land in southern Sweden. In: Boardman, J., Foster, I.D.L. and Dearing, J.A. (eds) *Soil Erosion on Agricultural Land.* John Wiley & Sons, Chichester, UK, pp. 87–105.

Dearing, J.A., Elner, J.K. and Happey-Wood, C.M. (1981) Recent sediment flux and erosional processes in a Welsh upland lake-catchment based on magnetic susceptibility measurements. *Quaternary Research* 16, 356–372.

Dillaha, T.A. and Inamdar, S.P. (1997) Buffer zones as sediment traps or sources. In: Haycock, N.E., Burt, T., Goulding, K. and Pinay, G. (eds) *Buffer Zones: their Processes and Potential in Water Protection. The Proceedings of the International Conference on Buffer Zones, September 1996.* Quest Environmental, Harpenden, Hertfordshire, UK, pp. 33–42.

European Environment Agency (1995) *Europe's Environment – the Dobris Assessment.* Stanners, D. and Bourdeau, P. (eds) European Environment Agency, Copenhagen.

Foster, I.D.L., Grew, R. and Dearing, J.A. (1990) Magnitude and frequency of sediment transport in agricultural catchments: a paired lake-catchments study in Midlands England. In: Boardman, J., Foster, I.D.L. and Dearing, J.A. (eds) *Soil Erosion on Agricultural Land.* John Wiley & Sons, Chichester, UK, pp. 153–172.

Grant, R., Laubel, A., Kronvang, B., Andersen, H.E., Svendsen, L.M. and Fuglsang, A. (1996) Loss of dissolved and particulate phosphorus forms in drainage water from four arable catchments on structured soils in Denmark. *Water Research* 30(11), 2633–2642.

Heckrath, G., Brookes, P.C., Poulton, P.R. and Goulding, K.W.T. (1995) Phosphorus leaching from soils containing different P concentrations in the Broadbalk experiment. *Journal of Environmental Quality* 24, 904–910.

Haygarth, P.M., Hepworth, L., Jarvis, S.C. and Smith, R.V. (1998) Forms of phosphorus transfer in hydrological pathways from soil under grazed grassland. *European Journal of Soil Science* 49, 65–72.

Iversen, T.M., Kjeldsen, K., Kristensen, P., Haan, B., Oirschot, M., Parr, W. and Lack, T. (1997) *Integrated Environmental Assessment on Eutrophication – a Pilot Study,* Technical Report No. 207. National Environmental Research Institute, Silkeborg, Denmark.

Jeppesen, E. (1998) The ecology of shallow lakes – trophic interactions in the pelagial, Technical Report No. 247. Doctoral dissertation (DSc), National Environmental Research Institute, Silkeborg, Denmark.

Jeppesen, E., Søndergaard, M., Kronvang, B., Jensen, J.P., Svendsen, L.M. and Lauridsen, T.L. (1999) Lake and catchment management in Denmark. In: Harper, D., Brierley, B., Ferguson, A., Phillips, G. and Madgewick, J. (eds) The Ecological Bases for Lake and Reservoir Management. *Hydrobiologia* 395/396, 419–432.

Kristensen, P. and Hansen, O.H. (1994) *European Rivers and Lakes – Assessment of their Environmental State,* EEA Environmental Monographs 1. European Environment Agency, Copenhagen.

Kronvang, B. (1996) Suspended sediment and nutrient loads: diffuse sources, delivery pathways, fate and management. PhD thesis, The National Environmental Research Institute, Silkeborg, Denmark.

Kronvang, B. and Bruhn, A.J. (1996) Choice of sampling strategy and estimation method when calculating nitrogen and phosphorus transport in small lowland streams. *Hydrological Processes* 10, 1483–1501.

Kronvang, B., Ærtebjerg, G., Grant, R., Kristensen, P., Hovmand, M. and Kirkegaard, J. (1993) Nationwide Monitoring of Nutrients and their Ecological Effects: state of the Danish Aquatic Environment. *AMBIO* 22(4), 176–187.

Kronvang, B., Grant, R., Larsen, S.E., Svendsen, L.M. and Kristensen, P. (1995) Non-point source nutrient losses to the aquatic environment in Denmark. Impact of agriculture. *Marine and Freshwater Research* 46, 167–177.

Kronvang, B., Græsbøll, P., Larsen, S.E., Svendsen, L.M. and Andersen, H.E. (1996) Diffuse nutrient losses in Denmark. *Water Science and Technology* 33(4–5), 81–88.

Kronvang, B., Laubel, A. and Grant, R. (1997) Suspended sediment and particulate phosphorus transport and delivery pathways in an arable catchment, Gelbæk stream, Denmark. *Hydrological Processes* 11, 627–642.

Kronvang, B., Hoffmann, C.C., Svendsen, L.M., Windolf, J., Jensen, J.P. and Dørge, J. (1999) Retention of nutrients in river basins. *Aquatic Ecology* 33, 29–40.

Kronvang, B., Laubel, A., Larsen, S.E. and Iversen, H.L. (2000) Soil erosion and sediment delivery through buffer zones in 135 Danish slope units. In: Stone, M. (ed.) *The Role of Erosion and Sediment Transport in Nutrient and Contaminant Transfer.* IAHS Publication No. 263, IAHS, Wallingford, UK, 67–73.

Larsen, S.E., Kronvang, B., Windolf, J. and Svendsen, L.M. (1999) Trends in diffuse nutrient concentrations and loading in Denmark: statistical trend analysis of stream monitoring data. *Water Science and Technology* 39(12), 197–205.

Laubel, A., Jacobsen, O.H., Kronvang, B. and Grant, R. (1999a) Subsurface drainage loss of particles and phosphorus from field plot experiments and a tile-drained catchment. *Journal of Environmental Quality* 28, 576–584.

Laubel, A., Svendsen, L.M., Kronvang, B. and Larsen, S.E. (1999b) Bank erosion in a Danish lowland stream system. *Hydrobiologia* 410, 279–285.

Laubel, A., Kronvang, B., Larsen, S.E., Pedersen, M.L. and Svendsen, L.M. (2000) Bank erosion as a source of sediment and phosphorus input to small Danish streams. In: Stone, M. (ed.) *The Role of Erosion and Sediment Transport in Nutrient and Contaminant Transfer*. IAHS Publication No. 263, IAHS, Wallingford, UK, 75–82.

Pedersen, L.M. and Kronvang, B. (1998) Catchment analysis and modelling. In: Boegestrand, J. (ed.) *Streams and Springs – National Freshwater Monitoring Programme 1998*, Technical Report No. 292. National Environmental Research Institute, Silkeborg, Denmark, pp. 43–47 (in Danish).

Platts, W.S., Torquemada, R.J., McHenry, M.L. and Graham, C.K. (1989) Changes in salmon spawning and rearing habitat from increased delivery of fine sediment to the South Fork Salmon River, Idaho. *Transactions of the American Fisheries Society* 118, 274–283.

Smith, R.V., Stevens, R.J., Foy, R.H. and Gibson, C.E. (1982) Upward trend in nitrate concentrations in river discharging into Lough Neagh for the period 1969–1979. *Water Research* 16, 183–188.

Smith, K.A., Chalmers, A.G., Chambers, B.J. and Christie, P. (1998) Organic manure phosphorus accumulation, mobility and management. *Soil Use and Management* 14, 154–159.

Turner, R.E. and Rabalais, N.N. (1991) Changes in Mississippi river water quality this century. *BioScience* 41(3), 140–147.

Uhlen, G. and Lundekvam, H. (1988) *Avrenning av Nitrogen, Fosfor og Jord fra Jordbruk 1949–1979/88.* (Runoff of nitrogen, phosphorus and soil from arable land) (in Norwegian) SEFO-projekt under NTNF-programmet, Naturresurs og Samfund, Ås-NLH 11 November 1988, Ås, Norway.

Walling, D.E. (1988) Erosion and sediment yield research – some recent perspectives. *Journal of Hydrology* 100, 113–141.

Walling, D.E., Owens, P.N. and Leeks, G.J.L. (1999) Rates of contemporary overbank sedimentation and sediment storage on the floodplains of the main channel systems of the Yorkshire Ouse and River Tweed, UK. *Hydrological Processes* 13, 993–1009.

14 Development of Geographical Information Systems for Assessing Hydrological Aspects of Diffuse Nutrient and Sediment Transfer from Agriculture

A.I. Fraser
National Soil Resources Institute (Cranfield University), North Wyke Research Station, Okehampton, Devon EX20 2SB, UK

Key words: diffuse pollution, geographical information systems, hydrological pathways, land use classification modelling, nutrient transfer, remote sensing, sediment transfer, soil erosion, soil hydrology

Introduction

Farm managers, regulators and policy makers worldwide are faced with an increasingly complex decision making environment where sustainable agricultural production needs to be achieved within narrowing economic, social and environmental limits (Matthews *et al.*, 1999). These restrictions, driven by increasingly competitive markets, regulatory enforcement, increasing public awareness and subsequent pressures, have led to a growing demand for decision-support tools which simulate and predict environmental processes and their effect on water quality. These decision-support systems (DSS) are diverse and are used by a wide range of disciplines (Shepherd *et al.*, 1999). They provide appropriate information in a format that can be readily understood by non-scientists and allow land managers and regulators to explore the options and assess the impacts of strategies to meet given agricultural, environmental and social objectives (Oliver *et al.*, 1990; Shaw and Falco, 1990; Refsgaard *et al.*, 1996; Tim, 1996).

The recent proliferation of geographical information system(s) (GIS) technology, digital terrain or digital elevation models (DTM/DEM), spatial data-sets, digitized maps (previously available in hard copy form only) and powerful desktop computers has enabled rapid progress in the development of quantitative research tools (Garen *et al.*, 1999). Through the manipulation and extrapolation of extensive data-sets, which would otherwise be difficult using conventional database software (Adinarayana *et al.*, 1999; De Roo and Jetten, 1999a), GIS can assist in understanding complex land and water interactions (e.g. Moore *et al.*, 1983; Lee *et al.*, 1987; Skidmore *et al.*, 1991; Harris and Boardman, 1998). Figure 14.1 illustrates a decision flow chart for determining whether development of a GIS would improve the understanding of a given system. By coupling expert-systems to GIS through the integration of multi-source environmental data, spatial and temporal variations in process-controlling factors of diffuse nutrient and sediment pollution may be expressed. This approach to environmental modelling using GIS has been outlined by Hutchinson and Dowling (1991) and Hallett *et al.* (1996). This chapter reviews recent GIS techniques developed to simulate hydrological aspects of diffuse agricultural pollution and provides an appraisal of the components that constitute the main input data sources therein.

Geographical Information Systems and Knowledge-based System Linkage

Although different GIS vary in their capabilities, Martin (1991) found that most contain the following components:

1. Data acquisition and input: concerned with receiving data for processing in the GIS, including manual digitizing, scanning, manual entry of attribute information and online retrieval of data from other database systems.
2. Data storage and retrieval: organization of the spatial data into a format that permits efficient retrieval by the user for subsequent analysis, in addition to the need for rapid and accurate updates and correction of the data.

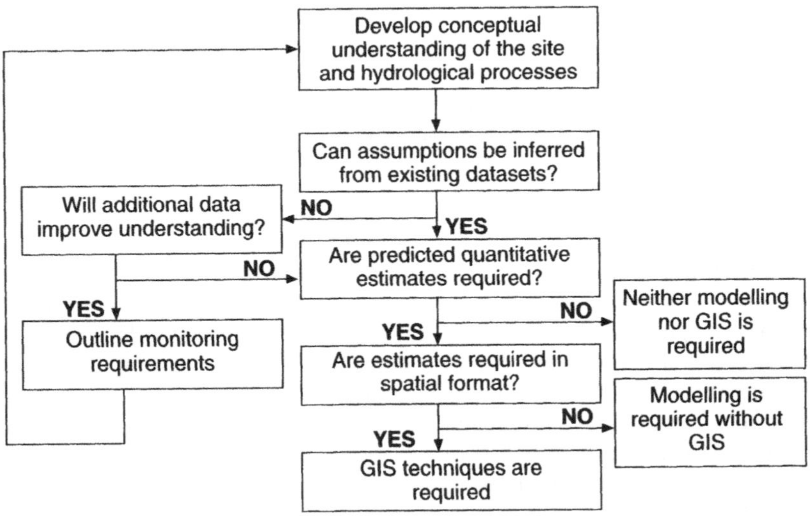

Fig. 14.1. Decision flow chart to determine if GIS and/or modelling is required (adapted from Bontwell *et al.*, 1985).

3. Data manipulation and analysis: representing the application of techniques which convert data through user-defined rules, or which produce estimates of parameters and constraints for various spatial and temporal optimization and simulation models.
4. Data reporting: covering a range of techniques to display all or part of the original database, as well as manipulated data, and the output from spatial models in tabular, graph or map form.

The integration of models with a GIS has a number of advantages such as short running time and quick production of results (Baban, 1999). For example, Smith *et al.* (1992) used a GIS to calibrate a continuous simulation model in addition to an event-driven model. The GIS approach had advantages in that model output was enhanced because of increased accuracy and minimization of human errors and time input. Furthermore, model output may be validated using GIS. For example, a temporally and spatially distributed model to predict total P sources and loads to Bassenthwaite Lake in the UK was validated successfully using GIS (Penning-Rowsell, 1996).

The term 'knowledge-based system(s)' (KBS) describes the numerous potential techniques which may be adopted by researchers when conducting environmental analysis. KBS includes models implemented as object-oriented or procedural software, declarative rules based (e.g. expert) systems and others such as neural/Baysian nets (Matthews *et al.*, 1999). GIS can be regarded as a distinct class of KBS with specialized spatial data handling capabilities. In a review of the strategies used for linking GIS to environmental models, Fedra (1996) identifies a continuum of linkage intimacy from 'loose' to 'deep' coupling. Loose coupling is accomplished via the exchange of data files between the KBS and GIS interface. The required model parameters are created in a database, the models are run and their outputs are then applied to a GIS for mapping. The interface between the GIS and KBS is thus defined by common file formats (Matthews *et al.*, 1999). A loosely coupled system has disadvantages in that data processing is typically slow, as memory usage, required for file exchange, is considerable. However, this technique requires little investment in software development and therefore provides scientists with a convenient tool for spatial data extrapolation. Nevertheless, loosely coupled GIS are vulnerable to error where manual data handling and manipulation are required. In contrast, deep coupling merges the GIS and KBS to the extent that they form a single system (Fedra, 1996). A deeply coupled system has transparent exchange of data via common systems memory and a single user interface. Implementation of such techniques, however, requires open GIS architectures and considerable amounts of software development (Matthews *et al.*, 1999). To date, loosely coupled GIS have been the most widely developed approaches, although these systems often only provide qualitative output such as vulnerability maps (Hiscock *et al.*, 1995). Only a few studies, such as those of Gilliland and Baxter-Potter (1987), Oloufa *et al.* (1995), McKinney and Tsai (1996) have demonstrated models, developed within a GIS framework, which produce quantitative results.

Data Acquisition for Integration with Geographical Information Systems

The process of data acquisition is critical to the development of any GIS. It is a labour-intensive component of GIS work and can account for between 60 and 90% of the time needed for construction of the system. A number of data acquisition options for hydrologically oriented GIS have been suggested by van Blargan and Ragan (1991), and include manual digitizing, commercial digitizing, satellite imaging and automated scanning. Subsequent data storage within the GIS is established by using either raster or vector data structuring. A raster structure represents the geographic space in terms of a two-dimensional grid of rows and columns, while a vector structure represents the space in terms of lines, points and areas (otherwise known as polygons), typically constituting sets of X and Y coordinates representing the cartographic outline of geographical objects. In diffuse pollution studies, a vector-based GIS has advantages over a raster-based system, as manipulation and processing of the input data are typically much faster.

The collation of base-line data for monitoring and managing purposes, and its subsequent storage within a GIS is referred to as an 'inventory' exercise (Baban, 1999). Figure 14.2 illustrates typical base-line data which may be used to build a complex GIS for quantifying diffuse pollution transfers. Prior to conducting analysis of data, there may be a requirement for lengthy and time-consuming preparations, including collecting data such as catchment topography, source areas of pollution, soil characteristics and land use, followed by compilation, storage and retrieval of data. There is an advantage of using GIS in these exercises as highly sophisticated database management systems which can store voluminous spatial data typically required in hydrological modelling (Vieux *et al.*, 1988; Bhaskar *et al.*, 1992).

The Role of Remote Sensing in the Development of Geographical Information Systems

The development of remote sensing techniques has been a crucial factor in the growth of GIS and is particularly relevant for the development of KBS focusing on diffuse pollution. Remote sensing imaging provides an alternative to conventional data capture, overcoming constraints which are inherent in time-consuming and expensive sampling and monitoring programmes. A number of studies have illustrated that remote sensing can meet the demands of data acquisition for many samples over a large spatial area and also provide a temporal view of parameters such as water quality (Yang *et al.*, 1999), land use (Higgitt and Lu, 1999) and climatic variations (Legates and Willmott, 1990).

The main objective of remote sensing is to characterize surface areas of land or water by using electromagnetic radiation energy sensors mounted on either aircraft or orbiting satellites. Wavelengths of most use in determining terrestrial coverages such as vegetation and soil are those of the near-infrared range (0.7–2.0 μm). For geological features, wavelengths of the middle infrared (2.0–2.35 μm) are most useful (Curran, 1985) (Table 14.1). The measurement of electromagnetic energy and subsequent dis-

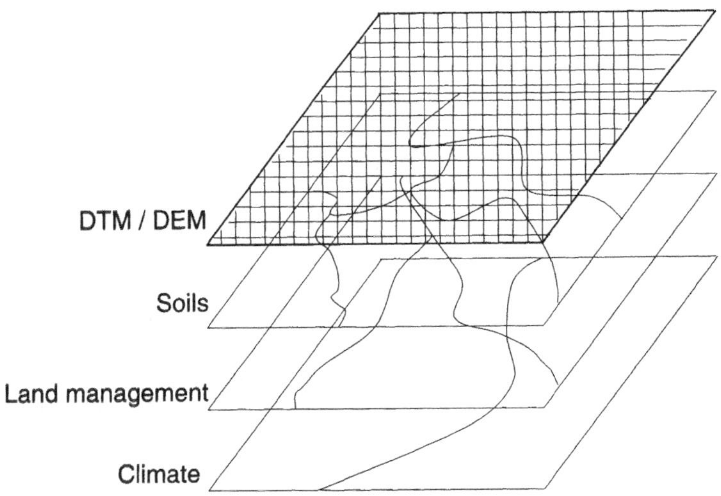

Fig. 14.2. Major components of Geographical Information Systems developed for quantifying diffuse pollution transfers from agricultural land.

Table 14.1. Wavebands recorded by the Thematic Mapper sensor (after Curran, 1985).

Band number	Band Name	Bandwidth (μm)	Characteristics
TM1	Blue/green	0.45–0.52	Strong vegetation absorbance
TM2	Green	0.52–0.6	Strong vegetation reflectance
TM3	Red	0.63–0.69	Very strong vegetation absorbance
TM4	Near infrared	0.76–0.90	High land contrasts, very strong vegetation reflectance
TM5	Near middle infrared	1.55–1.75	Very moisture sensitive
TM6	Thermal infrared	10.4–12.5	Very sensitive to soil moisture and vegetation
TM7	Middle infrared	2.08–2.35	Good geological discrimination

tinction of surface characteristics is an expression of an energy balance relationship (Baban, 1999):

$$E_I(\lambda) = E_R(\lambda) + E_A(\lambda) + E_T(\lambda),$$

where the incident energy (E_I) is a function of the reflected (E_R), absorbed (E_A) and transmitted (E_T) energy.

The most widely used remotely sensed information is derived from satellite-mounted sensors which operate at different wavelengths. The most popular instruments are the Landsat Thematic Mapper (TM), Système pour l'Observation de la Terre (SPOT), and the Advanced Very High Resolution Radiometer (AVHRR). Recently, the use of Synthetic Aperture Radar (SAR) imaging devices has become popular, as the visible and infrared atmospheric windows are shuttered when cloud obscures conventional imaging. Microwave wavelengths from the SAR system penetrate cloud, bounce off surface features and return to the satellite.

Barrett and Curtis (1992) identified the following criteria for the selection of appropriate remotely sensed data:

1. Spatial resolution: the smallest feature size that can be studied in an image. Essentially, this is a measure of the smallest ground object than can be distinguished as separate entity, typically a single pixel.
2. Spectral resolution: the capability of the sensor to measure spectral information within a range of wavelengths. Some sensors operate on a broad-band basis, while others can be finely adjusted to record discrete reflectance.
3. Radiometric resolution: the amount of electromagnetic energy collected by a sensor varies more or less continuously but the actual range of energy that can be recorded and the precision of the measurement depend on the operating range and radiometric resolution of the instrument.
4. Temporal resolution: satellites in regular orbit around the Earth revisit a given location. The time between visits depends on the angular field of view of the sensor, the latitude and longitude of the area of interest, and the characteristics of the satellite orbit.

Sensor design is typically a compromise between these various types of resolution, since an increase in performance in one area generally means a deterioration in another. The balance is struck by optimizing the sensor design to meet the specific objectives of its expected use.

The final stages of data acquisition through remote sensing are concerned with the processing of the raw image. This processing includes correction of the image for any distortion and degradation, increasing the distinction between features to optimize visual interpretation, and maximizing the contrast between features of interest. Finally, for accurate interpretation of surface features, the image needs to be calibrated with measured characteristics on the ground. This calibration is known as 'ground-truthing'.

Data Types for Hydrologically Oriented Geographical Information Systems

Precipitation inputs

The first prerequisite for hydrological modelling and extrapolation with GIS is an accurate estimate of precipitation (Garen *et al.*, 1999). While hydrological pathways of diffuse nutrient, sediment and pesticide transfer are subsequently determined by land use, soil characteristics and topography (Beven and Kirkby, 1979; Beven, 1986; Beven, 1995), rainfall determines the overall magnitude of water flux through a catchment. Several characteristics of a rainfall event are important, including rainfall intensity and duration, raindrop size and antecedent soil moisture condition (i.e. time lapsed following previous rainfall) (Basnyat *et al.*, 2000). While only a small number of studies have focused on the specific effects of rainfall on nutrient and pesticide transfer (e.g. Meyers, 1981; Kinnell, 1983; Rose and Dalal, 1988; Fraser *et al.*, 1999), there has been a considerable amount of research concentrating on the relationship between rainfall and sediment transfer, principally in an attempt to evaluate soil detachment processes (e.g. Massey and Jackson, 1952; Oxley, 1974; Fox and Le Bissonnais, 1998; Kuznetsov *et al.*, 1998).

At the field or small catchment scale, it is not particularly difficult to accurately quantify rainfall for GIS extrapolation, as long as precise rain gauges are properly installed and maintained. At larger regional and national scales, however, the spatial and temporal variability of rainfall must be realized if erroneous GIS output is to be avoided. It is not adequate to assume that precipitation is uniform over the entire area, particularly where convective, orographic or other spatial variability of rainfall occurs (Garen *et al.*, 1999). This problem is compounded by the fact that larger-scale GIS extrapolation is typically accompanied by the use of larger grid sizes, and this, in itself, may introduce greater error in hydrological model predictions (Vieux, 1993). Subsequently, studies have attempted to resolve such difficulties, by resorting to grouping strategies which reduce the effects of scattering (e.g. Milliman and Syvitski, 1992; Lu and Higgitt, 1998). Recent spatial interpolation techniques, such as the Parameter-elevation Regressions on Independent Slopes Model (PRISM) (Daly *et al.*, 1994) (www.ocs.orst.edu/prism) or the daily time series model developed by Garen *et al.* (1994) may be used to obtain better estimates of spatial and temporal ranges in precipitation over more extensive areas. If sub-daily input variables are required, these may be derived by disaggregating daily amounts on the basis of high time resolution rain gauges positioned within the area in question, as described by Garen *et al.* (1999).

Following the identification of a number of problems by Favis-Mortlock (1995) and Johnson *et al.* (1996) associated with stochastic weather generator models, applications utilizing data derived from these models, as opposed to historically observed data, have benefited from some improvements in statistical behaviour, weather variables and time resolution (Johnson *et al.*, 1996). These improvements should improve inputs, thereby enhancing predictive accuracy (Garen *et al.*, 1999).

Land use and land cover classification

The role of land use in determining non-point source pollution has been well established in the literature (Perry and Vanderklein, 1996; Basnyat *et al.*, 2000). Agricultural activities have been identified as major sources of diffuse pollutants, including soil nutrients, sediment, liquid farm waste, crop residues, inorganic salts and minerals, and pesticides (Viessman and Hammer, 1993), all of which may lead to a deterioration in water quality. Indeed, in many cases, in the process of sediment generation and transfer, land use is more critical than climate (Morgan, 1995; Evans, 1996). However, the integration of land use parameters into any modelling and GIS framework, in order to forecast impacts on water quality, is problematic. For instance, it has been demonstrated that the prediction of future land use is difficult, in many ways more than prediction of climate change (Favis-Mortlock and Guerra, 1999). Furthermore, few GIS have incorporated accurate land cover information because of the difficulty in obtaining detailed land use data, particularly at the regional and catchment scales (Higgitt and Lu, 1999).

To date, remotely sensed Landsat imagery, which has been widely available for over 20 years, has been by far the most widely used land classification system in GIS. An early GIS approach, developed by Smith and Blackwell (1980) incorporated digital Landsat data, DTM, conventional maps and ground-truthing techniques for Lake Tahoe and its catchment. The GIS allowed correlation of the remotely sensed information and topographic data relating to parameters such as surface runoff, drainage basin area and terrain configuration (Baban, 1999). More recently, other sources of remotely sensed imagery have been used to extract land use information (Table 14.2), including the SPOT and AVHRR systems, previously discussed. There has also been an increase in the use of aerial photography to validate this information.

Having classified land use for the study areas used in a GIS, distinct land cover groups can be incorporated into the spatial analysis. As the influence of agriculture is of major concern in diffuse pollution processes, the analysis typically considers a number of issues, as identified by Higgitt and Lu (1999). For example, the relationship between the proportion of arable, grassland and other land use and overland flow is typically considered. This form of analysis is particularly evident in studies focusing on sediment yield estimations, which also incorporate a slope index calculation. However, GIS output following such procedures requires some calibration with observed data, as a number of studies have demonstrated (e.g. De Roo and Jetten, 1999a,b; Funke, 1999; Lee *et al.*, 2000). In a study aimed at estimating sediment yields from arable land in the Upper Yangtze, China, Higgitt and Lu (1999) stated that the distribution of sediment yields of a particular magnitude from a grid coverage of cultivated land contrasted with observed amounts. Nevertheless, there was generalized grouping of data within the model output, coinciding with broad patterns observed within the catchment. For example, the lowest specific sediment yields occurred in areas of low overland flow and limited arable land. Higgitt and Lu (1999) concluded that percentage cropland may be used as a crude indicator of erosion potential, the assumption being that it is the extension of arable land on to steep slopes that is influential in the apparent increase in land areas affected by soil erosion.

Furthermore, the effects of land use on non-point pollution at the local scale and interaction with land use impacts at the larger catchment scale are complex. Land use is only one of a number of factors that influence the transfer and delivery of diffuse pollutants, and attempts to segregate the impact of distinct land uses from other controlling processes are confounded by the interaction of variables and scale effects. Indeed, accurate classification of land use alone can be problematic, particularly at the small catchment scale. Typically, the increase in spatial resolution, necessary for detailed small-scale GIS, is accompanied by variability within remotely sensed land cover grids (Aplin *et al.*, 1999). This inevitably leads to deterioration in the classification accuracy at the per-pixel level (Townshend, 1981, 1992; Cushnie, 1987). Several studies have aimed to resolve this difficulty through analysis of groups of pixels within land cover parcels (Pedley and Curran, 1991; Johnsson, 1994; Harris and Ventura, 1995; Janssen and Molenaar, 1995).

Occasionally, it is not possible to test whether GIS output is representative of real events, as appropriate observation data are not always available. This has been demonstrated by Leek and Olsen (2000) in a GIS-based assessment of temporal trends in soil erosion events in Denmark. Moreover, in a GIS-based erosion assessment using the Revised Universal Soil Loss Equation (RUSLE) within the Walnut Gulch experimental watershed in south-east Arizona, USA, estimations were lower than field sediment yields observed in normal and high rainfall years (Yitayew *et al.*, 1999). In concluding the assessment, Yitayew *et al.* (1999) postulated that utilizing RUSLE with GIS to predict erosion still requires quantification of channel erosion and deposition to achieve accurate results.

Table 14.2. A number of land use classification methods for integration with diffuse pollution-oriented GIS.

GIS objective	Study area/scale	Land use classification method	Validated using	Reference
Establish NO_3^- concentrations from agricultural and urban catchment	Fish River basin, Alabama, USA, 138 km^2	Landsat Thematic Mapper and SPOT panchromatic data	National aerial photography programme (NAPP)	Basnyat *et al.*, 2000
Examine the relationship between land use and sediment yield	Upper Yangtze, China, not specified	AVHRR derived DTM, ecosystem interpretation and county level land use data	No validation undertaken	Higgitt and Lu, 1999
Quantification of groundwater NO_3^- transport from agricultural catchment	Poitou-Charantes, France, 20 km^2	Crop maps and farm information and Agricultural Department data on cultivation practices	No validation undertaken	Lasserre *et al.*, 1999
Estimate potential eroded sediment delivery to streams within a wetland catchment	Beaver Dam River, Wisconsin, USA, 410 km^2	Wisconsin Wetland Inventory and Natural Resources Conservation Service wetland inventory	Landsat Thematic Mapper	Richardson and Gatti, 1999
Calibration and validation of the catchment-based hydrochemical (dissolved N and P) transport model, 'GWLF'	Choptank River Basin, Chesapeake Bay, USA, 2057 km^2	Derived from digital coverages of Maryland Dept of Planning and Delaware Dept of Agriculture	No validation undertaken	Lee *et al.*, 2000
Estimation of dissolved N and P delivery to surface waters	Windrush and Slapton catchments, UK, 363 km^2, 46 km^2	UK Ministry of Agriculture, Fisheries and Food Annual Agricultural Census and Population census	No validation undertaken	Johnes *et al.*, 1996
Estimation of total P transfer from agricultural land	England and Wales, national	UK Ministry of Agriculture, Fisheries and Food Annual Agricultural Census	No validation undertaken	Fraser and Harrod, 1998
Development of GIS for catchment management	Namoi Basin, New South Wales, Australia, 41,998 km^2	Aerial photography and the Genamap data-set (spatial data of terrain, geology, land use, soil)	Field study ground-truthing	Zhang and Beavis, 1999
Regional assessment of soil erosion using the SEMMED model and remotely sensed data	Southern area of the Ardèche province, France, 12 km^2	Landsat Thematic Mapper imaging on four occasions	Functions between field data and spectral indices	De Jong *et al.*, 1999
Application and calibration of the INCA model	River Tywi, South Wales, UK, 1090 km^2	Institute of Terrestrial Ecology land cover classes	No validation undertaken	Whitehead *et al.*, 1999

The need for improvement of the methods of extrapolating RUSLE with GIS was also demonstrated in the findings of Formaggio *et al.* (1998), where soil erosion losses were estimated in a catchment in Sao Paulo State, Brazil.

Nevertheless, a number of studies have isolated the relationship between the potential for diffuse pollutant transfer and land use. For example, Meals and Budd (1998) demonstrated the relationship between land use and P loads in the Lake Champlain basin, USA. This was achieved by developing a loading function model that combined P concentration coefficients with regional hydrology data. In addition, Tufford *et al.* (1998) developed multiple regression models for eight catchments on the coastal plain of South Carolina, USA, by integrating land use and in-stream concentrations of total N and total P. Land use data were obtained from archive sources and provided a means of delineating diffuse source effects (Sweeney, 1999).

The precise effects of land use on diffuse pollution processes remain, therefore, inconclusive. Future improvements in the resolution and availability of land use datasets, and their integration with other spatial information such as soil and climate would enable more accurate predictions of the form and magnitude of diffuse pollutants.

Soil hydrology and pathways from agricultural land to water

As an expression of the interrelationships between geology, geomorphology, topography, climate and land management, the inclusion of soils data within numerous environmental-driven GIS has proved to be particularly important in identifying a range of characteristics within given study areas. In the construction of GIS-driven diffuse pollution tools, by far the most critical factors are soil hydrology and the quantification of dominant pathways. An understanding of variability in the physical and chemical properties of soil types is essential if estimates of the amounts of nutrients and sediment are to be calculated (Basnyat *et al.*, 2000).

The simulation of water movement through catchments has been the most fundamental requirement in the modelling of diffuse pollution from agriculture and has been developed more than other aspects of nutrient and sediment processes. In order to accurately depict water flux incorporating the interrelated concepts of partial contributing areas (Gburek and Sharpley, 1997, 1998) (or 'contributing zone' (Basnyat *et al.*, 2000)), terrain control of water flow paths and soil hydrology (Pionke *et al.*, 1997), a number of GIS approaches have been developed using DTM/DEM and spatial data-sets. An example is the Water Erosion Prediction Project (WEPP) (Cochrane and Flanagan, 1999), a process-based continuous simulation erosion model (Flanagan and Nearing, 1995). The WEPP model has been integrated with a GIS to estimate catchment runoff and sediment yield in order to improve the application of the model. Within the WEPP approach DTM/DEM coverages are used to represent topography within catchments via conversion from triangular irregular networks (TIN). The subsequent DEM is grid-based; each grid point represents a cell of a certain size or resolution.

Using a similar grid-based approach, Fraser (2000) constructed a GIS to estimate diffuse P transfer from agricultural land in England and Wales through integration of a KBS which incorporated soil information, DTM and land use data. The approach provides annual estimated P transfers along a number of hydrological pathways which can be subsequently summed to estimate a total P transfer rate (Fig. 14.3).

Mapping of features such as drainage networks, slope form and steepness can be achieved through flow-routing algorithms that determine the steepest descent direction and gradient between cells (Cochrane and Flanagan, 1999). Such flow-routing algorithms are described in O'Callaghan and Mark (1984), Zevenbergen and Thorne (1987), Jenson and Dominique (1988) and Martz and Garbrecht (1992). These algorithms have been used to integrate GIS and sediment generation models, such as the Universal Soil Loss Equation (USLE)

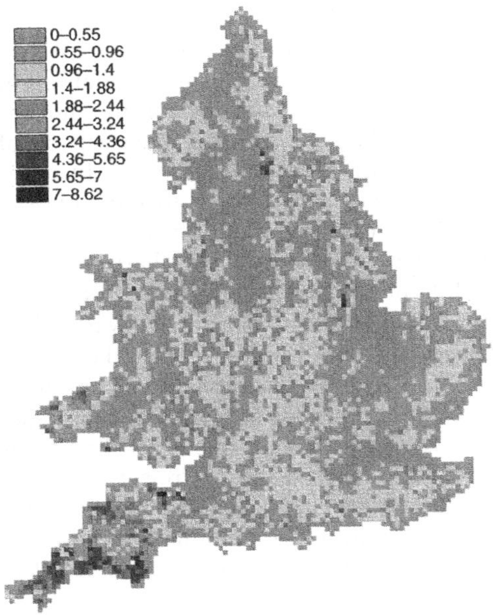

Fig. 14.3. Estimated diffuse annual total P transfer (kg ha^{-1}) from agricultural land in England and Wales, 1988 (after Fraser, 2000).

(Desmet and Govers, 1996). In addition, models such as ANSWERS (Areal Non-point Watershed Response Simulation), AGNPS (AGricultural Non-Point Source) and SWAT (Soil and Water Assessment Tool) (Srinivasan and Arnold, 1994) all incorporate flow-routing algorithms to estimate runoff amounts. Through these approaches, GIS analysis using DEMs/DTMs has proved to be a useful tool for the parametrization of hydrological drainage networks at a number of scales.

The progress made in the understanding of topographic control and subsurface dynamics of water movement represents an important step in the integration of modelling approaches with GIS. This has been achieved with models such as TOPMODEL, developed by Beven and Kirkby (1979), which attempt to simulate the saturation excess overland flow pathway. While hydrologists have recognized the importance of these pathways in the generation of streamflow, their significance in the area of nutrient and sediment transfer from land to water is less well understood (Garen *et al.*, 1999). It has been demonstrated, however, that sediment losses and associated nutrient transfers, caused by both saturation excess and infiltration excess overland flow during winter periods can be considerable (Kwaad, 1991; Catt *et al.*, 1994, 1998). Furthermore, it has been suggested by Moore *et al.* (1988), Baade *et al.* (1993), Huang and Laflen (1996) and Poesen *et al.* (1996) that topographic convergence and subsurface hydrology are related to the amount of erosion and nutrient transfer due to their contribution to the formation of ephemeral rills and gullies. There is still a need within this area to develop and refine existing gully-erosion models, such as those of Thorne and Zevenbergen (1990) and the Soil Conservation Service (1992), to account for these contributions.

Spatial analysis techniques which incorporate GIS and algorithms for DTM/DEM interpolation can be used to assist in the description of hydrological processes within sediment, nutrient and pesticide transfer models (Quinn *et al.*, 1991; Costa-Cabral and Burges, 1994; Garbrecht and Martz, 1996; Tarboton, 1997). Such techniques may be used to identify catchment boundaries, determine extended drainage networks, and establish overland flow routes (Garen *et al.*, 1999). Through these approaches, detailed information may be provided, contributing to the development of either fully spatially distributed models, such as the Système Hydrologique Européan (SHE) model (Wicks and Bathurst, 1996), the LImburg Soil Erosion Model (LISEM) (De Roo *et al.*, 1996; De Roo and Jetten, 1999a) and the hydrology–vegetation model for complex terrain (Wigmosta *et al.*, 1994); or semi-distributed models such as TOPMODEL (Beven and Kirkby, 1979) and the approaches described by Schumann and Funke (1996) and Schultz (1996). By comparing runoff data from two small, adjacent agricultural catchments, Bonta (1998) demonstrated that annual runoff amounts were larger in one catchment than the other. These results indicated that deterministic and random spatial variability of soil information with soil map units, at small scales, would be necessary for adequate modelling of catchment hydrology.

The Application and Integration of Geographical Information Systems with Appropriate Models

While GIS can be powerful tools for DSS and spatial analysis, their inherent capabilities for complex and dynamic analysis are limited (Sweeney, 1999). Simulation models, however, are primarily designed for extensive analysis of large data-sets, but lack the visualization and spatial-analysis functions that GIS provides. The integration of GIS and models that seek to simulate agricultural and environmental interactions enables more accessible and functional systems to be developed.

Selecting an appropriate model for combination with a GIS, can be a complicated and confusing task, particularly for the non-expert (Garen *et al.*, 1999). The problem arises because many current models have been developed independently by individual scientists or working groups. Each model may incorporate a number of components borrowed from a number of approaches, but each will display unique properties and algorithms. The result is that there is overlap in the capability of models. For example, in recognizing this problem, the United States Department of Agriculture has conducted a study to compare and evaluate the components of two catchment-scale models, SWAT and AGNPS. SWAT is based on spatially lumped techniques, where a single spatial unit represents a sub-catchment. AGNPS, on the other hand, is grid-based and does not represent sub-catchment variability. The aim of this comparative exercise is to reduce the duplication of effort when constructing models and GIS. Several studies have described the integration of GIS and SWAT. Yildirim *et al.* (1997) demonstrated the integration of SWAT with GIS to examine the most appropriate management practices for a region in central Nebraska, where the concentration of NO_3^--N in the groundwater exceeds maximum permitted concentrations. Srinivasan *et al.* (1997) applied SWAT to the entire Rio Grande/Rio Bravo river basin following a compilation of hydrographic and geographic maps and databases developed to simulate the hydrology of the river basin. Similarly, ArcView©, a desktop mapping software, was used by Di Luzio *et al.* (1997) to construct a user friendly GIS tool with SWAT. The resulting software assisted water resource managers to improve the efficiency of analyses for diffuse- and point-source pollution assessment and management plans. In addition, Qiu and Prato (1998) determined the most cost-effective spatial pattern of farms for improving water quality and evaluated the economic value of riparian buffers in reducing agricultural diffuse source pollution in a catchment in the US mid-west by modifying the SWAT model using GIS techniques.

Feijtel *et al.* (1998) developed a novel model and integrated GIS to estimate the distribution of predicted environmental concentrations (PEC), both spatially and temporally, of drain-delivered chemicals in European surface waters on river and catchment scales. The approach used a GIS for data storage and visualization, combined with established mathematical models for prediction of chemical fate. Output from the GIS incorporates an interactive graphical user interface (GUI) and animated map coverages to identify vulnerable areas and critical sources of diffuse pollution.

In an attempt to evaluate the effectiveness of P management practices on Lake Okeechobee, Florida, a GIS-driven model (GIS-PLAT) was developed by Zhang *et al.* (1997). This, and similar GIS, such as the approach described by Cluis *et al.* (1997), distinguish between point and diffuse sources of pollution and therefore improve the tools available for the management of water quality. These GIS enable the targeting of potentially critical areas where BMPs should be implemented as a priority.

Therefore, movements towards reducing the impacts of diffuse pollution of surface waters are being addressed through the application of GIS/model integrations. For example, Bachhaus and Braun (1998) developed a GIS based on a land use classification derived from Landsat data, a DEM and a calibrated relief-dependent water distribution model to provide an assessment of sustainable land management. However, in some circumstances, diffuse pollution models and GIS are applied to areas where no hydrological and chemical loading information is available, making calibration impossible. Under these conditions, it is crucial that a suitable model is selected, which can produce reasonably accurate results in the absence of calibration data. It must be stressed, however, that calibration and validation of model output is always necessary as simulation error is far greater without calibration stages (Grunwald and Frede, 1999).

Conclusions

Ultimately, GIS provide a useful and accessible means of introducing information and knowledge which can assist in the decision making process, in terms of targeting management strategies for the remediation of diffuse pollution. Linked with modelling capabilities, GIS can provide a valuable tool for identifying and quantifying diffuse pollution within agricultural catchments. Moreover, the use of GIS integrated with a transport model facilitates the determination of the effects of implementing alternative management and remediation strategies on water quality. Consequently, one of the most important outcomes of the next generation of diffuse pollution GIS will be the inclusion and precise representation of remediation measures, reflecting the contribution of specific hydrological and erosional processes occurring. With respect to water quality issues specifically, the methods with which flow pathways of water are calculated could make significant improvements in the accuracy of prediction of diffuse nutrient, pesticide and sediment loads delivered to receiving waters.

References

Adinarayana, J., Gopal Roa, K., Rama Krishna, N., Venkatachalam, P. and Suri, J.K. (1999) A rule-based soil erosion model for a hilly catchment. *Catena* 37, 309–318.

Aplin, P., Atkinson, P.M. and Curran, P.J. (1999) Fine spatial resolution simulated satellite sensor imagery for land cover mapping in the UK. *Remote Sensing of Environment* 68, 206–216.

Baade, J., Barsch, D., Mäusbacher, R. and Schukraft, G. (1993) Sediment yield and sediment retention in a small loess-covered catchment in SW Germany. *Zeitschrift für Geomorphologie* 92, 217–230.

Baban, S.M.J. (1999) Use of remote sensing and geographical information systems in developing lake management strategies. *Hydrobiologia* 395/396, 211–226.

Bachhaus, R. and Braun, G. (1998) Integration of remotely sensed and model data to provide the spatial information basis for sustainable land use. *Acta Astronautica* 45, 541.

Barrett, E.C. and Curtis, L.F. (1992) *Introduction to Environmental Remote Sensing*, 3rd edn. Chapman & Hall, London.

Basnyat, P., Teeter, L.D., Lockaby, B.G. and Flynn, K.M. (2000) The use of remote sensing and GIS in watershed level analyses of non-point source pollution problems. *Forest Ecology and Management* 128, 65–73.

Beven, K.J. (1986) Hillslope runoff processes and flood frequency characteristics. In: Abrahams, A.D. (ed.) *Hillslope Processes*. Allen and Unwin, Winchester, Massachusetts, pp. 187–202.

Beven, K.J. (1995) Linking parameters across scales – Subgrid parameterizations and scale dependent hydrological models. *Hydrological Processes* 9, 507–525.

Beven, K.J. and Kirkby, M.J. (1979) A physically-based variable contributing area model of basin hydrology. *Hydrolological Sciences Bulletin* 24, 43–69.

Bhaskar, N.R., James, W. and Devulapalli, R.K. (1992) Hydrologic parameter estimation using geographic information systems. *Journal of Water Resources Planning and Management ASCE* 118, 492–513.

Bonta, J.V. (1998) Spatial variability of runoff and soil properties on small watersheds in similar soil-map units. *Transactions of the American Society of Agricultural Engineers* 41, 575.

Boutwell, S.H., Brown, S.M., Roberts, B.R. and Attwood, D.F. (1985) *Modelling Remedial Actions at Uncontrolled Hazardous Waste Sites*, Report No. EPA/540/2–85/001. Anderson-Nichols and Co. Inc., Prepared for Environmental Protection Agency, Hazardous Waste Engineering Research Laboratory, Cincinnati, Ohio.

Catt, J.A., Quinton, J.N., Rickson, R.J. and Styles, P. (1994) Nutrient losses and crop yields in the Woburn erosion reference experiment. In: Rickson, R.J. (ed.) *Conserving Soil Resources: European Perspectives.* CAB International, Wallingford, UK, pp. 94–104.

Catt, J.A., Howse, K.R., Brockie, D., Harris, G.L., Pepper, T.J., and Goss, M.J. (1998) Effects of tillage practices on nitrate and phosphorus losses in the clay soil of the Brimstone Experiment, UK. In: *16th World Congress of Soil Science. Montpellier, France*, 20 pp.

Cluis, D., Quentin, E. and Gangbazo, G. (1997) Targeting critical areas vulnerable to agricultural pollution using a Geographic Information System. *Transactions of the American Society of Agricultural Engineers* 2, 5.

Cochrane, T.A. and Flanagan, D.C. (1999) Assessing water erosion in small watersheds using WEPP with GIS and digital elevation models. *Journal of Soil and Water Conservation* 54.

Costa-Cabral, M.C. and Burges, S.J. (1994) Digital elevation model networks (DEMON): a model of flow over hillslopes for computation and dispersal areas. *Water Resources Research* 30, 1681–1692.

Curran, P.L. (1985) *Principles of Remote Sensing.* Longman, London.

Cushnie, J.L. (1987) The interactive effect of spatial resolution and degree of internal variability within land-cover types on classification accuracies. *International Journal of Remote Sensing* 8, 15–29.

Daly, C., Neilson, R.P. and Phillips, D.L. (1994) A statistical-topographical model for mapping climatological precipitation over mountainous terrain. *Journal of Applied Meteorology* 33, 140–158.

De Jong, S.M., Paracchini, M.L., Bertolo, F., Folving, S., Megier, J. and de Roo, A.P.J. (1999) Regional assessment of soil erosion using the distributed model SEMMED and remotely sensed data. *Catena* 37, 291–308.

De Roo, A.P.J. and Jetten, V.G. (1999a) Calibrating and validating the LISEM model for two data-sets from The Netherlands and South Africa. *Catena* 37, 477–494.

De Roo, A.P.J. and Jetten, V.G. (1999b) Modelling soil erosion by water at the catchment scale. *Catena* 37, 275–276.

De Roo, A.P.J., Wesseling, C.G., Jetten, V.G. and Ritsema, C.J. (1996) LISEM: a physically-based hydrological and soil erosion model incorporated in a GIS. In: Kovar, K. and Nactnebel, H.P. (eds) *Application of Geographic Information Systems in Hydrology and Water Resources Management*, IAHS Publication No. 235. International Association of Hydrological Sciences, Wallingford, UK, pp. 395–403.

Desmet, P.J.J. and Govers, G. (1996) A GIS procedure for automatically calculating the USLE-LS factor on topographically complex landscape units. *Journal of Soil and Water Conservation* 51, 427–433.

Di Luzio, M., Srinivasan, R. and Arnold, J.G. (1997) Integrated user interface for SWAT using ArcView and AVENUE. *Transactions of the American Society of Agricultural Engineers* 2, 16.

Evans, R. (1996) *Soil Erosion and its Impacts in England and Wales.* Friends of the Earth, London, UK.

Favis-Mortlock, D. (1995) The use of synthetic weather for soil erosion modelling. In: McGregor, D.F.M. and Thompson, D.A. (eds) *Geomorphology and Land Management in a Changing Environment.* John Wiley & Sons, Chichester, UK, pp. 265–282.

Favis-Mortlock, D.T. and Guerra, A.J.T. (1999) The implications of general circulation model estimates of rainfall for future erosion: a case study from Brazil. *Catena* 37, 329–354.

Fedra, K. (1996) Distributed models and embedded GIS. In: Goodchild, M.F. (ed.) *GIS and Environmental Modelling: Progress and Research Issues.* GIS World Books, Fort Collins, Colorado, pp. 413–417.

Feijtel, T., Boeije, G., Matthies, M., Young, A., Morris, G., Gandolfi, C., Hansen, B., Fox, K., Matthijs, E., Koch, V., Schroder, R., Cassani, G., Schowanek, D., Rosenblom, J. and Holt, M. (1998) Development of a geography-referenced regional exposure assessment. *Journal of Hazardous Materials* 61, 59.

Flanagan, D.C. and Nearing, M.A. (1995) *USDA-Water Erosion Prediction Project: Hillslope Profile and Watershed Model Documentation,* NSERL Report No. 10. USDA-ARS National Soil Erosion Laboratory, West Lafayette, Indiana.

Formaggio, A.R., Gameiro, M.G. and Epiphanio, J.C.N. (1998) Soil erosion modeling using 'USLE': Two approaches for evaluating the parameters 'L' and 'S'. *International Geosciences Remote Sensing Symposium* 2, 856.

Fox, D.M. and Le Bissonnais, Y. (1998) Process-based analysis of aggregate stability effects on sealing, infiltration, and interill erosion. *Soil Science Society of America Journal* 62, 717–724.

Fraser, A.I. (2000) Towards a systematic framework for quantifying and extrapolating diffuse phosphorus transfer from agricultural fields to the national scale. PhD thesis, Cranfield University, Silsoe, UK.

Fraser, A.I. and Harrod, T.R. (1998) A systematic approach to predict phosphorus transfer from agriculture using GIS. In: *Practical and Innovative Measures for the Control of Agricultural Phosphorus Losses to Water.* Greenmount College of Agriculture and Horticulture, Belfast, Northern Ireland, pp. 182–183.

Fraser, A.I., Harrod, T.R. and Haygarth, P.M. (1999) The effect of rainfall intensity on soil erosion and particulate phosphorus transfer from arable soils. *Water Science and Technology* 39, 41–45.

Funke, R. (1999) Parameterization and regionalization of a runoff generation model for heterogeneous catchments. *Physics and Chemistry of the Earth Part B – Hydrology, Oceans and Atmosphere* 24, 49–54.

Garbrecht, J. and Martz, L.W. (1996) Digital landscape parameterization for hydrological applications. In: Kovar, K. and Nachtnebel, H.P. (eds) *Application of Geographic Information Systems in Hydrology and Water Resources Management,* IAHS Publication No. 235. International Association of Hydrological Sciences, Wallingford, UK, pp. 169–173.

Garen, D., Woodward, D. and Geter, F. (1999) A user agency's view of hydrologic, soil erosion and water quality modelling. *Catena* 37, 277–290.

Garen, D.C., Johnson, G.L. and Hanson, C.L. (1994) Mean areal precipitation for daily hydrologic modelling in mountainous regions. *Water Resources Bulletin* 30, 481–491.

Gburek, W.J. and Sharpley, A.N. (1997) Delineating sources of phosphorus export from agricultural watersheds. In: *ASAE Annual International Meeting.* American Society of Agricultural Engineers, Minneapolis, Minnesota.

Gburek, W.J. and Sharpley, A.N. (1998) Hydrologic controls on phosphorus loss from upland agricultural watersheds. *Journal of Environmental Quality* 27, 267–277.

Gilliland, M.W. and Baxter-Potter, W. (1987) A geographic information system to predict non-point source pollution potential. *Water Resources Bulletin* 23, 281–291.

Grunwald, S. and Frede, H.G. (1999) Using the modified agricultural non-point source pollution model in German watersheds. *Catena* 37, 319–328.

Hallett, S.H., Jones, R.J.A. and Keay, C.A. (1996) Environmental information systems developments for planning sustainable land use. *International Journal of Geographic Information Systems* 10, 47–64.

Harris, P.M. and Ventura, S.J. (1995) The integration of geographic data with remotely sensed imagery to improve classification in the urban area. *Photogrammetric Engineering and Remote Sensing* 61, 993–998.

Harris, T.M. and Boardman, J. (1998) Alternative approaches to soil erosion prediction and conservation using Expert Systems and Artificial Neural Networks. In: Boardman, J. and Favis-Mortlock, D.T. (eds) *Modelling Soil Erosion by Water,* NATO-ASI Series I-55. Springer, Berlin, pp. 461–478.

Higgitt, D.L. and Lu, X.X. (1999) Challenges in relating land use to sediment yield in the Upper Yangtze. *Hydrobiologia* 410, 269–277.

Hiscock, K.M., Lovett, A.A., Brainard, J.J. and Parfitt, J.P. (1995) Groundwater vulnerability assessment: two case studies using GIS methodology. *Quarterly Journal of Engineering Geology* 28, 179–194.

Huang, C. and Laflen, J.M. (1996) Seepage and soil erosion for a clay loam soil. *Soil Science Society of America Journal* 60, 408–416.

Hutchinson, M.F. and Dowling, T.I. (1991) A continental hydrological assessment of a new grid-based digital elevation model of Australia. *Hydrological Processes* 5, 45–58.

Janssen, L.L.F. and Molenaar, M. (1995) Terrain objects, their dynamics and their monitoring by the integration of GIS and remote sensing. *IEEE Transactions on Geoscience and Remote Sensing* 33, 749–758.

Jenson, S.K. and Dominique, J.O. (1988) Extracting topographic structure from digital elevation data for geographical information system analysis. *Photogrammetric Engineering and Remote Sensing* 54, 1593–1600.

Johnes, P., Moss, B. and Phillips, G. (1996) The determination of total nitrogen load and total phosphorus concentrations in freshwaters from land use, stock headage and population data: testing of a model for use in conservation and water quality management. *Freshwater Biology* 36, 451–473.

Johnson, G.L., Hanson, C.L., Hardegree, S.P. and Balard, E.B. (1996) Stochastic weather simulation: overview and analysis of two commonly used models. *Journal of Applied Meteorology* 35, 1878–1896.

Johnsson, K. (1994) Segment-based land-use classification from SPOT satellite data. *Photogrammetric Engineering and Remote Sensing* 60, 47–53.

Kinnell, P.L.A. (1983) The effect of kinetic energy of excess rainfall on soil loss from non-vegetated plots. *Australian Journal of Soil Research* 21, 445–453.

Kuznetsov, M.S., Gendugov, V.M., Khalilov, M.S. and Ivanuta, A.A. (1998) An equation of soil detachment by flow. *Soil and Tillage Research* 46, 97–102.

Kwaad, F.J.P.M. (1991) Summer and winter regimes of runoff generation and soil erosion on cultivated soils (The Netherlands). *Earth Surface Processes and Landforms* 16, 653–662.

Lasserre, F., Razack, M. and Banton, O. (1999) A GIS-linked model for the assessment of nitrate contamination in groundwater. *Journal of Hydrology* 224, 81–90.

Lee, K.Y., Fisher, T.R., Jordan, T.E., Correll, D.L. and Weller, D.E. (2000) Modelling the hydrochemistry of the Choptank River Basin using GWLF and Arc/Info: 1. Model calibration and validation. *Biogeochemistry* 49, 143–173.

Lee, T., Richargs, J.A. and Swain, P.H. (1987) Probabilistic and evidential approaches for multisource data analysis. *IEEE Transactions on Geoscience and Remote Sensing* 25, 283–293.

Leek, R. and Olsen, P. (2000) Modelling climatic erosivity as a factor for soil erosion in Denmark: changes and temporal trends. *Soil Use and Management* 16, 61–65.

Legates, R. and Willmott, C.J. (1990) Mean seasonal and spatial variability in gauge-corrected, global precipitation. *International Journal of Climatology* 10, 111–128.

Lu, X.X. and Higgitt, D.L. (1998) Recent changes of sediment yield in the Upper Yangtze, China. *Environmental Management* 22, 697–709.

Martin, D. (1991) *Geographic Information Systems and their Socioeconomic Applications*. Routledge, London.

Martz, L.W. and Garbrecht, J. (1992) Numerical definition of drainage network and subcatchment areas from digital elevation models. *Computers and Geoscience* 18, 747–761.

Massey, H.F. and Jackson, M.L. (1952) Selective erosion of soil fertility constituents. *Soil Science Society of America Proceedings* 16, 353–456.

Matthews, K.B., Sibbald, A.R. and Craw, S. (1999) Implementation of a spatial decision support system for rural land use planning: integrating geographic information system and environmental models with search and optimisation algorithms. *Computers and Electronics in Agriculture* 23, 9–26.

McKinney, D.C. and Tsai, H.L. (1996) Multigrid methods in GIS gridcell-based modelling environment. *Journal of Computers in Civil Engineering* 10, 25–30.

Meals, D.W. and Budd, L.F. (1998) Lake Champlain basin nonpoint source phosphorus assessment. *Journal of American Water Resources Association* 34, 251.

Meyers, L.D. (1981) How rainfall intensity affects interill erosion. *Transactions of the American Society of Agricultural Engineers* 22, 639–644.

Milliman, J.D. and Syvitski, J.P.M. (1992) Geomorphic/tectonic control of sediment discharge to the ocean: the importance of small mountainous rivers. *Journal of Geology* 100, 252–544.

Moore, D.M., Lees, B.G. and Davey, S.M. (1983) A new method for predicting vegetation distributions using decision tree analysis in a GIS. *Journal of Environmental Management* 15, 59–71.

Moore, I.D., Burch, G.J. and Mackenzie, D.H. (1988) Topographic effects on the distribution of surface soil water and the location of ephemeral gullies. *Transactions of the American Society of Agricultural Engineers* 30, 1403–1418.

Morgan, R.P.C. (1995) *Soil Erosion and Conservation*. Longman, Harlow, UK.

O'Callaghan, J.F. and Mark, D.M. (1984) The extraction of drainage networks from digital elevation data. *Computer Vision, Graphics and Image Processing* 28, 323–344.

Oliver, G., Burt, J. and Solomon, R. (1990) The use of surface runoff models for water quality decisions – a user's perspective. In: Decoursey, D.G. (ed.) *Proceedings of an International Symposium on Water Quality Modelling of Agricultural Non-Point Source Pollution*, ARS-81. USDA-ARS, Washington, DC.

Oloufa, A.A., El-Kadi, A.I., Eltahan, A.A. and Malik, H.U. (1995) An analytical groundwater approach for natural resources utilizing GIS. *Civil Engineering Systems* 12, 49–65.

Oxley, N.C. (1974) Suspended sediment delivery rates and the solute concentration of stream discharge in two Welsh catchments. In: Gregory, K.J. and Walling, D.E. (eds) *Fluvial Processes in Instrumented Watersheds.* Institute of British Geographers Special Publication No. 6, London, UK, pp. 141–154.

Pedley, M.I. and Curran, P.J. (1991) Per-field classification: an example using SPOT HRV imagery. *International Journal of Remote Sensing* 12, 2181–2192.

Penning-Rowsell, K.R. (1996) Creation and testing of a model to predict phosphorus loads in streams flowing into Bassenthwaite Lake. MSc thesis, University of Edinburgh, UK.

Perry, J. and Vanderklein, E. (1996) *Water Quality Management of a Natural Resource.* Blackwell Science, Cambridge, Massachusetts.

Pionke, H.B., Gburek, W.J., Sharpley, A.N. and Zollweg, J.A. (1997) Hydrologic and chemical controls on phosphorus losses from catchments. In: Tunney, H., Carton, O.T., Brookes, P.C. and Johnston, A.E. (eds) *Phosphorus Loss to Water from Agriculture.* CAB International, Wallingford, UK, pp. 225–242.

Poesen, J.W., Vandaele, K. and van Wesemael, B. (1996) Contribution of gully erosion to sediment production on cultivated lands and rangelands. In: Walling, D.E. and Webb, B.W. (eds) *Erosion and Sediment Yield: Global and Regional Perspectives,* IAHS Publication No. 236. International Association of Hydrological Sciences, Wallingford, UK, pp. 251–266.

Qiu, Z. and Prato, T. (1998) Economic evaluation of riparian buffers in an agricultural watershed. *Journal of the American Water Resources Association* 34, 877.

Quinn, P., Beven, K., Chevalier, P. and Planchon, O. (1991) The prediction of hillslope flow paths for distributed hydrological modelling using digital terrain models. *Hydrological Processes* 5, 59–79.

Refsgaard, J.C., Storm, B. and Abbott, M.B. (1996) Comment on 'A discussion of distributed hydrological modelling' by K. Beven. In: Abbott, M.B. and Refsgaard, J.C. (eds) *Distributed Hydrological Modelling.* Kluwer Academic Publishers, Dordrecht, The Netherlands.

Richardson, M.S. and Gatti, R.C. (1999) Prioritizing wetland restoration activity within a Wisconsin watershed using GIS modelling. *Journal of Soil and Water Conservation* 54, 537–542.

Rose, C.W. and Dalal, R.C. (1988) Erosion and runoff of nitrogen. In: Wilson, J.R. (ed.) *Advances in Nutrient Cycling in Agricultural Ecosystems.* CAB International, Wallingford, UK, pp. 212–235.

Schultz, G.A. (1996) Remote sensing applications to hydrology: runoff. *Hydrological Sciences Journal* 41, 453–475.

Schumann, A.H. and Funke, R. (1996) GIS-based components for rainfall-runoff models. In: Kovar, K. and Nachtnebel, H.P (eds) *Application of Geographic Information Systems in Hydrology and Water Resources Management,* IAHS Publication No. 235. International Association of Hydrological Sciences, Wallingford, UK, pp. 477–484.

Shaw, R.R. and Falco, J.W. (1990) A management perspective on water quality models for agricultural non-point sources of pollution. In: Decoursey, D.G. (ed.) *Proceedings of an International Symposium on Water Quality Modelling of Agricultural Non-Point Source Pollution,* ARS-81. USDA-ARS, Washington, DC.

Shepherd, B., Harper, D. and Millington, A. (1999) Modelling catchment-scale nutrient transport to watercourses in the UK. *Hydrobiologia* 395/396, 227–237.

Skidmore, A.K., Ryan, P.J., Dawes, W., Short, D. and O'Loughlin, E. (1991) Use of expert system to map forest soils from a GIS. *International Journal of Geographical Information Systems* 5, 431–445.

Smith, A.Y. and Blackwell, R.J. (1980) Development of an information data base for watershed monitoring. *Photogrammetric Engineering and Remote Sensing* 46, 154–165.

Smith, R.H., Sahoo, S.N. and Moore, L.W. (1992) A GIS-based synthetic watershed sediment routing model. In: *Proceedings of the Water Resources Planning and Management Conference, Baltimore.* USDA-ARS, Washington, DC, pp. 200–207.

Soil Conservation Service (1992) *Ephemeral Gully Erosion Model (EGEM) User Manual.* United States Department of Agriculture, Washington, DC.

Srinivasan, R. and Arnold, J.G. (1994) Integration of a basin-scale water quality model with GIS. *Water Resources Bulletin* 30, 453–462.

Srinivasan, R., Ramanarayanan, T.S., Jayakrishnan, R. and Wang, H. (1997) Hydrological modeling of Rio Grande/Rio Bravo basin. *Transactions of the American Society of Agricultural Engineers* 2, 18.

Sweeney, M.W. (1999) Geographic information systems. *Water Environment Research* 71, 551–556.
Tarboton, D.G. (1997) A new method for the determination of flow directions and upslope areas in grid digital elevation models. *Water Resources Research* 33, 309–319.
Thorne, C.R. and Zevenbergen, L.W. (1990) Prediction of ephemeral gully erosion on cropland in the south-eastern United States. In: Boardman, J., Foster, I.D.L. and Dearing, J.A. (eds) *Soil Erosion on Agricultural Land.* John Wiley & Sons, Chichester, UK, pp. 447–460.
Tim, U.S. (1996) Coupling vadose zone models with GIS: emerging trends and potential bottlenecks. *Journal of Environmental Quality* 25, 535–544.
Townshend, J.R.G. (1981) The spatial resolving power of Earth resources satellites. *Progress in Physical Geography* 5, 32–55.
Townshend, J.R G. (1992) Land cover. *International Journal of Remote Sensing* 13, 1319–1328.
Tufford, D.L., McKellar, H.N J. and Hussey, J.R. (1998) In-stream non-point source nutrient prediction with land-use proximity and seasonality. *Journal of Environmental Quality* 27, 100.
van Blargan, J.E. and Ragan, R.M. (1991) Automated estimation of hydrologic parameters. In: *Proceedings of the ASCE National Conference Civil Engineering Applications of Remote Sensing and Geographic Information Systems*, Washington, DC, pp. 278–285.
Viessman, W., Jr. and Hammer, M.J. (1993) *Water Supply and Pollution Control.* Harper Collins College Publications, New York.
Vieux, B.E. (1993) Aggregation and smoothing effects on surface runoff modelling. *Journal Computing in Civil Engineering* 7, 310–338.
Vieux, D.E., Bralts, V.F. and Segerlind, L.J. (1988) Finite element analysis of hydrologic response areas using geographic information systems. In: *Proceedings of the ASCE International Symposium*, Chicago, pp. 437–447.
Whitehead, P.G., Wilson, E. and Butterfield, D. (1999) Water quality processes in catchments: an integrated modelling approach for scenario analysis. In: Trudgill, S.T., Walling, D.E. and Webb, B.W. (eds) *Water Quality: Processes and Policy.* John Wiley & Sons, Chichester, UK, pp. 85–100.
Wicks, J.M. and Bathurst, J.C. (1996) SHESED: a physically based, distributed erosion and sediment yield component for the SHE hydrological modelling system. *Journal of Hydrology* 175, 213–238.
Wigmosta, M.S., Vail, L.W. and Lettenmaier, D.P. (1994) A distributed hydrology-vegetation model for complex terrain. *Water Resources Research* 30, 1665–1679.
Yang, M.D., Merry, C.J. and Sykes, R.M. (1999) Integration of water quality modelling, remote sensing, and GIS. *Journal of the American Water Resources Association* 35, 253–263.
Yildirim, Y.E., Skonard, C.J., Arumi, J., Martin, D.L. and Watts, D.G. (1997) Evaluation of best management practices using an integrated GIS and SWAT model for field sized areas. *Transactions of the American Society of Agricultural Engineers* 2, 9.
Yitayew, M., Pokrzywka, S.J. and Renard, K.G. (1999) Using GIS for facilitating erosion estimation. *Applied Engineering in Agriculture* 15, 295–301.
Zevenbergen, L.W. and Thorne, C.R. (1987) Quantitative analysis of land surface topography. *Earth Surface Processes and Landforms* 12, 47–56.
Zhang, L. and Beavis, S.G. (1999) Development of a spatial database for large-scale catchment management: Geology, soils and land use in the Namoi Basin, Australia. *Environment International* 25, 853–860.
Zhang, J., Guan, W. and Essex, A. (1997) GIS-based model for analyzing phosphorus loading and transport in the Lake Okeechobee watershed. *Transactions of the American Society of Agricultural Engineers* 2, 15.

15 Wetlands as Regulators of Pollutant Transport

Martin S.A. Blackwell, David V. Hogan and Edward Maltby
Royal Holloway Institute for Environmental Research, Department of Geography, Royal Holloway University of London, Surrey GU25 4LN, UK

Key words: buffer zone, constructed wetland, denitrification, ditches, nutrient transformation processes, wetland

Introduction

What are wetlands?

The term wetland is used to describe a wide range of often complex ecosystems, which occupy the transitional zone between aquatic and terrestrial environments. They are characteristically waterlogged for substantial periods by ground or surface water and can experience periodic or long-term flooding. Wetlands occupy *c.* 6% of the world's land surface (Maltby and Turner, 1983) and occur in all continents. They include systems influenced by both freshwater and saltwater and may comprise herbaceous (e.g. marshes and fens) or woody (e.g. cypress and mangrove swamps) vegetation, or may lack higher plant cover altogether (e.g. mudflats). Because of their wide geographical distribution, they include some of the most and the least productive ecosystems in the world. For example, the average annual primary production of peat bogs ranges from 150 t ha^{-1} in the humid tropics to only 1.7 t ha^{-1} in polar regions (Maltby and Turner, 1983). More than half of the global wetland resource is characterized by the accumulation of peat, comprising variably decomposed plant remains which, under waterlogged conditions, can accumulate progressively to as much as tens of metres in thickness. Among the many definitions of wetlands there is general agreement that they possess one or more of the following attributes (Mitsch and Gosselink, 1993): (i) the frequent presence of water either at the surface or in the root zone; (ii) characteristic (hydric) soil conditions different from adjacent non-wetland areas; (iii) vegetation adapted to permanently or seasonally wet conditions.

The wetlands most commonly associated with agricultural land and affected by its impacts are river marginal types and have been defined as follows:

> River marginal wetlands include all currently active and historical floodplain ecosystems in which inundation by surface water and/or the prevalence of an elevated water table are regular phenomena. Also included may be adjacent valley slopes which are regularly affected by overland flow, groundwater seepage, or otherwise have an elevated water table. Waterlogging should normally be of sufficient duration to support hydrophytic vegetation and to enable the development of hydromorphic soils. There may or may not be a hydrological connection between the floodplain and the valley slopes.
>
> (Maltby *et al.*, 1996)

Loss and degradation of wetlands

The interaction among physical, chemical and biological processes operating in wetlands can result in the performance of a variety of environmentally beneficial functions, some of which result from the regulation of both water flow and quality. Cycles of periodic wetting and drying and the consequent shift between anaerobic and aerobic conditions enable biogeochemical transformations to take place in wetland soils and sediments which, together with physical processes, can influence both the chemical and the biological quality of water. However, historically, wetlands have been seen as undesirable, as sources of flooding or disease, and generally posing a threat to society. In consequence, substantial areas have been drained for agricultural use, and more recently for urban expansion (Tiner, 1984; Hollis and Jones, 1991).

In the UK, changes to flooding regimes of rivers through impoundment, the building of flood defence structures and extensive land drainage, along with continued requirements for river and groundwater abstraction have led to concerns over the progressive degradation of wetlands through desiccation, particularly in the relatively dry south and east of England (Hogan and Maltby, 1992). Other factors impacting on wetlands in the UK have been afforestation, poor burning practices of semi-natural vegetation and the extraction of peat. However, wetland loss and degradation have resulted principally from agricultural activities, and in particular drainage for conversion to cropped land.

Increasing inputs of nutrients, both directly by fertilizer application and indirectly by the discharge of nutrient-rich agricultural runoff, have altered the composition of vegetation in many wetlands (English Nature, 1992). This, combined with adverse impacts from herbicides and the cessation of traditional grazing practices, has left many wetlands seriously degraded and under threat of complete loss (Williams, 1990). Though most of the wetlands that remain in the UK have been altered through the effects of management or other environmental impacts, they are usually still capable of performing important functions, though often to a lesser degree than if they had remained in a pristine condition. Many impacts such as nutrient enrichment are insidious in their effects, and become evident only through gradual declines in observable factors such as bird and fish populations, or changes in plant assemblages.

Kidneys of the landscape

The ability of wetlands to regulate both water quality and quantity has resulted in them being referred to as 'the kidneys of the landscape' (Mitsch and Gosselink, 1993). Of all wetlands, it is the destruction of those located on significant hydrological pathways (e.g. those that intercept and process nutrient-rich runoff originating from agricultural land) that has had the most deleterious effect on water quality. The effects of modern farming have been to reduce the capacity of wetlands to buffer the aquatic environment against damaging impacts. This ability of wetlands to act as buffer zones has motivated many investigations of the potential of using both natural and constructed wetlands for the removal of pollutants, especially nutrients, from both diffuse (Cooper *et al.*, 1986; Pinay and Décamps, 1988; Lowrance, 1992; Haycock and Pinay, 1993) and point sources (Godfrey *et al.*, 1985; Richardson, 1985; Hammer, 1989; Kadlec and Knight, 1996).

How Wetlands can Affect Water Quality

As already noted, although modern farming practices can often lead to the pollution of ground- and/or surface waters, natural processes taking place in wetlands can ameliorate the environmental impact of these pollutants (Table 15.1). A generalized diagram showing the most important pollutant control processes acting in wetlands is presented in Fig. 15.1.

Denitrification

Denitrification involves the reduction of oxidized forms of N, in particular NO_3^-, to gaseous forms, primarily dinitrogen (N_2) and nitrousoxide (N_2O). This reduction is carried out by facultative anaerobic bacteria which inhabit nearly all known environments, but are particularly abundant in wetlands (Groffman, 1994). An essential requirement of the process is anoxic conditions that stimulate denitrifying organisms to use NO_3^- as an electron acceptor in respiration in the absence of free O_2, causing NO_3^- to be progressively reduced to N_2. The characteristically high water tables found in wetlands result in a prevalence of anaerobic soil conditions which, in combination with the abundance of denitrifying bacteria, promote denitrification. The particular importance of this process in terms of nutrient removal is that it involves total export of N from the system to the atmosphere, rather than temporary storage by other processes such as plant uptake or organic matter accumulation in the soil. It is responsible for major losses of soil N with some estimates suggesting that up to 30% of applied fertilizer N can be lost via this process (Averill and Tiedje, 1981). Reported rates of N removal by denitrification in freshwater wetlands vary both temporally and spatially, ranging from < 0.1 to 1428 mg N m^{-2} day^{-1} (Johnston *et al.*, 1997). Some examples of rates measured in different wetlands are shown in Table 15.2.

Table 15.1. Wetland processes for amelioration of impacts of agriculturally derived pollutants.

Agriculturally derived pollutant	Associated water quality problems	Wetland process for amelioration
Nutrients (especially N and P)	Eutrophication Toxicity	Denitrification Precipitation Plant uptake Adsorption Sediment deposition/retention
Pesticides/herbicides	Toxicity	Adsorption Plant uptake
Sediment	Eutrophication Silting of gravels	Sediment deposition/retention Hydrological regulation
Pathogens	Disease	Sediment deposition/retention Adsorption Predation
Heavy metals	Toxicity	Plant uptake Sediment deposition/retention Precipitation Sorption
BOD and COD	De-oxygenation	Adsorption Sediment deposition/retention Oxidation/mineralization

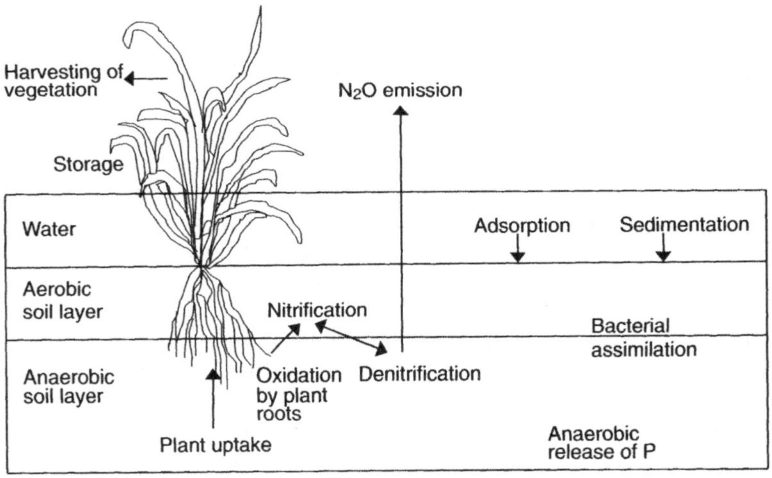

Fig. 15.1. Main processes acting in wetlands that contribute to their ability to remove nutrients from agricultural runoff.

Sedimentation

Land management can lead to erosion of soil and sediment which, together with scouring from river banks, can result in large sediment loads in rivers. Deposition of this material into river bed gravels can cause deoxygenation with subsequent loss of invertebrate habitats, and failure of spawning by salmonids and other fish. Nutrients, especially P, as well as pesticides, heavy metals and pathogens can all be sorbed on to, or associated with, sediment particles, while high BOD is often related to the presence of organic particles. Wetlands can help to reduce the sediment load in water by the following mechanisms.

Reducing flow velocity

This can result from reduction in slope gradient, discharge into ponded water bodies or increased friction and detention of surface water by vegetation, causing surface deposition of particulate matter from suspension (Dillaha *et al.*, 1989; Baker, 1992). Densely packed, fine vegetation such as grass offers the most resistance to flow at shallow depths (i.e. has the greatest roughness coefficient (Mitsch and Gosselink, 1993) though its impact can be reduced in deeper flows (Hammer, 1992). The greater the reduction in flow velocity or the longer the period of ponding, the greater the amount of sediment deposited. The lower the energy of an environment the greater the capability to deposit finer particles. Ponded water and large vegetation roughness coefficients often associated with wetlands mean that they are well suited to the removal of sediment in this way. The importance of this process for P retention was demonstrated by Mitsch *et al.* (1979) who reported P retention by sedimentation of 3.6 g P m^{-2} $year^{-1}$ in a riverine wetland. This was estimated as being 18 times that of all other P retention mechanisms.

Filtration by vegetation

Vegetation can act as a filter, depending on its structure and density. Larger soil particles can become trapped within litter layers or among stems and leaves (Dillaha and Inamdar, 1997). The usually abundant vegetation in wetlands and the presence of typically thick litter layers mean that they have a high capacity for sediment removal by this process.

Oxidation

Many wetland plants can release large amounts of O_2 from their roots, and consequently provide aerobic pockets within an otherwise anaerobic soil environment. This can be important for the coupling of nitrification (the transformation of NH_4^+ to NO_3^-) and denitrification processes (Reddy *et al.*, 1989; Lloyd, 1993), optimizing the removal of N from the wetland system. Additionally, the O_2 produced in this process can help to reduce BOD and COD levels in surface waters.

Chemical precipitation

Under certain conditions, P can be transformed from soluble to insoluble forms and precipitate out of solution. In low energy environments this precipitate will settle and become stored, again showing the value of sediment retention in reducing nutrient inputs to aquatic systems. Precipitates with Al and Fe readily form under acidic conditions, while combinations with Ca and Mg are more usual under alkaline conditions. However, the anaerobic conditions in wetlands can have the reverse effect and result in the mobilization of previously precipitated P, as can alterations in pH (Richardson, 1989). The relative importance of precipitation

Table 15.2. Examples of the range of denitrification rates measured in wetlands.

Wetland type and location	Denitrification rate (g N m^{-2} $year^{-1}$)	Reference
Alder and ash swamps, Denmark	Alder 4.8, Ash 2.9	Struwe and Kjoller (1989)
Organic, riparian wetland, New Zealand	408.8	Schipper *et al.* (1993)
Blanket bog, Pennines, UK	0.1	Martin and Holding (1978)
Bottomland hardwood swamps, Georgia, USA	31.5	Lowrance *et al.* (1984)
Peat fens, Somerset, UK	156.2	Kirkham and Wilkins (1993)

processes in preventing P transfer is not clear because particulate matter of this kind has been shown to be transferred along various subsoil pathways (Haygarth *et al.*, 1998), and therefore may be simply transported in a different form.

Plant uptake

Plant roots remove nutrients from soils by uptake of dissolved solutes in soil water, although direct exchange between nutrient ions adsorbed on to soil particles and H^+ ions on root surfaces can also take place (Reddy *et al.*, 1989). The biomass productivity of many wetlands is high, with consequent large capacities for nutrient removal by plant uptake. However, where plant growth is strongly seasonal, the process of nutrient removal and storage by this means is correspondingly limited to the growing season. During dormant periods, the senescence and decomposition of above ground plant material results in the release of nutrients which are recycled into the system. To avoid this, harvesting and removal of the vegetation is required to achieve nutrient removal. The amount of plant uptake depends on the growth rates of plants and the period of storage, which varies among plant types, but generally is longest in woody vegetation. However, nutrients taken up by herbaceous plants can enter into long-term storage if the vegetation becomes incorporated into peat, as often occurs in wetlands. Plant age and stage of development are also important determinants of nutrient uptake. In addition, heavy metals can become incorporated into plant material, as either short- or long-term storage, depending on plant types and conditions. A study of freshwater riverine wetlands in Wisconsin, USA, reported nutrient uptake rates by *Scirpus fluviatalis* of 20.8 g N m^{-2} $year^{-1}$ and 5.3 g P m^{-2} $year^{-1}$, but retention of only 26% of the N and 38% of the P. The rest was returned to the wetland through litterfall and leaching (Klopatek, 1978).

Adsorption

Adsorption of dissolved materials on to soil particles through ionic bonding can be a significant process for the reduction of concentrations of various nutrients and heavy metals, and of dissolved BOD and COD. Since the capacity for adsorption depends on the amount of soil surface available, the small size and relatively large surface area of clay particles means that clay soils generally have the highest adsorption capacity of mineral soils. The type and amount of ions adsorbed also depends on the anion and cation exchange capacities of a soil which are influenced largely by the mineralogy of the clay particles (greatest in swelling clays such as smectite) and the pH of the soil. Cations are adsorbed more strongly than anions, but at low pH anion adsorption capacity increases. In addition to clays, humified soil organic material also has a high adsorptive capacity and, typically, both of these soil components are found in wetland soils. Mander *et al.* (1991) reported removal rates in riparian wetlands of 100% of incoming P, and attributed much of the removal to adsorption.

Bacterial assimilation

Soil bacterial populations can greatly increase under favourable conditions, resulting in the assimilation of large quantities of nutrients. However, the storage time can vary considerably, and unfavourable conditions can result in rapid reductions in bacterial numbers with the release of large quantities of nutrients.

Hydrological regulation

The ability of wetlands to reduce flood peaks in river discharge and contribute to the maintenance of base-flows can play a part in the improvement of water quality. Flood peak attenuation arises from any feature or process that results in the retention of water and can reduce flood discharge peaks in rivers, thereby reducing the likelihood of flooding downstream. The location and morphological configuration of wetlands determine how effectively they can intercept and detain both runoff from adjacent valley slopes and overbank flooding from a river channel (Novitzki, 1979; Verry and Boelter, 1979). During periods of hydrological recharge when water tables are at shallow depths, water storage capacity will be reduced, though detention can still occur as wetland vegetation often has a high roughness coefficient, with the consequent high friction acting to reduce flow rates and detain water. Flood peak attenuation tends to be reduced during winter and spring, when water tables are high and the aerial parts of herbaceous plants have died. Although flooding is a natural process in river dynamics and floodplain development, in the absence of wetlands, floods can result in severe erosion and riverbed scouring. Novitzki (1985) reported that floodflow in the Chesapeake Bay drainage basin, USA, was only 50% of that in similar catchments with no wetland storage. There can also be significant economic benefits from reducing flood risk to property. The US Army Corps of Engineers (1972) demonstrated that the 3400 ha of wetlands in the Charles River catchment prevent approximately US$17 million of flood damage year^{-1} because of their ability to regulate flooding.

While most wetlands make poor aquifers (Burt, 1995), river base-flow maintenance can arise through the ability of some wetlands to act like sponges, both by soaking up water, which can help to attenuate flood peaks (above), and by subsequently slowly releasing it, which can be a particularly useful process during drier periods. This slow release of often high quality water from wetlands enables river base-flows to be maintained and can also reduce the risk of damaging siltation occurring by maintaining sediment in suspension (Novitzki, 1985; Roulet, 1990).

Additional significant processes

Improvements in the quality of water draining from wetlands can result from the inactivation and predation of pathogenic bacteria and viruses. Protozoa, which are often abundant in wetlands, can consume large quantities of bacteria, while bacteriophages (viruses that infect bacteria) can reduce numbers of susceptible bacteria by fatal infection (Kadlec and Knight, 1996; Nuttall *et al.*, 1997).

Wetlands as Buffer Zones

Wetland functioning and buffer zones

Many of the functions performed by wetlands, such as the provision of cleaner water or a reduction in the frequency of flooding, can be of value to society. A current important research area is the linking of wetland functional assessment to the many techniques available for calculating the economic values of different wetland functions (Barbier *et al.*, 1997). Wetland functions arise from the performance of natural processes as described above (e.g. water storage resulting in flood attenuation and plant nutrient uptake resulting in water quality improvement: Maltby *et al.*, 1996). As a consequence of these functions, some wetlands are capable of protecting or 'buffering' rivers and lakes from potentially harmful impacts including those arising from modern farming practices. The concept of buffer zones is now widely accepted and their establishment and use is increasing in most developed countries (Haycock *et al.*, 1997). However, the terminology used in reference to buffer zones and wetlands can be confusing; some definitions are given here.

A buffer zone is a generic term and is defined as 'a vegetated area lying between agricultural land and a surface water body, and acting to protect the water body from harmful impacts such as high nutrient, pesticide or sediment loadings that might otherwise result from land use practices' (Blackwell *et al.*, 1999). It may or may not be a wetland. A buffer strip is an area of land acting as a buffer zone, but which usually is a narrow, linear feature, often situated along a field boundary or following a contour. A riparian buffer zone (or strip) is a buffer zone situated adjacent to a river. A wetland buffer zone is a buffer zone that comprises wetland, and may also be a buffer strip or a riparian buffer zone. From these definitions, it can be seen that the terms buffer zone and wetland are not synonymous. Whether or not a wetland is acting as a buffer zone will depend on the functions it is performing and the extent to which they are performed. Both of these will depend to some extent on position in the landscape and hydrological pathways.

It is a common misconception that buffer zones must be situated adjacent to water bodies in order to be most effective. Hydrological pathways are the routes of most pollutant transport into surface water bodies, and consequently buffer zones will only be capable of improving water quality if they are located in a position to intercept and process water draining from agricultural land. However, much agricultural drainage is by means of subsurface pipes or by ditches that often discharge directly into main watercourses, effectively by passing buffer zones. Whipple (1991) argues that large quantities of nutrient-rich water enter main channels via this route, rendering buffer zones in many areas largely ineffective. While riparian buffer zones usually offer some protection, supplementation with alternatively located buffer zones can be an effective approach to the protection of water quality.

Pollutant removal can be optimized by targeting areas as buffer zones to address specific problems. Locations selected can include ditch systems or areas of nutrient-rich groundwater discharge, which may be sited at some distance from main channels (Mockler *et al.*, 1994). Reduction of pollutant loadings in surface waters may also be achieved by diverting drainage from ditches into river marginal wetlands.

It is becoming increasingly common to use both natural and constructed wetlands for the treatment of polluted waters, especially those derived from agriculture. Such practices have been reported to be highly effective (Tilton and Kadlec, 1979; De Busk and Reddy, 1987; Knight *et al.*, 1987), although on natural sites degradation of habitat through changes in vegetation can occur as a result of increased nutrient inputs (Kadlec and Bevis, 1990). In consequence, it is important to identify areas suitable for such practices and assess the implications of using a wetland in this way in order to maximize the buffer zone potential while minimizing degradation.

Implementation of wetland buffer zones

In considering the establishment of wetland buffer zones, it may be that existing wetlands are already performing water quality functions to some extent, but their potential to carry these out may be limited by their location and configuration. These factors determine both the amount of runoff intercepted and the detention time of water in the wetland. The amount of runoff intercepted by a wetland will depend largely on its catchment area, that is, the source area of the water that drains into that wetland. The boundary is effectively delineated by the watershed. The catchment can be considered as the area within which a drop of water falling at any given point would actually enter the wetland before discharging into a surface water body. The point at which this fails to occur represents the watershed. The delineation of the catchment area in this way is often complicated by the presence of artificial land drainage structures, designed to intercept water that would otherwise pass through a wetland.

Whether or not a wetland can remove pollutants from water flowing into it will depend largely on the pollutant concentration and the residence time of water within the wetland. If a large volume of water discharges rapidly into a small wetland it is likely that most of the water will flow rapidly through it, with little improvement in quality. This is an example of hydrological short-circuiting, which may also occur when discharges into a wetland become focused or channelized, rather than operating as sheet or diffuse flow. Another means of hydrological short-circuiting is through artificial land drainage systems including pipes and via secondary treatments of moling and subsoiling. Their presence can severely compromise the opportunities for wetlands to remove pollutants from runoff.

The alteration of hydrological pathways to direct water through existing wetlands can usefully target natural features such as oxbow lakes in order to increase their nutrient removal functioning. These adjustments typically involve simple solutions such as the diversion of ditches and drains. However, the questions of size, nutrient status and wildlife conservation value of the potential receiving wetlands should be given careful consideration in order to target the required functions most effectively.

The capacity of an existing wetland to remove nutrients from runoff will depend largely on its size in relation to the quantity of water passing through it. If a source of high discharge is diverted into too small a wetland then not only will little nutrient removal occur due to hydrological short-circuiting, but undesirable changes in the wetland vegetation and loss of the soil and substrate through erosion may result. The relationships between wetland size, stream discharge, stream catchment area and potential of wetland for improving water quality can be complex, and other factors including catchment soil type, vegetation, land use, topography and climate will affect the quality and quantity of runoff that a wetland is able to process. Various techniques have been devised for determining minimal or optimal buffer zone widths and areas, some of which may be applicable to wetlands. For example, simple ratio methods based on comparisons of the area of wetland to the area of stream catchment exist; ratios ranging from 1:10 to 1:50 are most common (Williams and Nicks, 1993; Uusi-Kämppä and Yläranta, 1996). This method does not use site-specific data and is simple to apply once the ratio has been decided. Another simple approach is that devised by Trimble and Sartz (1957), which relates wetland buffer zone area to catchment slope and catchment area. More complex models have been devised such as the Riparian Ecosystem Management Model (REMM) (Lowrance *et al.*, 1998), Chemicals, Runoff and Erosion from Agricultural Management Systems Model (CREAMS) (Knisel, 1980), and The Water Erosion Prediction Project (WEPP) model (Flanagan and Nearing, 1995). All of these require considerable amounts of data relating to fundamental hydrological and erosion mechanics, making them inappropriate in many cases.

Consideration must be given to the potential impact of polluted water on the conservation value of a wetland to be used for nutrient removal. For example, oligotrophic wetlands containing plants adapted to a limited amount of available nutrients will be adversely affected by inputs of nutrient-rich waters, unlike a wetland of higher trophic status that may already have been impacted by pollutants. Examples of the efficiency of natural wetlands for the removal of N and P are included in Table 15.3.

Table 15.3. Examples of the efficiency of natural and created wetlands for N and P removal.

Function	Removal efficiency (%)	Wetland type	Reference
N-removal	89	Natural wetland, USA	Peterjohn and Correll (1984)
	56–100	Natural wetland, New Zealand	Cooper (1990)
	80–90	Created wetland, USA	Kadlec and Knight (1996)
P-removal	65	Natural wetland, Switzerland and France	Pommel and Dorioz (1995)
	27–88	Natural wetland, Estonia	Mander *et al.* (1991)
	98	Created wetland, Norway	Jenssen *et al.* (1993)
	20–42	Created wetland, Norway	Braskerud (1994)

The role of ditches

While wetlands can offer a means of reducing pollutant inputs to aquatic systems, there may be reluctance from land managers to maintain wetlands for a variety of reasons, not least the lack of financial incentive. However, one area where wetland functions can be delivered and which may be more acceptable to land managers under current circumstances is in the management of drainage ditches, which are one of the major pathways by which nutrients are transported to aquatic systems (Newbold *et al.*, 1989). It is possible to manage ditches to act, in effect, as linear wetlands, performing the functions and delivering the services normally provided by wetlands, but with little impact on surrounding land (Mander *et al.*, 1991). This overcomes the problems of flows being diverted away from wetlands since functional areas are actually located along the key hydrological pathways. Three main management techniques can be applied to ditches in order to increase their capacity to remove nutrients from runoff.

Modification of ditch clearance practice

It is common practice to clear and dredge drainage ditches regularly in order to maintain them for the purpose for which they were created; to move water rapidly from the land to a convenient surface water body such as a river or lake. Commonly, long stretches of ditch are cleared at one time, allowing the free and rapid movement of water and any pollutants it contains. During periods of high discharge, ditches may themselves become sources of sediment by the erosion of unprotected soils from their sides resulting from high velocity flows. These problems can be alleviated by simple changes to their management, such as clearing only short stretches of individual ditches at any one time, or delaying clearance until necessary.

Insertion of baffles

Baffles are simple barriers, usually made from wood, which can be inserted into ditches, particularly those that are deep and/or on moderate to steep slopes. This creates a series of pools in which sediment can be deposited, and eventually wetland vegetation may develop (Newbold *et al.*, 1989). This will offer the benefits of particulate pollutant removal and provide opportunities for the development of small wetland areas capable of improving the quality of water draining from them. Deep ditches are preferred for these practices so that the ponded water does not affect outfalls discharging drainage waters from adjacent fields.

Establishment of wetland vegetation

Ditches on gently sloping land may be widened and planted with wetland species such as common reed (*Phragmites australis*), creating, in effect, a linear wetland capable of performing wetland functions (Newbold *et al.*, 1989). Although individual wetlands created within ditches in this way will only be small in area, their combined impact on water quality may be significant.

Constructed Wetlands

Design and applications

Constructed wetlands are engineered systems designed to simulate natural wetlands for the purpose of water quality enhancement and/or habitat creation. They provide a low cost, low energy, robust, passive (requiring no chemical additions and only low maintenance) method for the treatment of wastewater. They can be applied to a wide range of situations and are able to treat episodic or continuous, low or high flows. Essentially, constructed wetlands are biological systems in which chemical and microbial interactions occur, although physical processes such as sediment deposition can be important. Basically they comprise vegetation and a support medium, over or through which water is directed in order to enable purification processes to take place. Applications include the treatment of primary, secondary or tertiary sewage effluent, surface water and stormwater from both agricultural and urban environments, acid mine drainage, landfill leachate and sludge dewatering. They compare well with other, more conventional types of wastewater treatment systems in terms of economics and water quality (Nuttall *et al.*, 1997). Some examples of the efficiency of constructed wetlands for the removal of N and P are included in Table 15.3.

One of the plant species most often used is the common reed (*Phragmites australis*), which can be harvested and utilized for a variety of purposes providing some economic return, including thatching and weaving. Species of willow (*Salix* spp.) are also used in many constructed wetlands as they have a high capacity for nutrient uptake (Brix, 1994). If managed by coppicing, there can be economic returns from the timber produced, which can be used for fuel, construction or shredded for mulch or compost. Sometimes, especially in developing countries, vegetation may be harvested and used in bio-gas production (Wiloso *et al.*, 1995).

Large constructed wetlands can provide a source of recreation, amenity or passive enjoyment. The constructed wetland at Barn Elms, in the heart of London, has not only attracted a wealth of wildlife, but also has enabled the development of luxury homes which have provided financial support for the project. The site provides both a popular attraction for birdwatchers and a valuable educational resource (K. Peberdy, London, 1999, personal communication).

Size is a major consideration when selecting the most suitable type of constructed wetland for a specific purpose. In the UK and more widely in Europe, where population densities are generally high and land is at a premium, subsurface flow wetlands (SSFW), the most compact type, are most popular. In the USA, where land is generally more available, surface flow wetlands (SFW) are more common. The main characteristics of these two types of constructed wetland are detailed in Table 15.4.

Table 15.4. Characteristics of constructed wetland types.

Wetland type	Sub-type	Size	Hydrology	Vegetation	Efficiency
Surface flow wetlands (SFW)	N/A	Usually large because of restricted interaction of water with support medium. Typically range from 100 m^2 to over 10,000 m^2	Water flows over the surface of the support medium, which restricts interaction with microbes. Flow is usually continuous	Can support emergent, submergent and floating vegetation	Highly efficient for lowering BOD, COD and particulate pollutants. Less efficient for dissolved pollutants, particularly N and P compounds
Subsurface flow wetlands (SSFW)	Horizontal flow	Usually much smaller than SFWs, typically ranging from 10 to 3500 m^2	Water flows through the support medium predominantly in a horizontal direction. Flow is usually continuous	Supports emergent vegetation only	Highly efficient for reducing BOD and COD, even at low temperatures, and more efficient than SFWs
	Vertical flow	Typically the smallest of all constructed wetlands ranging from a few square metres and not normally exceeding 100 m^2	Water flows through the support medium in a vertical direction. Flow is usually periodic	Supports emergent vegetation only	Highly efficient for removing dissolved N and P, but low efficiency for reducing BOD and particulate pollution

Key components of constructed wetlands

The main components that control the types of processes operating within constructed wetlands are the hydrology, support medium, vegetation and temperature.

Hydrology

The important hydrological factors are water depth, precipitation, evapotranspiration, flow rate, retention time (which may vary from a few hours to several weeks) and hydraulic loading.

Support medium

The support medium acts as a sediment sink for deposited suspended solids and particulate-bound pollutants, including a proportion of BOD and COD. It also acts as a rooting medium for the vegetation and a habitat for microscopic and macroscopic animals and plants. Additionally, it provides sites for chemical ionic exchange for adsorption and micro-zones for aerobic and anaerobic processes.

Vegetation

The vegetation affects the processes operating within a constructed wetland and gives rise to habitat value and economic returns. The influence of vegetation in a SFW is substantial. Roots and stems stabilize the support medium, reducing scouring. Shading by the vegetation canopy limits the risk of algal growth in the water column (which would detrimentally affect performance). Plants take up nutrients, heavy metals and other pollutants. Roots release small amounts of O_2 to provide support for microbial populations within the support medium. Vegetation also reduces water velocity giving enhanced retention time, regulates temperature by shading (in summer) and insulation (in winter), and provides wildlife habitat and aesthetic quality to the landscape. The three main genera considered to be the most useful in constructed wetlands are *Phragmites*, *Typha* and *Scirpus*, because of their rapid growth rates, high biomass production, large root development and tolerance to a variety of water conditions. *Phragmites* also provides a type of habitat (reedbed) that is valuable in terms of biodiversity, particularly in the UK, with a resource that can be harvested for economic purposes. Bullrushes (*Typha* spp.) are often used where heavy metals are a problem as they have a high capacity to remove these. However, many other species have been used in constructed wetlands with varying degrees of success depending on the nature of the pollutants and the water regimes employed (Nuttall *et al.*, 1997).

Temperature

Temperature is a key factor controlling types of processes and the rates at which they take place. Many of the processes important for water quality improvement are dependent on biological media such as uptake by plants and denitrification. Consequently, at low temperatures these important processes experience reduced rates, or may even cease.

Case Studies

Case study 1: Footslope–floodplain boundary wetland for nitrate removal

Tetcott Barton is situated on the eastern bank of the River Tamar, south-west England, at a point where the floodplain is approximately 250 m wide. The first 125 m of floodplain from the river carries deep, well drained, silty soils (entric and gleyic cambisols; FAO-UNESCO, 1974) and is used for pasture, while the next 125 m comprises poorly drained, silty fluvisols, which had been recently cultivated for flax production. The adjacent footslopes have dystric gleysols of low permeability, supporting willow scrub and tall herb fen. The backslope contains, mainly, more permeable dystric cambisols and lies immediately below flatter land used for arable cropping. Figure 15.2 represents a cross-

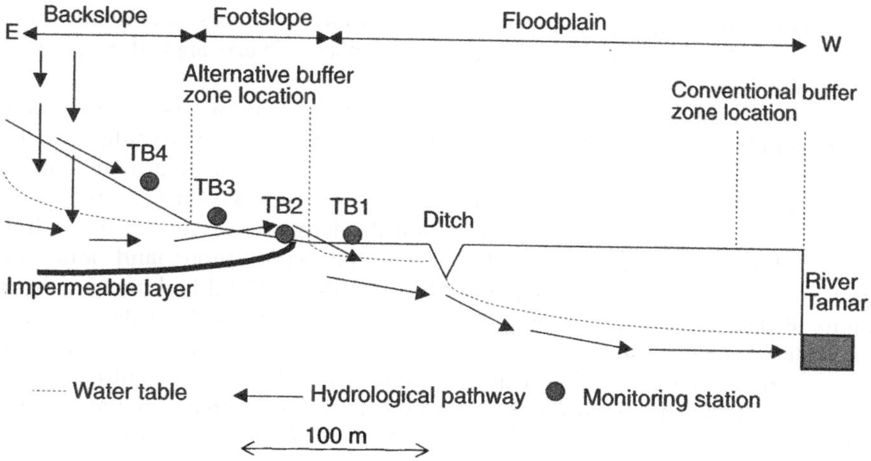

Fig. 15.2. Cross-section of Tetcott Barton study site showing generalized hydrology and relative locations of buffer zones and monitoring stations.

section of the study transect showing the locations of the four monitoring and sampling stations, the wetland buffer zone location and the generalized hydrology of the site. The sampling stations are named TB (Tetcott Barton) and numbered 1–4; station 4 is at the upslope end of the transect and station 1 at the downslope end.

Data from Tetcott Barton are presented in Figs 15.3–15.4. Results for soil water NO_3^--N concentrations (Fig. 15.3) represent a mean value for monthly samples collected from September 1997 to January 1998. At station TB4, the boundary between the buffer zone and the arable field upslope, NO_3^--N concentrations in soil water are highest, ranging from 75 to 13 mg NO_3^--N l^{-1}. Mean concentrations at TB3 (16 mg NO_3^--N l^{-1}) are less than half those at TB4 (43 mg NO_3^--N l^{-1}), and at TB2 and TB1 concentrations are almost zero, demonstrating a large decrease in NO_3^- concentration downslope through the buffer zone.

Denitrification rates were measured from soil cores taken at 0–10 cm depth, collected during a single assay in December 1997. They have a general increasing trend downslope through the buffer zone (Fig. 15.4), with a minimum rate of 0.03 mg N_2O-N m^{-2} h^{-1} at TB4 and

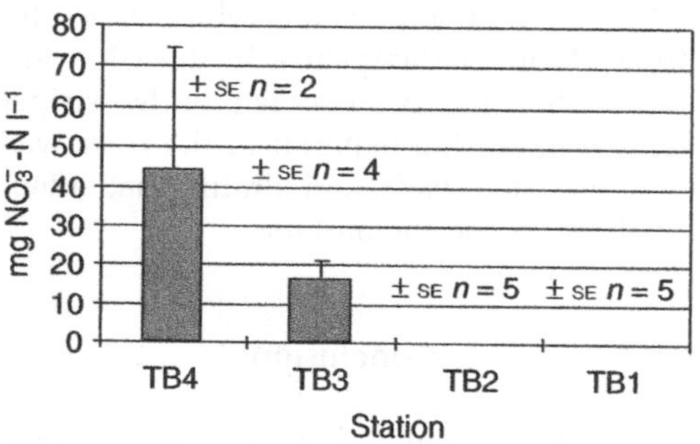

Fig. 15.3. Decrease in mean soil water NO_3^--N concentrations from upslope edge of buffer zone (TB4) to floodplain (TB1) at Tetcott Barton.

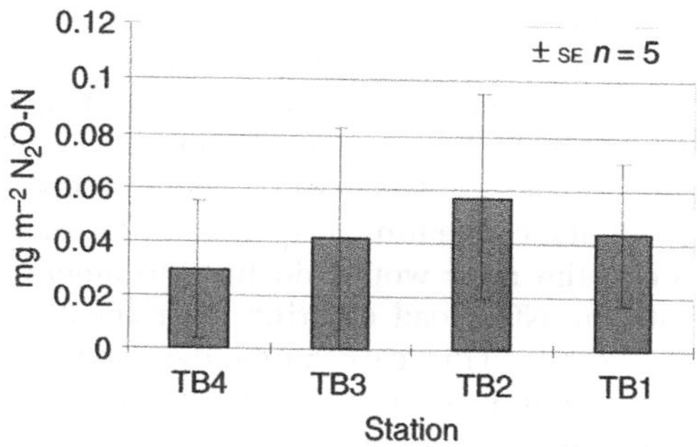

Fig. 15.4. Differences in denitrification rates through a wetland buffer zone from upslope edge of buffer zone (TB4) to floodplain (TB1) at Tetcott Barton.

a maximum rate of 0.06 mg N_2O-N m^{-2} h^{-1} at TB2. Rates at TB4 are probably lowest (despite the abundance of NO_3^- in the soil and soil water) because of the drier and hence more aerobic nature of the soil at this station, which would promote nitrification and restrict denitrification. The increase in denitrification rates down the transect at TB3 and TB2 reflects the increasing soil moisture and consequent anaerobicity at these stations, with optimal conditions for denitrification occurring at TB2. The subsequent decline in denitrification activity at TB1, despite the presence of anaerobic soil conditions, probably reflects the decrease in availability of NO_3^-.

While it is possible that some of the reductions in NO_3^- concentrations in the soil water downslope through the buffer zone may be the result of dilution by groundwater of low NO_3^- concentration, the trend in denitrification rates tends to support the view that much of the NO_3^- is lost through denitrification. At the time of year when the studies were carried out it is unlikely that much NO_3^- would have been removed by plant uptake and therefore most of the losses can be assumed to have been caused by dilution, microbial denitrification activity, or re-ammonification, demonstrating that even in winter such buffer zones can be effective. Whatever the process, the net result is a dramatic reduction in NO_3^- concentration through the wetland buffer zone, reflected in the low concentrations measured in water samples collected monthly from the drainage ditch located at the footslope/floodplain boundary. These range from 0.08 to 4.05 mg NO_3^--N l^{-1} but could have been much higher if the buffer zone were not present. Water from this ditch flows directly into the River Tamar some 200 m away. The establishment of a conventional riparian buffer zone along the river would do little to ameliorate the NO_3^- load entering the river from this ditch. This case study demonstrates the importance of considering landform and hydrological pathways when establishing buffer zones, and that it is not always most efficient to establish or preserve buffers adjacent to large rivers.

Case study 2: Ditches and the role they can play in improving water quality

Kismeldon Meadows is a river marginal wetland in the River Torridge catchment, south-west England. It provides an example of how ditch systems can reduce pollutant discharge into rivers. A ditch draining improved agricultural land and passing through the wetland had become degraded at several locations along its course, forming areas of overland flow (Fig. 15.5). Monitoring of NO_3^- concentrations in the ditches indicated that a reduction in concentration of $> 90\%$ occurred regularly in the ditch discharging from these zones; the lowest observed monthly reduction was 60% (Blackwell, 1997). These overland flow areas, which occupy only a small percentage of the total wetland area, are effectively acting as buffer zones to the River Torridge for the removal of NO_3^-. The establishment of a conventional riparian buffer zone along the banks of the Torridge would have no impact on the quality of the water in the ditch, which bypasses any potential conventional buffer zone area, discharging directly into the river. Optimizing the potential of these alternative buffer zone locations along ditch systems is an important component of a strategy to minimize NO_3^- pollution. Targeting buffer zones along ditch systems in addition to establishing them in riparian strips is particularly effective for water quality maintenance and improvement. Similar conclusions have been drawn by Haycock and Muscutt (1995), who reported that while 85% of some sub-catchments of the River Avon in Hampshire in the UK were served by effective riparian buffer strips, 60% of polluting material in the river was delivered by roads and drains that effectively bypassed the buffers.

Conclusion

Research has shown that wetlands can provide a low maintenance method for reducing the input of nutrients and other agriculturally derived pollutants to surface

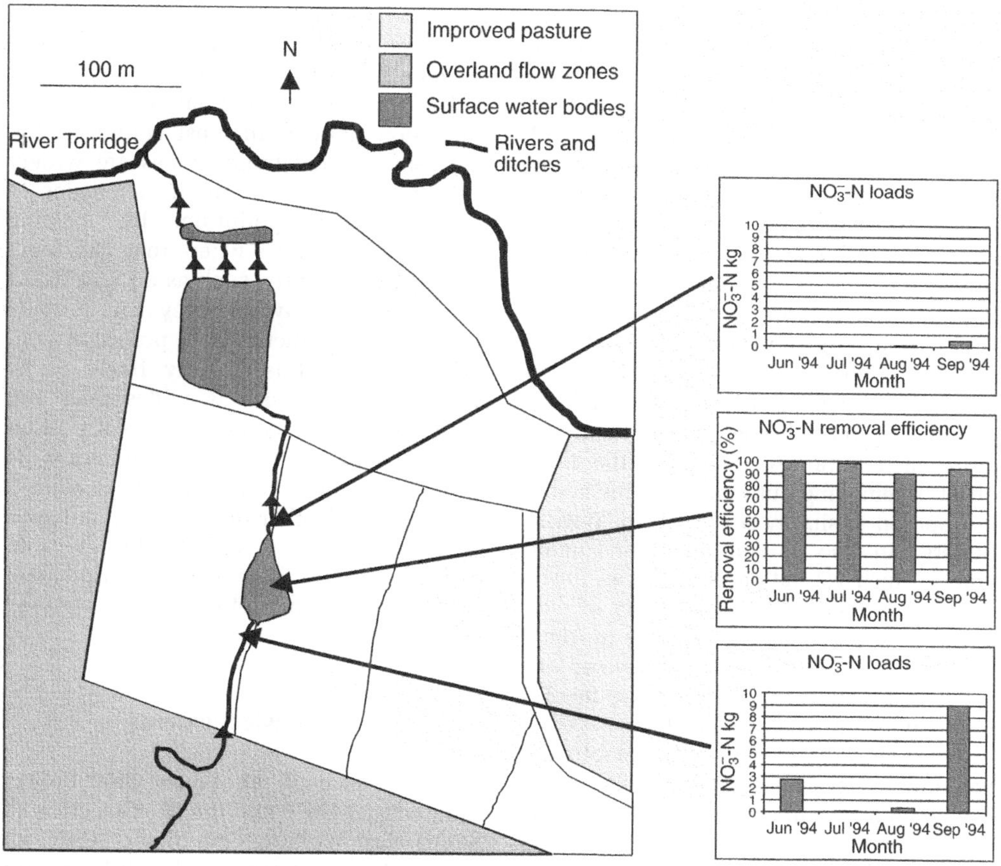

Fig. 15.5. Kismeldon Meadows: an example of alternative wetland buffer zone locations. The diagram illustrates the high efficiency of NO_3^- removal by small wetlands on a ditch system.

and shallow groundwaters. However, their performance is variable depending on the nature, quantity and timing (i.e. seasonality) of pollution inputs, as well as their size and ability to interact with the pollutants.

Point source pollution, where a discrete source of discharge can be identified, is generally straightforward to control or regulate, and although its environmental impacts can be serious, it can often be dealt with and action taken to limit or reduce its deleterious effects. On the other hand, diffuse sources, where no single discharge point occurs, have been identified as a significant cause of pollution, but are much harder to counteract because they result from complex interactions among soil processes, land management factors and water movement. They are difficult to both locate and regulate. Often there can be little incentive to farmers and land managers to carry out BMPs which can reduce levels of diffuse pollution, especially as many of these measures involve capital outlay or loss of income. In some countries, prosecution can result from causing pollution from a point source, but with diffuse pollution where there is no risk of prosecution it has been difficult to convince farmers of the need to adopt BMPs to prevent this.

However, there are several economic incentives that can be used to encourage BMPs. Savings can be made from more efficient fertilizer usage, improved livestock waste and feed management and better cultivation practices, all of which can lead to the reduction of diffuse pollution. Buffer zones can be established through subsidies and grants available from a variety of sources, the main one in the UK being MAFF's Countryside Stewardship Scheme (MAFF, 1997). In New Zealand, a voluntary option has been adopted, encouraging the implementation of wetland buffer zones through education and community consultation in preference to compensatory approaches (Cooper *et al.*, 1997). It is economic constraints that largely restrict the adoption of wetland buffer zone schemes. Their establishment on farms can be made viable if income is gained through the production of crops such as timber, which has the added effect of removing nutrients from the system as opposed to simply storing them. Wetland buffers can also provide areas for summer grazing; lush vegetation is produced during dry periods when the productivity of well-drained pastures may be severely restricted by drought. At the same time they can also promote biodiversity by providing increasingly limited areas of wet habitat.

One of the problems with establishing wetland buffer zones is that it is not an exact science. Numerous models exist for their design, although they require large amounts of site-specific data that are not usually readily available. In most cases, the general approach to wetland buffer zone implementation developed from the results of a large number of individual case studies may be more appropriate. Despite the lack of strategic wetland buffer zone implementation policies in most countries, the fact remains that buffer zones are widely recommended as one of the strategies to combat diffuse pollution. By locating buffer zones in areas where they can function most efficiently, such as on significant hydrological pathways, they can provide very important methods of pollution control. However, this is only likely to be effective if their use is widespread and implemented in conjunction with other BMPs as part of a holistic, catchment-wide approach to environmental management. It is likely that further research will enhance our knowledge of this relatively new 'soft' technology, and how best to implement both constructed wetlands and wetland buffer zones.

Acknowledgements

The assistance of the Tamar 2000 Project (funded by MAFF and the EU Objective 5b South West Working Group) and NERC for funding the work presented in this chapter is acknowledged. Thanks to Victoria Tainton for assistance with figure construction.

References

Averill, B.A. and Tiedje, J.M. (1981) The chemical mechanism of microbial denitrification. *FEBS Letters* 138(1), 8–12.

Baker, L.A. (1992) Introduction to non-point source pollution in the United States, and prospects for wetland use. *Ecological Engineering* 1, 1–26.

Barbier, E.B., Acreman, M. and Knowler, D. (1997) *Economic Valuation of Wetlands: a Guide for Policy Makers and Planners.* Ramsar Convention Bureau, Gland, Switzerland.

Blackwell, M.S.A. (1997) Zones of enhanced denitrification in river marginal wetlands. PhD thesis, Department of Geography, Exeter University, UK.

Blackwell, M.S.A., Hogan, D.V. and Maltby, E. (1999) The use of conventionally and alternatively located buffer zones for the removal of nitrate from diffuse agricultural run-off. *Water Science and Technology* 39 (12) 157–164.

Braskerud, B. (1994) Sedimentation of phosphorus and soil particles in constructed wetlands. In: Persson, R. (ed.) *Agrohydrology and Nutrient Balances.* Proceedings of NJF-seminar 247, Communications 94 (5). Division of Agricultural Hydrotechnics, Swedish University of Agricultural Sciences, Uppsala, Sweden, pp. 66–69.

Brix, H. (1994) Functions of macrophytes in constructed wetlands. *Water Science and Technology* 29, 4, 71–78.

Burt, T.P. (1995) The role of wetlands in runoff generation from headwater catchments. In: Hughes, J.M.R. and Heathwaite, L. (eds) *Hydrology and Hydrochemistry of British Wetlands.* John Wiley & Sons, Chichester, UK, pp. 21–38.

Cooper, A.B. (1990) Nitrate depletion in the riparian zone and stream channel of a small headwater catchment. *Hydrobiologia* 202, 13–26.

Cooper, A.B., Ngapo, N.I., Parminter, T.G. and Stroud, M.J. (1997) Encouraging implementation of riparian buffer schemes – the New Zealand experience. In: Haycock, N.E., Burt, T.P., Goulding, K.W.T. and Pinay, G. (eds) *Buffer Zones: Their Processes and Potential in Water Protection. Proceedings of the International Conference on Buffer Zones, September 1996.* Quest Environmental, Harpenden, UK, pp. 295–304.

Cooper, J.R., Gilliam, J.W. and Jacobs, T.C. (1986) Riparian areas as a control of non-point pollutants. In: Correll, D.L. (ed.) *Watershed Research Perspectives.* Smithsonian Institute Press, Washington, DC, pp. 166–192.

De Busk, W.F. and Reddy, K.R. (1987) Removal of floodwater nitrogen in a cypress swamp receiving primary wastewater effluent. *Hydrobiologia* 153, 79–86.

Dillaha, T.A. and Inamdar, S.P. (1997) Buffer zones as sediment traps or sources. In: Haycock, N.E., Burt, T.P., Goulding, K.W.T. and Pinay, G. (eds) *Buffer Zones: Their Processes and Potential in Water Protection. Proceedings of the International Conference on Buffer Zones, September 1996.* Quest Environmental, Harpenden, UK, pp. 33–42.

Dillaha, T.A., Reneau, R.B., Mostaghimi, S. and Lee, D. (1989) Vegetative filter strips for agricultural non-point source pollution control. *Transactions of the American Society of Agricultural Engineers* 32, 513–519.

English Nature (1992) How nutrient enriching pollution destroys wetland habitats. *English Nature* 3, 6–7.

FAO-UNESCO (1974) *Soil Map of the World*, 1:5,000,000, Vol. 1 Legend. UNESCO, Paris.

Flanagan, D.C. and Nearing, M.A. (1995) *USDA – Water Erosion Prediction Project: Hillslope Profile and Watershed Model Documentation*, NSERL Report No.10. USDA-ARS National Soil Erosion Research Laboratory, West Lafayette, Indiana.

Godfrey, P.J., Kaynor, E.R., Pelczarski, S. and Benforado, J. (1985) *Ecological Considerations in Wetlands Treatment of Municipal Wastewaters.* Van Nostrand Reinhold, New York.

Groffman, P.M. (1994) Denitrification in freshwater wetlands. *Current Topics in Wetland Biogeochemistry* 1, 15–35.

Hammer, D.A. (1989) *Constructed Wetlands for Wastewater Treatment: Municipal, Industrial and Agricultural.* Lewis Chelsea, Michigan.

Hammer, D.A. (1992) Designing constructed wetland systems to treat agricultural non-point source pollution. *Ecological Engineering* 1, 49–82.

Haycock, N.E. and Muscutt, A.D. (1995) Landscape management strategies for the control of diffuse pollution. *Landscape and Urban Planning* 31, 313–321.

Haycock, N.E. and Pinay, G. (1993) Groundwater nitrate dynamics in grass and poplar vegetated riparian buffer strips during the winter. *Journal of Environmental Quality* 22, 273–278.

Haycock, N.E., Burt, T.P., Goulding, K.W.T. and Pinay, G. (1997) *Buffer Zones: Their Processes and Potential in Water Protection. Proceedings of the International Conference on Buffer Zones, September 1996.* Quest Environmental, Harpenden, UK.

Haygarth, P.M., Hepworth, L. and Jarvis, S.C. (1998) Forms of phosphorus transfer in hydrological pathways from soil under grazed grassland. *European Journal of Soil Science* 49, 65–72.

Hogan, D.V. and Maltby, E. (1992) Water Resource Management and the Protection of Wetland Functioning: Research and Development Priorities. Report for NRA, R & D Note 114, Bristol, UK.

Hollis, G.E. and Jones, T.A. (1991) Europe and the Mediterranean Basin. In: Finlayson, M. and Moser, M. (eds) *Wetlands.* International Waterfowl and Wetlands Research Bureau, Facts on File, Oxford, UK, pp. 27–56.

Jenssen, P.D., Mæhlum, T. and Krogstad, T. (1993) Potential use of constructed wetlands for wastewater treatment in northern environments. *Water Science and Technology* 28(10), 149–157.

Johnston, C.A., Schubauer Berigan, J.P. and Bridgham, S.D. (1997) Riverine wetlands as buffer zones. In: Haycock, N.E., Burt, T.P., Goulding, K.W.T. and Pinay, G. (eds) *Buffer Zones: Their Processes and Potential in Water Protection. Proceedings of the International Conference on Buffer Zones, September 1996.* Quest Environmental, Harpenden, UK, pp. 33–42.

Kadlec, R.H. and Bevis, F.B. (1990) Wetlands and wastewater: Kinross, Michigan. *Wetlands* 10 (1), 77–92.

Kadlec, R.H. and Knight, R.L. (1996) *Treatment Wetlands*. Lewis Publishers, CRC Press, Boca Raton, Florida.
Knisel, W.G. (1980) *CREAMS: Field Scale Model for Chemicals, Runoff, and Erosion from Agricultural Management Systems*, Conservation Report No. 26. US Department of Agriculture, Tifton, Georgia.
Kirkham, F.W. and Wilkins, R.J. (1993) Seasonal fluctuations in the mineral nitrogen content of an undrained wetland peat soil following differing rates of fertilizer nitrogen application. *Agriculture, Ecosystems and Environment* 43, 11–29.
Klopatek, J.M. (1978) Nutrient dynamics of freshwater riverine marshes and the role of emergent macrophytes. In: Good, R.E., Whigham, D.F. and Simpson, R.L. (eds) *Freshwater Wetlands: Ecological Processes and Management Potential*. Academic Press, New York, pp. 195–216.
Knight, R.L., McKim, T.W. and Kohl, H.R. (1987) Performance of a natural wetland treatment system for wastewater management. *Journal of Water Pollution Control Federation* 59, 746–754.
Lloyd, D. (1993) Aerobic denitrification in soils and sediments: from fallacies to facts. *TREE* 8, 352–356.
Lowrance, R.R. (1992) Groundwater nitrate and denitrification in a coastal plain riparian forest. *Journal of Environmental Quality* 21, 401–405.
Lowrance, R.R., Todd, R.L. and Asmussen, L.E. (1984) Nutrient cycling in an agricultural watershed: I. Phreatic movement. *Journal of Environmental Quality* 13, 22–27.
Lowrance, R., Altier, L.S., Williams, R.G., Inamdar, S.P., Bosch, D.D., Sheridian, J.M., Thomas, D.L. and Hubbard, R.K. (1998) The Riparian Ecosystem Management Model: simulator for ecological processes in riparian zones. *Proceedings of the First Federal Interagency Hydrologic Modelling Conference, Las Vegas, Nevada, April 1998*. Interagency Advisory Committee on Water Data, pp. 1.81–1.88.
MAFF (1997) *The Habitat Scheme, Water Fringe Areas, Information Pack*. MAFF Publications, London.
Maltby, E. and Turner, R.E. (1983) Wetlands are not wastelands. *Geographical Magazine* LV, 92–97.
Maltby, E., Hogan, D.V. and McInnes, R.J. (1996) *Functional Analysis of European Wetland Ecosystems – Phase I (FAEWE)*, Ecosystems Research Report 18. Office for Official Publications of the European Communities, Luxembourg.
Mander, Ü., Matt, O. and Nugin, U. (1991) Perspectives on vegetated shoals, ponds and ditches as extensive outdoor systems of wastewater treatment in Estonia. In: Etnier, C. and Guterstam, B. (eds) *Ecological Engineering for Wastewater Treatment. Proceedings of the International Conference at Stensund Folk College, Sweden*. Lewis Publishers, pp. 271–282.
Martin, N.J. and Holding, A.J. (1978) Nutrient availability and other factors limiting microbial activity in the blanket peat. In: Heal, O.W. and Perkins, D.F. (eds) *Production Ecology of British Moors and Montane Grasslands*, Ecological Studies No. 27. Springer-Verlag, New York, pp. 113–116.
Mitsch, W.J. and Gosselink, J.G. (1993) *Wetlands*, 2nd edn. van Nostrand Rheinhold, New York.
Mitsch, W.J., Dorge, C.L. and Wiemhoff, J.R. (1979) Ecosystem dynamics and a phosphorus budget of an alluvial cypress swampi n southern Illinois. *Ecology* 60, 1116–1124.
Mockler, N.J., Blackwell, M.S.A. and Maltby, E. (1994) Buffer zones for nitrate removal. Poster paper, RHIER, Royal Holloway University, London.
Newbold, C., Honnor, J. and Buckley, K. (1989) *Nature Conservation and the Management of Drainage Channels*. Nature Conservancy Council, Peterborough, UK.
Novitzki, R.P. (1979) Hydrologic characteristics of Wisconsin's wetlands and their influence on floods, stream flow and sediment. In: Greeson, P.E., Clark, J.R. and Clark, J.E. (eds) *Wetland Functions and Values: the State of Our Understanding*. American Water Resource Association, Minneapolis, Minnesota, pp. 377–388.
Novitzki, R.P. (1985) The effects of lakes and wetlands on flood flows and base flows in selected northern and eastern states. In: Groman, H.A., Meyers, E., Burke, D. and Kusher, J. (eds.), *Proceedings of a Conference – Wetlands of the Chesapeake*. Environmental Law Institute, Washington, DC, pp. 143–154.
Nuttall, P.M., Boon, A.G. and Rowell, M.R. (1997) *Review of the Design and Management of Constructed Wetlands*, Report 180. Construction Industry Research and Information Association, Westminster, London, UK.
Peterjohn, W.T. and Correll, D.L. (1984) Nutrient dynamics in an agricultural watershed: observations on the role of a riparian forest. *Ecology* 65, 1466–1475.

Pinay, G. and Décamps, H. (1988) The role of riparian woods in regulating nitrogen fluxes between the alluvial aquifer and surface water: a conceptual model. *Regulated Rivers: Research Management* 2, 507–516.

Pommel, B. and Dorioz, J.M. (1995) Movement of phosphorus from agricultural soil to water. Paper presented at international workshop *Phosphorus Loss to Water from Agriculture*, 27–29 September, Teagasc, Johnstown Castle, Wexford, Ireland.

Reddy, K.R., Patrick, W.H. and Lindau, C.W. (1989) Nitrification–denitrification at the plant root sediment interface in wetlands. *Limnologici Oceanographica* 34, 1004–1013.

Richardson, C.J. (1985) Mechanisms controlling phosphorus retention capacity in freshwater wetlands. *Science* 228, 1424–1427.

Richardson, C.J. (1989) Wetlands as transformers, filters and sinks for nutrients. *Freshwater Wetlands: Perspectives on Natural, Managed and Degraded Ecosystems*, University of Georgia, Savannah River Ecology Laboratory, Ninth Symposium, Charleston, South Carolina.

Roulet, N. (1990) The hydrological role of peat covered-wetlands. *Canadian Geographer* 34, 82–83.

Schipper, L.A., Cooper, A.B., Harfoot, C.G. and Dyck, W.J. (1993) Regulators of denitrification in an organic riparian soil. *Soil Biology and Biochemistry* 25(7), 925–933.

Struwe, S. and Kjoller, A. (1989) Field determination of denitrification in waterlogged forest soils. *FEMS Microbiological Ecology* 62, 71–78.

Tilton, D.L. and Kadlec, R.H. (1979) The utilisation of a fresh-water wetland for nutrient removal from secondarily treated wastewater effluent. *Journal of Environmental Quality* 8, 328–334.

Tiner, R.W. (1984) *Wetlands of the United States: Current Status and Trends.* US Department of the Interior, Fish and Wildlife Service, US Government Printing Office , Washington, DC.

Trimble, G.R. and Sartz, R.S. (1957) How far from a stream should a logging road be located? *Journal of Forestry* 55, 339–341.

US Army Corps of Engineers (1972) *Charles River Watershed, Massachusetts.* New England Division, Waltham, Massachusetts, p. 65.

Uusi-Kämppä, J. and Yläranta, T. (1996) Effect of buffer strips on controlling soil erosion and nutrient losses in southern Finland. In: Mulamoottil, G., Warner, B.G. and McBean, E.A. (eds) *Wetlands: Environmental Gradients, Boundaries and Buffers.* CRC, Lewis Publishers, New York, pp. 221–235.

Verry, E.S. and Boelter, D.H. (1979) Peatland hydrology. In: Greeson, P.E., Clark, J.R. and Clark, J.E. (eds) *Wetland Functions and Values: the State of Our Understanding.* American Water Resource Association, Minneapolis, Minnesota, pp. 389–402.

Whipple, W. Jr (1991) Discussion. *Water Resources Bulletin* 27 (3) 553.

Williams, M. (1990) *Wetlands: a Threatened Landscape,* Special publications series, Institute of British Geographers, Blackwell, Oxford.

Williams, R.D. and Nicks, A.D. (1993) A modelling approach to evaluate best management practices. *Water Science and Technology* 28, 675–678.

Wiloso, E.I., Basuki, T. and Aiman, S. (1995) Utilization of agricultural wastes for biogas production in Indonesia. In: Ishizuka, K., Hisajima, S. and Macer, D.R.J. (eds) *Traditional Technology for Environmental Conservation and Sustainable Development in the Asian-Pacific Region. Proceedings of the UNESCO – University of Tsukuba International Seminar on Traditional Technology for Environmental Conservation and Sustainable Development in the Asian-Pacific Region, Tsukuba Science City, Japan, 11–14 December 1995.* University of Tsukuba, pp. 134–138.

Section 3

Water Quality: Impacts and Case Studies from Around the World

Introduction: Impacts of Agriculture on Water Quality Around the World

Graham Harris
CSIRO Land and Water, Canberra, Australian Capital Territory, Australia

After many years of focus on point sources of pollution, and considerable success in their control, the emphasis has now shifted back to the diffuse sources of nutrients and pesticides that degrade water quality (Carpenter *et al.*, 1998). This change in emphasis arises from examples of what might be called 'landscape failure' where the entire landscape starts to leak and the degradation of water quality consequently becomes very difficult to manage. Two examples, among others, illustrate landscape failure from this section: the first is the leakage of P from the soils of Northern Ireland after many years of animal manure application (Foy and O'Connor, Chapter 19) and the second is dryland salinity problems of Australia, where clearing of the native Australian bush has altered the entire hydrological balance of the landscape and mobilized large amounts of salt (Williams *et al.*, Chapter 21). These and other examples should be set in the context of the disruption of global biogeochemical cycles of key elements, also leading to increased fluxes and catchment loadings (Vitousek *et al.*, 1997).

One recent discovery that affects these problems is the observation that there is a much closer linkage between water quality and soil chemistry (and its changes over time) than we had previously realized (Engstrom *et al.*, 2000). There has, in my view, been too much of an intellectual divide between limnologists and freshwater scientists on the one hand, and agronomists and soil scientists on the other. There has been a great tendency to collect water samples while not looking too closely at the condition of the catchment, or to maximize production with little regard for the off-site impacts. This is now changing rapidly, and not before time.

Two major features of land use appear to control the linkage between catchment condition and water quality. One is the large-scale distribution of landscape features, such as forest and agricultural lands and the presence or absence of riparian vegetation. The clearing of forest and deep-rooted vegetation and its replacement with pasture changes the hydrology of catchments and the routes of water and sediment flow to rivers and streams. This is particularly true if animal densities are high and manure and fertilizer loadings are also high. The second feature is that, at the microscopic scale, microbial processes and soil physical properties, such as the presence or absence of macropores, also determine the forms and mobility of nutrients and pesticides through the soil profile. Such scales are, in reality, interlinked and a good example of this is where deep-rooted plants provide preferential flow pathways through the soil profile (e.g. Burgess *et al.*, 2001).

Rivers and water bodies are the ultimate integrators of our land use decisions, and water quality degradation often results from policies that control the intensive use of agricultural lands without considering the downstream impacts. Some of our biggest problems lie as much with institutional failures as they do with lack of understanding of key processes or the consequences of our actions. The more we can integrate our knowledge and actions across scales and between disciplines the better able we will be to control diffuse sources of pollution and their impact on water quality.

References

Burgess, S.S.O., Adams, M.A., Turner, N.C., White, D.A. and Ong, C.K. (2001) Tree roots: conduits for deep recharge of soil water. *Oecologia* 126, 158–165.

Carpenter, S.R., Caraco, N.F., Correll, D.L., Howart, R.W., Sharpley, A.N. and Smith, V.H. (1998) Non point pollution of surface waters with phosphorus and nitrogen. *Ecological Applications* 8, 559–568.

Engstrom, D.R., Fritz, S.C., Almendinger, J.E. and Juggins, S. (2000) Chemical and biological trends during lake evolution in recently deglaciated terrain. *Nature* 408, 161–166.

Vitousek, P.M., Aber, J.D., Howarth, R.W., Likens, G.E., Matson, P.A., Schindler, D.W., Schlesinger, W.H. and Tilman, D.G. (1997) Human alteration of the global nitrogen cycle: sources and consequences. *Ecological Applications* 7, 737–750.

16 Solutions to Nutrient Management Problems in the Chesapeake Bay Watershed, USA

J. Thomas Sims[1] and Frank J. Coale[2]

[1]*Department of Plant and Soil Sciences, 163 Townsend Hall, University of Delaware, Newark, DE 19717-1303, USA;* [2]*Department of Natural Resource Sciences and Landscape Architecture, University of Maryland, College Park, MD 20742, USA*

Keywords: agriculture, animal manures, estuaries, nitrogen, non-point source pollution, nutrient management, phosphorus, regional approaches, watershed management, water quality

Introduction: the Chesapeake Bay Watershed

Overview

The Chesapeake Bay is the largest and most important estuary in North America. The bay is an amazingly diverse ecosystem, supporting more than 3600 species of plants, fish and animals. It has been a vital aspect of the economy of mid-Atlantic USA since European settlers arrived in the region in the early 1600s, particularly because of its abundant supply of fish and shellfish (Chesapeake Bay Program, 2000). The word Chesapeake itself is derived from the Alongquin Indian term 'Chesepioc', which means 'great shellfish bay'. Today, despite numerous water quality problems that have plagued the bay for decades, it continues to be prized for its beauty, biological diversity and contribution to the regional economy. As an example, approximately 70 M kg of fish and shellfish are harvested annually from the bay, which also contains *c.* 25% of the total US waters approved for commercial harvest of oysters and clams. The bay is also a year-round habitat for many important species of commercial and sport fish (i.e. white perch, bay anchovy, channel and white catfish) and is the nursery for early life stages of several migratory species such as Atlantic menhaden, American eel, weakfish, spotted sea trout and striped bass. Protecting the Chesapeake Bay by reducing both point and non-point source pollution has been, and continues to be, a central goal of the six states within Chesapeake Bay's watershed (Delaware, Maryland, New York, Pennsylvania, Virginia and West Virginia).

The Chesapeake Bay watershed encompasses about one-sixth of the eastern seaboard of the USA and has a total land area of *c.* 180,000 km^2 (Fig. 16.1). The total surface area of the Chesapeake Bay is 9900 km^2 with 400 km^2 of tidal fresh waters, 9200 km^2 in the mixing zone of fresh and salt waters, and 300 km^2 of salt waters. There are also 4000 km^2 of fresh and tidal wetlands in the bay's watershed. The Chesapeake Bay, in common with many estuaries on the US Atlantic coast, is rather shallow (mean depth of 10 m); the total water volume in the bay is only 75 km^3. An important point with regard to non-point source pollution impacts on water quality in the Chesapeake Bay is that its land area:water volume ratio (2410 km^2 land km^{-3} water) is much higher than any other riverine estuary in the world (e.g. the ratio for the next largest coastal water, the Gulf of Finland, is *c.* 400 km^2 land km^{-3} water). Thus there is great potential for non-point source pollution of the bay from pollutants discharged from the land within its watershed.

Five major US rivers (the Potomac, Susquehanna, Rappahannock, York and James rivers) and dozens of smaller rivers and streams drain into the bay, providing a substantial freshwater inflow of *c.* 8.7×10^6 m^3 h^{-1}. In contrast, seawater moves into the bay in the opposite direction and at greater depths than the seaward moving fresh water. This bi-directional and stratified flow creates a complex water mixing and circulation pattern that affects not only the biota of the bay but the fate and transport of any pollutants entering the bay from either point or non-point sources (Boynton, 2000). Stratification of the water column is particularly important for water quality because deeper waters in the bay are often not exposed to the atmosphere for weeks, or even months, and thus can become depleted in dissolved O_2. Another consequence of this bi-directional, stratified flow pattern, and the relatively weak tides entering the bay, is the long residence time of pollutants in the bay. Many important characteristics of the bay that affect water quality also vary temporally and spatially, changing with season and as a function of their proximity to the fresh and saline waters entering the bay. Two examples illustrate this: (i) inputs of fresh water, sediments, nutrients and organic matter to the bay are about tenfold greater in the spring than in the autumn; and (ii) the limiting nutrient for eutrophication in the winter and early spring changes from N at the southern, seaward end of the bay to P at the northern, freshwater end of the

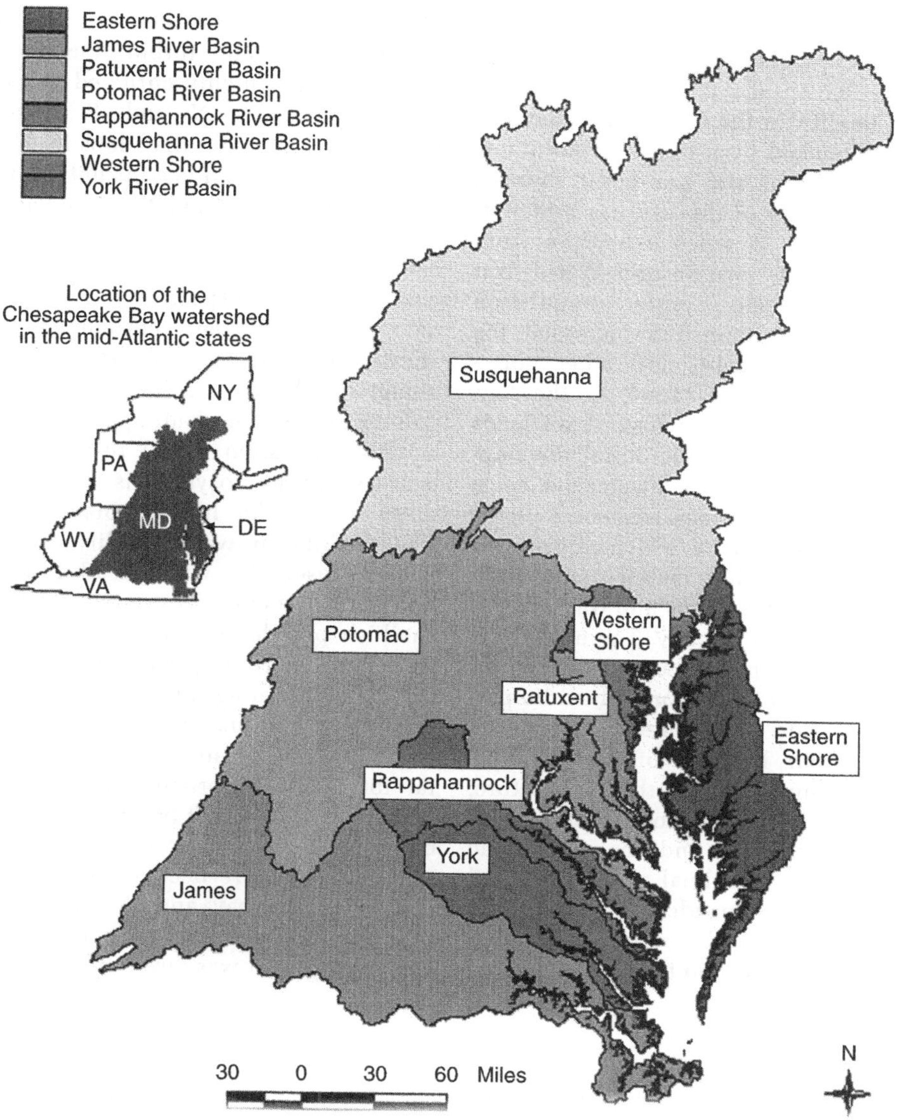

Fig. 16.1. Location of the Chesapeake Bay watershed in eastern USA, indicating the location of the six US states and eight major basins in the watershed (Chesapeake Bay Program, Annapolis, Maryland).

bay, while in the summer and early autumn N is the limiting nutrient throughout the bay (Boynton, 2000).

Much of the land in the Chesapeake Bay watershed is forested (58%); the remainder is used for agriculture (33% in total, 15% in cropland, 7% in hayland and 11% in pasture) and urban areas (8%). Several areas of concentrated animal production (i.e. with large numbers of animal units per hectare of land available for manure application) are found in the bay's watershed. Most notable are the southern part of the Delmarva (Delaware–Maryland–Virginia) peninsula (poultry) on the bay's eastern shore, southeastern Pennsylvania (dairy and poultry), north-central Virginia (poultry, beef, dairy) and western Maryland (dairy). It is widely accepted today that the most difficult nutrient management problems existing in the Chesapeake Bay watershed are associated with animal-based agricultural systems.

Major water quality problems in the Chesapeake Bay watershed

Water quality in the Chesapeake Bay has been influenced by a range of pollutants from both point and non-point sources. While pollution of the bay has undoubtedly occurred to some extent for hundreds of years, post-Second World War urbanization and rapid population growth in the watershed increased the delivery of sediments and other pollutants in stormwater runoff to the bay. Added to this was the loss of wetlands through urban and agricultural development and increased wastewater discharge from municipal sewage treatment plants and industrial sources, which contributed nutrients, organic matter (increased BOD), organic chemicals and trace metals to all portions of the bay. The expanding urban population also contributed pollutants to the bay in the form of acidic precipitation and atmospheric deposition (e.g. N from burning of fossil fuels). Increased nutrient use by agriculture, encouraged by the advent of commercial chemical fertilizers and the geographic intensification of animal agriculture, and soil loss from expanded areas of tilled cropland also contributed to the degradation of water quality in the bay.

With time, the waters of the bay became murky and turbid, low in dissolved O_2, and highly enriched in nutrients, particularly N and P. Nutrient budgets developed in the 1980s for N and P (Table 16.1) provided insight into the sources of these pollutants to the bay and indicated that N and P loading rates were 5–8 and 13–24 times higher than in pre-colonial times, respectively (Boynton *et al.*, 1995). Nutrient enrichment was shown to stimulate algal growth which, along with higher turbidity in the water column, increased shading of submerged aquatic vegetation (SAV) resulting in a decrease in photosynthesis rates below those needed for healthy plant growth. The decline in water quality was followed closely by widespread losses of SAV and dramatic decreases in fish and shellfish populations. The losses of SAV were particularly important because of its role in maintaining water clarity (through binding of sediments), nutrient cycling, and supplying organic matter and a habitat for many important benthic species (Kemp *et al.*, 1983). It became clear in the early 1980s that substantial, watershed-wide actions were needed to restore the quality of the bay and that controlling nutrient inputs would be a key part of any strategy that was developed.

Table 16.1. Nitrogen and phosphorus budgets (10^6 kg year^{-1}) for the Chesapeake Bay (Boynton *et al.*, 1995).

	Nitrogen	Phosphorus
Nutrient inputs to the bay		
Point sources	42.9 (28%)[a]	4.0 (35%)
Diffuse sources	90.5 (60%)	6.6 (58%)
Atmospheric deposition	18.2 (12%)	0.7 (7%)
Import from coastal ocean	—	4.1 (37%)
Nutrient storage and export		
Burial in sediments	52.8 (35%)	14.5 (128%)
Denitrification	40.0 (26%)	—
Export to coastal ocean	45.9 (30%)	—
Fisheries harvest	13.5 (9%)	0.6 (5%)

[a] Values in parentheses are percentages of total inputs of either N or P.

History of water quality protection efforts in the Chesapeake Bay watershed

In 1975, the Chesapeake Bay became the first US estuary to be specifically targeted for protection and restoration when the US Congress directed the US Environmental Protection Agency (USEPA) to conduct studies to determine the causes of the decline in water quality in the bay. By the early 1980s, sufficient research had been conducted to formulate a working hypothesis that the primary pollution problem for the bay was nutrient enrichment with N and P. It was also clear that all states in the watershed were contributing to non-point source nutrient pollution of the bay. In 1983, in response to growing environmental and ecological concerns and mounting public pressure, the first Chesapeake Bay Agreement was signed by the Governors of Maryland, Virginia, Pennsylvania, the government of the District of Columbia, the Chesapeake Bay Commission (which represented the legislatures of these states) and USEPA (USEPA, 1983). Delaware, New York and West Virginia were non-signatory states to this initial agreement; however, these three states have regularly contributed to regional efforts to develop policies and practices that will reduce pollution of the bay. This historic agreement also established the Chesapeake Bay Program, a regional, federal–state– local partnership which has since coordinated efforts from all parties to restore water quality in the bay. A Chesapeake Executive Council, consisting of the governors of the bay states, the mayor of the District of Columbia and the chairperson of the Chesapeake Bay Commission, was established to lead the effort to restore the bay and to be accountable to the public for progress towards that end. In 1987, the agreement was renewed and specific goals were established, the most important of which was the recommendation that a watershed-wide strategy be developed and implemented to reduce N and P inputs to the main stem of the Chesapeake Bay by 40% by the year 2000, relative to 1985 point source loads and non-point source inputs under average rainfall conditions. In 1992, the Chesapeake Bay Agreement was amended to re-affirm the importance of the 40% nutrient reduction and to devote more of the Chesapeake Bay Program's efforts and resources to reducing upstream sources of nutrients (i.e. those entering the bay through its smaller tributaries). Throughout this process these agreements have focused on the bay as an ecosystem and have established measures of success related to ecosystem restoration, not just to quantitative changes in water quality parameters.

In June of 2000, The Chesapeake Bay Council signed 'Chesapeake 2000' a comprehensive and far reaching Bay agreement that will guide Maryland, Pennsylvania, Virginia, The District of Columbia. Delaware, the Chesapeake Bay Commission and the US Environmental Protection Agency (EPA) in their combined watershed-wide agreement again re-affirmed the commitment of the Chesapeake Bay Program to

> engage everyone – individuals, businesses, communities, and governments – in our effort; to commit all citizens of the Chesapeake Bay Watershed in a shared vision – a system with abundant, diverse populations of living resources, fed by healthy streams and rivers, sustaining strong, local and regional economies, and our unique quality of life.

The agreement delineated a number of specific goals (Table 16.2) and emphasized the importance of individual responsibility and community engagement to the overall success of the programme. It stressed promotion of individual stewardship and assisting individuals, community-based organizations, local governments, and schools in undertaking initiatives that enhance public outreach and education programmes and that foster 'government by example' (i.e. that lands and resources owned and managed by governments adhere to the goals delineated in Table 16.2).

The importance of this renewed commitment was emphasized by the Chesapeake Bay Foundation's (CBF, a private foundation formed in 1967 with more than 80,000 members that is dedicated to restoring the Chesapeake Bay) 1999 'State of the Bay Assessment' which rated the overall health of the bay at an index of 28 on a scale of 100

Table 16.2. Summary of the long-term commitments and specific goals of the Chesapeake Bay Program.[a]

Commitment and goals	Key targets for each goal
Living resource protection and restoration: Restore, enhance, and protect the finfish, shellfish, and other living resources, their habitats and ecological relationships to sustain all fisheries and provide a balanced ecosystem	*Oysters*: By 2010 achieve a tenfold increase in oysters relative to a 1994 baseline *Exotic species*: Identify those exotic species that have a negative effect on the bay by 2002 and by 2004 implement management programmes to control these species *Fish passage and migratory fish and resident fish*: By 2002 identify steps needed to restore fish passage for migratory fish by 2003. For priority species, assess trends in populations, determine desired, tributary-specific population sizes based on projected fish passage and available habitat and provide recommendations to achieve these populations *Multi-species management*: By 2005 develop multi-species management plans for selected species; by 2007 implement multi-species fisheries management plans that include ecological and socio-economic considerations *Crabs*: Restore a healthy biomass, size and age structure for blue crabs; set target harvest by 2001 and implement bay-wide fisheries management strategies to meet this target
Vital habitat protection and restoration: Preserve, protect and restore those habitats and natural areas vital to the survival and diversity of the living resources of the bay and its rivers	*Submerged aquatic vegetation (SAV)*: Recommit to restoring 46,000 ha of SAV and revise goals in 2002 to reflect historic abundance; set specific targets for water clarity to be met by 2010. Accelerate SAV restoration in areas of critical importance *Wetlands*: Achieve no net loss of jurisdictional wetlands through use of regulatory programmes; restore 10,000 ha of tidal and non-tidal wetlands by 2010. Support local, watershed-based wetland restoration programmes *Forests and stream corridors*: Ensure that measures are in place by 2003 to restore 3200 km of riparian forest buffers. Promote practices that link contiguous forests and support the adoption of local stream corridor programmes that conserve and restore riparian forests. Initiate pilot projects to promote stream corridor protection and restoration and support the efforts of local governments to develop plans that protect these corridors
Water quality restoration and protection: Achieve and maintain the water quality necessary to support the aquatic living resources of the bay and its tributaries and to protect human health	*Nutrients*: Achieve and maintain a 40% reduction in nutrient loading to the bay. By 2010 correct a sufficient number of nutrient-related problems to remove the bay and its tidal tributaries from the list of US impaired waters. By 2001 define water quality conditions needed to protect aquatic resources and the load reductions in N and P required to meet these conditions for each major tributary of the bay *Sediment*: Correct all sediment-related problems in the bay and tidal portions of its tributaries by 2010; establish, by 2001, sediment load reductions needed to achieve desired water quality conditions and assign load reductions to all major bay tributaries *Chemical contaminants*: Commit to the 1994 goal of reducing input of chemical contaminants from controllable sources to levels that result in no toxic or bioaccumulative impact on aquatic organisms or human health *Air pollution*: Assess by 2003 effects of airborne N compounds and chemical contaminants on the bay; strengthen air pollution prevention programmes in the airshed *Boat discharge*: Establish, by 2003, no discharge zones in the bay for human waste from boats; by 2010, expand by 50% number of waste pump-out facilities

Sound land use: Develop, promote, and achieve sound land use practices which protect and restore watershed resources and water quality, maintain reduced pollutant loadings for the bay and its tributaries, and restore and preserve aquatic living resources	*Land conservation*: Expand, by 2002, use of voluntary, market-based and governmental mechanisms to preserve natural resource lands (e.g. purchasing development rights). Develop and maintain strong local GIS systems to aid in land use planning *Public access*: Expand public access points to the bay by 30% by 2003; increase the length of water trails in the region by 800 km; enhance outreach efforts that promote public awareness of the value of the natural, recreational and historical resources of the bay. *Development*: Reduce, by 2010, rate of conversion of forest and agricultural lands to development by at least 30%. Support state and local efforts to manage growth (e.g. tax policies, investment opportunities) and implement policies that limit impervious cover, control stormwater erosion, improve septic system designs and rehabilitate brownfields *Transport*: Promote coordination of transport and land use planning between state and local governments to reduce dependence on automobiles, enhance compact development patterns and revitalize existing communities

[a] The full text of the Chesapeake Bay Program's Watershed Partnership plan is available at: http://www.chesapeakebay.net/pubs/agree99.htm

(an index of 100 refers to the bay's status at the beginning of European settlement in the area in the early 1600s). In comparison, the first index developed by the CBF in 1983 was 23. This assessment of the bay's overall health was based on three broad criteria: (i) a pollution factor (what enters the bay); (ii) a harvest factor (what is taken from the bay); and (iii) a habitat/resilience factor (measure of the bay's ability to filter out pollution) (see Table 16.3 for specific aspects of this assessment). In making this assessment, the CBF stated that 'the bay will never again reach the pristine levels of the past, but a index of 70 is achievable by 2050' and should be the agreed-upon target for a 'saved bay'.

Restoring the health of the Chesapeake Bay is a long-term challenge that will require significant changes of virtually every citizen and every group, business, organization, agency and government in the watershed. The breadth of this challenge is clearly illustrated by the complex series of criteria used to judge the health of the bay (Table 16.3) and the action plans outlined for the decades ahead (Table 16.2). The focus of this chapter is on current and planned efforts to reduce nutrient inputs, specifically N and P, from agricultural non-point sources. Agriculture's contribution is significant, estimated at one-third of the total N and one-half of the total P entering the Chesapeake Bay. Reducing agricultural nutrient inputs will require a sound understanding of the processes controlling nutrient cycling and transport in the watershed, the development and, more importantly, the implementation of nutrient management plans at all scales (from regional to individual farms), and innovation to identify best management practices (BMPs) for nutrient use that will be more effective than those in widespread use today.

Agriculture and Nutrient Management in the Chesapeake Bay Watershed

Overview of agriculture in the Chesapeake Bay watershed

Approximately one-third of the land in the Chesapeake Bay watershed is devoted to agriculture. Row crop production accounts for 15% of the land use in the watershed. The major row crops grown in the watershed include maize (for grain and silage), soybeans (for grain) and small grains (barley, oats, wheat). The remainder of the agricultural land is used for hay production, pasture (7 and 11% of the total land in the watershed, respectively) and the production of high-value speciality crops such as fruits, vegetables, tobacco and turf. The broad relationship between the various types of agricultural land use and other land uses (e.g. urban and forests) is shown for each sub-basin of the Chesapeake Bay watershed in Table 16.4. Note, however, that agricultural land use varies between states and within sub-basins. For example, in Delaware and Maryland most agricultural land (> 80%) is used for the production of row crops for grain; conversely, in New York, Pennsylvania, Virginia and West Virginia, more than 50% of the land is used for hay production (e.g. lucerne and grass) (NASS, 1999). These differences in cropping systems are largely associated with the different types of animal agriculture in these regions of the bay watershed.

The nature of agriculture in the Chesapeake Bay is somewhat different from many other regions in the USA. Farms are small to medium in size (average of *c.* 100 ha compared with a national average of *c.* 200 ha) and tend to be more diversified in terms of the types of crops grown and animals produced. Irrigated agriculture is uncommon because there is plentiful rainfall (3% of the watershed is irrigated compared with 13% in the USA). Animal production is a mainstay of the agricultural economy in most of the watershed. Farm income is primarily derived from the sale of animals and animal products, particularly milk, beef and pigs, poultry (both meat and eggs), vegetables and orchard fruits. Much of the grain and forage produced is consumed locally by livestock and poultry operations, some of which are large and as geographically intensified as any in the USA.

As with crop production, the type of animal agriculture practised varies within the

Table 16.3. The Chesapeake Bay Foundation's 'State of the Bay' report for 1999.[a]

Category, assessment parameter and observations for 1998–1999	Goal for 2010
Habitat (rating)[b]	
Wetlands (42): Wetlands now cover 43% of their land area before European settlement. In the past year ditching and draining resulted in a loss of thousands of hectares of wetlands in Virginia, erasing wetland restoration gains	Reduce loss of existing wetlands by 75%. Restore 50,000 ha of wetlands, relative to 1995 baseline
Forested buffers (53): The CBF estimates 53% of the streams and shorelines in the watershed have riparian forests. Restoration of buffers continues but so does destruction by development; additional efforts are needed to protect riparian forests	Reduce loss of streamside forest buffers by 75%, vs. 1995 baseline. Restore 8000 km of buffers
Underwater grasses (12): Underwater grasses cover about 11% of their historical area; strong increases in grasses were observed in some areas. Continued reductions in nutrient and sediment loading are needed to restore these vital habitats	Underwater grasses should cover 90,000 ha of the bay and its tributaries
Resource lands (33): The CBF estimates that about 40,000 ha of farmland and open space are lost annually to development. The rate of development is alarming and greater efforts are needed to manage growth in the watershed	Current rate of land development should be reduced by 50%. An additional 200,000 ha of farms and forests should be permanently protected from development
Pollution	
Toxics (30): Industrial discharges to the bay have declined during the past 15 years, but there is no evidence of a bay-wide decline in toxic levels. The CBF advocates zero discharge goals and increases in regulatory and monitoring programmes for toxic compounds	Amount of toxics entering the bay should be reduced by 50% vs. 1995 levels
Nutrients (N: 16, P: 16): The CBF estimates nutrient loading to the bay is seven times the values of pre-colonial times. No bay-wide trend for reduced levels of N and P is evident, although new nutrient management laws in MD, PA and VA may help in this regard	Total N and P loading (including atmospheric deposition) should be permanently reduced by 50% from 1985 levels
Water clarity (16): Water clarity in the bay is poor and indicates excessive nutrient and sediment loading. Losses of wetlands, forested buffers, and underwater grasses have reduced the bay's ability to filter nutrients and sediments. Greater enforcement of sediment control programmes and reductions in nutrient runoff are needed	Water should be clear enough to support underwater grasses at a depth of 2 m
Dissolved oxygen (15): There are widespread areas of anoxia and hypoxia in the bay. No evidence of bay-wide improvements in dissolved oxygen is available, stressing the importance of further nutrient reductions	Dissolved oxygen levels in waters of the bay should be at least 5 ppm in spawning and nursery areas and 3 p.p.m. elsewhere

Continued

Table 16.3. *Continued.*

Category, assessment parameter and observations for 1998–1999	Goal for 2010
Fish and shellfish	
Crabs (48): Heavy fishing and highly reduced levels of underwater grasses have depleted crab populations well below pre-colonial values. A bay-wide effort to reduce the level of excessive harvesting is needed	The crab population should be restored to its 1960 size and composition
Rockfish (75): Rockfish (striped bass) numbers are the highest since the 1960s. Regional controls on harvesting have helped and must be maintained; the major concern is a decline in the menhaden population the main food source for the rockfish	Current populations should be maintained and the number of older, larger rockfish increased
Oysters (2): The bay's oyster population continues to suffer from disease and a lack of suitable habitat. Some progress has been made in restoring oyster populations by transplanting oysters on to reefs throughout the bay	10% of the bay's traditional oyster grounds should be established as sanctuaries that support 400 ha of rebuilt reefs
Shad (3): The shad population is also severely depleted, but efforts to restore fish passages and phase out ocean fisheries are encouraging	The 10 largest bay tributaries should have spring shad runs sufficient to support fisheries and sustain continued population growth

[a] The full text of the Chesapeake Bay Foundation's 'State of the Bay Report' is available at: http://www.savethebay.cbf.org

[b] Value in parentheses is the rating for that habitat based on a value of 0–100; the rating is based on the best professional judgement of scientists at the Chesapeake Bay Foundation, using historical and currently available information.

Table 16.4. Major land use categories (% of total land in the watershed) in the main basins of the Chesapeake Bay watershed (Source: Chesapeake Bay Program, 1999a).

Basin	Forest	Non-agric. herbaceous[a]	Agriculture					Urban[c]
			Hi-till[b]	Lo-till	Pasture	Hay	Total agric.	
Maryland Eastern Shore	42	12	16	22	2	1	42	5
Virginia Eastern Shore	46	11	34	2	1	0	38	6
Maryland Western Shore	39	10	4	8	5	3	20	31
James River	63	10	2	2	10	3	17	10
Patuxent River	44	11	7	6	5	3	21	25
Potomac River	52	8	3	6	14	6	29	11
Rappahannock River	53	11	7	4	13	6	30	5
York River	63	14	7	3	5	3	18	5
Susquehanna River	61	10	7	4	6	6	23	5
Total watershed	57	10	6	5	8	5	25	8

[a] Non-agricultural herbaceous represents the difference between total, non-forested herbaceous land and total agricultural land.

[b] Hi-till refers to conventionally tilled cropland and lo-till to cropland with conservation tillage (> 30% plant residue cover at time of planting).

[c] Urban includes pervious and impervious land areas.

Table 16.5. Animal inventories (no. of animals $\times 10^3$), cropland area, and animal density in the major basins of the Chesapeake Bay watershed.

Basin	Beef	Dairy	Swine	Layers	Broilers	Cropland (10^3 ha)	Animal density (AEU ha^{-1})
Maryland and Delaware Eastern Shore	19	24	43	986	66,822	433	0.50
Eastern Shore of Virginia	< 1	< 1	1	< 1	2,447	3	0.22
Maryland Western Shore	7	10	18	265	105	79	0.30
James River	158	21	115	69	6,109	438	0.54
Patuxent River	4	2	8	8	29	47	0.16
Potomac River	245	199	195	6,024	42,554	931	0.71
Rappahannock River	73	11	1	1	< 1	207	0.47
York River	23	3	1	2	244	136	0.22
Susquehanna River	101	577	911	20,725	20,670	1,640	0.62
Total watershed	630	848	1,294	28,080	138,980	3,914	0.58

Data obtained from the US Department of Agriculture National Agricultural Statistics Service (NASS, 1999). Number for broilers is the housing capacity on the farm; five to six flocks of broilers are typically grown each year.

watershed (Table 16.5). Poultry (broiler) production predominates on the eastern shore of the bay and in some counties in central Virginia, while beef and dairy are the main animal industries in western Maryland and Virginia and most of Pennsylvania and New York. Animal densities range from a low of 0.16 AEU ha^{-1} (AEU = animal equivalent unit, defined as 450 kg of live weight of any animal) in the Patuxent river basin to 0.71 AEU ha^{-1} in the Potomac river basin, averaging 0.58 AEU ha^{-1} on a watershed basis. In comparison, animal densities in Europe range from 0.5 (Spain) to 4.0 AEU ha^{-1} (Netherlands), averaging 0.9 AEUs ha^{-1} for the European Union (Gassman and Bouzaher, 1995). Some counties within the watershed have much higher animal densities than the watershed average, which has led to local or regional problems with nutrient imbalances, manure surpluses, and increased nutrient losses in runoff and by leaching (Fig. 16.2). One of the areas facing the most significant problems in this regard is the Delmarva Peninsula where geographic concentration of poultry production has led to problems with NO_3^--N leaching, soil P saturation and emerging concerns about NH_3 emissions from animal agriculture. A case study for the Delmarva Peninsula is presented by Beegle *et al.* (Chapter 8, this volume).

Agricultural nutrient inputs to the Chesapeake Bay

Agriculture's role in non-point source pollution of the Chesapeake Bay has been

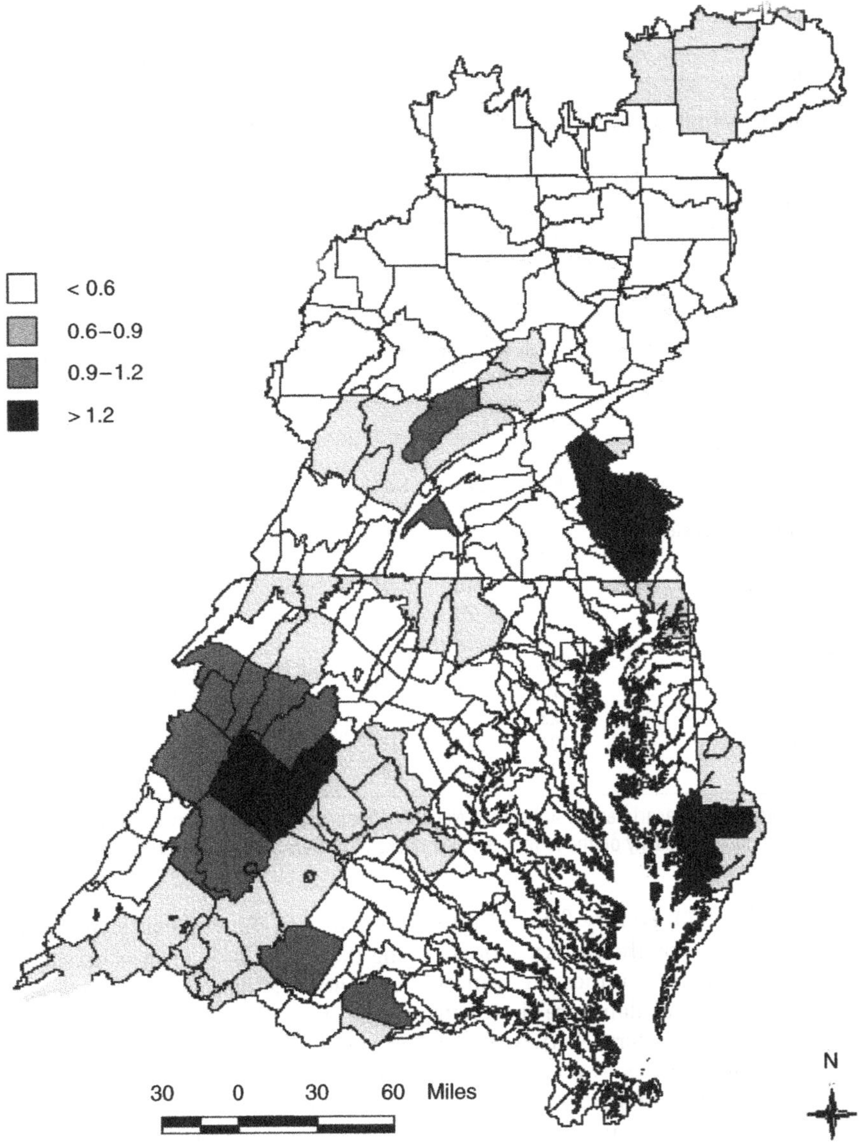

Fig. 16.2. Distribution of estimated animal densities (AEU ha^{-1}) in the Chesapeake Bay watershed (source: Chesapeake Bay Program and USDA NASS, 1999).

studied intensively for years and particularly since the signing of the Chesapeake Bay Agreement in 1983. It is currently estimated that agriculture is the single largest contributor to nutrient pollution of the bay, accounting for 39% of the N loading and 49% of the P loading. Some of the bay's sub-watersheds rank in the top 10% in the USA in terms of N loadings and soil erosion.

N and P from agricultural sources enter the bay's nine major tributaries (Fig. 16.1) as a result of soil erosion, surface or subsurface flows of soluble forms of these nutrients, by groundwater discharge, and by atmospheric deposition of NH_3 emitted from confined animal production facilities, manure storage areas and the land application of manures and N fertilizers. However, most of the approximately 145 M kg of total N (97%) and 9 M kg of total P (90%) input to the bay each year from point and non-point sources is delivered by the three rivers (Susquehanna, Potomac and James) that drain the bulk of the bay's cropland. It has been estimated that agriculture contributes 40% of the N and 60% of the P to the Susquehanna, the largest of the bay's tributaries, and that 70% of the N and 82% of the P input to the bay from the southern Delmarva Peninsula originates from agriculture.

Nutrient inputs to the bay are known to vary according to the type of agriculture practised and the drainage patterns present in each basin. For example, Taylor and Pionke (2000) estimated that 80% of the agricultural contribution of P came from 25% of the watershed that is used for the production of row crops and hay. Intensive and sophisticated efforts are now underway to spatially locate the main sources of N and P to the bay in order to prioritize the implementation of improved nutrient management practices. One example is the SPARROW (SPAtially Referenced Regressions On Watershed attributes) model, which provides statistically based relations between stream water quality and environmental factors such as contaminant sources in the watershed, land surface characteristics that affect contaminant delivery to streams, and instream contaminant losses (Smith *et al.*, 1997). An example of the application of SPARROW to delineate potential sources of N loading to the bay is shown in Fig. 16.3. Information such as this will be extremely useful in future efforts to target research, education and technical assistance programmes designed to further reduce nutrient loading to the Chesapeake Bay.

In summary, it has become apparent that wider implementation of currently recommended best management practices for nutrient use and innovations in nutrient management will be required in the future to reduce agricultural nutrient inputs to the bay. In the past few years, state and federal governments and regulatory agencies have stepped up their efforts to reduce agricultural non-point source pollution of the bay. The remainder of this chapter briefly reviews the current BMPs recommended for N and P management by agriculture, summarizes key changes in policy and legislation that will affect agricultural nutrient use, and outlines some of the new ideas being considered to improve the efficiency of agricultural N and P use in the watershed.

Current Best Management Practices for Nitrogen and Phosphorus in the Chesapeake Bay Watershed

The agricultural BMPs currently recommended in the Chesapeake Bay watershed for N and P reflect more than 40 years of effort by universities, technical advisory agencies, private consultants, agribusiness and farmers to design and implement nutrient management practices that are profitable and protective of the environment. It is beyond the scope of this chapter to review the basic and applied research that underlies and supports these BMPs, or to present crop-specific recommendations for N and P use. Rather, an overview of the basic principles of N and P management is presented, empha-

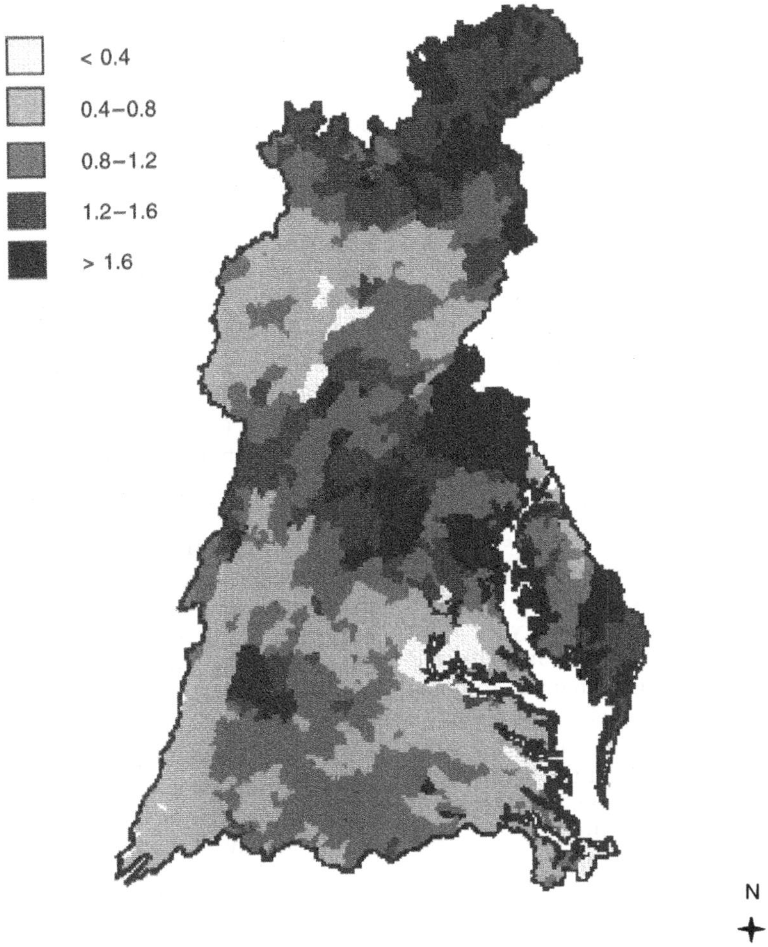

Fig. 16.3. Incremental (local) total N yield (gm^{-2} year^{-1}) to the stream reaches in the Chesapeake Bay Watershed, as estimated by the SPARROW model (Smith *et al.*, 1997)

sizing some of the reasons that additional steps will be needed in the future if we are to further reduce nutrient loading to the Chesapeake Bay.

Agricultural nitrogen management

Nitrogen recommendations for agricultural crops are based on empirical relationships established in multi-site, multi-year field studies that investigated the rate and application method for N needed to achieve economically optimum crop yield goals. This is because, despite years of research, an accurate pre-plant soil test for N is still not available. Crop yield goals reflect both natural conditions (soil type, climate, drainage, topography) and management practices (irrigation, tillage method, crop rotation). Typical N rates in the watershed are 100–200 kg N ha^{-1} for maize, 60–120 kg N ha^{-1} for wheat, barley and oats, and 125–175 kg N ha^{-1} for grass pastures (Sims and Gartley, 1996; Pennsylvania State University, 2000). Nitrogen applications are not recommended for legumes (e.g. soybeans, lucerne) or for pastures with > 50% legumes present in the sward.

Once a yield goal has been established

and the appropriate rate of N application identified, credits are assigned for current or past manure use and/or the inclusion of legumes in a crop rotation. Manure credits can be based on actual manure tests or on manure N availability factors and manure rate applied (e.g. *c.* 75% of the N in poultry manure and 50% of the N in dairy manure will be available in the first year after application if incorporated immediately but only 30–50% will be available if incorporated after 5 days through losses of NH_3), or based on the frequency of manure use. For example, in Pennsylvania, recommended rates of N application for small grains in fields that receive manure every 1–2 years are 0–30 kg N ha^{-1} compared with 60–120 kg N ha^{-1} for non-manured fields. If supplemental fertilizer N is still required after these other N credits are accounted for, the most efficient and economic N source and application method for the crop to be grown are selected. For row crops, multiple applications of N are recommended during the growing season because this approach is known to increase crop N recovery and thus N use efficiency.

For many crops, in-season and postharvest monitoring are recommended to assess the efficiency of the N management programme and make ongoing adjustments where possible. In-season monitoring practices include soil testing (for some crops) and diagnostic plant tests that provide an indication of crop N status. Examples of these N tests include the pre-sidedress soil NO_3^- test (PSNT) (Magdoff *et al.*, 1984; Fox *et al.*, 1992; Sims *et al.*, 1995), the leaf chlorophyll meter (LCM) (Binford *et al.*, 1990) and the stalk NO_3^- test for maize (Binford and Hansen, 2000).

Despite these and other advances in N management, losses of N from agricultural soils continue to be a water quality problem in the Chesapeake Bay watershed. The major reasons for this are as follows:

1. The dynamic nature of N and the variable, unpredictable humid climate in the watershed make it extremely difficult to avoid environmentally significant N losses. For example, on the eastern shore of the bay many soils are sandy and crops are highly susceptible to summer droughts, which occur regularly. Thus, even if farmers follow all available advice on N management, significant, drought-induced reductions in yields can prevent N uptake and result in large amounts of residual N in the soil after maize harvest. Much of this NO_3^--N is then lost by leaching during the autumn and winter. Use of autumn cover crops to 'trap' residual N, while recommended, is not a widespread practice in this region. Indeed, the main winter crops grown (small grains) regularly receive autumn applications of fertilizer and manure N because a reliable autumn test for soil N availability has not been developed.

2. Farmers are unaware of, or do not follow, recommended nutrient management BMPs because of time and economic constraints. A recent survey of states in the mid-Atlantic and north-east indicated that less than 5% of the farmers in these states used the PSNT regularly and that many were unaware of the LCM or stalk NO_3^- test (Fox *et al.*, 1999).

3. The use of animal manures in the watershed is widespread and there are difficulties in estimating N availability from manures, which depends on soil moisture and temperature; there are also logistical constraints which prevent the timely application of manure N relative to crop N uptake patterns. As an example, many farmers must apply manure in the winter or early spring because of inadequate storage or insufficient time to apply all the manure that is generated close to the planting of summer annual crops.

Agricultural phosphorus management

Best management practices for P are more straightforward than those required for N because soil P availability can be easily and accurately assessed by routine soil testing methods. Also, because P is less susceptible to leaching losses and there are no gaseous pathways for P loss from soils, the management of native and applied P is less problematic than N. The situations where crops respond to applied P are clearly understood and the economically optimum P rates and methods of application are well established. Thus, in general, barring serious soil erosion problems, following the recommended BMPs for P management should minimize P transport to surface and groundwaters. Unfortunately, erosion does occur in the watershed and in some settings the recommended BMPs cannot be followed because of the type of agriculture that is being practised. Understanding where the BMPs for P management fail is the first step in developing new practices that will be more successful in reducing P loss to water.

The general approach for P management recommended throughout the Chesapeake Bay watershed is conceptually simple (Sims and Vadas, 1997). Soils should be tested for P on a regular basis using the soil test method developed for the region of interest. If soils test below the optimum soil P value, fertilizer or manure P should be applied in a timely and efficient manner. Typical P application rates for crops grown in the watershed when soils test in the low range are from 40 to 70 kg P ha^{-1}. If soils are above the optimum range, no P should be applied except for a small amount (5–10 kg P ha^{-1}) as starter fertilizer; however, in some cases a maintenance recommendation of 10–20 kg P ha^{-1} will be made to ensure that soil P does not decrease from optimum to below optimum concentrations. Soil and water conservation practices should be adopted throughout the farm to minimize the potential for soil erosion (e.g. conservation tillage, buffer strips, strip cropping, terraces). Following these practices should minimize, to the best of the farmer's ability, the transport of particulate and soluble P to nearby surface waters and the leaching of P to shallow groundwaters.

The main nutrient management issues related to P in the Chesapeake Bay are:

1. Soil P build up to very high or excessive levels due to the long-term over-application of manures and fertilizers. This results in soil particles that are highly enriched in bioavailable P which may be lost and also increases the potential for the losses of soluble P in runoff.

2. The occurrence of soil erosion, most commonly from conventionally tilled row crops where conservation practices are not installed or not maintained.

A concerted effort has been made in the watershed to promote and provide funding to support the use of conservation tillage and other soil conservation practices. Today, about half the cropland in the watershed that is not in pasture or hayland receives some form of conservation tillage. Incentive programmes to promote the use of buffer strips and riparian forests are increasing (see Table 16.2) as are cost-sharing programmes for winter annual cover crops. Preserving and restoring wetlands, which act as filters to remove sediments from runoff is also a high priority in the watershed.

The application of fertilizer and manure P to soils in the economically optimum (or higher) range is another matter. Taylor and Pionke (2000) reported that, on average, 30 kg P ha^{-1} were applied to cropland in the watershed each year; 73% of this was applied in commercial fertilizer and 27% in manure. This is despite the fact that soil test summaries from these states show that most soils are in the optimal or higher range of soil test P (Sims *et al.*, 2000). For example, Sims (2000) reported that most (> 60–70%) of the agricultural soil samples tested in Pennsylvania, Delaware and Maryland were in the optimum or excessive categories, where P applications would not be recommended. In counties with high animal densities, > 90% of the soils were in these categories. If soil testing recommendations were being followed, most farmers would not purchase and apply fertilizer P (other than starter P) to soils such as these. However, they have no option but to regularly apply large amounts of manure P because there are no economically viable alternative uses for manures available today. An increasing body of research clearly shows that soluble P losses from soils increase in proportion to increases in soil test P. Thus, unless there are other uses for animal manures in the watershed (or export from the watershed) and reductions in fertilizer P sales, it seems likely that soil P concentrations and thus P losses to water will increase in the future, particularly in areas where animal agriculture predominates.

It is apparent that greater efforts to reduce soil loss and prevent the unnecessary application of fertilizer and manure P to soils will be needed in the future. Since P loss to water is a complex function of climatic conditions, field-scale properties (e.g. topography, drainage, proximity to water) and P source management, a more holistic approach to agricultural P management will be needed in the future if P loading to the bay is to be reduced.

Environmental Policy and Legislation in the Chesapeake Bay Watershed

Agricultural nutrient management planning and practices in the USA have historically been voluntary, with the general caveat that farmers are not allowed to directly pollute the waters of the USA. The long-standing model for nutrient management planning has been the approach historically used by land-grant universities for most agricultural problems: (i) learn of the problems from extension educators and specialists or farmers themselves; (ii) conduct basic and/or applied research under well-controlled conditions (e.g. small research plots) to investigate potential solutions to the problem; (iii) develop practical and economically feasible, farm-scale management practices based on the results of the controlled experiments; (iv) test these practices out through on-farm demonstrations for several years and at different locations that vary in soil types, crop rotations, hydrology and climate; and (v) use extension education programmes to encourage widespread adoption of practices proved to be effective. This approach has worked well for production-related problems where implementation of a problem-solving practice usually increases profitability. However, for environmental problems, where the goal is often to manage nutrients more carefully, thus increasing the costs of production but not the value of agricultural products, this model has been less successful. Practices that are acknowledged by farmers to increase the efficiency of nutrient use have often not been widely adopted because they require more time and effort to perform and/or the assumption of more economic risk on the part of the producer. In recent years, public concerns about declining water quality in the Chesapeake Bay watershed have led to a different, more structured approach, which involves more stringent agricultural policies and even, in some states, regulations for nutrient management. The approximate chronology of these actions is described below.

Pennsylvania Nutrient Management Act

In 1993, Pennsylvania passed the first nutrient management law in the Chesapeake Bay watershed. The Pennsylvania Nutrient Management Act required regulatory oversight of nutrient management on some farms, specifically concentrated animal operations (CAOs). By law, CAOs were defined as those agricultural operations where the 'animal density exceeds 5 AEU ha^{-1} on an annual basis. An AEU is *c.* 450 kg of liveweight of any animal' (Beegle, 1997). According to the regulations developed subsequent to passage of this law, the land that can be used to calculate animal density on a farm includes 'cropland, hay land or pastureland (owned or rented) that is an integral part of the operation and land that is, or will be, used for the application of manure from the operation'. Farmsteads and forests cannot be included in the calculation of animal density.

The regulations to enforce the Pennsylvania Nutrient Management Act of 1993 took 4 years to develop and became effective on 1 October 1997. All CAOs were required to develop a nutrient management plan and submit it for review to either the local conservation district or the State Conservation Commission within 1 year of the passage of the regulations. Farms with lower

Table 16.6. Summary of nutrient management legislation in Pennsylvania, Delaware, Maryland and Virginia.

State	Overview of key components of the legislation
Pennsylvania (Nutrient Management Act 1993)	• The act is designed to: (i) establish the criteria, planning requirements and an implementation schedule for nutrient management measures on certain agricultural operations that produce or use manure; (ii) develop educational programmes on nutrient management to prevent the pollution of surface and ground waters; (iii) provide for technical and financial assistance for nutrient management and alternative uses of animal manure; and (iv) require the state Department of Environmental Protection to assess the extent of other nonpoint sources of nutrient pollution • The act establishes a Nutrient Management Advisory Board that will review and comment on all regulations developed by the state's Conservation Commission. The board includes five farmers, an animal nutrition specialist, representatives of the feed and fertilizer industries, commercial lenders and local government, two citizens, a university agronomist, a hydrologist and an environmental representative • The act requires that all concentrated animal operations (CAOs, those operations where 'the animal density exceeds two animal equivalent units per acre') develop and implement nutrient management plans. These plans must include six components: (i) farm identification information; (ii) summary document that provides an overview of the plan, including the amount of manure generated, used on the operation and exported from the farm; (iii) nutrients available on the farm, crop nutrient needs, and nutrient application rates and procedures; (iv) alternative uses for excess manure; (v) manure and barnyard management; and (vi) storm water runoff control. The act also provides for plan review, documents record keeping requirements and enforcement procedures, and establishes a procedure for obtaining technical and financial assistance with plan implementation. For operations with lower animal densities, the act encourages voluntary nutrient management planning and offers opportunities for financial assistance in the form of grants, loans and loan guarantees. It also requires the state Conservation Commission to re-evaluate, after 5 years, the criteria for defining a CAO
Delaware (Nutrient Management Act 1999)	• Establishes a 19-member, politically appointed Delaware Nutrient Management Commission (DNMC) to develop and implement a state nutrient management programme. Anyone with > 8 animal units or who applies nutrients to > 4 ha of land must have and implement a nutrient management plan. Plans will be reviewed beginning in 2003 and the state nutrient management programme must be implemented by 2007. For agronomic plans N applications will be limited to that required for a 'realistic yield' (defined as best four yields from the past 7 years); application of P to 'high P' soils (defined in 2002 by the DNMC as soils with mehlich 3P7 150 mg kg^{-1}) shall not exceed a 3-year crop removal rate. The act also proposes that the state's NPDES programme for CAFOs be delegated to the Department of Agriculture who will rely on the act to the greatest extent practical in the development of CNMPs for CAFOs. Municipalities applying biosolids to cropland are exempt from the act (except for reporting) because they are already permitted by the state through other programmes. Financial penalties are included for those who do not comply with the act • The DNMC will also: (i) consider the establishment of critical areas for targeting voluntary and regulatory programmes; (ii) establish BMPs to reduce nutrient losses to the environment; (iii) develop educational and awareness programmes; (iv) consider the need for a transport and alternative use incentive programme to move nutrients from areas of overabundance to areas where they are needed; (v) establish a state certification programme for four classes of nutrient users and a method to evaluate an applicant's suitability for certification; (vi) cooperate with state agencies to provide cost-share funds for NMP development and BMP implementation; (vii) work with 'commercial processors' (e.g. poultry integrating companies) to provided technical assistance and educational programmes for contract growers to improve the storage and management of animal wastes; and (viii) keep records and provide annual reports to the legislature on the progress of the DNMC in implementing the requirements of the act. The DNMC is now developing the regulations required by the act

Maryland (Water Quality Improvement Act 1998)	• Requires that any agricultural operation with > US$2500 in gross annual income or > 8 animal units must develop and implement a NMP. The law clearly includes municipal biosolids application to cropland and impacts non-agricultural nutrient users (e.g. commercial lawn care, nurseries, turfgrass producers) who apply nutrients to more than 1.25 ha of land. The WQIA requires that anyone who only uses commercial fertilizer must 'develop and implement a N and P based plan by 31/12/02; those using biosolids or animal manures must comply by 31/12/05. The act includes financial penalties for those who do not comply. The addition of restrictions on the amount of P that can be applied is a major change from nutrient management planning efforts in Maryland • All NMPs must be developed and approved through the Maryland Department of Agriculture; cost-sharing is provided for plan development. A state Nutrient Management Advisory Committee with wide-ranging representation developed and recently released (May 2000) the final regulations that will implement this act. These proposed regulations are now under review and may be subject to additional public hearings
Virginia (Poultry Waste Management Bill 1999)	• Requires the development and implementation of nutrient management plans for 'any person owning or operating a confined poultry feeding operation'. These NMPs will govern the storage, treatment and management of poultry waste. Also provides for poultry waste 'tracking and accounting' • States that N application rates in mandated NMPs shall not exceed crop nutrient needs as determined by the Virginia Department of Conservation and Recreation (DCR). Poultry wastes shall also 'be managed to minimize runoff, leaching, and volatilization losses and reduce adverse water quality impacts from nitrogen' • After 1/10/01 P application rates in mandated NMPs shall not exceed the greater of crop nutrient needs or crop nutrient removal as determined by DCR. Poultry wastes shall *'be managed to minimize runoff and leaching and reduce adverse water quality impacts from P'*. Prior to 1/10/01 comprehensive soil conservation plans consistent with USDA-NRCS guidelines will be required for soils with a soil test P (Mehlich 1) value > 55 mg kg^{-1} (ppm)

animal densities were encouraged to develop nutrient management plans voluntarily with economic incentives from the state government. The specific requirements of the Pennsylvania act are provided by Beegle *et al.* (1997) and summarized in Table 16.6. It is important to note, however, that the regulatory threshold set by this act (*c.* 5 AEU ha^{-1}) is very high compared with the situation in most of the Chesapeake Bay watershed (average animal density = 0.58 AEU ha^{-1}; Table 16.5), higher even than The Netherlands (4.0 AEU ha^{-1}; Gassman and Bouzaher, 1995) which has a long history of serious environmental problems with high density animal agriculture. Thus, for the majority of farms, the Pennsylvania approach continues to rely on voluntary efforts to reduce non-point nutrient pollution.

Delaware and Virginia's TMDL programmes

The next significant step towards a more regulatory approach to nutrient management in the Chesapeake Bay watershed was a lawsuit filed against USEPA in 1996 by a consortium of environmental groups for 'failure to perform its mandatory duties under the Clean Water Act to identify and then improve water quality' in Delaware. The Federal Clean Water Act requires states to develop a list of water bodies for which existing pollution control activities are not sufficient to attain water quality standards and to develop total maximum daily loads (TMDLs) for pollutants of concern. A TMDL sets a limit on the amount of a pollutant that can be discharged into a water body such that water quality standards are met.

In 1997 the state of Delaware negotiated a TMDL agreement with USEPA and established a 10-year schedule to develop TMDLs for affected water bodies and then to promulgate pollution control strategies to ensure that pollutant loadings are below the TMDL values. Since 1997, TMDLs for N and P have been established for a number of important waterways in Delaware. The reductions in N and P loading required to meet these TMDLs are significant and will require intensive efforts by both point and non-point sources. For example, elimination of all point source discharge of N and P and major reductions in non-point loading of N (85%) and P (65%) will be necessary to meet the TMDLs for the Indian River in southern Delaware. The TMDL rule also requires a 20% reduction in atmospheric deposition of N, which is of relevance to agriculture because of concerns about NH_3 emissions from the poultry industry located in close proximity to this river.

Virginia entered into a TMDL agreement with USEPA in 1998, adopting a 12-year schedule to set TMDLs and, subsequent to this, to implement plans to reduce pollutant inputs to levels needed to meet the desired water quality. Maryland does not have a TMDL agreement but a lawsuit has been filed to compel USEPA to establish TMDLs for impaired water bodies in that state.

The Maryland Water Quality Improvement Act

The next piece of water quality legislation passed in the Chesapeake Bay watershed was Maryland's Water Quality Improvement Act of 1998. Passage of this act was stimulated by public concerns over fish kills in rivers on the eastern shore of Maryland in the summer of 1997. These fish kills were reportedly caused by *Pfiesteria* spp., a toxic dinoflagellate that had been implicated in earlier, massive fish kills in North Carolina and also in human health problems. In response to a public outcry and intense media pressure, the Governor of Maryland appointed the 'Citizens Pfiesteria Action Commission' in the autumn of 1997. The report of this commission led to legislation in early 1998 requiring all agricultural operations with annual incomes greater than US$2500 or more than eight animal units to develop and implement a N and P-based nutrient management plan, phasing the plans in over a 7-year period, depending on the type of agricultural operation. The legislation is wide-ranging in scope, going beyond farming to include municipalities that apply biosolids to land, vegetable growers, nurseries, greenhouses, turfgrass producers and some horse farms in its requirements. It was also the first legislation in the USA to mandate P-based nutrient management for agriculture. A 35-member Nutrient Management Advisory Committee was established to develop the regulations required by this act. The regulations were completed in May 2000 and will be effective in the near future. More detailed information on the events that led to Maryland's law is provided by Simpson (1998) and key components of the law are summarized in Table 16.6. However, the Maryland law, which passed in a politically charged atmosphere, stimulated similar efforts in Virginia and Delaware, under pressure from the USEPA, to move away from the voluntary nutrient management practices advocated in the past and in the direction of regulated programmes, especially for large confined animal feeding operations (CAFOs), usually those operations with $>$ 1000 AEU.

The Virginia Poultry Waste Management Bill

In January 1999, the Virginia legislature passed the Poultry Waste Management Bill (Table 16.6). This bill gave regulatory authority over poultry operations with $>$ 200 AEU (20,000 chickens) to the Virginia Department of Environmental Quality (DEQ) and charged the Department of Conservation and Recreation with setting the standards for nutrient management plans. These plans restrict poultry litter application based on crop nutrient needs (including limits based on soil P). The bill also requires commercial poultry companies to detail to DEQ how they will assist poultry growers with whom they contract to ensure that poultry waste is properly stored, managed and transported to areas where it can be used most effectively.

The Delaware Nutrient Management Act

In 1998, in response to ongoing public concerns about water quality and fish kills, and under pressure from USEPA, the Governor of Delaware appointed an Agricultural Industry Advisory Committee on Nutrient Management, consisting of ten farmers, to develop recommendations for state actions. The efforts of this committee led to the passage in June 1999 of Delaware's state Nutrient Management Act. A 15-member Delaware Nutrient Management Commission (DNMC) was established to develop and implement a state nutrient management programme to protect and improve water quality. As in Maryland, the Delaware act requires nutrient management plans by anyone with $>$ 8 AEU or who applies nutrients to $>$ 4 ha of land and includes a wide range of non-farming operations; however, it specifically excludes municipalities that apply biosolids to land. It also requires the DNMC to establish a state certification programme for four classes of nutrient users (nutrient generator, private nutrient applicator, commercial nutrient applicator and nutrient consultant) and to develop and implement a method to evaluate an applicant's suitability for certification (e.g. an examination programme). The DNMC is now in the process of drafting regulations that fulfil the requirements of the law. Specific information on this act is provided by Sims (1999) and summarized in Table 16.6.

USDA and USEPA national policies

National policy initiatives are also underway that impact agricultural nutrient management in the Chesapeake Bay watershed. By far the most significant is the USEPA-USDA Unified National Strategy for Animal Feeding Operations (AFOs), adopted in March 1999 after lengthy discussion and public review. This document has nine 'guiding principles' for the joint effort between the nation's lead regulatory agency (USEPA) and its lead technical agency for agriculture (USDA) to 'address the water quality and public health impacts associated with AFOs' (USDA and USEPA, 1999):

1. Minimize water quality and public health impacts from AFOs.
2. Focus on AFOs that represent the greatest risk to the environment and public health.
3. Ensure that measures to protect the environment and public health complement the long-term sustainability of livestock production in the USA.
4. Establish a national goal and performance expectations for AFOs.
5. Promote, support and provide incentives for the use of sustainable practices and systems.
6. Build on strengths of federal agencies and their state/local partners and make appropriate use of diverse tools including voluntary, regulatory and incentive-based approaches.
7. Foster public confidence that AFOs meet performance expectations and that federal and state/local governments are ensuring the protection of water quality and public health.
8. Coordinate activities among USDA, USEPA and state agencies and other organizations that affect or influence the management and operation of AFOs.
9. Focus technical and financial assistance to support AFOs in meeting national goal and performance expectations established in the Unified National Strategy.

Following this strategy, in summer 1999 the USDA Natural Resources Conservation Service (USDA-NRCS) released a national nutrient policy which will require more comprehensive nutrient management planning and implementation of plans for farmers receiving technical assistance and cost-sharing funds. In September 1999, the USEPA released for public comment the Guidance Manual and Example NPDES Permit for Concentrated Animal Feeding Operations, and in December 1999 USDA-NRCS released for comment the Technical Guidance for Developing Comprehensive Nutrient Management Plans (CNMPs) for both the USDA and USEPA (see websites http://www.nhq/nrcs/usda/gov and http://www.epa.gov/owm, respectively). As described by USDA-NRCS guidance, the objective of a CNMP is 'to combine management activities and conservation practices into a system that, when implemented, will minimize the adverse impacts of animal feeding operations on water quality'. Six elements are proposed for a CNMP: (i) animal outputs: manure and wastewater collection, handling, storage, treatment and transfer; (ii) evaluation and treatment of sites proposed for land application; (iii) land application; (iv) records of CNMP implementation; (v) inputs to animals; and (vi) other manure uses (e.g. power generation, composting, pelletization). In the summer of 2000, USEPA released for comment the 'National Management Measures to Control Nonpoint Source Pollution from Agriculture'. This document was designed to provide technical assistance to state programme managers and others on the best available, economically achievable means of reducing nonpoint-source pollution of surface and ground water from agriculture. Finally, in 2001 USEPA proposed new national rules for CAFO's, to be adopted by 2003, that will markedly increase the regulation of these operations (see http://cfpub.epa.gov/npdes/ for detailed information on these newly proposed regulations. State agencies, farm organizations, universities and environmental advocacy groups are now working to understand and then to integrate federal guidance and regulations with those mandated by new state laws. Most anticipate that this process will take from 5 to 10 years.

Future Directions in Nutrient Management in the Chesapeake Bay Watershed

Clearly, in the past 5 years dramatic changes have occurred in the approach that will be used in the future for agricultural nutrient management in the Chesapeake Bay watershed. Most changes are still in progress, or are still being debated, and are not likely to be finalized for several years. Given this, it is somewhat difficult to predict, or describe, what the future holds for nutrient management and water quality in this vitally important watershed. Three major changes, however, are likely to occur that will permanently alter the way agriculture is practised in the watershed.

Nutrient management planning

Nutrient management planning is not a new agricultural activity; farmers and their advisers have developed and implemented plans for the profitable use of nutrients for decades. Environmental considerations (e.g. preventing soil erosion, reducing NO_3^- leaching) have always played a role in these voluntary plans. It is apparent, however, that nutrient management plans developed in the future will be more formal, even regulatory, in nature. The passage of the state laws described in Table 16.6 and the development of regulations to implement these laws will require most nutrient users (agricultural and non-agricultural) to implement nutrient management plans for their operations within the next few years. Record-keeping will be required and penalties assessed for failure to comply with requirements of the laws. These state and federal actions are likely to lead to some changes, as follows.

Expanded nutrient management education and planning

Expansions in funding and activities of government, university and private sector advisers involved in educational efforts on nutrient management plans, the design and implementation of these plans and associated BMPs, and cost-sharing to facilitate plan implementation can be expected in the future. This should result in more efficient use of nutrients by agriculture which should in turn contribute to the long-term improvements in water quality hoped for by those that passed the state legislation and the national agencies that designed the CNMPs for AFOs.

Intensified and integrated nitrogen and phosphorus management

Perhaps the most significant specific change that will occur will be the redesign of nutrient management plans to include more detailed N and P management practices that are more protective of the environment. For example, state laws and programmes will provide more funds for soil testing which should result in more widespread use of the PSNT, LCM and stalk NO_3^- tests, thus overcoming some of the obstacles to the use of these widely accepted, but poorly implemented, N tests. Manure testing for N and P will expand, as state and private laboratories increase their capability in these areas. Soil testing for P will also increase as will the use of new soil P tests (e.g. soil P saturation, water soluble P). Other, more intensive BMPs for P will be more widely adopted. Steinhilber and Weismiller (2000) reviewed these and identified several promising options, including:

- use of chemical amendments (e.g. alum) as manure treatments to stabilize manure P in less soluble forms;
- amending soils in buffer strips with by-products (e.g. alum water treatment residuals, fly ash) to immobilize soil P;
- altering animal diets to decrease P excretion, for example by using phytase enzymes as a feed supplement while decreasing mineral P supplements in the diets;
- phytoremediation of high P soils, provided that plants capable of high P removal can be identified or 'bio-engineered';
- more incorporation of manures, in conjunction with wider use of buffer strips and cover crops;
- periodic deep ploughing to incorporate high P topsoils thus reducing the loss of soluble P and highly enriched soil particles;
- the removal of P-enriched sediments from field ditches.

A regional effort has been underway for several years to develop a holistic approach to P management that integrates site properties related to P transport, water body sensitivity to P, soil P status, and P management of fertilizers and manures into a field-scale risk assessment tool (the Phosphorus Site Index; Leytem *et al.*,

2000). In 2001 the state nutrient management commissions in Delaware and Maryland formally adopted *P Site Index* assessment tools that will be used by nutrient management planners to identify areas requiring more intensive P management. Similar actions are anticipated in Pennsylvania and Virginia in the near future.

Increasing emphasis on strategic nutrient management planning

Wide-scale development of nutrient management plans will probably confirm the magnitude and geographic locations of nutrient surpluses in the watershed. Site-specific or precision management technology is evolving which will make agronomically and environmentally sound, field-by-field nutrient management decisions commonplace. Farm-scale, field-scale, and intrafield-scale plans are designed to use the optimum amount of nutrients needed to achieve realistic yields; not to dispose economically of surplus nutrients not needed on the farm. Individual farmers do not have the economic resources to design and construct the infrastructure needed to economically use nutrient surpluses of the magnitude estimated to be present in some regions, such as the Delmarva Peninsula. State or regional efforts, involving private industry, as described below, will be required for long-term resolution of the nutrient imbalance problems facing animal agriculture.

Alternative uses for animal wastes

For the past 2 years a concerted effort has been under way in the Chesapeake Bay watershed to critically analyse alternative uses for animal wastes based on the premise that application to agricultural cropland cannot remain the sole end use for the by-products of animal production. Most recently a task force was established by the Chesapeake Bay Executive Council with the charge to 'recommend appropriate procedures pertaining to the interstate distribution of animal wastes'. This task force conducted a thorough review of all proposed alternatives to land application of animal wastes and made the following list of recommendations (Chesapeake Bay Program, 1999b):

1. A memorandum of understanding should be signed by the six states in the Chesapeake Bay watershed (Delaware, Maryland, Pennsylvania, New York, Virginia and West Virginia) that will formalize a long-term commitment to ensure the proper land application of animal wastes regardless of their final destination.
2. The six states should adopt technical guidelines for animal waste transport, nutrient management plans, storage of wastes, biosecurity and the monitoring and tracking of animal waste transfers within and out of the watershed.
3. Potentially feasible alternatives to land application of animal wastes include expanded land application, composting, pelletization or granulation for fertilizers, use in animal feed, and energy generation. The task force provided detailed analyses of each option and stated that:

- The most appropriate mix of alternative uses and incentive solutions will vary by area. Areas with slight oversupply should expand the use of more distant land application of animal wastes and secondary uses such as composting for the landscape industry. Areas with significant oversupply should pursue waste reuse options such as pelletization and shipment to nutrient deficit areas, energy generation and minor use options.
- Nutrient reduction through more effective feeding strategies should be encouraged for liquid wastes which are more difficult to transport for alternative uses.
- Incentive programmes should be developed to promote alternative uses, with preferences given to recurring incentives; incentives with a phase-out period; start-up incen-

tives such as grants and low-interest loans; and insurance mechanisms to reduce the risk of adopting a new practice. Market-based solutions and research on alternative uses should be encouraged.

- Long-term success will require good faith cooperation, funding, risk assumption, and personnel resources from the integration of animal production companies, livestock growers, manure brokers, state and federal agencies, utilities involved in energy generation from wastes and groups such as the Chesapeake Bay Program.

Nutrient trading

The decision by the Chesapeake Bay Program to reduce nutrient inputs to the bay from point and non-point sources by 40% and then to maintain that degree of loading thereafter has caused an interest in nutrient trading between point and non-point sectors. Nutrient trading means that, if expansion in one sector (e.g. municipal wastewater plant) results in an increase in nutrient discharge, then nutrient reductions from another sector (e.g. agriculture) must occur to offset that increase. Presumably, the sector desiring the increase would help to subsidize the efforts of the sector that must provide the decrease and this trade would be economically beneficial to both. There is a market for trading in the bay's watershed. About 350 large wastewater treatment plants are operating in the watershed and new nutrient reduction technologies (e.g. biological nutrient removal) may be able to offer nutrient credits for sale in the future. These credits may be useful to non-point sources that have difficulty implementing BMPs to reduce nutrient losses, or to other municipalities where population growth requires expansion of existing wastewater treatment plants.

In 1999, the Chesapeake Bay Program recognized that for nutrient trading to be successful, innovative and creative measures would be required. Consequently, a 'Nutrient Trading Negotiation Team' was established, composed of representatives of all states in the watershed, the District of Columbia, USEPA and individuals representing the interests of local watersheds, the public, the point sources (municipal and industrial), the rural non-point sources, the urban non-point sources and the Chesapeake Bay Commission. This team identified six elements for a successful trading framework: (i) nutrient reduction goals must be delineated for major tributaries and the effects of changes in these caps defined; (ii) eligibility – who will be eligible to trade and how will the number and types of nutrient credits be determined; (iii) trade administration, or establishment of a marketplace structure for the trading process, including recording keeping and brokerage practices; (iv) accountability, to ensure that the effectiveness of the trade is monitored, including but not limited to effects of the trade on water quality; (v) assessment, at both tributary and bay-wide scales of a process to assess the overall effectiveness of nutrient trading; and (vi) stakeholder involvement, to permit interested parties to observe the trading programme, monitor it and offer comments for improvement. It is anticipated that a set of Nutrient Trading Guidelines were developed in 2000 and released for public comment in 2001. A series of public workshops were held throughout the watershed in 2001 to discuss these guidelines and explain the fundamental principles involved in nutrient trading. The current goal is to have these guidelines endorsed by the Chesapeake Bay Program in 2002 and subsequently to be available for use by localities and states as they consider the need to develop voluntary local programmes for nutrient trading.

Conclusions

Agriculture in the Chesapeake Bay watershed is at a crossroads. Fundamental questions are being asked about the sustainability of the animal agriculture that dominates the economy of this region. The voluntary approaches to nutrient management used in the past are being replaced by regulatory programmes that will require farmers, and other nutrient users, to follow more intensive, and thus costly, approaches to nutrient management planning. Many in agriculture question the need and scientific validity of these changes. Others ask for delays in their implementation to allow for more research to justify the actions required in state laws and more time for an economic evaluation of their impact on agricultural profitability. At the same time, federal agencies, environmental advocacy groups and the media are pressing states to move forward more rapidly to protect and improve water quality.

Given the rather contentious nature of this issue, what steps should be taken today to address the need to sustain agriculture and protect environmental quality? First, there is a need for all parties involved to recognize that the issue of nutrient management and water quality requires serious attention, in the short and long term. Sufficient information is available from regional research and research elsewhere to justify the immediate need for improved nutrient management practices, particularly for animal agriculture. Consequently, there is a need for reasoned public debate on the issue to identify the most appropriate changes that should be required to reduce agriculture's impact on water quality. Second, some changes should occur now, specifically concerted efforts to implement nutrient management practices that are accepted by agriculture, but not widely used due to economic constraints. Examples include the use of diagnostic N tests to increase fertilizer use efficiency, more extensive efforts to identify soils that are high enough in P to be of environmental concern, use of modified animal diets and manure treatment technologies to reduce the potential for nutrient losses, wider implementation of soil conservation practices (e.g. buffer strips) and avoiding poorly timed or badly located applications of animal wastes. Finally, most scientists agree that more information is needed on nutrient cycling and management, particularly for P. A significant research effort is under way today and all indications are that support for nutrient management research (basic and applied) will continue to be strong for the next decade. This argues for patience and flexibility in the implementation of regulations until the research is completed and clearly understood by all interested individuals.

References

Beegle, D. (1997) *Nutrient Management Legislation in Pennsylvania: Who will be affected?* Agronomy Facts 40. Pennsylvania State University, University Park, Pennsylvania.

Beegle, D., Lanyon, L.E. and Lingenfelter, D. (1997) *Nutrient Management Legislation in Pennsylvania: a Summary of the Final Regulations*, Agronomy Facts 40. Pennsylvania State University, University Park, Pennsylvania.

Binford, G.D. and Hansen, D.J. (2000) Diagnostic nitrogen tests for manured soils: current status and future outlook. *Proceedings Conference Managing Nutrients and Pathogens from Animal Agriculture*, NRAES-130. Natural Resource, Agriculture, and Engineering Service, Ithaca, New York, pp. 355–368.

Binford, G.D., Blackmer, A.M. and El-Hout, N.M. (1990) Tissue test for excess nitrogen during corn production. *Agronomy Journal* 82, 124–129.

Boynton, W.R. (2000) Impact of nutrient inflows on the Chesapeake Bay. In: Sharpley, A.N. (ed.) *Agriculture and Phosphorus Management: the Chesapeake Bay*. Lewis Publishers, Boca Raton, Florida, pp. 23–40.

Boynton, W.R., Garber, J.H., Summers, R. and Kemp, W.M. (1995) Inputs, transformations, and transport of nitrogen and phosphorus in the Chesapeake Bay and selected tributaries. *Estuaries* 18, 285–314.

Chesapeake Bay Program (1999a) *The State of the Chesapeake Bay*, EPA 903-R99–013 and CBP/TRS 222/108. Chesapeake Bay Program, Annapolis, Maryland.

Chesapeake Bay Program (1999b) *Task Force Report on Interstate Animal Waste Distribution and Use Technology*. Chesapeake Bay Program, Annapolis, Maryland.

Chesapeake Bay Program (2000) 410 Severn Avenue, Suite 109, Annapolis, MD 21403, http://www.chesapeakebay.net

Fox, R.H., Meisinger, J.J. and Sims, J.T. (1992) Predicting N fertilizer needs for corn in humid regions: Advances in the Mid-Atlantic States In: *Predicting N Fertilizer Needs for Corn*, Bulletin Y-226. National Fertilizer and Environmental Research Center, Tennessee Valley Authority, Muscle Shoals, Alabama, pp. 43–57.

Fox, R.H., Piekielek, W.P. and Bundy, L.G. (1999) Status of diagnostic tests for nitrogen availability to corn. *Agronomy Abstracts*, ASA, Madison, Wisconsin.

Gassman, P.W. and Bouzaher, A. (1995) Livestock pollution: lessons from the European Union. In: Steele, K. (ed.) *Animal Waste and the Land Water Interface*. Lewis Publishers, CRC, New York, pp. 515–522.

Kemp, W.M., Twilley, R.R., Stevenson, J.C., Boynton, W.R. and Means, J.C. (1983) The decline of submerged vascular plants in the upper Chesapeake Bay: summary of results concerning possible causes. *Marine Technology Society* 17, 78–89.

Leytem, A.B., Sims, J.T. and Coale, F.J. (2000) Developing a reliable site index for the Delmarva peninsula. *Proceedings Conference, Managing Nutrients and Pathogens from Animal Agriculture*, NRAES-130. Natural Resource, Agriculture, and Engineering Service, Ithaca, New York, pp. 282–302.

Magdoff, F.R., Ross, D. and Amadon, J. (1984) A soil test for nitrogen availability to corn. *Soil Science Society America Journal* 48, 1301–1304.

NASS (National Agricultural Statistics Service) (1999) *Census of Agriculture, 1997*. US Department of Agriculture, Washington, DC, http://www.usda.gov/nass

Pennsylvania State University (2000) *Agronomy Guide*. College of Agricultural Science, Pennsylvania State University, University Park, Pennsylvania.

Simpson, T.W. (1998) *A Citizen's Guide to Maryland's Water Quality Improvement Act*. University of Maryland Cooperative Extension, College Park, Maryland.

Sims, J.T. (1999) Overview of Delaware's 1999 nutrient management act. Fact Sheet NM-02. College of Agricultural Natural Resources, University of Delaware, Newark.

Sims, J.T. (2000) The role of soil testing in environmental risk assessment for phosphorus. In: Sharpley, A.N. (ed.) *Agriculture and Phosphorus Management: the Chesapeake Bay*. Lewis Publishers, Boca Raton, Florida, pp. 57–81.

Sims, J.T. and Gartley, K.L. (1996) *Nutrient Management Handbook for Delaware*, Coop. Bull. No. 59. University of Delaware, Newark.

Sims, J.T. and Vadas, P.A. (1997) Nutrient management strategies for the profitable, environmentally sound use of phosphorus, Fact Sheet ST_08. College of Agricultural Sciences and Cooperative Extension, University of Delaware, Newark.

Sims, J.T., Vasilas, B.L., Gartley, K.L., Milliken, B. and Green, V. (1995) Evaluation of soil and plant nitrogen tests for maize on manured soils of the Atlantic Coastal Plain. *Agronomy Journal* 87, 213–222.

Sims, J.T., Edwards, A.C., Schoumans, O.F. and Simard, R.R. (2000) Integrating environmental soil phosphorus testing into environmentally-based agricultural management practices. *Journal of Environmental Quality* 29, 60–72.

Smith, R.A., Schwarz, G.E. and Alexander, R.B. (1997) Regional interpretation of water quality monitoring data. *Water Resource Research* 33, 12.

Steinhilber, P. and Weismiller, R. (2000) On-farm management options for controlling phosphorus inputs to the Chesapeake Bay. In: Sharpley, A.N. (ed.) *Agriculture and Phosphorus Management: the Chesapeake Bay*. Lewis Publishers, Boca Raton, Florida, pp. 169–178.

Taylor, A.W. and Pionke, H.B. (2000) Inputs of phosphorus to the Chesapeake Bay watershed. In: Sharpley, A.N. (ed.) *Agriculture and Phosphorus Management: the Chesapeake Bay*. Lewis Publishers, Boca Raton, Florida, pp. 7–23.

USDA (US Department of Agriculture) and USEPA (US Environmental Protection Agency) (1999) *Unified National Strategy for Animal Feeding Operations*. USDA Natural Resources Conservation Service, Washington, DC.

USEPA (US Environmental Protection Agency) (1983) *Chesapeake Bay Program: a Framework for Action*. USEPA Region III, Philadelphia, Pennsylvania.

17 Nutrient and Pesticide Transfer from Agricultural Soils to Water in New Zealand

K.C. Cameron, H.J. Di and L.M. Condron
Centre for Soil and Environmental Quality, PO Box 84, Lincoln University, Canterbury, New Zealand

Key words: agriculture, leaching, New Zealand, nitrogen, overland flow, pesticides, phosphorus, soil

Introduction

Land-based primary production industries are a major contributor to New Zealand's economy, from which more than half of the country's export earnings is derived. Over half of New Zealand's land area is used for farming and plantation forestry (*c.* 16 Mha). Pastoral farming, based primarily on grass–legume pasture associations, is the dominant form of land use (*c.* 13 Mha). Over the past decade, there has been a decline in sheep farming and an increase in dairy and deer farming and plantation forestry, largely driven by declining returns for meat and wool. Figures from Statistics New Zealand (1996) show that the sheep population has declined from over 70 million in 1982 to *c.* 46 million in 1996. The dairy cattle population increased from *c.* 3

(eds P.M. Haygarth and S.C. Jarvis)

million to 4.4 million from 1982 to 1998. In 1996, beef cattle and farmed deer herds were estimated to be 4.4 million and 1.2 million, respectively. The land areas used for horticulture and plantation forestry have also increased significantly in recent years, with horticulture occupying *c.* 123,000 ha and plantation forestry occupying *c.* 1.7 Mha (Statistics New Zealand, 1996). Increases in fertilizer use and pesticides are coupled with selected land use changes. The total amount of solid fertilizers used in 1996 was more than 3 Mt, including 122,000 t of urea and 1.4 Mt of superphosphate. Pesticide use reached a peak of 3700 t of active ingredients in 1994.

New Zealand's environment is generally regarded as 'clean and green' partly because of the lack of industries that have been associated with causing environmental pollution (Roberts *et al.*, 1996). However, there has been increasing concern in recent years about the potential impact of intensified agricultural activities on the quality of surface and groundwater. A recent report on the state of New Zealand's environment identified NO_3^--N and pesticide leaching from agricultural soils as posing major potential threats to groundwater quality (Ministry for the Environment, 1997). This report also highlighted that 40% of 177 New Zealand lakes examined contained high to very high levels of total N (> 0.30 mg l^{-1}) and total P (> 0.02 mg l^{-1}) and were considered to be either eutrophic or hypereutrophic. As high concentrations of NO_3^--N in drinking water are deemed to be harmful to human health, the New Zealand health authorities have established drinking water guidelines limiting NO_3^--N concentration to 11.3 mg l^{-1}, that is, the level also adopted by the EU. Guidelines have also been established for a number of pesticides based on toxicological data (Ministry of Health, 1995). It is therefore critically important that rivers, lakes and aquifers are adequately protected from contamination by nutrients (principally N and P) and pesticides.

This chapter reviews the evidence of water contamination by NO_3^--N, P and pesticides in New Zealand, discusses the major factors that affect their transfer from agricultural land to ground and surface waters, and outlines strategies which can be used to minimize the transfers.

Nitrogen

Nitrate in New Zealand groundwater

Burden (1982), Dillon *et al.* (1989) and Painter *et al.* (1997) have reviewed NO_3^--N concentrations in New Zealand groundwater. The concentration of NO_3^--N in some shallow aquifers exceeds the limit for drinking water set by the Ministry of Health (Table 17.1). For example, in the Hamilton basin (Waikato), over 30% of shallow wells (i.e. < 15 m) have > 10 mg NO_3^--N l^{-1}, and it has been recommended that no shallow bore water in this area should be used for feeding infants (Hoare, 1986). In mid-Canterbury, NO_3^--N concentrations in inland shallow aquifers and the uppermost confined aquifer were also found to exceed 10 mg l^{-1}. However, deep aquifers generally have very low NO_3^--N concentrations (Bowden, 1986). The locations of selected regions within New Zealand are shown in Fig. 17.1.

Grazed pasture and cropland are the major sources of NO_3^--N contamination in most of the studies reported in Table 17.1. In some localized areas, point sources (e.g. dairy factory, piggery) and market gardens have also been identified as contributing to the NO_3^--N load.

Factors affecting the transfer of nitrate from soil to water

Soil

New Zealand soils commonly contain between 0.1 and 0.6% N in the top 15 cm, which represents approximately 2000–12,000 kg N ha^{-1}. As most (> 95%) of this N is present in soil organic matter, it is not available for plant uptake or leaching unless mineralized. Only 1–2% of the total soil N pool is present in the mineral forms

Table 17.1. Nitrate concentrations reported for New Zealand groundwater.

Location	Nitrate concentration (mg NO_3^--N l^{-1})	Suggested major sources of NO_3^--N	Reference
Ashburton, Canterbury	Shallow + unconfined aquifers 4–8 and in places > 10	Grazed pasture, cropland	Burden (1982)
Ashley catchment, North Canterbury	4% wells > 10 32% wells > 5	Grazed pasture, cropland	Bowden *et al.* (1982)
Canterbury	Shallow wells (24 m): 5 Deep wells (178 m): 0.1	Grazed pastures, crop land	CRC (1995)
Christchurch, Canterbury	Shallow + unconfined often > 10. Deep < 1	Grazed pasture, cropland some point sources	Bowden (1986)
Clutha Valley, Otago	Shallow and unconfined 0.2–5.3	Grazed pastures, crop land, some point sources	Smith *et al.* (1993)
Hamilton Basin, Waikato	30–50 % shallow (< 15 m) > 10	Grazed pasture, dairy factory effluent	Hoare (1986); Selvarajah *et al.* (1994)
Hauraki Plain, Waikato	Most wells < 10 (range up to 37)	Grazed pasture	Roberts *et al.* (1996)
Heretaunga Plains, Wellington	Unconfined aquifer 20–57		Burden (1982)
Manawatu	Wells 0–30 m: 3–13 Wells > 30 m: < 2		Brougham *et al.* (1985)
Waimea Plain, Nelson	Unconfined aquifer > 10	Market gardening (N fertilizer), piggery manure	Fenemor (1987)
Wairarapa	Deep confined aquifers < 7.5		O'Dea (1980)
Pukekohe, Auckland	Upper unconfined aquifer 17–26	Intensive market gardens	Cathcart (1994)

of NH_4^+-N and NO_3^--N simultaneously (McLaren and Cameron, 1996). Ammonium ions may be held by cation exchange reactions or fixed by certain clay minerals in New Zealand soils and are therefore not generally prone to leaching. However, the NO_3^--N retention capacity of New Zealand soils is generally low and NO_3^--N is therefore prone to leaching.

Sandy soils contain less water at field capacity than clay or silt loam soils and leaching is therefore induced more quickly in lighter soils. The hydraulic conductivity of a sandy soil is also greater than that of a clay soil and therefore the rate of NO_3^--N leaching is faster through light textured soils. The more rapid the rate of leaching the less opportunity there is for plant uptake, denitrification and/or immobilization to remove NO_3^--N from the soil solution.

In poorly drained soils, denitrification may reduce the NO_3^--N concentration in the soil solution and thus result in a low leaching loss or transfer from soil to surface water bodies (Cameron and Haynes, 1986). For example, denitrification in riparian zones near New Zealand rivers and lakes has been found to be effective in removing over 90% of NO_3^--N from incoming drainage water (Schipper *et al.*, 1993). Denitrification can also have a significant effect on leaching losses from irrigated soils.

Mineralization of N from soil organic matter and/or plant residues increases the concentration of NO_3^--N in soil solution and therefore increases the potential for leaching to occur. The amount of N mineralized each year may vary between 10 and 500 kg N ha^{-1}, depending on soil, environmental and management conditions. Nitrate leaching after mineralization of

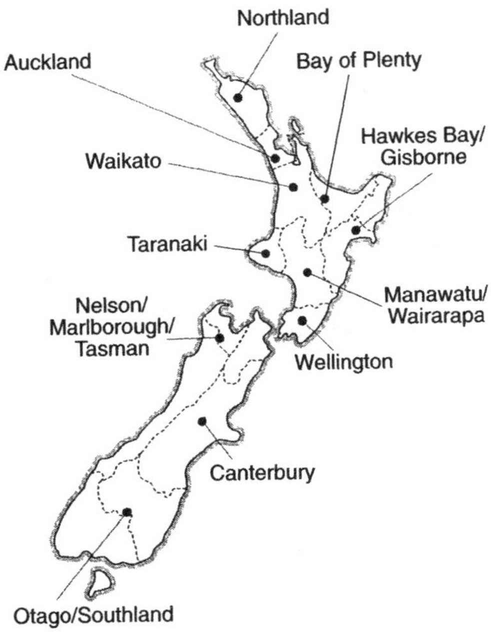

Fig. 17.1. Map of New Zealand showing selected regions.

ploughed pasture is a major source of NO_3^--N contamination of New Zealand groundwater (Francis *et al.*, 1992). Over 90 kg N ha^{-1} $year^{-1}$ were leached after ploughing of clover residues as part of the normal crop rotation in Canterbury (Adams and Pattinson, 1985; Francis *et al.*, 1995).

Season and climate

Most of mainland New Zealand lies in the temperate zone of the southern hemisphere. During the summer months (December–March), significant amounts of soil moisture are lost by evapotranspiration, and irrigation is often necessary to meet the moisture deficit and ensure plant growth. Drainage during this period is generally low and mostly occurs during the cool winter to early spring period (July–September), when evapotranspiration losses are low. Nitrogen uptake by plants during this period is also slow and therefore this is the time of year when most NO_3^--N leaching occurs (Cameron and Haynes, 1986; Ledgard *et al.*, 1996; Di *et al.*, 1999). However, leaching can also occur at other times of the year if the soil is close to 'field capacity' and heavy rain or irrigation is received. A recent study by Di *et al.* (1999) on flood irrigated (600 mm $year^{-1}$) ryegrass (*Lolium perenne*)/clover (*Trifolium* sp.) pastures showed that between 15 and 20% of the N applied in the autumn (May) as ammonium chloride was lost in leaching in 1 year, compared with an equivalent of 8–10% losses of N applied in late spring (November). The amount of drainage in spring is particularly critical as it often determines the amount of leaching loss from freshly applied fertilizers and excreta from grazing animals. The intensity of rainfall is also important because, when combined with appropriate soil and moisture conditions, it may induce macropore flow and rapid leaching. A dry summer can cause an accumulation of NO_3^--N in soil through poor crop uptake and this may result in higher than average leaching losses from soils over the subsequent winter (Scholefield *et al.*, 1993).

Plant uptake

Plant uptake is a primary factor affecting the amount of NO_3^--N leached from soils and fertilizers. The greater the uptake, the lower the concentration of NO_3^--N in the soil solution and thus the lower potential for leaching. Plant root morphology can influence the NO_3^--N leaching loss since deep and extensively rooted plants have a greater opportunity to capture NO_3^--N from the soil. For example, lucerne (*Medicago sativa*) can obtain NO_3^--N from the subsoil through its deep rooting system, while the extensive root system of most grasses enables them to capture efficiently NO_3^- present in the topsoil.

Extensive pasture systems and pasture cut for hay or silage generally have low leaching losses (Di and Cameron, 2000a). Relatively large leaching losses can, however, occur under intensively managed pasture particularly when high stocking rates are combined with high fertilizer inputs (Table 17.2). This is because animal urine patches contain large amounts of N, equiv-

Table 17.2. Nitrate leaching from grazed pasture in New Zealand.

Region	Stock	Pasture	Fertilizer (kg N ha^{-1} $year^{-1}$)	NO_3^--N leached (kg N ha^{-1} $year^{-1}$)
Manawatu (Heng *et al.*, 1991)	Sheep	Grass + clover	50	17
Palmerston North, Manawatu (Ruz-Jerez *et al.*, 1995)	Sheep	Grass + clover	220–270	6–7
Canterbury (Adams and Pattinson, 1985)	Sheep	Grass + clover	0	< 10
Waikato (Ledgard *et al.*, 1996)	Dairy	Grass + clover	225	57
			400	110
Canterbury (Silva *et al.*, 1999)	Dairy	Grass + clover	200	54

alent to approximately 500 kg N ha^{-1} for sheep and 1000 kg N ha^{-1} for cattle (Steele, 1982; Jarvis *et al.*, 1995; Silva *et al.*, 1999). These levels are greater than the plant assimilative capacity and therefore significant leaching losses may occur. Studies in New Zealand have found that between 8 and 20% of the N applied in animal urine patches may be leached (Field *et al.*, 1985; Fraser *et al.*, 1994; Silva *et al.*, 1999). It has been estimated that N leached from intensive dairy farms in the Waikato area is contributing at least 60 kg N ha^{-1} $year^{-1}$ to the shallow aquifers in that region (Selvarajah *et al.*, 1994).

Fertilizer applications in different production systems

Until relatively recently, New Zealand pastoral farmers relied almost exclusively on N fixation by clover plants in the pasture sward. A well-established clover-based pasture, consisting of 20% clover and 80% ryegrass, could contribute about 200 kg N ha^{-1} $year^{-1}$ to pasture nutrition. However, in recent years the amount of N fertilizers used in New Zealand has increased considerably (Roberts *et al.*, 1992). The greatest increase has been on dairy farms where applications of 100 kg N ha^{-1} $year^{-1}$ are common, with some farmers using up to 400 kg N ha^{-1} $year^{-1}$. Recent studies showed that NO_3^--N leaching losses were generally low when the application rate of N was less than 200 kg N ha^{-1}, but increased significantly when the application rate of fertilizer N reached 400 kg N ha^{-1} (Ledgard *et al.*, 1996; Di *et al.*, 1998b,c; Silva *et al.*, 1999). Nitrate leaching losses increased quadratically as the amount of potentially leachable N (mineral and mineralizable N) in the soil increased (Fig. 17.2; Di and Cameron, 2000b).

Cereals generally use fertilizer N efficiently and leaching losses are therefore small at optimum fertilizer rates during the plant-growing season (Prins *et al.*, 1988; Cookson *et al.* 2000). For example, *c.* 90 kg N ha^{-1} can be removed in the harvested grain of a winter wheat (*Triticum aestivum* L.) (Haynes, 1997). Leaching losses in an arable cropping system mainly occur in the autumn after the harvest of the crop when there is no vegetation to take up NO_3^--N from the soil. Leaching losses of 30–60 kg N ha^{-1} can occur, and could increase significantly when fertilizer N is applied at more than 200 kg N ha^{-1} $year^{-1}$ (Haynes, 1997; Di and Cameron, 2000a).

Fertilizer N recovery by field vegetables is generally low (e.g. lettuce (*Lactuca sativa* L.) 11%; spinach (*Spinacea oleracea* L.) 31%; red beet (*Beta vulgaris* L.) 52%; potatoes (*Solanum tuberosum* L.) 50%) because of their shallow rooting depth and the high rates of irrigation (Prins *et al.*, 1988). This and the high N application rates in most market gardens often result in high NO_3^--N leaching losses (70–300 kg N ha^{-1}) (Prins *et al.*, 1988; Spiers *et al.*, 1996).

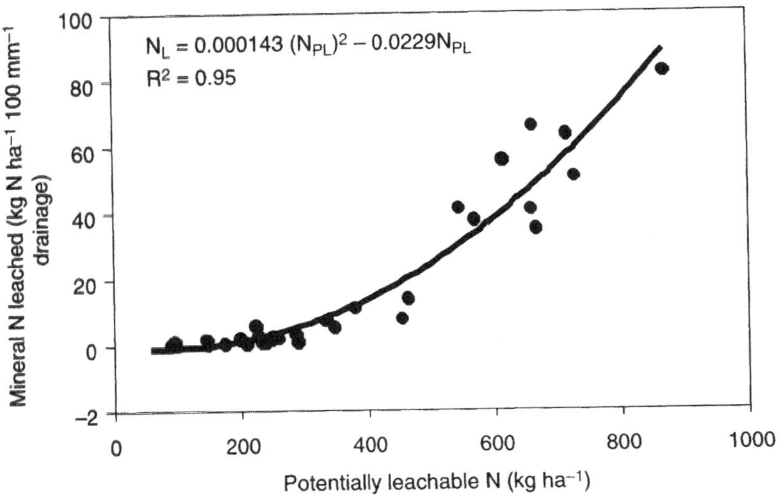

Fig. 17.2. Relationship between annual mineral leaching loss and the amount of potentially leachable nitrogen in the soil determined using soil lysimeters (Di and Cameron, 2000b).

Organic manure applications

The application of organic manure (e.g. dairy shed effluent, pig slurry) to soil has increased recently (Hart and Speir, 1992; Cameron *et al.*, 1997). Land application of dairy farm effluent and pig slurry at rates that can be utilized by a pasture or crop (e.g. < 200 kg N ha^{-1}) appear to present no greater threat to groundwater quality than many other agricultural practices such as fertilizer applications and ploughing-up pasture (Francis *et al.*, 1995; Cameron *et al.*, 1996; Di *et al.*, 1998b,c, 1999; Silva *et al.*, 1999). Recent studies using large undisturbed soil lysimeters showed that NO_3^--N leaching losses from organic manure were generally lower than those from N fertilizers (e.g. urea or ammonium chloride) when applied at the same rates (Table 17.3). Only a fraction of the N in organic manure is in mineral N forms (mainly NH_4^+-N) and the rest is in organic forms. It takes time for the organic N to be mineralized and thus for leaching to occur. The potential for NO_3^--N leaching from organic manure is thus not related to the total N content of the manure, but to the content of mineral N and the rate at which the organic N is mineralized (Cameron *et al.*, 1999). While 60–85% of the N in pig slurry may be present as NH_4^+-N, only about 15% of the N in dairy pond sludge (the sludge under the anaerobic treatment pond) is present as mineral N (Cameron *et al.*, 1996, 1999). For dairy shed effluent (the effluent from the milking shed) about 25% is mineral N (mainly NH_4^+-N) (Di *et al.*, 1998b,c). Because the origins and the composition of various effluents vary widely, the rate at which the organic N is mineralized is also likely to differ significantly. Fresh effluents with simpler organic N compounds are likely to release N more readily than effluents that contain predominantly recalcitrant organic N compounds. Recent studies using isotopic dilution techniques (Barraclough, 1995; Di *et al.*, 2000a) have shown that the application of organic manure can have a significant impact on the overall N mineralization rates in the soil (Zaman *et al.*, 1999a,b). There is a need for more studies to improve our understanding of N transformation rates as affected by the application of different forms of manure.

Most regional authorities in New Zealand have now established rules to limit the rate of application of organic manure to less than 150–200 kg N h^{-1} year^{-1}, unless special circumstances apply. A recent study by Di and Cameron (2000b) showed that these limitations were appro-

Table 17.3. Nitrate nitrogen leaching losses from different organic manures and wastes compared with those from N fertilizers applied to the same Canterbury soil (Templeton fine sandy loam, Udic Ustochrept) under cut grassland in New Zealand.

N source and rate (kg N ha^{-1} $year^{-1}$)	No. of applications	Drainage (mm)	N leaching loss (kg N ha^{-1} $year^{-1}$)	Reference
Control	—	300–600	2.6–2.8	Di *et al.* (1998b)
Dairy shed effluent 400	2 × 200	300–600	13–25	Di *et al.* (1998b)
Dairy shed effluent 400	4 × 100	350	10	Silva *et al.* (1999)
Dairy pond sludge 300	1 × 300	89–150	1–4	Cameron *et al.* (1996)
Pig slurry 200–400	1 × 200 or	207	30	Carey *et al.* (1997)
	1 × 400	207	105	
Sewage sludge 455	1 × 455	84	5	McLaren *et al.* (1999)
Urea 400	4 × 100	350	17	Silva *et al.* (1999)
Ammonium chloride 400	2 × 200	300–600	47–49	Di *et al.* (1998b,c)

priate for the application of dairy shed effluent to grazed dairy pastures, but were too restrictive for systems where the pasture is cut for hay or silage.

Irrigation

Irrigation applied at optimum rates for plant growth can enhance N uptake and reduce NO_3^--N leaching, although excessive irrigation rates can increase the loss. Irrigation of pasture permits an increased stocking rate and thus there is potential for higher leaching losses caused by greater urine returns. Irrigation water can, nevertheless, dilute the drainage water NO_3^--N concentration and thus reduce the peak concentrations entering groundwater. It may also increase denitrification losses (N_2O and N_2 emissions) and thus reduce the concentration of NO_3^--N in solution. This is supported by recent lysimeter studies which showed that NO_3^--N concentrations in the drainage water were lower under flood irrigation (600 mm $year^{-1}$) than under spray irrigation (300 mm $year^{-1}$) (Di *et al.*, 1998b,c). However, more studies are required to understand the changes in the amount of NO_3^--N leached and NO_3^--N concentrations in the drainage water as affected by different irrigation conditions.

Minimizing nitrate leaching

Di and Cameron (2000a) have provided a detailed review of strategies that can be adopted to minimize NO_3^--N leaching from agricultural land to water. The strategies that have been shown to be effective in reducing NO_3^--N leaching include minimizing N fertilizer application rates, synchronizing fertilizer N supply to plant N demand, balancing the input of different nutrients, use of buffer zones and growing a cover crop. Ryegrass (*Lolium* sp.) and rye (*Secale cereale* L.) were found to be effective winter cover crops in New Zealand that could reduce NO_3^--N leaching by up to 30 kg N ha^{-1} compared with fallowed land (McLenaghen *et al.*, 1996). In mixed cropping systems, delaying the cultivation of pasture leys until late autumn can also reduce NO_3^--N leaching by minimizing the amount of N released from mineralization (Francis *et al.*, 1995). Other strategies that are under consideration and are being researched include the use of slow release fertilizers and nitrification inhibitors.

Computer models can be useful in improving our understanding of various processes and factors of NO_3^- leaching, identifying knowledge gaps and providing guidance for the development of best management practices or regulatory guidelines (e.g. Bergstrom and Jarvis, 1991; Scholefield *et al.*, 1991; Hutson and Wagenet, 1992; Di and Cameron, 2000b). There is great potential in the use of computer modelling to mitigate NO_3^--N leaching in the form of a decision support system, incorporating NO_3^--N leaching, farm management practices, plant N requirements, and soil and environmental conditions (Smith *et al.*, 1996). As an example, the nitrogen leaching estimation (NLE) model of Di and Cameron (2000b) is a semi-empirical model for estimating NO_3^--N leaching losses and critical N application rates (in order not to exceed the drinking water standard of 11.3 mg NO_3^--N l^{-1}) for dairy pasture (ryegrass/clover) systems. It is based on the assumption that the amount of annual N leaching loss is determined by the amount of potentially leachable N in the soil (including mineral and mineralizable N) and the amount of drainage. The major N cycling processes (all in kg N ha^{-1} $year^{-1}$) that determine the amount of potentially leachable N in the soil are:

$$N_{PL} = N_{AD}+N_F+N_B+N_M+N_A-N_P-N_V-N_D \quad (17.1)$$

where:

N_{PL} = potentially leachable N;
N_{AD} = atmospheric deposition;
N_F = annual fertilizer or effluent N application rate;
N_B = annual biological N fixation;
N_M = annual net N mineralization;
N_A = annual animal N returns (mainly urine) to the pastures;
N_P = annual pasture N uptake;
N_V = volatilization losses after the application of N fertilizers or effluents;
N_D = denitrification loss.

A quadratic relationship was used to relate the amount of N leached to the amount of potentially leachable N in the soil (Fig. 17.2), which was derived from lysimeter studies conducted over a number of years. The model takes into account the effect of urine-N returns by the grazing animal on NO_3^--N leaching, by calculating a weighted average of N leaching losses from the grazed paddock on the basis of the fractional areas occupied by urine and non-urine areas:

$$N_L = N_{L1} \times P_1 + N_{L2} \times P_2 \quad (17.2)$$

where N_L is the NO_3^--N leaching losses from the grazed paddock, N_{L1} and N_{L2} are the leaching losses from the urine and non-urine patch areas, respectively, and P_1 and P_2 are the proportions of areas occupied by urine and non-urine patch areas, respectively. The value of P_1 varies with stocking rates, and probably averages between 20 and 30% $year^{-1}$ under New Zealand conditions.

Based on data from Canterbury, New Zealand, the model estimated that the critical N application rates (above which the NO_3^--N concentration will exceed 11.3 mg NO_3^--N l^{-1}) for grazed pasture systems were 160–200 kg N ha^{-1} $year^{-1}$ if urea was applied, or 250–300 kg N ha^{-1} if dairy shed effluent was applied. For cut pastures, the critical application rates were *c.* 400 kg N ha^{-1} for urea and 600 kg N ha^{-1} for dairy shed effluent. These values compare with the guideline rates of 150–200 kg N ha^{-1} $year^{-1}$ as adopted by most regional authorities in New Zealand. The different NO_3^--N leaching potential between the different forms of N and management systems should ideally be recognized in guidelines regulating land application of manure and fertilizers.

Phosphorus

The ongoing P requirements of agroecosystems are determined by crop demand and a combination of factors, including soil chemical and physical properties and the history and intensity of land use (Hedley *et al.*, 1995). In effect, P transfer from soil (via overland flow or subsurface flow) to surface water and (via saturated flow or preferential flow) groundwater represents an inefficient and potentially harmful use of a finite natural resource. It is important to balance agronomic requirements with the need to limit P loss to natural waters (Haygarth *et al.*, 1998; Frossard *et al.*, 2000).

Phosphorus status of natural waters and transfer from soil in New Zealand

Available survey data collected from 100 New Zealand rivers have shown that dissolved reactive P (< 1.00 μm) levels ranged from 0.003 to 0.01 mg P l^{-1} (median 0.005 mg P l^{-1}) (Fig. 17.3). These levels are well below the established 'suitability threshold' of 0.015–0.030 mg P l^{-1}, and indicate that, as yet, P enrichment is not a serious problem in most New Zealand rivers. A recent detailed study of P levels in streamwater in a South Island high-country catchment (Lake Hayes, Otago) found that median annual concentrations of P determined in five tributaries were 0.003–0.007 mg P l^{-1} for dissolved reactive P (< 0.45 μm) and 0.009–0.064 mg P l^{-1} for total P (Caruso, 2000). The median inflow to Lake Hayes was 0.7 kg total P per day, although this increased to 81.5 kg total P per day during a storm event. Overall, the findings of this study showed that base flow accounted for 47% of annual total P loading to the lake, while the remainder originated from snowmelt (30%) and storm flow (23%).

Data on the trophic status of New Zealand lakes indicate that approximately 10% of shallow lakes can be classified as eutrophic (0.020–0.050 mg total P l^{-1}) or hypereutrophic (> 0.050 mg total P l^{-1}) (Ministry for the Environment, 1997). Furthermore, pastoral farming was the dominant land use in the catchments of nutrient enriched lakes, which were found mainly in the North Island.

Most research carried out on the extent of P transfer from soil to water in New Zealand to date has concentrated on P loss via overland flow, mainly from hill country. For example, Cooper and Thomsen (1988) found that the annual transfer of dissolved reactive P (< 0.45 μm) was only 0.020–0.040 kg P ha^{-1} under native or exotic pine forest compared with 0.37 kg P ha^{-1} under grazed pasture. Gillingham and Thorrold (2000) recently reviewed avail-

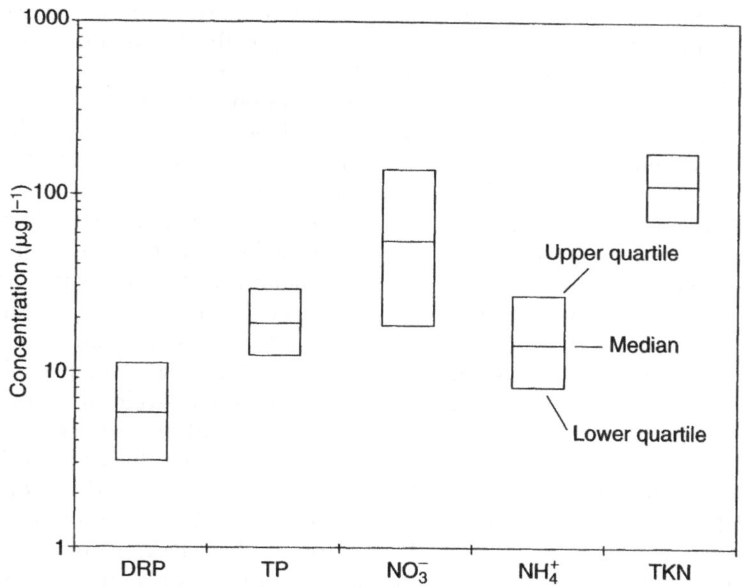

Fig. 17.3. Selected nutrient concentrations in water of the '100 Rivers' study in New Zealand; DRP = inorganic P < 1μm, TP = total P, TKN = dissolved Kjeldahl N (Hoare and Rowe, 1992).

able data on P loss from land under grazed pasture and showed that annual total P loss by overland flow ranged from 0.1 to 1.7 kg P ha^{-1}, with a nominal national average of 1.3 kg P ha^{-1}.

Several studies have attempted to describe quantitatively the P balance and associated P loss in New Zealand pasture ecosystems at field or catchment scale (Haynes and Williams, 1993; Thorrold *et al.*, 1999). For example, Williams and Haynes (1992) collected and examined soil, plant and animal P data from a long-term irrigated field trial at Winchmore in Canterbury. They calculated total gains and losses of P for unfertilized ('wilderness') pasture areas compared with plots which had received approximately 19 and 38 kg P ha^{-1} $year^{-1}$ as single superphosphate for 38 years (1952–1990). Results from this study revealed that 35–37% of the total fertilizer P applied over the period could not be accounted for in animal produce removed or accumulation in residual soil and plant P pools. The authors concluded that most of the P not accounted for may have been transferred in overland flow as suspended soil particles and animal manure during flood irrigation, although some P loss by saturated and/or preferential flow may have occurred from the shallow, free draining stony soil. It is interesting to note that Nguyen and Goh (1992) used a mass-balance modelling approach to examine the P budget on the same Winchmore trial which included some assessment of P transfers via animals (stock camp) and overland flow within the experimental plots. Using this approach, they found that 83–94% of the fertilizer P applied over 35 years could be accounted for by a combination of animal produce removed, P transfer and residual P in various soil and plant pools. This, in turn, suggested that net P loss from grazed pasture at Winchmore was substantially lower than that indicated by Williams and Haynes (1992). The contrasting findings of these studies conducted on the same trial strongly suggests that more direct, *in-situ* measurement of P transfer is required to improve our understanding of the fate of P in these systems.

Factors affecting the transfer of phosphorus from soil to water

Gillingham and Thorrold (2000) demonstrated that the amount and pattern of P loss by overland flow from land under pasture were influenced by a combination of factors including soil P status, the form and timing of fertilizer P application (particularly on hill country) and the presence of grazing animals. Most of the P in overland flow was in particulate forms (> 0.45 μm) and included sparingly soluble P fertilizer residues.

Phosphorus loss via subsurface flow (drainage) may be important in heavy textured soils under intensive land use such as dairying, although very little is known about the amounts and dynamics of P transfer by this pathway under New Zealand conditions (Gillingham and Thorrold, 2000). Similarly, comparatively little is known about the extent of P loss by saturated and preferential flow from free-draining, light textured soils. This may be particularly relevant in areas of the Canterbury Plains in the eastern South Island where in recent years there has been large-scale conversion from sheep and mixed cropping to dairy production on irrigated land with corresponding increases in P inputs. Typically, the quantity of fertilizer P required annually to maintain pasture production on irrigated land in Canterbury is 40–50 kg P ha^{-1} for dairying compared with *c.* 20 kg P ha^{-1} for intensive sheep production.

The potential for P transfer from soil may be directly related to the quantity of plant available P in topsoil, which in turn is determined by the amounts, forms and duration of P inputs. A study on arable soils in the UK indicated that a relationship exists between levels of plant available P in topsoil as determined by extraction with sodium bicarbonate (Olsen P) and P loss via subsurface drainage (Heckrath *et al.*, 1995). Initial findings from laboratory studies carried out on New Zealand pasture soils also tentatively indicate a direct relationship between Olsen P and potentially mobile P (as determined by

0.01 M calcium chloride extraction) (McDowell and Condron, 1999; McDowell and Condron, 2000) (see also discussion in McDowell *et al.*, Chapter 9, this volume). None the less, it is important to note that further detailed research is required to elucidate the precise nature of the relationship between soil P status and actual P transfer under New Zealand conditions and the consequent implications for P use and management. This includes assessment of the influence of different forms of P input (mineral fertilizer, animal manure), the role of biological processes in determining the amounts and chemical nature of mobile P, and the influence of hydrological conditions (e.g. preferential flow).

In summary, our understanding of the extent of P transfer from New Zealand soils and the influence of various biological, chemical and physical processes is limited. Accordingly, continued research is required to address this important environmental issue which, in turn, is an important factor in determining the long-term sustainability of New Zealand's land-based industries.

Pesticides

Pesticide use in New Zealand

The use of pesticides in New Zealand, as in many other developed countries, has become an integral part of the land-based primary production system. New Zealand's mild to warm temperatures combined with adequate moisture conditions are often favourable for fungal infections of plants, and a wide range of weed species also flourish under these conditions. It is therefore often necessary to use pesticides for weed and disease control for efficient and economic production. In general, for every NZ$1 spent on pesticides, a return of NZ$3–4 is expected from reduced crop losses. The use of pesticides is often necessary so that some produce destined for export meets the quality and quarantine standards set by overseas trading partners (e.g. Smith *et al.*, 1992).

Organochlorine pesticides were used for pest control in the 1950s and 1960s but the use was stopped in the 1970s. The main types included DDT, lindane, aldrin, dieldrin, chlordane, hexachlorobenzene (HCB), heptachlor and pentachlorophenol (PCP). Residues of some of these pesticides remain in soils and water today. It is estimated that *c.* 3300 t of active ingredients are used currently in the agricultural, horticultural and forestry production systems (Holland and Rahman, 1999). These pesticides comprise *c.* 68% herbicides, 24% fungicides and 8% insecticides. Approximately 34% of the pesticides are used for horticultural purposes (e.g. orchards, vegetables), 25% for pastoral agriculture, 16% in cropping systems and 25% in forestry. There has been an increase in pesticide sales for horticultural purposes in the past decade as more land is used for horticulture. The use of pesticides in New Zealand is not evenly distributed geographically. Horticulture, which involves the heavy use of pesticides, is concentrated more in the North Island, for example in the Waikato, Bay of Plenty, Hawke's Bay and Auckland regions (Fig. 17.1). However, vegetable and field crops are also widely grown in Canterbury in the South Island.

Pesticide residues in soil and water

In New Zealand, use of organochlorine insecticides such as DDT has resulted in the contamination of soils and water bodies, contamination that persists long after the cessation of their use. Recent surveys of organochlorine pesticides in the New Zealand environment found low concentrations of the pesticides in the air, soil, rivers and estuaries surveyed throughout the country (Ministry for the Environment, 1998). The concentrations of organochlorine pesticides in soils were found to be generally low, although they are variable. Dieldrin and DDT were found in most urban soils, and these and HCB were also detected in forest and grassland soils. The concentrations of DDT and its metabolic products were found to be geographically variable, reflecting the variations in the amounts of the pesticide used in different regions. The concentrations ranged from a low of 1.5 μg kg^{-1} in urban Hamilton soils to a high of 431 μg kg^{-1} in some urban Christchurch (Canterbury) soils. The presence of DDT in the soil at high concentrations can limit the use of land for intensive grazing as the DDT residues can be transferred to meat and milk products through the ingestion of contaminated soil by the grazing animal.

The most frequently detected organochlorine pesticides in river fish were HCB, dieldrin and DDT (Ministry for the Environment, 1998). Dieldrin and DDT were also most frequently found in sediment, fish and shellfish in the estuaries that were studied. Concentrations of dieldrin were generally below 0.05–0.38 μg kg^{-1} (dry weight) in the estuary sediments and were below 0.02–0.56 μg kg^{-1} in the estuary shellfish (fresh weight). DDT concentrations (including metabolic products) were mostly < 0.01–3.29 μg kg^{-1} (dry weight) in estuary sediments and < 0.01–2.77 μg kg^{-1} in estuary shellfish (fresh weight) (Scobie *et al.*, 1999). These concentrations are generally low compared with those reported in other countries. For example, DDT concentrations were *c.* 1–100 μg kg^{-1} in Florida estuary sediments and *c.* 0.1–1000 in Australian estuary sediments.

While the more water-soluble pesticides are less persistent compared with the organochlorine insecticides, they are more prone to leaching. A number of pesticides have been detected in groundwater surveys around the globe (e.g. Cohen *et al.*, 1986; Beitz *et al.*, 1994). Groundwater surveys have also been carried out throughout New Zealand and more detailed studies have been conducted in regions where pesticide use is widespread (Close, 1996; Hadfield and Smith, 1997). A list of pesticides and their concentrations measured in well water is summarized in Table 17.4. Although a total of 21 pesticides were measured, the

Table 17.4. Pesticides detected in two groundwater surveys, one carried out in the Waikato region and the other throughout New Zealand (Close, 1996; Hadfield and Smith, 1997).

Pesticides detected	Concentrations (μg l^{-1})	MAV[a] (μg l^{-1})	Regions concerned
Alachlor	0.02–3.67	20	Waikato, Gisborne, Wellington
Atrazine	0.02–0.9	2	Bay of Plenty, Gisborne, Waikato
Bromacil	0.06–0.29	300	Waikato
2,4-D	0.09	30	Waikato
Dieldrin	0.03–0.09	0.03	Waikato
Diuron	0.03–9.5	20	Waikato
Hexazinone	0.08–0.16	300	Waikato
Mecoprop	2.4–3.2	10	Marlborough
Metolachlor	0.1–4.5	10	Gisborne, Waikato
Metribuzine	0.02–0.21	50	Waikato
Oryzalin	0.19	300	Waikato
Oxydiazon	0.21	—[b]	Waikato
Phenylphenol	0.1	—[b]	Marlborough
Picloram	0.04–2.8	300	Waikato
Procymidone	0.02–3.04	700	Waikato
Pirimisulfuron	2.7	—[b]	Waikato
Propazine	0.2	10	Southland
Simazine	0.02–1.6	2	Taranaki, Tasman, Marlborough, Waikato, Canterbury, Otago, Southland
Terbuthylazine	0.02–0.3	20	Otago, Waikato, Canterbury
Triclopyr	0.02	100	Tasman, Waikato
Trifluralin	0.3	30	Otago

[a] Maximum acceptable value in drinking water in New Zealand.
[b] Not available.

concentrations were, in most cases, below the drinking water standards established by the New Zealand health authorities. However, some were above the drinking water standard of 0.1 $\mu g\ kg^{-1}$ set by the European Union. There seemed to be a greater number of pesticides detected in those regions with the highest pesticide use, although the results may also reflect the greater intensity of groundwater surveys carried out in some of those regions compared with others.

Factors affecting the transfer of pesticides from soil to water

The fate of pesticides in the soil and their transfer to ground and surface waters are governed by a number of processes and factors (Table 17.5). Pesticides may be sorbed by soil components (particularly soil organic matter), degraded to simpler organic or inorganic compounds, volatilized into the atmosphere, accumulated in soil biota, leached down the soil profile to groundwater and lost to surface waters via overland flow pathways (Jury *et al.*, 1987b). The amount of pesticides that may be transferred to ground or surface waters not only depends on the amount of the pesticides that are applied to agroecosystems, but also on the persistence of the pesticides in the soil, which is determined by the rates of dissipation. The ultimate process that leads to the detoxification of pesticides from the environment is the breakdown or degradation of the pesticides into simple inorganic compounds (water, carbon dioxide and other inorganic compounds). The dominant process of degradation in most agricultural soils is breakdown by microorganisms through mineralization or co-metabolic reactions. The rate of degradation varies widely depending on the

Table 17.5. Major factors that affect leaching and groundwater contamination potential of pesticides in New Zealand.

Number and amounts of pesticides applied
Pesticide properties:
Chemical structure
Water solubility
Vapour pressure
Sorption coefficient
Persistence
Soil properties that affect sorption coefficient, degradation rate and other dissipation processes:
Organic matter content
Microbial activities
Number of pesticide-degrading microorganisms
Soil texture
Soil pH
Soil redox potential
Macroporosity
Soil moisture content
Depth to groundwater
Topography
Environmental conditions:
Temperature
Rainfall and recharge rate
Sunlight
Management practices:
Irrigation
Previous exposure of soil to pesticides
Cultivation
Land application of organic manure
Mode of application

chemical composition and structure of the pesticide, as well as on the soil and environmental conditions; for example microbial activity, soil organic matter content, pH, temperature, moisture and redox potential (Hamaker, 1972; Rao and Davidson, 1980; Bollag and Liu, 1990). Organochlorine pesticides have been found to be particularly persistent in the soil, with half-lives (years) ranging as follows: aldrin 1–9, dieldrin 3–7, chlordane 1–8, heptachlor 1–4, and DDT 3–10 (Nash and Woolson, 1967; Augustijn-Beckers *et al.*, 1994). Other, more water-soluble, pesticides such as organophosphates, triazines, carbamates and substituted ureas are less persistent, with half-lives often shorter than 100–200 days. Soil conditions that are favourable for microbial activity usually increase the rate of pesticide degradation. Therefore, pesticide degradation may be accelerated by providing substrates and nutrients, or by providing optimum soil moisture and temperature conditions to stimulate microbial activity. Anaerobic soil conditions, as generated by flooding and the addition of organic substrates, are conducive to the degradation of organochlorine pesticides such as DDT (Boul *et al.*, 1994).

The groundwater contamination potential of pesticides is affected by the rates of pesticide degradation in both the surface soil and the subsurface soils (Di and Aylmore, 1997). Pesticide degradation rates in subsurface soils can be significantly different from those in the surface soils, and they are not always slower in subsurface soils than in surface soils (Di *et al.*, 1998a; Sparling *et al.*, 1998). For instance, a recent study showed that the half-life of atrazine was 150 days in a subsoil compared with 250 days in the surface soil (Di *et al.*, 2000b). The rates at which pesticides are degraded in subsoil layers in the unsaturated soil zone can have a major impact on the potential of pesticides to reach groundwater in significant quantities (Bergstrom, 1996; Di *et al.*, 2000b). One of the major limiting factors in our ability to estimate the potential for groundwater contamination by pesticides is the lack of information on the degradation rates of pesticides in subsurface soils (Sparling *et al.*, 1998; Di *et al.*, 2000b).

In addition to the persistence of pesticides in the soil, another important factor that strongly affects the mobility and thus the leaching potential of pesticides is the sorption coefficient (Di and Aylmore, 1997). Pesticides that are strongly sorbed by soil components are slow to leach and are less likely to reach the groundwater in significant quantities. The most important soil component responsible for pesticide sorption is soil organic matter. For ionic pesticides, Fe and Al oxides are also important for sorption. Organochlorine pesticides are usually strongly sorbed by soil components and are thus very immobile. For instance, the sorption coefficient normalized by the fractional weight of soil organic carbon (K_{OC}) is 240,000 for DDT. The newer generation pesticides such as triazines are less strongly sorbed and more water soluble and thus have a greater potential for groundwater contamination. The sorption coefficient of a particular pesticide, however, can vary significantly between soils and between different soil layers. For example, a recent study of five pesticides on a group of New Zealand soils showed seven- to 12-fold variations in sorption coefficients, where higher coefficients indicate a higher propensity for sorption: atrazine 16–148, 2,4-D 24–197, metsulfuron 4–29, phorate 46–558, and terbufos 62–584 (Baskaran *et al.*, 1996). Many of New Zealand's fine-textured pastoral soils have high organic carbon contents and therefore have high sorption capacities for pesticides. However, some of the coarse-textured soils contain low concentrations of organic carbon making them prone to pesticide leaching (Di *et al.*, 1995). It must be noted that pesticides, which are strongly sorbed to organic matter or mineral surfaces, may also leach when carried by organic or inorganic colloids, although the significance of this process is not well understood. This process may become particularly significant where organic manure effluents are applied because these tend to increase the concentration of soluble organic matter in the soil solution (Cameron *et al.*, 1997).

Other factors that affect the leaching and groundwater contamination potential, include soil depth to groundwater level and rainfall or groundwater recharge rate (Table 17.5). Pesticides are likely to reach the groundwater in greater amounts where the groundwater levels are high and recharge rates are high through high precipitation. In addition, pastoral soils in New Zealand are populated with earthworms, and pesticide leaching by preferential flow through earthworm channels can be significant, particularly under flood irrigation or heavy rainfall conditions. However, there have been few studies in New Zealand to quantify the magnitude of pesticide leaching through earthworm channels and macropores in general.

There are few studies of pesticide transfer by overland flow in New Zealand. Studies in other parts of the world show that overland flow losses of pesticides to surface waters are mainly related to the pesticide concentrations in the surface 10 mm of soil (Leonard, 1990). Pesticides may be lost in solution or on suspended soil particles. Overland flow losses are particularly significant in a storm event shortly after the application of pesticides. However, pesticide losses by overland flow from agricultural land, on average, constitute only a small fraction (1–2%) of the total amount of pesticides applied, although higher losses have been reported (Baker *et al.*, 1978). The actual amount of loss by overland flow may vary, depending on the pesticide properties, the amount of pesticide applied, method of application, the timing of application in relation to a storm and the topography. More research is needed to quantify these losses under a range of typical farming situations.

Another significant pathway by which pesticides can enter surface waters is spray drift during application (Holland *et al.*, 1997). Studies in New Zealand kiwifruit (*Actinidia deliciosa* L.) orchards showed that, on average, 79% of the pesticides sprayed ended up on target crops, 19% on the ground or shelter and *c.* 2% left the orchard as drift. The distance for 50% decline in pesticide deposition from drift was 26 m. Pesticide type, windspeed, shelter density, temperature and humidity affected the amount lost in spray drift. Volatility of the pesticides and density of shelter were two of the most important factors affecting drift.

Minimizing transfer of pesticides from soil to water

Minimizing the use of pesticides will obviously reduce the risks of water contamination. Where possible, pest control should be achieved by integrated pest management with minimal use of pesticides. In New Zealand, there is a growing consumer demand for agricultural produce that is grown organically without the use of pesticides. This consumer pressure for organically grown food has resulted in some farmers switching from conventional farming into organic farming. However, the demand for more food by the growing global population and the need for some produce to meet the quality and quarantine standards as required by trading partners will probably necessitate the use of pesticides by many conventional farmers.

Education of pesticide users on the proper handling and application methods is important in minimizing environmental impacts. The amount, location and timing of application are all critical factors in determining pesticide transfer from the location of application to non-target areas. As significant pesticide transfer by overland flow, saturated flow and/or preferential flow can occur immediately after a storm, application of pesticides should be avoided just before a heavy rain event, especially in combination with high antecedent soil moisture conditions. In addition, some buffer zones or shelters between cropping land and waterways can significantly reduce losses by overland flow and spray drift to waterways (Holland *et al.*, 1997).

Computer models have been developed to assess the leaching behaviour of pesticides in the soil, and to assist management decisions (e.g. Jury *et al.*, 1987a; Hutson and

Wagenet, 1992; Di *et al.*, 1995; Di and Aylmore, 1997). Models can be useful for evaluating the impacts of management practices, soil and environmental conditions and pesticide properties on the leaching and groundwater contamination potential. They can be used to help identify the pesticides that are most likely to cause water contamination in a given environment, or locate areas where significant leaching is likely to occur for particular pesticides. Modelling can also help to select the best pesticide to use where more than one pesticide is available with similar functions.

Many organochlorine pesticides can no longer be used in New Zealand because of the recognized environmental problems associated with persistence, bioaccumulation and adverse impacts on wildlife and other organisms. In addition, a number of pesticides have been found to have high mobility and thus the potential to leach (e.g. atrazine, simazine). The use of these pesticides in groundwater recharge zones needs to be managed with particular care.

Conclusions

Transfer of nutrients and pesticides from agricultural soils to ground and surface water bodies can cause contamination of water resources in some parts of New Zealand. The extent of water contamination by nutrients and pesticides from agricultural sources varies significantly between different regions and is strongly affected by land use practices, soil properties and climatic conditions. The transfer of nutrients and pesticides to water is particularly significant in areas where land use is intensive, involving the application of N and P fertilizers, organic manure and pesticides, frequent cultivation, or where the land is used for grazing by large animals such as cattle. The amounts of nutrient and pesticide transfers are also affected by a number of soil and climatic conditions. Adverse impacts may have a long lifetime. For example, DDT residues in the soil and aquatic environment, in some parts of the country, persist 30 years after its use was stopped. This clearly points to the importance of carrying out extensive testing for potential environmental impacts before a chemical is released into the environment, a practice that is now adopted in many countries, including New Zealand.

The management of New Zealand's natural resources and environment is governed by the Resource Management Act that was passed into law in 1991. A number of measures have also been put in place to regulate land use practices with regard to land application and disposal of manure, wastes and pesticides. The fertilizer industry has also developed codes of practice for the agricultural use of fertilizers. Major progress has been made in the past decade in improving our understanding of the factors and processes that affect the transfer of nutrients and pesticides from soil to water. The key to minimizing the environmental impacts of agricultural activities lies in developing best management practices that are suited to different sectors and different regions, so that the nutrients and chemicals are best utilized for efficient production with minimum impacts on the wider environment. A number of these practices have been proposed and implemented in the past few years such as guidelines for the amount, timing and location of fertilizer and pesticide applications, the timing of cultivation and the appropriate amount and location of manure application to land. Computer models have also been developed which can be used to provide practical management guidance with regard to fertilizer and pesticide use (e.g. Di and Cameron, 2000b).

However, land use patterns are changing rapidly in New Zealand. The increasing conversion of sheep farms to dairy farms, particularly in the South Island, and the increasing expansion of horticultural activities may have significant environmental implications. An increased effort is required in research, development and education to define the environmental issues associated with these land use changes and develop appropriate management strategies to ensure that New Zealand's environment remains 'clean and green'.

References

Adams, J.A. and Pattinson, J.M. (1985) Nitrate leaching losses under a legume-based crop rotation in Central Canterbury, New Zealand. *New Zealand Journal of Agricultural Research* 28, 101–107.

Augustijn-Beckers, P.W.M., Hornsby, A.G. and Waushope, R.D. (1994) The SCS/ARS/CES pesticides database for environmental decision making. II. Additional compounds. *Reviews in Environmental Contamination and Toxicology* 137, 1–82.

Baker, L.J., Laflen, J.N. and Johnson, H.P. (1978) Effect of tillage systems on surface runoff losses of pesticides. A rainfall simulation study. *Journal of Environmental Quality* 21, 886–892.

Barraclough, D. (1995) ^{15}N isotope dilution techniques to study soil nitrogen transformations and plant uptake. *Fertilizer Research* 42, 185–192.

Baskaran, S., Bolan, N.S., Rahman, A. and Tillman, R.W. (1996) Pesticide sorption by allophanic and non-allophanic soil of New Zealand. *New Zealand Journal of Agricultural Research* 39, 297–310.

Beitz, H., Schmidt, H. and Herzel, F. (1994) Occurrence, toxicological and ecotoxicological significance of pesticides in groundwater and surface water. In: Börner, H. (ed.) *Pesticides in Groundwater and Surface Water*, Chemistry of Plant Protection 9. Springer-Verlag, Berlin, pp. 1–56.

Bergstrom, L. (1996) Model prediction and field measurements of chlorsulfuron leaching under non-steady-state flow conditions. *Pesticide Science* 48, 37–45.

Bergstrom, L. and Jarvis, N.J. (1991) Prediction of nitrate leaching losses from arable land under different fertilization intensities using the SOIL-SOIL N models. *Soil Use and Management* 7, 79–85.

Bollag, J.M. and Liu, S.Y. (1990) Biological transformation processes of pesticides. In: Cheng, H.H. (ed.) *Pesticides in the Soil Environment: Processes, Impacts, and Modelling.* Soil Science Society of America, Madison, Wisconsin, pp. 169–211.

Boul, H.L., Garnham, M.L., Hucker, D., Baird, D. and Aislabie, J. (1994) Influence of agricultural practices on the levels of DDT and its residues in soils. *Environmental Science and Technology* 28, 1397–1402.

Bowden, M.J. (1986) *The Christchurch Artesian Aquifers.* North Canterbury Catchment Board, Christchurch, New Zealand.

Bowden, M.J., Ayrey, R.B., Duffield, D.N., Glennie, J.M., Harrison, N., Hurd, S.B., Mason, C.R., Talbot, J.D. and Weeber, J.H. (1982) *The Water Resources of the Ashley Catchment, North Canterbury.* North Canterbury Catchment Board and Regional Water Board, Christchurch, New Zealand.

Brougham, G.H., Kampman, I. and Caldwell, K.J. (1985) *Groundwater Resources of Waitohu, Otati and Mangaone*, Report No. 65. Manawatu Catchment Board and Regional Water Board, New Zealand.

Burden, R.J. (1982) Nitrate contamination of New Zealand aquifers. *New Zealand Journal of Science* 25, 205–220.

Cameron, K.C. and Haynes, R.J. (1986) Retention and movement of nitrogen in soils. In: Haynes, R.J. (ed.) *Mineral Nitrogen in the Plant–Soil System.* Academic Press, Orlando, Florida, pp. 166–241.

Cameron, K.C., Rate, A.W., Noonan, M.J., Moore, S., Smith, N.S. and Kerr, L.E. (1996) Lysimeter study of the fate of nutrients following subsurface injection and surface application of dairy pond sludge to pasture. *Agriculture, Ecosystems and Environment* 58, 187–197.

Cameron, K.C., Di, H.J. and McLaren, R.G. (1997) Is soil an appropriate dumping ground for our wastes? *Australian Journal of Soil Research* 35, 995–1035.

Cameron, K.C., Di, H.J. and McLaren, R.G. (1999) Comparison of nitrogen leaching losses from different forms and rates of organic wastes, fertilizers and animal urine applied to Templeton soil lysimeters. In: Currie, L.D., Hedley, M.J., Horne, D.J. and Loganathan, P. (eds) *Best Soil Management Practices for Production.* Occasional Report No. 12. Fertilizer and Lime Research Centre, Massey University, Palmerston North, New Zealand, pp. 241–249.

Carey, P.L., Rate, A.W. and Cameron, K.C. (1997) Fate of nitrogen in pig slurry applied to a New Zealand pasture soil. *Australian Journal of Soil Research* 35, 941–959.

Caruso, B.S. (2000) Spatial and temporal variability of stream phosphorus in a New Zealand high-country agricultural catchment. *New Zealand Journal of Agricultural Research* 43, 235–249.

Cathcart, S.N. (1994) Fertilizer and its role in the contamination of Pukekohe groundwater. In: Currie, L.D. and Loganathan, P. (eds) *The Effective Use of Fertilisers in a Changing Environment:*

Reconciling Productivity and Sustainability, Fertiliser and Lime Research Centre Occasional Report No 7. Massey University, Palmerston North, New Zealand, pp. 374–381.

Close, M.E. (1996) Survey of pesticides in New Zealand groundwaters, 1994. *New Zealand Journal of Marine and Freshwater Research* 30, 455–461.

Cohen, S.Z., Eiden, C. and Lorber, M.N. (1986) Monitoring groundwater for pesticides. In: Garner, W.Y., Honeycutt, R.C. and Nigg, H.N. (eds) *Evaluation of Pesticides in Ground Water*, ACS Symposium Series, Vol. 315. American Chemical Society, Washington, DC, pp. 170–196.

Cookson, W.R., Rowarth, J.S. and Cameron, K.C. (2000) The fate of autumn, winter and spring applied nitrogen fertilizer in a perennial ryegrass seed crop. *Agriculture, Ecosystems and Environment* 56, 99–107.

Cooper, A.B. and Thomsen, C.E. (1988) Nitrogen and phosphorus in streamwaters from adjacent pasture, pine and native forest catchments. *New Zealand Journal of Marine and Freshwater Research* 22, 279–291.

CRC (1995) *Canterbury Regional Council Regional Environmental Report 1993–1994*, Report 95 (5). Christchurch, New Zealand.

Di, H.J. and Aylmore, L.A.G. (1997) Modelling the probabilities of groundwater contamination by pesticides. *Soil Science Society of America of Journal* 61, 17–23.

Di, H.J. and Cameron, K.C. (2000a) Nitrate leaching in agroecosystems: sources, factors and mitigating strategies. In: *Nutrient Cycling in Agroecosystems* (in press).

Di, H.J. and Cameron, K.C. (2000b) Calculating nitrogen leaching losses and critical nitrogen application rates in dairy pasture systems using a semi-empirical model. *New Zealand Journal of Agricultural Research* 43, 139–147.

Di, H.J., Kookana, R.S. and Aylmore, L.A.G. (1995) Application of a simple model to assess the groundwater contamination potential of pesticides. *Australian Journal of Soil Research* 33, 1031–1040.

Di, H.J., Aylmore, L.A.G. and Kookana, R.S. (1998a) Degradation rates of eight pesticides in surface and subsurface soils under laboratory and field conditions. *Soil Science* 163, 404–411.

Di, H.J., Cameron, K.C., Moore, S. and Smith, N.P. (1998b) Nitrate leaching from dairy shed effluent and ammonium fertilizer applied to a free-draining soil under spray or flood irrigation. *New Zealand Journal of Agricultural Research* 41, 263–270.

Di, H.J., Cameron, K.C., Moore, S. and Smith, N.P. (1998c) Nitrate leaching and pasture yields following the application of dairy shed effluent or ammonium fertilizer under spray or flood irrigation: Results of a lysimeter study. *Soil Use and Management* 14, 209–214.

Di, H.J., Cameron, K.C., Moore, S. and Smith, N.P. (1999) Contributions to nitrogen leaching and pasture uptake by autumn-applied dairy effluent and ammonium fertilizer labelled with ^{15}N isotope. *Plant and Soil* 210, 189–198.

Di, H.J., Cameron, K.C. and McLaren, R.G. (2000a) Isotopic dilution methods to determine the gross transformation rates of nitrogen, phosphorus and sulphur: A review of the theory, methodologies and limitations. *Australian Journal of Soil Research* 38, 213–230.

Di, H.J., Sparling, G., Lee, B. and Magesan, G. (2000b) The effect of atrazine mineralization rates in surface and subsurface soils on its groundwater contamination potential. *Australian Journal of Soil Research* 39 (in press).

Dillon, P.J., Close, M.E. and Scott, R.J. (1989) Diffuse source nitrate contamination of groundwater in New Zealand and Australia. *Hydrology and Water Resources Symposium, Christchurch.* Institution of Engineers, Barton, ACT, Australia, pp. 351–355.

Fenemor, A.D. (1987) *Water Quality of the Waimea Plains Aquifers, 1971–86.* Nelson Regional Water Board, Nelson, New Zealand.

Field, T.R.O., Theobald, P.W., Ball, P.R. and Clothier, B.E. (1985) Leaching losses of nitrate from cattle urine applied to a lysimeter. *Proceedings of Agronomy Society of New Zealand* 15, 137–141.

Francis, G.S., Haynes, R.J., Sparling, G.P., Ross, D.J. and Williams, P.H. (1992) Nitrogen mineralization, nitrate leaching and crop growth following cultivation of a temporary leguminous pasture in autumn and winter. *Fertilizer Research* 33, 59–70.

Francis, G.S., Haynes, R.J. and Williams, P.H. (1995) Effects of the timing of ploughing-in temporary leguminous pastures and two winter cover crops on nitrogen mineralization, nitrate leaching and spring wheat growth. *Journal of Agricultural Science (Cambridge)* 124, 1–9.

Fraser, P.M., Cameron, K.C. and Sherlock, R.R. (1994) Lysimeter study of the fate of nitrogen in animal urine returns to irrigated pasture. *European Journal of Soil Science* 45, 439–448.

Frossard, E., Condron, L.M., Oberson, A., Sinaj, S. and Fardeau, J.C. (2000) Processes governing phosphorus bioavailability in temperate soils. *Journal of Environmental Quality* 29, 15–23.

Gillingham, A.G. and Thorrold, B.S. (2000) A review of New Zealand research measuring phosphorus in runoff from pasture. *Journal of Environmental Quality* 29, 88–96.

Hadfield, J.H. and Smith, D. (1997) Pesticide contamination of groundwater: an investigation in the Waikato region, New Zealand. *24th Hydrology Water Resources Symposium Proceedings*, pp. 246–251.

Hamaker, J.W. (1972) Decomposition: quantitative aspects. In: Goring, C.A.I. and Hamaker, J.W. (eds) *Organic Chemicals in the Soil Environment*, Vol. 1. Marcel Dekker, New York, pp. 253–340.

Hart, P.B.S. and Speir, T.W. (1992) Agricultural and industrial effluents and wastes as fertilizers and soil amendments in New Zealand. In: Gregg, P.E.H. and Currie, L.D. (eds) *The Use of Wastes and Byproducts as Fertiliser and Soil Amendments for Pastures and Crops*, Occasional Report No. 6. Fertiliser and Lime Research Centre, Massey University, Palmerston North, New Zealand, pp. 69–90.

Haygarth, P.M., Chapman, P.J., Jarvis, S.C. and Smith, R.V. (1998) Phosphorus budgets for two contrasting grassland farming systems in the UK. *Soil Use and Management* 14, 160–167.

Haynes, R.J. (1997) *Leaching of Nitrate from New Zealand Agricultural Soils.* A report prepared for MAF Policy, Wellington, New Zealand.

Haynes, R.J. and Williams, P.H. (1993) Nutrient cycling and soil fertility in the grazed pasture ecosystem. *Advances in Agronomy* 49, 119–199.

Heckrath, G., Brookes, P.C., Poulton, P.R. and Goulding, K.W.T. (1995) Phosphorus leaching from soils containing different phosphorus concentrations in the Broadbalk experiment. *Journal of Environmental Quality* 24, 904–910.

Hedley, M.J., Mortvedt, J.J., Bolan, N.S. and Syers, J.K. (1995) Phosphorus fertility management in agroecosystems. In: Tiessen, H. (ed.) *Phosphorus in the Global Environment.* John Wiley & Sons, Chichester, UK, pp. 43–58.

Heng, L.K., White, R.E., Bolan, N.S. and Scotter, D.R. (1991) Leaching losses of major nutrients from a mole-drained soil under pasture. *New Zealand Journal of Agricultural Research* 34, 325–334.

Hoare, R.A. (1986) Groundwater nitrate in the Hamilton Basin. Waikato Valley Authority, Tech. Rep. 1986/16.

Hoare, R.A. and Rowe, L.K. (1992) Water quality in NZ. In: Mosley, M.P. (ed.) *Waters of New Zealand.* New Zealand Hydrological Society, Caxton Press, Christchurch, New Zealand.

Holland, P. and Rahman, A. (1999) *Review of Trends in Agricultural Pesticide Use in New Zealand.* MAF Policy Technical Paper 99/11. Ministry of Agriculture, Wellington, New Zealand.

Holland, P.T., Maber, J.F., May, W.A. and Malcolm, C.P. (1997) Drift from orchard spraying. *Proceedings of the 50th New Zealand Plant Protection Conference*, pp. 112–118.

Hutson, J.L. and Wagenet, R.J. (1992) *Leaching Estimation and Chemistry Model, a Process-Based Model of Water and Solute Movement, Transformation, Plant Uptake and Chemical Reactions in the Unsaturated Zone,* Version 3. Department of Soil, Crop and Atmospheric Sciences Research Series No. 92–3. Cornell University, Ithaca, New York.

Jarvis, S.C., Scholefield, D. and Pain, B. (1995) Nitrogen cycling in grazing systems. In: Bacon, P.E. (ed.) *Nitrogen Fertilization in the Environment.* Marcel Dekker, New York, pp. 381–419.

Jury, W.A., Focht, D.D. and Farmer, W.J. (1987a) Evaluation of pesticide groundwater pollution potential from standard indices of soil-chemical adsorption and biodegradation. *Journal of Environmental Quality* 16, 422–428.

Jury, W.A., Winer, A.M., Spencer, W.F. and Focht, D.D. (1987b) Transport and transformations of organic chemicals in the soil-air-water ecosystem. *Reviews of Environmental Contamination and Toxicology* 99, 119–164.

Ledgard, S.F., Clark, D.A., Sprosen, M.S., Brier, G.J. and Nemaia, E.K.K. (1996) Nitrogen losses from grazed dairy pasture, as affected by nitrogen fertilizer application. *Proceedings of the New Zealand Grassland Association* 57, 21–25.

Leonard, R.A. (1990) Movement of pesticides into surface waters. In: Cheng, H.H. (ed.) *Pesticides in the Soil Environment: Processes, Impacts, and Modeling.* Soil Science Society of America, Madison, Wisconsin, pp. 303–349.

McDowell, R.W. and Condron, L.M. (1999) Developing a predictor for phosphorus loss from soil. In: Currie, L.D., Hedley, M.J., Horne, D.J. and Loganathan, P. (eds) *Best Soil Management Practices for Production.* Occasional Report No. 12. Fertilizer and Lime Research Centre, Massey

University, Palmerston North, New Zealand, pp. 153–164.

McDowell, R.W. and Condron, L.M. (2000) Chemical nature and potential mobility of phosphorus in fertilized grassland soils. *Nutrient Cycling in Agroecosystems* 57, 225–233.

McLaren, R.G. and Cameron, K.C. (1996) *Soil Science: Sustainable Production and Environmental Protection*. Oxford University Press, Auckland, New Zealand,.

McLaren, R.G., Taylor, M., Hendry, T. and Clucas, L. (1999) Leaching of metals and nutrients from soils treated with metal-amended sewage sludge. In: Currie, L.D., Hedley, M.J., Horne, D.J. and Loganathan, P. (eds) *Best Soil Management Practices for Production*. Occasional Report No. 12. Fertilizer and Lime Research Centre, Massey University, Palmerston North, New Zealand, pp. 251–260.

McLenaghen, R.D., Cameron, K.C., Lampkin, N.H., Daly, M.L. and Deo, B. (1996) Nitrate leaching from ploughed pasture and the effectiveness of winter catch crops in reducing leaching losses. *New Zealand Journal of Agricultural Research* 39, 413–420.

Ministry for the Environment (1997) *The State of New Zealand's Environment*. Ministry for the Environment, Wellington, New Zealand.

Ministry for the Environment (1998) *Reporting on Persistent Organochlorines in New Zealand*. Ministry for the Environment, Wellington, New Zealand.

Ministry of Health (1995) *Drinking Water Standards for New Zealand 1995*. Ministry of Health, Wellington, New Zealand.

Nash, R.G. and Woolson, E.A. (1967) Persistence of chlorinated hydrocarbon insecticides in soils. *Science* 157, 556–565.

Nguyen, M.L. and Goh, K.M. (1992) Nutrient cycling and losses based on a mass-balance model in grazed pasture receiving long-term superphosphate applications in New Zealand. 1. Phosphorus. *Journal of Agricultural Science (Cambridge)* 119, 89–106.

O'Dea, T.F. (1980) Wairarapa Groundwater: Study Report by Scientific Advisory Group. Wairarapa Catchment Board.

Painter, D., Williams, P., Francis, G., Bidwell, V., Cameron, K. and Bright, J. (1997) *Nitrogen Inputs at Landsurfaces and Groundwater Quality. Review of Science, Policy, and Management: International and New Zealand*. Lincoln Environmental, Canterbury, New Zealand.

Prins, W.H., Dilz, K. and Neetsen, J.J. (1988) Current recommendations for nitrogen fertilisation within the EEC in relation to nitrate leaching. *Proceedings of the Fertiliser Society, London* 276.

Rao, P.S.C. and Davidson, J.M. (1980) Estimation of pesticide retention and transformation parameters required in non-point source pollution models. In: Overcash, M.R. and Davidson, J.M. (eds) *Environmental Impact of Nonpoint Source Pollution*. Ann Arbor Publishers, Ann Arbor, Michigan, pp. 23–67.

Roberts, A.H.C., Ledgard, S.F., O'Connor, M.B. and Thomson, N.A. (1992) Effective use of N fertilizer, research and practice. *Proceedings of Ruakura Dairy Farmers' Conference, Hamilton, New Zealand*, pp. 77–82.

Roberts, A.H.C., Cameron, K.C., Bolan, N.S., Ellis, H.K. and Hunt, S. (1996) Contaminants and the soil environment in New Zealand. In: Naidu, R., Kookana, R.S., Oliver, D.P., Rogers, S. and McLaughlin, M.J. (eds) *Contaminants and Soil Environment in the Australasia-Pacific Region, Proceedings of the First Australasia-Pacific Conference on Contaminants and Soil Environment in the Australasia-Pacific Region, Adelaide, Australia*. Kluwer Academic Publishers, Dordrecht, The Netherlands, pp. 579–628.

Ruz-Jerez, B.E., White, R.E. and Ball, P.R. (1995) A comparison of nitrate leaching under clover-based pastures and nitrogen-fertilized grass grazed by sheep. *Journal of Agricultural Science (Cambridge)* 125, 361–369.

Schipper, L.A., Cooper, A.B., Harfoot, C.G. and Dyck, W.J. (1993) Regulators of denitrification in an organic riparian soil. *Soil Biology and Biochemistry* 25, 925–933.

Scholefield, D., Lockyer, D.R., Whitehead, D.C. and Tyson, K.C. (1991) A model to predict transformation and losses of nitrogen in UK pastures grazed by beef cattle. *Plant and Soil* 132, 165–177.

Scholefield, D., Tyson, K.C., Garwood, E.A., Armstrong, A.C., Hawkins, J. and Stone, A.C. (1993) Nitrate leaching from grazed grassland lysimeters: effects of fertilizer input, field drainage, age of sward, and patterns of weather. *Journal of Soil Science* 44, 601–613.

Scobie, S., Buckland, S.J., Ellis, H.K. and Salter, R.T. (1999) *Organochlorines in New Zealand: Ambient Concentrations of Selected Organochlorines in Estuaries*. Ministry for the Environment, Wellington, New Zealand.

Selvarajah, N., Maggs, G.R., Crush, J.R. and Ledgard, S.F. (1994) Nitrate in groundwater in the Waikato region. In: Currie, L.D. and Loganathan, P. (eds) *The Efficient Use of Fertilizers in a Changing Environment*, Occasional Report No. 7. Fertilizer and Lime Research Centre, Massey University, Palmerston North, New Zealand, pp. 160–185.

Silva, R.G., Cameron, K.C., Di, H.J. and Hendry, T. (1999) A lysimeter study of the impact of cow urine, dairy shed effluent and nitrogen fertilizer on drainage water quality. *Australian Journal of Soil Research* 37, 357–369.

Smith, C.M., Wilcock, R.J., Vant, W.N., Smith, D.J. and Cooper, A.B. (1993) Towards sustainable agriculture: freshwater quality in New Zealand and the influence of agriculture. MAF Policy Technical paper 93/10. Ministry of Agriculture, Wellington, New Zealand.

Smith, I.M., McNamara, D.G., Scott, P.R. and Harris, K.M. (1992) *Quarantine Pests for Europe*. CAB International, Wallingford, UK.

Smith, J.U., Bradbury, N.J. and Addiscott, T.M. (1996) SUNDIAL: a PC-based systems for simulating nitrogen dynamics in arable land. *Agronomy Journal* 88, 38–43.

Sparling, G., Dragten, R., Aislabie, J. and Fraser, R. (1998) Atrazine mineralization in New Zealand topsoils and subsoils: influence of edaphic factors and numbers of atrazine-degrading microbes. *Australian Journal of Soil Research* 36, 557–570.

Spiers, T.M., Francis, G.S. and Cant, G. (1996) The impact of the land disposal of organic wastes on water quality. In: Currie, L.D. and Loganathan, P. (eds) *Recent Developments in Understanding Chemical Movements in Soils Significance in Relation to Water Quality and Efficiency of Fertiliser Use*. Occasional Report No. 9, Fertiliser and Lime Research Centre, Massey University, New Zealand, pp. 248–252.

Statistics New Zealand (1996) *Agriculture Statistics*. Statistics New Zealand, Wellington, New Zealand.

Steele, K.W. (1982) Nitrogen in grassland soils. In: Lynch, P.B. (ed.) *Nitrogen Fertilisers in New Zealand Agriculture*. New Zealand Institute of Agricultural Research, Wellington, New Zealand, pp. 29–44.

Thorrold, B.S., Wilcock, R., Rodda, H., Monaghan, R.M., Stroud, M. and Hamill, K. (1999) Catchment scale effects of land use and management on nutrient loss. In: Currie, L.D., Hedley, M.J., Horne, D.J. and Loganathan, P. (eds) *Best Soil Management Practices for Production*, Occasional Report No. 12. Fertilizer and Lime Research Centre, Massey University, Palmerston North, New Zealand, pp. 223–230.

Williams, P.H. and Haynes, R.J. (1992) Balance sheets for phosphorus, sulphur and potassium in a long-term grazed pasture supplied with superphosphate. *Fertilizer Research* 31, 51–60.

Zaman, M., Di, H.J. and Cameron, K.C. (1999a) A field study of gross rates of N mineralization and nitrification and their relationships to microbial biomass and enzyme activities in soils treated with dairy effluent and ammonium fertilizer. *Soil Use and Management* 15, 188–194.

Zaman, M., Di, H.J., Cameron, K.C. and Frampton, C.M. (1999b) Gross nitrogen mineralization and nitrification rates and their relationships to enzyme activities and the soil microbial biomass in soils treated with dairy shed effluent and ammonium fertilizer at different water potentials. *Biology and Fertility of Soils* 29, 178–186.

18 Land, Water and People: Complex Interactions in the Murrumbidgee River Catchment, New South Wales, Australia

S. Pengelly[1] and G. Fishburn[2]
[1]*NSW Department of Land and Water Conservation, PO Box 365, Queanbeyan, NSW 2620, Australia;* [2]*NSW Department of Land and Water Conservation, PO Box 10, Wagga Wagga, NSW 2650, Australia*

Key words: algal blooms, biodiversity, irrigation, Murrumbidgee, New South Wales, salinity, soil acidity, soil erosion

Introduction

Australia is a land of natural variation and unpredictability and the Murrumbidgee River catchment is no exception. Droughts are expected and may be broken by floods. This variability, and the stress it produces, has played a vital role in shaping the landscape and the selection of flora and fauna within it. European settlers and their farming practices were not adapted to this extreme variability. Consequently, agricultural, industrial and urban developments necessitated numerous changes to the catchment. In particular, developing infrastructure to provide a dependable and secure water supply has been a high priority. The success of these modifications allowed colonization and agricultural development across the whole catchment. Valuable industries developed which are often reliant on the availability of large volumes of water.

It is now recognized that large-scale catchment modifications have not been achieved without unbalancing natural processes. These unforeseen changes are serious and threaten to undermine not only the agricultural systems, but also many natural components of the environment. Developing a balance between the required catchment changes, and avoiding others, is the crucial concept of sustainability. So far, we are yet to demonstrate that sustainability can be converted from this concept to reality.

This chapter initially provides an impression of the Murrumbidgee River catchment based on its physical, economic and social profiles. Key environmental issues affecting both land and water resources that have resulted from the complex interactions of land, water and people in the Murrumbidgee catchment are then listed and described in detail.

Most land-based issues have intimate links to water quality. The links between land and water health and their biological inhabitants are also direct and clear. The scale of catchment degradation is recognized as a community problem; one that spans social, economic and environmental issues. Consequently, solutions have been sought at local, catchment, basin-wide and national levels. Some of these solutions are discussed and reviewed. Some ideas for future improvements are also presented.

Physical profile

The Murrumbidgee Valley is one of the major catchments contributing to the larger Murray–Darling Basin which covers 14% of Australia's total area. The Murray–Darling is Australia's largest river system. The Murrumbidgee River catchment is an inland catchment situated in southern New South Wales (Fig. 18.1). The 84,000 km^2 of the Murrumbidgee River catchment start in the alpine Snowy Mountains and finish about 500 km further west in the semi-arid Hay Plains. The size of the catchment has necessitated that, for management purposes, it is divided into three zones. These are called the 'upper', 'mid' and 'lower' catchments which loosely reflect varying landscapes and land uses.

Although the east–west distance of the Murrumbidgee catchment is only 80 km, surface runoff falling in the catchment must travel far west inland towards Adelaide in South Australia. The quickest path to the sea is blocked by the Great Dividing Range which follows much of the eastern Australian coast. This range forces much of the east coast rainfall on a long western route. It was this westerly flow of many of the large rivers which convinced the early explorers there must be a large inland sea in the heart of Australia. In fact, Charles Sturt was so convinced of this that in 1844 his exploration party travelled with a sailing boat through what turned out to be numerous deserts in central Australia (Flannery, 1998).

The flows of the Murrumbidgee River and its tributaries are regulated by numerous water structures to provide a secure supply of water to people and agriculture throughout the year. These structures include 12 major dams, five minor dams, eight large weirs and over 10,000 km of irrigation canals in the catchment.

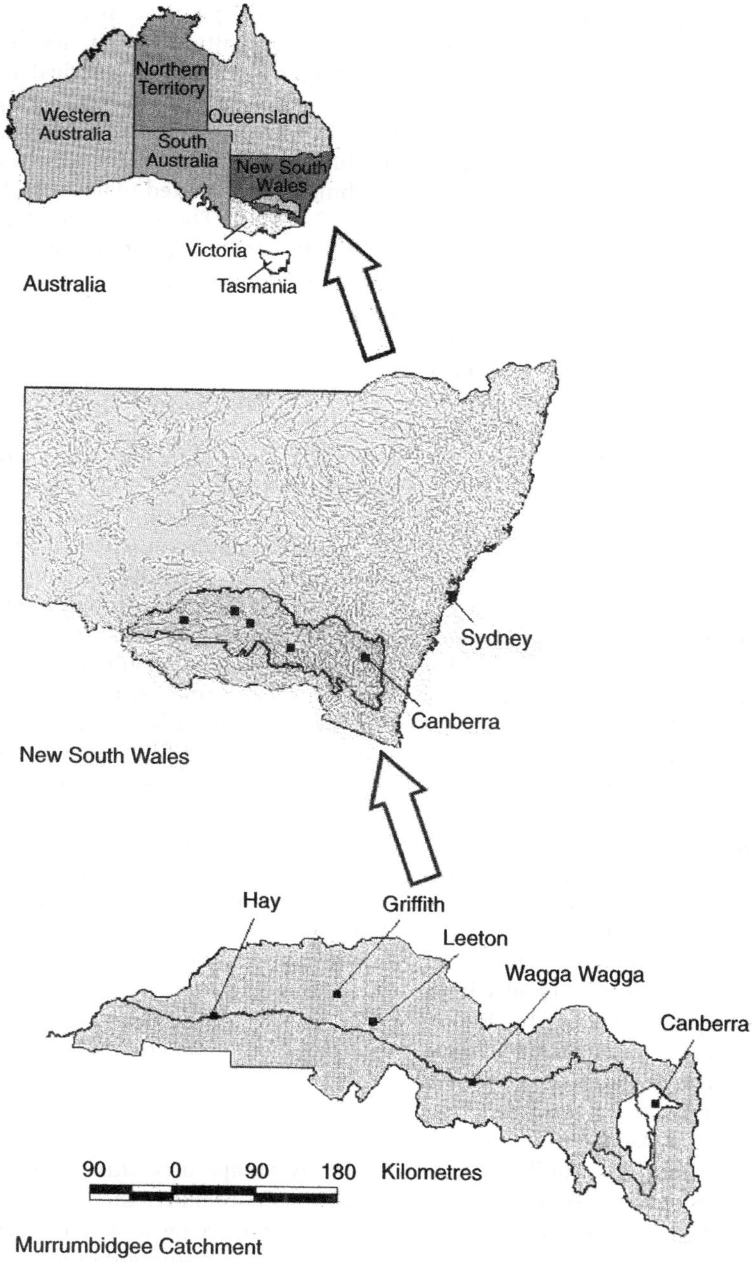

Fig. 18.1. Location of the Murrumbidgee catchment within New South Wales and Australia.

While the upper reaches of the Murrumbidgee River have steep gradients with rapids and topography typical of upland streams, the gradients further west are quite different. In 10 km of travel there may be altitude variation of as little as 1 m. Consequently, weir structures only 5 m high (such as Maude weir) can create a weir pool up to 75 km long (Bek, 1985). In the lower Murrumbidgee catchment, weir pools influence flows within up to 280 km of river channel (Webster and Pengelly, 1998).

After significant water volume reductions through evaporation and human diversions, the remaining Murrumbidgee flows join the larger Murray River for a convoluted journey further west. Water falling at the top of the Murrumbidgee River catchment may travel up to 1600 km before ending at its confluence with the Murray River. In total, the water may travel up to 2800 km before finally reaching the sea. If a major flood peak could be sustained throughout the length of only the Murrumbidgee River, it would still take approximately 28 days for the flood to travel from one end of the catchment to the other. Current regulated flows along the river take considerably longer.

Social profile

The Murrumbidgee catchment not only encompasses Canberra, but also includes numerous regional centres and towns, and with a total population of approximately 520,000, it is one of the most densely populated regions in rural Australia. Despite a static population, irrigated agriculture has continued to develop, mostly within irrigation areas and along Murrumbidgee waterways provided with regulated flows from upstream dams (Buchan, 1995).

Community groups play an important role in natural resource management within the Murrumbidgee catchment. Representation and input from varied groups including agricultural industries, government agencies, research organizations and environmentalists is keenly sought. All these groups have the opportunity to contribute to the development and implementation of environmental management.

Economic profile

The Murrumbidgee catchment has extensive irrigation areas responsible for large-scale food production for both domestic and export markets. These areas account for 25% of fruit and vegetable production for New South Wales and 50% of Australia's rice production (MCMC, 1998). Dryland agriculture (such as livestock, cropping and timber industries) is also a major contributor to agricultural production. The total annual value of agricultural production within the Murrumbidgee catchment is estimated to be AUS$1 billion (MCMC, 1998). The economic importance of the Murrumbidgee catchment as either a source of produce or revenue extends to the entire country.

Issues and Impacts

Few catchments have their origins in alpine terrain and their conclusion in a contrasting semiarid environment. The diversity of these natural environments leads to a similar contrast in agriculture, environmental issues and attitudes. Among such differences there are, however, some consistencies across the catchment. Few areas have escaped some form of environmental degradation. Land degradation alone has been estimated by the CSIRO to cost Australia about AUS$1 billion annually arising from lost production and ongoing nutrient losses (Environment Australia, 2000). Some of the key natural resource management issues in the Murrumbidgee catchment are listed in Table 18.1. Each of these issues is discussed below in greater detail.

Soil erosion

Soil erosion creates many problems beyond the site at which it occurs. Eroded materials rarely escape a catchment in one swift storm event. Usually, there is a staggered pace of downstream movement, which en route causes sedimentation, excessive flooding and partial stimulation of algal blooms. Each of these problems is significant, as is the initial loss of the land that was once relied on by farmers or wildlife.

Historical research has shown that current sources of sediment are quite different from those in the past. Existing fluvial geomorphic features, broad-scale aerial photography conducted since 1944 and anecdotal

Table 18.1. Some of the major issues in the Murrumbidgee catchment and their causes.

Issue (not in order of importance)	Causes
Soil erosion	Vegetation destruction Overstocking and ploughing Changed flow regimes In-stream willow invasion Salinity Soil acidity Soil structure decline
Noxious weeds and feral animals	Introduction of invasive pest species, e.g. willows, rabbits, grasses and European carp
Water supply and flow manipulation	Dam/weir construction and storage of winter flows for summer irrigation release
Water sharing	Diverse and excessive consumptive water requirements Extreme natural variability in supply (both drought and flood are 'normal')
Algal blooms	Anthropogenic eutrophication Ponding of water (from farm dams to large reservoirs)
Dryland and irrigation salinity	Vegetation clearing Irrigated agriculture Urban salinity
Soil acidity	Nitrogenous fertilizers Growing monoculture legume pastures
Biodiversity decline	Land degradation Native vegetation clearing
Declining water quality	All issues listed above

evidence (photos, sketches or passages from early explorer diaries) have been used by Starr (1995, 1996) to understand local catchment changes. Early vegetation decline and overstocking caused erosion of hill-slopes and gullies on a catastrophic scale. Recent studies show that the Murrumbidgee's erosion story is now well into a new phase, one that is influenced by waterway dominated erosion contributions.

Studies by Walbrink *et al.* (1996, 1998) have used radioactive isotopes to determine whether river sediments are derived from surface or subsoils. As a result of atmospheric atomic weapons testing in South Australia in the 1950s, the entire surface of the continent was labelled with ^{137}caesium radionuclide fallout. This labelling affected only the top 10–20 cm of the soil profile. By testing soil samples for this radioactive isotope, it was determined whether the origins were from the surface soil layer or a deeper, unlabelled layer. Results indicated that in the upland catchment areas up to 87% of sediment and P loads are generated from in-stream sources (i.e. the erosion of stream-banks, stream-beds and gullies) (Walbrink *et al.*, 1996, 1998). The implication of these studies is that, within the Murrumbidgee River catchment, the single most important way to reduce sediment and P inputs into waterways is to focus on the stabilization and protection of riparian environments (Pengelly, 1998). To achieve this, many community groups in the Murrumbidgee catchment are now actively focused on revegetation and stock control along waterways.

Noxious weeds and feral animals

The last 212 years of white settlement have provided numerous examples where seemingly innocent introductions of plants and animals have developed into environmental and economic disasters. No region of

Australia, whether alpine or arid, has escaped the deleterious impacts of invaders into fragile ecologies. The introduction of rabbits is one of the classic examples. In this case, just 24 rabbits imported in 1859 developed into a plague covering the entire continent. The problem of this feral is yet to be resolved. A recently emerged pest, sometimes described as the rabbit of the waterways, is the European carp (*Cyprinus carpio*). This fish seems to tolerate extremely varied environments and has a very high reproductive capacity. Carp are now estimated to account for between 70 and 90% of fish numbers in not only the Murrumbidgee catchment, but the entire Murray–Darling Basin (Crabb, 1997). In some areas, destructive carp feeding habits are thought to make a significant contribution to bank erosion and water turbidity. No control methods to date have made any impact on their dominance of inland fish stocks.

Introduced plants have also had incredible success at not only establishing in the Australian environment, but even out-competing the native species. The willow (*Salix*) is one of these examples and the negative impacts are only now being recognized. No woody shrubs or trees native to Australia grow within the low flow channel of waterways or on transient sand-bars. This zone was once restricted to only macrophytes, but is now shared by some willows. Here they manage to grow in stands, or 'galleries', which cause channel constriction. Consequently, they cause the diversion of water into adjacent stream banks. As the blockage created by the willows (and the flood debris they collect) grows, so does the bank erosion. In some situations the risk of complete stream diversion is high. Willows that can reproduce by seed pose the greatest risk. Along the Murrumbidgee River up to 400,000 willow seedlings per river kilometre have been observed. In such cases the cost of eradication is high and completely dependent on chemicals. The threat of such issues is extreme and immediate action is essential. The impacts of in-stream willows extends far beyond the immediate bank erosion they generate. Eroded sediment also fills downstream fish breeding holes and smothers aquatic life. Dense stands of willows do not even provide substantial habitat for native wildlife.

Even plants as small as grasses have generated a significant environmental and agricultural issue in the Murrumbidgee. Serrated tussock (*Nassella trichotoma*) is a prime example. A native to South America, it has flourished in the upper Murrumbidgee catchment (and across south-eastern Australia) to become one of the greatest threats to livestock production (Lamp and Collet, 1989). This aggressive weed is able to dominate pastures, yet provides no sustenance.

Many other species, whether introduced accidentally or deliberately, have also become integrated into the Murrumbidgee and the wider Australian environment. They alter natural balances and affect local, regional and national economies. While some control efforts have worked, some will require a significant boost in attention before results can be achieved. At the other end of the scale, some pests have had no control methods applied as the methods simply do not exist.

Water supply and flow manipulation

There is some evidence that Aboriginals in far northern Australia used natural rock constrictions and sheets of bark to dam water for later use (Gunn, 1954). However, it is clear that only since European settlement has there been a large impact on Australia's waterways from river regulation and dam construction. The scale of the flow changes that have occurred within the Murrumbidgee catchment are so significant that the volume of summer and winter flows for some of the Murrumbidgee River have been completely reversed. Winter dominated runoff is now converted into high summer flows.

Two of the 17 reservoirs in the Murrumbidgee River catchment store over 1,000,000 Ml. The largest of these, Blowering Dam (situated on the Tumut

River which joins the Murrumbidgee River between Canberra and Wagga Wagga), holds about three times the capacity of Sydney Harbour (Jeffcoat, 1995). Such large reservoirs not only provide the water requirements of towns and industries further west during the summer months, but also supply the water requirements of thirsty irrigated agriculture.

Water supply for irrigation areas was long believed to be unlimited. Only in the last decade has there been universal acceptance that water is a finite resource. Debate continues on whether we are at, or past, the limit whereby further water extractions increase environmental stress, resulting in declining riverine health. Initially, the consequences of the changed flow regime for the downstream riverine environments was not considered. Both the reservoirs and the water management legislation were constructed during a time when water supply for irrigation was the driving force. Today, the impacts of water storages are still significant, but their operating protocols now dictate that these impacts are minimized. Some consideration is now given to the timing and volume of water releases specifically for the benefit of downstream environments, especially wetlands. Thermal water pollution, generated by the release of cold, low level dam water, is also of concern, particularly with respect to fish breeding. For decades summer spawning below large reservoirs has been influenced by the 'cold shower' treatment. While it may not be possible to achieve natural river temperatures, increases in water temperatures can usually be reached by releasing warmer surface waters.

Water sharing

The conflict between competing water users in Australia is intense and increasing, for example between people upstream and downstream, or between people and the natural environment. After focusing on only the needs of humans, current efforts are giving consideration to the water requirements for a healthy environment. Current legislation (established in 1912) does not provide for recognition of water for the environment, but this is being addressed through a proposed Water Management Act (DLWC, 2000).

Engineering has allowed impressive feats of water storage and diversion. It has also severely complicated the issue of water sharing. Not only are we able to impound vast quantities of water along the Murrumbidgee River's natural path, but major diversions are also possible between rivers, and between basins. Inter-river and inter-basin transfers of water from the Murrumbidgee River have been made possible through the Snowy Mountains Hydro Electric Scheme. With this scheme, water is diverted along and through the granitic mountains; 75% of the diversions end up in the Murrumbidgee River catchment and the remaining 25% in the Murray River catchment (SMHEA, 2000). Such a project has not been constructed without significant environmental effects. Some rivers affected by the Snowy Mountains Hydro Electric Scheme have had 99% of their flows diverted away from their natural basins. The remaining 1% is obviously not sufficient to maintain ecological health.

Pressure from environmental lobby groups to return some portion of diverted flows back to their source has strengthened in recent years with a growing awareness of the environmental costs of the scheme. Such intentions, while seeming clear and logical, quickly sink into a difficult debate shrouded in conflict. The diverted water now has new uses for the power generation and irrigation industries. The resulting electricity and produce in turn generate a significant income. Losses from these industries are not suffered without a fight at all political levels.

Rivers that receive the diverted flows are also affected. Delivery of large volumes of water during the summer months to the irrigation areas requires supply rivers to flow at very high rates. Both the timing and duration of these releases are quite unnatural. The Tumut River below Blowering Dam in the Murrumbidgee catchment is an example of a river heavily affected in this way. The result of such high and consistent summer discharges is manifested most obviously in excessive bank erosion. Local Tumut communities see their river being used as only a conduit for irrigation water and hydroelectric power generation. They perceive that the benefits are delivered elsewhere, while the problems have landed on their doorstep. Nevertheless, Blowering Dam and the Snowy Mountains Hydro Electric Scheme have effectively removed flooding concerns along the Tumut River valley.

The uses of water are varied, and clear priorities for access have been set. Such prioritization becomes paramount during dry periods. Human consumption and domestic use take precedence over domestic stock needs, which in turn have priority over irrigation, industrial uses and recreational uses. It is now finally recognized that water in Australia is not an unlimited resource. Supply does not exceed demand. Sharing this valuable resource between water users, including the environment, is complex. It requires clear rules and a method for continual evaluation to determine if the diverse rules are in fact suitable. Monitoring and evaluation must also be used to highlight non-compliance with these rules. Balanced allocation of water resources is a relatively new field for the community and governments. A certain amount of flexibility must remain so the 'rules' guiding water use can be modified as new information comes to light, or the needs of water-based industries change. Solutions to water-based conflict are at best difficult, and some may remain invisible. The interaction of land and water resources with people is undeniably complex.

Algal blooms

Unlike the pest species that generated massive problems after their introduction into Australia, algal blooms are a native problem. Alteration to the natural flow of water through the catchment has proved to provide periodically ideal conditions for algal blooms. Regardless of the scale of these flow alterations (whether they are the size of a small farm dam or a large reservoir), these modifications seem to favour algal growth.

Although there are many bloom-forming algae, blue-green algae (or cyanobacteria) are the group of greatest concern because of their ability to produce various toxins. Within the Murrumbidgee catchment there are two main cyanobacterial species which commonly emerge during warmer months to create human health risks. They are *Microcystis aeruginosa* and *Anabaena circinalis*.

Exposure to algal toxins can occur through ingestion or skin contact. Drinking water supplies sourced from standing water bodies must be monitored for unacceptable cyanobacterial concentrations. If these are found, the local authority must have in place a contingency for either treating the water (e.g. carbon filtration) or finding an alternative safe water supply (e.g. importing water). Some local communities, even in the upper reaches of the Murrumbidgee River where the catchment is in relatively good condition, have been caught unawares.

Open water created by storages acts as a magnet for recreational use. Unfortunately, storages also seem to be a magnet for potentially toxic algae species. The combination of these algae and recreational users is unwise, so in the absence of algal control methods, the restriction of human access must be exercised. Besides the obvious environmental concerns created by a potentially toxic bloom, such situations also lead to social and economic impacts on local communities. There is a strong connection between the water conditions, surrounding land use and the people within the local environment. The status of one does have an impact on the other two.

Blooms along the river channel itself are infrequent as river flows generate sufficient turbulence to prevent thermal layering or stratification. Without this stratification, the algae are distributed through the entire water column and are not restricted to the upper warm and light layer which provides preferential growing conditions. River and canal reaches below large storages usually have high flows during the summer period when algal blooms are most likely to occur. Consequently, the risk of a river bloom is minimal.

Despite the frequency of algal blooms, their environmental impact is poorly understood. When a bloom occurs, immediate concerns relate to human water users, leaving the impact on natural communities a mystery. Anecdotal reports from the Murrumbidgee catchment suggest that the impacts vary. For example, no native fish deaths have been observed in Burrinjuck Dam, despite the fact that surface algal concentrations reached as high as 80.5 million cells ml^{-1} and dissolved O_2 concentrations were considered to be critically low. In contrast, veterinarians have reported sickness in wildlife and stock that have been drinking from a cyanobacteria contaminated water (Pengelly, 1999). These anecdotal reports are, however, a long way from representing sound ecological knowledge.

Dryland and irrigation salinity

This subject is discussed in detail in Williams *et al.*, Chapter 21, this volume, but a brief summary is also given here. Land salinization is the accumulation of salt on or near the ground surface. Salinization in Australia is the result of rising groundwater levels mobilizing large quantities of salt previously stored deep within the soil. The source of this salt is ancient oceanic sediments, rock weathering and the deposition of sea spray (Wooldridge, 1999). Salinization is an issue because, at high levels, it reduces agricultural production, damages or destroys vegetation and increases the salinity of surface waters (Buchan, 1995). There are two main forms, 'dryland' and 'irrigation' salinity. They are differentiated by their cause, and consequently occur across areas of quite different land use. Vegetation removal is the primary cause of dryland salinity. The replacement of deep-rooted native vegetation (such as eucalypts) and perennial native grasses and shrubs with shallow-rooted, annual pastures and crops has significantly altered the hydrological cycle. The introduced shallow-rooted plants no longer act as natural pumps to lower the water table through evapotranspiration. With continued infiltration of water into the groundwater, but reduced losses, the water table rises, bringing with it tonnes of salt. Most agricultural pastures and crops are unable to access the groundwater until it is at least 2 m from the surface. At this depth, capillary rise can bring salt to the surface where it concentrates. When the water table reaches the surface the ground becomes waterlogged as well as highly saline.

Dryland salinity can no longer be considered as solely a rural issue. Numerous towns and cities in the Murray–Darling Basin now recognize that their urban areas are also under attack from this issue. The city of Wagga Wagga, located in the centre of the Murrumbidgee catchment, suffers considerably from urban salinity problems. Currently, 600 houses require immediate remedial action and significant damage could be caused to a further 7500 properties by 2020 if there is no change in land management and residential practices. In the absence of any preventive measures, the cost of urban salinity to Wagga Wagga over the next 30 years is estimated at AUS$183 million $year^{-1}$. Nationally it has been estimated that dryland salinity costs AUS$250 million $year^{-1}$ (Murray–Darling Basin Ministerial Council, 1999).

Irrigation salinity results in similar impacts but the cause is quite different. While dryland salinity is created by interrupting one of the natural paths through which groundwater is removed from the water table, irrigation salinity is caused by increasing groundwater inputs. When irrigation water is surplus to plant needs, or there is leakage from supply canals and water storages, it percolates into the groundwater. These water losses from many irrigation areas are significant enough to raise the water table, and with it the salt.

A recent report has indicated that accessions to groundwater in the Murrumbidgee Irrigation Area will have to be cut by more than 25% (i.e. approximately 25,000 Ml) to prevent more land going out of use (Crabb, 1997). It has also been predicted that, without intervention, all irrigation areas in the southern Murray–Darling Basin (including the Murrumbidgee catchment) will have water tables within 2 m of the surface by 2010 (Murray–Darling Basin Ministerial Council, 1999).

In some localities, vegetation loss from salinity occurs on a large scale. In some cases, one environmental issue may generate many others. For example, land salinization can trigger other serious issues including soil erosion, eutrophication, increased water turbidity and the smothering of riparian habitats.

Land salinization inevitably leads to increasing water salinization. A large portion of the salt brought to the land surface is exported into the waterways and river systems. Some lower reaches of major rivers have natural groundwater intrusions of extremely high salinity (reaching as high as 50,000 EC). These natural contributions are, however, frequently aggravated by drainage flows from irrigated agriculture areas and rising groundwater tables (Crabb, 1997).

While the Murrumbidge has lower salinity than other major New South Wales rivers, dryland salinity has intensified, affecting downstream waters (Buchan, 1995). The amount of salt mobilized to the land surface in the Murrumbidgee catchment has been estimated to be 2,170,000 t year^{-1}. At the current rate of increase, this is likely to become almost 4,000,000 t year^{-1} by 2100 (Murray–Darling Basin Ministerial Council, 1999). Within 50–100 years, it expected that 4.5% of Australia's cultivated land will be affected by salinity (Wooldridge, 1999).

Soil acidity

Far less spectacular than salinity is the problem of soil acidity. Despite the enormous scale of this issue, it is known to few people beyond a limited number of specialists and those who are directly affected. Acid soils, or those suffering induced acidity, are soils which have a pH $\leq$ 5.6. When a soil pH is 5.0 or less the yields of most plants begin to decline (Crabb, 1997). New South Wales has the largest area of land considered highly acidic. Figure 18.2 shows the extent of acid soils in New South Wales (Crabb, 1997). Australia-wide, some 90 Mha of land are affected by acid soils (Crabb, 1997). Induced soil acidity is now seen as one of the most serious and most neglected of land degradation issues. Soil acidity is becoming more widely recognized as a serious restriction to agricultural production in much of the Murrumbidgee River catchment.

There are numerous causes of soil acidity. Humans, however, are considered to be the main cause through the use of inappropriate agricultural practices. These include the repeated application of NH_3/NH_4^+-based fertilizers without liming, and the growing of pastures based on legumes such as subterranean clover. Such pastures tend to die off during summer leaving N unused in the soil which then leaches and contributes to soil acidification instead of plant growth.

Current management of soil acidity is conducted mainly on a property basis. It involves soil testing and the addition of lime or the use of crops such as canola in a wheat growing rotation. Liming is, however, only an option for pastures accessible by ground vehicles. Also, the additional cost this imposes on production may not translate into productivity improvements. Such significant additional costs increase the pressures on farm viability. In a time when rural economies are low, such an issue could be the difference between business survival and bankruptcy.

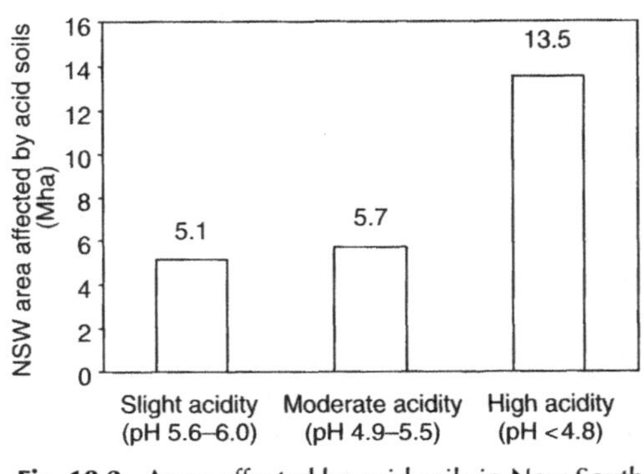

Fig. 18.2. Areas affected by acid soils in New South Wales (NSW) (Crabb, 1997).

Not treating the problem of soil acidity results in lower productivity and the generation of other forms of land degradation. These can include soil erosion and increased salinity which occur as a result of lost vegetation. To add to the multitude of problems generated by soil acidity, as the soil becomes increasingly acidic, the availability of elements such as Al and Mn increases and they may become toxic to plants. Some farms may not be able to afford the necessary remedial measures, but neither can they afford not to adopt them. Such complex situations require consideration and assistance from the broader community *before* the land degradation issue generates into more serious environmental, economic and social issues.

Biodiversity decline

Biological diversity refers to the diversity of all life forms. This includes the animals, plants, microorganisms, the genes that they contain and the ecosystems of which they form a part. Biological diversity, or biodiversity, is not static. It is increased by genetic change and evolutionary processes, and is reduced by processes of habitat degradation, population decline and extinction (Environment Australia, 2000). An important concept of biodiversity is the interrelatedness of the components of the biological world.

Australia's continental isolation has resulted in a high percentage of native species. The uniqueness of the biota is demonstrated by the fact that 82% of mammals, 85% of flowering plants, 89% of reptiles and 93% of frogs are found only in Australia (Environment Australia, 2000). Human activities have altered Australian ecosystems for approximately 50,000 years, but the scale and pace of these alterations has accelerated during the last 212 years of European settlement. Most of Australia's natural ecosystems have been extensively modified and degraded. The result has been dramatic declines in the distribution and abundance of many of the species (Environment Australia, 2000).

Despite the snowy origins of the Murrumbidgee River, much of its catchment is temperate and has thus been a prime target for human habitation, and all the impacts that closely follow. Loss of species, or serious reductions in populations, has been the result of direct causes such as habitat destruction, or indirect causes such as competition. Irrespective of cause, both terrestrial and aquatic habitats are being purged of species sensitive to change. Ecologically speaking, large shifts or losses in populations are not unusual. However, the latest wave of rapid changes in the Australian environment have been engineered by humans in only the last 212 years. Globally, current rates of decline and loss of biological diversity are the highest for at least 60 million years (Environment Australia, 2000). The speed of these recent and continuing ecological changes are significant because they are occurring in the same environment in which we wish to live.

The patchy statistics that we have on biodiversity decline are unambiguous. For example, in the western Murray River (which the Murrumbidgee River joins), there has been a 94% decline in silver perch and a 74% decline in golden perch (Crabb, 1997). Nationally, since European settlement, there has been extinction of 20 mammal, 20 bird and 76 plant species (Environment Australia, 2000). The list of endangered and vulnerable species is far longer. Introduced species now constitute up to 15% of the Australian flora (Environment Australia, 2000).

Single pest species, such as the European carp, are frequently targeted as the major causes of environmental woes. For example, 'there is growing evidence of the detrimental impact of carp on native species of fish and the overall aquatic environment' (MCMC, 1994). Single land degradation issues, such as dryland salinity, are also seen in some areas as the foundation of environmental and economic troubles. In reality, while some pests and environmental issues undeniably cause greater impacts than others, it is likely to be the overwhelming combination of many, or all, of the problems that is responsible for the downward spiral in the state of the catchments.

There is not likely to be one single reason why species become endangered or extinct. Their decline or disappearance is likely to be complex and the result of numerous factors. While some efforts should be directed to the specific requirements of endangered species to maximize their chance of survival, we should not lose sight of the broader issues underlying their decline.

The maintenance of biological diversity requires more than just protecting wildlife and their habitats in nature conservation reserves. Ecologically sustainable management of all Australia's terrestrial and marine environments is deemed essential for the conservation of biological diversity (Environment Australia, 2000). All over the world, individuals and community groups have an increasingly important role in the conservation of biological diversity. Only through their activities will there be a chance to protect all ecosystems, and not just those within the boundaries of designated reserves.

While it is a very difficult task (if at all possible) to quantify the relative economic costs for the conservation of biological diversity, the benefits become obvious when compared with the scenario of not applying such efforts. Significant economic benefits can be gained from immediate action to conserve biological diversity. Among the benefits are the future opportunities for resource use and the substantial future savings in the cost of rehabilitating species and ecosystems (Environment Australia, 2000). Biodiversity loss is a serious threat. It would be dangerous to wait and learn the true value of species only through the impact their loss creates.

Declining water quality

In 1830 the explorer Charles Sturt reached the far inland confluence of the Murray and Darling Rivers. As he and his men turned their boat into the Darling River and rowed upstream, a distinct difference in the quality of the water was noticed. In relation to the Darling River, Sturt wrote 'its water, though sweet, were turbid and had a taste of vegetable decay, as well as a slight tinge of green' (Flannery, 1998). This description is now believed to be referring to salinity, turbidity and an algal bloom. These three issues are only a sample of the overload of water quality concerns which today are see as serious threats.

Some water quality issues can be considered as 'natural', such as those observed by Sturt. Others, such as pesticides, are generated solely by humans, and their influences are new and unfortunate experiences for natural ecosystems. Extreme water quality changes in the Australian environment are clearly natural. Numerous large-scale modifications, such as creating stagnant reservoirs, point sources of pollution and rising water tables, have generated large-scale problems. Monitoring within the Murrumbidgee River catchment has indicated some disturbing water quality trends. When these trends are extrapolated into the future, it can be seen that there is often little time left before many properties exceed the Australian and New Zealand drinking water quality guidelines (ARMCANZ, 1996). Some already consistently exceed these recommended levels. For example, in South Australia, over 20% of surface water resources have salinity levels above those recommended for human consumption (Wooldridge, 1999).

At different times, and in different places, it is not unusual for some water quality factors to be outside the recommended guidelines for their use (whether that use is for recreation or consumption). Sometimes these exceedences are natural; however, increasingly, we are observing that some properties do not return to what is considered their 'normal' levels.

The term 'water quality' is extremely broad, but is effectively the sum of many different physical, chemical and biological properties. The properties of concern often vary, depending on the purpose for which the water is used, and, frequently, it is more than one that causes concern. Almost certainly these water quality concerns, whether relating to surface or groundwater, are generated by varied causes.

Solutions

Clearly, the future prospects in the Murrumbidgee are grim if inaction or delay are the paths to be chosen over immediate environmental reparations. Numerous and varied efforts show that slowing, halting or, ideally, reversing many of the critical issues are the intentions of many people. Levels of continuing support for environmental improvements and detailed monitoring will indicate whether these efforts are successful. The range of efforts is detailed below.

The scale of many of the issues requiring urgent attention is enormous. Community efforts and private and public investment are essential because minority interest in environmental restoration will not be able to achieve the needed results. Improvements must also be conducted at an appropriate scale. At this stage, it is difficult to pinpoint the magnitude of the efforts required. For many issues we have limited and poor information on not only the existing scale of these issues, but also the rate at which they are changing. Few trials or studies have been conducted at a catchment scale, or over a sufficiently long period, to truly understand the rate at which we can implement the needed environmental improvements.

The need for environmental improvements is, fortunately, recognized and being enacted at a variety of levels. Grassroots action is being stimulated by institutional and policy change at local, state and federal government levels. Correcting so many problems over such a large area is a challenge without a clear prescription for remedy.

Existing strategies

Below are some of the strategies adopted to date (either catchment, state or basin-wide). These strategies cover efforts focused on political, institutional, legislative and on-ground changes for the benefit of the environment and are described in order from those with the broadest influence, down to those with the influence at the most local level (Table 18.2).

Council of Australian Governments (COAG) water reforms

Finding solutions to water-based issues is seen as a critical national concern. Consequently, all states, territories and the commonwealth governments agreed that there was an urgent need to take action to improve the management of water resources, and so formed the Council of Australian Governments (COAG). The agreement that they formed, provides a national framework for water reform in each state and catchment. COAG has adopted a plan of water reforms stemming from a recognition that an important part of the solution lies in significant policy and institutional change. Such changes are complementary, even if seemingly distant, from the grassroots changes occurring at the local community level.

The explicit links between economic and environmental objectives are important features of the package of reforms. They seek to improve both the efficiency of water use and the environmental management of the nation's river systems.

The main elements of the COAG water reforms package include:

Table 18.2. Existing natural resource management strategies and their level of influence.

Strategy	Geographic level of influence
Council of Australian Governments (COAG) water reforms	National
Murray–Darling Basin Initiative	Basin
NSW water and vegetation reforms	State
Catchment management	Catchment
Landcare	Local

- pricing water for full cost recovery;
- establishing secure access to water;
- providing for permanent trading in water entitlements;
- provision of water for the environment;
- the operation of water services on a commercial basis;
- improved institutional arrangements; and
- public consultation and education.

Institutional reforms have been introduced, such as the corporatization and privatization of irrigation areas. Such reforms are starting to provide the incentive for water use efficiency improvements. Essentially, the COAG water reforms provide a framework at a national scale within which the states can operate consistently. Ultimately, however, it is the states and territories which have the responsibility to implement these reforms in line with COAG objectives. To achieve this in New South Wales, more specific water reform initiatives (described in detail below) are the driving influence behind changes in management of water-based natural resources.

The Murray–Darling Basin Initiative

The Murray–Darling Basin Initiative is the largest integrated catchment management programme in the world (Crabb, 1997). The Murray–Darling Basin Initiative refers to an agreement and the institutions formed for its facilitation. The agreement puts into effect a recommendation made by a 1902 Royal Commission on the Murray River which stated that 'the river and its tributaries should be looked on as one' (Crabb, 1997). The agreement on which the initiative is based provides a solid foundation for integrated catchment management which did not previously exist. The purpose of the agreement is to 'promote and coordinate effective planning and management for the sustainable use of water, land and other environmental resources of the Murray–Darling Basin' (Murray–Darling Basin Agreement, 1992, Part I, clause 1, cited in Crabb, 1997). To achieve its purpose, the agreement established new institutions at the political, bureaucratic and community levels.

During its short existence, the agreement has made a significant contribution to the management of the Murray–Darling Basin. For example, in 1997 the Murray–Darling Basin Ministerial Council imposed a permanent 'cap' on water diversions, which has since limited these to the volume of water which was diverted under 1993/94 levels of development. Whether this decision will, however, go far enough to protect water resources is questionable. It has been noted by Whittington *et al.* (2000) that the 'cap' is set at 'the level of diversions which contributed to the current degradation' and therefore there should be 'no expectation that the Cap ... will improve the riverine environment'; essentially it is designed to halt the growth in water diversions.

New South Wales water and vegetation reforms

The Murray–Darling Basin, of which the Murrumbidgee catchment is a major component, covers portions of Queensland, New South Wales, Victoria and South Australia and includes the Australian Capital Territory (ACT). These political divisions at the state level have led to considerable legislative and natural resource management inconsistency across the basin. Consequently, natural resource management progress across the states and territory has been varied and staggered.

The progress of government reforms relating to natural resources has varied across other states in the Murray–Darling Basin. Political boundaries which are not closely aligned with catchment management boundaries confuse natural resource management issues. However, since the Murrumbidgee catchment is mostly within New South Wales, this section focuses on natural resource management relating to this state.

Water reforms

Australia is the driest inhabited continent. There are few streams and rivers with constant flows. Large areas of New South Wales west of the Great Dividing Range, including the Murrumbidgee catchment, are semi-arid, so rainfall is low. Even in areas receiving high rainfall it is not reliable. New South Wales cannot sustain an ever-increasing use of water and so has introduced a number of water reform initiatives.

The New South Wales government's water reform agenda is based on three key goals (DLWC, 1998a): (i) to share the available water better, (ii) to enhance support to the rural water sector; and (iii) to reshape water management delivery in New South Wales. The package of water reforms will ensure greater security and equity of water for all users, including the environment.

The implementation of these water reforms involves an active community and government approach. The government involvement extends to all government natural resource management agencies. The community input into this process relies largely on the existing catchment management boards (which are discussed in detail below) and specific water management committees. These committees consider issues relating to surface and groundwater within both regulated and unregulated waterways. A key function of these committees is to develop water management plans that will recommend equitable sharing between all water users, including the environment. These recommendations will eventually have statutory backing.

Vegetation reforms

Native vegetation is a valuable resource and an important key to maintaining the health of the land and waterways. The vegetation reforms provide legislative protection for existing vegetation as well as incentives for landholders to improve the quality or amount of the vegetation that they manage. Regional vegetation management plans are being developed. These aim to balance social, economic and environmental concerns of the region through sound vegetation management. To enable this, the government acknowledges the need for landholders to identify areas for agricultural development in their region as well as the need to provide for the long-term management of native vegetation (DLWC, 1998b).

Property agreements and management contracts with individuals or groups of landholders also help to provide financial incentives to improve native vegetation management. The incentives can contribute to the cost of fencing areas, enhancing existing native vegetation, controlling feral animals and reducing weeds.

With *c.* 70% of Australia's land area under the control of private landholders (Environment Australia, 2000), their cooperation is essential for the success of all such flora and fauna conservation activities.

Catchment management

Catchment management uses watersheds as the most logical and natural unit on which integrated natural resource management should be based. These distinct natural areas become the focus for a holistic approach to balancing the management of natural resources and is based on an understanding of their interrelatedness. This philosophy of management, which has been adopted by the New South Wales Government, is described as either Total Catchment Management (TCM) or Integrated Catchment Management (ICM). Essentially, they are the same in their focus. Within the Murrumbidgee catchment the most frequently adopted term is TCM.

TCM can be defined as: 'the coordinated and sustainable use and management of land, water, vegetation and other natural resources on a water catchment basis so as to balance resource use and conservation' (MCMC, 1998).

The management of natural resources on a catchment basis in Australia is at a scale which seems to be comprehensible. The clear natural boundaries are logical and allow a sufficiently broad perspective while maintaining a local relevance. The understanding of local impacts spreading either up or down a slope or stream is fundamental. Such concepts assist individuals in understanding cumulative or off-site impacts. Bio-regional or global perspectives are equally valid; however, they are generally too broad for people to conceptualize the importance of small actions on a broader geographic scale.

To drive the process of fully integrated catchment management, a group with a purely Murrumbidgee focus has been established. The 18 constituents of this group are representatives of a diverse range of interests and locations including those of local government, state and territory agencies, nature conservation, Aboriginals, primary producers and irrigation-based water users (Table 18.3). The membership of this group is also organized so that nine of the representatives are land holders/users. This group, called the Murrumbidgee Catchment Management Board, considers natural resource management issues not only from an environmental perspective, but also with consideration of their economic and social impacts.

One of the key roles of the Board is to identify and set priorities for natural resource management issues within the Murrumbidgee River catchment. In this way the Board steers natural resource management. While the Board does not directly do, or fund, large-scale on-ground works projects, it does have an important role in supporting and directing any applications for government funding. The main tools used for promoting improved management of natural resources are education and awareness. Enforcement of legislation is seen as the least preferable option. The diversity of membership interests on this Board reflects the diversity of environments, industries and issues within the Murrumbidgee catchment. Such diversity inevitably means that decisions are unlikely to be made without serious debate and compromise. The decisions made by this Board are consensus-based.

Landcare

Landcare is a community-based approach to solving environmental problems and protecting the future of natural resources. The 'Decade of Landcare' initiative originated in 1989 from an unusual coalition of the National Farmers Federation and the Australian Conservation Foundation. Their proposal to the federal government for action on land degradation was well received and initiated a unique programme of community and government partnerships.

The Landcare initiative recognizes that Australia's future is inextricably linked to better management of its natural resources. Deterioration of our natural

Table 18.3. Board membership of the Murrumbidgee Catchment Management Board (18 members made up of 9 landholder/users and 9 others).

Interests represented	Membership category	Selection method	Number of people
Local government	Local government	Nominated by local government in the trust area	2
	Land holder/users	As above	1
Nature conservation	Environmental interest	Nominated by Nature Conservation Council	2
	Land holder/users	Identified through public advertisement	2
Primary producers, water user natural resource user[a]	Land holder/users	Nominated by industry groups e.g. NSW irrigators, NSW Farmers Association[a]	2
	Land holder/users	Identified through public advertisement	2
Aboriginal	Land holder/users	Nominated by appropriate process, local and state	2
State government	Government department or authority	Nominated by minister	4
ACT government	Government agency	Nominated by ACT Government	1

[a] In urban areas farmer/irrigator representation may be replaced by other natural resource users.

resource base will result in costs to the Australian public through the loss of economic production, disruption to ecological processes, loss of biological diversity and decline in rural communities.

Financial investment in the programme by the government has been high. Both government and corporate investments have increased each year. Nationally, government support for Landcare and similar environment and community-based projects for the period from 1996/97 to 2001/02 is expected to be AUS$1.5 billion (Natural Heritage Trust, 1999). These figures are dwarfed compared with the value of the matching community contributions, estimated to be over three times this amount. Whilst it is expected that the government provides information on best management practices and limited financial assistance, it is widely accepted that the community bears the responsibility for actually conducting most of the physical restoration work.

With 10 years of Landcare completed and considerable expenditure, clear achievements and measurable improvements in both land and water resources should be noticeable. Unfortunately, the success of the Landcare programme is not clear and currently depends upon how 'success' is judged. For example, awareness and involvement in Landcare has steadily increased (Fig. 18.3), but this does not translate directly into an improvement in the status of natural resources. With the large number of people involved in Landcare it could be called a success; however, these people still represent a minority of the Australian population.

While tangible and measurable environmental improvements are the desired outcomes to which efforts are directed, it is still worth considering other benefits of the Landcare programme. These benefits include the awareness through education that has stimulated recognition and ownership of local environmental problems. Large investments of time and money are required to reach this basic but essential understanding. Although the scale and speed of improvements through Landcare may need a boost, it is considered to be a good start. A survey of Landcare groups by Curtis (1999) showed that 'Landcare participation has increased awareness of issues and enhanced landholder skills and knowledge and contributed to increased adoption of best management practices'.

Currently, there are 4250 Landcare groups across Australia (Agriculture, Fisheries and Foresty, 2000). This far exceeds the target of 2000 established groups set for the 'Decade of Landcare'. In the Murrumbidgee catchment there are 165 groups with approximately 4200 members. The top four issues dealt with by these Murrumbidgee Landcare groups (in order of priority) are: (i) revegetation; (ii) soil erosion by water; (iii) dryland salinization; and (iv) weed control.

The significant effort required by the community either to follow the local planning process, or conduct the on-ground improvements, should not be underestimated. In some areas, a small number of motivated individuals may be the driving force for a programme which is trying hard to find acceptance among traditionally conservative rural communities. The whole process is still overwhelmingly positive, but now in the second decade of Landcare there are some additional key issues to be addressed, including community 'burnout'. Other key issues include: (i) monitoring and evaluation to quantify (any) environmental changes; (ii) prioritizing and targeting expenditure; (iii) financial accountability (at government and community levels); and (iv) cost–benefit analyses.

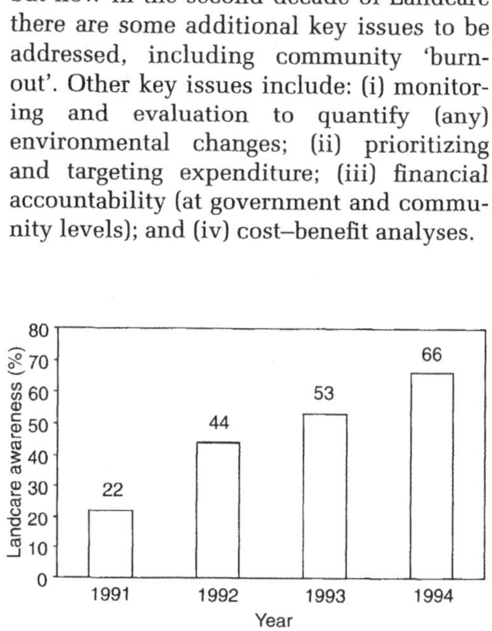

Fig. 18.3. Percentage of Australians aware of Landcare (source: Agriculture, Fisheries and Forestry, 2000).

Restoration projects must be sensibly targeted so that limited funds bring maximum benefit. Part of the marketing for such projects must involve demonstration of the commercial benefits of sustainable land management. This requires a multidisciplinary approach of integrating soil, water and nature conservation principles with farm production.

Facilitating change

The 'carrot' versus the 'stick'

There is no debate that current efforts towards natural resource management are significant across the whole community. Uncertainty does, however, become evident when considering whether improvements are being adopted with the swiftness that is needed. The scale of some issues and speed at which they are expanding is daunting. It is not time to try some remedial options, but time to employ them all. Half-hearted efforts will only delay our arrival at an unenviable destination. The range of 'solutions' discussed in this chapter is broad. Solutions can, however, be simply categorized as either 'carrot' or 'stick' approaches, or a combination.

The Landcare model employs the 'carrot' approach to implementing change. That is, it aims to motivate people to adopt best management practices through understanding and incentive. This method may be as simple as providing good advice, or it may require financial incentives. Either way, positive change is adopted by people with their own free will.

The 'stick' approach relies on laws to prevent people from adopting unacceptable land or water management practices. If undesirable activities do still occur, then the stick can be used to enforce remedial action as well as penalties. While the stick is often not seen as the best ethos for natural resource management, it does have an important role. Even when an environment of community partnership is being fostered, it should still remain as a valid tool for environmental protection.

Dynamic environmental management

Assessing the state of various natural resources is essential to understanding whether the chosen methods of natural resource management are working. This information can later be reassessed and an indication of trends determined. A cycle of assessment and action, linked by research and a clear understanding of what actions are required (Fig. 18.4), will allow informed judgement on the success of any model of natural resource management that is adopted. Such a cycle accepts the need for adaptation in efforts as new information comes to light. It also highlights where significant resources are required.

The components within this cycle have clear links and if any is removed, the process will not continue. Assessment, information and action are equally important in understanding and facilitating change. Assessment is purely academic if it is not linked to on-ground action and actions are likely to be misdirected if not supported by current information. In the Murrumbidgee River catchment, the importance of all these components is understood and this model of dynamic environmental management is adopted.

Conclusions

Improvements

There is ample space for improvement in the way natural resource management in the Murrumbidgee River catchment is conducted. Genuine community partnerships and catchment management are relatively new concepts, even in the areas in which they have been adopted. Mistakes are made, faults are highlighted and lessons are learned.

From Murrumbidgee experiences, some brief ideas for improved integrated catchment management are given below.

- Red tape should be cut to facilitate more on-ground improvements.
- Current reforms should not be considered in isolation, but considered in the context of the whole landscape (water, vegetation and soil).

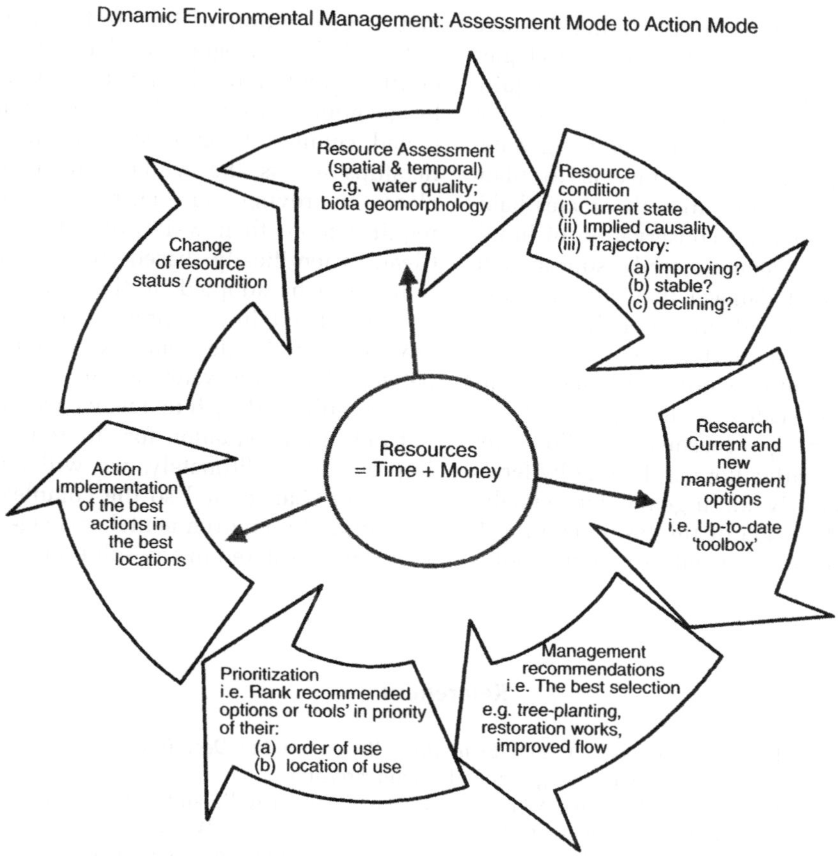

Fig. 18.4. The cycle of assessment and action.

- All players (not just the major ones) should be involved in decision making.
- Research should be directed to plug major knowledge gaps.
- Social, economic and environmental factors should be evaluated in considering optimum solutions.
- Natural resource management should be based on government and user partnerships with a realistic amount of investment.
- Marketing and awareness of issues and solutions should be improved.

General conclusion

There are numerous gaps in our knowledge. Ultimately, it may not be the lack of information, but our inability to absorb that which exists, that creates the greatest obstacle to environmental problem solving. This inability to learn from past mistakes and current research may be a critical barrier to the prevention of natural resource degradation.

The Murrumbidgee River catchment, like most Australian catchments, has numerous serious issues of natural resource degradation to contend with. With the strong dependence that the Australian economy and society still have on the rural sector, it is unfortunately unlikely that scales of the current and impending problems have been overestimated. The strong ties that these environmental issues have with social and economic health are direct. For this reason they are taken into consideration in the management of the Murrumbidgee River catchment.

One consolation in attempting to solve so many problems is that many strategies for improving environmental does require similar and complementary efforts. For example, a key method for reducing water tables and therefore salinity is to plant trees. Such an action directly creates habitat and benefits the conservation of biodiversity. Being able to apply solutions to environmental damage that address more than one issue will undoubtedly help the community to reach its goals.

With lists of rare and threatened species growing, salt scalds extending across fertile lands, reversing of seasonal river flows and eviction of native flora and fauna by ferals, there is clearly an urgency for remedial actions. The scale of the tasks required to improve on the current situation is enormous. Recognition of this fact is closely followed by recognition that remedial efforts cannot, and will not, be conducted by governments alone. The need for integrated community and government partnerships across the whole catchment is clear. Improving our current situation requires more than well-directed science. Clear, scientifically based recommendations must be adopted by the community and translated into improvements in our environment. While complex interactions between land and water resources are further complicated by humans, we cannot be left out of any equation for environmental improvement. Ultimately, we will be the agents of change, and so the complexities we bring to environmental management must be considered in any solutions.

References

Agriculture, Fisheries and Forestry (2000) *Evaluation Report on the Decade of Landcare Plan – National Overview.* http://landcare.gov.au/agfor/landcare/pub

ARMCANZ (Agriculture and Resource Management Council of Australia and New Zealand) (1996) *National Water Quality Management Strategy – Australian Drinking Water Guidelines.* National Health and Medical Research Council and Agriculture and Resource Management Council of Australia and New Zealand.

Bek, P. (1985) *Management Options for the Control of Algal Blooms.* New South Wales Department of Water Resources.

Buchan, A. (1995) *Murrumbidgee Catchment State of the Rivers Report 1994–1995.* New South Wales Department of Land and Water Conservation (DLWC).

Crabb, P. (1997) *Murray–Darling Basin Resources.* Murray–Darling Basin Commission.

Curtis, A. (1999) Landcare – beyond onground work. *Natural Resource Management* September 1999.

DLWC (Department of Land and Water Conservation) (1998a) *Water Sharing the Way Forward – Overview of the New South Wales Water Initiatives.* New South Wales Department of Land and Water Conservation.

DLWC (Department of Land and Water Conservation) (1998b) *Rural Production and Native Vegetation Conservation – a Partnership for the Future.* New South Wales Department of Land and Water Conservation.

DLWC (Department of Land and Water Conservation) (1999) *A White Paper – a Proposal for Up-Dated and Consolidated Water Management Legislation in New South Wales.* New South Wales Department of Land and Water Conservation.

DLWC (Department of Land and Water Conservation) (2000) *New South Wales Salinity Strategy – Taking on the Challenge.* New South Wales Department of Land and Water Conservation.

Environment Australia (2000) *National Strategy for the Conservation of Australia's Biological Diversity.* http://www.environment.gov.au/portfolio/esd/biodiv/strategey

Flannery, T. (1998) *The Explorers.* The Text Publishing Company, Melbourne.

Gunn, A. (1954) *We of the Never-Never.* Robertson and Mullens, Melbourne.

Jeffcoat, K. (1995) *Major Rural Dams in New South Wales.* New South Wales Department of Land and Water Conservation.

Lamp, C. and Collet, F. (1989) *A Field Guide to Weeds in Australia.* Inkarta Press, Sydney.

MCMC (Murrumbidgee Catchment Management Committee) (1994) *Proceedings of the Forum on European Carp, Wagga Wagga, New South Wales, Australia, 20 June 1994.* Murrumbidgee Catchment Management Committee.

MCMC (Murrumbidgee Catchment Management Committee) (1998) *Murrumbidgee Catchment Action Plan – for Integrated Natural Resources Management.* Murrumbidgee Catchment Management Committee.

Murray–Darling Basin Ministerial Council (1999) *The Salinity Audit of the Murray–Darling Basin – a 100-year Perspective.* Murray–Darling Basin Commission.

Natural Heritage Trust (1999) *Annual Report 1998–99. Helping Communities Helping Australia.* Environment Australia, Canberra.

Pengelly, S. (1998) *Nutrients in the 'Bidgee – Current Knowledge and Recommended Future Directions.* New South Wales Department of Land and Water Conservation.

Pengelly, S. (1999) *The Impacts of Algal Blooms on Stock – Murrumbidgee Veterinarian Survey.* New South Wales Department of Land and Water Conservation.

SMHEA (Snowy Mountains Hydro Electric Authority) (2000) *The Snowy Mountains Scheme.* www.snowyhydro.com.au

Starr, B. (1995) *The Numeralla – River of Change.* Numeralla and District Landcare Group.

Starr, B. (1996) *Bredbo and the 'Bidgee – Management Strategies for the Murrumbidgee River and it's Tributaries in the Bredbo District.* Bredbo Community Landcare Group.

Walbrink, P.J., Olley, J.M., Murray, A.S. and Olive, L.J. (1996) *The contribution of channel banks and gully walls to total phosphorus loads in the Murrumbidgee River.* In: Rutherford, I.D. and Walker, M. (eds) *First National Conference on Stream Management, Merrijig, 19–23 February*, pp 1–6.

Walbrink, P.J., Murray, A.S. and Olley, J.M. (1998) Determining source and transit times of suspended sediment in the Murrumbidgee River, New South Wales, Australia, using fall-out ^{137}Cs and ^{210}Pb. *Water Resources Research* 34/4, 879–887.

Webster, I. and Pengelly, S. (1998) *Algae, Flow and Weirs – a Summary for the Murrumbidgee River Management Committee.* CSIRO and New South Wales Department of Land and Water Conservation.

Whittington, J., Cottingham, P., Gawne, P., Hillman, T., Thoms, M. and Walker, K. (2000) *Ecological Sustainability of Rivers of the Murray–Darling Basin – Review of the Operation of the Cap.* Cooperative Research Centre for Freshwater Ecology.

Wooldridge, A.C. (1999) Salinity – We are all in this together. In: Lantry, J.O. (ed.) *Rising Water Tables and Salinity in the Yass River Valley.* Murrumbidgee Landcare Association, Wagga Wagga, New South Wales.

19 Managing the Effects of Agriculture on Water Quality in Northern Ireland

R.H. Foy and W.C.K. O'Connor

Agricultural and Environmental Science Division, Department of Agriculture and Rural Development, Newforge Lane, Belfast BT9 5PX, UK

Key words: diffuse loadings, eutrophication, farm pollution, lakes, Lough Neagh, phosphorus, River Bush, rivers, salmon, suspended solids

Introduction

Visitors to Northern Ireland often remark on three aspects of the local countryside and, indeed, these often strike the returning resident as unusual and characteristic of Ireland. The first is the verdure of the countryside, reflecting a temperate maritime climate with abundant rainfall distributed evenly throughout the year and low frequencies of snow and prolonged frosts. A second and related factor is the small area devoted to arable cropping, which accounts for less than 5% of the farmed area leaving grassland as the major land use. Livestock rearing is the dominant farming type with beef and dairy enterprises most in evidence but there are also significant poultry and pig sectors that rely on imported foodstuffs. Although cattle utilize grass, a significant proportion of their energy requirements is obtained from imported foodstuffs. The result is that agriculture operates under conditions of nutrient surpluses, where more nutrients are imported to the system in the form of fertilizers and animal foodstuffs than are exported in the form of agricultural product (Heaney *et al.*, 2001). The third feature of the countryside is a paucity of forests and woodlands, so that forestry accounts for no more than 5% of the land area, and that area is mostly coniferous plantations planted on the uplands (> 200 m), often on blanket peat. In total, therefore, agriculture occupies most of the landscape and it is not surprising that the major water quality issues encountered in Northern Ireland arise from agricultural activities. Before describing three of these, it is necessary to outline the soils and associated hydrology of Northern Ireland.

Soils with impeded drainage predominate; gleyed soils make up 54% of the land area and a further 16% of the land area is peat (Cruickshank, 1997). A number of consequences arise from this, beyond the obvious one that most soils do not favour arable cropping. The first is that, as downward water movement through the soil profile is restricted, water tends to move from the land via overland flow or via shallow lateral flow-pathways. Rivers therefore respond rapidly to rainfall events, especially in the winter when potential evaporation is low and soils are at field capacity. Flow duration curves for Northern Ireland rivers show high ratios of peak flows to base flows, suggesting a small contribution to river flow from groundwater transfers (Wilcock, 1997). Second, farmers have traditionally attempted to improve drainage by installing field drains, and fields with mole and/or tiled drains may be considered the normal condition in lowland areas (Wilcock, 1979). As rivers respond rapidly to rainfall and overtop their banks, the tendency has been to regard them as conduits for drainage water and river management has been towards reducing the flooding of farmland. By the mid-1980s, arterial river drainage schemes had impacted on most major waterways in Northern Ireland and on many minor watercourses, often to the extreme detriment of the associated riparian and river habitats (Wilcock, 1997; Sear *et al.*, 2000).

The last major drainage scheme in Northern Ireland was on the River Blackwater in the 1980s. The impact of this scheme on the river and associated wetlands was considered by environmentalists to be almost an exemplar of what should not be done, although all of the features of this scheme have been replicated elsewhere in the British Isles (Purseglove, 1988; Sear *et al.*, 2000). Arterial drainage generally involves lowering the river bed and removing bends, with associated losses of pools, riffles and overhanging riparian vegetation. Constructed berms provide further protection against flooding. The new river channel has diminished ability to sort gravels to provide a diverse habitat resource for aquatic invertebrates (Brookes, 1988). Because the river is confined within an enlarged and straightened channel, water velocities will be higher under flood conditions than were experienced under pre-drainage conditions and may be beyond the tolerance of some aquatic life forms. Sediment, formerly deposited over the flood plain, is constrained within a narrower channel area. Bank instability is often encountered post-drainage, leading to further engineering solutions where the bank is armoured to protect it against erosion. The converse is the case under low flows, which, because they occur in a wider channel, result in lower water velocities and reduced water depths. This potentially favours the colonization of the river bed by aquatic macrophytes and may adversely impact the habitat types required for juvenile salmonids (Elliot *et al.*, 1999). The importance of the riparian zone as a habitat worth preserving and, if possible, restoring is now recognized, and the River Habitat Quality Survey documented the physical characteristics of rivers and streams in the UK (Raven *et al.*, 1998). It is a measure of the historic importance of river drainage, that river channels in Northern Ireland score a higher proportion exhibiting heavy and extensive modification than England, Scotland or Wales (Sear *et al.*, 2000).

This chapter examines three specific issues that are affecting water resources in Northern Ireland: (i) point source farm pollution and the associated damage to tributary or headwater streams; (ii) diffuse P losses and associated eutrophication; and (iii) sedimentation of spawning redds of the Atlantic salmon (*Salmo salar* L.). Trends in their impact on water quality are assessed, as is the degree to which they are interacting with each other.

Study areas

Farm pollution studies were undertaken from 1990 to 1998 on small tributaries of two of the Colebrooke and the Upper Bann rivers (Foy and Kirk, 1995; Foy *et al.*, 2001). These rivers, shown in Fig. 19.1, flow into Lough Erne and Lough Neagh, respectively. Phosphorus exports from the Upper Bann and the other five major rivers that drain into Lough Neagh have been determined since October 1974 and have been used to assess the changing impact of diffuse P loading to Lough Neagh (Heaney *et al.*, 2001). The final river system studied is the River Bush, which drains into the Atlantic Ocean (Fig. 19.1). A fish trap, installed close to the seaward end of the river in the early 1970s, has allowed a census to be taken of adult salmon migrating upstream and juvenile salmon smolts descending to the sea. From these data, the success of salmon reproduction in fresh water can be determined (Kennedy and Crozier, 1998).

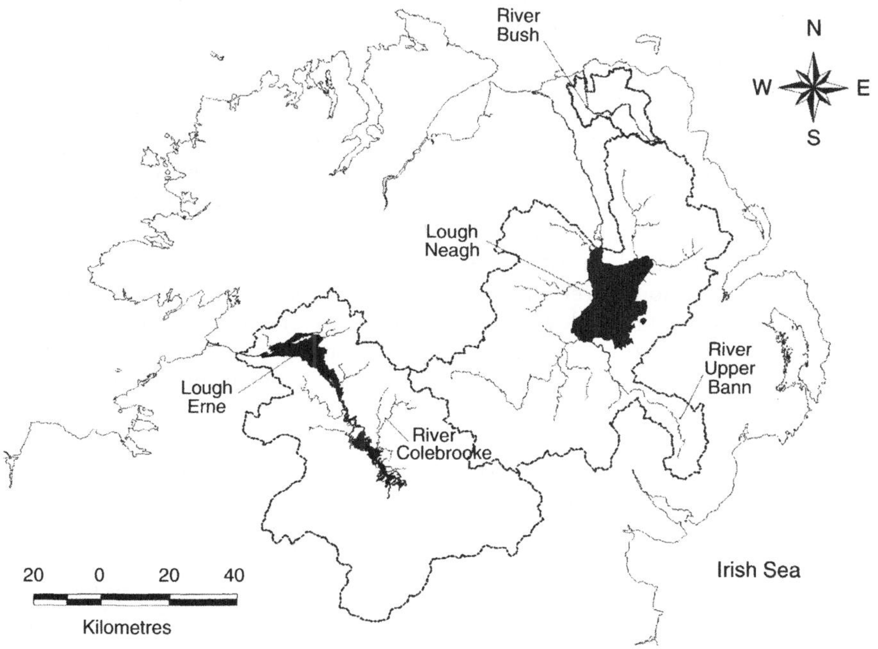

Fig. 19.1. Location map showing the Colebrooke, Upper Bann, Lough Neagh and River Bush catchments. The six major rivers of the Neagh system shown on the map are monitored at points close to where they enter Lough Neagh.

Water quality assessment

Depending on the water resource under consideration, water quality may be determined by a variety of means. For rivers in Northern Ireland, fisheries protection has been an important goal and water quality assessments combine a chemically based definition of suitability for salmonid and cyprinid fish species with a biological determination based on the sensitivity of aquatic macroinvertebrates to organic pollution. The chemical quality assessment scheme follows the criteria of the Environment Agency of England and Wales, which defines six classes (Foy and Kirk, 1995). In the farm pollution study, these have been reduced to three broader categories based on their suitability for salmonid and cyprinid fish species, with a 'no fish' class representing streams that were degraded to such a degree that self-sustaining fish populations are unlikely (Table 19.1). The boundary values that define these classes are the 90th percentile values of DO percentage saturation (DO %sat), NH_4^+ and BOD concentrations; the default water quality class is the worst quality class for these three variables (Table 19.1). Methods employed for determining DO %sat, NH_4^+ or BOD are given in Foy *et al.* (2001)

The biological assessment used in the farm pollution study is that described by the UK Biological Monitoring Working Party (Chesters, 1980). Based on their sensitivity to organic pollution, invertebrates, identified to family or taxon level, are scored on a scale of 1 (tolerant) to 10 (highly sensitive) (Armitage *et al.*, 1983). In this chapter, the composite annual average score per taxon (ASPT) is used, which is the sum of the family-taxon scores recorded at a site over a year (three samples per year) divided by the number of taxa recorded (Armitage *et al.*, 1983; Foy *et al.* 2001).

Because it is a driver of eutrophication, P is of great interest in lakes and special emphasis has been placed on relating changes observed in the P status of Lough Neagh to catchment P inputs (Gibson, 1998; Gibson *et al.*, 2001, and references therein). Lough Neagh annual TP data presented here are based on composite samples (0–10 m depth) obtained from a mid-point location in the Lough, which was sampled at a weekly frequency from 1974 to 1990 and every 14 days from 1991 to 1999. Phosphorus inputs to Lough Neagh are based on a weekly sampling programme of the six major inflowing rivers shown in Fig. 19.1. Samples were analysed for P according to the methodology given by Gibson and Stevens (1979). Unfiltered samples are analysed for TP by persulphate digestion. Filtered (0.45 μm) samples are analysed, either after persulphate digestion to give total soluble P (TSP), or without digestion to give SRP. Particulate P is calculated by difference (TP–TSP) as is soluble unreactive P (SUP = TSP–SRP). Annual estimates of P loadings are derived from P concentration versus flow relationships for each river (for summary of the methodology see Lennox *et al.*, 1997). Annual TP and SRP point source loads from sewage treatment works in the Lough Neagh catchment were calculated as described by Foy *et al.* (1995) and Heaney *et al.* (2001).

Table 19.1. Percentile concentrations employed in the determination of chemical water quality classes (DO %sat = dissolved oxygen percentage saturation).

	Fishery status and water quality class					
	Salmonid		Cyprinid		No fish	
	1	2	3	4	5	6
DO %sat (%)	> 80	> 70	> 60	> 50	> 20	< 20
NH_4^+ (mg N l^{-1})	< 0.25	< 0.6	< 1.3	< 2.5	< 9.0	> 9.0
BOD (mg O l^{-1})	< 2.5	< 4.0	< 6.0	< 8.0	< 15.0	> 15.0

Point Source Farm Pollution

Farm pollution as a water quality issue

Degradation of the aquatic resource in Northern Ireland can be strikingly evident from farmyard discharges of manures and silage effluent. The rapid adoption of silage making from the 1970s was accompanied by an increase in silage-related pollution, while discharges of manures and silage effluent have historically formed the largest category of recorded water pollution in Northern Ireland (Foy *et al.*, 1994; Lennox *et al.*, 1998). Government response has been to: (i) provide cost-sharing grants to farmers for new or upgraded on-farm facilities, (ii) provide free pollution control advice; (iii) promote codes of good practice for minimizing pollution; and (iv) introduce legislative measures which increased maximum fines for causing pollution. Assessing the effectiveness of these initiatives, or even the impact of farm pollution on water quality, is problematic. Pollution statistics may be unreliable indicators of changes in farming impact if they are influenced by variable resources devoted to pollution detection and, as many recorded incidents originate from the general public, changes in public awareness. Furthermore, statutory water quality monitoring networks may not be sensitive to farm pollution trends as monitoring sites are located on the larger rivers and so reflect many pollutant sources and, in particular, the performance of sewage treatment works (Royal Commission, 1992). This contrasts with point source farm pollution, which is most apparent in small tributaries that are intimately connected with farmsteads.

Upper Bann and Colebrooke pollution study area

In response to these concerns, both chemical and biological water quality were determined in rural tributary streams in the Upper Bann and Colebrooke catchments during 1990. From each catchment, 21 streams were selected to include all lowland (< 150 m) tributaries marked on the 1:50,000 scale maps. Details of site geolocation, land use and animal stocking rates or the stream catchments are given in Foy and Kirk (1995) and McGuckin *et al.* (1999). A relational database, linked to geo-location, of farm pollution incidents was also established from which trends in pollution types and causes could be determined (Lennox *et al.*, 1998). It was found that less than half of streams merited a 'salmonid' water quality assessment and 26% were so polluted that they were unlikely to support fish life (Table 19.2). With respect to the entire study area, this value inflates the observed impact, as polluted streams

Table 19.2. Chemical classifications of streams in farm pollution study area in 1990, and from 1996 to 1998. Catchment changes in water quality from 1996 to 1998 assessed relative to 1990 values using the Wilcoxon signed ranks test (Snedecor and Cochran, 1967).

			Fishery status (numbers of streams)					
		Change from	Salmonid		Cyprinid		No fish	
Year	River	1990[a]	1	2	3	4	5	6
1990	Colebrooke		3	10	6	0	2	0
	Upper Bann		1	3	4	4	6	3
1990	Combined		4	13	10	4	8	3
1996[b]	Combined	NS	0	15	10	5	11	1
1997	Combined	*	2	15	17	4	4	0
1998	Combined	*	1	25	8	4	3	1

[a]Statistical significance of difference from 1990 baseline study assessed using Wilcoxon signed ranks test: NS = not significant at $P < 0.05$, *significant at $P < 0.05$ and ** $P < 0.01$.
[b]Twelve months from March 1996 to February 1997.

tended to be smaller than average. However, when weighted by stream drainage areas, streams with a 'no fish' classification drained 15.8% of the study area compared with only 1.3% of the main river channel length in Northern Ireland of a similar poor quality. This level of rural water pollution is not unprecedented as, of 49 tributary and headwater streams surveyed in England, over half showed signs of environmental degradation (Furse *et al.*, 1993). Similarly, point source pollution was severe in small streams in a dairy farming catchment in Wales (Schofield *et al.*, 1990). Tributary streams are generally too small to be of direct interest for angling, but they do provide important habitats for juvenile fish, particularly migratory salmonids. In addition, the study of Furse *et al.* (1993) demonstrated that headwater streams had a potentially high conservation status due to their invertebrate fauna.

Recorded pollution statistics as an index of farm pollution

Recorded farm pollution in Northern Ireland declined by 35% between 1987 and 1997 but this decline was almost entirely attributable to a reduction in silage pollution (Fig. 19.2). From 1990 onwards, when water quality was measured in the Colebrooke and Upper Bann tributary streams, pollution numbers in the Colebrooke + Upper Bann study area and in Northern Ireland as a whole remained relatively unchanged. Despite the extent of poor water quality in the study area, incidents recorded there in any one year were low, averaging 19 incidents year^{-1}, suggesting that a substantial under-recording of pollution incidents occurred. None the less the number of incidents in the study area was consistent with Northern Ireland as a whole: the study area contained 5.0% of the land area and 5.3% of recorded incidents between 1987 and 1997 (Foy *et al.*, 2001). A breakdown of pollution types found that, not only had silage pollution declined, but numbers of incidents attributed to manures had also decreased. This decrease did not translate into a reduction in non-silage related pollution because less polluting farm pollution categories, such as yard runoff and waste from dairies, became more prevalent (Lennox *et al.*, 1998). The increasing prevalence of these pollution categories suggests that a shift in detection had taken place; pollution totals were more likely to encompass less polluting discharges. This shift had, in turn, masked a decline in the number of incidents caused by the most polluting farm effluents.

Chemical water quality trends

Can the above conclusion be substantiated from the water quality monitoring results of the Upper Bann and Colebrooke streams? Pollution was substantially a summer phenomenon in 1990 and monitoring from 1991 to 1995 was confined to the summer, year-round monitoring not resuming until 1996. The chemical water quality classification of 1996 was marginally worse than in 1990, suggesting that a decrease in farm pollution had not taken place, but in 1997 and 1998 water quality improved (Table 19.2). A detailed breakdown of the summer (May–September) monitoring data provided evidence that

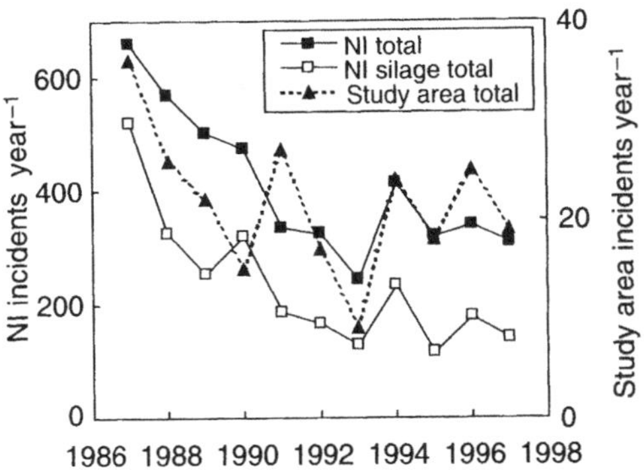

Fig. 19.2. Total farm pollution and silage pollution incidents recorded in Northern Ireland (NI) from 1987 to 1997. Also shown are farm pollution incidents recorded in the study area over the same period. In 1994 the maximum fines for causing pollution increased tenfold.

improvements evident in 1997 and 1998 were not the result of a major reduction in farm pollution in these years, but reflected the cumulative improvements which were evident from 1990 (Foy *et al.*, 2001). This is most clearly seen from the improving trends from 1990 to 1998 in percentages of samples with the high DO (> 70% saturation) levels required for a salmonid classification (Fig. 19.3). Streams are categorized according to their water quality status in 1990 as salmonid, cyprinid or no-fish (Fig. 19.3). Although failing to attain the high DO status of the salmonid streams, which maintained their high status throughout the study, steady improvements are evident in the O_2 status of samples recorded from the cyprinid and no-fish streams.

Why these improvements were not reflected in improved water quality assessments for 1996, in part reflects the ability of a relatively small number of poor quality samples to determine the 90th percentile concentrations. A second feature of the 1996 monitoring was that the percentage of samples which failed BOD salmonid (< 4 mg O l^{-1}) status was higher than in 1990, in contrast to improvements in both DO and NH_4^+ (Table 19.3). Even in 1997 and 1998, the numbers of samples failing the salmonid BOD standard were not lower than in 1990. In general, BOD behaved in an anomalous fashion to DO and NH_4^+. Low DO levels and polluting concentrations of NH_4^+ occurred in combination with high BOD concentrations, with visible pollution evident from sewage fungus colonizing the stream-bed. For BOD, the converse was not the case as, during summer freshets, elevated BOD concentrations could occur despite high DO and low NH_4^+ concentrations. A likely source of these elevated BODs was field runoff rather than polluting point discharges from farms. Peaks in BOD and NH_4^+ are evident in flood events immediately after the application of manures to grassland but tend not to persist (Kiely, 1981; Sherwood and Fanning, 1981). That elevated BOD levels during floods were not accompanied by increases in NH_4^+ suggests that they were not a residual effect of manure applications to fields.

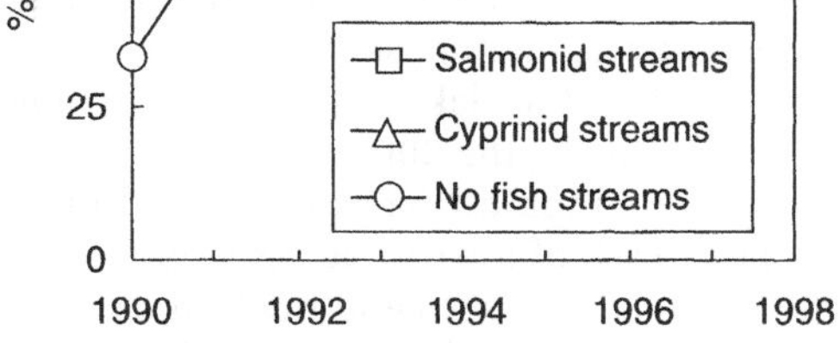

Fig. 19.3. Trends in the percentages of samples attaining dissolved oxygen > 70% saturation during May–September 1990 to 1998. Streams are grouped according to their water quality status in 1990 (from Foy *et al.*, 2001).

Biological water quality trends

Improvements in the chemical indices of water quality in 1997 and 1998 suggest that farm pollution had been reduced, but the biological dimension of water quality failed to improve at all (Fig. 19.4). Time plots of mean ASPT values for the salmonid, cyprinid and no fish streams demonstrate a remarkable stability with no evidence that they responded to the improved overall water quality determined in 1997 and 1998. This may simply reflect the limited time-span available for biological recovery and re-colonization during 1997 and 1998. As noted above, the most polluted streams tended to be small and, as pollution was extensive, they contained limited refuge areas from which pollution intolerant species could re-colonize the stream. Alternatively, the failure of biological quality to improve may, fundamentally, reflect the nature of farm pollution and the nature of the chemically derived water quality index classification which permits 10% of samples to 'fail', while still maintaining a good water quality index value.

Table 19.3. Percentages of water samples failing the standards for salmonid waters as defined in Table 19.1. Differences assessed by χ^2 test with respect to 1990 results.

		1990	1996[a]	1997	1998
Salmonid fails	DO %sat (< 70% sat)	14.3	6.5***	6.1***	3.5***
	NH_4^+ (> 0.6 mg N l^{-1})	10.6	8.0*	4.6***	4.2***
	BOD (> 4 mg O l^{-1})	11.5	17.0***	11.1^{NS}	10.9^{NS}

[a] Twelve months from March 1996 to February 1997.
NS = not significant at $P < 0.05$, *significant at $P < 0.05$ and ***$P < 0.01$.

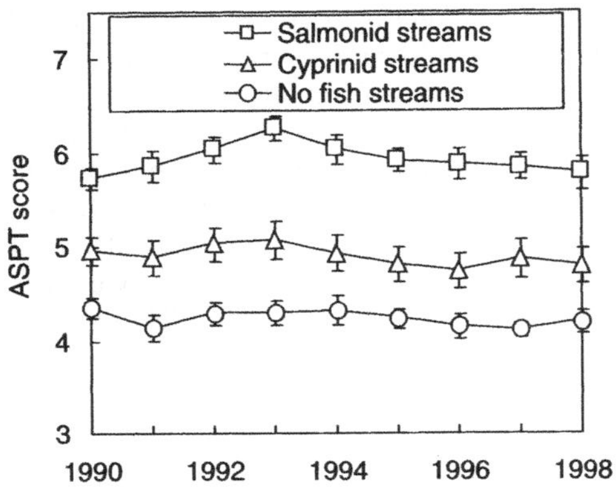

Fig. 19.4. Trends in stream average score per taxon (ASPT) values from 1990 to 1998. Streams are grouped according to their water quality status in 1990. Error bars denote standard errors (from Foy *et al.*, 2001).

Farm effluent discharges occur intermittently and pollution events evident from one sample often did not persist to the subsequent sample, an interval of 14 days; while streams could be visibly polluted when sampled for invertebrates, but not on either the preceding or subsequent routine chemical monitoring dates. Therefore, not only is there a high likelihood that intermittent pollution from agriculture may go undetected, but this is exacerbated by the highly polluting nature of even the lower impact farm wastes such as yard runoff. BOD concentrations in excess of 1000 mg O l^{-1} have been routinely determined for these 'dirty water' effluents (Bloxham, 1999; Cumby *et al.*, 1999). Although lower than recorded for manures and silage liquor, they are over 200 times higher than the required standard for salmonid waters. Short-lived farm pollution incidents therefore have the capacity to severely perturb stream ecosystems but, even if detected by a monitoring programme, may fail to breach the annual 90th percentile limits used to determine the annual water quality of the impacted stream. A conclusion from the study is that, despite progressive reductions in farm pollution, the final indicator to improve will be the biotic index score, as the persistence of a few polluting farms will be sufficient to prevent the recovery of stream biology.

Diffuse Phosphorus Loadings

Lough Neagh

The catchment of Lough Neagh shown in Fig. 19.1 drains 4450 km^2. The lake is shallow (mean depth 8.9 m), with a water turnover time of 1.25 years and is the largest potable water source in Northern Ireland. It is a multipurpose resource supporting the largest eel fishery in Europe and is an internationally important refuge for migratory birds. During the late 1960s, the eutrophic state of the lake first became apparent and this stimulated extensive research on the lake and associated catchment processes, most notably on nutrient exports (Wood, 1999). Palaeo-limnological investigations in the early 1970s demonstrated that the lake had only become enriched during the 20th century. By inference the eutrophication process was linked to the growth of towns, the development of urban sewage systems and the introduction of P-containing household detergents (Battarbee and Carter, 1993). This interpretation appeared to be confirmed by a strong correlation between the inflowing river SRP load and catchment urban population density (Smith, 1977). A P budget for the late 1970s attributed 55% of the SRP loading to point sources and, in order to reduce lake P inputs, a catchment programme to reduce the P loading of major sewage treatment works (STWs) commenced in 1981 (Foy *et al.*, 1982, 1996). Much of the budgetary work has focused on the SRP fraction of the loadings to Lough Neagh, as it is highly available for algal growth, in contrast to either the SUP or PP fractions. The influence of the PP faction on lake TP concentrations appears to be small and it is thought to be rapidly lost to the sediments (Gibson *et al.*, 1988, 2001).

Time series of diffuse phosphorus loadings

Reducing point source P inputs has not brought about the expected long-term recovery and lake TP concentrations have steadily increased since 1988 (Fig. 19.5). As sediment P has not shown any marked trend since 1974, changes in the release from sediment are judged to be an unlikely source of this increase (Gibson *et al.*, 2001). Rather, the increase in lake TP can be related to a clear upward trend in diffuse SRP inputs from 1974 (Fig. 19.6). Diffuse SRP loadings are highest in the winter when river flows are greatest and it is during the winter that a positive trend in SRP river concentrations has occurred (Smith *et al.*, 1994; Foy *et al.*, 1995). The positive effect of discharge on diffuse SRP losses is evident from Fig. 19.6. Although annual loadings of PP and SUP are also highly flow dependent, increases in the loading of these P fractions have not been noted (Foy and Bailey-Watts, 1998; Foy and Lennox, 2000). The SRP loading trend line in Fig. 19.6 is obtained from a multiple regression of SRP loading versus flow and time (Heaney *et al.*, 2001). The trend line regression slope of 6.25 t P $year^{-1}$, with 95% confidence limits of $\pm$ 1.20 t P $year^{-1}$, represents the rate at which annual loadings increased from 1974 to 1997 and equates to an area loss rate increase of 0.016 $\pm$ 0.003 kg P ha^{-1} $year^{-1}$. These values refer to the drainage area (3894 km^2) of the six monitored rivers, which have an average annual flow of 2.471×10^9 m^3 $year^{-1}$. Dividing the annual loading increase of 6.25 t P $year^{-1}$ by this mean flow, gives an estimate of the underlying increase in river SRP concentrations of 2.5 μg P l^{-1} $year^{-1}$. Over a year, this is a trivial increase but, over the river monitoring period of 23 years, the accumulated increase is a limnologically large concentration of 58 μg P l^{-1}.

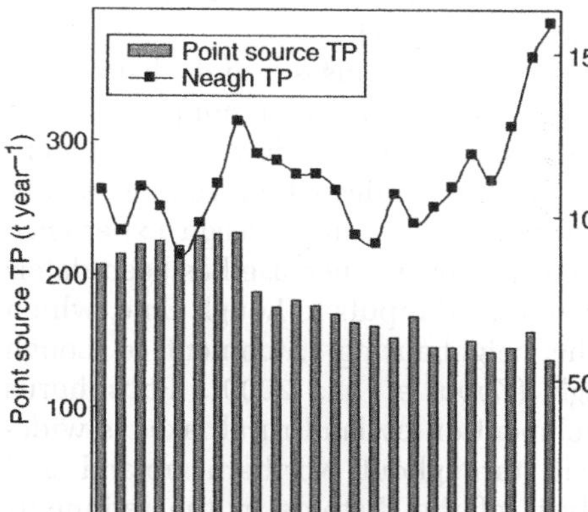

Fig. 19.5. Time series of annual Lough Neagh total phosphorus (TP) and lake inputs of TP from sewage treatment works and industrial point sources (from Heaney *et al.*, 2001).

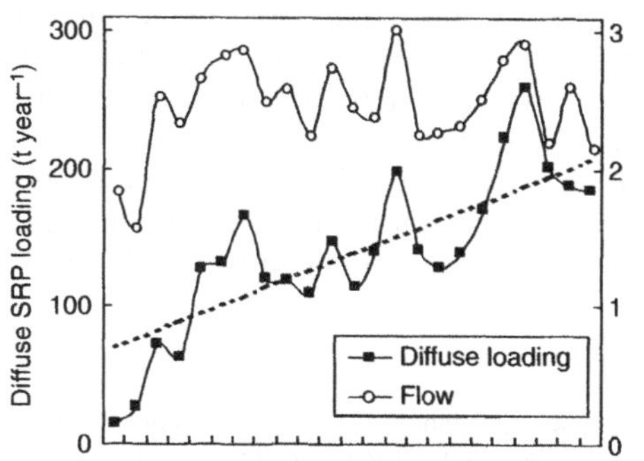

Fig. 19.6. Diffuse soluble reactive phosphorus (SRP) loadings and river flows from the six major rivers entering Lough Neagh. Diffuse SRP loads are calculated as the difference between the river SRP loads and load to rivers from sewage treatment works and industrial point sources. The linear trend line equates to the trend line for diffuse loads under average (1974–1997) flow conditions.

Phosphorus use in agriculture

Agriculture in Northern Ireland operates a P surplus. More P enters than leaves in products, with an average surplus of *c.* 12,700 t P year^{-1} (Heaney *et al.*, 2001). For the farmed area, excluding upland rough grazing, the average P surplus is equivalent to 15 kg P ha^{-1} year^{-1} and this surplus has remained approximately constant since 1950. As P accumulates in the soil, it has resulted in a reduction of the proportion of agronomically deficient soils encountered in Northern Ireland (Table 19.4). Although diffuse P loss rates to the Lough Neagh rivers have increased by an environmentally large amount, the increase is small in proportion to the amount of P that has accumulated in the soil. Over the 22 years of river monitoring for P, the soils in the Lough Neagh catchment accumulated in the region of 90,000 t of P compared with a cumulative increase in the annual loss rate of SRP from diffuse sources of just 143 t P.

Linking diffuse phosphorus losses to agriculture

The above analysis should not be taken to mean that soil P is the dominant or only source for soluble P loads to Lough Neagh, but it does appear that soil P is an important driver of change. A number of studies have shown that increasing soil P is accompanied by increasing loss. Thus, drainage P from experimental plots in England presented a non-linear relationship, with negligible drainage P evident until soil Olsen P exceeded 60 mg P kg^{-1} (Heckrath *et al.*, 1995). Recent results from the same research group suggest that the change or break-point between drainage and soil P is variable and can be substantially lower than 60 mg P kg^{-1} (Hesketh and Brookes, 2000). Overland flow P concentrations from grassland plots in the USA provide less evidence for a break-point in the soil P vs. P loss relationship, but nevertheless showed strong positive correlations with soil P (Pote *et al.*, 1999). McDowel *et al.* (Chapter 9, this volume) provide a discussion on the nature of the soil P vs. P loss relationship. A strong positive relationship has been observed between soil test P and the water export of P from grassland fields in the Republic of Ireland (Tunney *et al.*, 2000). In these studies, SRP was the principal P fraction that increased in drainage or overland flow water, rather than PP or SUP. From a monitored farm drain in Northern Ireland, where grassland management has remained constant, drainage SRP increased at a rate of 1 μg P l^{-1} year^{-1} from 1980 to 1990 (Smith *et al.*, 1995). The predominance of SRP in the computed diffuse P inputs to Lough Neagh is consistent with data from grass-grazed mini or drain catchments in Northern Ireland, which show that SRP accounts for about 50% of the flow-weighted mean TP concentrations (Jordan and Smith, 1985; Foy and Withers, 1995; Lennox *et al.*, 1997).

Although surpluses, and hence an increase in soil P, are commonplace in the British Isles, there are few similar catchment loading studies which can be used to test whether the Lough Neagh experience is widespread. An increase has been determined in SRP inputs to Lough Erne, which is the neighbouring catchment to Lough Neagh (Zhou *et al.*, 2000). Phosphorus enrichment of lowland rural lakes is widespread throughout Northern Ireland and analysis of the diatoms in the sediments from a sample of these lakes has shown that enrichment is largely a phenomenon of the previous 50 years (Anderson, 1997; Foy and Bailey-Watts, 1998). In contrast, no evidence was found for an increase in P export over a 10-year period from the

Table 19.4. Phosphorus status of soils (% of soils) in Northern Ireland in the 1940s and 1990s assessed for grass production and based on samples taken for advisory purposes.

Phosphorus status	1940s	1990s
Deficient	37.2	5
Low	35.6	16
Adequate	23.4	31
Excess	4.2	48

Ythan River in Scotland (McDonald *et al.*, 1995). Similarly, a 22-year time series of TP concentrations in Lough Ennell in the Irish midlands provided no evidence that diffuse P inputs were increasing over time, although lake TP did respond to changes in the TP loading from a STW (Foy *et al.*, 1996). In the introduction to this chapter, the high proportion of soils with impeded drainage and the extensive field drainage network in Northern Ireland were noted. Each of these factors favours flow pathways that bypass the capacity of the soil profile to fix P (Haygarth *et al.*, 2000). In this respect, it is significant that these pathways are less prevalent in the River Ythan and Lough Ennell catchments, which contain relatively free draining soils.

Manure P production has increased by less than 25% in Northern Ireland since 1974, so that the time series for manure P cannot be easily linked to the increasing diffuse P loads (Foy and Lennox, 2000; Heaney *et al.*, 2001). None the less, it is certain that manure is an important source of diffuse P to Lough Neagh, for the flow pathways that facilitate the transfer of a soluble P pool located in the surface soil profiles will also ensure losses of P from manure applications. An analysis of the seasonality of slurry applications demonstrated that winter applications are commonplace in Northern Ireland and indeed a necessity on many farms which often have limited storage capacity (Tunney *et al.*, 1998). Following manure applications, high P losses via overland flow have been demonstrated under conditions found in the British Isles (Kiely, 1981; Sherwood and Fanning, 1981; Smith *et al.*, 2001). Sherwood and Fanning (1981) found that the P signal persisted in runoff for 6 weeks following manure applications to grassland in south-east Ireland, in contrast to the BOD and N signals in runoff which lasted for less than a week. Kiely (1981) reported a similar response from another Irish site, where losses of P via overland flow were in the range of 6 to 16% of the P applied in manures. Phosphorus losses were strongly influenced by discharge volumes that followed manure application and were also related to application rate, although the relationship was not necessarily linear (Sherwood and Fanning, 1981). This is consistent with losses of P measured following winter applications of manures to arable land in England where losses increased exponentially at application rates above 50 m^3 ha^{-1} (Smith *et al.*, 2001). In these experiments, interflow was found not to be a significant source of loss.

However, many other studies have shown that contamination of drain flow can occur within 24 hours of manure applications (McAllister, 1976; O'Callaghan and Pollock, 1976; Stevens, 1980; Jordan and Smith, 1985; Williams and Nicholson, 1995; Parkes *et al.*, 1997). McAllister (1976) observed a pulse of severely contaminated water after manure applications, while similar responses to slurry applications have been found at another grassland site in Northern Ireland (Stevens, 1980; Jordan and Smith, 1985). Stevens (1980) observed that that the P signal in drain flow, evident during slurry application, persisted over subsequent flood events. Cracks or fissures in the soil have been implicated in this type of contamination and cracking of clay soils by drought can provide a pathway for direct contamination of drains. The results of Williams and Nicholson (1995) on field drain contamination caused by the disposal of dirty water via field irrigation systems indicate the importance of choosing application rates that did not initiate drain flow.

The diffuse P losses plotted in Fig. 19.6 include losses associated with discharges of manures and silage from farmyards. As these discharges have declined with time, it is unlikely they are the cause of the increasing loadings of diffuse P to Lough Neagh. Nevertheless, they must contribute to the overall P losses from agriculture. A plot of TP loss rate versus cattle stocking rates of tributary streams in the Colebrooke catchment shows a positive correlation between the two (Fig. 19.7). As stocking rate is also correlated with Olsen soil P in these catchments, a strong correlation can also be established between soil P and P loss (Foy, unpublished results). However,

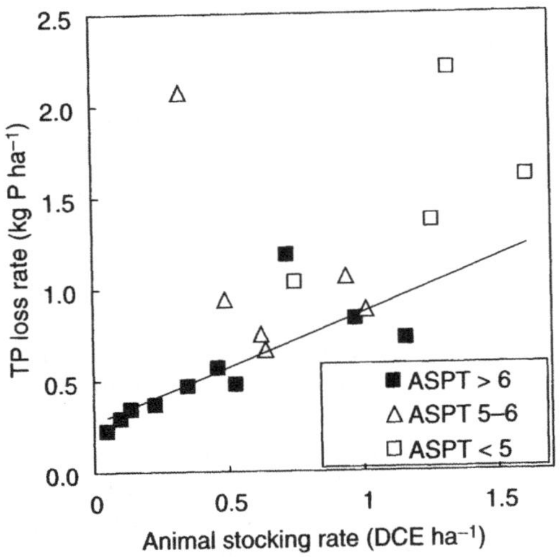

Fig. 19.7. Impact of stream average score per taxon (ASPT) values on the relationship between animal stocking rates, expressed as dairy cow equivalents (DCE) per hectare vs. total phosphorous (TP) loss rates in 1997. The regression line is drawn for TP loss rates from Colebrooke streams vs. DCE stocking rates for streams with ASPT values of > 6.0.

the highest loss rates were associated with depressed ASPT biotic index scores, which provide an indication of the impact of farmyard discharges on P loss.

River Bush Sediment Studies

River Bush

The River Bush is a well-known salmon river and the International Council for the Exploration of the Sea has designated it as an index river for salmonid research. It drains a catchment of 340 km^2 and, for most of its length, flows through a fertile valley of grassland-based agriculture with limited arable crops. Drainage of the main river channel was completed by the mid-1960s but there is a programme of regular follow-up channel maintenance. Flows are in the range of 1 to 90 m^3 s^{-1}, with a mean discharge of 5 m^3 s^{-1} (Kennedy and Crozier, 1998). The trapping facilities operating on the river since 1973 enable assessments of changes in the survival of the marine and freshwater stages of salmon. From the mid-1980s, there has been a reduction in the ratio of salmon smolts leaving the river to the number of ova in the spawning salmon, indicating a decline in survival during the juvenile–freshwater stages of the salmon life cycle (Heaney *et al.*, 2001). While a number of factors have been implicated in this decline, including cormorant predation, Kennedy and Crozier (1995, 1998) concluded that freshwater survival has been adversely impacted by a reduction in ova to fry survival. This section examines the role of sediment on ova survival and a possible interaction with nutrient enrichment.

Sediment impacts on River Bush

After returning to their native river, Atlantic salmon spawn between the months of November and February in areas of suitable gravel, termed redds, normally in fast flowing, well oxygenated riffles at the downstream ends of pools (White, 1942). Redd selection is probably influenced by a combination of water depth, velocity and intra-gravel flow (Stuart, 1953a; Beland *et al.*, 1982). It has long been recognized that high levels of fine sediment can inhibit the permeability of redds and, by restricting the supply of well-oxygenated water to the ova, prove detrimental to spawning salmonid success (Harrison, 1923; Stuart, 1953b; Shelton, 1955; Alabaster and Lloyd, 1980). This issue is also considered by Harrod and Theurer, Chapter 7 in this volume. A variety of thresholds have been proposed with regard to the sediment content and particle size classes that should be designated as fine sediment. Hausle and Coble (1976) proposed that, for brook trout (*Salvelinus fontinalis*), fry emergence was retarded in spawning substrates with > 20% of fine sediments. McNeil and Ahnell (1964) found that fry emergence of pink salmon (*Oncorhynchus gorbuscha*) was inversely related to the proportion of spawning substrate < 0.833 mm in diameter. Peterson (1978) demonstrated that gravel permeability was negatively correlated with the amount of fine sediment < 2 mm in diameter, concluding that permeability was reduced when the level of fines was in excess of 8%, and almost zero when the fines content reached 20%. Others have reported declines in salmonid embryo survival related to the abundance of fine sediments of < 2 mm in diameter (Crouse *et al.*, 1981; Petts *et al.*, 1989). Ottaway *et al.* (1981) argued that spawning substrates used by salmonids in the British Isles should be composed of < 10% of fine material < 2 mm in diameter.

A freeze coring methodology, developed by Petts *et al.* (1989) to sample spawning gravels, has been used to study the fine sediment content of spawning gravels on the River Bush (O'Connor, 1997, 1998; O'Connor and Andrew, 1998). Potentially adverse levels of sediment occurred in salmon spawning redds; a mean content of 16.6% for fine material (< 2 mm in diameter) and 67% of spawning gravels with a fine sediment content of > 10%. Incubation experiments carried out under both river and laboratory conditions demonstrated that water flow through gravel, and therefore the availability of O_2 to the deposited ova, was considerably higher in the absence of sediment pollution (O'Connor and Andrew, 1998). Mean fry survival in the control incubators, loaded with clean, homogeneous gravel, was 38%, but survival decreased to 9.3% with the addition of 10% fine material. When fines were added at a 15% level, mean fry survival was 2.7% and only 1% in replicates with 20% fines. At the highest loading of fine material (25%), fry survival was zero. Thus, although the many spawning reaches on the River Bush possess an excellent morphology for salmon spawning, with abundant shoals of gravel, the degree of sedimentation will invariably lead to gravel compaction and result in loss of spawning terrain (Gustafson-Greenwood and Moring, 1991).

Many studies emphasize the influence of land use in the sedimentation of spawning grounds. Cordone and Kelley (1961) suggested that the danger to salmonid habitat lay not so much in the occasional incident of sedimentary pollution, but in long-term degradation caused by erosion associated with land use. Many land-use activities can result in erosion of fine sediment but some, such as logging and tillage of arable soils, do not appear to be major sources in the River Bush (Meehan, 1991). Alternatively, drinking points on the river and the ability of cattle and sheep to graze up to the riverbank edge has resulted in areas of bank erosion (O'Connor, 1998). Naismith *et al.* (1996) observed a similar situation for bank erosion in the River Torridge catchment in Devon. An ongoing programme of channel maintenance work also exacerbates the amount of sediment entering the Bush system. This involves re-sectioning the riverbank by mechanical excavation and removal of riparian vegetation. It may take a year before vegetation is fully re-established on the resulting areas of exposed bank and, in the intervening period, flood events can wash sediment into the river (Lisle and Lewis, 1992). As indicated in Fig. 19.8, gravel samples extracted downstream from areas of bank disruption have a significantly higher level ($P < 0.001$) of sediment pollution than the substrates sampled from an undisturbed control area (O'Connor, 1998). However, while undoubtedly damaging, it is unclear why the channel maintenance programme has caused the precipitous decline in spawning success from only the mid-1980s as it has continued for almost 40 years, while cattle grazing pressures are not dramatically greater than in the 1970s.

Impacts of nutrients on the River Bush

A contributing factor to habitat decline in the river may be eutrophication. The ready availability of nutrients is evident from large quantities of the aquatic macrophyte *Ranunculus penicillatus* (Dumort.) Bab. var. *penicillatus* growing in many reaches of the river and, especially, on salmon spawning fords. This species has the capacity to arrest flows

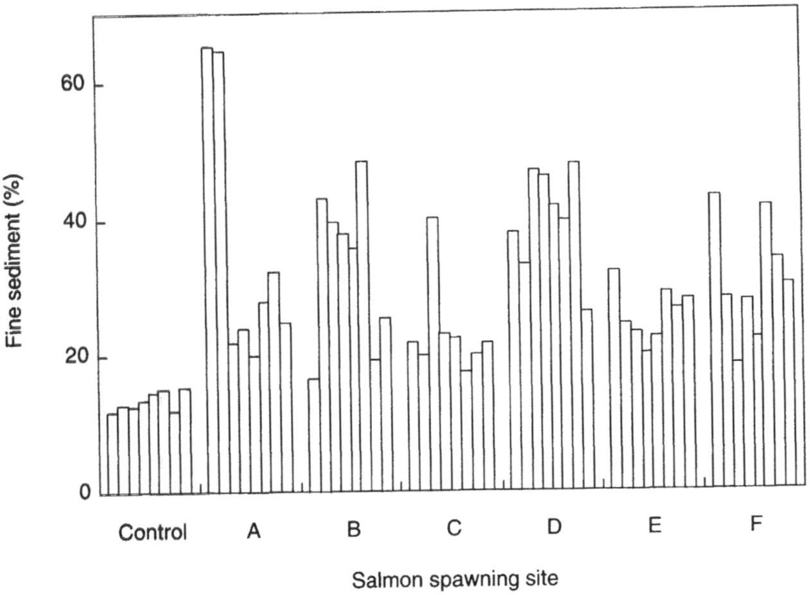

Fig. 19.8. Percentages of fine material (< 2 mm diameter) in gravel taken from a control reach of the river and at six sites downstream of bank disturbance on the River Bush (from O'Connor, 1998).

and accumulate sediment in the vicinity of the roots (Ladle and Casey, 1971). Suspended solids, chlorophyll *a*, SRP, NO_3^- and soluble silica (SiO_2) data from Livery Bridge (Irish grid reference D 023301), which is in the main salmon spawning area of the river, are presented in Table 19.5. A time series for SS and chlorophyll *a* from 1995 to 1998 shows periodic peaks of both, with an exceptionally large chlorophyll *a* concentration of 254 μg l^{-1} at the end of May 1995 which was accompanied by the maximum SS concentration of 58 mg l^{-1} (Fig. 19.9). However, if this observation is excluded, the correlation between the two variables was low at only $R^2 = 0.045$ ($N = 201$), suggesting that SS was not markedly associated with autochthonous production in the river. Even when the maximum chlorophyll *a* value is discounted, the peaks in chlorophyll *a* were surprisingly frequent for a river that is both small and subjected to regular freshets.

These observations relate to material suspended in the flowing water. The poten-

Table 19.5. Summary statistics for River Bush at Livery Bridge: suspended solids, chlorophyll *a*, soluble reactive phosporus (SRP), SiO_2 and NO_3^- concentrations from April 1994 to December 1998, rounded to nearest whole number.

	Suspended solids (mg l^{-1})	Chlorophyll *a* (μg l^{-1})	SRP (μg P l^{-1})	SiO_2 (mg SiO_2 l^{-1})	NO_3^- (mg N l^{-1})
Mean	4	8	46	13	2
Median	2	5	43	13	2
Minimum	1	1	5	3	1
Maximum	58	254	154	24	3
SD	5	17	24	4	1
Observations	208	231	237	228	230

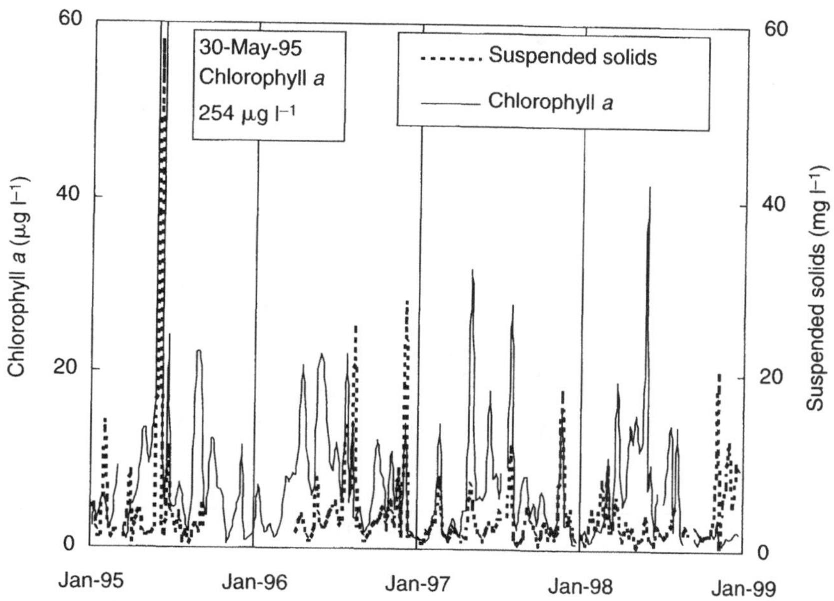

Fig. 19.9. Suspended solids and chlorophyll *a* concentrations in the River Bush at Livery Bridge 1995–1998. Note that the maximum chlorophyll *a* concentration of 254 μg l^{-1} on 30 May 1995 is off-scale (from Heaney *et al*., 2001).

tial for benthic algal production was assessed by measuring epiphytic chlorophyll *a* colonization on unglazed ceramic tiles placed in the river during 1996 (O'Connor, 1998). Epiphytic colonization and production were at a maximum during March and April with minima in both winter and July (Fig. 19.10). The winter minimum is consistent with the growth limiting effects of low irradiances and temperatures. The impact of grazers is the most probable cause of the precipitous decline in epiphytic colonization in July and August, which cannot be attributed to depletion of SRP, NO_3^- or SiO_2. The annual cycles of nutrients during 1996 show that SRP was the only nutrient to be markedly reduced by the spring epiphytic growth (Fig. 19.10). However, although the lowest concentrations occurred from March to May, they did not decline below 11 μg P l^{-1}. Over 4 years (1995–1998) of monitoring, only one SRP concentration was < 10 μg P l^{-1} and summary statistics for NO_3^- and SiO_2 demonstrate that these nutrients also did not decline to trace or growth limiting concentrations (Table 19.5).

Linking eutrophication to salmon decline

Eutrophication may therefore be a factor in sediment accumulation in the River Bush but, without the benefit of long-term data, it is difficult to place the current P status of the Bush, or of any other water quality variable currently monitored in the river, into a historical context. The absence of SRP at trace concentration (< 5 μg P l^{-1}) suggests that the Bush is now adequately supplied with P but, if SRP has increased in the river in a manner similar to the Lough Neagh rivers, this P sufficiency may only be a comparatively recent phenomenon. Using P concentrations to define trophic state is much better developed for lakes than for river systems, but experience in Ireland is that adverse impacts on salmonids occur when mean unfiltered MRP concentrations are above the range of 20–30 μg P l^{-1} (OECD, 1982; McGarrigle, 1998). As unfiltered MRP in river water exceeds SRP, the mean SRP concentration of 46 μg P l^{-1} of the River Bush substantially exceeds this criterion and points to a potentially deleterious impact of P. The

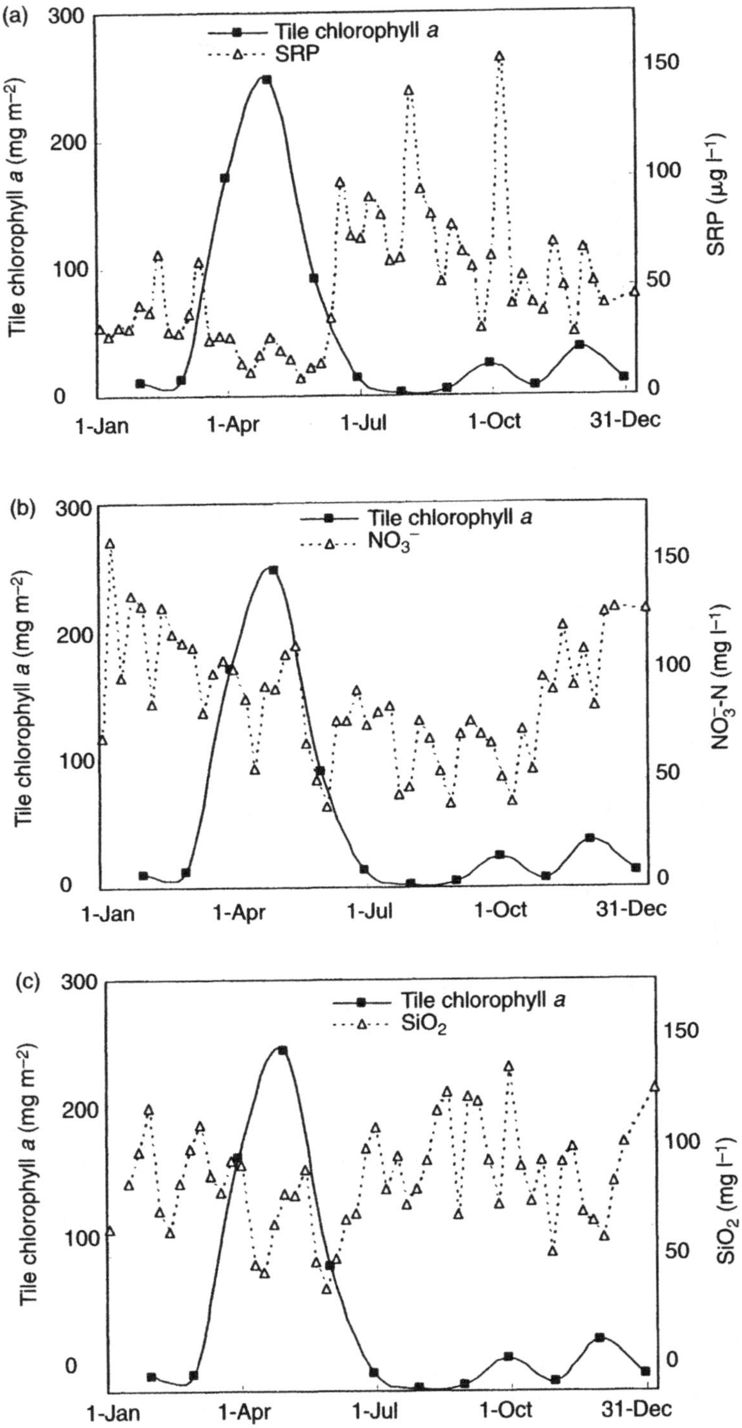

Fig. 19.10. Monthly accumulation (■) of epiphytic chlorophyll *a* on unglazed ceramic tiles in the River Bush 1996. Also plotted are weekly concentrations of: (a) soluble reactive phosphorus (SRP), (b) NO_3^--N and (c) SiO_2 at Livery Bridge on the River Bush (from Heaney *et al.*, 2001).

importance of P as a limiting nutrient is also suggested by the high N:P ratio (147:1 by atoms) for NO_3^- and SRP, which comfortably exceeds the critical Redfield atomic ratio of 15:1 and is indicative of diffuse P sources rather than point sources for P in the River Bush (Caraco and Cole, 1999).

Suspended solids in the River Bush do not seem to be excessive, as salmon are capable of tolerating much higher concentrations (> 1000 mg l^{-1}) than the maximum recorded (Alabaster and Lloyd, 1980). Moreover, SS concentrations in the River Bush are lower than in the River Test, an English river, where their impact on sediments has been observed and are comparable to those in the River Moy, a major Irish salmon river, where sediment pollution is not considered a problem (McGarrigle *et al.*, 1998; Acornley and Sear, 1999). Using floodplain profiles of radionuclides, Walling and He (1999) concluded that there was no evidence to suggest that overbank sedimentation rates had increased at 21 English and Welsh sites during the 20th century. Acornley and Sear (1999) showed that sediment deposition on to river gravels varied with season, and that a higher deposition did not necessarily require high concentrations of sediment *per se*, but flows sufficiently low to permit sedimentation to take place. As these conditions occurred after salmonid spawning, it was argued that this interaction was important in considering the impact of SS on spawning.

Similarly, Walling and Amos (1999) showed that elevated SS levels in an English trout stream with a low SS yield reflected remobilization of sediment within the stream channel and significant instream sediment re-deposition. For the River Bush, the low correlation between river SS concentrations and chlorophyll *a* suggests that algal material is not the primary source of fine sediment. Riverbank erosion by grazing animals and channel maintenance are proposed as important sediment sources and the sediment 'fingerprinting' techniques, which have been developed in recent years, may provide quantitative assessments of their importance (e.g. Walling *et al.*, 1999; Olley and Caitcheon, 2000). The mechanisms by which eutrophication affects salmonids are not well understood. Macrophyte growth will modify the river-bed and impede flows and, while this is often ecologically beneficial in creating a more diverse habitat, it may contribute to sedimentation of spawning gravels (Newall, 1995; O'Connor, 1998). For the River Bush, the possibility remains that an interaction between sediment load and eutrophication is taking place, exacerbated by the altered river channel, where stimulated primary production has altered sediment retention in the river, rather than an absolute increase in river sediment load.

Conclusions

It is evident from the above that, not only does agriculture adversely affect water quality in Northern Ireland, but it is proving difficult to reverse this. Farm pollution has proved to be relatively intractable in obtaining a resolution, and there is no indication that even the rate of increase of diffuse P losses has reduced. With respect to the decline in salmon in the River Bush, there is a considerable degree of scientific uncertainty about what is required to solve the problem or even agreement on what the problem is. Based on a brief fluvial geomorphological appraisal of the river, Newson (2000) raised the possibility that the river may be suffering from a shortage of the coarse gravel sizes required for salmon spawning redds rather than an excessive loading of fine silt. While prolific growths of *Ranunculus* were noted in the river, it was argued that armouring of the banks, which was done to prevent post-drainage bank erosion, could be effectively preventing erosion of glacial gravel deposits. These are common in the central reaches of the river and limitations on erosive activity on them were therefore preventing the renewal of salmon spawning redds. In the future, it will be necessary to link fluvial geomorphological approaches and hydrological insights to ecological investigations in the manner suggested by Petts *et al.* (1995). Methods for modifying the post-drainage physical habitat of rivers have been attempted with varying success but it remains uncertain whether these could be effective within the constraints of the existing modified river channel (O'Grady, 1991; Brookes, 1992; Kelly and Bracken, 1998). Specific techniques to restore spawning areas have been developed, which involve removing aquatic macrophytes and flushing fine sediment from spawning redds with high-powered water pumps or mechanical excavator scouring (Reeves *et al.*, 1991; Shackle *et al.*, 1999). On the River Bush, fresh gravel was added to a stretch of river following removal of macrophytes, but within 2 years there had been a marked increase in fine sediment in the clean gravel and recolonization by *R. penicillatus* (Heaney *et al.*, 2001). This suggests that, without tackling the underlying causes, such restoration methods can be only regarded as short-term, palliative solutions. If, within the framework of conservation options given by Boon (1998), a radical channel realignment of the River Bush is required, the question remains whether the river could ever return to the ecological status of the 1950s, if it is now enriched with P.

Increasing SRP loadings from diffuse sources in the Lough Neagh catchment are steadily increasing P concentrations in the Lough Neagh. There is some knowledge of the causes of the problem, if not the relative magnitude of the specific P sources. Waiting for an ever-more detailed understanding of the problem is not an attractive option, given the forthcoming European Union Water Framework Directive which has the potential to radically alter the way water quality is assessed (Pollard and Huxham, 1998). The issue of high soil P should be a comparatively straightforward one on most farms for, if existing levels of P are adequate, there is little justification for adding more P from either agronomic or environmental perspectives. Attempts to reduce the P surplus in Northern Ireland involve a variety of voluntary measures encouraging farms take into account the P available in manures within a nutrient management planning framework (Withers *et al.*, 2000). For poultry and pig production, however, even a cessation of P fertilizer use will not reverse the build up of P as, on these farms, a surplus is inevitable given the reliance on imported foodstuffs. Evidence to date shows that farms may be reluctant to reduce P use, suggesting either a lack of confidence in the plans or in their ability to utilize efficiently the P present in manures (Dils *et al.*, 1998; Beegle *et al.*, 2000). At best, many nutrient management plans will only halt the rise in soil P. For the rise in river SRP to be reversed will also require a focus on the manure P management on farms in Northern Ireland. Here, modifying existing practices by applying manures only when the potential for P transport is low is likely to lead to extra capital costs to provide extra manure storage facilities.

Although, as presented here, there are reasons to suppose that discharges of manures and silage effluent have become less prevalent, progress towards elimination of farm pollution has been painfully slow. It has an impact on overall P transfer rates from agriculture, although eliminating point source pollution may be best considered as part of the solution to reducing high P loss rates. A number of authors have commented on the failure of current regulatory measures to appreciably reduce farm pollution numbers in the UK (Lowe *et al.*, 1992; Seager *et al.*, 1992; Bloxham, 1999). In Northern Ireland, detected pollution incidents have been traced to farm facilities constructed with grant-aid for pollution control measures, while increasing fines did not reduce the numbers of incidents recorded (Northern Ireland Audit Office, 1998; Lennox *et al.*, 1998). Their survey of English dairy farmers' attitudes to pollution led Lowe and Ward (1997) to claim that existing regulatory approaches were ineffective and that a re-orientation was required by the local community as to what was perceived as right and wrong. Watson (1997) also argued that current farm advisory approaches were ineffective in the UK as they did not interact with or actively involve farmers in defining why the problem of diminished water quality occurred.

Guerin (1999), in examining why Australian landowners failed to accept technology transfer related to environmental protection, suggested that it was necessary to involve them in the technology development process. This participation of farmers, representing a 'bottom-up' approach to environmental management, is regarded as a key constituent in an Australian programme to mitigate diffuse pollution from grassland farming (Walker, 1999; see also Pengelly and Fishburn, Chapter 18 this volume). Participatory action research has been defined as involvement and collaboration in the research process leading to problem resolution and relies in part on the incorporation of local knowledge (Greenwood *et al.*, 1993). Re-orientating relationships between farmer, extension officer and scientific researcher presents real challenges, particularly for the latter group who avoid multidisciplinary and eclectic research. However, the scale of the impacts resulting from farming actions suggests that gaining an understanding of the processes involved and devising appropriate mitigation approaches can only emerge from within a collaborative framework.

References

Acornley, R.M. and Sear, D.A. (1999) Sediment transport and siltation of brown trout (*Salmo trutta* L.) spawning gravels in chalk streams. *Hydrological Processes* 13, 447–458.

Alabaster, J.S. and Lloyd, R. (1980) *Water Quality Criteria for Freshwater Fish*. Butterworths, London.

Anderson, N.J. (1997) Reconstructing historical phosphorus concentration in rural lakes using diatom models. In: Tunney, H., Carton, O.T., Brookes P.C. and Johnston A.E. (eds) *Phosphorus Loss from Soil to Water*. CAB International, Wallingford, UK, pp. 95–118.

Armitage, P.D., Moss, D., Wright, J.F. and Furse, M.T. (1983) The performance of a new biological water quality score system based on macro invertebrates over a wide range of unpolluted running-water sites. *Water Research* 17, 333–347.

Battarbee, R.W. and Carter, C. (1993) The recent sediments of Lough Neagh, Part B: Diatom and chironomid analysis. In: Wood, R.B. and Smith, R.V. (eds) *Lough Neagh – the Ecology of a Multipurpose Water Resource*. Kluwer Academic, Dordrecht, pp. 133–148.

Beegle, D.B., Carton, O.T. and Bailey, J.S. (2000) Nutrient management planning: justification, theory, practice. *Journal of Environmental Quality* 29, 72–79.

Beland, K.F., Jordan, R.M. and Meister, A.L. (1982) Water depth and velocity preferences of spawning Atlantic salmon in Maine rivers. *North American Journal of Fisheries Management* 2, 11–13.

Bloxham, P.F. (1999) The management of agricultural manures. *Journal of the Royal Agricultural Society of England* 160, 138–155.

Boon, P.J. (1998) River restoration in five dimensions. *Aquatic Conservation: Marine and Freshwater Ecosystems* 8, 257–264.

Brookes, A. (1988) *Channelized Rivers: Perspectives for Environmental Management*. John Wiley & Sons, Chichester, UK.

Brookes, A. (1992) Recovery and restoration of some engineered British river channels. In: Boon, P.J., Calow, P. and Petts, G.E. (eds) *River Conservation and Management*. John Wiley & Sons, Chichester, UK, pp. 337–352.

Caraco, N. and Cole, J. (1999) Regional scale export of C, N, P and sediment: what river data tell us about key controlling variables. In: Tenhunen, J.D. and Kabat, P. (eds) *Integrating Hydrology, Ecosystem Dynamics and Biogeochemistry in Complex Landscapes*. John Wiley & Sons, Chichester, UK, pp. 239–254.

Chesters, R.K. (1980) *Biological Monitoring Working Party. The 1978 National Testing Exercise*, Technical Memorandum 19. Department of the Environment, London.

Cordone, A.J. and Kelley, D.E. (1961) The influence of inorganic sediments on the aquatic life of streams. *California Fish and Game* 47, 189–228.

Crouse, M.R., Callahan, C.A., Malueg, K.W. and Dominguez, S.E. (1981) Effects of fine sediments on growth of juvenile coho salmon in laboratory streams. *Transactions of the American Fisheries Society* 110, 281–286.

Cruickshank, J.G. (1997) *Soil and Environment: Northern Ireland.* Department of Agriculture for Northern Ireland and the Queen's University Belfast, Belfast.

Cumby, T.R., Brewer, A.J. and Dimmock, S.J. (1999) Dirty water from dairy farms, I: biochemical characteristics. *Bioresource Technology* 67, 155–160.

Dils, R., Coyle, N.M., Bailey, J.S. and Foy, R.H. (1998) Nutrient management planning: Lough Erne, Northern Ireland. In: Foy, R.H. and Dils, R. (eds) *Practical and Innovative Measures for the Control of Agricultural Phosphorus Losses to Water. Workshop Papers and Poster Abstracts.* Department of Agriculture for Northern Ireland, Belfast, pp. 60–61.

Elliot, C.R.N., Dunbar, M.J., Gowing, I. and Acreman, M.C. (1999) A habitat assessment approach to the management of groundwater dominated rivers. *Hydrological Processes* 13, 459–475.

Foy, R.H. and Bailey-Watts, A.E. (1998) Observations on the temporal variation in the phosphorus status of lakes in the British Isles. *Soil Use and Management* 14, 131–138.

Foy, R.H. and Kirk, M. (1995) Agriculture and water quality: a regional study. *Journal of Institute of Water and Environmental Management* 9, 247–256.

Foy, R.H. and Lennox, S.D. (2000) Contributions of diffuse and point sources to the phosphorus loads in the River Main over a 22 year period. *Boreal and Environmental Research* 5, 27–37.

Foy, R.H. and Withers, P.J.A. (1995) The contribution of agricultural phosphorus to eutrophication. *Proceedings of the Fertiliser Society* 365.

Foy, R.H., Smith, R.V., Stevens, R.J. and Stewart, D.A. (1982) Identification of factors affecting nitrogen and phosphorus loadings to Lough Neagh. *Journal of Environmental Management* 15, 109–129.

Foy, R.H., Smith, R.V., Smyth, D., Lennox, S.D. and Unsworth, E.F. (1994) The impact of climatic and agricultural variables on the frequency of silage pollution incidents. *Journal of Environmental Management* 41, 105–121.

Foy, R.H., Smith, R.V., Jordan, C. and Lennox, S.D. (1995) Upward trend in soluble phosphorus loadings to Lough Neagh despite phosphorus reduction at STWs. *Water Research* 29, 1051–1063.

Foy, R.H., Champ, W.S.T. and Gibson, C.E. (1996) The effectiveness of restricting phosphorus loadings from sewage treatment works as a means of controlling eutrophication in Irish lakes. In: Giller, P.S. and Myers, A.A. (eds) *Disturbance and Recovery in Ecological Systems.* Royal Irish Academy, Dublin, pp. 136–154.

Foy, R.H., Lennox, S.D. and Smith, R.V. (2001) Chemical and biological water quality indices and pollution statistics as measures of the impact of agricultural pollution control measures. *Water Research* 35, 3004–3012.

Furse, M.T., Winder, J.M., Symes, K.L., Clarke, R.T., Gunn, R.J.M., Blackburn, J.H. and Fuller, R.M. (1993) *The Faunal Richness of Headwater Streams: Stage 2 Catchment Studies* Volume 1 Main Report. National Rivers Authority, Bristol.

Gibson, C.E. (1998) *Freshwater Research in the Department of Agriculture for Northern Ireland.* Department of Agriculture for Northern Ireland, Belfast.

Gibson, C.E. and Stevens, R.J. (1979) Changes in phytoplankton physiology and morphology in response to dissolved nutrients in Lough Neagh, Northern Ireland. *Freshwater Biology* 9, 105–109.

Gibson, C.E., Smith, R.V. and Stewart, D.A. (1988) A long term study of the phosphorus cycle in Lough Neagh, Northern Ireland. *International Revue de gesamten Hydrobiologie* 73, 159–170.

Gibson, C.E., Wang, G., Foy, R.H. and Lennox, S.D. (2001) The importance of catchment and lake processes in the phosphorus budget of a large lake. *Chemosphere* 42, 215–220.

Greenwood, D.J., Whyte, W.F. and Harkavy, I. (1993) Participatory action research as a process and as a goal. *Human Relations* 46, 175–192.

Guerin, T.F. (1999) An Australian perspective on the constraints to the transfer and adoption of innovations in land management. *Environmental Conservation* 26, 289–304.

Gustafson-Greenwood, K.I. and Moring, J.R. (1991) Gravel compaction and permeabilities in redds of Atlantic salmon, *Salmo salar* L. *Aquaculture and Fisheries Management* 22, 537–540.

Harrison, C.W. (1923) Planting eyed salmon and trout eggs. *Transactions of the American Fisheries Society* 105, 57–63.

Hausle, D.A. and Coble, D.W. (1976) Influence of sand in redds on survival and emergence of brook trout (*Salvelinus fontinalis*). *Transactions of the American Fisheries Society* 105, 57–63.

Haygarth, P.M., Heathwaite, A.L., Jarvis, S.C. and Harrod, T.R. (2000) Hydrological factors for phosphorus transfer from agricultural soils. *Advances in Agronomy* 69, 153–178.

Heaney, S.I., Foy, R.H., Kennedy, G.J.A., Crozier, W.W. and O'Connor, W.C.K. (2001) Impacts of agriculture on aquatic systems: lessons learnt and new unknowns in Northern Ireland. *Marine and Freshwater Research* 52, 151–163.

Heckrath, G., Brookes, P.C., Poulton, P.R. and Goulding, K.W.T. (1995) Phosphorus leaching from soils containing different phosphorus concentrations in the Broadbalk Experiment. *Journal of Environmental Quality* 24, 904–910.

Hesketh, N. and Brookes, P.C. (2000) Development of an indicator for risk of phosphorus leaching. *Journal of Environmental Quality* 29, 105–100.

Jordan, C. and Smith, R.V. (1985) Factors affecting leaching of nutrients from an intensively managed grassland in County Antrim, Northern Ireland. *Journal of Environmental Management* 20, 1–15.

Kelly, F.L. and Bracken, J.J. (1998) Fisheries enhancement of the Rye Water, lowland river in Ireland. *Aquatic Conservation: Marine and Freshwater Ecosystems* 8, 131–143.

Kennedy, G.J.A. and Crozier, W.W. (1995) Factors affecting recruitment success in salmonids. In: Harper, D. and Ferguson, A. (eds) *The Ecological Basis for River Management.* John Wiley & Sons, Chichester, UK, pp. 349–362.

Kennedy, G.J.A. and Crozier, W.W. (1998) The River Bush. In: Moriarty, C. (ed.) *Studies of Irish Rivers and Lakes* (Congress of Societies Internationalis Limnologiae (SIL) Dublin). Marine Institute, Dublin, pp. 257–273.

Kiely, P.V. (1981) Catchment pollution. In: Brogan, J.C. (ed.) *Nitrogen Losses and Surface Run-off.* Martinus Nijhoff, The Hague, pp. 87–91.

Ladle, M. and Casey, H. (1971) Growth and nutrient relationships of *Ranunculus penicillatus* var. *calcareus* in a small chalk stream. *Proceedings of the European Weed Research Council, 3rd International Symposium on Aquatic Weeds*, pp. 53–64.

Lennox, S.D., Foy, R.H., Smith, R.V. and Jordan, C. (1997) Estimating the contribution from agriculture to the phosphorus load in surface water concentration. In: Tunney, H., Carton, O.T., Brookes, P.C. and Johnston A.E. (eds) *Phosphorus Loss from Soil to Water.* CAB International, Wallingford, UK, pp. 55–75.

Lennox, S.D., Foy, R.H., Smith, R.V., Unsworth, E. and Smyth, D.R. (1998) A comparison of agricultural pollution incidents in Northern Ireland and England and Wales. *Water Research* 32, 649–656.

Lisle, T.E. and Lewis, J. (1992) Effects of sediment on survival of salmonid embryos in a natural stream: a simulation approach. *Canadian Journal of Fisheries and Aquatic Sciences* 49, 2337–2344.

Lowe, P. and Ward, N. (1997) The moral authority of regulation: the case of agricultural pollution. In: Romstad, E., Simonsen, J. and Vatn, A. (eds) *Controlling Mineral Emissions in European Agriculture.* CAB International, Wallingford, UK, pp 59–71.

Lowe, P., Clark, J., Seymour, S. and Ward, N. (1992) *Pollution Control on Dairy Farms, An Evaluation of Current Policy and Practice.* SAFE Alliance, London.

McAllister, J.S.V. (1976) Studies in Northern Ireland on problems relating to the disposal of slurry. In: *Agriculture and Water Quality.* Ministry of Agriculture, Fisheries and Food Technical Bulletin 32. HMSO, London, pp. 418–431.

McDonald, A.M., Edwards, A.C., Pugh, K.B. and Balls, P.W. (1995) Soluble nitrogen and phosphorus in the River Ythan system, UK: annual and seasonal trends. *Water Research* 29, 837–846.

McGarrigle, M.L. (1998) Impact of eutrophication on Irish river quality. In: Wilson, J. (ed.) *Eutrophication in Irish Waters.* Royal Irish Academy, Dublin, pp. 82–90.

McGarrigle, M., O'Grady, M., Champ, W.S. and McCarthy, T.K. (1998) The Moy catchment. In: Moriarty, C. (ed.) *Studies of Irish Rivers and Lakes* (Congress of Societies Internationalis Limnologiae (SIL) Dublin). Marine Institute, Dublin, pp. 213–236.

McGuckin, S.O., Jordan, C. and Smith, R.V. (1999) Deriving phosphorus export coefficients for CORINE land cover types. *Water Science and Technology* 12, 47–53.

McNeil, W.J. and Ahnell, W.H. (1964) Success of pink salmon spawning relative to size of spawning bed materials. *US Fish and Wildlife Service Special Scientific Report, Fisheries* 469.

Meehan, W.R. (1991) Introduction and overview. In: Meehan, W.R. (ed.) *Influences of Forest and Rangeland Management on Salmonid Fishes and their Habitats.* American Fisheries Society Special Publication 19, pp. 1–15.

Naismith, I., Wyatt, R., Gulson, J. and Mainstone, C.P. (1996) *The Impacts of Land Use on Salmonids. A Study of the River Torridge Catchment*, Research and Development Report 30. National Rivers Authority, Bristol.

Newall, A. (1995) The microflow environments of aquatic plants – an ecological perspective. In: Harper, D.M. and Ferguson, A.J.D. (eds) *The Ecological Basis for River Management.* John Wiley & Sons, Chichester, UK, pp. 79–92.

Newson, M.D. (2000) *Rehabilitation of Salmon Fisheries Production Areas in the River Bush, Co. Antrim, Northern Ireland. Towards Strategies and Designs Informed by Fluvial Geomorphology.* Technical Report for Rivers Agency, through River Restoration Centre, Silsoe UK. Available from Rivers Agency, Department of Agriculture and Rural Development, Belfast, Northern Ireland.

Northern Ireland Audit Office (1998) *Control of River Pollution Report by the Comptroller and Auditor General for Northern Ireland,* HC 693 Session 1997–98. The Stationery Office, London.

O'Callaghan, J. and Pollock, K.A. (1976) A land spreading trial with pig slurry. In: *Agriculture and Water Quality.* Ministry of Agriculture, Fisheries and Food Technical Bulletin 32. HMSO, London, pp. 371–399.

O'Connor, W.C.K. (1997) A freeze sampling assessment of spawning gravel siltation in the River Bush, Co. Antrim. *Irish Naturalists' Journal* 25, 337–341.

O'Connor, W.C.K. (1998) An investigation of Atlantic salmon (*Salmo salar* L.) spawning habitat in Northern Ireland rivers. DPhil thesis, University of Ulster, Faculty of Science.

O'Connor, W.C.K. and Andrew, T.E. (1998) The effects of siltation on Atlantic salmon, *Salmo salar* L., embryos in the River Bush. *Fisheries Management and Ecology* 5, 393–401.

OECD (1982) *Eutrophication of Waters, Monitoring, Assessment and Control.* Organisation for Economic Co-operation and Development, Paris.

O'Grady, M. (1991) Rehabilitation of salmonid habitats in a drained Irish river system. In: Steer, M. (ed.) *Irish Rivers: Biology and Management.* Royal Irish Academy, Dublin, pp. 187–204.

Olley, J. and Caitcheon, G. (2000) Major element chemistry of sediments from the Darling-Barwon River and its tributaries: implication for sediment and phosphorus sources. *Hydrological Processes* 14, 1159–1175.

Ottaway, E.M., Carling, P.A., Clarke, A. and Reader, N.A. (1981) Observations on the structure of brown trout, *Salmo trutta* Linnaeus, redds. *Journal of Fish Biology* 19, 593–607.

Parkes, M.E., Campbell, M.E. and Vinten, A.J.A. (1997) Practice to avoid contamination of drainflow and runoff from slurry from slurry spreading in spring. *Soil Use and Management,* 13, 36–42.

Peterson, R.H. (1978) *Physical Characteristics of Atlantic Salmon Spawning Gravel in Some New Brunswick Streams,* Technical Report No. 785. Canadian Fisheries and Marine Service, St Andrew's, New Brunswick, Canada.

Petts, G.E., Thoms, M.C., Brittain, K. and Atkin, B. (1989) A freeze-coring technique applied to pollution by fine sediment in gravel-bed rivers. *The Science of the Total Environment* 84, 259–272.

Petts, G.E., Maddock, I., Bickerton, M. and Ferguson, A.J.D. (1995) Linking hydrology and ecology: the scientific basis for river management. In: Harper, D.M. and Ferguson, A.J.D. (eds) *The Ecological Basis for River Management.* John Wiley & Sons, Chichester, UK, pp. 1–16.

Pollard, P. and Huxham, M. (1998) The European Water Framework Directive: a new era in the management of aquatic ecosystem health. *Aquatic Conservation: Marine and Freshwater Ecosystems* 8, 773–792.

Pote, D.H., Daniel, T.V., Sharpley, A.N., Moore, P.A. Jr, Miller, D.M. and Nichols, D.J. (1999) Relationship between phosphorus levels in three ultisols and phosphorus concentrations in runoff. *Journal of Environmental Quality* 28, 170–175.

Purseglove, J. (1988) *Taming the Flood: a History and Natural History of Rivers and Wetlands.* Oxford University Press in association with Channel Four Television, Oxford.

Raven, P.J., Holmes, N.T.H., Dawson, F.H., Fox, P.J.A., Everard, M., Fozzard, I.R. and Rouen, K.J. (1998) *River Habitat Quality: the Physical Character of Rivers and Streams in the UK and Isle of Man.* Environment Agency, Bristol.

Reeves, G.H., Hall, J.D., Roelofs, T.D., Hickman, T.L. and Baker, C.O. (1991) Rehabilitating and modifying stream habitats. In: Meehan, W.R. (ed.) Influence of Forest and rangeland management on Salmonid fishes and their habitats. *American Fisheries Society Special Publication* 19, 519–557.

Royal Commission (1992) *Sixteenth Report, Freshwater Quality,* Cmnd 1166. Royal Commission on Environmental Pollution. HMSO, London.

Schofield, K., Seager, J. and Meriman, R.P. (1990) The impact of intensive dairy farming activities on river quality: the Eastern Cleddau catchment study. *Journal of Institute of Water and Environmental Management* 4, 176–186.

Seager, J., Jones, J. and Rutt, G. (1992) Assessment and control of farm pollution. *Journal of the Institute of Water and Environmental Management* 6, 48–54.

Sear, D., Wilcock, D., Robinson, M. and Fisher, K. (2000) River channel modification in the UK. In: Acreman, M. (ed.) *The Hydrology of the UK. A Study of Change*. Routledge, London, pp. 55–81.

Shackle, V.J., Hughes, S. and Lewis, V.T. (1999) The influence of three methods of gravel cleaning on brown trout, *Salmo trutta*, egg survival. *Hydrological Processes* 13, 477–486.

Shelton, J.M. (1955) The hatching of chinook salmon eggs under simulated stream conditions. *Progressive Fish-Culturist* 17, 20–35.

Sherwood, M. and Fanning, A. (1981) Nutrient content of surface runoff from land treated with animal wastes In: Brogan, J.C. (ed.) *Nitrogen Losses and Surface Run-off*. Martinus Nijhoff, The Hague, pp. 87–91.

Smith, K.A., Jackson, D.R. and Withers, P.J.A. (2001) Nutrient losses by surface runoff following the application of organic manures to arable land. 2. Phosphorus. *Environmental Pollution* 112, 53–60.

Smith, R.V. (1977) Domestic and agricultural contributions to the inputs of phosphorus and nitrogen to Lough Neagh. *Water Research* 11, 453–459.

Smith, R.V., Foy, R.H. and Lennox, S.D. (1994) The application of mathematical modelling techniques to evaluate the impact of phosphorus reduction on phosphate loads to Lough Neagh. In: Barnett, V. and Turkman, K.F. (eds) *Statistics for the Environment 2: Water Related Issues*. John Wiley & Sons, Chichester, UK, pp. 271–284.

Smith, R.V., Lennox, S.D., Jordan, C., Foy, R.H. and McHale, E. (1995) Increase in soluble phosphorus transported in drainflow from a grassland catchment in response to soil phosphorus accumulation. *Soil Use and Management* 11, 204–209.

Snedecor, G.W. and Cochran, W.G. (1967) *Statistical Methods*, 6th edn. Iowa State, Ames.

Stevens, R.J. (1980) Some factors affecting N and P losses in drainage water. *Annual Report of Research and Technical Work (1979)*. Department of Agriculture for Northern Ireland, Belfast, pp. 128–130.

Stuart, T.A. (1953a) Water currents through permeable gravels and their significance to spawning salmonids. *Nature* 172, 407–408.

Stuart, T.A. (1953b) Spawning, migration, reproduction and young stages of loch trout (*Salmo trutta* L.). Freshwater and Salmon Fisheries Research 5, Scottish Home Department, Edinburgh.

Tunney, H., Foy, R.H. and Carton, O.T. (1998) Phosphorus inputs to water from diffuse agricultural sources. In: Wilson, J.G. (ed.) *Eutrophication in Irish Waters*. Royal Irish Academy, Dublin, pp. 25–39.

Tunney, H., Coulter, B., Daly, K., Kurz, I., Coxon, C., Jeffrey, D., Mills, P., Kiely, P.V. and Morgan, G. (2000) *Quantification of Phosphorus Loss from Soil to Water*. Environmental Protection Agency, Wexford.

Walker, J. (1999) Environmental indicators of catchment and farm health. In: Williams, J., Hook, R.A. and Gascoigne, H.L. (eds) *Farming Action Catchment Reaction*. CSIRO Publishing, Collingwood, Victoria, pp. 99–117.

Walling, D.E. and Amos, C.M. (1999) Source, storage and mobilization of fine sediment in a chalk stream. *Hydrological Processes* 13, 323–340.

Walling, D.E. and He, Q. (1999) Changing rates of overbank sedimentation on the flood plains of British rivers during the past 100 years. In: Brown, A.G. and Quine, T.A. (eds) *Fluvial Processes and Environmental Change*. John Wiley & Sons, Chichester, UK, pp. 207–222.

Walling, D.E., Owens, P. and Leeks, G.J.L. (1999) Fingerprinting suspended sediment sources in the catchment of the River Ouse, Yorkshire, UK. *Hydrological Processes* 13, 955–975.

Watson, R.D. (1997) Integrating catchment management: the human dimension. In: Cresser, M. and Pugh, K. (eds) *Multiple Land Use and Catchment Management*. Macaulay Land Use Research Centre, Aberdeen, pp. 125–137.

White, H.C. (1942) Atlantic salmon redds and artificial spawning bed. *Journal of the Fisheries Research Board of Canada* 6, 37–44.

Wilcock, D.N. (1979) Post-war land drainage, fertilizer use and environmental impact in Northern Ireland. *Journal of Environmental Management* 8, 137–149.

Wilcock, D.N. (1997) Rivers, drainage basins and soils. In: Cruickshank, J.G. (ed.) *Soil and Environment: Northern Ireland*. Department of Agriculture for Northern Ireland and The Queen's University Belfast, Belfast, pp. 85–98.

Williams, J.R. and Nicholson, R.J. (1995) Low rate irrigation of dirty water and its effect on drainage water quality. *Agricultural Engineer* 20(2), 24–28.

Withers, P.J.A., Davidson, I.A. and Foy, R.H. (2000) Prospects for controlling non-point phosphorus loss to water. *Journal of Environmental Quality* 29, 167–175.
Wood, R.B. (1999) Lough Neagh. In: Moriarty, C. (ed.) *Studies of Irish Rivers and Lakes* (Congress of Societies Internationalis Limnologiae (SIL) Dublin). Marine Institute, Dublin, pp. 257–273.
Zhou, Q., Gibson, C.E. and Foy, R.H. (2000) Long-term changes of nitrogen and phosphorus loadings to a large lake in North west Ireland. *Water Research* 34, 922–926.

20 Conflicts and Problems with Water Quality in the Upper Catchment of the Manyame River, Zimbabwe

Kudakwashe E. Motsi, Esther Mangwayana and Ken E. Giller
Department of Soil Science and Agricultural Engineering, University of Zimbabwe, PO Box MP167, Mount Pleasant, Harare, Zimbabwe

---

Key words: agriculture, irrigation, livestock, Manyame, pesticides, water quality, Zimbabwe

Introduction

The population of Harare, the capital city of Zimbabwe, and associated settlements is increasing at a rate of 6.5–8.1% each year, much faster than the national growth rate of 2.5% per annum (CSO, 1997). In 1997 the population was estimated to be approximately 2.35 million, having risen from < 0.4 million in 1969 and 1.5 million in 1992. Among the demands of the increasing urban population, the supply of adequate water of sufficient quality for drinking is an acute one.

Some of the features raised above are common to many cities in Africa – rapidly increasing urbanization and unpredictable water supplies – but Harare has a particular and peculiar problem. The major water supply for the city comes from Lake Chivero, which was created in 1952 by damming the Manyame River. Harare, and its satellite towns, are situated in the upper Manyame catchment resulting in problems of water quality control of a particularly acute nature. The problem of recycling water, against a backdrop of restricted financial investment, is the subject of this chapter. We review the current status and history of water use and quality in Harare and environs, in relation to the very rapid increase in urbanization, industrialization and urban cultivation for food production. Possible options for the future development of sufficient supplies of clean water are then discussed. It is clear that major investments will be required to avoid critical water shortages, or massive pollution events.

The Upper Manyame Catchment

The Upper Manyame catchment area is the most urbanized in Zimbabwe; it contains two cities, Harare itself and Chitungwiza (a 'dormitory' satellite for the majority of the labour force), and the towns Norton and Ruwa (Fig. 20.1). The Upper Manyame River, one of the tributaries of the Zambezi River, is defined as the upstream reach from its source to the Manyame Dam wall, and its basin covers about 3900 km^2. Within the catchment there are several streams, most of them seasonal, that are tributaries of the Manyame River. There are four major impoundments: Lake Chivero, Lake Manyame, Seke and Harava Dams. Three of the urban centres, Harare, Chitungwiza and Ruwa, sit directly above Lake Chivero. Norton is below Lake Chivero but above Lake Manyame. During the dry season these impoundments provide a reliable source of water.

The main activities that affect water quality in the catchment are urbanization, industrialization, agriculture and rural settlements. The city of Harare and surrounding towns draw their water for industrial and domestic use from the catchment. All the towns and the city discharge most of their treated/untreated sewage and industrial effluent back into the catchment rivers and impoundments. The agricultural sector mainly draws untreated water for irrigation and livestock rearing from the rivers and impoundments. Poor agricultural management by some members of the farming community has resulted in some nutrients and chemicals, intended for the crops, contaminating the water within the catchment.

Problems within the catchment are mainly pollution, siltation of the main water bodies and adequacy of the water supply. Eutrophication of the catchment water impoundments has resulted in problems of water purification, unpleasant tastes and odours in the water and toxins that cannot be removed by normal water purification methods. Inappropriate agricultural activities in some parts of the catchment have resulted in the problem of heavy silt loads. The lake in the catchment became eutrophic as a result of discharge of N and P-rich effluent causing excessive growth of algae, which then died and decomposed, depleting dissolved O_2, causing massive fish kills and making the water expensive and difficult to treat. Conflicts are arising from the fact that the beneficiaries are competing for utilization of the water and at the same time polluting the very same scarce resource. These conflicts arise in the areas of environmental control, industrial growth and development, residential expansion of human settlements and water purification costs.

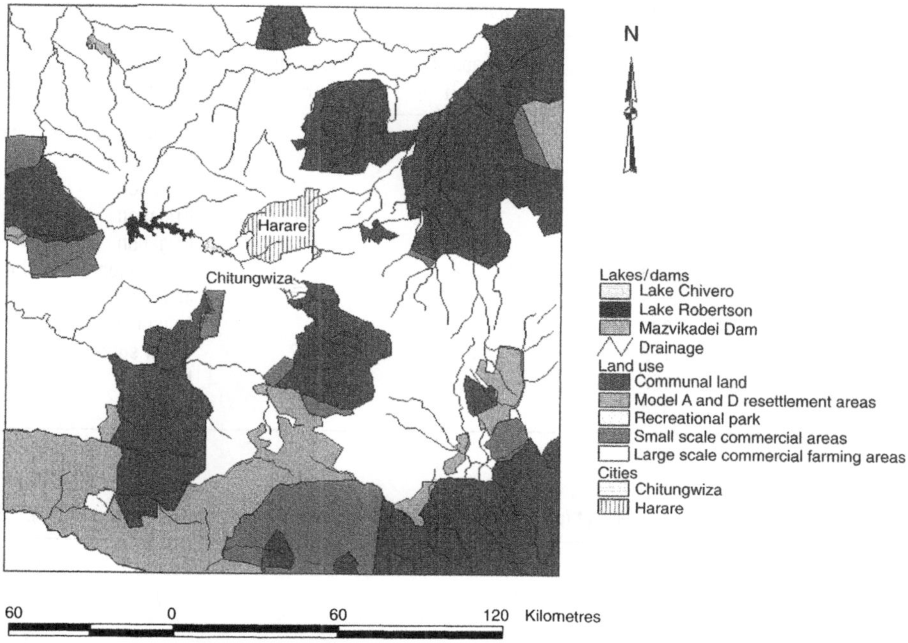

Fig. 20.1. The Manyame River catchment and surroundings showing the drainage and the major land use patterns.

Characteristics of the catchment

River Manyame's source of origin is just outside Marondera, 80 km to the east of Harare. The river stretches for more than 400 km discharging into the Zambezi River, but this account will be limited to a section from the source to the end of Lake Manyame. To the north of the river are the Seke communal area, Chitungwiza town and Harare South District commercial farm. This section of the catchment discharges both into the Nyatsime River and directly into the Manyame River (Fig. 20.1). On the western bank are the intensive commercial farms under Ruwa, Bromely, Marondera and Harare South districts and the City of Harare. This section of the catchment discharges into the Ruwa, Mukuvisi and Marimba Rivers which are all tributaries of the Manyame River. There are approximately 50 agricultural rights holders within the catchment. The main water-using agricultural activities within this area are intensive horticulture and tobacco production.

In the Upper Manyame river basin, the major impoundments are Lake Chivero, Seke Dam and Harava Dam (Table 20.1). Several rivers flow into these water bodies. Their general dimensions and water capacity use are shown in Table 20.2. The mean annual runoff of the Manyame River is approximately 307.7×10^6 m^3, while the total annual runoff varies from 207×10^6 m^3 to 796.5×10^6 m^3. The Manyame occupies as much as 80% of the total gauged flow, while Mukuvisi, Nyatsime and Marimba rivers contribute to the remainder of the flow rate. Groundwater inflows and outflows are considered to be minimal in comparison with the surface flows. The catchment area is generally a gently undulating, featureless plateau with a few hills around the Lake Chivero area. Most of the catchment area lies at an altitude ranging from 1400 to 1500 m above sea level and the lowest altitude is approximately 1000 m above sea level.

Climate in the catchment area is very seasonal with three distinct, wet or dry, sea-

Table 20.1. Study basin characteristics in relation to water use.

Water body	Catchment area (km^2)	Rated capacity (10^3 m^3)	Flow rate (10^3 m^3 day^{-1})	Water use
Lake Chivero	421	247,181		Water supply, recreational and fisheries
Marimba River	215		63	Irrigation
Mukuvisi River	230		114	Irrigation
Nyatsime River	280		125	Irrigation
Lake Manyame	3,930	480,236		Water supply, recreational and fisheries
Gwebi	680			Irrigation
Seke and Harava Dam	115	12,406	42	Water supply
Ruwa River	195		102	Irrigation
Manyame River	474			Irrigation

Table 20.2. Morphometric characteristics, at full supply level, of the Manyame lakes (source: Thornton, 1980).

	Date of construction	Area (ha)	Volume (10^{-6} m^3)	Max. depth Z_{max} (m)	Mean depth Z (m)
Cleveland	1913	30	1.00	—	3.3
Seke (Prince Edward)	1929	109	3.64	10	3.3
Harava (Henry Hallam)	1973	215	9.25	17.5	4.3
Chivero (McIlwaine)	1952	2630	250	227.4	9.5
Manyame (Robertson)	1976	8100	490	22.6	6

NB The pre-1990 names of these lakes are inserted in parentheses as they are used in earlier references.

sons. 'Spring' is a hot and dry season from September to November. The average daily temperature is approximately 22°C ± 6°C. 'Summer' is classified as the rainy season with hot and wet conditions from December to April. The average daily temperature is 20°C ± 6°C. The remaining period of the year is a cold and generally dry season, called 'winter'. The average temperature is approximately 16°C ± 6°C. The rainfall in Zimbabwe is strongly seasonal, normally falling between November and March with a dry season of six months (Fig. 20.2). Rainfall in Harare is 817 mm on average, but can be highly variable, ranging from 440 to 1220 mm. Approximately 80% of the total annual rainfall occurs during the summer season. The early 1990s witnessed a series of droughts which led to acute water shortages and rationing; in the past few years Zimbabwe has received heavy rains, and most dams filled to capacity.

Population of the catchment

The catchment area encompasses the four urban centres of Harare (1997 population: 1.9 million), Chitungwiza (1992 population: 274,912), Ruwa (1992 population: 1447) and Norton (1992 population: 20,405), as well as the communal tenure areas of Seke and Chihota and privately owned commercial farms. The catchment area is 2700 km^2 with a rural population in excess of 1.5 million. Settlement densities are among the highest in the country; the Harare/Chitungwiza urban areas account for 47% of the national urban population. Harare's urban area has a settlement density of 215 people km^{-2}. This figure seems to be growing by the day as more and more people are migrating from the communal areas, towns and cities within Zimbabwe and far afield in search of

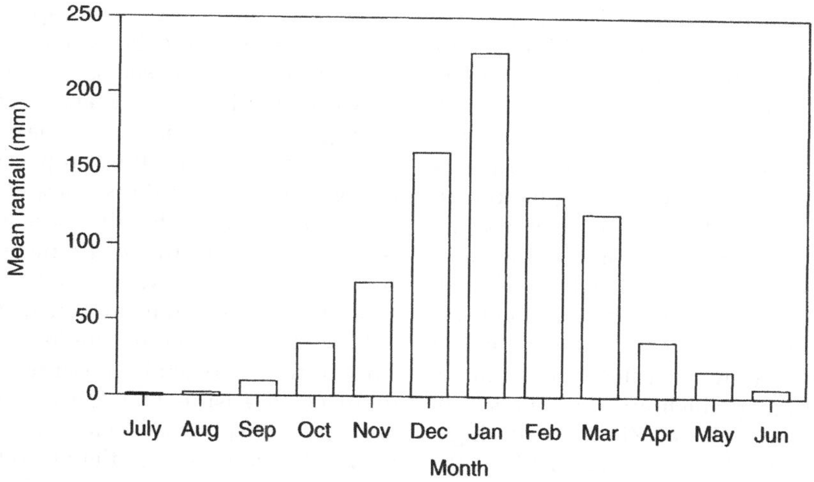

Fig. 20.2. Mean seasonal rainfall at Kutsaga (Harare Airport) over the 20 year period, 1980–2000.

employment. Settlement densities in the rural parts of the catchment area have a lower average of 27 people km^{-2}.

The City of Harare encompasses an area of 872 km^2 and the first municipality authority was formed as a Board of Management in 1891. It was granted city status in 1935 and has been growing since that time. Most of the industrial, social and economic development is taking place in the capital city. The municipality of Chitungwiza, currently the third largest in the country, lies approximately 9 km south of the Harare City boundary and covers an area of more than 42 km^2. It was formed in 1978 and was granted town council status in 1981. Since then a strong emphasis has been placed on attracting various industries and commercial developers to the town. Chitungwiza was proclaimed a municipality in February in 1996. This high-density area of Chitungwiza is densely populated with a housing density of 17 units per hectare.

About 8 km^2 of the Ruwa farming area was designated as a growth point in 1996. The centre is located in a large-scale commercial farming area about 23 km to the east of Harare City. The area has been formally excised from the rural district council's authority subsequent to the formation of the Ruwa Local Board. The three urban centres of Chitungwiza, Ruwa and Norton are all designated growth points, enjoying tax concessions for newly locating industries. This, coupled with Harare, makes the catchment area one of the fastest growing areas in the country. The area's population growth rates have been estimated to range from 6.5 to 8.1% per annum.

Agricultural Activities within the Catchment

Urban agriculture

Under normal circumstances cultivation in open spaces (green corridors) is not allowed in Harare and the surrounding towns as this leads to a multitude of problems which include water pollution and siltation of dams. In the past, Harare City Council has tried to promote and provide a framework for urban agriculture as an interim use of future development land. This was with the provision that before cultivation, residents should: identify pieces of land, organize themselves into cooperatives and then approach the city council's Department of Housing and Community Services to obtain permission and assistance for cultivation where necessary.

The practice of urban agriculture has been on the increase in major cities such as Harare and Chitungwiza. The practice has expanded over the years, dating back to as early as the 1950s and has gained momentum ever since, as economic hardships forced people to turn to urban agriculture to supplement their livelihoods. This has led to stream bank cultivation as all other suitable pieces of land became fully exploited. By 1994, 36% of public land (9300 ha) was under cultivation, and 67% was cultivated in vlei areas (low-lying flood plains prone to waterlogging which are characteristic of the landscape of the Zimbabwean highlands) (Bowyer-Bower *et al.*, 1995). In the last cropping season (1999–2000), almost all public land was cultivated within the city, except in the low-density suburbs. The general recommendation in Zimbabwe is that all cultivation activities must be at least 30 m from the stream. As the pressure for land increases, this recommendation is no longer followed and there is no longer close monitoring. This has led to rapid siltation of the rivers, dams and lakes within the catchment. Although technically illegal, cultivation is a short-term solution to providing food for the city poor while the long-term problems of water provision are of concern to the city authorities. There is clearly a conflict of interest between the city authorities and the residents. This problem cannot be solved easily unless the economic situation in the country improves.

On average the urban agriculture plot holders have plots that vary in size between 20 and 200 m^2 (Drakakis-Smith, 1995). As vegetable production is highly profitable, many urban agriculture plot holders apply fertilizers and pesticides in excess, often designing their own cocktails from commercially available agrochemicals. This applies to all income groups. Considering the average size of individual plots, this may appear to have little impact on the ecology of the catchment in terms of agrochemical use. However, as an aggregate group, their impact is likely to be equivalent to, if not higher than, those of commercial farmers (Bowyer-Bower and Tengeh, 1995).

Apart from using open spaces and vleis, areas near the large sewage treatment works are also used as plots for market gardening by some vegetable vendors. They grow tomatoes, onions, beans, rape and cabbage among other crops near the sewage ponds using the sewage water for irrigation through the dry season. This acts as a cheap source of liquid fertilizer since the sewage water contains high quantities of essential nutrients. Produce is harvested and sold on the local market. No control or hygienic measures are put in place to ensure that these vegetables are thoroughly rinsed with clean water before they are taken to the market. Some residents who consume vegetables from the open market sometimes complain of gastrointestinal problems, and circumstantial evidence exists which links these cultivators at sewage farms to intense outbreaks of diarrhoea and cholera in some suburbs. Effluent from the Firle Sewage Works was found to be highly contaminated with *Salmonella* bacteria (Gopo and Chingobe, 1995). Apart from the potential danger of bacterial disease the vegetables could also contain heavy metals from the sewage effluent. Contamination of soils and vegetables with heavy metals and links between sewage irrigation waters and occurrence of gastrointestinal and other health problems are a focus for our current research.

Commercial agriculture

There are commercial farms in Ruwa, Bromely, Marondera and Harare South District that are within the Manyame River catchment. Apart from the Manyame River, the Gwebi River and the Muzururu River also have catchments that drain into Lake Manyame. The main activities within the Gwebi River subcatchment are agricultural. These include cattle rearing, crop production and horticulture. Commercial farming is suspected to be a source of pollution from runoff containing nutrients and various noxious substances from the use of fertilizer, herbicides and pesticides. As yet no research has been conducted to assess the magnitude of this source of nutrient loading in the catchment. A study into pesticide residues in Lake Chivero (Mhlanga and Madziva, 1990) found traces of aldrin, dieldrin, DDE, DDD and DDT in water, soil, fish and sediment samples (Table 20.3). DDT and dieldrin levels in water and sediments appear to have risen slightly over the years, however; the concentrations are well within the permissible drinking water standards (Mhlanga and Madziva, 1990). The concentration of DDT fell considerably in fish in the 1980s, which may be explained by the ban on the use of DDT.

Direct use of river water for irrigation purposes is minimal because of limited flow available during the dry season. This has resulted in competition for water with other economic sectors within the catchment, especially during drought years.

Communal/subsistence agriculture

Within the catchment there are communal farmers who were settled in infertile pieces of land, which were called native 'reserves' during the colonial era and are now referred to as communal areas. In the catchment area, these include Seke, Chihota and Dema districts. These areas are characterized by overcrowding and agriculture is the major activity. The farmers tend to utilize all available land to the extent that soil conservation techniques are compromised. Communal grazing lands are also often over-utilized resulting in high rates of soil erosion. All this has resulted in high silt loads being carried to the rivers and water impoundments within the catchment.

Farmers in the periphery of towns supplement their pastures by taking their livestock to graze in the open spaces. A classic example was in the 1991/1992 grazing season when there was a severe drought such that there was no water for cattle from communal areas surrounding Chitungwiza Municipality. As a result, the cattle owners or the herd boys blocked main sewers so that the sewage effluent flooded storm drains, some residential properties and eventually the valleys. The results were health hazards to the urban dwellers whose housing was flooded with raw sewage. During a survey conducted by Mbiba (1996) vast expanses of land were flooded by this effluent. Cattle drank some of the sewage effluent, which is known to be heavily contaminated with heavy metals and meat from these cattle may in turn have created health hazards for consumers (see below). To the municipality of Chitungwiza the blockage of main sewer pipes is a management problem as it results in time wastage detecting and repairing these blocked pipes.

Table 20.3. The concentration (mg l^{-1} in water or mg kg^{-1} dry weight) of some pesticides in water, sediments and fish from Lake Chivero. The green-headed bream (*Orechromis macrochir*) is herbivorous and the African catfish (*Clarius gariepinus*) is an omnivore (compiled from Greichus *et al.*, 1978; Greichus, 1982; and Mhlanga and Madziva, 1990).

Pesticide	Sample	1974	1979	1987–1988
Total DDT	Water	< 0.1	—	0.4
	Sediments	57	—	76
	O. macrochir > 500 g	450	1270	210
	C. gariepinus	—	1510	180
Dieldrin	Water	< 0.1	—	0.2
	Sediments	4	—	5
	O. macrochir > 500 g	120	25	
	C. gariepinus	—	990	—

Industrial Activities within the Catchment

Cities within the catchment

Harare, like most major cities of Zimbabwe, is located on the high plateau. At this elevation most streams are in their formative stages of flow stream order. The progression of the Harare water supply dams downstream illustrates a fundamental problem: in order to obtain sufficient water supply, water has to be drawn from elevations lower than the city. Thus the city's wastewaters flow either into the water supply facilities or into water supplies for other settlements. In general, water pollution sources are categorized into those of natural origin and those from human activities, comprising point and distributed sources in terms of the manner of the effluent discharge into public water bodies. The sources of pollution have been identified as including sewage discharge from urban areas, treated/untreated industrial wastewater, effluent from sewage treatment plants, agricultural chemicals, fertilizer, livestock from commercial/communal farmland, urban runoff and sediments (Munzwa, 1982). The relative importance of the different sources becomes clearer when the sources of pollution are zoned by sub-catchments as discussed below.

Industrial activities are limited to urban centres within the catchment. The urban areas of Harare City, Chitungwiza Municipality, Norton Town Council and Ruwa Local Board cover more than 3151 ha. There are three main types of industrial activities, which are categorized by the nature of work that goes on within the factories or places of work (Table 20.4). These are the heavy, light and informal/backyard industries which include, for example, motor vehicle assembly, soft drink manufacturing and servicing of vehicles, respectively. Most of these industries require water for the manufacture of the final end product. Others also require water for cooling purposes during the manufacturing processes. It is not possible to discuss all the industries within the catchment. Therefore, a few have been selected per sub-catchment to highlight some of the problem areas as outlined below.

The Mukuvisi sub-catchment

Studies of riverine nutrient loadings to the lake conducted by Marshall and Falconer in the late 1960s and by Thornton in the late 1970s showed that the Mukuvisi was contributing the bulk of the nutrient loading to the lake (see Thornton, 1980; Marshall, 1997). The Mukuvisi Rivers catchment is almost wholly urban (90%), except for the lower part of the river that is occupied by commercial agriculture (intensive livestock and crop farming). Within the Mukuvisi River the main activities that are sources of pollution are outlined below.

Table 20.4. Features in each industrial area (1995) (from JICA, 1997).

Local authority	Industrial area (ha)	Main type of industry	No. of employees
Harare City	2,352	Food processing Textile products Metal products	85,845
Chitungwiza Municipality	135	Food processing Transportation equipment	2,089
Norton Town Council	273	Pulp & paper Metal products	2,960
Ruwa Local Board	211	Plastic products Metal products	2,321
Total	3,151		93,215

The Mukuvisi River, within its upper reaches, receives effluent from a PO_4^{3-} fertilizer manufacturing plant and several other sundry industries within Masasa Industrial Area. It also runs past a tobacco processing plant. After this, the river runs through the residential areas of Hillside and Braeside, the City's railway station and the main commercial area of the city. There is an area along the river as it approaches the middle reaches that is set aside for landfill programmes dating back to the 1950s. Within its middle reaches, the river passes through the residential areas of Mbare, Parktown, Houghton Park, Waterfalls, Highfields and Glen Norah. Between Houghton Park and Highfields, it receives effluent from the Southerton Industrial Area. In the Mbare residential area there is an informal industrial area and some of the waste products from here are dumped in the Mukuvisi River. In addition to pollution from the Mbare informal industry the banks of the river are used as latrines and for bathing and washing. After Glen Norah, the river receives effluent from Firle sewage Works, the major sewage treatment plant for the city serving the eastern, southern and south-western suburbs, the city centre and industrial areas. The plant was designed to treat 72,000 m^3 of wastewater day^{-1}. Currently the plant is receiving 100,000 m^3 of wastewater day^{-1}. Expansion of the sewage treatment plant is under way.

Lake Manyame sub-catchment

Lake Manyame is at risk of water pollution from two major sources: the nutrient-rich water which comes from Lake Chivero and the growth of industry and population in Norton, which was initially built as a dormitory town of Harare. The earliest work on this lake dates from 1976 (Cotterill and Thornton, 1985) when it was still in the relatively eutrophic, post-filling phase that is characteristic of new reservoirs on the Zimbabwean highveld. The loadings of N and P were relatively high but by 1979 they had fallen considerably and the lake was considered to be mesotrophic (Watts, 1982). However, by 1991 the concentrations of P and N (especially as NH_4^+) had risen markedly and it is clear that the process of eutrophication in this lake had begun.

Marimba River sub-catchment

The Marimba River, which originates in Sherwood Park and on the University of Zimbabwe campus, passes through the residential suburbs of Avondale, Belvedere, West Wood, between Kambuzuma and Warren Park I, Marimba Park and finally near Mufakose before reaching Lake Chivero. Around Belvedere suburb the river passes near the industrial area and in the Mufakose suburb it receives effluent from the Crowborough Sewage Works. There are a number of companies located in the western industrial sites of Harare including food processing companies and the Harare Power Station. The industrial activities of all companies result in wastewaters enriched in heavy metals, oils, detergents and other pollutants. The bulk of these find their way to the Crowborough Sewage Works but some of them are washed and carried to the Marimba River.

Wastewater Management within the Catchment

The urban areas in the catchment area, Harare City, Chitungwiza, Norton and Ruwa, have public sewage systems. In contrast, the rural areas such as the Rural Districts do not have such facilities and use on-site facilities (septic tanks and latrines) and the 'bush toilet' (literally means what it says). The bush toilet is a potential health hazard with regard to the spread of water-borne diseases.

Harare

There are five sewage treatment works within Harare, namely Firle, Crowborough, Hatcliff, Marlborough and Donnybrook sewage treatment works. However, Hatcliff is outside the drainage basin. Firle is the largest of the sewage treatment works, servicing more than 560,000 people plus industrial wastewater. The design capacity of this sewage treatment works was 72,000 m^3 day^{-1} but it was overloaded with about 90,000 m^3 day^{-1} and a new phase with a capacity of 72,000 m^3 day^{-1} was completed by 1997. The Crowborough Sewage Treatment Works is the second largest, having a design capacity of 54,000 m^3 day^{-1} and presently serves more than 470,000 people. The current sewage inflow comprises domestic sewage and industrial wastewater that is almost equal to the design capacity. A study carried out at the Crowborough Sewage Works during 1961–1970 showed that the works had been discharging P into Lake Chivero at an alarming rate of 70 t per annum (Watts, 1982). This has since been rectified. The Marlborough Sewage Treatment Works mainly serves Marlborough and has a design capacity of 2000 m^3 day^{-1}. This was enough to serve the old residential suburb, but is overloaded by the development of a new residential suburb almost equal in size to the old one. In Mabvuku and Tafara residential suburbs, there are four sets of stabilization pond systems that were designed to serve a total population of 79,000, of which 55,000 people were to be served by Donnybrook 1 and Donnybrook 2 and 24,000 by Donnybrook 3 and 4.

Wastewater from industry is controlled in compliance with the Urban Council Act (1995), the requirements of which are listed in Table 20.5. The purpose of this law is to prevent damage to the sewer itself and protect the health of the sewage workers who enter the sewer. There are no BOD or COD regulations, and wastewaters containing high concentrations of organic pollutants from factories are accepted into public sewers. Wastewater quality and amount is periodically monitored at the Central Laboratory in the Department of Works. The water quality check analyses the following items: pH, heavy metals, permanganate value and settleable solids, out of the ten items that are controlled by the law. Sewage tariffs are collected based on the amount and the strength (permanganate value) of the industrial wastewater. Most of the established factories have pretreatment facilities where the effluent is treated before discharging to the sewer.

Chitungwiza

In Chitungwiza town the sewage treatment works is based at Zengeza with a design capacity of 21,750 mm^3 day^{-1} serving a population of more than 400,000 people. The current sewage inflow comprises domestic sewage and industrial wastewater, and is more than the design capacity. Tilcor Industrial Area has a pretreatment facility to reduce wastewater BOD from 6000 to 1000 mg l^{-1} by anaerobic ponds. The treated effluent is then sent to the inlet works of the Zengeza Sewage Treatment Works from the pump station via a pipeline. By mid-1995 the sewage plant was heavily overloaded such that it discharged > 16,000 m^3 of partially treated sewage into Nyatsime River, which is a tributary of Manyame River on which Lake Manyame and Lake Chivero are located. During the dry season, this outflow would have constituted the main flow of River Nyatsime. The treatment process at this sewage treatment works is a combination of aerated ponds and trickling filters. The reasons for using aerated ponds instead of trickling filters are that the sewage handled is of high strength, and this simplifies

Table 20.5. Industrial wastewater standards based on the Urban Councils Act (1995).

Characteristics	Limits
pH	6.8–9.0
Settleable solids (cm^3 l^{-1})	< 10
Fats (mg l^{-1})	< 400
Mineral oils	Nil
Organic solvents	Nil
Individual heavy metals (mg l^{-1})	< 50
Calcium carbide	Nil
Bitumen	Nil
Cyanides (mg l^{-1})	Nil
Temperature (°C)	< 60

sludge treatment. The sludge from the trickling filter system is not removed within the treatment facility but is sent, along with treated water, to irrigated land.

Norton

Norton's sewage treatment works handles domestic sewage and industrial wastewater, mainly from a pulping industry, with a design capacity of 3400 m^3 day^{-1}. The treatment process employed at this sewage treatment works is trickling filters. Sewage treated by this method does not meet the effluent standards for discharge into Lake Manyame, so the whole treated sewage is kept in a storage pond until it is pumped to irrigation land along with the treated water from a nearby pulp factory.

Ruwa

Ruwa sewage treatment works has a design capacity of 5300 m^3 day^{-1} and handles mainly domestic sewage. The treatment process employed is a wastewater stabilization pond. As the sewage treated in this way does not meet the effluent standards for discharge into rivers, the treated water is sent to irrigation land. At this point, Ruwa Local Board did not have a clear policy for sludge disposal, but it is probable that the dried sludge is reused for pasture/agricultural land.

Problems of overloading

The general conditions of the existing public sewage systems of the urban areas within the catchment are summarized in Table 20.6. With the current urban population growth rates exceeding 6.5% per annum the quantity of sewerage effluent is rising rapidly and its disposal will become increasingly problematic. In March 2000, the City Council declared a moratorium on new housing developments because of chronic overloading of the sewage treatment plants. Most of the sewage effluent finds its way into the catchment's water impoundments, which have been subjected

Table 20.6. Sewered areas and population in the Manyame River catchment (Source: JICA, 1997).

Item	Harare City	Chitungwiza Municipality	Norton Town	Ruwa Local Board	Total
Municipality area (ha)	55,745	4,200	1,950	3,140	65,035
Sewerage area (ha)	Crowborough 15,200 Firle 17,777 Marlborough 378 Donnybrook 1690	2,730	970	765	39,510
Municipality population	1,264,999	400,000	30,000	26,000	1,637,399
Served population	Crowborough 470,000 Firle 590,000 Marlborough 9,000 Donnybrook 79,000	400,000	20,000	1,500	1,545,500

to the problem of eutrophication for more than 30 years. The first major blooms of blue-green algae appeared in the winter of 1960 when the dry season flow of sewage effluent was at a rate of 30,000 m^3 day^{-1} (McKendrick and Williams, 1968). By 1964 the flow had risen to 40,000 m^3 day^{-1} and algal blooms had become a permanent feature of Lake Chivero, causing serious difficulties in water purification. This may have been responsible for seasonal outbreaks of gastroenteritis among children in the city (Zilberg, 1966). A study of the performance of three large sewage works indicated excessive release of PO_4^{3-}, NO_3^-, SS and BOD from the Chitungwiza sewage works, whereas the Crowborough works achieved most standards (Zaranyika *et al.*, 1999).

Because of eutrophication problems which resulted from sewage discharge into the rivers, the sewage waters were diverted to irrigation of land. The large sewage works now rely on irrigation of large areas of pastures on municipal farms for disposal of the wastewaters leaving the settling ponds. At times, particularly during the rainy season when heavy storms occur, these tanks and pastures flood and sewage is washed directly into the rivers. Many of the pastures used for grazing cattle on these municipal farms have been treated with sewage effluents for more than 20 years. The soils are heavily loaded with P. Concentrations of resin-extractable P are up to 76–114 mg kg^{-1} in the surface 10 cm horizons compared with concentrations < 4 mg kg^{-1} in a comparable area which has not received sludge (Nyamangara and Mzezewa, 2000). It is highly likely that there is substantial leakage of P from the pasture soils into the rivers and lakes, both because of surface flow during storms and subsurface flow as the soils are saturated during irrigation periods.

Mangwayana (1995) found elevated concentrations of Cd, Pb and Ni in soils and grasses at the Crowborough works. Of particular concern were soil concentrations of Cd as high as 28 mg kg^{-1} in soil where sewage sludge had been applied; almost ten times the permitted concentration in sewage sludge-treated soils of 3 mg kg^{-1} in the European Union and the UK (CEC, 1986; DoE, 1989). Nyamangara and Mzezewa (1999) also found major contamination of the soils with Zn, Cu, Ni and Pb. Where only the effluent waters from the sewage works were used (i.e. where the majority of the suspended particulate matter was removed), the concentrations of heavy metals in soil were much smaller, reaching for example 1.6 mg Cd kg^{-1} (Mangwayana, 1995). This indicates that the effluent waters could be managed as a resource for irrigation without much further purification. A detailed study of flows of water and sewage loading rates to land supported this conclusion (Manjonjo, 1999).

The cattle grazing this land are likely to suffer acute toxicity problems from the heavy metals in the pastures and soil. They graze land currently under irrigation and can be seen drinking directly from the irrigation channels, so they are likely to ingest substantial amounts of contaminated sewage directly. Meat from the animals is sold for human consumption but potential health risks have not been investigated to our knowledge.

The Problem of Multiplicities of Responsibility

The water hyacinth problem in Lake Chivero is a prime example of the complexity of environmental management in Zimbabwe. In 1984 the Natural Resources Board identified the reappearance of water hyacinth in Lake Chivero as an urgent problem. However, the control of the weed is the responsibility of Department of Agricultural and Extension Services (AGRITEX) through the Noxious Weeds Control Act. The effluent control responsibilities are entrusted to the Water Pollution Control Advisory Board, in the Ministry of Lands and Water Resources, which administers the Water Act. Within the catchment, there are urban authorities that have corporate status and communal areas where there are no identifiable responsible authorities for water resources and pollution management. In the latter case, any Act is practically unenforceable as it is targeted at individual or corporate respondents. To address this problem of multiplicity of responsibility there is urgent need to produce an internally cohesive institutional framework for the management of water resources and especially wastewater management.

Interference of Water Quality with Commercial and Recreational Activities

The quality of raw water has been deteriorating over the years; the first signs of serious eutrophication were noticed at Lake Chivero in 1963. There is sparse published data on nutrient loadings but concentrations of MRP up to 15 mg l^{-1} and NO_3^--N up to 47 mg l^{-1} were recorded in 1995 (Zaranyika *et al.*, 1996). The eutrophication of the catchment's river system and water impoundments, Lake Chivero in particular, has led to large amounts of algae and floating aquatic plants, particularly water hyacinth (*Eichornia crassipes*), water lettuce (*Salvinia molesta*) and water fern (*Azolla filiculoides*). Rooted macrophytes such as *Lagarosiphon major* are also very common in lakes and these aquatic systems have diverse populations of aquatic fauna. About 28 species of fish are known from the Upper Manyame impoundments (Marshall, 1982), though few occur in the most polluted sections of the rivers within the catchment (Marshall, 1999). The productivity of the different fish species has tended to increase with the increase in the eutrophic nature of the impoundments, though local extinctions have occurred as a result of the combined effects of periodic drought and pollution (Marshall, 1999). Periodic fish kills occurred in the 1960s and 1970s and actually led to diversion of sewage effluents to irrigation of pastures, so that the problem of eutrophication of the waters was considered to be solved (Thornton, 1981). However, in 1996 thousands of green-headed tilapia (*Oreochromis macrochir*) died as a result of deoxygenation of the water triggered by a seasonal turnover of the water (Moyo, 1997). Thus as the sewage loadings have increased, the problem of eutrophication of the waters has become increasingly acute.

The floating aquatic plants interfere with commercial fishing and recreational use of Lake Chivero in particular. There is a need to control the levels of the aquatic fauna as there are negative effects of competition with fish for O_2 and, in some cases, sunlight. The fisheries industry is negatively affected as the water hyacinth interferes with nets. Also the aquatic flora hampers the navigation of boats and canoes within the lakes. Methods of controlling *E. crassipes* have included spraying with herbicides, manual removal and more recently, biological control using a weevil, *Neochetina eichorniae*. The first method was effective but also interfered with the harmonious interdependence of the flora and fauna and some species were removed. As an alternative, manual removal was adopted but this method of *E. crassipes* destruction was found to be prohibitively expensive. To save on costs and maintain the balance in the aquatic ecology while at the same time eliminating the *E. crassipes*, biological control methods were introduced whose results, so far, are promising.

Other problems include algal blooms, poor sedimentation, filter clogging, colour, smell and taste. The dominant blue-green algae in the lakes are *Mycrocystis aeruginosa* and *Anabaena flos-aquae*, both cosmopolitan species in eutrophic waters. These are reported to occur in dense blooms with up to several million colonies per litre (Falconer, 1973). The algal blooms cause problems in water purification, leave unpleasant tastes and odours in the water and may secrete toxins that cannot be removed by normal water purification methods. However, these algal blooms are responsible for very high levels of primary productivity that, in turn, promotes secondary production. High pH, alkalinity and low turbidity have made the water treatment process very difficult, hence the need for flocculant aids. This also has implications for the treatment costs of the raw water. Jarawaza (1997) showed marked increases in dissolved and suspended solids and hardness in the raw water from Lake Chivero between 1988 and 1995 (Table 20.7).

Table 20.7. Lake Chivero raw water quality from 1988 to 1995 (source: Jarawaza, 1997).

	1988	1989	1990	1991	1992	1993	1994	1995
Total solids (mg 1^{-1})	209	131	135	158	359	254	364	430
Dissolved solids (mg 1^{-1})	207	129	132	154	356	251	359	420
Suspended solids (mg 1^{-1})	1	1	3	4	32.6	3	5	10
Turbidity	1.4	0.9	1.5	2.2	—	—	—	—
Total hardness (mg 1^{-1} $CaCO_3$)	50	50	62	75	87	94	116	132
BOD (mg 1^{-1})	0.8	0.7	0.9	1.5	3.8	—	6.1	5.5

The Future

It is not clear at present what can be done to solve the chronic problems of pollution in the Upper Manyame catchment. Effective treatment of the ever-increasing volume of sewage will require massive investment. Even if a means of diverting the polluted waters and sludges from the lakes is found, an alternative disposal route will have to be found. Underfunding of the City Council also means that there is no effective policing or control of industrial pollution of the waterways within the catchment.

A proposal for construction of a dam to the east of Harare, outside the Manyame catchment, to provide a source of drinking water for the city has been suspended and awaits finance. This would obviously help enormously in terms of provision of a clean supply of drinking water, but does not address the underlying problems of largely uncontrolled pollution because of a lack of investment in environmental protection. Although there is widespread awareness of the problems, a general lack of funds and competing priorities means that little is being done to assess or address what is undoubtedly an increasingly serious impending ecological disaster.

Acknowledgements

We thank the UK Department for International Development (DFID) Crop Post Harvest Programme for partly financing this review. Any views expressed are the sole responsibility of the authors.

References

Bowyer-Bower, T.A.S. and Tengeh, G. (1995) The environmental implications of (illegal) urban agriculture in Harare, Zimbabwe. Paper presented to the *Workshop on the Environmental, Social and Economic Impacts of (Illegal) Urban Agriculture in Harare, Zimbabwe. University of Zimbabwe, Harare.*

Bowyer-Bower, T.A.S., Mapaure, I. and Drummond, R.B. (1995) Effects on the natural vegetation of cultivation of the vleis: case study of urban agriculture in Harare, Zimbabwe. Paper presented to the *Workshop on the Environmental, Social and Economic Impacts of (Illegal) Urban Agriculture in Harare, Zimbabwe. University of Zimbabwe, Harare.*

CEC (1986) Council Directive on the protection of the environment, and in particular of soil, when sewage sludge is used in agriculture, Directive 86/278/EEC. *Official Journal of the European Communities* No L181/6–12.

Cotterill, N.G. and Thornton, J.A. (1985) Hydroclimate development following impoundment in a tropical African man-made lake Lake Robertson, Zimbabwe. *Journal of the Limnological Society of Southern Africa* 11, 54–61.

CSO (1997) Central Statistical Office, Harare.

DoE (1989) *Code of Practice for Agricultural Use of Sewage Sludge.* HMSO, London.

Drakakis-Smith, D. (1995) Socio-economic aspects of urban agriculture in Harare: a preliminary report of field research in 1994/95. Paper presented to the *Workshop on the Environmental, Social and Economic Impacts of Illegal Urban Agriculture in Harare, Zimbabwe. University of Zimbabwe, Harare.*

Falconer, A.C. (1973) The phytoplankton ecology of Lake Mcllwaine, Rhodesia. MPhil thesis, University of London.

Gopo, J.M. and Chingobe, N. (1995) Salmonella contamination of recycled effluent of treated sewage and urban waste-water. *Water SA* 21, 245–250.

Greichus, Y.A. (1982) Insecticides in Lake Mcllwaine, Zimbabwe. In: Thornton, J.A. (ed.) *Lake Mcllwaine: the Eutrophication and Recovery of a Tropical African Lake.* Dr W. Junk, The Hague, pp. 94–99.

Greichus, Y.A., Greichus, A., Draayer, H.A. and Marshall, B.E. (1978) Insecticides, polychlorinated biphenyls and metals in African Lake Ecosystems II. Lake Macllawaine, Rhodesia. *Bulletin of Environmental Contamination and Toxicology* 19, 444–453.

Jarawaza, M. (1997) Water quality monitoring in Harare. In: Moyo, N.A.G. (ed.) *Lake Chivero: a Polluted Lake.* University of Zimbabwe Publications, Harare, Zimbabwe, pp. 27–34.

JICA (1997) *The Study on Water Pollution Control in the Upper Manyame River Basin in the Republic of Zimbabwe.* Japan International Cooperation Agency/Ministry of Local Government, Rural and Urban Development, Harare, Zimbabwe.

Mangwayana, E. (1995) Heavy metal pollution from sewage sludge and effluent of soils and grasses at Crowborough Farm. Unpublished BSc thesis, University of Zimbabwe, Harare, Zimbabwe.

Manjonjo, M. (1999) Impact of land disposal of sewage sludge and effluent on surface water – a case study of the impact of Crowborough Farm on Marimba River in Harare, Zimbabwe. Unpublished MSc thesis, University of Zimbabwe, Harare, Zimbabwe.

Marshall, B.E. (1982) The fish of Lake Mcllwaine. In: Thornton, J.A. (ed.) *Lake Mcllwaine: the Eutrophication and Recovery of a Tropical African Lake.* Dr W. Junk, The Hague, pp. 156–188.

Marshall, B.E. (1997) Lake Chivero after forty years: the impact of eutrophication. In: Moyo, N.A.G. (ed.) *Lake Chivero: a Polluted Lake.* University of Zimbabwe Publications, Harare, Zimbabwe, pp. 1–12.

Marshall, B.E. (1999) The fishes of Zimbabwe: a century of change. *The Zimbabwe Science News* 33, 45–52.

Mbiba, B. (ed.) (1996) *Urban Agriculture in Zimbabwe Implications for Urban Management and Poverty.* Ipswich Book Company Ltd, Avebury, UK.

McKendrick, J. and Williams, R.K. (1968) The effects of urban drainage on Lake Mcllwaine, Rhodesia, and subsequent water purification difficulties. Paper presented to the *Conference of the Institute of Water Pollution Control, East London.*

Mhlanga, A.T. and Madziva, T.J. (1990) Pesticide residues in Lake Macllawaine, Zimbabwe. *Ambio* 19, 368–372.

Moyo, N.A.G. (1997) Causes of massive fish deaths in Lake Chivero. In: Moyo, N.A.G. (ed.) *Lake Chivero: a Polluted Lake.* University of Zimbabwe Publications, Harare, Zimbabwe, pp. 98–104.

Munzwa, K. (1982) Land use survey of the Upper Hunyani Catchment. In: Thornton, J.A. (ed.) *Lake Mcllwaine: the Eutrophication and Recovery of a Tropical African Lake.* Dr W. Junk, The Hague, pp. 43–59.

Nyamangara, J. and Mzezewa, J. (1999) The effect of long-term sewage application on Zn, Cu, Ni and Pb levels in a clay loam soil under pasture grass in Zimbabwe. *Agriculture, Ecosystems and Environment* 73, 199–204.

Nyamangara, J. and Mzezewa, J. (2000) Effect of long-term application of sewage sludge to a grass pasture on organic carbon and nutrients of a clay soil in Zimbabwe. *Nutrient Cycling in Agroecosystems* (in press).

Thornton, J.A. (1980) *A Review of Limnology in Zimbabwe: 1959–1979. National Water Quality Report Series* No. 1. Ministry of Water Development and Department of National Parks and Wildlife Management, Harare.

Thornton, J.A. (1981) Lake Mcllwaine: an ecological disaster averted. *Zimbabwe Science News* 15, 87–88.

Watts, C.J. (1982) An examination of phytoplankton nutrient limitation in Lake Macllwaine and the Hunyani River system. In: Thornton, J.A. (ed.) *Lake Mcllwaine: the Eutrophication and Recovery of a Tropical African Lake.* Dr W. Junk, The Hague, pp. 117–133.

Zaranyika, M.F., Mtetwa, L., Zvomuya, S., Gongoraad, G. and Mathuthu, A.S. (1996) The effect of industrial effluent and leachate from landfills on the levels of selected trace heavy metals in the waters of Upper and Middle Mukuvisi in Harare, Zimbabwe. *Bulletin of the Chemical Society of Ethiopia* 7, 1–10.

Zaranyika, M.F., Mahamadi, C. and Mathuthu, A.S. (1999) A comparative study of the performance of Chitungwiza, Marondera and Crowborough sewage treatment plants in Zimbabwe. *Journal of Applied Science in Southern Africa* 5, 1–7.

Zilberg, B. (1966) Gastroenteritis in Salisbury European children. *Central African Journal of Medicine* 12, 164–166.

21 Dryland Salinization: a Challenge for Land and Water Management in the Australian Landscape

John Williams, Glen R. Walker and Tom J. Hatton
CSIRO Land and Water and Perth Laboratories, GPO Box 1666, Canberra, ACT 2601, Australia

Key words: discharge, drainage, groundwater, leakage, Murray–Darling Basin, recharge, runoff, salinity, vegetation (native, remnant) water table

Why Do We Need to Worry about Dryland Salinity?

Dryland salinity is undoubtedly the greatest and most intractable threat to the health and utility of Australia's rivers, soil and vegetation. Saving the waters and land of the Murray–Darling Basin (MDB) from salinization is one of the biggest environmental challenges facing Australians, both in the scale of the problem and in the time needed to turn it around. The 1998 report to the Prime Minister's Science Engineering and Innovation Council (PMSEIC) on dryland salinity noted that 'the time scales over which salinity establishes itself, spreads, and has its effects can be long, but once established it can be very difficult or impossible to contain or reverse. As a consequence salinity must inevitably continue to get worse in Australia as a result of land use decisions already made.'

In 1998, PMSEIC estimated that the costs of dryland salinity include AUS$700 million in lost land and AUS$130 million annually in lost production. The effects of dryland salinity include increasing stream salinity, and losses of remnant vegetation, riparian zones and wetland areas. Salinity is degrading rural towns and infrastructure, and crumbling building foundations, roads and sporting grounds.

The problem is not new (Wood, 1924; Martin, 1971; Talsma and Philip, 1971) and is certainly not under control. We can expect the effects of dryland salinity to increase dramatically (see Fig. 21.1 for the estimated upper limit of the extent of the salinity hazard). For example, if we do not find and implement effective solutions, the area of land affected by dryland salinity is likely to rise to between 10 and 12 Mha over the next 50 years.

To place this in context, the area of dryland cropping is *c.* 45 Mha, of which *c.* 18 Mha are reliable cropping land. Thus, the area of land (15 Mha) that could be salinized in the next 50 years could be similar to the area of quality cropping land in Australia. Western Australia has the greatest area of dryland salinity at present (1.8 Mha) with the potential to rise to 6.1 Mha; all the rivers of south-western Western Australia are salinized or salinizing. A similar picture is emerging for South Australia, where Jolly *et al.* (2000a) showed that all surface waters are either salinized or at risk of serious salinization. New South Wales is of critical concern, with 7.5 Mha potentially at risk, much of which will endanger MDB rivers, groundwaters (Leaney and Herczeg, 1999) and the related ecological systems. While land lost to production poses a significant threat to agricultural production, its primary impact is in the salinization of previously fresh rivers. If this is not controlled, there will be a marked decline in the viability of much of the MDB. Salinity will damage the supply of drinking water and water suitable for irrigation, with serious economic, social and environmental consequences for rural and urban communities.

Stream salinity is therefore a major concern. Projections for the town of Morgan, a key location used to monitor the effect of salinity in the lower part of the Basin, and in particular, the water supply for Adelaide, illustrate the problem. Here, the salinity of the River Murray is expected to increase by a further 240 EC units ($\mu s\ cm^{-1}$) over the next 50 years. This will bring salinity in this part of the river close to the World Health Organization's (WHO) limit of 800 EC units for desirable drinking water, and create concern for its long-term sustainability as a source of water for urban and agricultural use. In most northern tributaries of the Darling River in Queensland and New South Wales, it is expected that river salinity will rise to levels that seriously constrain the use of river water for irrigation. Not only is irrigated agriculture in jeopardy, but salinity rises will seriously threaten the biodiversity and ecosystem function of the floodplains, riverine habitat and wetlands. Many of these riverine environments have very high biodiversity values.

Salinization occurs in both dryland and irrigation farming (MDBMC, 1987). Replacing native vegetation or applying irrigation causes more water to enter the landscape than can drain from it. This increased amount of water moving in the landscape mobilizes and relocates salt

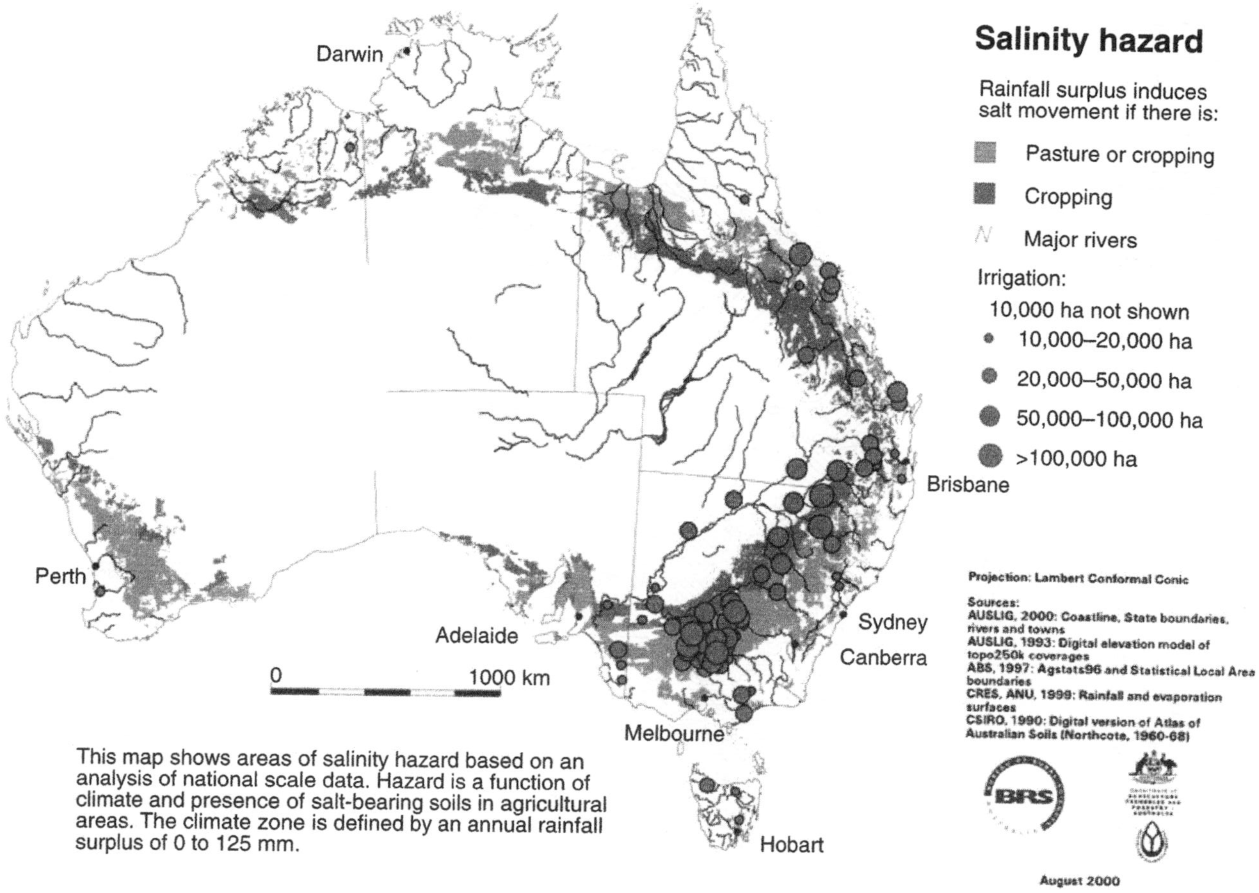

Fig. 21.1. Estimate of the areas of the Australian continent in which there exists a potential dryland salinity hazard or salinity hazard as consequence of irrigation (published by Bureau of Rural Sciences at http://www.affa.gov.au/content/output.cfm?).

stored in the soil, regolith or groundwater. Once water moves near the soil surface or discharges there, the salinization process is promoted by the evaporation of soil water containing small quantities of soluble salts. This concentrates the salts and deposits them on the soil surface, where they are readily washed to surface streams and rivers. In other geological and hydrological circumstances, the mobilized salt is carried in the groundwater to the streams and rivers along the discharge pathways in the landscape. It is possible, therefore, to have salinization of streams and rivers with very little salinization of land.

Salt in the Australian Landscape

Australia's geological history makes it different from most other parts of the world. The very ancient, flat, continental landmass has been stable through enormous lengths of geological time (see Fig. 21.2 for

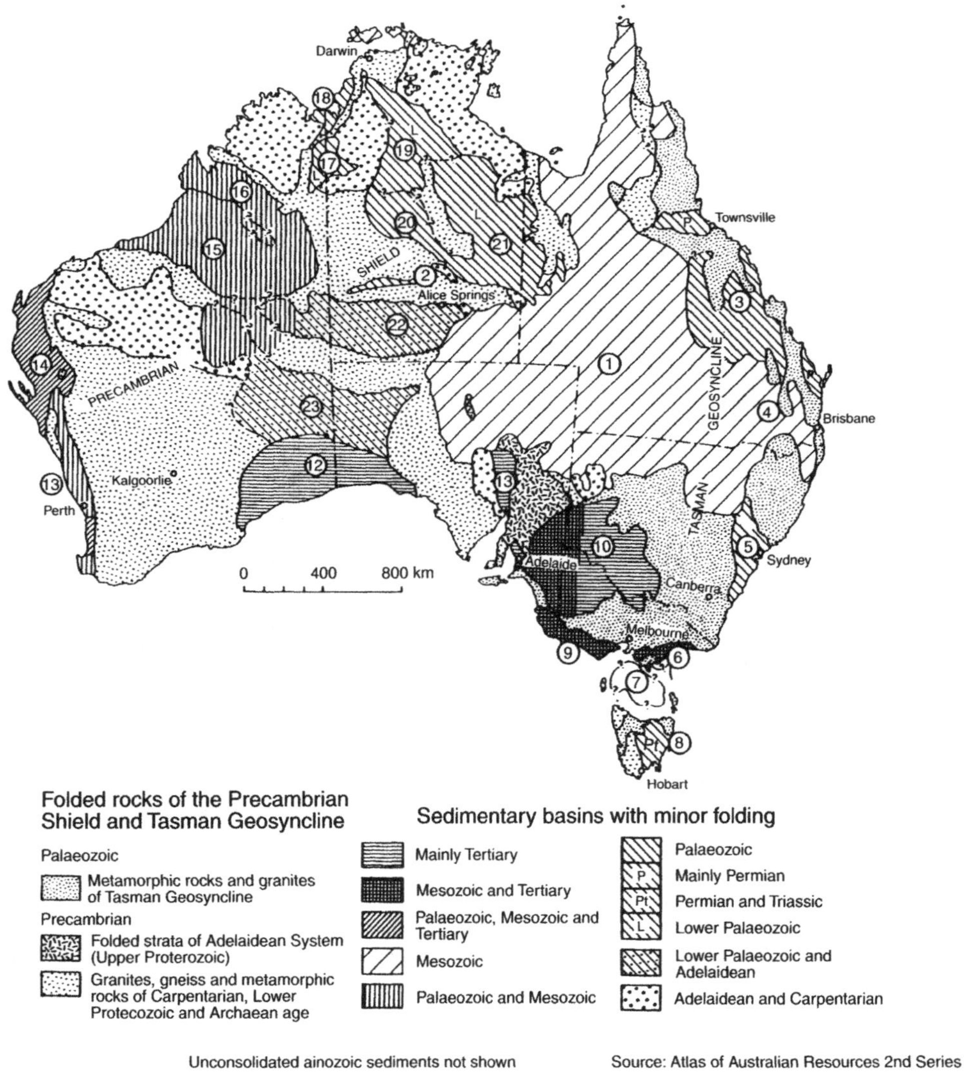

Fig. 21.2. The geological framework of Australia (redrawn from AWRC, 1975).

the geological framework). Land surfaces and rocks have weathered, eroded, mobilized, accumulated and redistributed sediments and salts over this long, stable history (Beckmann, 1983; Isbell *et al.*, 1983; White, 1997).

In addition, Australia has existed as a lone island floating in a vast expanse of ocean for many millions of years. While some of the salts in the landscape are released from weathering rocks, particularly marine sediments, most are carried from the surrounding oceans in rain to be deposited, trapped and accumulated in the soils, regolith, lakes and groundwater. This has been the pattern for millennia (Holmes, 1971; Isbell *et al.*, 1983; Simpson and Herczeg, 1994). The accumulated salts were blown and redistributed across the landscape during the extremely dry periods of geological time. Enormous amounts have remained in the landscape because there is little capacity to drain the continent of water and its salt burden. The continent is very flat, and dominated by a gentle fall towards the interior, so most of our rivers and groundwater systems are sluggish, with only a small capacity to move salt from the continent. Because these arteries and veins of the Australian landscape are not in good condition, they have limited capacity to drain the continent.

By its very nature, the continent must accumulate salt weathered from its ancient rocks and deposited from the surrounding oceans. Much of it amassed in the arid central regions as saline lakes, seeps and depressions, of which Lake Eyre is a good example. This stored salt is distributed widely across the semiarid and arid landscapes of Australia, frequently occurring in patchy, complex patterns that reflect remnant features of ancient sequences of climate and geological events that characterize the evolution of the continent. These salt storages stretch in a huge arc from north Queensland, extending south and adjacent to the Great Dividing Range, then broadening and sweeping south-west across the Murray–Darling Basin to encompass the Riverina and Mallee regions of New South Wales, Victoria and South Australia. In the Western Australian landscape, massive amounts of salt are stored in a large arc that sweeps south and east across the semiarid and arid landscapes of south Western Australia (see Fig. 21.1 and Holmes, 1971).

A combination of natural factors has led to Australia's significant problems with dryland salinity. The continent is geologically old and stable. The climate is very dry and highly variable, with low rainfall compared with potential evaporation. This low, highly variable rainfall pattern means we have one of the lowest amounts of runoff (see Fig. 21.3 for definition) to rivers and deep drainage to groundwater in the world (McMahon, 1992). As a result, most rainfall is used by native plants and only small amounts leak to the groundwater (Holmes and Colville, 1970; Holmes, 1971; Allison and Hughes, 1972; Allison *et al.*, 1985). This combination of ancient, flat, low-permeability landscapes results in conditions conducive to the accumulation of salt, which is not flushed from the landscape by leaching. The consequence is that saline lakes, streams and land are a natural part of the Australian landscape, and native vegetation has adapted to these unique conditions.

The Landscape before Clearing

In most cases, native vegetation in semiarid Australia is dominated by trees or woody shrubs, with only limited areas of natural perennial grasslands (Cocks, 1992). This perennial vegetation, with its relatively deep roots, has become effective at taking full advantage of any available water (see Fig. 21.3 for the terms in the hydrological cycle). As a result, native vegetation can use most of the rainfall in ways that have minimized the amount of water that leaks past the root zone (Holmes, 1971; Allison and Hughes, 1972; Allison *et al.*, 1985; Williams *et al.*, 1997; Williams *et al.*, 1998). The interception of rainfall on leaves, branches and stems and its subsequent evaporation before

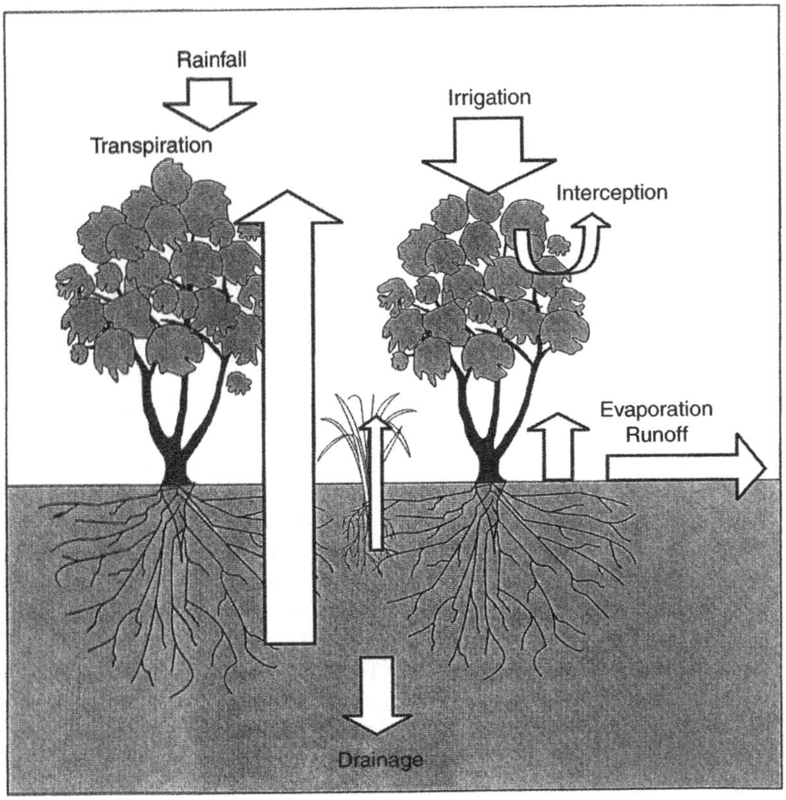

Fig. 21.3. The terms of the hydrological cycle. An increase in the drainage term through vegetation management drives the salinization process. The change can be brought about by decreasing the magnitude of any one or any combination of the other terms (diagram is courtesy of Dr Val Snow, CSIRO Land and Water, Canberra).

reaching the soil was an important strategy (Hillel, 1990) in the hydrological cycle in native vegetation. Further, some native vegetation, including native perennial grasslands, sheds rain as overland flow (Dunin *et al.*, 1999), while most use strategies of storage in soil and subsequent evapotranspiration (Nulsen *et al.*, 1986).

Whatever the dominant combination of these mechanisms in our diverse native vegetation, the 'leakage' of excess water (the 'drainage' term in Fig. 21.3) into the deeper soil below the roots is usually small, if it happens at all. Various studies have shown that over most of Australia's current dryland grazing and cropping areas, this leakage was commonly between 1 and 5 mm year^{-1} (Allison and Hughes, 1972; Allison *et al.*, 1985; Williamson, 1998). Over thousands of years, the minimal leakage has allowed the salts introduced through rainfall or rock weathering to build up in the soil below the depth of plant roots. The native vegetation evolved over a long time so that the deep drainage or leakage beneath the plant roots into the landscape's internal drainage systems was approximately the same as the drainage rates or discharge rates from the landscape. It is well documented that healthy native ecosystems within catchments are in hydraulic and salt balance (Peck and Hurle, 1973; Peck, 1976). The input of salt to the catchment is balanced by the salt discharged from the catchment. Once clearing occurs and agriculture is introduced, the salt discharged from the catchment begins to greatly exceed the salt entering the catchment (Peck and Hurle, 1973; Peck, 1976; Simpson and Herczeg, 1994; Jolly *et al.*, 2000b).

What Happens after Clearing

European settlers have unintentionally changed the hydrology of the Australian landscape to a remarkable degree in a relatively short time. Large-scale clearing of native vegetation and its replacement with annual crops and pastures have substantially increased the amount of water leaking beneath the root zone and entering the internal drainage and groundwater systems of the landscape (Williamson, 1998). These increased amounts of water now entering the groundwater under land used for agriculture greatly exceed the capacity of the groundwater systems to discharge the additional water to rivers and streams. Since the input exceeds the output to the groundwater, the water table must rise (see Figs 21.4 and 21.5).

As the water table rises, more water is discharged to the land surface as seepage comes to the surface (usually at lower positions in the landscape). Wherever this groundwater contains or intercepts salt stored in the soils or regolith of the landscape, salt is mobilized to these seepage faces, and hence to the land surface, streams, rivers and wetlands. This simple salinization process operates at a number of scales in the landscape (see Fig. 21.6).

The process can occur in local, shallow groundwater on a hillslope stretching over less than a kilometre, where seepage faces develop as the slope flattens near the stream. Alternatively, it can occur in large, regional groundwater basins stretching over hundreds of kilometres, where the salinization is expressed on the lower parts of the basin and on the floodplains. The amount of water leaking into the groundwater system depends on various factors, including the climate (particularly the distribution and amount of rainfall), the depth, water storage capacity and permeability of soils and subsoil, and vegetation characteristics. For example, any water leaking beyond the root zone does not always end up in groundwater; in certain situations it can move laterally through the soils and end up in surface streams. In other situations, leakage can occur from the base of the streams into groundwater systems. Once the leakage beneath the root zone is increased and this water begins to move along pathways that connect salt stored in the landscape to land surfaces, and intersects with rivers and streams, the dryland salinization march has begun.

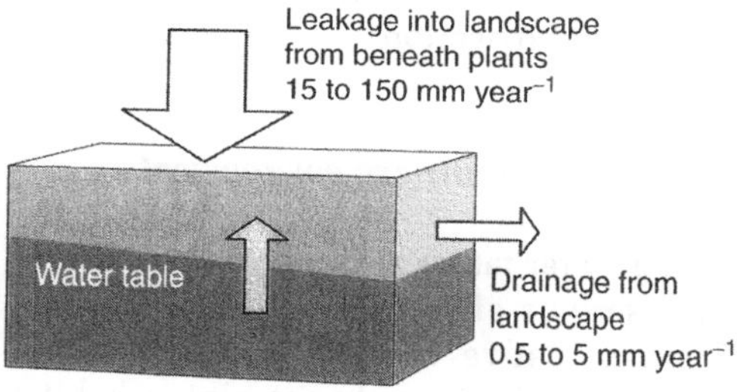

Fig. 21.4. Schematic diagram of the fundamental cause of dryland salinization. The leakage of water beneath the root zone of the replacement plants greatly exceeds the drainage capacity of the landscape. The drainage capacity of most Australian landscapes is low (0.5–5 mm year^{-1}), while the drainage beneath annual agricultural plants is high (15–150 mm year^{-1}). Hence, the rise in water table, interception of salt and the subsequent salinization of water and land.

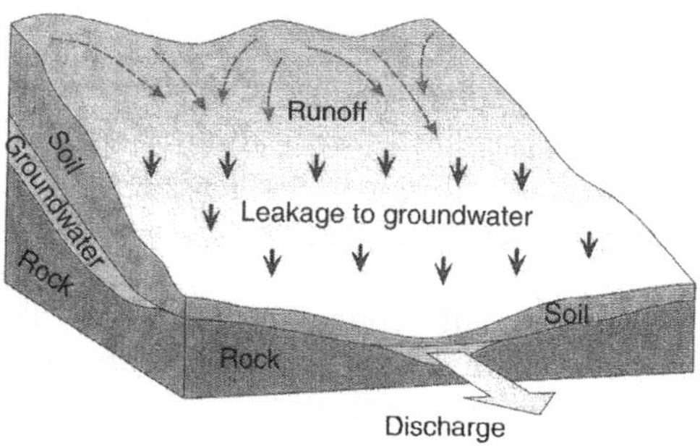

Fig. 21.5. Schematic diagram of groundwater flow with geologically constrained discharge that is common in the Australian landscape. This makes it easy for leakage to groundwater from agricultural crops and plants to exceed the discharge capacity and drive the rise in water tables and the salinization process. The native vegetation had leakage rates to groundwater that were similar, over time, to the discharge capacity.

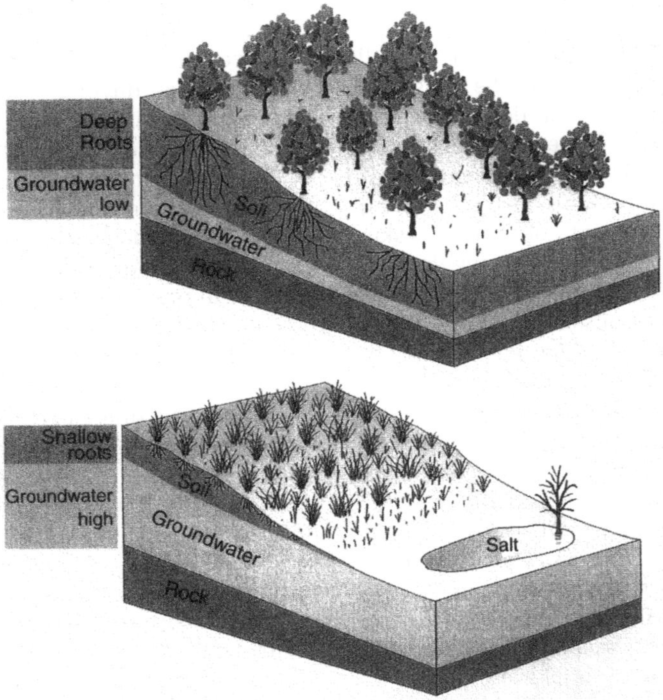

Fig. 21.6. Schematic diagram of the dryland salinization process following clearing of native deep-rooted perennial vegetation and its replacement with annual crops and pastures.

Effects of Salinity on Land and Water

In 1998, PMSEIC outlined the effects of dryland salinity. These can be summarized as follows:

Stream salinity

Increasing salt concentrations can be observed in many streams and rivers, particularly in the southern half of the MDB (see Figs 21.7 and 21.8). Rising groundwater in the Basin leads to saline discharges to streams and at the soil surface, where it affects runoff quality. Salinity levels in the Murrumbidgee River are increasing at between 0.8 and 15% per annum, depending on where measurements are made. Stream salinity in the Murray currently exceeds WHO levels for about 10% of the year. These changes have significant effects on aquatic ecosystems and all extractive uses, including drinking water, mining and irrigation.

Biodiversity and environment

Rising water tables and increasing salinity have serious effects on native vegetation. Remnant native vegetation is threatened, and as this is the only remaining habitat for a variety of important animal species, these are also under threat. Riparian vegetation, critical to the stability of stream banks, and wetland areas, are already damaged and under increasing threat. For example, the 1996 situation statement for salinity in south-western Western Australia identified the extensive impact of salinity on natural values. It found that

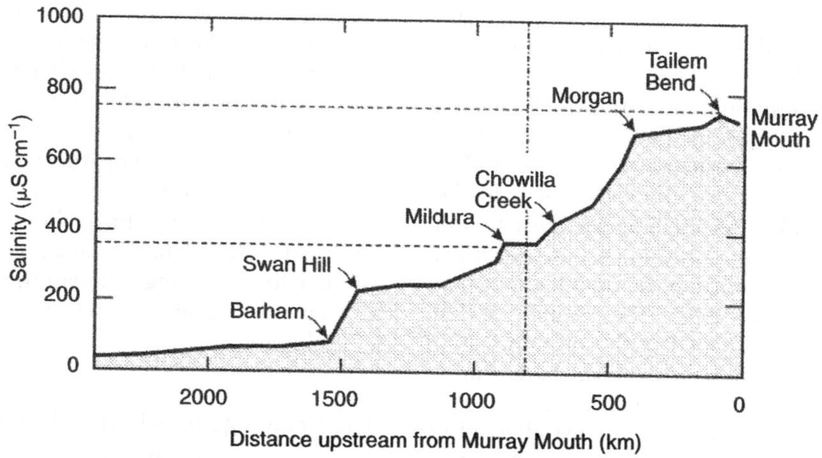

Fig. 21.7. The salt loading from the landscape to the Murray River, from the eastern highlands in New South Wales and Victoria to the estuary in South Australia (redrawn from Jolly *et al.*, 2000a).

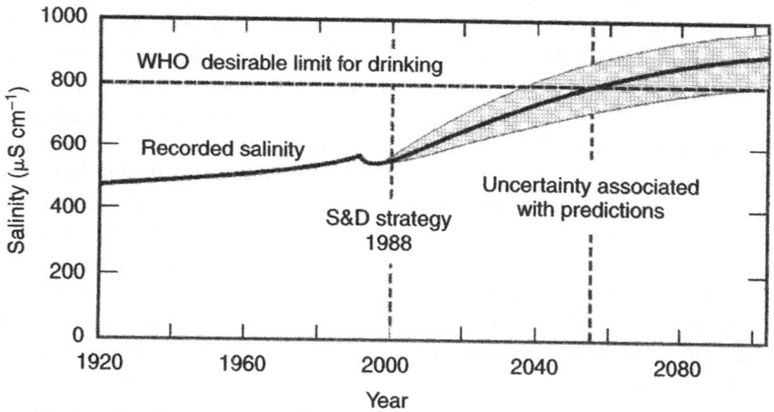

Fig. 21.8. The salinity trends over time in the Murray River at Morgan, South Australia (redrawn from Stirzaker *et al.*, 2000).

80% of the length of rivers and streams was degraded by salinity, and half the waterbird species had disappeared from the many wetlands that were once fresh or brackish. Under threat are 80% of remnant vegetation on private land, up to 50% of conservation reserves, at least one population of each of ten declared rare fauna and the entire population of a declared rare orchid. These consequences of salinity for biodiversity and nature conservation also affect tourism.

Public infrastructure

Road and bridge damage caused by shallow, saline groundwater is a major cost to local governments. It is believed that high water tables affect about 34% of state roads and 21% of national highways in south-western New South Wales, with damage costing AUS$9 million year^{-1}. Preliminary work by the MDB Commission has reliably estimated the annual cost of salinity in eight of the most salt-affected of the 22 tributary catchments at AUS$247 million $year^{-1}$ for all forms of damage to public infrastructure (K. Goss, Murray–Darling Basin Commission, Canberra, 1999, personal communication). A 1997 investigation of salinity and waterlogging on main roads in Western Australia showed that 230 km were already affected, a figure likely to double in 10–20 years and to cost an additional AUS$50–100 million in maintenance and reconstruction over the same period. Structures associated with communication and gas pipelines are subject to a similar fate. The town of Wagga Wagga in New South Wales is one of the worst affected by dryland salinity; it sustains costs of *c.* AUS$500,000 $year^{-1}$ from salinity-induced damage to roads, footpaths, parks, sewage pipes, housing and industry. Across Australia, many other towns and rural infrastructure are at risk of, and currently experience, rising water tables and consequent salinity problems (GHD, 1999).

Urban households

Saline water and high water tables affect households in various ways, from structural damage to houses and motor vehicles, through damage to hot water systems and household appliances, to increased use of soaps, detergents and water conditioners. Loss of property value is a major concern to householders and can lead them to deny the existence of problems until they are very obvious. Increased salinity in the Murray–Darling system is estimated to cost Adelaide AUS$55–65 million $year^{-1}$, largely in dealing with harder, more saline water.

Agriculture

Salinity leads to loss of productive land area and agricultural production. As noted earlier, PMSEIC estimated that the capital value of land lost is *c.* AUS$700 million, with production losses of around AUS$130 million annually, and increasing. In addition, the costs to farmers of protecting land and surface waters from salinization and/or of moving to alternative, more sustainable uses have not been determined, but are likely to be high (VCG and GNRM, 2000). The effects are being felt mainly in the grains, wool and grazing industries of the intensive land use-zone of Australia.

In summary, the high level of intervention needed to deal with dryland salinity, combined with the landscape's slow response to any changes, mean that now is the time to devise new ways to manage the problem.

The Challenge of Salinity: Options and Solutions

The recent synthesis of the nature and impact of salinity on Australia's water, soil and vegetation in the PMSEIC (1998) report, the MDBC Salinity Audit (MDBMC, 1999) and the Natural Resource Management Policy discussion paper (AFFA, 1999), has focused community attention on the economic, social and environmental costs of salinity. In a new development, governments, community leaders and scientific leaders have recognized the seriousness of the problem and the need for radical changes in land use if we are to bring salinity under control. Radical changes are necessary because farming systems that can control salinity while generating sufficient income for social and community well-being do not yet exist (Walker *et al.*, 1999; Stirzaker *et al.*, 2000) for most of the agricultural zones in Australia.

It can take a long time for salinity to establish itself, spread and have its effects, but once established, it can be difficult to contain or reverse (Fitzpatrick *et al.*, 2000). The result is that, inevitably, salinity must continue to get worse in Australia as a result of land-use decisions that have already been made. The repercussions of those decisions need not be local. In some places, the cause of the problem and the resultant salt scalds at the bottom of hill slopes are on the one property; in other cases, the cause of the problem may be hundreds of kilometres from where the symptoms become obvious.

Decisions taken now cannot prevent salinity and its consequences from becoming worse in the short term. While we should seek to slow the degradation and to learn to adapt to it, salinity has indirect, delayed and intractable affects. Australia's potential response to the salinity challenge entails serious trade-offs between short and long-term benefits, between existing primary industries and alternative land uses, between economic gain and the environment, and between rural and urban values. The management or response options are set out in Fig. 21.9 and discussed in the following paragraphs.

The complex combination of a heterogeneous landscape and socio-economic trade-offs makes a uniform national response inappropriate. The nature of the groundwater systems could be used as a basis for choosing salinity management options (Coram, 1998; Fig. 21.2). Some of the local options are to: (i) do nothing that will add to the problem; (ii) maintain current land use; (iii) adapt to a salinizing environment; (iv) adopt engineering approaches to mitigate salinization; and (v) reduce recharge.

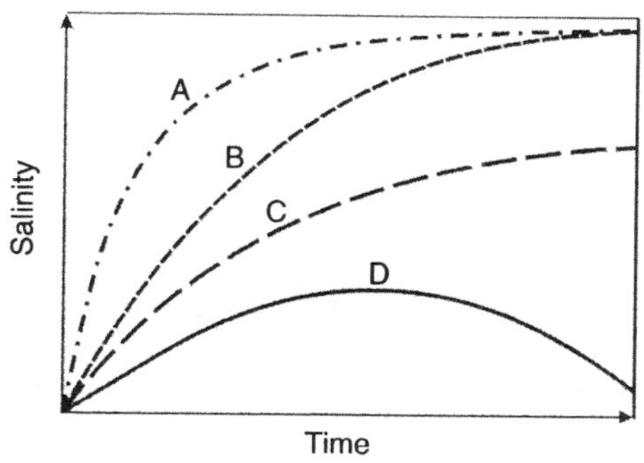

Fig. 21.9. Management options for salinity control. Option A is to do nothing; option B is to change land use and slow down the rate of salinization but adjust to living with the same extent and severity; option C is extensive and radical change in land use to slow the rate of salinization and reduce its extent and severity; option D seeks to reverse the extent and severity (redrawn from Walker *et al.*, 1999).

Do nothing that will add to the problem

It would be inaccurate to wholly ascribe the causes of salinization to past resource policy. Widespread land clearance for agriculture still occurs in places such as northern New South Wales and Queensland, posing a significant salinity hazard (Bui *et al.*, 1996; Williams *et al.*, 1997). While southern Australia has been the focus of salinity analysis, there is sound scientific evidence (Bui *et al.*, 1996; Williams *et al.*, 1997; Bui, 2000; Gordon *et al.*, 2000) that all the factors that contribute to salinity hazard exist over large areas of the semiarid zones of tropical Australia. The degree to which salinity will develop following land clearing, and the time delay for it to occur, will vary from catchment to catchment, but all the factors that contribute to salinity hazard exist throughout northern New South Wales. While there will be landscapes where salinity will not be a serious issue, the regions at greatest risk of salinization resulting from land clearing need to be established. There is an urgent need for robust analysis of salinity hazard for northern Australia as a foundation for implementing policy and good practice in vegetation management. Broad-scale land clearing with little or no regard for the salinity hazard is the recipe to repeat, over large areas of northern Australia, the huge problems that we now face in the southern and central parts of the MDB. A hazard analysis is essential to establish those areas of northern Australia at greatest risk of salinity from land clearing and vegetation. The technology and science to underpin such an analysis has been developed. An investment in this approach to avoid salinization in the first place is a lot more cost-effective than expensive action to combat active expression of salinization.

Clearly, a society committed to protecting land and water resources from salinization would take a more cautious approach to development. This precautionary approach extends to the development of new industries that, by virtue of their demand for water, will increase river salinity downstream. A more subtle precaution relates to initiatives, driven by commercial forestry, to extensively afforest the headwaters of major river systems. In these higher rainfall areas, historical clearing and associated decreases in evapotranspiration yielded greater runoff and little or no salt. Afforestation of these areas is attractive from a silvicultural point of view, but will be associated with decreased flows of fresh water downstream (Vertessy and Bessard, 1999; Zhang *et al.*, 1999). These flows are vital in diluting saline rivers.

Maintain current land use

In a number of situations, the amount of salt stored in the regolith is relatively small, the gradients are reasonable, and the rainfall is relatively high. While these areas are currently contributing salt loads downstream, there is a foreseeable horizon (perhaps only a few decades) at which discharge will begin to freshen. Eventually, farming systems yielding high groundwater recharge may contribute positively to regional water resources and dilution flows to salinity-affected river systems downstream.

Adapt to a salinizing environment

There is no question that whatever Australia's response, regional river and land salinity will continue to increase. Any national response must facilitate our adaptation to, and exploitation of, salinized resources. This may take the form of new saltland farming (Stirzaker *et al.*, 2000; Walker, 2000), inland mariculture, or other new extractive industries. Another form of adaptation is to find alternative sources of fresh water for towns and industries that currently rely on salinizing sources. In this context, aquifer storage and recovery of urban runoff and desalination are both being tried, but it is too early to be certain of their effectiveness.

Adopt engineering approaches to mitigate salinization

Enhancing drainage to control high-saline water tables, or intercepting saline groundwater by pumping to evaporation basins, can be effective strategies for controlling salinity, especially in cases of extreme necessity and urgency (see Fig. 21.10 for a brief description of a salt interception arrangement). Given the immediacy of salinity risk and the impact of salinity on important built environments (e.g. 80 towns) and natural assets (e.g. key Ramsar wetlands), no solution involving recharge control will afford timely protection, and Australia will have to look to engineering approaches to protect these assets. Similarly, large areas that are already affected, such as the regional valley systems in Western Australia, are in such an advanced state of salinization that no form of recharge control is likely to maintain current farming enterprises. Local and regional drainage schemes are in operation, under development, or planned as part of the mix of activities aimed at keeping farmers in business.

With any engineering solution, the crucial issue is the disposal of the saline

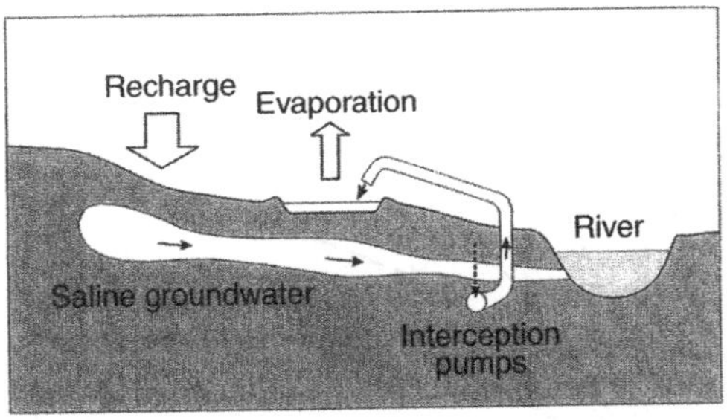

Fig. 21.10. Schematic diagram of salt interception scheme. To reduce salt loading to the Murray River, salt is intercepted by pumping saline groundwater as it moves towards the river and holding it in large evaporation basins where the water is evaporated and the salt concentrated. The opportunity to use the concentrated salts for industrial purposes is being explored and developed. Note that the water is essentially wasted as it originally leaked past the root zone of the agriculture, only to be evaporated without ever having passed through a plant.

drainage water. The management of the salt that must accumulate in all such schemes poses a significant environmental hazard. Careful feasibility studies, due diligence and understanding of environmental impact are essential. The degradation of rivers through disposal of salt can not only destroy fresh water ecosystems but, equally, it can damage natural saline river ecosystems. While many rivers in Western Australia and South Australia are already salinized, and further salinization may be acceptable, this is not the case in the MDB, where much care and attention will be required in the case of these engineering solutions. This inherently involves the trade-off between on-site benefits and off-site effects. In many out-salinizing catchments, however, the systems are either so thoroughly degraded already, or on such a steep trajectory towards degradation, that the consequences of drainage to further salinization of the rivers may be acceptable to protect high value assets.

Reduce recharge

The most widely promoted response to salinity control is to restore the original water balance (or best approximation) to ensure catchments are not leaking water in ways that mobilize salt. Leakage (see Fig. 21.11 for the influence of tree clearing) can be much greater under current agricultural systems than under natural vegetation (Holmes, 1971; Williams *et al.*, 1997; Williamson, 1998; Walker *et al.*, 1999). It occurs when the plant/soil system cannot cope with the amount of water that has fallen over a period of time. The magnitude and the periodic nature of deep drainage or

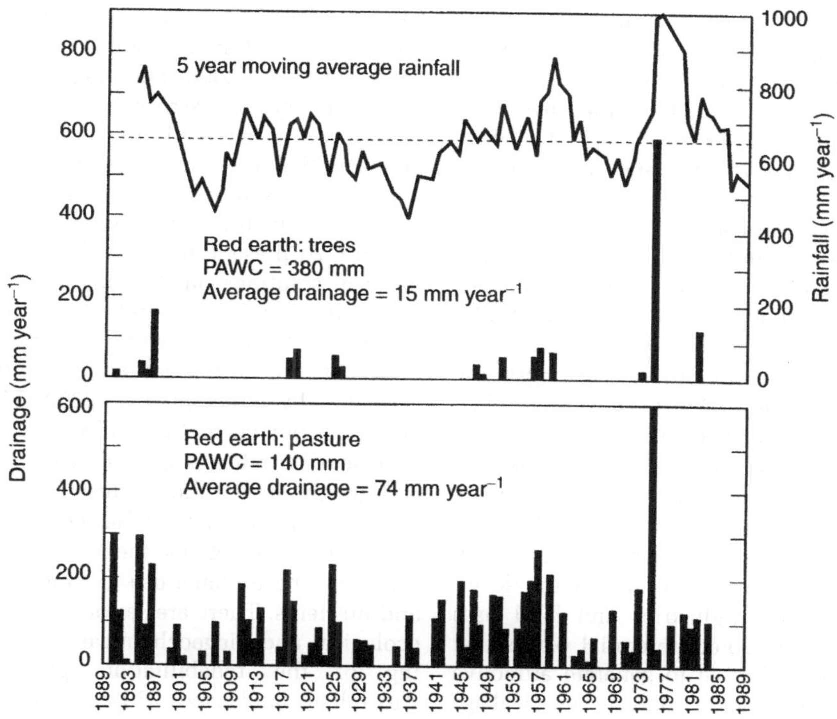

Fig. 21.11. One hundred years of simulated deep drainage (leakage) beneath trees and perennial native grass compared with perennial native grass pastures for a deep permeable soil in the semiarid tropics of North Australia near Charters Towers, Queensland. Note the greatly increased leakage beneath the grass and the strong periodicity of the leakage driven by the highly variable monsoon rainfall pattern (redrawn from Williams *et al.*, 1997). PAWC, plant-available water capacity.

leakage is caused by a complex interaction between soil, climate, amount and distribution of rainfall and vegetation characteristics such as canopy and root development.

Controlling salinity by reducing recharge requires a major shift in the water balance towards that existing in the vegetation of native ecosystems (Hatton and Nulsen, 1999). This is generally due to the extraordinarily low transmission (discharge) capacity of most of the aquifers at risk. It is so low that they rarely accommodate more recharge than that originally generated by the native ecosystems, without raising water tables (Salama and Hatton, 1999). Therefore, the approach must encompass and foster genuine commitment to change and improvement, so that our land uses and land management practices better match the capacity and capabilities of the land and meet performance standards for water balance.

The onerous recharge target implied above has two aspects. The first is the long-term average recharge rate under a particular land use; the second is the scale at which the land use must be adopted to effectively control groundwater discharge. The first aspect becomes a key issue when radical change in land use is advocated to address the cause of salinization. It is often assumed that there are land uses available that will control salinization and generate sufficient wealth to support a viable rural community. For most of the cereal cropping regions of Australia, this is not so. We have very few farming systems that can control salinization and support economically viable communities (Walker *et al.*, 1999; Stirzaker *et al.*, 2000). It seems that policy developers and most scientists remain reluctant to face this critical issue. The cause of salinity can only be brought under control by the development of new industries and land uses based on deep-rooted perennial plants that are commercial, able to generate attractive farm incomes and control the leakage beneath the root zone at levels similar to native vegetation. This is a most demanding task and will require a long-term, well-focused and funded strategy of research and development and on-farm innovation.

Recent reports about the effectiveness of current farming systems in the control of dryland salinity (Walker *et al.*, 1999; Stirzaker *et al.*, 2000) evaluated the role we can expect our current farming systems to play in controlling salinity in the future. The reports establish the prospects for new farming systems if they are to be part of strategies to control salinization by treating its cause. It is ironic that Australia is a continent that lacks both water and nutrients, yet land and water degradation, such as dryland salinity, result from an excess of water and nutrients leaking into parts of the environment where they can cause destruction (Williams, 1991, 1995, 1999). This immediately raises the prospect that if we can develop farming systems that make full use of available water and nutrients, they may be both more productive and ecologically sustainable. Unfortunately, in large areas of regional Australia we do not have farming systems that will do this at the moment (Walker *et al.*, 1999; Stirzaker *et al.*, 2000). This is demonstrated in Fig. 21.12 and in Petheram *et al.* (2000) for a range of crops and pasture systems across the MDB.

Discovering and building new land-use practices that meet these essential criteria will require solutions to scientific and technical problems that are many, complex and difficult. At present, there are few such biophysical solutions on the horizon as illustrated by Fig. 21.12 and the extensive review of field estimated recharge rates under annual, perennial and deep-rooted vegetation by Petheram *et al.* (2000). Most, if not all of our current agricultural systems, have leakage rates to groundwater systems that greatly exceed the drainage capacity of the landscape. Little work has been done on the use of native plants, their genes and the processes these plants use to capture water and nutrients. There are serious gaps in our ecological and biogeochemical understanding of the rehabilitation process in Australian landscapes (Fitzpatrick *et al.*, 2000). There is good evidence that there is very strong hysteresis in most biophysical processes associated with landscape rehabilitation and renewal. For instance, once the water tables rise following clearing, the

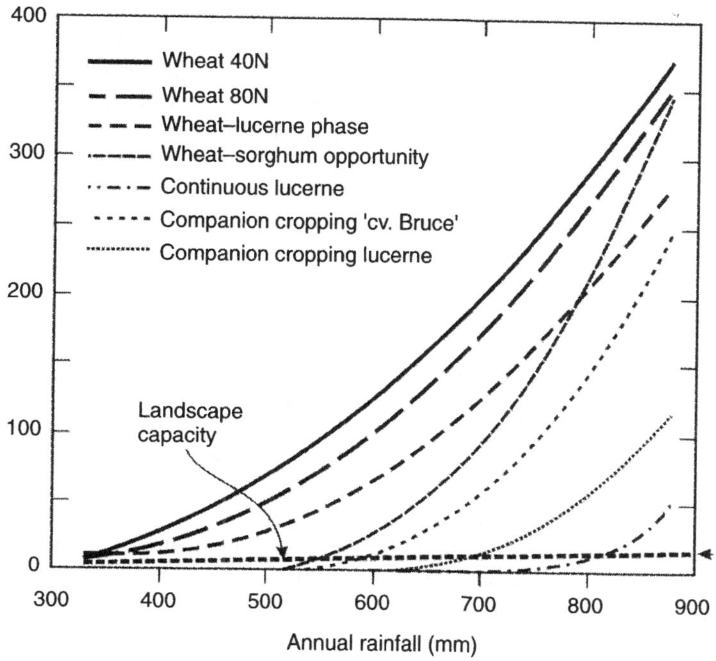

Fig. 21.12. Simulated average deep drainage (leakage) along a rainfall gradient across the Murray–Darling Basin for a series of cropping and pasture production systems compared with estimates of the landscape's drainage capacity. Most agricultural systems have leakage rates many times greater than the drainage capacity of the landscape (redrawn from Stirzaker *et al.*, 2000). 40N, 40 kg ha^{-1} of nitrogen fertilizer; 80N, 80 kg ha^{-1} of nitrogen fertilizer.

redox status of soils changes and subsequent reversal on rehabilitation releases a spectrum of chemical effects, ranging from acid sulphate conditions to mobility of Fe, Al and Si. The implications for water quality are significant. Re-afforestation to reduce groundwater recharge will usually result in reduction in water yield. These are just two illustrations of the strongly interacting biophysical processes that must be addressed in landscape rehabilitation and renewal.

With our present knowledge and capacity in system synthesis we may solve one problem only to generate others. There are encouraging prospects for new solutions arising from research, development and innovation (Stirzaker *et al.*, 2000). However, these innovative solutions, which may lead to revolutionary new ways to use the land, will need to be incorporated into the landscape not only to help deal with the growing problem of salinity, but also to maintain native biodiversity, water resources and community well-being.

One robust biophysical solution is to plant trees and expand remnant native vegetation. There is no question that establishing trees in most of the drier country at risk (< 750 mm of annual rainfall) will essentially eliminate groundwater recharge (George *et al.*, 1999; Hatton and Nulsen, 1999). However, the control that trees are likely to exert on water tables is only local to the plantings (George, 1990). Further, we lack tree crops that can compete economically with current farming systems in this dry end of the system (Hajkowicz and Young, 2000; Stauffacher *et al.*, 2000; Stirzaker *et al.*, 2000).

Even with attractive, alternative land uses that do not leak substantial amounts of water to saline aquifers, the issue of scale remains. The most widespread and pernicious impacts of salinity are, and will be, associated with more regional groundwater systems. In such systems, most of the landscape will require recharge control (Bari, 1998; Hatton and Salama, 1999).

Changing land use on this scale is a heroic challenge, economically and socially. A significant aspect of this challenge is the timescale of response. While it is possible for local catchments to respond to best management practices within several years (e.g. George, 1990; Bari and Schofield, 1991), the larger regional and intermediate systems may take upwards of 50–100 years to show any response at all. If we fail to achieve the required scale and intensity of intervention by changing land use, we will gain some time, but achieve little else.

Conclusion

Changes to the Australian landscape over the past 200 years or so have resulted in the widespread and rapidly growing problem of dryland salinity. Farmers were among the first to be affected, through salinization of rivers and agricultural land. It is becoming clear that regional and urban infrastructure, such as water supply, roads and buildings, as well as biodiversity, are at risk. It is perhaps our most serious natural resource management issue. While landholders will have a crucial role to play in reshaping the nature of agriculture and land use in Australia, landholders and rural communities cannot and should not bear the economic burden alone: scientists, politicians and society as a whole must work together to find solutions. The enormous level of intervention needed to deal with dryland salinity, and the landscape's slow response to any changes, mean that now is the time to devise new ways of managing the problem or of implementing the variety of existing responses.

Options must recognize the geographic scale of the problem and ensure that investigations and actions take place at the appropriate scale, typically regional or catchment-wide. They must address the diverse nature of the issue, and deal simultaneously with the biological, physical, social, economic and institutional factors involved. Where possible, they must ensure that proposed solutions address causes rather than merely ameliorate symptoms or transfer the salt damage to another segment of the landscape. For example, many of the engineering options essentially intercept the saline water and store it as safely as possible, while the water that was lost to productive purpose in the first place is evaporated. However, where local built and natural assets are at immediate risk, engineering interventions may be required and justified. In any event, we will be faced with the need to devise productive uses for salinized resources. This includes effective engineering solutions as well as new industries based on saline land and water.

Much of the activity currently associated with salinity control will not ultimately protect the built or natural environment. Rather, it gives us time to employ management strategies that make sufficient changes to slow down the onset and expansion of dryland salinity, but eventually allow it to reach a new, more saline equilibrium. A combination of 'buying time', devising appropriate agriculture and land-use patterns to control leakage, and developing strategies and production systems for adapting to a salinizing environment, seems the best overall approach, combined with a commitment to limit those activities that have contributed to salinization in the past. The magnitude of the required response is huge, but Australia faces little choice.

Acknowledgements

Our chapter draws on the work of many of our colleagues in CSIRO. We cannot mention them all, yet we are indebted for their important contributions to our thinking. In particular, we thank Brian Keating for access to his data and analysis; Margaret Bryant for her perspective and encouragement; Leanne Dempsey, Hester Gascoigne, Heinz Buettikofer and Greg Heath who assisted with editorial work and provided the graphics. Our colleagues Don Blackmore and Kevin Goss of the Murray Darling Basin Commission gave us financial support and the opportunity to attempt our synthesis.

References

AFFA (1999) *Managing Natural Resources in Rural Australia for a Sustainable Future – a Discussion Paper Developing a National Policy.* Agriculture, Fisheries and Forestry – Australia, Canberra, ACT.

Allison, G.B. and Hughes, M.W. (1972) Comparison of recharge to groundwater under pasture and forest using environmental tritium. *Journal of Hydrology* 17, 81–95.

Allison, G.B., Stone, W.J. and Hughes, M.W. (1985) Recharge in karst and dune elements of a semi-arid landscape as indicated by natural isotopes and chloride. *Journal of Hydrology* 176, 1–25.

AWRC (1975) *Groundwater Resources of Australia.* Australian Water Resources Council, Australian Government Printing Service, Canberra.

Bari, M.A. (1998) Reforestation reduces groundwater level and salt loads. *Proceedings of Productive Use of Saline Lands Conference, Tamworth, 9–13 March.* Land and Water, Canberra, ACT.

Bari, M.A. and Schofield, N.J. (1991) Effects of agroforestry-pasture associations on groundwater level and salinity. *Agroforestry Systems* 16, 13–31.

Beckmann, G.G. (1983) Development of old landscapes and soils. In: *Soils – an Australian Viewpoint.* CSIRO Division of Soils, CSIRO, Melbourne; Academic Press, London, pp. 51–72.

Bui, E.N. (2000) Environmental auditing-risk assessment in the face of controversy: tree clearing and salinization in North Queensland. *Environmental Management* 26, 447–456.

Bui, E.N., Smettem, K.R.J., Moran, C.J. and Williams, J. (1996) Use of soil survey information to assess regional salinization risk using geographical information systems. *Journal of Environmental Quality* 25, 433–439.

Cocks, K.D. (1992) *Use with Care: Managing Australia's Natural Resources in the Twenty-first Century.* NSW University Press, Kensington, New South Wales.

Coram, J. (ed.) (1998) *National Classification of Catchments for Land and River Salinity Control.* RIRDC Publication No. 98/78, Final Report AGS-1A. Rural Industries Research and Development Corporation, Canberra.

Dunin, F.X., Williams, J., Verburg, K. and Keating, B.A. (1999) Can agricultural management emulate natural ecosystems in recharge control in south-eastern Australia? *Agroforestry Systems* 45, 343–364.

Fitzpatrick, R.W., Merry, R.H. and Cox, J.W. (2000) What are saline soils? What happens when they are drained? *Journal of the Australian Association of Natural Resource Management (AANRM)* Special Issue (June 2000), 26–30.

George, R.J. (1990) Reclaiming sandplain seeps by intercepting perched groundwater with Eucalypts. *Land Degradation and Rehabilitation* 2, 13–25.

George, R.J., Nulsen, R.A., Ferdowsian, R. and Raper, G.P. (1999) Interactions between trees and groundwaters in recharge and discharge areas – a survey of Western Australian sites. *Agricultural Water Management* 39, 91–113.

GHD (1999) *Salinity Impact Study,* Final Report, Gutterige, Haskins and Davey Pty Ltd, Melbourne, Australia.

Gordon, I., Roth, C.H., Littleboy, M., Kalma, S. and Rogers, G. (2000) *Paddock Scale Guidelines for Salinity Management in the Balfes Creek Catchment.* LWRRDC Final Report QNR 12. Queensland Department of Natural Resources, Brisbane.

Hajkowicz, S.A. and Young, M.D. (2000) *An Economic Analysis and Cost Sharing Assessment for Dryland Salinity Management: a Case Study of the Lower Eyre Peninsula in South Australia.* CSIRO Consultancy Report to Primary Industries and Resources South Australia, CSIRO Land and Water, Policy and Economic Research Unit, Adelaide.

Hatton, T.J. and Nulsen, R.A. (1999) Towards achieving functional ecosystem mimicry with respect to water cycling in southern Australian agriculture. *Agroforestry Systems* 45, 203–214.

Hatton, T.J. and Salama, R.B. (1999) Is it feasible to restore the salinity-affected rivers of the Western Australian wheatbelt? In: Rutherfurd, I. and Bartley, R. (eds) *Proceedings of the 2nd Australian Stream Management Conference – the Challenge of Rehabilitating Australia's Streams, 8–11 February 1999, Adelaide, South Australia,* pp. 313–317.

Hillel, D.J. (1990) Saline seeps in Australia and North America. In: *Out of the Earth: Civilisation and the Life of the Soil.* The Free Press, New York.

Holmes, J.W. (1971) Salinity and the hydrological cycle. In: Talsma, T. and Philip, J.R. (eds) *Salinity and Water Use.* Australian Academy of Science, Macmillan Press, London, pp. 25–40.

Holmes, J.W. and Colville, J.S. (1970) Forest hydrology in a karstic region of southern Australia. *Journal of Hydrology* 10, 50–74.

Isbell, R.F., Reeve, R. and Hutton, J.T. (1983) Salt and sodicity. In: *Soils: an Australian Viewpoint.* CSIRO Publications, Melbourne; Academic Press, London, pp. 107–117.

Jolly, I.D., Walker, G.R., Stace, P., van der Wel, B. and Leaney, R. (2000a) *Assessing the Impacts of Dryland Salinity on South Australia's Water Resources.* CSIRO Land and Water Technical Report 00, Canberra.

Jolly, I.D., Williamson, D.R., Gilfedder, M., Walker, G.R., Morton, R., Zhang, L., Dowling, T.I., Dyce, P., Nathan, R.J., Nandakumar, N., Clarke, R., McNeill, V., Robinson, G. and Jones, H. (2000b) Historical stream salinity trends and catchment salt balances in the Murray–Darling Basin, Australia. *Marine and Freshwater Research* 52, 53–65.

Leaney, F.W.J. and Herczeg, A.L. (1999) *The Origin of Fresh Groundwater in the SW Murray Basin and its Potential for Salinisation*, Technical Report 7/99. CSIRO Land and Water, Adelaide.

Martin, A.E. (1971) Salinity and water use: future research directions. In: Talsma, T. and Philip, J.R. (eds) *Salinity and Water Use.* Australian Academy of Science, Macmillan Press, London, pp. 275–285.

McMahon, T.A. (1992) *Global Runoff: Continental Comparisons of Annual Flows and Peak Discharges.* Catena Verlag, Cremlingen-Destedt, Germany.

MDBMC (1987) *Murray–Darling Basin Environmental Resources Study.* Murray–Darling Basin Ministerial Council, Canberra, ACT.

MDBMC (1999) *The Salinity Audit of the Murray–Darling Basin.* Murray–Darling Basin Ministerial Council, Canberra, ACT.

Nulsen, R.A., Bligh, K.J., Baxter, I.N., Solin, E.J. and Imrie, D.H. (1986) The fate of rainfall in a mallee and heath vegetated catchment in southern Western Australia. *Australian Journal of Ecology* 11, 361–71.

Peck, A.J. (1976) Salinisation of non-irrigated soils and associated streams – a review. *Australian Journal of Soil Research* 16, 157–168.

Peck, A.J. and Hurle, D.H. (1973) Chloride balances of some farmed and forested catchments in southeastern Australia. *Water Resources Research* 9, 648–657.

Petheram, C., Zhang, L., Walker, G.R. and Grayson, R. (2000) *Towards a Framework for Predicting Impacts of Land-use on Recharge: a Review of Recharge Studies in Australia.* CSIRO Technical Report 28/00. CSIRO, Canberra, ACT.

PMSEIC (Prime Minister's Science, Engineering and Innovation Council) (1998) *Dryland Salinity and its Impacts on Rural Industries and the Landscape.* A paper prepared for the second meeting of the Prime Minister's Science, Engineering and Innovation Council, held in Canberra on 4 December 1998.

Salama, R.B. and Hatton, T.J. (1999) Predicting land use impacts on regional scale groundwater recharge and saline discharge. *Journal of Environmental Quality* 28, 446–460.

Simpson, H.J. and Herczeg, A.L. (1994) Delivery of marine chloride in precipitation and removal by rivers in the Murray–Darling Basin. *Australian Journal of Hydrology* 154, 323–350.

Stauffacher, M., Bond, W., Bradford, A., Coram, J., Cresswell, H., Dawes, W., Gilfedder, M., Huth, N., Keating, B., Moore, A., Paydar, Z., Probert, M., Simpson, R., Stefanski, A. and Walker, G.R. (2000) *Assessment of Salinity Management Options for Wanilla, Eyre Peninsula.* CSIRO Land and Water Technical Report 1/00, Canberra.

Stirzaker, R., Lefroy, T., Keating, B. and Williams, J. (2000) *A Revolution in Land Use: Emerging Land Use Systems for Managing Dryland Salinity.* CSIRO Land and Water, Canberra.

Talsma, T. and Philip, J.R. (eds) (1971) *Salinity and Water Use.* Australian Academy of Science, Macmillan Press, London.

Vertessy, R.A. and Bessard, Y. (1999) Conversion of grasslands to plantations: anticipating the negative hydrologic effects. *Proceedings of the VIth International Rangelands Congress, Townsville, 19–23 July*, 2, pp. 679–683.

The Virtual Consulting Group (VCG) and Griffin NRM (GRNM) Pty Ltd (2000) *National Investment in Rural Landscapes: an Investment Scenario for NFF and ACF.* CSIRO, Canberra, Australia.

Walker, G.R. (2000) Robust systems for saline land-challenges in agriculture and conservation. *Journal of the Australian Association of Natural Resource Management* Special Issue June 2000.

Walker, G.R., Gilfedder, M. and Williams, J. (1999) *Effectiveness of Current Farming Systems in the Control of Dryland Salinity.* CSIRO Land and Water, Canberra.

White, M.E. (1997) *Listen – Our Land is Crying: Australia's Environment: Problems and Solutions.* Kangaroo Press, Kenthurst, New South Wales, pp. 43–52.

Williams, J. (1991) Search for sustainability: agriculture and its place in the natural ecosystem. *Agricultural Science* 4, 32–39.

Williams, J. (1995) Farming without harming: how Australia made rural industries sustainable. In: Eckersley, R. and Jeans, K. (eds) *Challenge to Change: Australia in 2020.* CSIRO, East Melbourne, Australia, pp. 223–240.

Williams, J. (1999) Biophysical aspects of natural resource management. In: *Commodity Markets and Resource Management, Proceedings of the National Agricultural and Resources Outlook Conference, Canberra, 17–18 March.* Vol. 1, pp. 113–123.

Williams, J., Bui, E.N., Gardner, E.A., Littleboy, M. and Probert, M.E. (1997) Tree clearing and dryland salinity hazard in the upper Burdekin Catchment of North Queensland. *Australian Journal of Soil Resources* 35, 785–801.

Williams, J., Hook, R.A. and Gascoigne, H.L. (eds) (1998) *Farming Action – Catchment Reaction: the Effect of Dryland Farming on the Natural Environment.* CSIRO Publications, Collingwood, Victoria.

Williamson, D.R. (1998) Land degradation processes and water quality: waterlogging and salinization. In: Williams, J., Hook, R.A. and Gascoigne, H.L. (eds) *Farming Action – Catchment Reaction: the Effect of Dryland Farming on the Natural Environment.* CSIRO Publishing, Collingwood, Victoria, pp. 173–202.

Wood, W.E. (1924) Increase of salt in soil and streams following the destruction of native vegetation. *Journal of the Royal Society of Western Australia* 10(7), 35–47.

Zhang, L., Dawes, W.R. and Walker, G.R. (1999) *Predicting the Effect of Vegetation Changes on Catchment Average Water Balance.* Cooperative Research Centre for Catchment Hydrology Technical Report 9/99. CRC, Canberra, ACT.

22 Quantifying Nutrient Limiting Conditions in Temperate River Systems

A.C. Edwards[1] and P.A. Chambers[2]
[1]Macaulay Land Use Research Institute, Craigiebuckler, Aberdeen AB15 8QH, UK;
[2]National Water Research Institute, 867 Lakesheve Boulevard, PO Box 5050, Burlington, Ontario L7R 4S6, Canada

Key words: ecoregions, impacts, nitrogen, nutrient enrichment, nutrient guidelines, phosphorus, rivers, temperate, thresholds, trophic status

Introduction

Concentrations of N and P measured in surface and groundwaters have increased globally over the past 30 years as a direct consequence of human activities (Caraco, 1995; Carpenter *et al.*, 1998). Managing and restricting potential environmental impacts caused by nutrient enrichment of surface and groundwater is now a major scientific and political challenge. The situation is especially acute for NO_3^-; health concerns caused by consumption of contaminated drinking water supplies have resulted in stipulated maximum allowable concentrations for potable waters (CCME, 1999) and legislation to protect supplies (EEC, 1991). Despite the overwhelming evidence of increased productivity and reduced species diversity caused by nutrient enrichment, we have only a limited understanding of the relationship between nutrient concentrations and ecological response, particularly in the case of rivers. This situation also means that there is limited capacity to manipulate the key processes and transport mechanisms that control the flux and redistribution of nutrients at the landscape scale.

Effective management of nutrient transfers from land to water requires the estab-

(eds P.M. Haygarth and S.C. Jarvis)

lishment of the relationships between source, transport and impact, from which threshold concentrations can be formulated for the protection of ecosystems. Concerted efforts have resulted in progress being made on individual aspects of this problem, such as the quantification of nutrient cycling and losses (Heathwaite *et al.*, 1997), processes influencing biological availability and environmental impact (Fisher *et al.*, 1995), and predicting benthic algal abundance in rivers from nutrient concentrations (Biggs, 2000; Dodds and Welch, 2000). Fewer studies, however, have attempted to investigate the complete enrichment continuum by incorporating aspects of source, transport and aquatic ecological impact and, among those that have, the focus has been on standing waters. These studies have examined point source and, in some cases, non-point source loadings of N or P and have contributed to the development of N or P guidelines for lakes (e.g. OECD, 1982). Although certain aspects of the approach used for setting lentic nutrient guidelines may be transferable to flowing waters, the specific relationships used to develop actual threshold concentration ranges are probably not (Edwards *et al.*, 2000). The development of scientifically credible nutrient guidelines and management strategies for rivers remains an urgent research requirement.

In this chapter, we examine the linkages between terrestrial and aquatic landscapes, with respect to nutrient transfer and environmental impacts. We review nutrient sources and fluxes from terrestrial landscapes, particularly those of significance in agricultural catchments. We then examine the effects of added nutrients on aquatic systems, focusing on impacts on rivers and methodologies used to determine river trophic status. Finally, we provide advice on managing nutrient fluxes to rivers from agricultural landscapes, developing nutrient targets for rivers, and research priorities. By examining the continuum from terrestrial source to terrestrial and aquatic transport and, ultimately, to effects on aquatic biota, we argue that effective management of nutrient loading to rivers from diffuse sources requires a quantification of the linkages between nutrient source, transport and fate.

Nutrient Sources and Fluxes

Nutrients present in surface waters have a variety of origins (Fig. 22.1). Sources that contribute to the background nutrient status of rivers include runoff from unmanaged land, atmospheric deposition, groundwater discharge and tributary inputs from upstream. For any particular location and time, the residence period and potential biological availability of specific nutrients are controlled by a combination of abiotic and biotic instream processes (Munn and Meyer, 1990). In the majority of situations, these background contributions have been augmented by nutrient inputs from anthropogenic sources that include sewage treatment effluent, stormwater, industrial and drainage outfalls (point sources) or from runoff from agricultural fields and urban areas (non-point or diffuse sources). Although many of the general mechanisms responsible for redistributing nutrients at the landscape scale are similar for both N and P, important differences do exist. These differences have influenced the type and emphasis of legislation aimed at developing strategies for managing individual nutrients. The relative significance of individual transfer pathways remains open to debate because of the selectivity of individual processes involved, lag times and system inertia that influence the temporal and spatial dynamics of nutrient fluxes.

The redistribution of nutrients in food products and fertilizers which now occurs over local, regional and global scales, often crossing geo-political boundaries (Beaton *et al.*, 1995; Forsberg, 1998), has introduced a tremendous heterogeneity in the spatial distribution of nutrients. Prior to the recent intensification of agriculture with its heavy reliance on use of manufactured fertilizers, and permanent settlement, the N and P cycles were relatively ‘closed’ with very little of these elements ‘leaking’ to adjacent ecosystems. Human activity has disrupted the natural cycles of both N and P; approximately 60% of the additional anthropogenic N_2 fixation and 80% of the PO_4^{3-} produced by the world’s industries are used for fertilizer production (Steen,

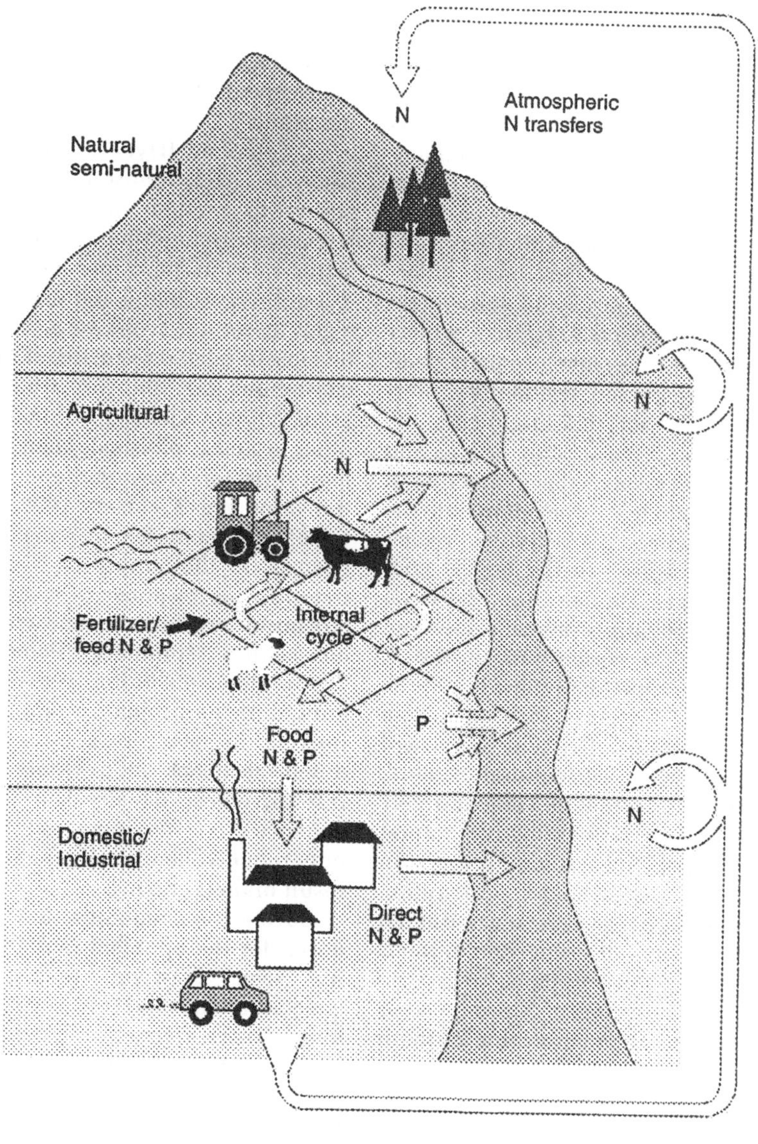

Fig. 22.1. Origins and sources of nutrients in temperate river systems.

1998; UNEP, 1999). This fixation and redistribution of N and P coupled with poor utilization efficiencies across most of the stages of food production and inadequate nutrient management have resulted in substantial nutrient surpluses (Smaling *et al.*, 1999). The cumulative effect of these surpluses means that concentrations of total and extractable N and P in many agricultural topsoils are now well above the native or background values, particularly evident for areas with high livestock density (Sims *et al.*, 2000). Thus, the drive towards intensification and specialization within agriculture together with increased urbanization of society has produced localized areas in which the soils are highly enriched in nutrients, thereby threatening the quality of surface and groundwaters. Because many of the changes to agronomic and management practices associated with intensification occurred simultaneously with the increased use of fertilizers, it is extremely difficult to attribute environmental impacts to specific individual causal agents (Forsberg, 1998).

There can be particular difficulties associated with making system level judgements for pollutant source and subsequent environmental impacts. It is widely accepted that no ecosystems remain unaffected by the increased availability of N (UNEP, 1999), a fact that highlights the need to understand the specific origins and transport mechanisms that operate at regional scales. This is evident in the example of N where the activities associated with agriculture are the largest contributors to the annual flux between terrestrial and aquatic systems. However, the significance of agriculture in the N cycle does not automatically mean that the most serious impacts arising from increased availability of N will be located downstream of agricultural land. Atmospheric deposition of comparatively small quantities of N to sensitive natural and semi-natural ecosystems areas can cause major ecosystem disruption (Vitousek *et al.*, 1997) yet receives limited attention. A substantial proportion of the N actually reaching these systems is derived from long-range transport of N produced as a by-product from the combustion of fossil fuels (notably car exhausts), thus demonstrating the need for a balanced appreciation of nutrient cycling.

Effects of Nutrients on Rivers

P and N have long been recognized as the agents driving the cultural eutrophication of lakes, rivers and coastal waters worldwide and as a basis on which responses to disturbances can be predicted (Dodds *et al.*, 1998; Williamson *et al.*, 1999). This has been demonstrated by reducing the inputs of P and N to lakes, which has been successful in decreasing phytoplankton abundance (Dillon and Rigler, 1974; Edmonson and Lehman, 1981). However, although broad relationships between particular landscape features and the structure and functioning of riverine communities have been described (Hynes, 1970; Delong and Brusven, 1992; Johnson *et al.*, 1997), our ability to predict site-specific biological responses arising from a change in nutrient status remains poor for river systems. For example, growth and abundance of benthic algae in certain rivers have been found to be closely related to instream nutrient concentrations (e.g. Bothwell, 1992; Chessman *et al.*, 1992; Kjeldsen, 1996) whereas in other systems, physical, chemical and biological factors may confound these relationships (e.g. Kjeldsen, 1996; Biggs *et al.*, 1998). The complex response of rivers to changes in the supply of N or P is illustrated in the response of a pristine tundra stream to fertilization conducted over four consecutive summers (Peterson *et al.*, 1993). Here, the initial nutrient-enhanced growth of the primary producers was suppressed in later years by a strong 'top-down' feedback loop where herbivorous insects, whose density had been initially enhanced by the increased food supply, now reduced benthic algal abundance as a result of their grazing.

Several explanations have been proposed for the apparent lack of predictability seen between instream nutrient concentrations and the response of river biota: (i) the dynamic nature of river systems which results in loss of biomass during high flows (Dodds and Welch, 2000); (ii) the downstream recycling of N or P in rivers such that a single atom taken up by an algal cell is subsequently released during decomposition and then taken up yet again by other cells further downstream (Paul and Duthie, 1988; Mulholland *et al.*, 1990); (iii) the control of plant production by higher trophic levels (i.e. 'top-down' control of plant biomass by invertebrates and fish) (Stevenson, 1997; Bourassa and Cattaneo, 1998); (iv) nutrient concentrations that far exceed physiological requirements and are thus independent of plant biomass (Bothwell, 1989); and (v) the temporal (usually seasonal) variability in responsiveness of aquatic biota to added nutrients. Despite these many factors that may confound attempts to relate the response of aquatic communities to nutrient enrichment, progress is being made in establishing quantitative models that predict benthic algal abundance from nutrient concentrations and developing nutrient targets for streams (e.g. Biggs, 2000; Dodds and Welch, 2000).

It will become increasingly important to be able to measure the environmental benefits accruing from reductions in nutrient fluxes against the economic cost of implementing nutrient management strategies. For this to be achieved, the factors responsible for variation in the response of stream ecosystems to reduced nutrient losses from terrestrial landscapes must be understood. However, our ability to relate terrestrial nutrient losses to stream response may be confounded by: (i) a non-linear relationship between reduced losses from the terrestrial system and the inputs to the aquatic system; (ii) a disconnection between the timing of nutrient release, its residence time in the terrestrial and riparian system, the timing of its entry into the aquatic system and temporal variation in the sensitivity of biota; and (iii) hysteresis within the aquatic system whereby a reduction in nutrient inputs may not have been sufficient to cross critical concentrations or threshold boundaries. The difficulties associated with making quantitative assessments across the various stages of source, transport and impact outlined for P in Table 22.1 have also recently been highlighted for the Baltic Sea. Here, action programmes initiated in 1980 to reduce N and P export by 50% over a 15-year period do not appear to have been able to meet the target (Grimvall *et al.*, 2000). These authors suggest that a serious under-estimation of the system's inertia, which controlled nutrient loss, was responsible for the discrepancy. The variability in the success of management measures that aim to improve stream ecosystems by reducing agricultural nutrient losses reinforces the need for research targeted at determining the timing of nutrient release, its residence time in the terrestrial system, the timing of its entry into the aquatic system and temporal variation in the sensitivity of biota.

Methods to Determine River Trophic Status

A wide range of methodologies is used to assess the current trophic status of rivers and, in many cases, to allow predictions of the effects caused by changes to the loading regime (Table 22.2). Water chemistry measurements are the most common form of assessment, largely because of the ease of collection and analysis. Water samples for trophic status determinations are usually analysed for reactive and unreactive forms of N and P, and sometimes dissolved O_2. Although changes in river nutrient concentrations may be used

Table 22.1. Stages that might be involved in a reduction in an environmental impact of phosphorus together with the main legislative mechanism and indicator.

Stages	Mechanism	Indicator
Agricultural P balance ↓	Farm/field nutrient planning	Surplus/deficit in relation to crop removal
Soil P status ↓	Soil testing	Critical concentration
P loss from land ↓	P index	Potential risk field scale, e.g. erosion risk
P content in river ↓	Maximum acceptable daily load	Critical concentration or load
Environmental impact	Ecological site index	Change in diversity and/or productivity

Table 22.2. A comparison of various approaches to determine the trophic status of rivers.

Approach	Advantages	Disadvantages	Examples
In situ *observations/monitoring*			
Water chemistry (N, P, DO)	Rapid, standardized methods	Represent a single point in time. No information on the effects of any chemical changes on biota	Muscutt and Withers (1996)
Aquatic plants			
Biomass	Rapid, standardized methods which provide abundance estimate for entire assemblage	No species information; any nutrient response may be masked by other factors (light limitation, herbivory, scour, etc.)	Dodds *et al.* (1997); Biggs (2000); Chetelat *et al.* (1999)
Tissue N and P	Relates changes in water nutrients directly to changes in plant nutrients	No species information; tissue nutrients may not be related to abundance	Biggs and Lowe (1994); Hillebrand and Sommer (1999)
Taxonomic composition	Robust analysis of entire assemblage	Time consuming; requires specialist knowledge in taxonomy and algal nutrient preferences	Rott (1995); Pan *et al.* (1996)
Productivity	Carbon assimilation or oxygen evolution measured *in situ* in respiration chambers; change in biomass over time measured under controlled conditions	Respiration chambers involve complex methodology and no standardized protocol	Bott *et al.* (1997); Bothwell (1988)
Enzyme activity (alkaline phosphatase activity)	Indicative of cellular phosphorus status	No species information	Hullar and Vestal (1989); Perrin *et al.* (1987)
Effects on higher trophic levels (benthic invertebrates, fish, fish habitat)	Varies with the parameter of interest May be more relevant to users	Varies with the parameter of interest; relationship to nutrients may be less clear than for aquatic plants	Wright *et al.* (1984); Bourassa and Cattaneo (1998)
In situ *experiments*			
Nutrient diffusing substrata	Measures algal biomass on replicated control and nutrient treatments. Allows comparison among reaches (if other factors such as scour, water depth, etc. are standardized)	Relies on natural colonization. Little or no control over nutrient diffusion rate. Location in river is critical as growth response is influenced by local site factors	Wold and Hershey (1999); Scrimgeour and Chambers (2000)
Immobilized algal systems	Relatively simple to use and highly reproducible. Can incorporate mixed species. Some control over nutrient diffusion rate. Uses a known algal growth/physiological state	Has only limited life once placed. Location in river is critical as growth response is influenced by local site factors	Twist *et al.* (1998)
Whole river nutrient manipulations	Provides information on direct effects of a nutrient pulse on biota. Allows whole community assessment and determination of kinetics of in-stream processes	Restricted to small streams and usually short experimental duration. Replication difficult. Difficult to extrapolate to other sites	Peterson *et al.* (1993); Perrin *et al.* (1987)

Controlled experiments			
Artificial streams	Allows determination of biomass, tissue nutrient content or taxonomic composition of the natural river assemblage or cultured species under controlled conditions in a laboratory or alongside the river	Can be expensive/complex to set up Does not identify site-specific factors that may be limiting growth	Carr and Chambers (1998);. Bothwell (1993)
Single-species laboratory tests	Determines critical concentration and provides an indication of the nutrient most likely to be limiting under controlled conditions	Often uses exotic species, which can make extrapolations to field difficult	Skulberg (1995); Nyholm and Peterson (1997)

to identify point and non-point sources of nutrient loading and patterns over time, water chemistry concentrations *per se* provide no information on biological responses. Elevated nutrient concentrations may indicate increased biological activity; conversely, other environmental factors (e.g. irradiance, current velocity) may mask the effects of nutrients, resulting in little or no change in trophic status (e.g. Biggs, 1995; Kjeldsen, 1996; Wellnitz *et al.*, 1996; Bourassa and Cattaneo, 1998). In addition, because of rapid cycling between particulate and dissolved forms, the measured concentration of a nutrient is not often indicative of availability over the longer term. There are also analytical issues associated with determining the concentration of a nutrient at the point where it actually is limiting growth (Hudson *et al.*, 2000). Low concentrations may also be indicative of either intrinsically oligotrophic conditions or the increased demand from an abundant algal population. Finally, chemical measurements typically represent a point in time, which may or may not be indicative of the average situation. Studies along a river system in Quebec, Canada, showed that up to 11 total P or N measurements were needed to obtain a mean value with an error *c.* 20% (Cattaneo and Prairie, 1995).

Biological indices are direct indicators of trophic status. Typically, plant population or assemblage characteristics are assessed; however, responses of higher trophic levels (notably benthic macroinvertebrates) may be considered. With a lifespan of weeks to months, algae and aquatic macrophytes can provide a longer-term assessment of nutrient loading to a river than single chemical measurements. Although benthic invertebrates and fish integrate environmental conditions over even longer scales (months to years), the effects of nutrient loading on organisms higher up the food web may be obscured owing to variable feeding behaviour (e.g. feeding at several trophic levels) or long distance movement (which may result in individual feeding in river reaches of different trophic status).

Measurements using aquatic plants include biomass (expressed as mass or, in the case of algae, chlorophyll *a* content per unit volume or area), taxonomic composition, tissue nutrient content, productivity (measured as C assimilation, O_2 evolution or change in biomass per unit time) or enzyme activity (such as alkaline phosphatase) (Table 22.2). These variables can be measured for either natural aquatic plant assemblages or plants grown in experiments conducted *in situ* or in the laboratory.

Natural plant assemblages provide direct evidence of the effects of nutrient loading. The strength of the relationship between river nutrient concentrations and biotic response may be greater for methods where the variable is not expressed as a function of surface area (i.e. measurements such as tissue nutrient composition, alkaline phosphatase activity or taxonomic composition). Responses that are expressed as a function of surface area (such as biomass) may be masked by other environmental factors, such as grazers that reduce total biomass, high current speeds that scour the standing crop, or increased turbidity or reduced temperatures that lower production rates (e.g. Biggs, 1995; Kjeldsen, 1996). Because natural aquatic plant assemblages exhibit temporal patterns in abundance and composition (Biggs, 1996), the timing of sampling will affect the interpretation of trophic status: most investigators attempt to control this by sampling at regular intervals (weekly or biweekly) or at the time of peak biomass (Steinman and Lamberti, 1996). Sampling programmes are usually designed to minimize spatial variability (caused by longitudinal or cross-channel variations in shading, current velocity, water depth and substrate size).

In situ experiments allow the investigator to minimize the effects of temporal and spatial variability and compare the effects of added nutrients with unamended control conditions (Table 22.2). These manipulative studies include:

- Nutrient diffusing substrata, in which algal biomass is measured on the diffusing surface of containers filled with N, P or N plus P stabilized in an agar matrix. This type of bioassay provides information on which nutrient is limiting community production and allows comparison between reaches. However, the presence of mixed species within a community could confound the results because of overlapping ranges in N and P optima.
- Immobilized algal systems, in which algae pre-immobilized in a permeable gel are deployed in the river to determine *in situ* growth rates. At present, only single species have been introduced; however, the technique could be expanded to include mixed species representative of the particular river.
- Whole river nutrient manipulations, in which nutrients are experimentally added to a stream for durations ranging from hours to months. This method provides a realistic assessment of the impacts of nutrient addition; however, logistical constraints limit such experiments to small streams and rarely with replication.

Controlled experiments range from single-species laboratory tests to those conducted in artificial streams. Laboratory growth/toxicity algal bioassays allow determination of N and P threshold values for single species and these results, when compared with instream nutrient concentrations, may allow identification of any limiting nutrient. Often the algal species used in bioassays are planktonic forms and thus the extrapolation of their responses to the benthic algae typically found in streams and rivers may be tenuous. Artificial streams (i.e. stream microcosms) have been used to investigate ecological interactions in running waters since the 1960s, including the examination of trophic relationships between algae, insects and fish (see Lamberti and Steinman, 1993, and papers therein). An important advantage of these systems is that they allow replication of treatments and controls and, when located alongside the stream, enable the use of natural river water under ambient water temperature and light regimes.

No single approach is generally adequate for assessing trophic status. Most investigators employ a combination of water chemistry and *in situ* biotic sampling. Intensive assessments that are specifically directed at predicting the effect that changes to the loading regime have on trophic status often rely on *in situ* experiments. Cumulative effects arising from multiple nutrient inputs or from inputs that may have interactive effects (e.g. a nutrient and a toxicant input) are often best evaluated using artificial streams. The need to incorporate an understanding of how the local biophysical properties of the receiving ecosystem are influenced by specific attributes inherited from upstream requires an appreciation of the site within a surrounding regional context (Roth *et al.*, 1996).

Managing Nutrient Fluxes to Rivers

Contrasting physico-chemical properties and dominant transport mechanisms between N and P (Heathwaite *et al.*, 2000) result in marked differences in their behaviour in a river system while also influencing the timing of their loss and potential biological availability. Recognition of these differences has led to separate management and legislation for each nutrient. In the case of N, management action plans have largely been directed at diffuse (or landscape scale) losses and tend to operate at the catchment scale, such as those proposed within the European Nitrates Directive (EEC, 1991). Under this legislation, surface and groundwaters that are or could be affected by pollution and vulnerable zones that contribute to pollution must be identified, and detailed farm nutrient plans and restrictions on timing and amounts of manure applications must be developed. The way in which this legislation has been implemented varies considerably among member states. For example, The Netherlands has designated the whole country as a nitrate vulnerable zone, whereas zones in the UK are smaller and targeted at protecting specific aquifers or water bodies. In the case of P, initial management strategies targeted point sources. However, because obvious improvements in water quality were not always forthcoming, the emphasis has recently switched to non-point source P control. This may take the form of limiting applications of livestock wastes such as in The Netherlands or setting maximum concentrations of extractable soil P. The development and implementation of management plans on a regional scale have been described for the Chesapeake Bay watershed by Sims and Coale (Chapter 16, this volume).

Separating the contribution that individual sources make to the total fluxes of N or P at any particular time or location in a river system and, at the same time, apportioning a direct biological impact remains a challenge. One management approach is simply to attempt a blanket reduction in the loadings of N, P or both to the receiving water. Although this may prove successful where the major inputs are derived from only a few point sources, strategies aimed at achieving blanket reductions in nutrient loading fail to fully appreciate the degree of synchronicity between timing of loss and a biological demand or response. Apportioning the total nutrient load between point and non-point sources or among individual sources represents another approach for prioritizing and targeting nutrient contributors. A frequent outcome of this approach is the focus of attention on reducing losses from readily identifiable point sources. This has been successful with respect to reducing the total load of nutrients to certain water bodies.

A more useful approach for distinguishing between nutrient sources might be the biological availability of the nutrient, the timing of its release, and its concentration and residence time, all of which influence the biological response to that nutrient. Differences in the rates of release, transport and biological availability result in temporal and spatial selectivity in the movement of nutrients within and across system boundaries. This approach necessitates action at the landscape level and ensuring that the timing of management of interventions matches the timing of losses and/or biological responses.

Establishing Nutrient Targets

Stressors of aquatic ecosystems can be broadly classified into those that have a toxic effect and those that have a non-toxic effect. Water quality guidelines for toxic stressors (e.g. organic contaminants, heavy metals, and NH_3/NH_4^+ and NO_3^- as toxicants) are derived from controlled dose-response experiments conducted on individuals or populations to determine concentrations resulting in lethality or a chronic condition. Many countries have developed water quality guidelines for toxic substances to protect human health, aquatic organisms, livestock or irrigated crops and many of these national guidelines have been developed using shared data on toxicological properties (e.g. CCME, 1999; US EPA, 1999).

In contrast, non-toxic stressors (e.g. P and N as nutrients, turbidity, flow, temperature) can change the structure or function of an ecosystem by affecting such properties as productivity, biodiversity or species numbers. For example, nutrients can increase algal growth and the frequency of occurrence of harmful algal blooms; turbidity can reduce primary production; and high discharge can affect habitats for many species. Protocols for establishing scientifically credible guidelines for non-toxic stressors are less well established largely because, in contrast to toxic stressors, there is no single number (e.g. a nutrient or turbidity concentration, a discharge or a temperature) that consistently signals ecosystem impairment. Because non-toxic stressors are not directly lethal and because their effects are determined by interactions with other components of the receiving environment (e.g. light availability, water residence time, etc.), water quality guidelines that are applicable worldwide cannot be developed for non-toxic stressors.

Nutrients (e.g. N and P) are a good example of the difficulty in developing scientifically credible water quality guidelines for a non-toxic stressor. Three factors contribute to a difficulty in ascribing nutrient guidelines:

- The lack of any widely applicable relationship between nutrient concentration and aquatic plant growth (or other relevant metric, e.g. DO and changes in higher trophic levels), particularly for rivers.
- Regional variation in nutrient concentrations and aquatic plant growth for sites with little or no anthropogenic impacts (i.e. 'background' or 'reference' sites).
- The fact that there is generally a wide range in nutrient concentrations between background conditions and unequivocal degradation in aquatic ecosystem health. Depending on public expectation as to which aquatic ecosystem uses (recreational sports; fisheries; water supply for drinking, irrigation and/or industrial use; industrial, municipal and/or agricultural wastewater dilution; aquatic ecosystem health) should be protected, anywhere along this continuum between reference conditions and an unequivocally degraded state may be deemed acceptable.

The development of water quality guidelines for nutrients, specifically P, is best developed for lakes. Here, empirical models relating P concentration to phytoplankton biomass (expressed as chlorophyll *a*) have been developed at both world and local scales (Dillon and Rigler, 1974; Jones and Bachmann, 1976; Smith and Shapiro, 1981). These models, in conjunction with knowledge of best achievable water quality for the region (i.e. reference conditions) and societal values for the water body, have been used to set loading limits into lakes in order to reduce phytoplankton biomass (e.g. Vollenweider, 1976).

The development of nutrient guidelines for rivers and streams is less advanced, largely because physical factors (current velocity, substrate size and stability, light penetration) assume greater importance and can obscure nutrient–biotic responses sporadically (either seasonally or on an unpredictable basis) or continuously (as a perpetual background stressor), thereby reducing the predictability of nutrient–periphyton relationships (e.g. Dodds *et al.*, 1997; Biggs, 2000). Because of the difficulty in confidently ascribing biotic changes to nutrients in systems that are controlled by physical forces, nutrient loading regulations for rivers continue to rely largely on the traditional approach of technology-based chemical guidelines (P, N, BOD loads). However, some jurisdictions are working towards the adoption of scientifically credible guidelines for rivers. This might result in extremely small target concentrations, as would be the case for an oligotrophic Western Canadian river, where an increase in the ortho-phosphate concentration of only 1 $\mu g\ P\ l^{-1}$ resulted in high accumulations of benthic algae (Bothwell, 1989). For many Irish rivers eutrophication becomes a potential problem when median annual PO_4^{3-} concentrations exceed 30 $\mu g\ P\ l^{-1}$ (McCarrigle, 1998).

A fundamental decision when initiating a programme to develop nutrient targets for rivers is whether to adopt a catchment or an ecoregion-based approach. Omernik and Bailey (1997) evaluated the advantages and disadvantages of both approaches (summarized in Table 22.3). The catchment approach, because it uses a physically defined source area, allows solute fluxes through the landscape to be expressed on a unit area basis and enables point sources of pollution to be identified and separated from the complicating effects arising from upstream influences. In contrast, the ecoregion approach groups sites on the basis of similarity in physico-chemical attributes (e.g. geology, climate, soils, terrestrial vegetation), irrespective of whether they are in the same or different catchments or their position in a catchment. Although the two frameworks have very different purposes, Omernik and Bailey (1997) argued that these approaches can be combined to improve the assessment and management of environmental resources. Comparisons among sites with similar physiographic

Table 22.3. A summary of benefits and limitations of catchment or ecosystem-based classifications for water quality management.

	Catchment		Ecoregion	
Characteristics	Benefits	Limitations	Benefits	Limitations
Regional context	Provides a strong longitudinal context, good for assessing point sources	Requires multiple sampling sites in order to characterise upstream reaches	Provides an indication of biological potential	Difficult to define so that it is acceptable to a wide range of users
Inter-site comparison	Allows focus on one (lowest) point of main river	Only relevant for points on main stem of river	Allows grouping of spatially separated areas either within individual catchments or between catchments	Choice of appropriate scale can be problematic
Geographic scale	Research catchments tend to be small and provide detailed process-based information	Not appropriate for large land areas where it is not possible to define hydrological boundaries	Provides a mechanism for selecting representative sites	Widescale availability of information can be a problem
Temporal scale	Monitoring sites often have a long run history of data for trend analysis	Often poor temporal resolution	Allows extrapolation of data across regions	Requires some additional information such as stream network in order to interpret trends
Extrapolation of information	Additional information is often collected such as river discharge	Sampling site selection may not be representative of the ecological resource making extrapolation difficult	Provides part of a framework for making vulnerability assessments	
Calculating loads	Possible to calculate nutrient loads and ascribe back to defined land area	Many of the solute properties and behaviour are influenced by catchment size		

characteristics (the ecoregion approach) will yield an improved understanding of reference conditions and the biological response to a particular nutrient concentration while measurements of source, transport and aquatic impacts within a catchment scale will improve basin management. Thus, combining the ecoregion and catchment approaches:

- allows the sensitivities of the ecological resources within an individual catchment to be considered on a regional scale;
- provides a means of designing and weighting monitoring programmes so that sampling sites reflect the distribution and density of ecosystems represented within individual catchments;
- provides a spatial framework within which the existing information on fluxes of nutrients from terrestrial ecosystems can be extrapolated or interpreted against an ecological effect.

The result is management plans that are unique to a catchment and, at the same time, the development of a database that allows comparison of the response to nutrient enrichment for similar sites with an ecoregion.

Having determined the appropriate geographic scale (catchment, ecoregion or a combined approach), the development of a series of nutrient concentration guidelines requires collection of relevant environmental data which may include other related factors such as siltation of spawning beds. Two broad types of sampling programmes can be identified: those involving a spatially extensive sampling regime conducted on a single date or within a single hydrological period (such that inter-sample variation reflects time-independent inter-site differences), versus those aimed at developing a long time series for a single site or limited number of sites (such that inter-sample variation is temporal and site specific) (Wiley *et al.*, 1997).

The identification of reference or 'background' nutrient concentrations and biotic conditions (e.g. biomass) requires information on both temporal and spatial variability within the catchment or ecoregion. Establishing reference conditions can be problematical because most ecosystems have been impacted to some degree and long records of water quality data are scarce. By adopting an ecoregion approach it is possible to include information for rivers located outside the catchment of interest. A database developed for sites experiencing minimal or no anthropogenic impact can be used to define 'background' conditions (mean and variance) for streams of similar hydrological characteristics (e.g. stream order, channel size or discharge characteristics). A database that includes both reference and impacted sites, sampled either as a spatially extensive survey or as an intensive long-term study of a few sites, is also needed to develop a relationship between stream nutrient concentration and the desired characteristic either for rivers of a given hydrological class or for all rivers within a catchment or ecoregion, by including hydrological or geomorphic characteristics in the model. From knowledge of reference conditions for a catchment or ecoregion as well as public expectation for water quality for various river reaches, it is possible to identify a threshold for the desired property (aquatic plant biomass, infilling of spawning beds, DO, etc.), define instream nutrient concentrations that correspond to the threshold condition and, in turn, prescribe nutrient loading limits to ensure that the threshold is not exceeded.

Conclusions

Effective management of nutrient loss from diffuse sources requires an unambiguous relationship between source, transport and impact. For this to be achieved a multidisciplinary scientific approach that involves terrestrial, agricultural and soil scientists, aquatic ecologists and hydrologists is required. Until recently, studies linking land management practices and soil processes with nutrient concentrations in aquatic ecosystems and effects on aquatic biota were rare. Physical and terrestrial scientists typically viewed nutrient transport out of the terrestrial environment as a loss from their system and tended to emphasize the relationship between this loss and management practices within the catchment, often with little consideration given to the sensitivities or attributes of the actual drainage network. In contrast, hydrologists and aquatic scientists viewed nutrient fluxes from the terrestrial environment as inputs and the starting point from which an assessment of aquatic environmental impact was made. In addition, aquatic environmental assessments tended to be undertaken at a catchment scale, typically in rivers receiving both point and non-point source inputs. We encourage further communication and collaborative research and planning among terrestrial, soil, agricultural, aquatic and hydrological scientists. This will aid in ensuring that data collected as part of monitoring and research programmes are appropriate for making impact assessments that link sources to effects. These collaborations should also facilitate determination of the relative importance of various transfer pathways that link terrestrial and aquatic systems.

The development of nutrient–biotic relationships and threshold concentrations requires data on nutrient sources, flux, transport, instream concentrations and biological impact. Ideally, this information should be collected from sites nested within a hierarchical framework of both catchment and ecoregion. Historically, the majority of monitoring sites in agricultural watersheds in North America and Europe were selected on the basis of their catchment size and accessibility. This often means that they are small in area and structurally relatively simple, consisting of a limited number of land uses, soil types and human activities. The problems have always been over whether these catchments are representative of the entire geo-political region and how to interpret and extrapolate information gained from them. We suggest that the development of nutrient management plans and the establishment of nutrient targets requires that rivers are examined from both the catchment and the ecoregion perspective with the result that:

- comparisons among rivers with similar physiographic characteristics (the ecoregion approach) and hydrological features (e.g. stream order, channel size, discharge characteristics, etc.) will yield an improved understanding of reference conditions (by examining only rivers with little or no anthropogenic impacts) and the biological response to nutrient enrichment (by examining a range of both reference and impact sites);
- measurements of nutrient sources, fluxes and aquatic impacts within a catchment scale will improve basin management.

An ecoregion approach to nutrient assessment will facilitate the development of a conceptual framework within which site-specific information on nutrient fluxes can be linked with aquatic ecosystem impacts. This framework will enable an evaluation and prioritization of the vulnerability of ecological resources to nutrient impairment and, in turn, facilitate the development of catchment management plans. We also advise that when developing catchment management plans, prioritizing of nutrient sources for control should not be based solely on numerical comparison of total loads. The effect of a given nutrient source will also relate to the timing of nutrient release, its residence time in the terrestrial system, the timing of its entry into the aquatic system and temporal variation in the sensitivity of biota. The synchronicity between timing of nutrient supply, residence times, biological availability and system sensitivity remains a high research priority.

Acknowledgements

The Macaulay Land Use Research Institute is funded by the Scottish Executive Rural Affairs Department. The National Water Research Institute is funded by Environment Canada.

References

Beaton, J.D., Roberts, T.L., Halstead, E.H. and Cowell, L.E. (1995) Global transfers of P in fertilizer materials and agricultural commodities. In: Tiessen, H. (ed.) *Phosphorus in the Global Environment, Transfers, Cycles and Management*, SCOPE 54. John Wiley & Sons, Chichester, UK, pp. 7–26.

Biggs, B.J.F. (1995) The contribution of flood disturbance, catchment geology and land use to the habitat template of periphyton in stream ecosystems. *Freshwater Biology* 33, 419–438.

Biggs, B.J.F. (1996) Patterns in benthic algae in streams. In: Stevenson, R.J., Bothwell, M.L. and Lowe, R.L. (eds) *Algal Ecology: Freshwater Benthic Ecosystems*. Academic Press, San Diego, pp. 31–56.

Biggs, B.J.F. (2000) Eutrophication of streams and rivers: dissolved nutrient-chlorophyll relationships for benthic algae. *Journal of the North American Benthological Society* 19, 17–31.

Biggs, B.J.F. and Lowe, R.L. (1994) Responses of two trophic levels to patch enrichment along a New Zealand stream continuum. *New Zealand Journal of Marine and Freshwater Research* 28, 119–134.

Biggs, B.J.F., Stevenson, R.J. and Lowe, R.L. (1998) A habitat matrix conceptual model for stream periphyton. *Archiv fur Hydrobiologia* 143, 21–56.

Bothwell, M.L. (1988) Growth rate responses of lotic periphyton diatoms to experimental phosphorus enrichment: the influence of temperature and light. *Canadian Journal of Fisheries and Aquatic Sciences* 45, 261–270.

Bothwell, M.L. (1989) Phosphorus-limited growth dynamics of lotic periphytic diatom communities: areal biomass and cellular growth rate responses. *Canadian Journal of Fisheries and Aquatic Sciences* 46, 1293–1301.

Bothwell, M.L. (1992) Eutrophication of rivers by nutrients in treated kraft pulp mill effluent. *Water Pollution Research Journal of Canada* 27, 447–472.

Bothwell, M.L. (1993) Artificial streams in the study of algal/nutrient dynamics. *Journal of the North American Benthological Society* 12, 327–333.

Bott, T.L., Brock, J.T., Battrup-Pedersen, A., Chambers, P.A., Dodds, W.K., Himbeault, K.T., Lawrence, J.R., Planas, D., Snyder, E. and Wolfaardt, G.M. (1997) An evaluation of techniques for measuring periphyton metabolism in chambers. *Canadian Journal of Fisheries and Aquatic Sciences* 54, 715–725.

Bourassa, N. and Cattaneo, A. (1998) Control of periphyton biomass in Laurentian streams (Quebec). *Journal of the North American Benthological Society* 17, 420–429.

Caraco, N.F. (1995) Influence of human populations on phosphorus transfers to aquatic systems: a regional scale study using large rivers. In: Tiessen, H. (ed.) *Phosphorus in the Global Environment, Transfers, Cycles and Management*, SCOPE 54. John Wiley & Sons, Chichester, UK, pp. 235–244.

Carpenter, S.R., Caraco, N.F., Correll, D.L., Howarth, R.W., Sharpley, A.N. and Smith, V.H. (1998) Nonpoint pollution of surface waters with phosphorus and nitrogen. *Ecological Applications* 8, 559–568.

Carr, G.M. and Chambers, P.A. (1998) Macrophyte growth and sediment phosphorus and nitrogen in a Canadian prairie river. *Freshwater Biology* 39, 525–536.

Cattaneo, A. and Prairie, Y.T. (1995) Temporal variability in the chemical characteristics along the Rivière de l'Achigan: how many samples are necessary to describe stream chemistry. *Canadian Journal of Fisheries and Aquatic Sciences* 52, 828–835.

CCME (Canadian Council of Ministers of the Environment) (1999) Environmental Quality Guidelines. CCME Winnipeg, Manitoba.

Chessman, B.C., Hutton, P.E. and Burch, J.M. (1992) Limiting nutrients for periphyton growth in subalpine forest, agricultural and urban streams. *Freshwater Biology* 28, 349–361.

Chetelat, J., Pick, F.R. and Morin, A. (1999) Periphyton biomass and community composition in rivers of different nutrient status. *Canadian Journal of Fisheries and Aquatic Sciences* 56, 560–569.

Delong, M.D. and Brusven, M.A. (1992) Patterns of periphyton chlorophyll *a* in an agricultural nonpoint source impacted stream. *Water Resources Bulletin* 28, 731–741.

Dillon, P.J. and Rigler, F.H. (1974) The phosphorus-chlorophyll relationship in lakes. *Limnology and Oceanography* 19, 767–773.

Dodds, W.K. and Welch, E.B. (2000) Establishing nutrient criteria in streams. *Journal of the North American Benthological Society* 19, 186–196.

Dodds, W.K., Smith, V.H. and Zander, B. (1997) Developing nutrient targets to control benthic chlorophyll levels in streams: a case study of the Clark Fork River. *Water Research* 31, 1738–1750.

Dodds, W.K., Jones, J.R. and Welch, E.B. (1998) Suggested classification of stream trophic status: distributions of temperate stream types by chlorophyll, total nitrogen, and phosphorus. *Water Research* 32, 1455–1462.

Edmondson, W.T. and Lehman, J.R. (1981) The effect of changes in nutrient income on the conditions of Lake Washington. *Limnology and Oceanography* 26, 1–28.

Edwards, A.C., Twist, H. and Codd, G.A. (2000) Assessing the impact of terrestrially derived phosphorus on flowing water systems. *Journal of Environmental Quality* 29, 117–124.

EEC (1991) Council Directive concerning the protection of waters against pollution caused by nitrates from agricultural sources. *Official Journal of the European Communities*, 91/676/EEC.

Fisher, T.R., Melack, J.M., Grobbelaar, J.U. and Howarth, R.W. (1995) Nutrient limitation of phytoplankton and eutrophication of inland, estuarine and marine waters. In: Tiessen, H. (ed.) *Phosphorus in the Global Environment, Transfers, Cycles and Management*, SCOPE 54. John Wiley & Sons, Chichester, UK, pp. 301–322.

Forsberg, C. (1998) Which policies can stop large scale eutrophication? *Water Science and Technology* 37, 193–200.
Grimvall, A., Stålnacke, P. and Tonderski, A. (2000) Time scales of nutrient losses from land to sea – a European perspective. *Ecological Engineering* 14, 363–371.
Heathwaite, A.L., Johnes, P.J. and Peters, N.E. (1997) Trends in nutrients. In: Peters, N.E., Bricker, O.P. and Kennedy, M.M. (eds) *Water Quality Trends and Geochemical Mass Balance.* John Wiley & Sons, Chichester, UK, pp. 139–169.
Heathwaite, A.L, Sharpley, A. and Gburek, W. (2000) A conceptual approach for integrating phosphorus and nitrogen management at watershed scales. *Journal of Environmental Quality* 29, 158–166.
Hillebrand, H. and Sommer, U. (1999) The nutrient stoichiometry of benthic microalgal growth: Redfield proportions are optimal. *Limnology and Oceanography* 44, 440–446.
Hudson, J.J., Taylor, W.D. and Schindler, D.W. (2000) Phosphate concentrations in lakes. *Nature* 406, 54–56.
Hullar, M.A. and Vestal, J.R. (1989) The effects of nutrient limitation and stream discharge on the epilithic microbial community in an oligotrophic Arctic stream. *Hydrobiologia* 172, 19–26.
Hynes, H.B.N. (1970) *The Ecology of Running Waters.* Liverpool University Press, Liverpool, UK.
Johnson, L.B., Richards, C., Host, G.E. and Arthur, J.W. (1997) Landscape influences on water chemistry in Midwestern stream ecosystems. *Freshwater Biology* 37, 193–208.
Jones, J.R. and Bachmann, R.W. (1976) Prediction of phosphorus and chlorophyll levels in lakes. *Journal of Water Pollution and Control Federation* 48, 2176–2182.
Kjeldsen, K. (1996) Regulation of algal biomass in a small lowland stream: field experiments on the role of invertebrate grazing, phosphorus and irradiance. *Freshwater Biology* 36, 535–546.
Lamberti, G.A. and Steinman, A.D. (1993) Research in artificial streams: applications, uses, and abuses. *Journal of the North American Benthological Society* 12, 313–384.
McCarrigle, M.L. (1998) Impact of eutrophication on Irish water Quality. In: Wilson, J.G. (ed.) *Eutrophication in Irish waters.* Royal Irish Academy, Dublin, pp. 82–90.
Mulholland, P.J., Steinman, A.D. and Elwood, J.W. (1990) Measurement of phosphorus uptake length in streams: comparison of radiotracer and stable PO_4 releases. *Canadian Journal of Fisheries and Aquatic Sciences* 47, 2351–2357.
Munn, N.L. and Meyer, J.L. (1990) Habitat-specific solute retention in two small streams: an intersite comparison. *Ecology* 71, 2069–2082.
Muscutt, A.D. and Withers, P.J.A. (1996) The phosphorus content of rivers in England and Wales. *Water Research* 30, 1258–1268.
Nyholm, N. and Peterson, H.G. (1997) Laboratory bioassays with microalgae. In: Wang, W., Lower, W.R., Gorsuch, J.W. and Hughes, J. (eds) *Plants for Environmental Studies.* CRC Lewis Publications, Boca Raton, Florida, pp. 225–276.
OECD, (1982) Eutrophication of Waters, Monitoring Assessment and Control. Organization for Economic Co-operation and Development, Paris.
Omernik, J.M. and Bailey, R.G. (1997) Distinguishing between watersheds and ecoregions. *Journal of the American Water Research Association* 33, 935–949.
Pan, Y., Stevenson, R.J., Hill, B.H., Herlihy, A.T. and Collins, G.B. (1996) Using diatoms as indicators of ecological conditions in lotic systems: a regional assessment. *Journal of the North American Benthological Society* 15, 481–495.
Paul, B.J. and Duthie, H.C. (1988) Nutrient cycling in the epilithon of running waters. *Canadian Journal of Botany* 67, 2302–2309.
Perrin, C.J., Bothwell, M.L. and Slaney, P.A. (1987) Experimental enrichment of a coastal stream in British Columbia: effects of organic and inorganic additions on autotrophic periphyton production. *Canadian Journal of Fisheries and Aquatic Sciences* 44, 1247–1256.
Peterson, B.J., Deegan, L., Helfrich, J., Hobbie, J.E., Hullar, M., Moller, B., Ford, T.E., Hershey, A., Hiltner, A., Kipphut, G., Lock, M.A., Fiebig, D.M., McKinley, V., Miller, M.C., Vestal, J.R., Ventullo, R. and Volk, G. (1993) Biological responses of a tundra river to fertilisation. *Ecology* 74, 653–672.
Roth, N.E., Allan, J.D. and Erickson, D.L. (1996) Landscape influences on stream biotic integrity assessed at multiple spatial scales. *Landscape Ecology* 11, 141–156.
Rott, E. (1995) Diatoms of the Grand River, Ontario, Canada restudied after 25 years. *Limnologica* 25, 165–192.

Scrimgeour, G.J. and Chambers, P.A. (2000) Cumulative effects of pulp mill and municipal effluents on epilithic biomass and nutrient limitation in a large northern river ecosystem. *Canadian Journal of Fisheries and Aquatic Sciences* 57, 1342–1354.
Sims, J.T., Edwards, A.C., Schoumans, O.F. and Simard, R.R. (2000) Integrating soil phosphorus testing into environmentally based agricultural management practices. *Journal of Environmental Quality* 29, 60–71.
Skulberg, O.M. (1995) Use of algae for testing water quality. In: Wiessner, W. (ed.) *Algae, Environment and Human Affairs*. Biopress Ltd, Bristol, UK, pp. 181–199.
Smaling, E.M.A., Oenema, O. and Fresco, L.O. (1999) *Nutrient Disequilibria in Agroecosystems and Case Studies*. CABI Publishing, Wallingford, UK.
Smith, V.H. and Shapiro, J. (1981) Chlorophyll-phosphorus relations in individual lakes: their importance to lake restoration strategies. *Environmental Science and Technology* 15, 444–451.
Steen, I. (1998) Phosphorus availability in the 21st century: management of a non-renewable resource. *Phosphorus & Potassium* 217, 25–31. www.nhm.ac.uk/mineralogy/phos/p&k217/steen.htm (06 November 2000)
Steinman, A.D. and Lamberti, G.A. (1996) Biomass and pigments of benthic algae. In: Hauer, F.R. and Lamberti, G.A. (eds) *Methods in Stream Ecology*. Academic Press, San Diego, California, pp. 295–313.
Stevenson, R.J. (1997) Resource thresholds and stream ecosystem sustainability. *Journal of the North American Benthological Society* 16, 410–424.
Twist, H., Edwards, A.C. and Codd, G.A. (1998) Algal growth responses to waters of contrasting tributaries of the River Dee, north-east Scotland. *Water Research* 32, 2471–2479.
UNEP (1999) *United Nations Environment Programme GEO-2000, Millennium Report on the Environment*. Earthscan, London.
US EPA (1999) *National Recommended Water Quality Criteria*. US EPA: 822/Z-99-001. EPA, Cincinnati, Ohio, USA.
Vitousek, P.M., Aber, J., Howarth, R.W., Likens, G.E., Matson, P.A., Schindler, D.W., Schlesinger, W.H. and Tilman, G.D. (1997) Human alteration of the global nitrogen cycle: causes and consequences. *Ecological Applications* 7, 737–750.
Vollenweider, R.A. (1976) Advances in defining critical loading levels for phosphorus in lake eutrophication. *Memorie dell'Istituto Italiano di Idrobiologia* 33, 53–83.
Wellnitz, T.A., Rader, R.B. and Ward, J.V. (1996) Importance of light and nutrients in structuring an algal community in a Rocky Mountain stream. *Journal of Freshwater Ecology* 11, 399–413.
Wiley, M.J., Kohler, S.L. and Seelbach, P.W. (1997) Reconciling landscape and local views of aquatic communities: lessons from Michigan trout streams. *Freshwater Biology* 37, 133–148.
Williamson, C.E., Morris, D.P., Pace, M.L. and Olson, O.G. (1999) Dissolved organic carbon and nutrients as regulators of lake ecosystems: resurrection of a more integrated paradigm. *Limnology and Oceanography* 44, 795–803.
Wold, A.P. and Hershey, A.E. (1999) Spatial and temporal variability of nutrient limitation in 6 North Shore tributaries to Lake Superior. *Journal of the North American Benthological Society* 18, 2–14.
Wright, J.F., Moss, D., Armitage, P.D. and Furse, M.T. (1984) A preliminary classification of running-water sites in Great Britain based on macro-invertebrate species and the prediction of community type using environmental data. *Freshwater Biology* 14, 221–256.

Index

Browse Read and Buy
www.cabi.org/bookshop
ANIMAL & VETERINARY SCIENCES
BIODIVERSITY CROP PROTECTION
HUMAN HEALTH NATURAL RESOURCES
ENVIRONMENT PLANT SCIENCES
SOCIAL SCIENCES
Subjects
Search
Reading Room
Bargains
New Titles
Forthcoming
CABI Publishing
A division of CAB International
Online BOOK SHOP
Crop Pollination by Bees
A DICTIONARY OF Entomology
G Gordh and D H Headrick
Principles of CATTLE PRODUCTION
C.J.C. Phillips
Seeds
THE ECOLOGY OF REGENERATION IN PLANT COMMUNITIES
2nd EDITION
Edited by Michael Fenner
Order & Pay Online!
MasterCard
VISA
AMERICAN EXPRESS
FULL DESCRIPTION
BUY THIS BOOK
BOOK OF THE MONTH
Tel: +44 (0)1491 832111 Fax: +44 (0)1491 829292

Printed and bound by CPI Group (UK) Ltd, Croydon, CR0 4YY
18/09/2026
14965013-0001